POWER SYSTEM
OPERATION AND CONTROL

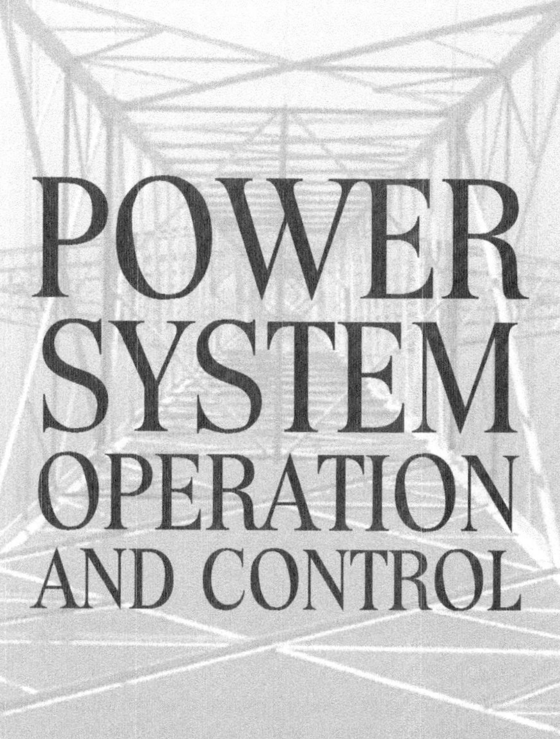

POWER SYSTEM
OPERATION
AND CONTROL

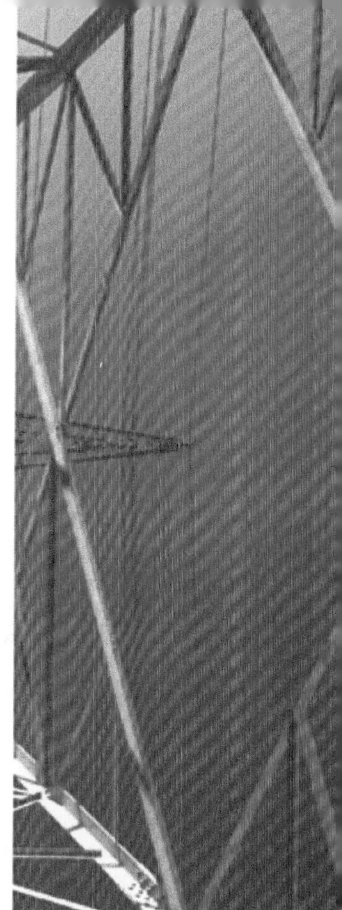

S. Sivanagaraju
Associate Professor
Department of Electrical and
Electronics Engineering
University College of Engineering
JNTU Kakinada
Kakinada, Andhra Pradesh

G. Sreenivasan
Associate Professor
Department of Electrical and
Electronics Engineering
INTELL Engineering College
Anantapur, Andhra Pradesh

 Pearson

Although the author and publisher have made every effort to ensure that the information in this book was correct at the time of editing and printing, the author and publisher do not assume and hereby disclaim any liability to any party for any loss or damage arising out of the use of this book caused by errors or omissions, whether such errors or omissions result from negligence, accident or any other cause. Further, names, pictures, images, characters, businesses, places, events and incidents are either the products of the author's imagination or used in a fictitious manner. Any resemblance to actual persons, living or dead or actual events is purely coincidental and do not intend to hurt sentiments of any individual, community, sect or religion.

In case of binding mistake, misprints or missing pages etc., the publisher's entire liability and your exclusive remedy is replacement of this book within reasonable time of purchase by similar edition/reprint of the book.

ISBN 978-81-317-2662-4

First Impression, 2009

Published by Pearson India Education Services Pvt. Ltd, CIN: U72200TN2005PTC057128.

Head Office: 1st Floor, Berger Tower, Plot No. C-001A/2, Sector 16B, Noida – 201 301, Uttar Pradesh, India

Registered Office: Featherlite, 'The Address' 5th Floor, Survey No 203/10B, 200 Ft MMRD Road, Zamin Pallavaram, Chennai - 600044
Phone: 044-66540100
Website: in.pearson.com, Email: companysecretary.india@pearson.com

Printer Manipal Technologies Limited, Manipal

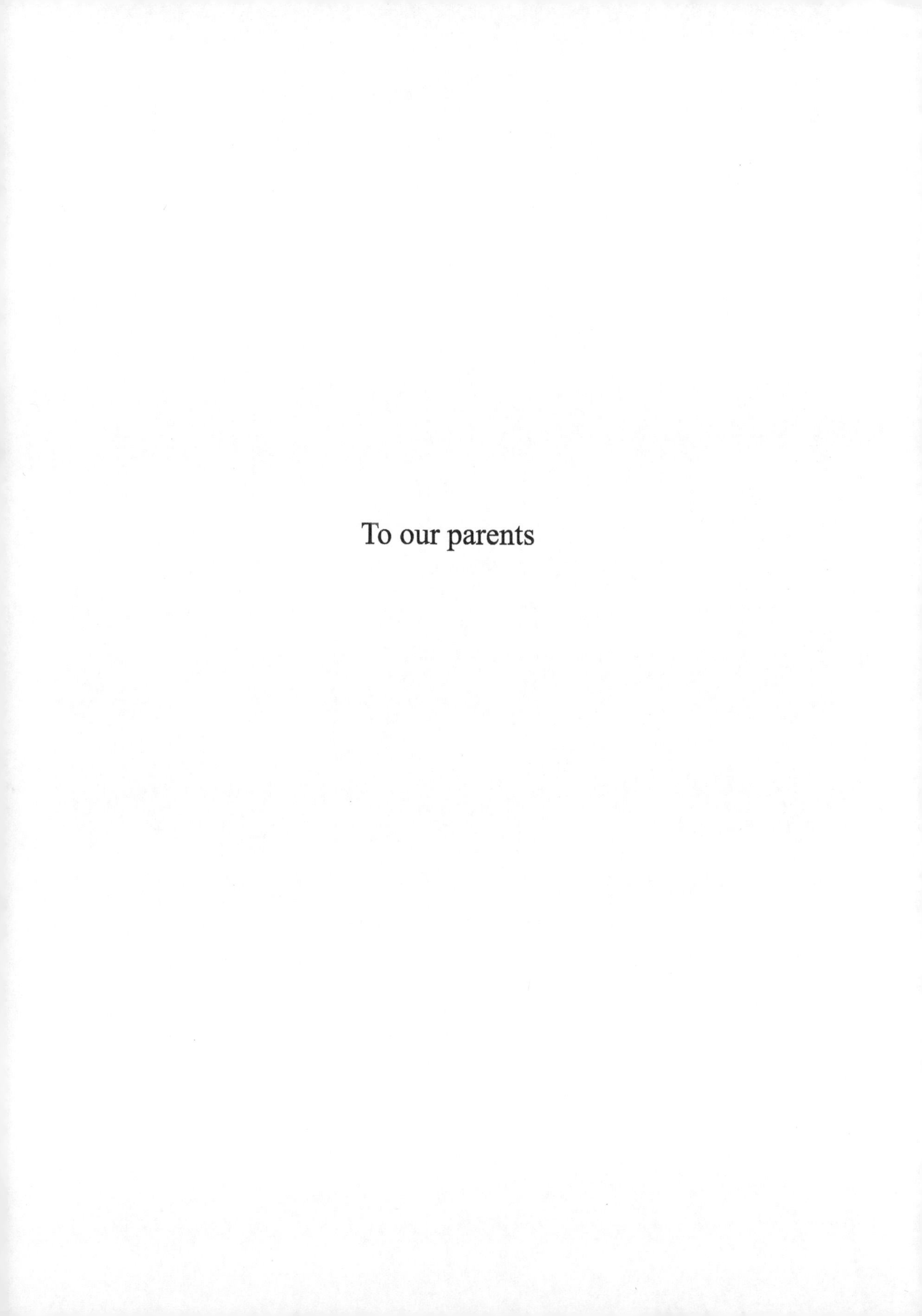

To our parents

Brief Contents

Contents

Preface

This book entitled *Power System Operation and Control* has been intended for use by undergraduate students in Indian universities. With a judicious mix of advanced topics, the book may also be useful for some institutions as a first course for postgraduates. The organization of this book reflects our desire to provide the reader with a thorough understanding of the basic principles and techniques of power system operation and control. Written to address the need for a text that clearly presents the concept of economic system operation in a manner that kindles interest, the topics are dealt with using a lucid approach that may benefit beginners as well as advanced learners of the subject. It has been designed as a functional aid to help students learn independently.

Chapter 1 introduces the economic aspects of power system and provides definitions for the various terms used in its analysis. It explains reserve requirements, the importance of load forecasting, and its classification.

Chapter 2 describes system variables and their functions. The characteristics of thermal and hydro-power units are illustrated in this unit. Non-smooth cost functions with multi-valve effect and with multi-fuel effect are briefly discussed. This chapter explains the mathematical formulation of economic load dispatch among various units by neglecting transmission losses, and it also gives an overview of the applications of various computational methods to solve the optimization problem. The flowchart required to obtain the optimal scheduling of generating units is also described here.

Chapter 3 looks at the derivation of the expression for transmission loss and explains the mathematical determination of economic load dispatch taking transmission loss into consideration. The theory of incremental transmission loss and penalty factor is clearly discussed. It also analyzes the optimal scheduling of generating units, determined with the help of a flowchart.

Chapter 4 expounds on the optimal unit commitment problem and its solution methods by taking a reliable example. Reliability and start-up considerations in optimal unit commitment problems are effectively discussed.

Chapter 5 explains the optimal power-flow problem and its solution techniques with and without inequality constraints. In this chapter, inequality constraints are considered first on control variables, and then on dependent variables. Kuhn–Tucker conditions for the solution of an optimal power flow are presented in this unit.

Chapter 6 spells out the important principle of hydro-thermal scheduling and its classification. It discusses the general mathematical formulations and methods of solving long-term and the short-term hydro-thermal scheduling problems.

Chapter 7 deals with single-area load frequency control. It describes the characteristics of the speed governor and its adjustment in case of parallel operating units. Generator controllers, namely, $P\text{–}f$ and $Q\text{–}V$ controllers, the speed-governing system model, the turbine model, and the generator–load model and their block diagram representations are clearly discussed. Steady- and dynamic-state analyses of a single-area load frequency control system are also explained. The chapter also discusses the analysis of integral control of a single-area load frequency control system.

Chapter 8 deals with the response of a two-area load frequency control for uncontrolled and controlled cases very effectively. A dynamic-state variable model for a two-area load frequency control and for a three-area load frequency control system is derived.

Chapter 9 delineates reactive-power compensation along with the objectives of load compensation. This chapter discusses uncompensated transmission lines under no-load and load conditions, and compensated transmission lines with the effects of series and shunt compensation using thyristor-controlled reactors and capacitors. It also elucidates the concept of voltage stability and makes clear how the analysis of voltage stability is carried out using P–V curves and Q–V curves.

The relationship among active power, reactive power, and voltage is derived in Chapter 10. This chapter also speaks about the methods of voltage control and the location of voltage-control equipments.

Chapter 11 deals with the principles of modeling hydro-turbines and steam turbines. It also looks at the modeling of synchronous machines including the simplified model with the effect of saliency. The determination of self-inductance and mutual inductance, and the development of general machine equations are discussed in this chapter. Park's transformation and its inverse, the derivations of flux linkage equations and voltage equations of synchronous machines, and the steady-state and dynamic-state model analysis are elucidated.

Chapter 12 offers an insight into the modeling of speed-governing systems for steam- and hydro-turbines. Mechanical–hydraulic-controlled speed-governing systems, electro–hydraulic-controlled speed-governing systems, and the general model for speed-governing systems for steam turbines are explained in detail. It throws light on excitation system modeling in various aspects such as methods of providing excitation, classification of excitation systems, and various components with their transfer functions. Standard block diagram representations for the different excitation systems are illustrated in this chapter.

Chapter 13 explains the steady-state security analysis and the transient security analysis of a power system. The concept of state estimation is developed in this chapter, and the method of least squares estimation of a system state has been clearly explained.

Acknowledgments

We express our gratitude to those who have helped us in many ways in making this book a reality:

- To our college administration for providing us a conducive atmosphere and extending immense support to carry out this work.
- To our department colleagues, especially those who, with their valuable inputs by way of suggestions, content-related guidance and discussions, helped us to successfully complete this book.
- To our students for their services.
- To all those who have either directly or indirectly helped us in bringing our efforts to fruition.
- To our families for their constant and favorable encouragement at every step of the way.

We are grateful to our publishers, Pearson Education, India, for making this book a reality. We specifically thank Sojan Jose, Jennifer Sargunar, M. E. Sethurajan, and Thomas Mathew Rajesh for their editorial inputs and for the successful completion of the project.

S. Sivanagaraju
G. Sreenivasan

1 Economic Aspects

OBJECTIVES

After reading this chapter, you should be able to:

- know the economic aspects of power systems
- analyze the various load curves of economic power generation
- define the various terms of economic power generation
- understand the importance of load forecasting

1.1 INTRODUCTION

A power system consists of several generating stations, where electrical energy is generated, and several consumers for whose use the electrical energy is generated. The objective of any power system is to generate electrical energy in sufficient quantities at the best-suited locations and to transmit it to the various load centers and then distribute it to the various consumers maintaining the quality and reliability at an economic price. Quality implies that the frequency be maintained constant at the specified value (50 Hz in our country; though 60-Hz systems are also prevailing in some countries) and that the voltage be maintained constant at the specified value. Further, the interruptions to the supply of energy should be as minimum as possible.

One important characteristic of electric energy is that it should be used as it is generated; otherwise it may be stated that the energy generated must be sufficient to meet the requirements of the consumers at all times. Because of the diversified nature of activities of the consumers (e.g., domestic, industrial, agricultural, etc.), the load on the system varies from instant to instant. However, the generating station must be in a 'state of readiness' to supply the load without any intimation from the consumer. This 'variable load problem' is to be tackled effectively ever since the inception of a power system. This necessitates a thorough understanding of the nature of the load to be supplied, which can be readily obtained from the load curve, load–duration curve, etc.

1.2 LOAD CURVE

A load curve is a plot of the load demand (on the y-axis) versus the time (on the x-axis) in the chronological order.

From out of the load connected, a consumer uses different fractions of the total load at various times of the day as per his/her requirements. Since a power system has to supply load to all such consumers, the load to be supplied varies continuously with time and does not remain constant. If the load is measured (in units of power) at regular intervals of time, say, once in an hour (or half-an-hour) and recorded, we can draw a curve known as the load curve.

A time period of only 24 hours is considered, and the resulting load curve, which is called a '**Daily load curve**', is shown in Fig. 1.1. However, to predict the annual requirements of energy, the occurrence of load at different hours and days in a year and in the power supply economics, '**Annual load curves**' are used.

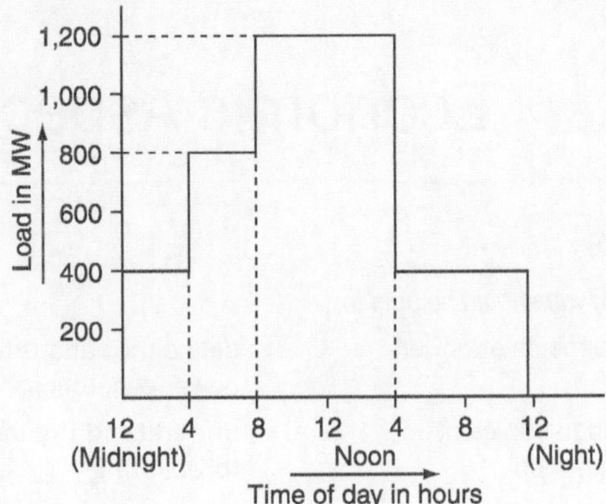

FIG. 1.1   *Daily load curve*

An annual load curve is a plot of the load demand of the consumer against time in hours of the year (1 year = 8,760 hours).

Significance: From the daily load curve shown in Fig. 1.1, the following information can be obtained:

- Observe the variation of load on the power system during different hours of the day.

- Area under this curve gives the number of units generated in a day.

- Highest point on that curve indicates the maximum demand on the power station on that day.

- The area of this curve divided by 24 hours gives the average load on the power station in the day.

- It helps in selection of the rating and number of generating units required.

1.3 LOAD–DURATION CURVE

The load–duration curve is a plot of the load demands (in units of power) arranged in a descending order of magnitude (on the y-axis) and the time in hours (on the x-axis). The load–duration curve can be drawn as shown in Fig. 1.2.

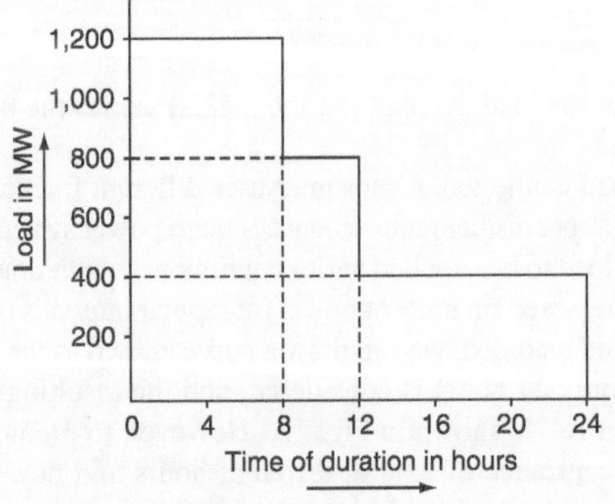

FIG. 1.2   *Load–duration curve*

1.4 INTEGRATED LOAD–DURATION CURVE

The integrated load–duration curve is a plot of the cumulative number of units of electrical energy (on the x-axis) and the load demand (on the y-axis).

In the operation of hydro-electric plants, it is necessary to know the amount of energy between different load levels. This information can be obtained from the load–duration curve. Thus, let the duration curve of a particular power station be as indicated in Fig. 1.3(a); obviously the area enclosed by the load–duration curve represents the daily energy generated (in MWh).

The minimum load on the station is d_1 (MW). The energy generated during the 24-hour period is $24\,d_1$ (MWh), i.e., the area of the rectangle $od_1b_1a_1$. So, we can assume that the energy generated varies linearly with the load demand from zero to d_1 to d_2 MW as indicated in Fig. 1.3(a). As the load demand increases from d_1 to d_2 MW, the total energy generated will be less than $24\,d_2$ MWh, since the load demand of d_2 MW persists for a duration of less than 24 hours. The total energy generated is given by the area $od_2\,b_2\,a_1$. So, the energy generated between the load demands of d_2 and d_1 is (area $od_2b_2a_1$ − area $od_1b_1a_1$) = area $d_1d_2b_1$ (shown cross-latched in Fig. 1.3(a)).

Now, if the total number of units generated was to be plotted as abscissa corresponding to a given load, we shall obtain what is called the integrated load–duration curve. Thus, if the area $od_2b_2a_1$ were designated as c_2 (MWh), then point p has the co-ordinates (e_2, d_2) on the integrated load–duration curve shown in Fig. 1.3(b).

The integrated load–duration curve is also the plot of the cumulative integration of area under the load curve starting at zero loads to the particular load. It exhibits an increasing slope upto the peak load.

1.4.1 Uses of integrated load–duration curve

   (i)   The amount of energy generated between different load levels can be obtained.

  (ii)   From acknowledgment of the daily energy requirements, the load that can be carried on the base or peak can be easily determined.

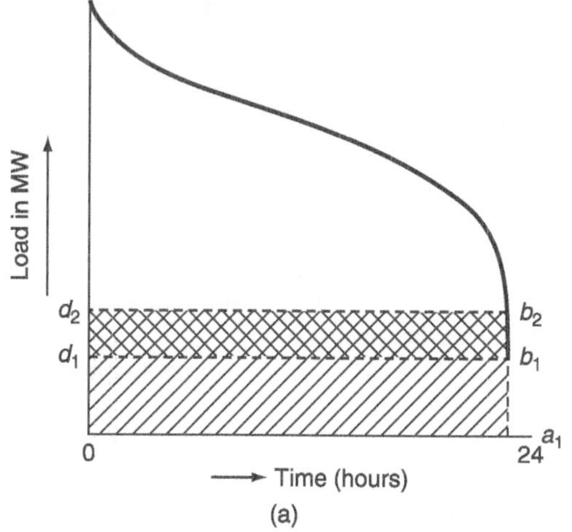

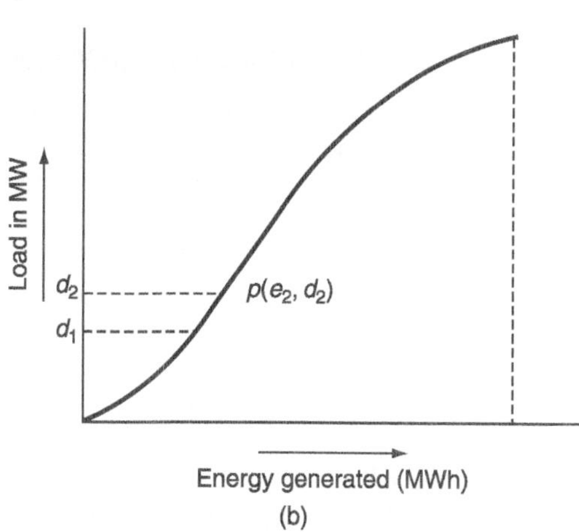

FIG. 1.3 *Integrated load–duration curve*

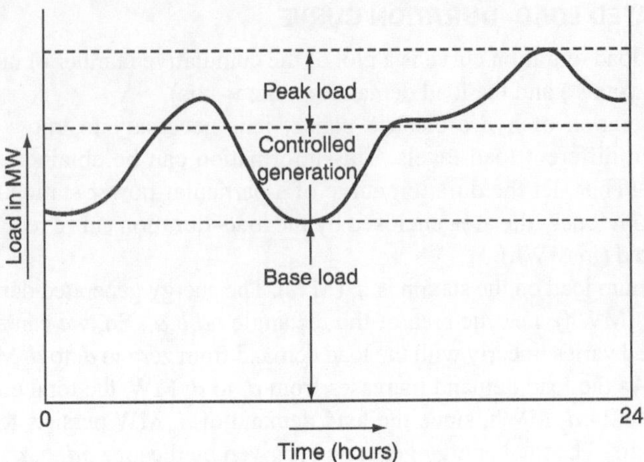

FIG. 1.4 *Daily load curve*

To have a clear idea of 'base-load' and 'peak load', let us consider a power system, the daily load curve of which is depicted in Fig. 1.4.

In a power system, there may be several types of generating stations such as hydro-electric stations, fossil-fuel-fired stations, nuclear stations, and gas-turbine-driven generating stations. Of these stations, some act as base-load stations, while others act as peak load stations.

Base-load stations run at 100% capacity on a 24-hour basis. Nuclear reactors are ideally suited for this purpose.

Intermediate or controlled-power generation stations normally are not fully loaded. Hydro-electric stations are the best choice for this purpose.

Peak load stations operate during the peak load hours only. Since the gas-turbine-driven generators can pick up the load very quickly, they are best suited to serve as peak load stations. Where available, pumped-storage hydro-electric plants can be operated as peak load stations.

A base-load station operates at a high-load factor, whereas the peak load plant operates at a low-load factor. So, the base-load station should have low operating costs.

1.5 DEFINITION OF TERMS AND FACTORS

Several terms are used in connection with power supply to an area, whether it be for the first time (as is the case when the area is being electrified for the first time) or subsequently (due to the load growth). These terms are explained below.

1.5.1 Connected load

A consumer, for example, a domestic consumer, may have several appliances rated at different wattages. The sum of these ratings is his/her connected load.

Connected load is the sum of the ratings (W, kW, or MW) of the apparatus installed on a consumer's premises.

1.5.2 Maximum demand

It is the maximum load used by a consumer at any time. It can be less than or equal to the connected load. If all the devices connected in the consumer's house run to their fullest extent simultaneously, then the maximum demand will be equal to the connected load.

But generally, the actual maximum demand will be less than the connected load since all the appliances are never used at full load at a time.

The maximum demand is usually measured in units of kilowatts (kW) or megawatts (MW) by a maximum demand indicator. (Usually, in the case of high-tension consumers, the maximum demand is measured in terms of kVA or MVA.)

1.5.3 Demand factor

The ratio of the maximum demand to the connected load is called the 'demand factor':

$$DF = \frac{\text{maximum demand}}{\text{connected load}}$$

Note: Maximum demand and the connected load are to be expressed in the same units (W, kW, or MW).

1.5.4 Average load

If the number of kWh supplied by a station in one day is divided by 24 hours, then the value obtained is known as the daily average load:

$$\text{Daily average load} = \frac{\text{kWh in one day}}{24}$$

$$\text{Monthly average load} = \frac{\text{kWh in one day}}{30 \times 24}$$

$$\text{Yearly average load} = \frac{\text{kWh in one day}}{365 \times 24}$$

1.5.5 Load factor

The ratio of the average demand to the maximum demand is called the load factor:

$$\text{Load factor (LF)} = \frac{\text{average demand}}{\text{max. demand}}$$

If the plant is in operation for a period T,

$$\text{Load factor} = \frac{\text{average demand} \times T}{\text{max. demand} \times T}$$

$$= \frac{\text{units generated in } T \text{ hours}}{\text{max. demand} \times T}$$

The load factor may be a daily load factor, a monthly load factor, or an annual load factor, if the time period is considered in a day or a month or a year, respectively. Load factor is always less than one because average load is smaller than the maximum demand. It plays a key role in determining the overall cost per unit generated. Higher the load factor of the power station, lesser will be the cost per unit generated.

1.5.6 Diversity factor

Diversity factor is the ratio of the sum of the maximum demands of a group of consumers to the simultaneous maximum demand of the group of consumers:

$$\text{Diversity factor} = \frac{\text{sum of individual max. demands}}{\text{max. demand on system}}$$

A power system supplies load to various types of consumers whose maximum demands generally do not occur at the same time. Therefore, the maximum demand on the power system is always less than the sum of individual maximum demands of the consumers.

A high diversity factor implied that with a smaller maximum demand on the station, it is possible to cater to the needs of several consumers with varying maximum demands occurring at different hours of the day. The lesser the maximum demand, the lesser will be the capital investment on the generators. This helps in reducing the overall cost of the units (kWh) generated.

Thus, a higher diversity factor and a higher load factor are the desirable characteristics of the load on a power station. The load factor can be improved by encouraging the consumers to use power during off-peak hours with certain incentives like offering a reduction in the cost of energy consumed during off-peak hours.

1.5.7 Plant capacity

It is the capacity or power for which a plant or station is designed. It should be slightly more than the maximum demand. It is equal to the sum of the ratings of all the generators in a power station:

1.5.8 Plant capacity factor

It is the ratio of the average demand on the station to the maximum installed capacity of the station.

$$\text{Plant capacity factor} = \frac{\text{average demand}}{\text{max. installed capacity}}$$

or capacity factor = (load factor) × (utilization factor).
Reserve capacity = plant capacity − maximum demand

1.5.9 Utilization factor (or plant-use factor)

It is the ratio of kWh generated to the product of the plant capacity and the number of hours for which the plant was in operation:

$$\text{Plant-use factor} = \frac{\text{station output in kWh}}{\text{plant capacity} \times \text{hours of use}}$$

1.5.10 Firm power

It is the power that should always be available even under emergency.

1.5.11 Prime power

It is the maximum power (may be thermal or hydraulic or mechanical) continuously available for conversion into electric power.

1.5.12 Dump power

This is the term usually used in hydro-electric plants and it represents the power in excess of the load requirements. It is made available by surplus water.

1.5.13 Spill power

It is the power that is produced during floods in a hydro-power station.

1.5.14 Cold reserve

It is the reserve-generating capacity that is not in operation, but can be made available for service.

1.5.15 Hot reserve

It is the reserve-generating capacity that is in operation, but not in service.

1.5.16 Spinning reserve

It is the reserve-generating capacity that is connected to bus bars and is ready to take the load.

1.6 BASE LOAD AND PEAK LOAD ON A POWER STATION

Base load: It is the unvarying load that occurs almost during the whole day on the station.

Peak load: It is the various peak demands of load over and above the base load of the station.

Example 1.1: A generating station has a maximum demand of 35 MW and has a connected load of 60 MW. The annual generation of units is 24×10^7 kWh. Calculate the load factor and the demand factor.

Solution:

$$\text{No. of units generated annually} = 24 \times 10^7 \text{ kWh}$$

$$\text{No. of hours in a year (assuming 365 days in a year)} = 365 \times 24$$
$$= 8,760 \text{ hours}$$

$$\therefore \text{Average load on the station} = \frac{24 \times 10^7}{8,760} = 27,397.26 \text{ kW} = 27.39726 \text{ MW}$$

$$\therefore \text{Load factor} = \frac{\text{average load}}{\text{max. demand}} = \frac{27.39726\,(\text{MW})}{35\,(\text{MW})} = 0.7828 \text{ or } 7$$

$$\text{Demand factor} = \frac{\text{max. demand}}{\text{connected load}} = \frac{35\,(\text{MW})}{60\,(\text{MW})} = 0.583 \text{ or } 58.3\%$$

Example 1.2: A generating station supplies four feeders with the maximum demands (in MW) of 16, 10, 12, and 7 MW. The overall maximum demand on the station is 20 MW and the annual load factor is 45%. Calculate the diversity factor and the number of units generated annually.

Solution:

$$\text{Sum of maximum demands} = 16 + 10 + 12 + 7 = 45 \text{ MW}$$
$$\text{Simultaneous maximum demand} = 20 \text{ MW}$$

$$\therefore \text{Diversity factor} = \frac{\text{sum of max. demands}}{\text{simultaneous max. demand}} = \frac{45}{20} = 2.25$$

$$\text{Average demand} = (\text{maximum demand}) \times (\text{load factor})$$
$$= 20 \times 0.45 = 9 \text{ MW}$$

$$\therefore \text{No. of units generated annually} = 9 \times 8,760 = 78,840 \text{ MWh}$$

Alternatively,

$$\text{Annual load factor} = \frac{\text{No. of units generated annually}}{(\text{Max. demand}) \times 8,760}$$

i.e.,

$$0.45 = \frac{\text{No. of units generated annually}}{20 \times 8,760}$$

so that the number of units generated annually $= 0.45 \times 20 \times 8,760$ MWh
$$= 78,840 \text{ MWh}$$

Example 1.3: The yearly load–duration curve of a power plant is a straight line (Fig. 1.5). The maximum load is 30 MW and the minimum load is 20 MW. The capacity of the plant is 35 MW. Calculate the plant capacity factor, the load factor, and the utilization factor.

Solution:

No. of units generated per year $=$ Area $OACD =$ Area $OBCD +$ Area BAC

$$= 20 \times 8,760 + \frac{1}{2}(30 - 20) \times 8,760$$

$$= 8,760 \left(20 + \frac{1}{2} \times 10 \right)$$

$$= 8,760 \times 25 = 2,19,000 \text{ MWh}$$

$$\therefore \text{Average annual load} = \frac{\text{no. of units generated per year}}{8,760} = \frac{2,19,000}{8,760} = 25 \text{ MW}$$

$$\therefore \text{Load factor} = \frac{\text{average annual load}}{\text{maximum load}} = \frac{25}{30} = 0.833$$

$$\text{Plant capacity factor} = \frac{\text{Average annual load}}{\text{Rated plant capacity}} = \frac{25}{35} = 0.714$$

$$\text{Utilization factor} = \frac{\text{max. demand}}{\text{rated capacity}} = \frac{30}{35} = 0.857$$

Alternatively,

$$\text{Utilization factor} = \frac{\text{capacity factor}}{\text{load factor}} = \frac{0.714}{0.833} = 0.857$$

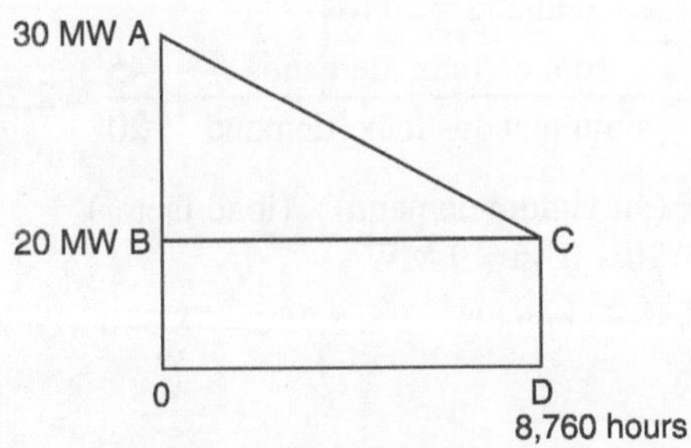

FIG. 1.5 *Load–duration curve*

Example 1.4: Calculate the total annual energy generated, if the maximum demand on a power station is 120 MW and the annual load factor is 50%.

Solution:

Maximum demand on a power station $= 120$ MW

Annual load factor $= 50\%$

$$\text{Load factor} = \frac{\text{energy generated/annum}}{\text{maximum demand} \times \text{hours in a year}}$$

$\therefore$ Energy generated/annum $=$ maximum demand $\times$ LF $\times$ hours in a year

$$= (120 \times 10^3) \times (0.5) \times (24 \times 365) \text{ kWh}$$

$$= 525.6 \times 10^6 \text{ kWh}$$

Example 1.5: Determine the demand factor and the load factor of a generating station, which has a connected load of 50 MW and a maximum demand of 25 MW, the units generated being 40×10^6/annum.

Solution:

Connected load $= 50$ MW

Maximum demand $= 25$ MW

Units generated $= 40 \times 10^6$/annum

$$\text{Demand factor} = \frac{\text{maximum demand}}{\text{connected load}} = \frac{25}{50} = 0.5$$

$$\text{Average demand} = \frac{\text{units generated/annum}}{\text{hours in a year}} = \frac{40 \times 10^6}{8,760} = 4,566.21 \text{ kW}$$

$$\text{Load factor} = \frac{\text{average demand}}{\text{maximum demand}} = \frac{4566.21}{25 \times 10^3} = 18.26\%$$

Example 1.6: Calculate the annual load factor of a 120 MW power station, which delivers 110 MW for 4 hours, 60 MW for 10 hours, and is shut down for the rest of each day. For general maintenance, it is shut down for 60 days per annum.

Solution:

Capacity of power station $= 120$ MW

Power delivered $= 110$ MW for 4 hours

$= 60$ MW for 10 hours

$= 0$ for the rest of each day

And for general maintenance, it is shut down for 60 days per annum.

Energy supplied in 1 day $= (110 \times 4) + (60 \times 10) = 1,040$ MWh

No. of working days in a year $= 365 - 60 = 305$

Energy supplied per year $= 1,040 \times 305 = 3,17,200$ MWh

$$\text{Annual load factor} = \frac{\text{MWh supplied/annum}}{\text{maximum demand in MW} \times \text{working hours}} \times 100$$

$$= \frac{3,17,200}{(120) \times (305 \times 24)} \times 100 = 36.11\%$$

Example 1.7: Customer-connected loads are 10 lamps of 60 W each and two heaters of 1,500 W each. His/her maximum demand is 2 kW. On average, he/she uses 10 lamps, 7 hours a day, and each heater for 5 hours a day. Determine his/her: (i) average load, (ii) monthly energy consumption, and (iii) load factor.

Solution:

Maximum demand $= 2$ kW

Connected load $= 10 \times 60 + 2 \times 1,500 = 3,600$ W

Daily energy consumption $=$ number of lamps used $\times$ wattage of each lamp $\times$ working hours per day $+$ number of heaters $\times$ wattage of each heater $\times$ working hours per day

$$= 10 \times 60 \times 7 + 2 \times 1,500 \times 5$$
$$= 19.2 \text{ kWh}$$

(i)   Average load $= \dfrac{\text{daily energy consumption in kWh}}{24} = \dfrac{19.2}{24} = 0.8$ kW

(ii)   Monthly energy consumption $=$ daily energy consumption $\times$ no. of days in a month

$$= 19.2 \times 30 = 576 \text{ kWh}$$
$$= 576 \text{ kWh}$$

(iii)   Monthly load factor $= \dfrac{\text{monthly energy consumption}}{\text{max. demand} \times 24 \times 30} = \dfrac{576}{2 \times 24 \times 30} = 40\%$

Example 1.8: The maximum demand on a generating station is 20 MW, a load factor of 75%, a plant capacity factor of 50%, and a plant-use factor of 80%. Calculate the following:

(i)   daily energy generated,

(ii)   reserve capacity of the plant,

(iii)   maximum energy that could be produced daily if the plant were in use all the time.

Solution:

Maximum demand, $\text{MD} = 20$ MW

Load factor, $\text{LF} = 75\%$

Power capacity factor $= 50\%$

Plant-use factor $= 80\%$

Average load $= \text{MD} \times \text{LF}$

$$= 20 \times 0.75 = 15 \text{ MW}$$

(i)   Daily energy generated $=$ average load $\times 24 = 15 \times 24 = 360$ MWh

(ii)   Power-station-installed capacity $= \dfrac{\text{average load}}{\text{plant capacity factor}} = \dfrac{15}{0.5} = 30$ MW

Plant reserve capacity $=$ installed capacity $-$ maximum demand

$$= 30 - 20$$
$$= 10 \text{ MW}$$

(iii) The maximum energy that can be produced daily if the plant is running all the time

$$= \frac{\text{actual energy generated}}{\text{plant capacity factor}} = \frac{360}{0.5} = 720 \text{ MWh}$$

Example 1.9: A certain power station's annual load–duration curve is a straight line from 25 to 5 MW (Fig. 1.6). To meet this load, three turbine-generator units, two rated at 15 MW each and one rated at 7.5 MW are installed. Calculate the following:

  (i) installed capacity;

  (ii) plant factor;

  (iii) units generated per annum;

  (iv) utilization factor.

Solution:

  (i) Installed capacity $= 2 \times 15 + 7.5$

$$= 37.5 \text{ MW}$$

  (ii) From the load–duration curve shown in Fig. 1.6,

$$\text{Average demand} = \frac{1}{2}[25 + 5]$$

$$= 15 \text{ MW}$$

$$\therefore \text{ Plant factor} = \frac{\text{average demand}}{\text{plant capacity}} = \frac{15}{37.5} = 0.4$$

  (iii) Units generated per annum $=$ area (in kWh) under load–duration curve

$$= \frac{1}{2}[5,000 + 25,000] \times 8,760 \text{ kWh}$$

$$= 131.4 \times 10^6 \text{ kWh}$$

  (iv) Utilization factor $= \dfrac{\text{maximum demand}}{\text{plant capacity}} = \dfrac{25,000}{37,500} = 66.7\%$

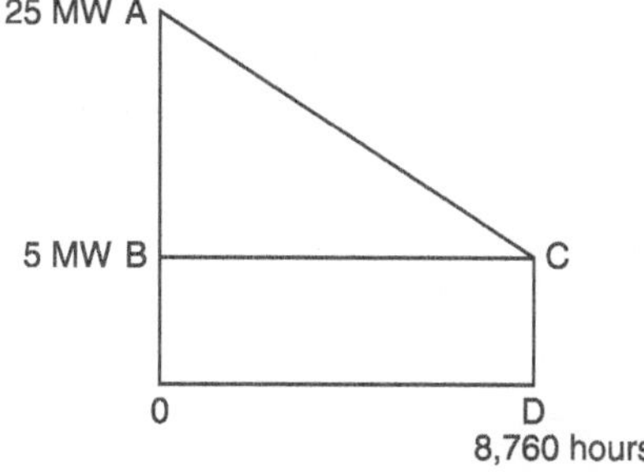

FIG. 1.6 *Load–duration curve*

Example 1.10: A consumer has a connected load of 12 lamps each of 100 W at his/her premises. His/her load demand is as follows:

From midnight to 5 A.M.: 200 W.

5 A.M. to 6 P.M.: no load.

6 P.M. to 7 P.M.: 700 W.

7 P.M. to 9 P.M.: 1,000 W.

9 P.M. to midnight: 500 W.

Draw the load curve and calculate the (i) energy consumption during 24 hours, (ii) demand factor, (iii) average load, (iv) maximum demand, and (v) load factor.

Solution:

From Fig. 1.7,

(i) Electrical energy consumption during the day = area of load curve

$$= 200 \times 5 + 700 \times 1 + 1,000 \times 2 + 500 \times 3$$

$$= 5,200 \text{ Wh}$$

$$= 5.2 \text{ kWh}$$

(ii) Average load $= \dfrac{\text{energy consumed during a day}}{24} = \dfrac{5,200}{24} = 216.7 \text{ W}$

(iii) Demand factor $= \dfrac{1,000}{12 \times 100} = 0.83$

(iv) Maximum demand $= 1,000 \text{ W}$

(v) Load factor $= \dfrac{\text{average load}}{\text{maximum demand}} = \dfrac{216.7}{1,000} = 0.2167 = 21.7\%$

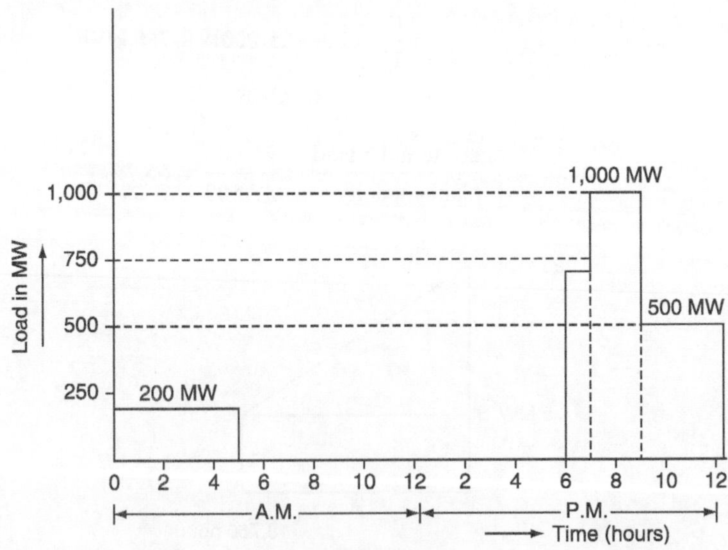

FIG. 1.7 *Load curve*

Example 1.11: Calculate the diversity factor and the annual load factor of a generating station, which supplies loads to various consumers as follows:

Industrial consumer = 2,000 kW;

Commercial establishment = 1,000 kW

Domestic power = 200 kW;

Domestic light = 500 kW

and assume that the maximum demand on the station is 3,000 kW, and the number of units produced per year is 50×10^5.

Solution:

$$\text{Load industrial consumer} = 2,000 \text{ kW}$$
$$\text{Load commercial establishment} = 1,000 \text{ kW}$$
$$\text{Domestic power load} = 200 \text{ kW}$$
$$\text{Domestic lighting load} = 500 \text{ kW}$$
$$\text{Maximum demand on the station} = 3,000 \text{ kW}$$
$$\text{Number of kWh generated per year} = 50 \times 10^5$$

$$\text{Diversity factor} = \frac{2,000 + 1,000 + 200 + 500}{3,000} = 1.233$$

$$\text{Average demand} = \frac{\text{kWh generated/annum}}{\text{hours in a year}}$$

$$= \frac{50 \times 10^5}{8,760} = 570.77 \text{ kW}$$

$$\text{Load factor} = \frac{\text{average load}}{\text{maximum demand}} = \frac{570.77}{3,000} = 19\%$$

Example 1.12: Calculate the reserve capacity of a generating station, which has a maximum demand of 20,000 kW, the annual load factor is 65%, and the capacity factor is 45%.

Solution:

$$\text{Maximum demand} = 20,000 \text{ kW}$$
$$\text{Annual load factor} = 65\%$$
$$\text{Capacity factor} = 45\%$$
$$\text{Energy generated/annum} = \text{maximum demand} \times \text{LF} \times \text{hours in a year}$$

$$= (20,000) \times (0.65) \times (8,760) \text{ kWh} = 113.88 \times 10^6 \text{ kWh}$$

$$\text{Capacity factor} = \frac{\text{units generated/annum}}{\text{plant capacity} \times \text{hours in a year}}$$

$$0.45 = \frac{113.88 \times 10^6}{\text{plant capacity} \times 8,760}$$

$$\therefore \text{Plant capacity} = \frac{113.88 \times 10^6}{0.45 \times 8,760} = 28,888.89 \text{ kW}$$

$$\text{Reserve capacity} = \text{plant capacity} - \text{maximum demand}$$
$$= 28{,}888.89 - 20{,}000 = 8{,}888.89 \text{ kW}$$

Example 1.13: The maximum demand on a power station is 600 MW, the annual load factor is 60%, and the capacity factor is 45%. Find the reserve capacity of the plant.

Solution:

$$\text{Utilization factor} = \frac{\text{capacity factor}}{\text{load factor}} = \frac{0.45}{0.6} = 0.75$$

$$\text{Plant capacity} = \frac{\text{maximum demand}}{\text{utilization factor}} = \frac{600}{0.75} = 800 \text{ MW}$$

$$\text{Reserve capacity} = \text{plant capacity} - \text{maximum demand}$$
$$= 800 - 600$$
$$= 200 \text{ MW}$$

Example 1.14: A power station's maximum demand is 50 MW, the capacity factor is 0.6, and the utilization factor is 0.85. Calculate the following: (i) reserve capacity and (ii) annual energy produced.

Solution:

$$\text{Energy generated/annum} = \text{maximum demand} \times \text{load factor} \times \text{hours in a year}$$
$$= (50 \times \text{LF} \times 8{,}760) \text{ MWh}$$

$$\text{Load factor} = \frac{\text{capacity factor}}{\text{utilization factor}} = \frac{0.6}{0.85} = 0.7058$$

$$\text{Energy generated/annum} = 50 \times 0.706 \times 8{,}760$$
$$= 3{,}09{,}228 \text{ MWh} = 0.3 \times 10^6 \text{ MWh}$$

$$\text{Plant capacity} = \frac{\text{maximum demand}}{\text{utilization factor}} = \frac{50}{0.85} = 58.82 \text{ MW}$$

$$\text{Reserve capacity} = \text{plant capacity} - \text{maximum demand}$$
$$= 58.82 - 50$$
$$= 8.82 \text{ MW}$$

Example 1.15: A power station is to feed four regions of load whose peak loads are 12, 7, 10, and 8 MW. The diversity factor at the station is 1.4 and the average annual load factor is 65%. Determine the following: (i) maximum demand on the station, (ii) annual energy supplied by the station, and (iii) suggest the installed capacity.

Solution:

$$\text{(i)} \quad \text{Maximum demand on station} = \frac{\text{sum of maximum demand of the regions}}{\text{diversity factor}}$$

$$= \frac{12 + 7 + 10 + 8}{1.4}$$

$$= 26.43 \text{ MW}$$

(ii) Units generated/annum = max. demand × LF × house in a year

$$= (26.43 \times 10^3) \times 0.65 \times 8{,}760 \text{ kWh}$$
$$= 150.49 \times 10^6 \text{ kMh}$$

(iii) The installed capacity of the station should be 15% to 20% more than the maximum demand in order to meet the future growth of load.

Taking the installed capacity to be 20% more than the maximum demand,

Installed capacity = 1.2 × max. demand

$$= 1.2 \times 26.43$$
$$= 31.716 \cong 32 \text{ MW}$$

1.7 LOAD FORECASTING

Electrical energy cannot be stored. It has to be generated whenever there is a demand for it. It is, therefore, imperative for the electric power utilities that the load on their systems should be estimated in advance. This estimation of load in advance is commonly known as load forecasting. It is necessary for power system planning.

Power system expansion planning starts with a forecast of anticipated future load requirements. The estimation of both demand and energy requirements is crucial to an effective system planning. Demand predictions are used for determining the generation capacity, transmission, and distribution system additions, etc. Load forecasts are also used to establish procurement policies for construction capital energy forecasts, which are needed to determine future fuel requirements. Thus, a good forecast, reflecting the present and future trends, is the key to all planning.

In general, the term forecast refers to projected load requirements determined using a systematic process of defining future loads in sufficient quantitative detail to permit important system expansion decisions to be made. Unfortunately, the consumer load is essentially uncontrollable although minor variations can be affected by frequency control and more drastically by load shedding. The variation in load does exhibit certain daily and yearly pattern repetitions, and an analysis of these forms the basis of several load-prediction techniques.

1.7.1 Purpose of load forecasting

  (i) For proper planning of power system;

 (ii) For proper planning of transmission and distribution facilities;

(iii) For proper power system operation;

(iv) For proper financing;

 (v) For proper manpower development;

(vi) For proper grid formation;

(vii) For proper electrical sales.

(i) *For proper planning of power system*

- To determine the potential need for additional new generating facilities;
- To determine the location of units;
- To determine the size of plants;

- To determine the year in which they are required;
- To determine that they should provide primary peaking capacity or energy or both;
- To determine whether they should be constructed and owned by the Central Government or State Government or Electricity Boards or by some other autonomous corporations.

(ii) *For proper planning of transmission and distribution facilities*

For planning the transmission and distribution facilities, the load forecasting is needed so that the right amount of power is available at the right place and at the right time. Wastage due to misplanning like purchase of equipment, which is not immediately required, can be avoided.

(iii) *For proper power system operation*

Load forecast based on correct values of demand and diversity factor will prevent overdesigning of conductor size, etc. as well as overloading of distribution transformers and feeders. Thus, they help to correct voltage, power factor, etc. and to reduce the losses in the distribution system.

(iv) *For proper financing*

The load forecasts help the Boards to estimate the future expenditure, earnings, and returns and to schedule its financing program accordingly.

(v) *For proper manpower development*

Accurate load forecasting annually reviewed will come to the aid of the Boards in their personnel and technical manpower planning on a long-term basis. Such a realistic forecast will reduce unnecessary expenditure and put the Boards' finances on a sound and profitable footing.

(vi) *For proper grid formation*

Interconnections between various state grids are now becoming more and more common and the aim is to have fully interconnected regional grids and ultimately even a super grid for the whole country. These expensive high-voltage interconnections must be based on reliable load data, otherwise the generators connected to the grid may frequently fall out of step causing power to be shut down.

(vii) *For proper electrical sales*

In countries, where spinning reserves are more, proper planning and the execution of electrical sales program are aided by proper load forecasting.

1.7.2 Classification of load forecasting

The load forecasting can be classified as: (i) demand forecast and (ii) energy forecast.

(i) *Demand forecast*

This is used to determine the capacity of the generation, transmission, and distribution system additions. Future demand can be predicted on the basis of fast rate of growth of demand from past history and government policy. This will give the expected rate of growth of load.

(II) *Energy forecast*

This is used to determine the type of facilities required, i.e., future fuel requirements.

1.7.3 Forecasting procedure

Depending on the time period of interest, a specific forecasting procedure may be classified as:

- Short-term.
- Medium (intermediate)-term.
- Long-term technique.

(1) *Short-term forecast*

For day-to-day operation, covering one day or a week, short-term forecasting is needed in order to commit enough generating capacity formatting the forecasting demand and for maintaining the required spinning reserve. Hence, it is usually done 24 hours ahead when the weather forecast for the following day becomes available from the meteorological office. This mostly consists of estimating the weather-dependent component and that due to any special event or festival because the base load for the day is already known.

The power supply authorities can build up a weather load model of the system for this purpose or can consult some tables. The final estimate is obviously done after accounting the transmission and distribution losses of the system. In addition to the prediction of hourly values, a short-term load forecasting (STLF) is also concerned with forecasting of daily peak-system load, system load at certain times of a day, hourly values of system energy, and daily and weekly system energy.

Applications of STLF are mainly:

- To drive the scheduling functions that decide the most economic commitment of generation sources.
- To access the power system security based on the information available to the dispatchers to prepare the necessary corrective actions.
- To provide the system dispatcher with the latest weather predictions so that the system can be operated both economically and reliably.

(2) *Long-term forecast*

This is done for 1–5 years in advance in order to prepare maintenance schedules of the generating units, planning future expansion of the generating capacity, enter into an agreement for energy interchange with the neighboring utilities, etc. Basically, two approaches are available for this purpose and are discussed as follows.

(a) *Peak load approach*

In this case, the simplest approach is to extrapolate the trend curve, which is obtained by plotting the past values of annual peaks against years of operation. The following analytical functions can be used to determine the trend curve.

(1)   Straight line, $Y = a + bx$

(2)   Parabola, $Y = a + bx + cx^2$

(3)   S-curve, $Y = a + bx + cx^2 + dx^3$

(4) Exponential, $Y = ce^{dx}$

(5) Gompertz, $\log_e Y = a + ce^{dx}$

In the above, Y represents peak loads and x represents time in years. The most common method of finding coefficients a, b, c, and d is the least squares curve-fitting technique.

The effect of weather conditions can be ignored on the basis that weather conditions, as in the past, are to be expected during the period under consideration but the effect of the change in the economic condition should be accommodated by including an economic variable when extrapolating the trend curve. The economic variable may be the predicted national income, gross domestic product, etc.

(b) *Energy approach*

Another method is to forecast annual energy sales to different classes of customers like residential, commercial, industrial, etc., which can then be converted to annual peak demand using the annual load factor. A detailed estimation of factors such as rate of house building, sale of electrical appliances, growth in industrial and commercial activities are required in this method. Forecasting the annual load factor also contributes critically to the success of the method. Both these methods, however, have been used by the utilities in estimating their long-term system load.

KEY NOTES

- A *load curve* is a plot of the load demand (on the y-axis) versus the time (on the x-axis) in the chronological order.

- The **load–duration curve** is a plot of the load demands (in units of power) arranged in a descending order of magnitude (on the y-axis) and the time in hours (on the x-axis).

- In the operation of hydro-electric plants, it is necessary to know the amount of energy between different load levels. This information can be obtained from the load–duration curve.

- The **integrated load–duration curve** is also the plot of the cumulative integration of area under the load curve starting at zero loads to the particular load.

- A base-load station operates at a high-load factor while the peak load plant operates at a-low load factor.

- **Demand factor** is the ratio of the maximum demand to the connected load.

- **Load factor** is the ratio of the average demand to the maximum demand. Higher the load factor of the power station, lesser will be the cost per unit generated.

- **Diversity factor** is the ratio of the sum of the maximum demands of a group of consumers and the simultaneous maximum demand of the group of consumers.

- **Base load** is the unvarying load that occurs almost the whole day on the station.

- **Peak load** is the various peak demands of load over and above the base load of the station.

SHORT QUESTIONS AND ANSWERS

(1) What is meant by connected load?

It is the sum of the ratings of the apparatus installed on a consumer's premises.

(2) Define the maximum demand.

It is the maximum load used by a consumer at any time.

(3) Define the demand factor.

The ratio of the maximum demand to the connected load is called the demand factor.

(4) Define the average load.

If the number of kWh supplied be a station in one day is divided by 24 hours, then the value obtained is known as the daily average load.

(5) Define the load factor.

It is the ratio of the average demand to the maximum demand.

(6) Define the diversity factor.

It is the ratio of the sum of the maximum demands of a group of consumers to the simultaneous maximum demand of the group of consumers.

(7) Define the plant capacity.

It is the capacity or power for which a plant or station is designed.

(8) Define the utilization factor.

It is the ratio of kWh generated to the product of the plant capacity and the number of hours for which the plant was in operation.

(9) What is meant by base load?

It is the unvarying load that occurs almost the whole day on the station.

(10) What is meant by peak load?

It is the various peak demands of load over and above the base load of the station.

(11) What is meant by load curve?

A load curve is a plot of the load demand versus the time in the chronological order.

(12) What is meant by load–duration curve?

The load–duration curve is a plot of the load demands arranged in a descending order of magnitude versus the time in hours.

MULTIPLE-CHOICE QUESTIONS

(1) In order to have a low cost of electrical generation,

   (a) The load factor and diversity factor are high.

   (b) The load factor should be low but the diversity factor should be high.

   (c) The load factor should be high but the diversity factor should be low.

   (d) The load factor and the diversity factor should be low.

(2) A power plant having maximum demand more than the installed capacity will have utilization factor:

   (a) Less than 100%.

   (b) Equal to 100%.

   (c) More than 100%.

   (d) None of these.

(3) The choice of number and size of units in a station are governed by best compromise between:

   (a) A plant load factor and capacity factor.

   (b) Plant capacity factor and plant-use factor.

   (c) Plant load factor and use factor.

   (d) None of these.

(4) If a plant has zero reserve capacity, the plant load factor always:

   (a) Equals plant capacity factor.

   (b) Is greater than plant capacity factor.

   (c) Is less than plant capacity factor.

   (d) None of these.

(5) If some reserve is available in a power plant,

   (a) Its use factor is always greater than its capacity factor.

   (b) Its use factor equals the capacity factor.

   (c) Its use factor is always less than its capacity factor.

   (d) None of these.

(6) A higher load factor means:

   (a) Cost per unit is less.

   (b) Less variation in load.

   (c) The number of units generated are more.

   (d) All of these.

(7) The maximum demand of two power stations is the same. If the daily load factors of the stations are 10 and 20%, then the units generated by them are in the ratio:

   (a) 2:1.

   (b) 1:2.

   (c) 3:3.

   (d) 1:4.

(8) A plant had an average load of 20 MW when the load factor is 50%. Its diversity factor is 20%. The sum of max. demands of all loads amounts to:

   (a) 12 MW.

   (b) 8 MW.

   (c) 6 MW.

   (d) 4 MW.

(9) A peak load station:

   (a) Should have a low operating cost.

   (b) Should have a low capital cost.

   (c) Can have a operating cost high.

   (d) (a) and (c).

   (e) (b) and (c).

(10) Two areas A and B have equal connected loads; however the load diversity in area A is more than in B, then:

   (a) Maximum demand of two areas is small.

   (b) Maximum demand of A is greater than the maximum demand of B.

   (c) The maximum demand of B is greater than the maximum demand of A.

   (d) The maximum demand of A more or less than that of B.

(11) The area under the daily load curve gives

   (a) The number of units generated in the day.

   (b) The average load of the day.

   (c) The load factor of the day.

   (d) The number of units generated in the year.

(12) The annual peak load on a 60-MW power station is 50 MW. The power station supplies loads having average demands of 9, 10, 17, and 20 MW. The annual load factor is 60%. The average load on the plant is:

   (a) 4,000 kW.

   (b) 30,000 kW.

   (c) 2,000 kW.

   (d) 1,000 kW.

(13) A generating station has a connected load of 40 MW and a maximum demand of 20 MW. The demand factor is:

   (a) 0.7.

   (b) 0.6.

   (c) 0.59.

   (d) 0.4.

(14) A 100 MW power plant has a load factor of 0.5 and a utilization factor of 0.2. Its average demand is:

   (a) 10 MW.

   (b) 5 MW.

   (c) 7 MW.

   (d) 6 MW.

(15) The value of the demand factor is always:

   (a) Less than one.

   (b) Equal to one.

   (c) Greater than one.

   (d) None of these.

(16) If capacity factor $=$ load factor, then:

   (a) Utilization factor is zero.

   (b) Utilization capacity is non-zero.

   (c) Utilization factor is equal to one.

   (d) None of these.

(17) If capacity factor $=$ load factor, then the plant's

   (a) Reserve capacity is maximum.

   (b) Reserve capacity is zero.

   (c) Reserves capacity is less.

   (d) None of these.

(18) Installed capacity of power plant is:

   (a) More than the maximum demand.

   (b) Less than the maximum demand.

   (c) Equal to the maximum demand.

   (d) Both (a) and (c).

(19) In an interconnected system, diversity factor determining:

   (a) Decreases.

   (b) Increases.

   (c) Zero.

   (d) None of these.

(20) The knowledge of diversity factor helps in determining:

(a) Plant capacity.

(b) Reserve capacity.

(c) Maximum demand.

(d) Average demand.

(21) A power station has an installed capacity of 300 MW. Its capacity factor is 50% and its load factor is 75%. Its maximum demand is:

(a) 100 MW.

(b) 150 MW.

(c) 200 MW.

(d) 250 MW.

(22) The connected load of a consumer is 2 kW and his/her maximum demand is 1.5 kW. The load factor of the consumer is:

(a) 0.75.

(b) 0.375.

(c) 1.33.

(d) none of these.

(23) The maximum demand of a consumer is 2 kW and his/her daily energy consumption is 20 units. His/her load factor is:

(a) 10.15%.

(b) 41.6%.

(c) 50%.

(d) 52.6%.

(24) In a power plant, a reserve-generating capacity, which is not in service but in operation is known as:

(a) Hot reserve.

(b) Spinning reserve.

(c) Cold reserve.

(d) Firm power.

(25) The power intended to be always available is known as:

(a) Hot reserve.

(b) Spinning reserve.

(c) Cold reserve.

(d) Firm power.

(26) In a power plant, a reserve-generating capacity, which is in service but not in operation is:

(a) Hot reserve.

(b) Spinning reserve.

(c) Cold reserve.

(d) Firm power.

(27) Which of the following is a correct factor?

(a) Load factor = capacity × utilization factor.

(b) Utilization factor = capacity factor × load factor.

(c) Utilization factor = load factor/utilization factor.

(d) Capacity factor = load factor × utilization factor.

(28) If the rated plant capacity and maximum load of generating station are equal, then:

(a) Load factor is 1.

(b) Capacity factor is 1.

(c) Load factor and capacity factor are equal.

(d) Utilization factor is poor.

(29) The capital cost of plant depends on:

(a) Total installed capacity only.

(b) Total number of units only.

(c) Both (a) and (b).

(d) None of these.

(30) The reserve capacity in a system is generally equal to:

(a) Capacity of the largest generating unit.

(b) Capacity of two largest generating units.

(c) The total generating capacity.

(d) None of the above.

(31) The maximum demand of a consumer is 5 kW and his/her daily energy consumption is 24 units. His/her % load factor is:

(a) 5.

(b) 20.

(c) 24.

(d) 48.

(32) If load factor is poor, then:

(a) Electric energy produced is small.

(b) Charge per kWh is high.

(c) Fixed charges per kWh is high.

(d) All of the above.

(33) If a generating station had maximum loads for a day at 100 kW and a load factor of 0.2, its generation in that day was:

(a) 8.64 MWh.

(b) 21.6 units.

(c) 21.6 units.

(d) 2,160 kWh.

(34) The knowledge of maximum demand is important as it helps in determining:

(a) Installed capacity of the plant.

(b) Connected load of the plant.

(c) Average demand of the plant.

(d) Either (a) or (b).

(35) A power station is connected to 4.5 and 6 kW. Its daily load factor was calculated as 0.2, where its generation on that day was 24 units. Calculate the demand factor.

(a) 2.6.

(b) 3.1.

(c) 3.0.

(d) 0.476.

(36) A 50-MW power station had produced 24 units in a day when its maximum demand was 50 MW. Its plant load factor and capacity factor that day in % were:

(a) 1 and 2.

(b) 2 and 3.

(c) 2 and 2.

(d) 4 and 3.

(37) Load curve of a power generation station is always:

(a) Negative.

(b) Zero slope.

(c) Positive.

(d) Any combination of (a), (b), and (c).

(38) Load curve helps in deciding the:

(a) Total installed capacity of the plant.

(b) Size of the generating units.

(c) Operating schedule of the generating units.

(d) All of the above.

(39) The load factor for domestic loads may be taken:

(a) about 85%.

(b) 50−60%.

(c) 25−50%.

(d) 20−15%.

REVIEW QUESTIONS

(1) Explain the significance of the daily load curve.

(2) Discuss the difference between the load curve and the load–duration curve.

(3) Explain the differences in operations of peak load and base-load stations.

(4) Explain the significance of the load factor and the diversity factor.

(5) Define the following:

(i) Load factor,

(ii) Demand factor,

(iii) Diversity factor,

(iv) Plant capacity factor, and

(v) Utilization factor.

(6) Explain the load forecasting procedures.

PROBLEMS

(1) Calculate diversity factor and annual load factor of a generating station that supplies loads to various consumers as follows:

Industrial consumer = 1,500 kW;
Commercial establishment = 7,500 kW
Domestic power = 100 kW;
Domestic light = 400 kW

In addition, assume that the maximum demand on the station is 2,500 kW and the number of units produced per year is 40×10^5 kWh.

(2) A power station is to feed four regions of load whose peak loads are 10, 5, 14, and 6 MW, respectively. The diversity factor at the station is 1.3 and the average annual load factor is 60%. Determine the following: (i) maximum demand on the station, (ii) annual energy supplied by the station, and (iii) suggest the installed capacity.

(3) A certain power station's annual load–duration curve is a straight line from 20 to 7 MW. To meet this load, three turbine-generator units, two rated at 12 MW each and one rated at 8 MW are installed. Calculate the following:

(i) installed capacity,

(ii) plant factor,

(iii) units generated per annum,

(iv) utilization factor.

2 Economic Load Dispatch-I

OBJECTIVES

After reading this chapter, you should be able to:

- study the different characteristics of steam and hydro-power generation units
- know the meaning of economical load dispatch
- develop the mathematical model for economical load dispatch
- discuss the different computational methods for optimization

2.1 INTRODUCTION

Power systems need to be operated economically to make electrical energy cost-effective to the consumer in the face of constantly rising prices of fuel, wages, salaries, etc. New generator-turbine units added to a steam power plant operate more efficiently than other older units. The contribution of newer units to the generation of power will have to be more. In the operation of power systems, the contribution from each load and from each unit within a plant must be such that the cost of electrical energy produced is a minimum.

2.2 CHARACTERISTICS OF POWER GENERATION (STEAM) UNIT

In analyzing the economic operation of a thermal unit, input–output modeling characteristics are significant. For this function, consider a single unit consisting of a boiler, a turbine, and a generator as shown in Fig. 2.1. This unit has to supply power not only to the load connected to the power system but also to the local needs for the auxiliaries in the station, which may vary from 2% to 5%. The power requirements for station auxiliaries are necessary to drive boiler feed pumps, fans and condenser circulating water pumps, etc.

The total input to the thermal unit could be British thermal unit (Btu)/hr or Cal/hr in terms of heat supplied or Rs./hr in terms of the cost of fuel (coal or gas). The total output of the unit at the generator bus will be either kW or MW.

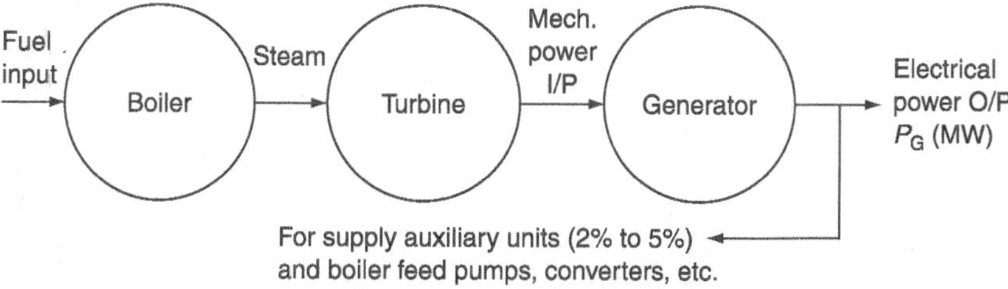

FIG. 2.1 *Thermal generation system*

2.3 SYSTEM VARIABLES

To analyze the power system network, there is a need of knowing the system variables. They are:

(i) Control variables.

(ii) Disturbance variables.

(iii) State variables.

2.3.1 Control variables (P_G and Q_G)

The real and reactive-power generations are called control variables since they are used to control the state of the system.

2.3.2 Disturbance variables (P_D and Q_D)

The real and reactive-power demands are called demand variables since they are beyond the system control and are hence considered as uncontrolled or disturbance variables.

2.3.3 State variables (V and δ)

The bus voltage magnitude V and its phase angle δ dispatch the state of the system. These are dependent variables that are being controlled by the control variables.

2.4 PROBLEM OF OPTIMUM DISPATCH—FORMULATION

Scheduling is the process of allocation of generation among different generating units. Economic scheduling is a cost-effective mode of allocation of generation among the different units in such a way that the overall cost of generation should be minimum. This can also be termed as an optimal dispatch.

Let the total load demand on the station $= P_D$ and the total number of generating units $= n$.

The optimization problem is to allocate the total load P_D among these 'n' units in an optimal way to reduce the overall cost of generation.

Let P_{G_1}, P_{G_2}, P_{G_3}, ..., P_{G_n} be the power generated by each individual unit to supply a load demand of P_D.

To formulate this problem, it is necessary to know the '*input–output characteristics of each unit*'.

2.5 INPUT–OUTPUT CHARACTERISTICS

The idealized form of input–output characteristics of a steam unit is shown in Fig. 2.2. It establishes the relationship between the energy input to the turbine and the energy output from the electrical generator. The input to the turbine shown on the ordinate may be either in terms of the heat energy requirement, which is generally measured in Btu/hr or kCal/hr or in terms of the total cost of fuel per hour in Rs./hr. The output is normally the net electrical power output of that steam unit in kW or MW.

In practice, the curve may not be very smooth, and from practical data, such an idealized curve may be interpolated. The steam turbine-generating unit curve consists of minimum and maximum limits in operation, which depend upon the steam cycle used, thermal characteristics of material, the operating temperature, etc.

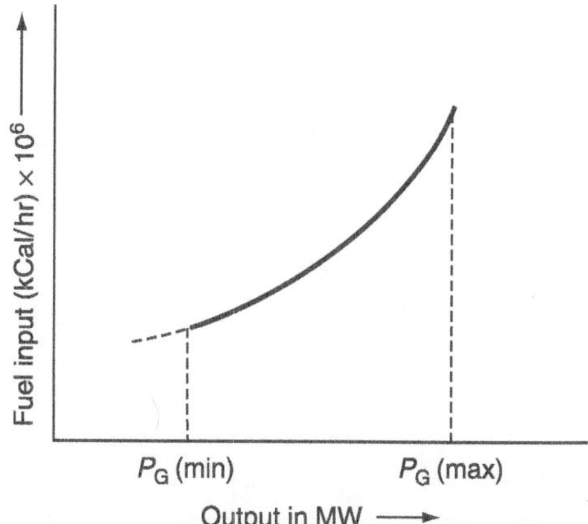

FIG. 2.2   *Input–output characteristic of a steam unit*

2.5.1 Units of turbine Input

In terms of heat, the unit is 10^6 kCal/hr (or) Btu/hr or in terms of the amount of fuel, the unit is tons of fuel/hr, which becomes millions of kCal/hr.

2.6 COST CURVES

To convert the input–output curves into cost curves, the fuel input per hour is multiplied with the cost of the fuel (expressed in Rs./million kCal).

$$\text{i.e.,} \quad \frac{\text{kCal} \times 10^6}{\text{hr}} \times \text{Rs./million kCal}$$

$$= \text{million kCal/hr} \times \text{Rs./million kCal}$$

$$= \text{Rs./hr}$$

2.7 INCREMENTAL FUEL COST CURVE

From the input–output curves, the incremental fuel cost (IFC) curve can be obtained.

The IFC is defined as the ratio of a small change in the input to the corresponding small change in the output.

$$\text{Incremental fuel cost} = \frac{\Delta \text{ input}}{\Delta \text{ output}} = \frac{\Delta F}{\Delta P_G}$$

where Δ represents small changes.

As the Δ quantities become progressively smaller, it is seen that the IFC is $\dfrac{d(\text{input})}{d(\text{output})}$ and is expressed in Rs./MWh. A typical plot of the IFC versus output power is shown in Fig. 2.3(a).

The incremental cost curve is obtained by considering the change in the cost of generation to the change in real-power generation at various points on the input–output curves, i.e., slope of the input–output curve as shown in Fig. 2.3(b).

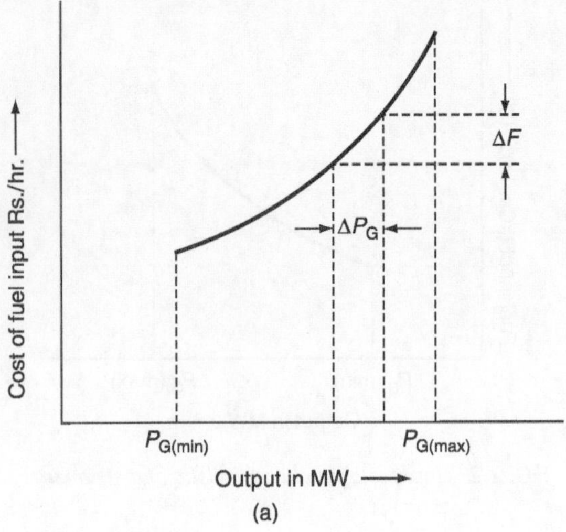

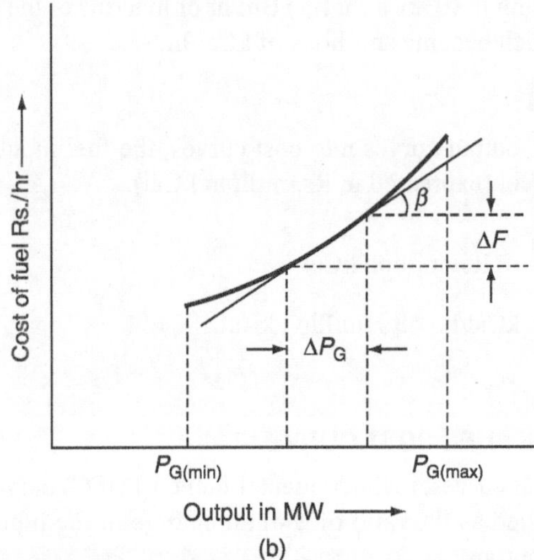

FIG. 2.3   *(a) Incremental cost curve; (b) Incremental fuel cost characteristic in terms of the slope of the input–output curve*

The IFC is now obtained as

$(IC)_i$ = slope of the fuel cost curve

i.e., $\tan \beta = \dfrac{\Delta F}{\Delta P_G}$ in Rs./MWh

The IFC (IC) of the ith thermal unit is defined, for a given power output, as the limit of the ratio of the increased cost of fuel input (Rs./hr) to the corresponding increase in power output (MW), as the increasing power output approaches zero.

$$\text{i.e.,} \quad (IC)_i = P_{G_i} \overset{Lt}{\underset{\rightarrow}{}} 0 \, \frac{\partial F_i}{\partial P_{G_i}}$$

$$= \frac{dF_i}{dP_{G_i}} \quad \text{(or)}$$

$$(IC)_i = \frac{dC_i}{dP_{G_i}} \left[\because \frac{dF_i}{dP_{G_i}} = \frac{dC_i}{dP_{G_i}} = \text{Incremental fuel cost of the } i^{\text{th}} \text{ unit} \right]$$

where C_i is the cost of fuel of the ith unit and P_{G_i} is the power generation output of that i^{th} unit.

Mathematically, the IFC curve expression can be obtained from the expression of the cost curve.

Cost-curve expression,

$$C_i = \frac{1}{2} a_i P_{G_i}^2 + b_i P_{G_i} + d_i \quad \text{(Second-degree polynomial)}$$

The IFC,

$$\frac{dC_i}{dP_{G_i}} = (IC)_i = a_i P_{G_i} + b_i \quad \text{(linear approximation)} \quad \text{for all } i = 1, 2, 3, \ldots, n$$

where $\dfrac{dC_i}{dP_{G_i}}$ is the ratio of incremental fuel energy input in Btu to the incremental energy output in kWh, which is called '*the incremental heat rate*'.

The fuel cost is the major component and the remaining costs such as maintenance, salaries, etc. will be of very small percentage of fuel cost; hence, the IFC is very significant in the economic loading of a generating unit.

2.8 HEAT RATE CURVE

The heat rate characteristic obtained from the plot of the net heat rate in Btu/kWh or kCal/kWh versus power output in kW is shown in Fig. 2.4.

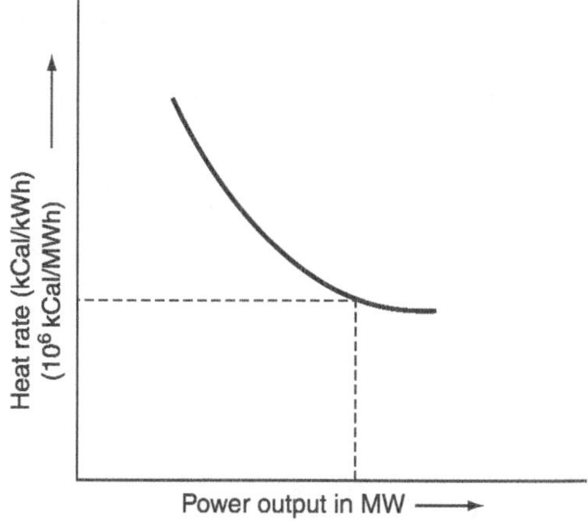

FIG. 2.4   *Heat rate curve*

The thermal unit is most efficient at a minimum heat rate, which corresponds to a particular generation P_G. The curve indicates an increase in heat rate at low and high power limits.

Thermal efficiency of the unit is affected by the following factors: condition of steam, steam cycle used, re-heat stages, condenser pressure, etc.

2.9 INCREMENTAL EFFICIENCY

The reciprocal of the incremental fuel rate or heat rate, which is defined as the ratio of output energy to input energy, gives a measure of fuel efficiency for the input.

$$\text{i.e., } \quad \text{Incremental efficiency} = \frac{\text{output}}{\text{input}} = \frac{dP_G}{dC}$$

2.10 NON-SMOOTH COST FUNCTIONS WITH MULTIVALVE EFFECT

For large steam turbine generators, the input–output characteristics are shown in Fig. 2.5(a).

Large steam turbine generators will have a number of steam admission valves that are opened in sequence to obtain an ever-increasing output of the unit. Figures 2.5(a) and (b) show input–output and incremental heat rate characteristics of a unit with four valves. As the unit loading increases, the input to the unit increases and thereby the incremental heat rate decreases between the opening points for any two valves. However, when a valve is first opened, the throttling losses increase rapidly and the incremental heat rate rises suddenly. This gives rise to the discontinuous type of characteristics in order to schedule the steam unit, although it is usually not done. These types of input–output characteristics are non-convex; hence, the optimization technique that requires convex characteristics may not be used with impunity.

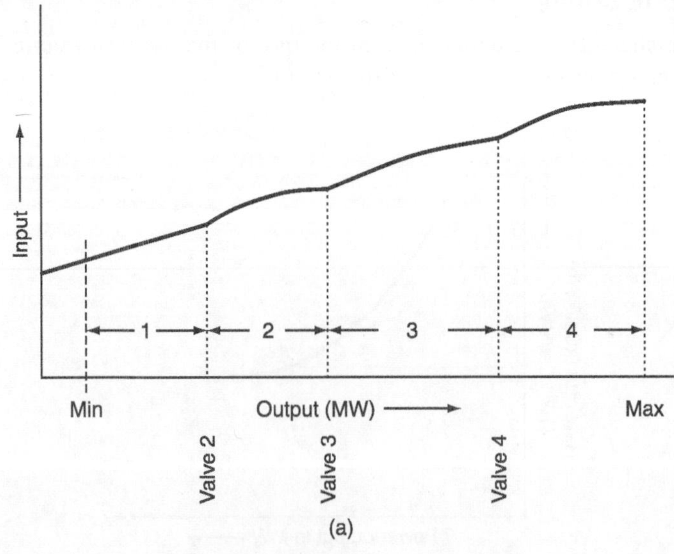

FIG. 2.5   (*Continued on next page*)

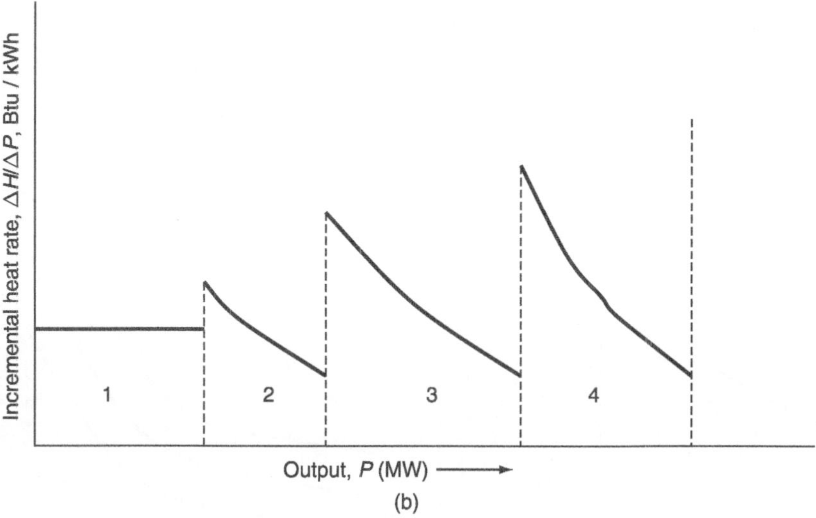

FIG. 2.5   *Characteristics of a steam generator unit with multivalve effect: (a) Input–output characteristic and (b) incremental heat rate characteristic*

2.11 NON-SMOOTH COST FUNCTIONS WITH MULTIPLE FUELS

Generally, a piece-wise quadratic function is used to represent the input–output curve of a generator with multiple fuels. Figure 2.6 represents the incremental heat rate characteristics of a generator with multiple fuels.

2.12 CHARACTERISTICS OF A HYDRO-POWER UNIT

A simple hydro-power plant is shown in Fig. 2.7(a).

The input–output characteristics of a hydro-power unit as shown in Fig. 2.7(b) can be obtained in the same way as for the steam units assuming the water head to be constant. The ordinates are water input or discharge (m^3/s) versus output power (kW or MW).

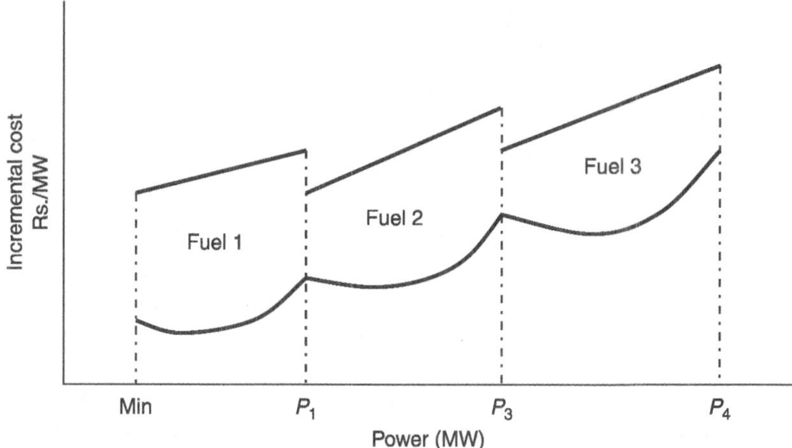

FIG. 2.6   *Incremental heat-rate characteristics of a steam generator with multiple fuels*

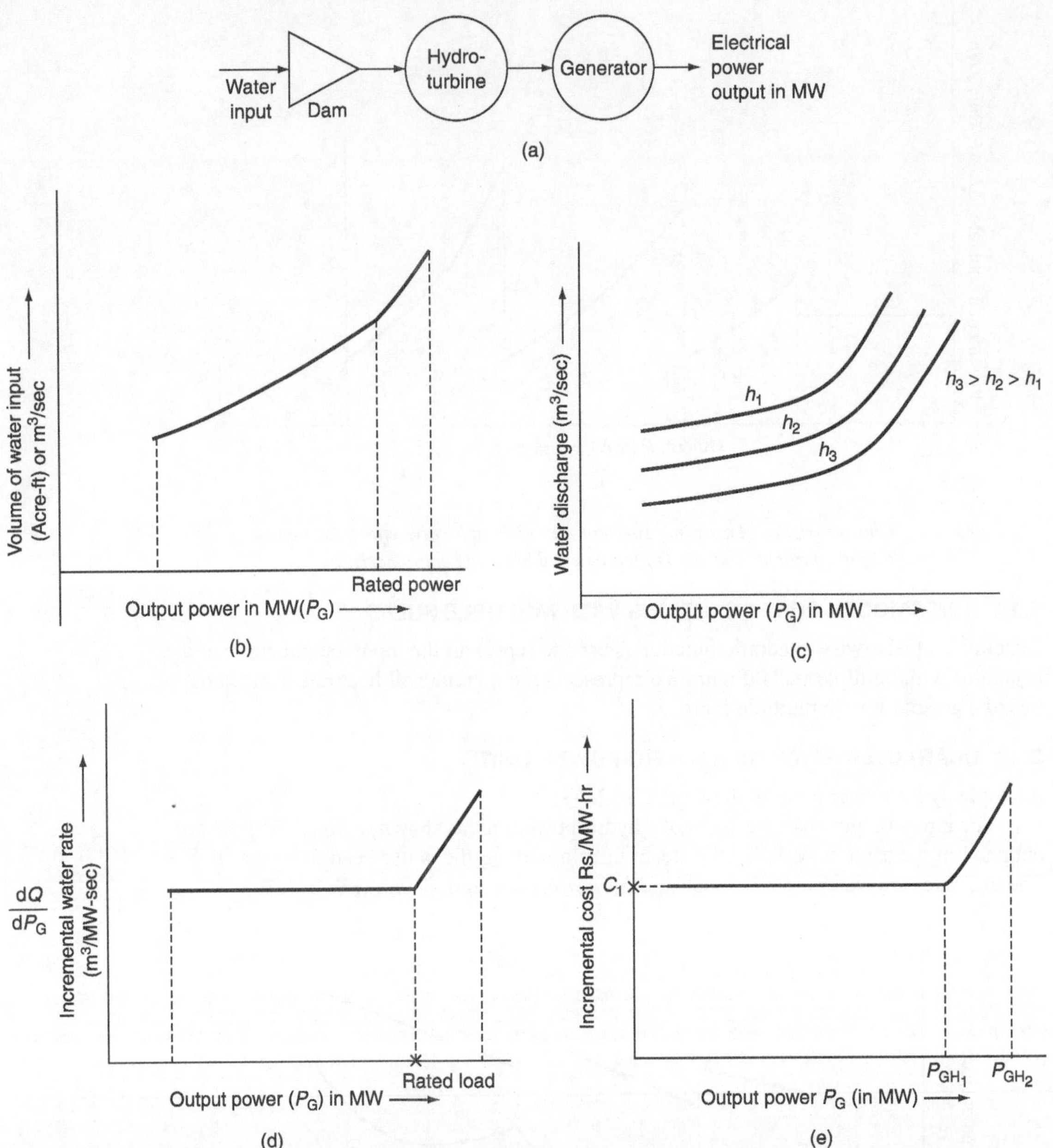

FIG. 2.7   *(a) A typical system of a hydro-power plant; (b) Input–output characteristics of a hydro-unit; (c) Effect of water head on water discharge; (d) Incremental water rate characteristic of a hydro-unit; (e) Incremental cost characteristic of a hydro-unit*

From Fig. 2.7(b), it is observed that there is a linear water requirement upto the rated load and after that greater discharge is needed to meet the increased load demand such that the efficiency of the unit decreases.

2.12.1 Effect of the water head on discharge of water for a hydro-unit

Figure 2.7(c) shows the effect of the water head on water discharge. It is observed that when the head of the water falls, the input–output characteristic of a hydro-power plant moves vertically upwards, such that a higher discharge of water is needed for the same power generation. The reverse will happen when the head rises.

2.12.2 Incremental water rate characteristics of hydro-units

A typical incremental water rate characteristic is shown in Fig. 2.7(d). It can be obtained from the input–output characteristic of a hydro-unit as shown in Fig. 2.7(b).

From Fig. 2.7(d), it is seen that the curve is a straight horizontal line upto the rated load indicating a constant slope and after that it rises rapidly. When the load increases more than the rated, more units will be put into operation (service).

2.12.3 Incremental cost characteristic of a hydro-unit

Actually, the input of a hydro-plant is not dependent on the cost. But the input water flow costs are due to the capacity of storage, requirement of water for the agricultural purpose, and running of the plant during off season (dry season). The artificial storage requirement (i.e., cost of construction of dams, canals, conduits, gates, penstocks, etc.) imposes a cost on the water input to the turbine as well as the cost of control on the water output from the turbine due to agricultural need.

The incremental cost characteristic can be obtained from the incremental water rate characteristic by multiplying it with cost of water in Rs./m^3.

$$\text{Incremental cost} = (\text{Incremental water rate}) \times \text{cost of water in Rs./m}^3$$
$$= \text{m}^3/\text{MWh} \times \text{Rs./m}^3$$
$$= \text{Rs./MWh}$$

The incremental cost characteristic (or) incremental production cost characteristic is shown in Fig. 2.7(e).

The analytical expression of an incremental cost characteristic is

$$(IC)_H = C_1, \, (0 \leq P_{GH} \leq P_{GH_1})$$

$$= m\,P_{GH} + C_1, \, (P_{GH_1} \leq P_{GH} \leq P_{GH_2})$$

where P_{GH} is the power generation of a hydro-unit and m is the slope of characteristic between P_{GH_1} and P_{GH_2}.

2.12.4 Constraints of hydro-power plants

The following constraints are generally used in hydro-power plants.

(i) *Water storage constraints*

Let γ_j be the storage volume at the end of interval j, $\gamma_{\min} \leq \gamma_j \leq \gamma_{\max}$.

(ii) *Water spillage constraints*

Even though there may be circumstances where allowing water spillage (S_{p_j}) > 0 for some interval j, prohibition of spillage is assumed so that all $S_{p_j} = 0$ might reduce the cost of operation of a thermal plant.

(iii) Water discharge flow constraints

The discharge flow may be constrained both in rate and in total as

$$q_{min} \leq q_j \geq q_{max} \quad \text{and} \quad \sum_{j=1}^{j_{max}} n_j q_j = q_{total}$$

2.13 INCREMENTAL PRODUCTION COSTS

The incremental production cost of a given unit is made up of the IFC plus the incremental cost of items such as labor, supplies, maintenance, and water.

It is necessary for a rigorous analysis to be able to express the costs of these production items as a function of output. However, no methods are presently available for expressing the cost of labor, supplies, or maintenance accurately as a function of output.

Arbitrary methods of determining the incremental costs of labor, supplies, and maintenance are used, the commonest of which is to assume these costs to be a fixed percentage of the IFCs.

In many systems, for purposes of scheduling generation, the incremental production cost is assumed to be equal to the IFC.

2.14 CLASSICAL METHODS FOR ECONOMIC OPERATION OF SYSTEM PLANTS

Previously, the following thumb rules were adopted for scheduling the generation among various units of generators in a power station:

(i) **Base loading to capacity:** The turbo-generators were successively loaded to their rated capacities in the order of their efficiencies.

(ii) **Base loading to most efficient load:** The turbo-generator units were successively loaded to their most efficient loads in the increasing order of their heat rates.

(iii) **Proportional loading to capacity:** The turbo-generator sets were loaded in proportion to their rated capacities without consideration to their performance characteristics.

If the incremental generation costs are substantially constant over the range of operation, then without considering reserve and transmission line limitations, the most economic way of scheduling generation is to load each unit in the system to its rated capacity in the order of the highest incremental efficiency. This method, known as the merit order approach to economic load dispatching, requires the preparation of the order of merit tables based upon the incremental efficiencies, which should be updated regularly to reflect the changes in fuel costs, plant cycle efficiency, plant availability, etc. Active power scheduling then involves looking into the tables without the need for any calculations.

2.15 OPTIMIZATION PROBLEM—MATHEMATICAL FORMULATION (NEGLECTING THE TRANSMISSION LOSSES)

An optimization problem consists of:

1. Objective function.

2. Constraint equations.

2.15.1 Objective function

The objective function is to minimize the overall cost of production of power generation.

Cost in thermal and nuclear stations is nothing but the cost of fuel. Let n be the number of units in the system and C_i the cost of power generation of unit 'i':

$\therefore$ Total cost $C = C_1 + C_2 + C_3 + \cdots + C_n$

i.e., $C = \sum\limits_{i=1}^{n} C_i$

The cost of generation of each unit in thermal power plants is mainly a fuel cost. The generation cost depends on the amount of real power generated, since the real-power generation is increased by increasing the fuel input.

The generation of reactive power has negligible influence on the cost of generation, since it is controlled by the field current.

Therefore, the generation cost of the ith unit is a function of real-power generation of that unit and hence the total cost is expressed as

$$C = \sum_{i=1}^{n} C_i(P_{G_i}) \tag{2.1}$$

i.e., $C = C_1(P_{G_1}) + C_2(P_{G_2}) + C_3(P_{G_3}) + \cdots + C_n(P_{G_n})$

This objective function consists of the summation of the terms in which each term is a function of separate independent variables. This type of objective function is called a **separable objective function.**

The optimization problem is to allocate the total load demand (P_D) among the various generating units, such that the cost of generation is minimized and satisfies the following constraints.

2.15.2 Constraint equations

The economic power system operation needs to satisfy the following types of constraints.

(1) Equality constraints

The sum of real-power generation of all the various units must always be equal to the total real-power demand on the system.

i.e., $P_\text{D} = \sum\limits_{i=1}^{n} P_{G_i}$

or $\sum\limits_{i=1}^{n} P_{G_i} - P_\text{D} = 0 \tag{2.2}$

where $\sum\limits_{i=1}^{n} P_{G_i} = $ total real-power generation and P_D is the total real-power demand. Equation (2.2) is known as the real-power balance equation when losses are neglected.

(2) Inequality constraints

These constraints are considered in an economic power system operation due to the physical and operational limitations of the units and components.

The inequality constraints are classified as:

(a) According to the nature

According to nature, the inequality constraints are classified further into the following constraints:

(i) Hard-type constraints: These constraints are definite and specific in nature. No *flexibility* will take place in violating these types of constraints.

e.g.,: The range of tapping of an on-load tap-changing transformer.

(ii) Soft-type constraints: These constraints have some flexibility with them in violating.

e.g.,: Magnitudes of node voltages and the phase angle between them.

Some penalties are introduced for the violations of these types of constraints.

(b) According to power system parameters

According to power system parameters, inequality constraints are classified further into the following categories.

(i) Output power constraints: Each generating unit should not operate above its rating or below some minimum generation. This minimum value of real-power generation is determined from the technical feasibility.

$$P_{G_i(\min)} \leq P_{G_i} \leq P_{G_i(\max)} \tag{2.3a}$$

Similarly, the limits may also have to be considered over the range of reactive-power capabilities of the generator unit requiring that:

$$Q_{G_i(\min)} \leq Q_{G_i} \leq Q_{G_i(\max)} \quad \text{for } i = 1, 2, 3, \ldots, n \tag{2.3b}$$

and the constraint $P_{G_i}^2 + Q_{G_i}^2 \leq (S_{\text{irated}})^2$ must be satisfied, where S_i is the rating of the generating unit for limiting the overheating of stator.

(ii) Voltage magnitude and phase-angle constraints: For maintaining better voltage profile and limiting overloadings, it is essential that the bus voltage magnitudes and phase angles at various buses should vary within the limits. These can be illustrated by imposing the inequality constraints on bus voltage magnitudes and their phase angles.

$$V_{i(\min)} \leq V_i \leq V_{i(\max)} \quad \text{for } i = 1, 2, \ldots, n$$

$$\delta_{ij(\min)} \leq \delta_{ij} \leq \delta_{ij(\max)} \quad \text{for } i = 1, 2, \ldots, n$$

where $j = 1, 2, \ldots, m, j \neq i$, n is the number of units, and m the number of loads connected to each unit.

(iii) Dynamic constraints: These constants may consider when fast changes in generation are required for picking up the shedding down or increasing of load demand. These constraints are of the form:

$$\left|\frac{dP_{G_i}(t)}{dt}\right|_{\min} \leq \left|\frac{dP_{G_i}(t)}{dt}\right| \leq \left|\frac{dP_{G_i}(t)}{dt}\right|_{\max}$$

In addition, in terms of reactive-power generation,

$$\left|\frac{dQ_{G_i}(t)}{dt}\right|_{\min} \le \left|\frac{dQ_{G_i}(t)}{dt}\right| \le \left|\frac{dQ_{G_i}(t)}{dt}\right|_{\max}$$

(iv) Spare capacity constraints: These constraints are required to meet the following criteria:

(a) Errors in load prediction.

(b) The unexpected and fast changes in load demand.

(c) Unplanned loss of scheduled generation, i.e., the forced outages of one or more units on the system.

The total power generation at any time must be more than the total load demand and system losses by an amount not less than a specified minimum spare capacity (P_{SP})

i.e., $P_G \ge (P_D + P_L) + P_{SP}$

where P_G is the total power generation, $P_D + P_L$ is the total load demand and system losses, and P_{SP} is the specified minimum spare power.

(v) Branch transfer capacity constraints: Thermal considerations may require that the transmission lines be subjected to branch transfer capacity constraints:

$$S_{i(\min)} \le S_{bi} \le S_{i(\max)} \quad \text{for } i = 1, 2, \ldots, n_b$$

where n_b is the number of branches and S_{bi} the ith branch transfer capacity in MVA.

(vi) Transformer tap position/settings constraints: The tap positions (or) settings of a transformer (T) must lie within the available range:

$$T_{(\min)} \le T \le T_{(\max)}$$

For an autotransformer, the tap setting constraints are:

$$0 \le T \le 1$$

where the minimum tap setting is zero and the maximum tap setting is 1.

For a 2-winding transformer, tap setting constraints are $0 \le T \le K$, where K is the transformation (turns) ratio.

For a phase-shifting transformer, the constraints are of the type:

$$\theta_{i(\min)} \le \theta_i \le \theta_{i(\max)}$$

where θ_i is the phase shift obtained from the ith transformer.

(vii) Transmission line constraints: The active and reactive power flowing through the transmission line is limited by the thermal capability of the circuit.

$$TC_i \le TC_{i(\max)}$$

where $TC_{i(\max)}$ is the maximum loading capacity of the ith line.

(viii) Security constraints: Power system security and power flows between certain important buses are also considered for the solution of an optimization problem.

If the system is operating satisfactorily, there is an outage that may be scheduled or forced, but some of the constraints are naturally violated. It may be mentioned that consideration of each and every possible branch for an outage will not be a feasible proportion. When a large system is under study, the network security is maintained such that computation is to be made with the outage of one branch at one time and then the computation of a group of branches or units at another time.

So, the optimization problem was stated earlier as minimizing the cost function (C) given by Equation (2.1), which is subjected to the equality and inequality constraint (Equations (2.2) and (2.3)).

2.16 MATHEMATICAL DETERMINATION OF OPTIMAL ALLOCATION OF TOTAL LOAD AMONG DIFFERENT UNITS

Consider a power station having 'n' number of units. Let us assume that each unit does not violate the inequality constraints and let the transmission losses be neglected.

The cost of production of electrical energy

$$C = \sum_{i=1}^{n} C_i(P_{G_i}) \tag{2.4}$$

where C_i is the cost function of the ith unit.

This cost is to be minimized subject to the equality constraint given by

$$P_D = \sum_{i=1}^{n} P_{G_i}$$

$$\text{(or)} \quad \sum_{i=1}^{n} P_{G_i} - P_D = 0 \tag{2.5}$$

where P_{G_i} is the real-power generation of the ith unit.

This is a constrained optimization problem.

To get the solution for the optimization problem, we will define an objective function by augmenting Equation (2.4) with an equality constraint (Equation (2.5)) through the Lagrangian multiplier (λ) as

$$C' = C - \lambda\left[\sum_{i=1}^{n} P_{G_i} - P_D\right]$$

$$\min\left[C'\right] = \min\left[C - \lambda\left[\sum_{i=1}^{n} P_{G_i} - P_D\right]\right] \tag{2.6}$$

The condition for optimality of such an augmented objective function is

$$\frac{\partial C'}{\partial P_{G_i}} = 0$$

From Equation (2.6),

$$\frac{\partial C'}{\partial P_{G_i}} = \frac{\partial C}{\partial P_{G_i}} - \frac{\partial}{\partial P_{G_i}}\left(\lambda\left[\sum_{i=1}^{n} P_{G_i} - P_D\right]\right) = 0$$

i.e., $\quad \dfrac{\partial C'}{\partial P_{G_i}} = \dfrac{\partial C}{\partial P_{G_i}} - \lambda(1-0) = 0$

Since P_D is a constant and is an uncontrolled variable,

$$\dfrac{\partial P_D}{\partial P_{G_i}} = 0$$

$$\therefore \dfrac{\partial C'}{\partial P_{G_i}} = \dfrac{\partial C}{\partial P_{G_i}} - \lambda = 0$$

(or) $\quad \dfrac{\partial C}{\partial P_{G_1}} - \lambda = 0$

$$\dfrac{\partial C}{\partial P_{G_2}} - \lambda = 0$$

$$\dfrac{\partial C}{\partial P_{G_3}} - \lambda = 0$$

$$\vdots$$

$$\dfrac{\partial C}{\partial P_{G_n}} - \lambda = 0$$

$$\therefore \dfrac{\partial C}{\partial P_{G_1}} = \dfrac{\partial C}{\partial P_{G_2}} = \dfrac{\partial C}{\partial P_{G_3}} = \cdots = \dfrac{\partial C}{\partial P_{G_n}} = \lambda \qquad (2.7)$$

Since the expression of C is in a variable separable form, i.e., the overall cost is the summation of cost of each generating unit, which is a function of real-power generation of that unit only:

i.e., $\quad \dfrac{\partial C}{\partial P_{G_1}} = \dfrac{\partial C_1}{\partial P_{G_1}}$

$$\dfrac{\partial C}{\partial P_{G_2}} = \dfrac{\partial C_2}{\partial P_{G_2}}$$

$$\vdots$$

$$\dfrac{\partial C}{\partial P_{G_n}} = \dfrac{\partial C_n}{\partial P_{G_n}}$$

$$\therefore \dfrac{\partial C_1}{\partial P_{G_1}} = \dfrac{\partial C_2}{\partial P_{G_2}} = \dfrac{\partial C_3}{\partial P_{G_3}} = \cdots = \dfrac{\partial C_n}{\partial P_{G_n}} = \lambda \qquad (2.8)$$

In Equation (2.8), each of these derivatives represents the individual incremental cost of every unit. Hence, the condition for the optimal allocation of the total load among the various units, when neglecting the transmission losses, is that the incremental costs of the individual units are equal. It a called a **co-ordination equation**.

Assume that one unit is operating at a higher incremental cost than the other units. If the output power of that unit is reduced and transferred to units with lower incremental operating costs, then the total operating cost decreases. That is, reducing the output of the unit with the higher incremental cost results in a more decrease in cost than the increase in cost of adding the same output reduction to units with lower incremental costs. Therefore, all units must run with same incremental operating costs.

After getting the optimal solution, in the case that the generation of any one unit is below its minimum capacity or above its maximum capacity, then its generation becomes the corresponding limit. For example, if the generation of any unit violates the minimum limit, then the generation of that unit is set at its minimum specified limit and vice versa. Then, the remaining demand is allocated among the remaining units as for the above criteria.

In the solution of an optimization problem without considering the transmission losses, we make use of equal incremental costs, i.e., the machines are so loaded that the incremental cost of production of each machine is the same.

It can be seen that this method does not sense the location of changes in the loads. As long as the total load is fixed, irrespective of the location of loads, the solution will always be the same and, in fact, for this reason the solution may be feasible in the sense that the load voltages may not be within specified limits. The reactive-power generation required may also not be within limits.

2.17 COMPUTATIONAL METHODS

Different types of computational methods for solving the above optimization problem are as follows:

(1) Analytical method.

(2) Graphical method.

(3) Using a digital computer.

The method to be adopted depends on the following:

(i) The mathematical equation representing the IFC of each unit, which can be determined from the cost of generation of that unit.

The cost of the ith unit is given by

$$C_i = \frac{1}{2} a_i P_{G_i}^2 + b_i P_{G_i} + d_i \quad \text{(Second-degree polynomial)} \tag{2.9}$$

$\therefore$ The IFC of the ith unit

$$(IC)_i = a_i P_{G_i} + b_i \quad \text{(Linear model)} \tag{2.10}$$

where a_i is the slope of the IFC curve and b_i the intercept of the IFC curve.

(ii) Number of units (n).

(iii) Need to represent the discontinuities (if any due to steam valve opening) in the IFC curve.

2.17.1 Analytical method

When the number of units are small (either 2 or 3), incremental cost curves are approximated as a linear or quadratic variation and no discontinuities are present in the incremental cost curves.

We know that the IFC of the ith unit

$$(\text{IC})_i = \frac{\partial C_i}{\partial P_{G_i}} = a_i P_{G_i} + b_i$$

For an optimal solution, the IFC of all the units must be the same (neglecting the transmission losses):

i.e., $\quad \dfrac{\partial C_1}{\partial P_{G_1}} = \dfrac{\partial C_2}{\partial P_{G_2}} = \cdots = \dfrac{\partial C_n}{\partial P_{G_n}} = \lambda$

The analytical method consists of the following steps:

(i)   Choose a particular value of λ.

   i.e., $\quad \lambda = a_i P_{G_i} + b_i$

(ii) Compute $P_{G_i} = \dfrac{\lambda - b_i}{a_i}$, for the ith unit.   (2.11)

(iii) Find total real-power generation $= \displaystyle\sum_{i=1}^{n} P_{G_i}$ for all $i = 1, 2, \ldots, n$.

(iv) Repeat the procedure from step (ii) for different values of λ.

(v)   Plot a graph between total power generation and λ.

(vi) For a given power demand (P_{D}), estimate the value of λ from Fig. 2.8.

That value of λ will be the optimal solution for optimization problem.

2.17.2 Graphical method

For obtaining the solution in this method, the following procedure is required:

(i)   Consider the incremental cost curves of all units:

   i.e., $\quad (\text{IC})_i = a_i P_{G_i} + b_i \quad$ for all $i = 1, 2, \ldots, n$

and the total load demand P_{D} is given.

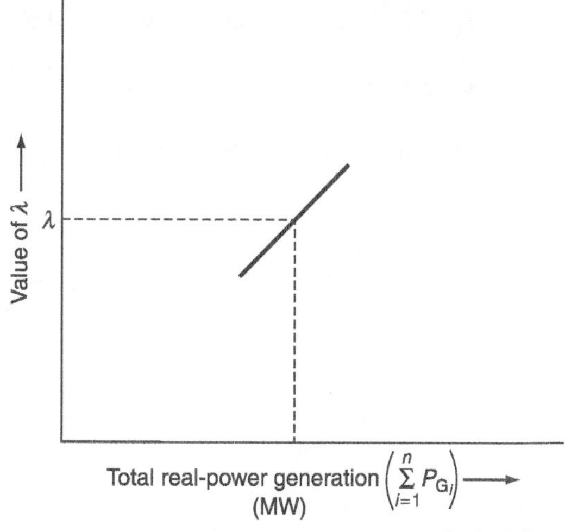

FIG. 2.8   *Estimation of optimum value of λ—analytical method*

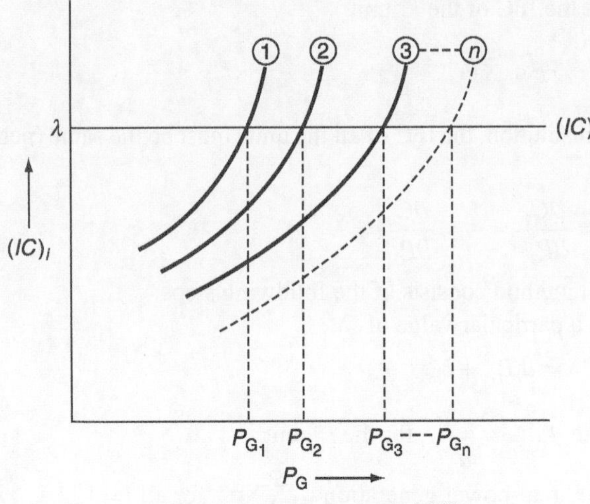

FIG. 2.9 *Graphical method*

(ii) For each unit, draw a graph between P_G and (IC) as shown in Fig. 2.9.

(iii) Choose a particular value of λ and $\Delta\lambda$.

(iv) Determine the corresponding real-power generations of all units:

i.e., $P_{G_1}, P_{G_2}, \ldots, P_{G_n}$

(v) Compute the total real-power generation $= \sum_{i=1}^{n} P_{G_i}$.

(vi) Check the real-power balance of Equation (2) as follows:

(a) If $\sum_{i=1}^{n} P_{G_i} - P_D = 0$, then the λ chosen will be the optimal solution and incremental costs of all units become equal.

(b) If $\sum_{i=1}^{n} P_{G_i} - P_D < 0$, increase λ by $\Delta\lambda$ and repeat the procedure from step (iv).

(c) If $\sum_{i=1}^{n} P_{G_i} - P_D > 0$, decrease λ by $\Delta\lambda$ and repeat the procedure from step (iv).

(vii) This process is repeated until $\sum_{i=1}^{n} P_{G_i} - P_D$ is within a specified tolerance (ε), say 1 MW.

i.e., $\sum_{i=1}^{n} P_{G_i} - P_D \leq \varepsilon$

2.17.3 Solution by using a digital computer

For more number of units, the λ-iterative method is more accurate and incremental cost curves of all units are to be stored in memory.

Information about the IFC curves is given for all units:

i.e., $\quad \lambda = (\text{IC})_i = a_i P_{G_i} + b_i$

or $\quad P_{G_i} = \dfrac{\lambda - b_i}{a_i} = \dfrac{\lambda}{a_i} - \dfrac{b_i}{a_i} \quad$ (when losses are neglected)

Let $\quad \alpha_i = -\dfrac{b_i}{a_i}$ $\hfill$ (2.12)

$\qquad \beta_i = \dfrac{1}{a_i}$ $\hfill$ (2.13)

and so on.

$$\therefore\; P_{G_i} = \alpha_i + \beta_i\,(\text{IC})_i + \gamma_i\,(\text{IC}_i)^2 + \ldots \tag{2.14}$$

for $i = 1, 2, \ldots, n$

The number of terms included depends on the degree of accuracy required and coefficients α_i, β_i, and γ_i are to be taken as input.

Algorithm for λ-iterative method

(i) Guess the initial value of λ° with the use of cost-curve equations.

(ii) Calculate $P_{G_i}^\circ$ according to Equation (2.14), i.e., $P_{G_i}^\circ = \alpha_i + \beta_i\,(\lambda^\circ)_i + \gamma_i\,(\lambda^\circ)_i^2 + \ldots$

(iii) Calculate $\sum\limits_{i=1}^{n} P_{G_i}^\circ$.

(iv) Check whether $\sum\limits_{i=1}^{n} P_{G_i}^\circ = P_D$:

$$\left(\sum\limits_{i=1}^{n} P_{G_i}^\circ - P_D \le \varepsilon \text{ (a tolerance value)} \right)$$

(v) If $\sum\limits_{i=1}^{n} P_{G_i}^\circ < P_D$, set a new value for λ, i.e., $\lambda' = \lambda^\circ + \Delta\lambda$ and repeat from step (ii) till the tolerance value is satisfied.

(vi) If $\sum\limits_{i=1}^{n} P_{G_i}^\circ > P_D$, set a new value for λ, i.e., $\lambda' = \lambda^\circ - \Delta\lambda$ and repeat from step (ii) till the tolerance value is satisfied.

(vii) Stop.

Example 2.1: The fuel cost functions in Rs./hr for three thermal plants are given by

$C_1 = 400 + 8.4 P_1 + 0.006 P_1^2$

$C_2 = 600 + 8.93 P_2 + 0.0042 P_2^2$

$C_3 = 650 + 6.78 P_3 + 0.004 P_3^2$

where P_1, P_2, and P_3 are in MW. Neglecting line losses and generator limits, determine the optimal scheduling of generation of each loading using the iterative method.

(i) $P_D = 550$ MW.

(ii) $P_D = 820$ MW.

(iii) $P_D = 1,500$ MW.

Solution:

For (i) $P_D = 550$ MW:

```
%MATLAB PROGRAM FOR ECONOMIC LOAD DISPATCH NEGLECTING LOSSES AND
GENERATOR LIMITS(dispatch1.m)
clc;
clear;
             % uno      d        b        a
Cost data=[1      400      8.4       0.006;
           2      600      8.93      0.0042;
           3      650      6.78      0.004];
Ng = length(cost data(:,1));
for i = 1:ng
    uno(i) = cost data(i,1);
     a(i) = cost data(i,2);
     b(i) = cost data(i,3);
     d(i) = cost data(i,4);
end
lambda = 9.0;
pd = 550;
delp = 0.1;
dellambda = 0;
iter = 0;
while(abs(delp)> = 0.001)
    iter = iter+1;
    lambda = lambda + dellambda;
    sum = 0;
    totgam = 0;
    for i = 1:ng
        p(i) = (lambda-b(i))/(2*d(i));
        sum = sum + p(i);
        totgam = totgam + 0.5*(1/d(i));
        ifc(i) = lambda;
    end
delp = pd-sum;
dellambda = delp/totgam;
end
totgencost = 0;
for i = 1:ng
totgencost = totgencost + (a(i)+ b(i)*p(i)+ d(i)*p(i)*p(i));
end
```

```
disp('OUTPUT OF MATLAB PROGRAM dispatch1.m');
lambda
disp('GENERATING UNIT        OPTIMAL GENERATION(MW)');
[uno;              p]'
disp('INCREMENTAL FUEL COST');
ifc(1)
```

OUTPUT OF MATLAB PROGRAM dispatch1.m

lambda = 9.6542

```
GENERATING UNIT              OPTIMAL GENERATION(MW)

       1                            104.5152
       2                             86.2121
       3                            359.2727
```

INCREMENTAL FUEL COST (Rs./MWhr)= 9.6542
TOTAL GENERATION COST(Rs./hr)= 6346.70

For (ii) P_d=820 MW:

OUTPUT OF MATLAB PROGRAM dispatch1.m

lambda = 10.4789

```
GENERATING UNIT              OPTIMAL GENERATION(MW)

    1.0000                          173.2424
    2.0000                          184.3939
    3.0000                          462.3636
```

INCREMENTAL FUEL COST(Rs./MWhr) = 10.4789
TOTAL GENERATION COST(Rs./hr) = 9064.70

For (iii) P_d = 1,500 MW:

OUTPUT OF MATLAB PROGRAM dispatch1.m

lambda = 12.5560

```
GENERATING UNIT              OPTIMAL GENERATION(MW)

    1.0000                          346.3333
    2.0000                          431.6667
    3.0000                          722.0000
```

INCREMENTAL FUEL COST(Rs./MWhr) = 12.5560
TOTAL GENERATION COST(Rs./hr) = 16897.00

2.18 ECONOMIC DISPATCH NEGLECTING LOSSES AND INCLUDING GENERATOR LIMITS

The output power of any generator should neither exceed its rating nor should it be below that necessary for the stable operation of a boiler. Thus, the generations are restricted to lie within given minimum and maximum limits. The problem is to find the active power generation of each plant such that the objective function (i.e., total production cost) is minimum, subject to the equality constraint, and the inequality constraints are

$$\sum_{i=1}^{n} P_{G_i} = P_D \quad \text{and} \quad P_{G_i(\min)} \leq P_{G_i} \leq P_{G_i(\max)},$$

respectively.

The solution algorithm for this case is the same as discussed in Section 2.17.3 with minor modifications. If any generating unit violates the above inequality constraints, set its generation at its respective limit as given below. In addition, the balance of the load is then shared between the remaining units on the basis of equal incremental cost.

The necessary conditions for optimal dispatch when losses are neglected:

$$\frac{\partial C_i}{\partial P_{G_i}} = \lambda \quad \text{for } P_{G_i(\min)} \leq P_{G_i} \leq P_{G_i(\max)}$$

$$\frac{\partial C_i}{\partial P_{G_i}} \leq \lambda \quad \text{for } P_{G_i} = P_{G_i(\max)}$$

$$\frac{\partial C_i}{\partial P_{G_i}} \geq \lambda \quad \text{for } P_{G_i} = P_{G_i(\min)}$$

Example 2.2: The fuel cost functions in Rs./hr. for three thermal plants are given by

$$C_1 = 400 + 8.4P_1 + 0.006P_1^2, \quad 100 \leq P_1 \leq 600$$

$$C_2 = 600 + 8.93P_2 + 0.006P_2^2, \quad 60 \leq P_2 \leq 300$$

$$C_3 = 650 + 6.78P_3 + 0.004P_3^2, \quad 300 \leq P_3 \leq 650$$

where P_1, P_2, and P_3 are in MW. Neglecting line losses and including generator limits, determine the optimal scheduling of generation of each loading using the iterative method.

(i) $P_D = 550$ MW.

(ii) $P_D = 820$ MW.

(iii) $P_D = 1,500$ MW.

Solution:

For (i) $P_D = 550$ MW:

```
%MATLAB PROGRAM FOR ECONOMIC LOAD DISPATCH NEGLECTING LOSSES AND
INCLUDING
%GENERATOR LIMITS(dispatch2.m)
clc;
clear;
               % uno      a       b       d       Pmin    Pmax
Cost data = [1      400     8.4     0.006   100     600;
             2      600     8.93    0.0042  60      300;
             3      650     6.78    0.004   300     650];
ng = length(cost data(:,1));
for i = 1:ng
    uno(i)= cost data(i,1);
    a(i)= cost data(i,2);
```

```matlab
        b(i)= cost data(i,3);
        d(i)= cost data(i,4);
        pmin(i)= cost data(i,5);
        pmax(i)= cost data(i,6);
end
lambda = 9.0;
pd = 550;
delp = 0.1;
dellambda = 0;
for i = 1:ng
    pv(i) = 0;
    pvfin(i) = 0;
end
while(abs(delp)>= 0.0001)
  lambda = lambda + dellambda;
  sum = 0;
  totgam = 0;
    for i = 1:ng
        p(i) = (lambda-b(i))/(2*d(i));
        sum = sum + p(i);
        totgam = totgam+0.5*(1/d(i));
    end
delp = pd - sum;
dellambda = delp/totgam;
ifc = lambda;
end
limvio = 0;
for i = 1:ng
    if(p(i) < pmin(i)|p(i) > pmax(i))
        limvio = 1;
        break;
    end
end
if limvio = = 0
    disp('GENERATION IS WITHIN THE LIMITS');
end
if (limvio = = 1)
sum = 0;
totgam = 0;
delp = 0.1;
loprep = 1;
while(abs(delp) >= 0.01 & loprep = 1)
    disp('GENERATION IS NOT WITHIN THE LIMITS');
    disp('VIOLATED GENERATOR NUMBER');
      i
            if p(i) < pmin(i)
                disp('GENERATION OF VIOLATED UNIT(MW)');
```

```
                p(i)
                disp('CORRESPONDING VOILATED LIMIT IS pmin');
            elseif p(i) > pmax(i)
                disp('GENERATION OF VIOLATED UNIT(MW)');
                p(i)
                disp('CORRESPONDING VIOLATED LIMIT IS pmax');
            end
        sum = 0;
        totgam = 0;
        for i = 1:ng
            pv(i) = 0;
        end
        for i = 1:ng
            if (p(i) < pmin(i)|p(i) > pmax(i))
                if p(i) < pmin(i)
                    p(i) = pmin(i);
                else
                    p(i) = pmax(i);
                end
                pv(i) = 1;
                pvfin(i) = 1;
                break;
            end
        end
        for i = 1:ng
            sum = sum + p(i);
            if (pvfin(i) ~= 1)
             totgam = totgam + 0.5*(1/d(i));
            end
        end
    delp = pd - sum;
    dellambda = delp/totgam;
    lambda = lambda+dellambda;
    ifc = lambda;
    for i = 1:ng
        if pvfin(i) ~ = 1
        p(i) = (lambda-b(i))/(2*d(i));
        end
        sum = sum + p(i);
    end
    delp = pd-sum;
    loprep = 0;

    for i = 1:ng
        if p(i) < pmin(i)|p(i) > pmax(i)
            loprep = 1;
            break;
        end
```

```
end
end
end
totgencost = 0;
for i = 1:ng
totgencost = totgencost+(a(i)+b(i)*p(i)+d(i)*p(i)*p(i));
end
disp('FINAL OUTPUT OF MATLAB PROGRAM dispatch2.m');
lambda
disp('GENERATING UNIT        OPTIMAL GENERATION(MW)');
[uno;              p]'
disp('INCREMENTAL FUEL COST(Rs./MWhr)');
ifc
disp('TOTAL GENERATION COST(Rs./hr.)');
totgencost
```

Results for (i) $P_d = 550$ MW:

```
GENERATION IS WITHIN THE LIMITS

FINAL OUTPUT OF MATLAB PROGRAM dispatch2.m

lambda = 9.6542

GENERATING UNIT           OPTIMAL GENERATION(MW)
    1.0000                     104.5152
    2.0000                      86.2121
    3.0000                     359.2727

INCREMENTAL FUEL COST(Rs./MWhr)= 9.6542

TOTAL GENERATION COST(Rs./hr) = 6346.70
```

Results for (ii) $P_d = 820$ MW:

```
GENERATION IS WITHIN THE LIMITS

FINAL OUTPUT OF MATLAB PROGRAM dispatch2.m

lambda = 10.4789

GENERATING UNIT    OPTIMAL GENERATION (MW)
    1.0000              173.2424
    2.0000              184.3939
    3.0000              462.3636
INCREMENTAL FUEL COST (Rs./MWhr)=10.4789
TOTAL GENERATION COST (Rs./hr) = 9064.70
```

Results for (iii) $P_d = 1,500$ MW:

```
GENERATION IS NOT WITHIN THE LIMITS
VIOLATED GENERATOR NUMBER = 2

GENERATION OF VIOLATED UNIT (MW) = 431.6667
```

```
CORRESPONDING VIOLATED LIMIT IS p_max

GENERATION IS NOT WITHIN THE LIMITS
VIOLATED GENERATOR NUMBER = 3

GENERATION OF VIOLATED UNIT (MW) = 801
CORRESPONDING VIOLATED LIMIT IS p_max

FINAL OUTPUT OF MATLAB PROGRAM dispatch2.m

lambda = 15

GENERATING UNIT          OPTIMAL GENERATION (MW)
       1                          550
       2                          300
       3                          650

INCREMENTAL FUEL COST (Rs./MWhr) = 15

TOTAL GENERATION COST (Rs./hr) = 17239
```

2.19 FLOWCHART FOR OBTAINING OPTIMAL SCHEDULING OF GENERATING UNITS BY NEGLECTING THE TRANSMISSION LOSSES

The optimal scheduling of generating units is represented by the flowchart as shown in Fig. 2.10.

2.20 ECONOMICAL LOAD DISPATCH—IN OTHER UNITS

The economical load dispatch problem has been solved for a power system area consisting of fossil fuel units. For an area consisting of a mix of different types of units, i.e.— fossil fuel units, nuclear units, pumped storage hydro-units, hydro-units, etc.—solving the economical load dispatch problem will become different.

2.20.1 Nuclear units

For these units, the fixed cost is high and operating costs are low. As such, nuclear units are generally base load plants at their rated outputs, i.e., the reference power setting of turbine governors for nuclear units is held constant at the rated output. Therefore, these units do not participate in economical load dispatch.

2.20.2 Pumped storage hydro-units

These units are operated as synchronous motors to pump water during off-peak hours. During peak load hours, the water is released and the units are operated as synchronous generators to supply power. The economic operation of the area is done by pumping during off-peak hours when the area incremental cost (λ) is low, and by generating during peak load hours when λ is high. Some techniques are available for incorporating pumped storage hydro-units into the economic dispatch of fossil units.

2.20.3 Hydro-plants

For an area consisting of hydro-plants located along a river, the objective of the economic dispatch problem becomes maximizing the power generation over the yearly water cycle rather than minimizing the total operating costs. For these types of plants, reservoirs are provided to store the water during rainy seasons. There are some constraints on the level of water such as flow of river, irrigation, etc. Optimal strategies are available for co-ordinating

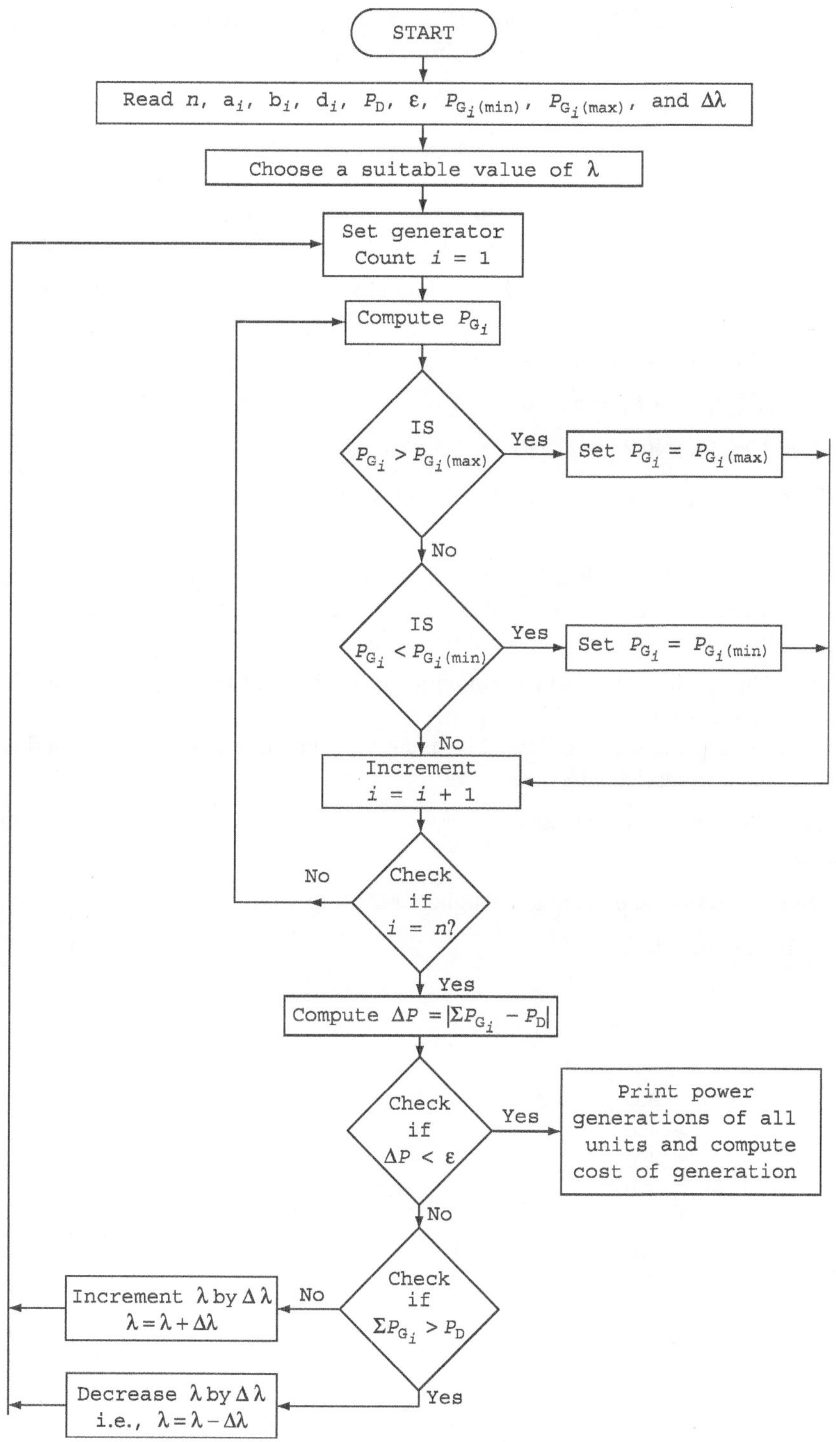

FIG. 2.10 *Flowchart*

the outputs of such plants along a river. There are also some economic dispatch strategies available for the mix of fossil fuel and hydro-systems.

2.20.4 Including reactive-power flows

In this case, both active and reactive powers are selected to minimize the operating costs. In particular, reactive-power injections from generators, switched capacitor banks, and static VAr systems along with transformer settings can be selected to minimize the transmission losses.

Example 2.3: A system consists of two units to meet the daily load cycle as shown in Fig. 2.11.

The cost curves of the two units are:

$$C_1 = 0.15P_{G_1}^2 + 60P_{G_1} + 135 \text{ Rs./hr}$$

$$C_2 = 0.25P_{G_2}^2 + 40P_{G_2} + 110 \text{ Rs./hr}$$

The maximum and minimum loads on a unit are to be 220 and 30 MW, respectively.

Find out:

    (i)   The economical distribution of a load during the light-load period of 7 hr and during the heavy-load periods of 17 hr. In addition, find the operation cost for this 24-hr period operation of two units.

    (ii)   The operation cost when removing one of the units from service during 7 hr of light-load period.

           Assume that a cost of Rs. 525 is incurred in taking a unit off the line and returning it to service after 7 hr.

    (iii) Comment on the results.

Solution:

(i) When both units are operating throughout a 24-hr period,

      Total time $= 24$ hr

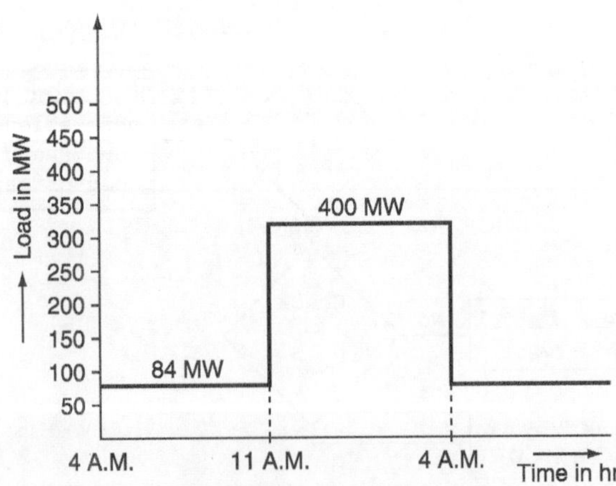

FIG. 2.11 *Daily load cycle*

Total load $= 84$ MW for 7 hr $\qquad + \qquad$ 400 MW for 17 hr
(from 4 A.M. to 11 A.M.) $\qquad$ (from 11 A.M. to 4 A.M.)

For a heavy load of 400 MW:

Heavy-load period, $t_h = 17$ hr
load, $P_{D_h} = 400$ MW
We have to find the optimal scheduling of two units with this load.
We have the cost curves of two units:

For Unit 1:

$$C_1 = 0.15 P_{G_1}^2 + 60 P_{G_1} + 135 \ \text{Rs./hr}$$

Incremental fuel cost , $\dfrac{dC_1}{dP_{G_1}} = 0.15 \times 2 P_{G_1} + 60$

$$= 0.3 P_{G_1} + 60 \ \text{Rs./MWh}$$

For Unit 2:

$$C_2 = 0.25 P_{G_2}^2 + 40 P_{G_2} + 110 \ \text{Rs./hr}$$

$$\Rightarrow \dfrac{dC_2}{dP_{G_2}} = 0.25 \times 2 P_{G_2} + 40$$

$$= 0.5 P_{G_2} + 40 \ \text{Rs./MWh}$$

For the optimal distribution of a load,

$$\dfrac{dC_1}{dP_{G_1}} = \dfrac{dC_2}{dP_{G_2}} = \lambda$$

$$0.3 P_{G_1} + 60 = 0.5 P_{G_2} + 40$$

$$0.3 P_{G_1} - 0.5 P_{G_2} = -20 \qquad\qquad (2.15)$$

$$P_{G_1} + P_{G_2} = 400 \ (\text{given}) \qquad\qquad (2.16)$$

From Equations (2.15) and (2.16), we have

$$0.3 P_{G_1} - 0.5 P_{G_2} = -20$$

$$0.5 P_{G_1} + 0.5 P_{G_2} = 200$$

$$\overline{\qquad\qquad\qquad\qquad\qquad}$$

$$0.8 P_{G_1} \qquad\qquad = 180$$

$$\therefore P_{G_1} \qquad\qquad = 225 \ \text{MW}$$

Substituting the P_{G_1} value in Equation (2.16), we get

$$P_{G_2} = 175 \ \text{MW}$$

Here, $P_{G_1} = 225$ MW and $P_{G_1} > P_{G_{max}}$; hence, set P_{G_1} at its maximum generation limit

i.e., $P_{G_1} = 220$ MW

$\therefore P_{G_2} = 400 - 220 = 180$ MW

The operation cost for a heavy-load period (i.e., from 11 A.M. to 4 A.M.) with this optimal distribution is

$$C = (C_1 + C_2) \times t_h$$
$$= [(0.15 \times 220^2 + 60 \times 220 + 135) + (0.25 \times 180^2 + 40 \times 180 + 110)] \times 17$$
$$= \text{Rs. } 6,12,085$$

For a light load of 84 MW:

Period, $t_l = 7$ hr

load, $P_D = 84$ MW

For optimal load sharing,

i.e., $\quad\quad 0.3\, P_{G_1} + 60 = 0.5\, P_{G_2} + 40$

$$0.3\, P_{G_1} - 0.5\, P_{G_2} = -20 \tag{2.17}$$

$$P_{G_1} + P_{G_2} = 84 \tag{2.18}$$

By solving Equations (2.17) and (2.18), we get

$$P_{G_1} = 27.5 \text{ MW}; \quad P_{G_2} = 56.5 \text{ MW}$$

Here, $\quad P_{G_1} = 27.5 \text{ MW} < P_{G_{min}} = 30 \text{ MW}$

Therefore, the load shared by Unit-1 is set to $P_{G_1} = 30$ MW and $P_{G_2} = 84 - 30 = 54$ MW.

The operation cost for a light-load period (i.e., from 4 A.M. to 11 A.M.) with this optimal distribution:

$$= [(0.15) \times (30)^2 + 60 \times 30 + 135) + (0.25 \times 54^2 + 40 \times 54 + 110)] \times 7$$

$$= \text{Rs. } 35,483$$

Hence, the total fuel cost when both the units are operating throughout the 24-hr period

$$= \text{Rs. } (6,12,085 + 35,483)$$
$$= \text{Rs. } 6,47,568$$

(ii)   If only one of the units is run during the light-load period,

i.e.,   Period, $t_l = 7$ hr

load, $P_D = 84$ MW

When Unit-1 is to be run,

Cost of operation $= C_1 \times t_1$
$$= [0.15 \times 84^2 + 60 \times 84 + 135] \times 7$$
$$= \text{Rs. } 43,633.80$$

When Unit-2 is to be run,

Cost of operation $= C_2 \times t_l$
$$= [0.25 \times 84^2 + 40 \times 84 + 110] \times 7$$
$$= \text{Rs. } 36,638$$

From the above, it is verified that it is economical to run Unit-2 during a light-load period and to put off Unit-1 from service.

The operating cost with only Unit-2 in operation $= \text{Rs. } 36,638$

The operating cost for the operation of both units in a heavy-load period and Unit-2 only in a light-load period = Rs. $(6,47,568 + 36,638)$ = Rs. 6,48,723

In addition, given that a cost of Rs. 525 is incurred in taking a unit off the line and returning it to service after 7 hr,

Total operating cost = operating cost + start-up cost = Rs. $(6,48,723 + 525)$ = Rs. 6,49,248.

(iii)　Total operating cost for (i) = Rs. 6,47,568

Total operating cost for (ii) = Rs. 6,49,248

It is concluded that the total operating cost when both units running throughout 24-hr periods is less than the operating cost when one of the units is put off from the line and returned to the service after a light-load period. Hence, it is economical to run both units throughout 24 hr.

Example 2.4: A constant load of 400 MW is supplied by two 210-MW generators 1 and 2, for which the fuel cost characteristics are given as below:

$$C_1 = 0.05\, P_{G_1}^2 + 20\, P_{G_1} + 30.0 \text{ Rs./hr}$$

$$C_2 = 0.06\, P_{G_2}^2 + 15\, P_{G_1} + 40.0 \text{ Rs./hr}$$

The real-power generations of units P_{G_1} and P_{G_2} are in MW. Determine: (i) the most economical load sharing between the generators. (ii) The saving in Rs./day thereby obtained compared to the equal load sharing between two generators.

Solution:

The IFCs are

$$\frac{dC_1}{dP_{G_1}} = 0.10 P_{G_1} + 20.0$$

$$\frac{dC_2}{dP_{G_2}} = 0.12 P_{G_2} + 15.0$$

(i)　For optimal sharing of load, the condition is

$$\frac{dC_1}{dP_{G_1}} = \frac{dC_2}{dP_{G_2}}$$

$$0.10 P_{G_1} + 20.0 = 0.12 P_{G_1} + 15.0$$

or　$0.10 P_{G_1} - 0.12 P_{G_1} = 15.0 - 20.0$

or　$0.10 P_{G_1} - 0.12 P_{G_1} = -5.0$ 　　　　　　(2.19)

Given:　$P_{G_1} + P_{G_2} = 400$ 　　　　　　(2.20)

Solving Equations (2.19) and (2.20), we have

$$0.10\, P_{G_1} - 0.12\, P_{G_2} = -5.0$$

Equation (2.20) $\times 0.12 \Rightarrow$ 　$0.12\, P_{G_1} + 0.12\, P_{G_2} = 48.0$

$$\overline{\phantom{0.12\, P_{G_1} + 0.12\, P_{G_2} = 48.0}}$$

$$0.22\, P_{G_1} = 43.0$$

$$\therefore P_{G_1} = 195.45 \text{ MW}$$

Substituting $P_{G_1} = 195.45$ MW in Equation (2.20), we get

$$P_{G_2} = 400 - 195.45 = 204.55 \text{ MW}$$

The load of 400 MW is economically shared by the two generators with $P_{G_1} = 195.45$ MW and $P_{G_2} = 204.55$ MW.

(ii) When the load is shared between the generators equally, then

$$P_{G_1} = 200 \text{ MW} \quad \text{and} \quad P_{G_2} = 200 \text{ MW}$$

With this equal sharing of load, the P_{G_1} value is increased from 195.45 with economical sharing to 200 MW.

$\therefore$ Increase in operation cost of Generator 1

$$= \int_{195.45}^{200} \left(\frac{dC_1}{dP_{G_1}} \right) dP_{G_1} = \int_{195.45}^{200} (0.10 P_{G_1} + 20.0) \, dP_{G_1}$$

$$= \left[0.05 P_{G_1}^2 + 20.0 P_{G_1} \right]_{195.45}^{200}$$

$$= 0.05(200^2 - 195.45^2) + 20.0 \, (200 - 195.45)$$

$$= 89.961323595 + 91.00$$

$$= 180.96 \text{ Rs./hr}$$

The P_{G_2} value is decreased from 204.55 to 200 MW.

$\therefore$ Decrease in operation cost of Generator 2

$$= \int_{204.55}^{200} (0.12 P_{G_2} + 15.0) \, dP_{G_2}$$

$$= \left[0.06 P_{G_2}^2 + 15.0 P_{G_2} \right]_{204.55}^{200}$$

$$= 0.06 \, (200^2 - 204.55^2) + 15.0 \, (200 - 204.55)$$

$$= -110.44 - 68.25 = -178.69 \text{ Rs./hr}$$

$\therefore$ Saving in cost $= 180.96 - 178.69 = 2.27$ Rs./hr

The saving in cost per day $= 2.27 \times 24$
$$= 56.75 \text{ Rs./day}$$

Example 2.5: Consider the following three IC curves:

$$P_{G_1} = -100 + 50(IC_1) + 2(IC_1)^2$$
$$P_{G_2} = -150 + 60(IC_2) - 2.5(IC_2)^2$$
$$P_{G_3} = -80 + 40(IC_3) - 1.8(IC_3)^2$$

where IC's are in Rs./MWh and P_G's are in MW.

The total load at a certain hour of the day is 400 MW. Neglect transmission losses and develop a computer program for optimum generation scheduling within the accuracy of ± 0.05 MW.

Note: All P_G's must be real and positive.

Solution:

$\alpha_1 = -100, \quad \beta_1 = 50, \quad \gamma_1 = 2 \quad (\because \text{Assume } d_1, d_2, d_3 \text{ are neglected})$

$\alpha_2 = -150, \quad \beta_2 = 60, \quad \gamma_2 = -2.5$

$\alpha_3 = -80, \quad \beta_3 = 40, \quad \gamma_3 = -1.8$

$$\alpha_i = \frac{b_i}{a_i}; \quad \beta_i = \frac{1}{a_i}$$

$\therefore a_1 = 0.02; \quad a_2 = 0.0166; \quad a_3 = 0.025$

$\quad b_1 = 2; \quad b_2 = 2.49; \quad b_3 = 2$

$$\frac{dC_1}{dP_{G_1}} = 0.02 P_{G_1} + 2 \, \text{Rs./MWh}$$

$$\frac{dC_2}{dP_{G_2}} = 0.0166 \, P_{G_2} + 2.49 \, \text{Rs./MWh}$$

$$\frac{dC_3}{dP_{G_3}} = 0.005 \, P_{G_3} + 2 \, \text{Rs./MWh}$$

For optimal load distribution among the various units,

$$\frac{dC_1}{dP_{G_1}} = \frac{dC_2}{dP_{G_2}} = \frac{dC_3}{dP_{G_3}}$$

$$\Rightarrow \frac{dC_1}{dP_{G_1}} = \frac{dP_{C_2}}{dP_{G_2}}$$

$$0.02 P_{G_1} + 2 = 0.0166 P_{G_2} + 2.49$$

$$\Rightarrow 0.02 \, P_{G_1} - 0.0166 P_{G_2} = 0.49 \tag{2.21}$$

$$\frac{dC_2}{dP_{G_2}} = \frac{dC_3}{dP_{G_3}}$$

$$0.0166 P_{G_2} + 2.49 = 0.025 P_{G_3} + 2$$

$$0.0166 P_{G_2} - 0.025 P_{G_3} = -0.49 \tag{2.22}$$

$$\frac{dC_3}{dP_{G_2}} = \frac{dC_1}{dP_{G_1}}$$

$$0.02 P_{G_1} + 2 = 0.025 P_{G_3} + 2$$

$$0.02 P_{G_1} - 0.025 P_{G_3} = 0 \tag{2.23}$$

Given: $\quad P_{G_1} + P_{G_2} + P_{G_3} = 400 \tag{2.24}$

or $\quad P_{G_2} + P_{G_3} = 400 - P_{G_1} \tag{2.25}$

Solving Equations (2.22) and (2.25), we have

$$0.0166P_{G_2} - 0.025P_{G_3} = -0.49$$

$$-(0.0166P_{G_2}) + (-0.0166P_{G_3}) = -(6.64 - 0.0166P_{G_1})$$

$$\overline{ -0.0416P_{G_3} = -0.49 - 6.64 + 0.0166P_{G_1}}$$

$$0.0166\,P_{G_1} + 0.0416P_{G_3} = 7.13 \tag{2.26}$$

Solving Equations (2.23) and (2.26), we get

Equation (2.23) $\times 0.0166 \Rightarrow \qquad 3.32 \times 10^{-4}P_{G_1} - 4.15 \times 10^{-4}P_{G_3} = 0$

Equation (2.26) $\times 0.02 \quad \Rightarrow \; -(3.32 \times 10^{-4}P_{G_1}) + (-8.32 \times 10^{-4}P_{G_3}) = -(0.1426)$

$$\overline{ -1.247 \times 10^{-3}P_{G_3} = -0.1426}$$

$$\therefore P_{G_3} = \frac{0.1426}{1.247 \times 10^{-3}} = 114.35 \text{ MW}$$

Substituting $P_{G_3} = 114.35$ MW in Equation (2.26), we get

$$P_{G_1} = 142.9375 \text{ MW}$$

Substituting $P_{G_1} = 142.9375$ MW and $P_{G_3} = 114.35$ MW in Equation (2.25), we get

$$P_{G_2} = 142.\,93175 \text{ MW}$$

Therefore, for optimal generation, the three units must share a total load of 400 MW as follows:

Cost of generation of 142.9375 MW by Unit-1

$$(C_1) = \frac{1}{2}\,(0.02)\,(142.9375)^2 + 2\,(142.9375)$$

$$C_1 = 490.186 \text{ Rs./MWh}$$

Similarly, $\quad C_2 = \left[\frac{1}{2} \times 0.0166 \times (142.91)^2 + (2.49 \times 142.91)\right]$

$$= \frac{1}{2} \times 0.0166 \times (142.91)^2 + (2.49 \times 142.91)$$

$$C_2 = 525.359 \text{ Rs./MWh}$$

and $\quad C_3 = \frac{1}{2} \times 0.025 \times (114.35)^2 + (2 \times 114.35)$

$$= 392.149 \text{ Rs./MWh}$$

Total cost of generation of 400 MW with economical load sharing

$$C = C_1 + C_2 + C_3$$

$$= 490.186 + 525.359 + 392.149$$

$$= 1{,}407.694 \text{ Rs./MWh}$$

Cost per day $= 1,407.694 \times 24$

$\qquad = $ Rs. $33,784.656/$day

Total cost per day with an equal distribution of load

$\qquad = 1,412.838 \times 24$

$\qquad = $ Rs. $33,908.112/$day

$\therefore$ Saving in cost $=$ Rs. $33,908.112 - 33,784.856 =$ Rs. $123.256/$day

Example 2.6: The fuel cost of two units are given by

$$C_1 = C_1(P_{G_1}) = 1.0 + 25P_{G_1} + 0.2P_{G_1}^2 \text{ Rs./hr}$$

$$C_2 = C_2(P_{G_2}) = 1.5 + 35P_{G_2} + 0.2P_{G_2}^2 \text{ Rs./hr}$$

If the total demand on the generators is 200 MW, find the economic load scheduling of the two units.

Solution:

The condition for economic load scheduling when neglecting the transmission losses is

$$\frac{dC_i}{dP_{G_i}} = \lambda \quad \text{for } i = 1, 2, \dots, n$$

i.e., $\quad \dfrac{dC_1}{dP_{G_1}} = \dfrac{dC_2}{dP_{G_2}} \dots \dfrac{dC_n}{dP_{G_n}} = \lambda$

$$C_1 = 1.0 + 25P_{G_1} + 0.12P_{G_1}^2$$

$$\frac{dC_1}{dP_{G_1}} = 25 + 0.4P_{G_1} \text{ Rs./MWh}$$

$$C_2 = 1.5 + 35P_{G_2} + 0.12P_{G_2}^2$$

$$\frac{dC_2}{dP_{G_2}} = 35 + 0.4P_{G_2}$$

For economical load dispatch,

$$\frac{dC_1}{dP_{G_1}} = \frac{dC_2}{dP_{G_2}} = \lambda$$

$$25 + 0.4P_{G_1} = 35 + 0.4P_{G_2}$$

or $\quad 0.4P_{G_1} - 0.4P_{G_2} = 10$ MW $\hfill (2.27)$

and $\qquad P_{G_1} + P_{G_2} = 200$ MW $\hfill (2.28)$

$$0.4P_{G_1} - 0.4P_{G_2} = 10$$

Multiplying both sides of Equation (2.28) by 0.4, we get

$$0.4P_{G_1} + 0.4P_{G_2} = 80 \hfill (2.29)$$

By adding Equations (2.27) and (2.29), we get

$$0.8P_{G_1} = 90$$

or $\quad P_{G_1} = \dfrac{90}{0.8} = 112.5 \text{ MW}$

Substituting $P_{G_1} = 112.5$ MW in Equation (2.28)

$$112.5 + P_{G_2} = 200$$

$$\therefore P_{G_2} = 87.5 \text{ MW}$$

Example 2.7: The incremental cost curves of three units are expressed in the form of polynomials:

$$P_{G_1} = -150 + 50(\text{IC}_1) - 2(\text{IC}_1)^2$$

$$P_{G_2} = -100 + 50(\text{IC}_2) - 2(\text{IC}_2)^2$$

$$P_{G_3} = -150 + 50(\text{IC}_3) - 2(\text{IC}_3)^2$$

The total demand at a certain hour of the day equals 200 MW. Develop a computer program that will render a solution for the optimum allocation of load among three units.

Solution:

Step 1: Assume $\lambda^o = 10$.

Step 2: Compute $P_{G_1}^{(o)}$ corresponding to $(\text{IC})_i = \lambda^o$, $i = 1, 2, 3$.

$$P_{G_1}^{(o)} = -150 + 50\lambda^o - 2(\lambda^o)^2 = -150 + 50(10) - 2(100) = 150 \text{ MW}$$

$$P_{G_2}^{(o)} = -100 + 50\lambda^o - 2(\lambda^o)^2 = -100 + 50(10) - 2(100) = 200 \text{ MW}$$

$$P_{G_3}^{(o)} = -150 + 50\lambda^o - 2(\lambda^o)^2 = -150 + 50(10) - 2(100) = 150 \text{ MW}$$

Step 3: Compute $\sum\limits_{i=1}^{n} P_{G_i}^{(o)}$:

i.e., $P_{G_1}^o + P_{G_2}^o + P_{G_3}^o = 500 \text{ MW}$

Step 4: Check if $\sum\limits_{i=1}^{3} P_{G_i}^{(o)} = P_D$:

We find $\sum\limits_{i=1}^{3} P_{G_i}^{(o)} > 200$

i.e., $500 > 200$

Step 5: Reduce λ by $\Delta\lambda = 3$:

i.e., $\lambda' = \lambda^o - \Delta\lambda = 10 - 3 = 7$

Step 6: Now find the generation $P_{G_1}^1, P_{G_2}^1,$ and $P_{G_3}^1$.

Step 7: Go to Step 4.

By repeating the above procedure, the following results are obtained and the above equations converge at $\lambda = 5$

λ	P_{G_1}	P_{G_2}	P_{G_3}	$\sum_{i=1}^{3} P_{G_i}$
10	150	200	150	500
7	102	152	102	356
6	78	121	78	284
5	50	100	50	200

Example 2.8: Two units each of 200 MW in a thermal power plant are operating all the time throughout the year. The maximum and minimum load on each unit is 200 and 50 MW, respectively. The incremental fuel characteristics for the two units are given as

$$\frac{dC_1}{dP_{G_1}} = 15 + 0.08 P_{G_1} \text{ Rs./MWh}$$

$$\frac{dC_2}{dP_{G_2}} = 13 + 0.1 P_{G_2} \text{ Rs./MWh}$$

If the total load varies between 100 and 400 MW, find the IFC and allocation of load between two units for minimum fuel cost for various total loads.

Solution:

For the minimum load of 100 MW,

$$P_{G_1} = 50 \text{ MW}, \quad P_{G_2} = 50 \text{ MW}$$

$$\frac{dC_1}{dP_{G_1}} = (\text{IC})_1 = 15 + 0.08 \times 50 = 19 \text{ Rs./hr} \tag{2.30}$$

$$\frac{dC_2}{dP_{G_2}} = (\text{IC})_2 = 13 + 0.1 \times 50 = 18 \text{ Rs./hr} \tag{2.31}$$

From Equations (2.30) and (2.31), it is noted that at a total minimum load of 100 MW, Unit-1 is operating at a higher IFC than Unit-2. Therefore, additional load on Unit-2 should be increased till $(\text{IC})_2 = \lambda = 19$ and at that point,

$$13 + 0.1 P_{G_2} = 19$$

$$\therefore P_{G_2} = 60$$

Hence, the total load being delivered at equal incremental costs of 19 Rs./MWh is 110 MW, i.e., $P_{G_1} = 50$ and $P_{G_2} = 60$.

Go on increasing the load on each unit so that the units operate at the same incremental cost, and these operating conditions are found by assuming various values of λ and by calculating the output for each unit.

λ	P_{G_1}	P_{G_2}	$\sum_{i=1}^{3} P_{G_i}$
18	50.0	50.0	100.00
19	50.0	60.0	110.00
20	62.5	70.0	132.50
22	87.5	90.0	167.50
25	125.0	120.0	245.0
28	163.0	150.0	313.0
30	188.0	170.0	358.0
31	200.0	180.0	380.0
32	200.0	200.0	400.0

Example 2.9: Determine the saving in fuel cost in Rs./hr for the economic distribution of the total load of 100 MW between two units of the plant as given in Example 2.8. Compare with equal distribution of the same total load.

Solution:

For the optimal distribution of the total load between the two units,

$$\frac{dC_1}{dP_{G_1}} = \frac{dC_2}{dP_{G_2}} = \lambda$$

$$\therefore 0.08P_{G_1} + 15 = 0.1P_{G_2} + 13$$

or $\quad 0.08P_{G_1} - 0.1P_{G_2} = 13 - 15 = -2$ $\hfill$ **(2.32)**

Given $\quad P_{G_1} + P_{G_2} = 110$ $\hfill$ **(2.33)**

By solving Equations (2.32) and (2.33), we get

$$0.08P_{G_1} - 0.1P_{G_2} = -2$$

Equation $(2.33) \times 0.1 \Rightarrow \quad 0.1P_{G_1} + 0.1P_{G_2} = 11$

$$0.18P_{G_1} = 9$$

or $\quad P_{G_1} = 50 \text{ MW}$

Substituting $P_{G_1} = 50$ MW in Equation (2.32), we get

$$P_{G_2} = 60 \text{ MW}$$

Operating cost of Unit-1,

$$C_1 = \int \frac{dC_1}{dP_{G_1}} = 0.04P_{G_1}^2 + 15P_{G_1}$$

Operating cost of Unit-2,

$$C_2 = \int \frac{dC_2}{dP_{G_2}} = 0.05P_{G_2}^2 + 13P_{G_2}$$

The operating costs of Unit-1 and Unit-2 are

$$C_1 = 0.04\,(50)^2 + 15(50) = 850 \text{ Rs./hr}$$
$$C_2 = 0.05(60)^2 + 13(60) = 960 \text{ Rs./hr}$$

For the equal distribution of load $\Rightarrow$ $P_{G_1} = 55$ MW and $P_{G_2} = 55$ MW.

The operating costs of Unit-1 and Unit-2 are

$$C_1 = 0.04(55)^2 + 15(55) = 946 \text{ Rs./hr}$$
$$C_2 = 0.05(55)^2 + 13(55) = 866.25 \text{ Rs./hr}$$

The increase in cost for Unit-1 when the delivering power increases from 50 to 55 MW is $946 - 850 = 96$ Rs./hr and for Unit-2 decreases in cost due to decrease in power generation from 60 to 55 MW is $960 - 866.25 = -93.75$ Rs./hr.

$\therefore$ Saving in cost $= 96 - 93.75 = 3.75$ Rs./hr.

Example 2.10: Three power plants of a total capacity of 500 MW are scheduled for operation to supply a total system load of 350 MW. Find the optimum load scheduling if the plants have the following incremental cost characteristics and the generator constraints:

$$\frac{dC_1}{dP_{G_1}} = 40 + 0.25P_{G_1}, \quad 30 \le P_{G_1} \le 150$$

$$\frac{dC_2}{dP_{G_2}} = 50 + 0.30P_{G_2}, \quad 40 \le P_{G_2} \le 125$$

$$\frac{dC_3}{dP_{G_3}} = 20 + 0.20P_{G_3}, \quad 50 \le P_{G_3} \le 225$$

Solution:

For economic load scheduling among the power plants, the necessary condition is

$$\frac{dC_i}{dP_{G_i}} = \lambda$$

For three plants,

$$\frac{dC_1}{dP_{G_1}} = \frac{dC_2}{dP_{G_2}} = \frac{dC_3}{dP_{G_3}} = \lambda$$

Given total load $= P_{G_1} + P_{G_2} + P_{G_3} = 350$ MW $\hfill(2.34)$

$$40 + 0.25P_{G_1} = 50 + 0.30P_{G_2} = 20 + 0.20P_{G_3} = \lambda \hfill(2.35)$$

$$\frac{dC_1}{dP_{G_1}} = \frac{dC_3}{dP_{G_3}}$$

$$\Rightarrow 40 + 0.25\,P_{G_1} = 50 + 0.30\,P_{G_2}$$

or $0.25 P_{G_1} - 0.30 P_{G_2} = 50 - 40 = 10$ **(2.36)**

and $40 + 0.25 P_{G_1} = 20 + 0.2 P_{G_3}$

or $0.25 P_{G_1} - 0.2 P_{G_3} = 20 - 40 = -20$ **(2.37)**

From Equation (2.36), we have

$$0.25 P_{G_1} - 10 = 0.30 P_{G_2}$$

$$\therefore P_{G_2} = \frac{0.25 P_{G_1} - 10}{0.3} = 0.833 P_{G_1} - 33.33$$ **(2.38)**

Substituting Equation (2.38) in Equation (2.34)

$$P_{G_1} + 0.833 P_{G_1} - 33.33 + P_{G_3} = 350$$

$$\text{or}\quad 1.833 P_{G_1} + P_{G_3} = 383.33$$ **(2.39)**

Solving Equations (2.37) and (2.39)

$$0.25 P_{G_1} - 0.2 P_{G_3} = -20$$

Equation (2.39) $\times\, 0.2 \Rightarrow\ 0.366 P_{G_1} + 0.2 P_{G_3} = 76.66$

$$\overline{\rule{0pt}{1em}\qquad\qquad\qquad\qquad\qquad}$$

$$0.616 P_{G_1} \qquad = 56.66$$

$$\text{or}\quad P_{G_1} \qquad = 91.98 \text{ MW}$$

Substituting the value of P_{G_1} in Equation (2.39),

$$1.833 \times 91.98 + P_{G_3} = 383.33$$

$$\text{or}\quad P_{G_3} = 214.73 \text{ MW}$$

Substituting the values of P_{G_1} and P_{G_2} in Equation (2.34), we get

$$91.98 + P_{G_2} + 214.73 = 350$$

$$\text{or}\quad P_{G_2} = 43.29 \text{ MW}$$

$\therefore$ For economic scheduling of the load, the generations of three plants must be

$$P_{G_1} = 91.98 \text{ MW}, \quad P_{G_2} = 43.29 \text{ MW}, \quad \text{and} \quad P_{G_3} = 214.73 \text{ MW}$$

Example 2.11: The fuel cost of two units are given by

$$C_1 = 0.1 P_{G_1}^2 = 25 P_{G_1} + 1.6 \text{ Rs./hr}$$

$$C_2 = 0.1 P_{G_2}^2 + 32 P_{G_2} + 2.1 \text{ Rs./hr}$$

If the total demand on the generators is 250 MW, find the economical load distribution of the two units.

Solution:

Given

$$\frac{dC_1}{dP_{G_1}} = 0.2P_{G_1} + 25 \text{ Rs./MWh}$$

$$\frac{dC_2}{dP_{G_2}} = 0.2P_{G_2} + 32 \text{ Rs./MWh}$$

Given the total load, $P_D = 250$ MW. For economical distribution of total load, the condition is

$$\frac{dC_1}{dP_{G_1}} = \frac{dC_2}{dP_{G_2}}$$

$$0.2P_{G_1} + 25 = 0.2P_{G_2} + 32$$

or $\quad 0.2P_{G_1} - 0.2P_{G_2} = 7$

$\quad$ or $\quad P_{G_1} - P_{G_2} = 35$ $\hspace{3cm}$ **(2.40)**

$\quad$ and $\quad P_{G_1} + P_{G_2} = 250$ $\quad$ (given) $\hspace{2cm}$ **(2.41)**

By solving Equations (2.40) and (2.41), we get

$$2P_{G_1} = 285$$

or $\quad P_{G_1} = 142.5$ MW

Substituting the P_{G_1} value in Equation (2.41), we get

$$P_{G_2} = 250 - P_{G_1} = 107.5 \text{ MW}$$

Example 2.12: A plant has two generators supplying the plant bus, and neither is to operate below 20 or above 125 MW. Incremental costs of the two units are

$$\frac{dC_1}{dP_{G_1}} = 0.15P_{G_1} + 20 \text{ Rs./MWh}$$

$$\frac{dC_2}{dP_{G_2}} = 0.225P_{G_2} + 17.5 \text{ Rs./MWh}$$

For economic dispatch, find the plant cost of the received power in Rs./MWh (λ) when $P_{G_1} + P_{G_2}$ equals: (a) 40 MW, (b) 100 MW, and (c) 225 MW.

Solution:

For economic operation,

$$\frac{dC_1}{dP_{G_1}} = \frac{dC_2}{dP_{G_2}} = \lambda$$

(a) When $P_{G_1} + P_{G_2} = 40$ MW $\hspace{4cm}$ **(2.42)**

$$\frac{dC_1}{dP_{G_1}} = \frac{dC_2}{dP_{G_2}}$$

$$0.15P_{G_1} + 20 = 0.225P_{G_2} + 17.5$$

or $\quad 0.15P_{G_1} - 0.225P_{G_2} = -2.5$ **(2.43)**

Equation (2.42) $\times 0.15 \Rightarrow 0.15P_{G_1} + 0.15P_{G_2} = 6.0$ **(2.44)**

Solving Equations (2.43) and (2.44), we get

$$-0.375P_{G_2} = -8.5$$

$$P_{G_2} = 22.66 \text{ MW}$$

Substituting $P_{G_2} = 22.66 \text{ MW}$ in Equation (2.42)

$$P_{G_1} = 40 - 22.666$$

$$= 17.34 \text{ MW}$$

$$\therefore \frac{dC_2}{dP_{G_2}} = \lambda$$

$$0.225P_{G_2} + 17.5 = \lambda$$

or $\quad 0.225(226) + 17.5 = \lambda$

$$\therefore \lambda = 22.59 \text{ Rs./MWh}$$

(b) $\quad$ When $P_{G_1} + P_{G_2} = 100 \text{ MW}$ **(2.45)**

Equation (2.45) $\times 0.15 \Rightarrow 0.15P_{G_1} + 0.15P_{G_2} = 15$ **(2.46)**

By solving Equations (2.43) and (2.46), we get

$$0.15P_{G_1} - 0.225P_{G_2} = -2.5$$

$$\underline{0.15P_{G_1} + 0.15P_{G_2} = 15}$$

$$-0.375P_{G_2} = -17.5$$

$$P_{G_2} = 46.66 \text{ MW}$$

Substituting the P_{G_2} value in Equation (2.45), we get

$$P_{G_1} = 53.34 \text{ MW}$$

$$\therefore \ 0.15P_{G_1} + 20 = \lambda \quad \text{or} \quad 0.225P_{G_2} + 17.5 = \lambda$$

$$0.15(53.34) + 20 = \lambda \quad \text{or} \quad \lambda = 0.225(46.66) + 17.5$$

$$\Rightarrow \lambda = 28 \text{ Rs./MWh}; \quad \lambda = 28 \text{ Rs./MWh}$$

(c) When $P_{G_1} + P_{G_2} = 225$ MW (2.47)

Equation $(2.47) \times 0.15 \Rightarrow -0.15 P_{G_1} + 0.15 P_{G_2} = 33.75$ (2.48)

By solving Equations (2.43) and (2.48), we get

$$0.15 P_{G_1} - 0.225 P_{G_2} = -2.5$$

$$-0.15 P_{G_1} + 0.15 P_{G_2} = 33.75$$

$$\overline{}$$

$$0.375 P_{G_2} = 36.25$$

$$\text{or} \quad P_{G_2} = 96.66 \text{ MW}$$

Substituting the P_{G_2} value in Equation (2.47), we get

$$P_{G_1} = 128.34 \text{ MW}$$

$$\therefore \ \lambda = 0.225 P_{G_2} + 17.5$$

$$= 0.225(96.66) + 17.5$$

$$= 39.24 \text{ Rs./MWh}$$

Example 2.13: The cost curves of two generators may be approximated by second-degree polynomials:

$$C_1 = 0.1 P_{G_1}^2 + 20 P_{G_1} + \alpha_1$$

$$C_2 = 0.1 P_{G_2}^2 + 30 P_{G_2} + \alpha_2$$

where α_1 and α_2 are constants

If the total demand on the generators is 200 MW, find the optimum generator settings. How many rupees per hour would you lose if the generators were operated about 15% of the optimum settings?

Solution:

For economic operation,

$$\frac{dC_1}{dP_{G_1}} = \frac{dC_2}{dP_{G_2}}$$

$$0.2 P_{G_1} + 20 = 0.2 P_{G_2} + 30$$

$$\text{or} \quad 0.2 P_{G_1} + 0.2 P_{G_2} = 10$$

$$\text{or} \quad P_{G_1} - P_{G_2} = 50 \tag{2.49}$$

and given that $P_{G_1} + P_{G_2} = 200$ (2.50)

Solving Equations (2.49) and (2.50), we get

$$2 P_{G_1} = 250$$

or $P_{G_1} = 125\,\text{MW}$

Substituting the P_{G_1} value in Equation (2.50), we get

$$P_{G_2} = 200 - 125 = 75\,\text{MW}$$

If the generators were operated about 15% of the optimum settings,

$$P_{G_1} = 125 - 125 \times \frac{15}{100} = 125 - 18.75 = 106.25\,\text{MW}$$

and $P_{G_2} = 75 \times \dfrac{15}{100} = 75 - 11.25 = 63.75\,\text{MW}$

The decrease in cost for Generator-1 is

$$\Delta C_1 = -\int_{125}^{106.25} (0.2P_{G_1} + 20)\,\mathrm{d}P_{G_1} = -\left[\frac{0.2P_{G_1}^2}{2} + 20P_{G_1}\right]_{125}^{106.25}$$

$$= -\left[0.1(106.25)^2 + 20(106.25) - 0.1(125)^2 - 20(125)\right]$$

$$= -\left[0.1(106.25)^2 - (125^2) + 20(106.25 - 125)\right]$$

$$= -\left[433.59375 - 375\right]$$

$$= -58.59\,\text{Rs./hr}$$

The decrease in cost for Generator-2 is

$$\Delta C_2 = -\int_{75}^{63.75} (0.2P_{G_2} + 30)\,\mathrm{d}P_{G_2} = -\left[0.1P_{G_2}^2 + 30P_{G_2}\right]_{75}^{63.75}$$

$$= -\left[0.1(63.75^2 - 75^2) + 30(63.75 - 75)\right]$$

$$= -[156.09375 - 337.5] = -181.40625\,\text{Rs./hr}$$

The loss of amount $= \Delta C_1 - \Delta C_2$

$$= -58.59 - (-181.40625)$$

$$= -122.81\,\text{Rs./hr}$$

Example 2.14: Determine the saving in fuel cost in Rs./hr for the economic distribution of a total load of 225 MW between the two units with IFCs:

$$\frac{\mathrm{d}C_1}{\mathrm{d}P_{G_1}} = 0.075P_{G_1} + 15$$

$$\frac{\mathrm{d}C_2}{\mathrm{d}P_{G_2}} = 0.085P_{G_1} + 12$$

Compare with equal distribution of the same total load.

Solution:

Given: $P_{G_1} + P_{G_2} = 225\,\text{MW}$ **(2.51)**

For optimal operation:

$$\frac{dC_1}{dP_{G_1}} = \frac{dC_2}{dP_{G_2}}$$

$$\Rightarrow 0.075 P_{G_1} + 15 = 0.085 P_{G_2} + 12$$

or $0.075 P_{G_1} - 0.085 P_{G_2} = -3$ **(2.52)**

Equation $(2.51) \times 0.085 \Rightarrow 0.085 P_{G_1} + 0.285 P_{G_2} = 225 \times 0.085 = 19.125$ **(2.53)**

By solving Equations (2.52) and (2.53), we get

$0.16 P_{G_1} = 16.125$

or $P_{G_1} = 100.78\,\text{MW}$

Substituting the P_{G_1} value in Equation (2.51), we get

$P_{G_2} = 225 - 100.78 = 124.218\,\text{MW}$

With equal distribution of the total load,

$\Rightarrow P_{G_1} = 112.5\,\text{MW}$ and $P_{G_2} = 112.5\,\text{MW}$

The increase in cost for Unit-1 is

$$\int_{100.78}^{112.5} [(0.075 P_{G_1} + 15)\,dP_{G_1}] = \left[0.0375 P_{G_2}^2 + 15 P_{G_1}\right]_{100.78}^{112.5}$$

$$= 0.0375[(112.5)^2 - (100.78)^2] + 15(112.5 - 100.78)$$

$$= 93.73656 + 175.8$$

$$= 269.53656\,\text{Rs./hr}$$

For Unit-2,

$$\int_{124.218}^{112.5} [(0.085 P_{G_2} + 12)\,dP_{G_2}] = \left[0.0425 P_{G_2}^2 + 12 P_{G_2}\right]_{124.218}^{112.5}$$

$$= 0.0425[(112.5)^2 - (124.218)^2] + 12(112.5 - 124.218)$$

$$= -117.8891148 - 140.616$$

$$= -258.505\,\text{Rs./hr}$$

The negative sign indicates a decrease in cost.

$\therefore$ Saving in fuel cost $=$ Rs. $269.53656 - 258.505$

$$= 11.03156\,\text{Rs./hr}$$

Example 2.15: Three plants of a total capacity of 500 MW are scheduled for operation to supply a total system load of 310 MW. Evaluate the optimum load scheduling if the plants have the following cost characteristics and the limitation:

$$C_1 = 0.06P_{G_1}^2 + 30P_{G_1} + 10, \quad 30 \leq P_{G_1} \leq 150$$

$$C_2 = 0.10P_{G_2}^2 + 40P_{G_2} + 15, \quad 20 \leq P_{G_2} \leq 100$$

$$C_3 = 0.075P_{G_3}^2 + 10P_{G_3} + 20, \quad 50 \leq P_{G_3} \leq 250$$

Solution:

The IFCs of the three plants are

$$\frac{dC_1}{dP_{G_1}} = 0.12P_{G_1} + 30 \text{ Rs./MWh}$$

$$\frac{dC_2}{dP_{G_2}} = 0.20P_{G_2} + 40 \text{ Rs./MWh}$$

$$\frac{dC_3}{dP_{G_3}} = 0.15P_{G_3} + 10 \text{ Rs./MWh}$$

and $\quad P_{G_1} + P_{G_2} + P_{G_3} = 310 \text{ MW}$

For optimum scheduling of units,

$$\frac{dC_1}{dP_{G_1}} = \frac{dC_2}{dP_{G_2}} = \frac{dC_3}{dP_{G_3}}$$

$$0.12P_{G_1} + 30 = 0.20P_{G_2} + 40 = 0.15P_{G_3} + 10$$

$$\frac{dC_1}{dP_{G_1}} = \frac{dC_3}{dP_{G_3}}$$

$$\Rightarrow 0.12P_{G_1} + 30 = 0.15P_{G_3} + 10$$

or $\quad 0.12P_{G_1} - 0.15P_{G_2} = -20$ (2.54)

and given that $\quad P_{G_1} + P_{G_3} = 310 - P_{G_2}$ (2.55)

Solving Equations (2.54) and (2.55), we have

$$0.12P_{G_1} - 0.15P_{G_3} = -20$$

Equation (2.55) $\times 0.15 \Rightarrow 0.15P_{G_1} + 0.15P_{G_3} = 46.5 - 0.15P_{G_2}$

$$\overline{\qquad\qquad\qquad\qquad\qquad\qquad\qquad\qquad}$$

$$0.27P_{G_1} = -20 + 46.5 - 0.15P_{G_2}$$

or $\quad 0.27P_{G_1} + 0.15P_{G_2} = 26.5$ (2.56)

and $\quad \dfrac{dC_1}{dP_{G_1}} = \dfrac{dC_2}{dP_{G_2}}$

$$0.12P_{G_1} + 30 = 0.2P_{G_2} + 40$$

$$0.12P_{G_1} - 0.2P_{G_2} = 10 \tag{2.57}$$

Solving Equations (2.56) and (2.57), we get

Equation $(2.56) \times 0.2 \quad \Rightarrow 0.054P_{G_1} + 0.03P_{G_2} = 5.3$

Equation $(2.57) \times 0.15 \quad \Rightarrow 0.018P_{G_1} - 0.03P_{G_2} = 1.5$

$$\overline{\qquad\qquad\qquad 0.072P_{G_1} = 6.8 \qquad}$$

$$\text{or} \quad P_{G_1} = 94.444 \text{ MW}$$

Substituting the P_{G_1} value in Equation (2.54), we get

$$0.12\,(94.444) - 0.15P_{G_3} = -20$$

$$11.33 - 0.15P_{G_3} = -20$$

$$31.33 = 0.15P_{G_3}$$

$$\text{or} \quad P_{G_3} = 208.86 \text{ MW}$$

Substituting the P_{G_1} and P_{G_3} values in Equation (2.55), we get

$$94.44 + 208.86 + P_{G_2} = 310$$

$$\therefore \quad P_{G_2} = 6.7 \text{ MW}$$

The optimal power generation is

$$P_{G_1} = 94.44 \text{ MW}$$

$$P_{G_2} = 6.7 \text{ MW}$$

$$\text{and} \quad P_{G_3} = 208.86 \text{ MW}$$

It is observed that the real-power generation of Unit-2 is 6.7 MW and it is violating its minimum generation limit. Hence, we have to fix its value at its minimum generation, i.e., $P_{G_2} = 20$ MW.

$$\text{Given:} \quad P_{G_1} + P_{G_2} + P_{G_3} = 310 \text{ MW}$$

$$P_1 + P_{G_3} = 310 - 20 = 290 \text{ MW}$$

The remaining load of 290 MW is to be distributed optimally between Unit-1 and Unit-3 as follows:

$$\frac{dC_1}{dP_{G_1}} = \frac{dC_3}{dP_{G_3}}$$

$$0.12P_{G_1} + 30 = 0.15P_{G_3} + 10$$

$$\text{or} \quad 0.12P_{G_1} - 0.15P_{G_3} = -20 \tag{2.58}$$

$$\text{and} \quad P_{G_1} + P_{G_3} = 290 \tag{2.59}$$

Solving Equations (2.58) and (2.59), we get:

$$0.12P_{G_1} - 0.5P_{G_3} = -20$$

Equation $(2.59) \times 0.15 \Rightarrow 0.15P_{G_1} + 0.15P_{G_3} = 43.5$

$$0.27P_{G_1} = 23.5$$

$$\text{or} \quad P_{G_1} = 87.03 \text{ MW}$$

Substituting the P_{G_1} value in Equation (2.59), we get

$$P_{G_3} = 290 - 67.14 = 202.96 \text{ MW}$$

The total load of 310 MW is distributed optimally among the units as

$$P_{G_1} = 87.03 \text{ MW}$$

$$P_{G_2} = 20 \text{ MW}$$

$$\text{and} \quad P_{G_3} = 202.96 \text{ MW}$$

Example 2.16: The incremental cost characteristics of two thermal plants are given by

$$\frac{dC_1}{dP_{G_1}} = 0.2P_{G_1} + 60 \text{ Rs./MWh}$$

$$\frac{dC_2}{dP_{G_2}} = 0.3P_{G_2} + 40 \text{ Rs./MWh}$$

Calculate the sharing of a load of 200 MW for most economic operations. If the plants are rated 150 and 250 MW, respectively, what will be the saving in cost in Rs./hr in comparison to the loading in the same proportion to rating.

Solution:

For economic operation,

$$\frac{dC_1}{dP_{G_1}} = \frac{dC_2}{dP_{G_2}}$$

$$0.2P_{G_1} + 60 = 0.3P_{G_2} + 40$$

or $0.2P_{G_1} - 0.3P_{G_2} = -20$ (2.60)

$$P_{G_1} + P_{G_2} = 200 \quad \text{(given)} \tag{2.61}$$

Solving Equations (2.60) and (2.61), we get

$$0.2P_{G_1} - 0.3P_{G_2} = -20$$

Equation $(2.61) \times 0.3 \Rightarrow 0.3P_{G_1} - 0.3P_{G_2} = 60$

$$0.5P_{G_1} = 40$$

$$\therefore \quad P_{G_1} = 80 \text{ MW}$$

Substituting the P_{G_1} value in Equation (2.61), $P_{G_2} = 120$ MW. If the plants are loaded in the same proportion to the rating,

i.e., $\quad P_{G_1} = 150$ MW, $\quad P_{G_2} = 250$ MW

Increase in the operation cost for Plant-1 is

$$\int_{80}^{150} (0.2P_{G_1} + 60)\, dP_{G_1} = \left[0.2\, P_{G_1}^2 + 60P_{G_1}\right]_{80}^{150}$$

$$= 0.1(150^2 - 80^2) + 60(150 - 80)$$
$$= 1{,}610 + 4{,}200$$
$$= 5{,}810 \text{ Rs./hr}$$

Increase in the operation cost for Plant-2 is

$$\int_{120}^{250} (0.3P_{G_2} + 40)\, dP_{G_2} = \left[0.15\, P_{G_2}^2 + 40P_{G_2}\right]_{120}^{250}$$

$$= 0.15(250^2 - 120^2) + 40(250 - 120)$$
$$= 7{,}215 + 5{,}200$$
$$= 12{,}415 \text{ Rs./hr}$$

$\therefore$ Saving in operation cost $= 12{,}415 - 5{,}810 = 66$ Rs./hr

Example 2.17: The IFCs of two units in a generating station are as follows:

$$\frac{dC_1}{dP_{G_1}} = 0.15P_{G_1} + 35$$

$$\frac{dC_2}{dP_{G_2}} = 0.20P_{G_2} + 28$$

Assuming continuous running with a total load of 150 MW, calculate the saving per hour obtained by using the most economical division of load between the units as compared with loading each equally. The maximum and minimum operational loadings are the same for each unit and are 125 and 20 MW, respectively.

Solution:

Given: $\quad \dfrac{dC_1}{dP_{G_1}} = 0.15P_{G_1} + 35$

$$\frac{dC_2}{dP_{G_2}} = 0.20P_{G_2} + 28$$

Total load $= P_{G_1} + P_{G_2} = 150$ MW $\hfill$ (2.62)

For optimality, $\dfrac{dC_1}{dP_{G_1}} = \dfrac{dC_2}{dP_{G_2}}$

$$0.15P_{G_1} + 35 = 0.20P_{G_2} + 28$$

or $0.15P_{G_1} - 0.20P_{G_2} = -7$ (2.63)

Solving Equations (2.62) and (2.63), we get

Equation (2.62) × 0.3 $\Rightarrow 0.2P_{G_1} + 0.2P_{G_2} = 30$

$$0.15P_{G_1} - 0.20P_{G_2} = -7$$

$$\overline{}$$

$$0.35P_{G_1} = 23$$

$$\therefore \quad P_{G_1} = 65.714 \text{ MW}$$

Substituting the P_{G_1} value in Equation (2.62), we get

$$P_{G_2} = 84.286 \text{ MW}$$

With an equal sharing of load, $P_{G_1} = 75$ MW and $P_{G_2} = 75$ MW.

With an equal distribution of load, the load on Plant-1 is increased from 65.714 to 75 MW.

The increase in cost of operation for Plant-1 is

$$\Delta C_1 = \int\limits_{65.714}^{75} (0.15P_{G_1} + 35)\, dP_{G_1}$$

$$= \left[0.075P_{G_1}^2 + 35P_{G_1} \right]_{65.714}^{75}$$

$$= 0.075(75^2 - 65.714)^2 + 35(75 - 65.714)$$

$$= (1{,}306.67)0.075 + 35(9.286)$$

$$= 98 + 325.01 = 423.01 \text{ Rs./hr}$$

The load on Plant-2 is decreased from 84.286 to 75 MW.

$$\Delta C_2 = -\int\limits_{84.286}^{75} (0.2P_{G_2} + 28P_{G_2}) = -\left[0.2P_{G_2}^2 + 28P_{G_2} \right]_{84.386}^{75}$$

$$= - \left[0.2(75^2 - 84.286^2) + 28(75 - 84.286) \right]$$

$$= -[-147.913 - 260]$$

$$= 407.921$$

$$\therefore \text{The solving in cost} = 423.01 - 407.921$$

$$= 15.089 \text{ Rs./hr}$$

Example 2.18: If two plants having cost characteristics as given

$$C_1 = 0.1P_{G_1}^2 + 60P_{G_1} + 135 \text{ Rs./hr}$$

$$C_2 = 0.15P_{G_2}^2 + 40P_{G_2} + 100 \text{ Rs./hr}$$

have to meet the following daily load cycle:

    0 to 6 hrs   – 7 MW
    18 to 24 hrs – 70 MW

find the economic schedule for the different load conditions. If a cost of Rs. 450 is involved in taking either plant out of services or to return to service, find whether it is more economical to keep both plants in service for the whole day or to remove one of them during light-load service.

Solution:

$$C_1 = 0.1P_{G_1}^2 + 60P_{G_1} + 135$$

$$\frac{dC_1}{dP_{G_1}} = 0.2P_{G_1} + 60 \text{ Rs./MWh}$$

$$C_2 = 0.15P_{G_2}^2 + 40P_{G_2} + 100$$

$$\frac{dC_2}{dP_{G_2}} = 0.3P_{G_2} + 40 \text{ Rs./MWh}$$

For 0–6 hr: Total load $= 7$ MW

i.e., $P_{G_1} + P_{G_2} = 7$ MW   (2.64)

The condition for the optimal distribution of load is

$$\frac{dC_1}{dP_{G_1}} = \frac{dC_2}{dP_{G_2}}$$

$$0.2P_{G_1} + 60 = 0.3P_{G_2} + 40$$

$$0.2P_{G_1} - 0.3P_{G_2} = -20 \qquad\qquad\qquad (2.65)$$

Solving Equations (2.64) and (2.65), we get

Equation (2.64) $\times 0.3 \;\Rightarrow\; 0.3P_{G_1} + 0.3P_{G_2} = 2.1$

$$0.2P_{G_1} - 0.3P_{G_2} = -20$$

$$\overline{\phantom{0.2P_{G_1} - 0.3P_{G_2} = -20}}$$

$$0.5P_{G_1} = -17.9$$

$$P_{G_1} = -35.8 \text{ MW}$$

Since the real-power generation of Plant-1 is $P_{G_1} = -35.8$ MW, it violates the minimum generation limit. Hence, to meet the load demand of 7 MW, it is necessary to run Unit-2 only with generation of 7 MW.

Operation cost of Unit-2 during 0–6 hr is

$$C_2 = 0.15(7)^2 + 40(7) + 100$$
$$= 7.35 + 280 + 100$$
$$= 387.35 \text{ Rs./hr}$$

For 18–24 hr:

Total load $= 70$ MW

i.e., $P_{G_1} + P_{G_2} = 70$ MW   **(2.66)**

Solving Equations (2.66) and (2.65), we get

Equation $(2.66) \times 0.3 \Rightarrow 0.3P_{G_1} + 0.3P_{G_2} = 21$

$$0.2P_{G_1} + 0.3P_{G_2} = -20$$

$$\overline{\qquad\qquad\qquad\qquad}$$

$$0.5P_{G_1} = 1$$

$$\therefore P_{G_1} = 2 \text{ MW} \quad \text{and} \quad P_{G_2} = 70 - 2 = 68 \text{ MW}$$

The cost of operation of Plant-1 with 2-MW generation is

$$C_1 = 0.1P_{G_1} + 60P_{G_1} + 135$$
$$= 0.1(2)^2 + 60(2) + 135 = 255.4 \text{ Rs./hr}$$

The cost of operation of Pant-2 with 68-MW generation is

$$C_2 = 0.15(68)^2 + 40(68) + 100 = 3{,}513.6 \text{ Rs./hr}$$

The operating cost during 18–24 hr $= 255.4 + 3{,}513.6 = 3{,}769$ Rs./hr
The total operating cost during an entire 24-hr period is

$$387.35 \times 6 + 3{,}769 \times 6 = \text{Rs. } 24{,}938.10$$

A cost of Rs. 450 is incurred as the start-up cost.

$$\therefore \text{ Total operating cost} = 24{,}938.1 + 450 = \text{Rs. } 25{,}388.10$$

Example 2.19: The IFCs in rupees per MWh for a plant consisting of two units are

$$\frac{dC_1}{dP_{G_1}} = 0.20P_{G_1} + 40.0$$

$$\frac{dC_2}{dP_{G_2}} = 0.25P_{G_2} + 30.0$$

Calculate the extra cost increased in Rs./hr, if a load of 220 MW is scheduled as $P_{G_1} = P_{G_2} = 110$ MW.

Solution:

For optimal scheduling of units,

$$\frac{dC_1}{dP_{G_1}} = \frac{dC_2}{dP_{G_2}}$$

$$0.20P_{G_1} + 40.0 = 0.25P_{G_2} + 30$$

or $\quad 0.20P_{G_1} - 0.25P_{G_2} = -10$ **(2.67)**

Given: $\quad P_{G_1} + P_{G_2} = 220$ **(2.68)**

Solving Equations (2.67) and (2.68), we get

$$0.20P_{G_1} - 0.25P_{G_2} = -10$$

Equation $(2.68) \times 0.25 \quad \Rightarrow 0.25P_{G_1} + 0.25P_{G_2} = 55$

$$\overline{\hspace{3cm}}$$

$$0.45P_{G_1} = 45$$

$$\text{or} \quad P_{G_1} = 100 \text{ MW}$$

Substituting the P_{G_1} value in Equation (2.68), we get

$$\therefore P_{G_2} = 220 - P_{G_1} = 120 \text{ MW}$$

For an equal distribution of load, $P_{G_1} = 110$ MW and $P_{G_2} = 110$ MW. The operation cost of Unit-1 is increased as the load shared by it is increased from 100 to 110 MW.

$\therefore$ Increase in operation cost of Unit-1

$$= \int_{100}^{110} (0.2P_{G_1} + 40.0) \ dP_{G_1} = \left[0.1P_{G_1}^2 + 40P_{G_1}\right]_{100}^{110}$$

$$= 0.1(110^2 - 100)^2 + 40(110 - 100)$$

$$= 210 + 400 = 610 \text{ Rs./hr}$$

The operation cost of Unit-2 is decreased as the load shared by it is decreased from 120 to 110 MW.

$\therefore$ Decrease in operation cost of Unit-2

$$= \int_{120}^{110} (0.25P_{G_2} + 30.0) \ dP_{G_2} = \left[0.125(P_{G_2})^2 + 30P_{G_2}\right]_{120}^{110}$$

$$= 0.125(110^2 - 120^2) + 30(110 - 120)$$

$$= -287.5 - 300$$

$$= -587.5 \text{ Rs/hr}$$

The extra cost incurred in Rs./hr if the load is equally shared by Unit-1 and Unit-2 is

$$610 - 587.5 = 22.5 \text{ Rs./hr}$$

Example 2.20: The fuel cost characteristics of two generators are obtained as under:

$$C_1(P_{G_1}) = 1,000 + 50P_{G_1} + 0.01P_{G_1}^2 \text{ Rs./hr}$$

$$C_2(P_{G_2}) = 2,500 + 45P_{G_2} + 0.005P_{G_2}^2 \text{ Rs./hr}$$

If the total load supplied is 1,000 MW, find the optimal load division between two generators.

Solution:

$$C_1(P_{G_1}) = 1,000 + 50P_{G_1} + 0.01P_{G_1}^2 \text{ Rs./hr}$$

$$C_2(P_{G_2}) = 2,500 + 45P_{G_2} + 0.005P_{G1}^2 \text{ Rs./hr}$$

The IFC characteristics are

$$\frac{dC_1}{dP_{G_1}} = 50 + 0.02P_{G_1}$$

$$\frac{dC_2}{dP_{G_2}} = 45 + 0.01P_{G_2}$$

The condition for optimal load division is

$$\frac{dC_1}{dP_{G_1}} = \frac{dC_1}{dP_{G_1}} = \lambda$$

$$50 + 0.02P_{G_1} = 45 + 0.01P_{G_2}$$

or $0.02P_{G_1} + 0.01P_{G_2} = -5.0$ **(2.69)**

$$P_{G_1} + P_{G_2} = 1,000 \quad \text{(given)} \tag{2.70}$$

Solving Equations (2.69) and (2.70), we get

$$0.02P_{G_1} + 0.01P_{G_2} = -5.0$$

Equation (2.70) $\times 0.01 \Rightarrow$ $\underline{\quad 0.01P_{G_1} + 0.01P_{G_2} = 10 \quad}$

$$0.03P_{G_1} = 5$$

$$\text{or} \quad P_{G_1} = 166 \text{ MW}$$

Substituting the P_{G_1} value in Equation (2.70), we get

$$P_{G_2} = 833 \text{ MW}$$

Substituting the P_{G_1} and P_{G_2} values in $\dfrac{dC_1}{dP_{G_1}}$ or $\dfrac{dC_2}{dP_{G_2}}$ equation, we get

$$\lambda = 53.33 \text{ Rs./MWh}$$

$\therefore$ The total load of 1,000 MW optimally divided in between the two generators is

$$P_{G_1} = 166 \text{ MW}$$

$$P_{G_2} = 833 \text{ MW}$$

And IFC, $\lambda = 53.33$ Rs./MWh

Example 2.21: Determine the economic operation point for the three thermal units when delivering a total of 1,000 MW:

Unit A: $P_{max} = 600$ MW, $\quad P_{min} = 150$ MW

$$C_A = 500 + 7\,P_{G_A} + 0.0015\,P_{G_A}^2$$

Unit B: $P_{max} = 500$ MW, $\quad P_{min} = 125$ MW

$$C_B = 300 + 7.88\,P_{G_B} + 0.002\,P_{G_B}^2$$

Unit C: $P_{max} = 300$ MW, $\quad P_{min} = 75$ MW

$$C_C = 80 + 7.99\,P_{G_C} + 0.05\,P_{G_C}^2$$

Fuel costs:

Unit A = 1.1 unit of price/MBtu

Unit B = 1.0 unit of price/MBtu

Unit C = 1.0 unit of price/MBtu

Find the values of P_{GA}, P_{GB}, and P_{GC} for optimal operation.

Solution:

Cost curves are:

$$C_A(P_{GA}) = H_A \times 1.1 = 550 + 7.7\,P_A + 0.00165\,P_A^2$$

$$C_B(P_{GB}) = H_B \times 1.0 = 300 + 7.88\,P_B + 0.002\,P_B^2$$

$$C_C(P_{GC}) = H_C \times 1.0 = 80 + 7.799\,P_C + 0.005\,P_C^2$$

Now IFCs are:

$$\frac{dC_A}{dP_{G_A}} = 7.7 + 0.0033\,P_{G_A}$$

$$\frac{dC_B}{dP_{G_B}} = 7.88 + 0.004\,P_{G_B}$$

$$\frac{dC_C}{dP_{G_C}} = 7.99 + 0.01\,P_{G_C}$$

For an economic system operation,

$$\frac{dC_A}{dP_{G_A}} = \frac{dC_B}{dP_{G_B}} = \frac{dC_C}{dP_{G_C}} = \lambda$$

$$\Rightarrow \frac{dC_A}{dP_{G_A}} = \frac{dC_C}{dP_{G_C}}$$

$$7.7 + 0.0033 P_{G_A} = 7.99 + 0.001 P_{G_C}$$

$$\text{or } \quad 0.0033 P_{G_A} - 0.01 P_{G_C} = 0.29 \tag{2.71}$$

$$P_{G_A} + P_{G_B} + P_{G_C} = 1,000 \quad \text{(given)}$$

$$\text{or } \quad P_{G_A} = 1,000 - (P_{G_B} + P_{G_C}) \tag{2.72}$$

Substituting P_{G_A} from Equation (2.72) in Equation (2.71), we get

$$0.0033\left[1,000 - (P_{G_B} + P_{G_C})\right] - 0.01 P_{G_C} = 0.29$$

$$\text{or } \quad 0.0033 P_{G_B} + 0.0133 P_{G_C} = 3.01 \tag{2.73}$$

$$\text{and } \quad \frac{dC_B}{dP_{G_B}} = \frac{dC_C}{dP_{G_C}}$$

$$7.88 + 0.004 P_{G_B} = 7.99 + 0.01 P_{G_C}$$

$$\text{or } \quad 0.004 P_{G_B} - 0.01 P_{G_C} = 0.11 \tag{2.74}$$

Solving Equations (2.73) and (2.74), we get

Equation $(2.73) \times 0.01 \qquad \Rightarrow \quad 0.000033 P_{G_B} + 0.000133 P_{G_C} = 0.0301$

Equation $(2.74) \times 0.0133 \qquad \Rightarrow \quad 0.0000532 P_{G_B} + 0.000133 P_{G_C} = 0.001463$

$$\overline{\qquad \qquad 8.62 \times 10^{-5} P_{G_B} = 0.031563}$$

$$\text{or } \quad P_{G_B} = 366.16 \text{ MW}$$

Substituting the P_{G_B} value in Equation (2.73), we get

$$0.0033(366.16) + 0.0133 P_{G_C} = 3.01$$

$$\text{or } \quad P_{G_C} = 135.464 \text{ MW}$$

Substituting P_{G_B} and P_{G_C} values in Equation (2.72), we get

$$P_{G_A} = 498.376 \text{ MW}$$

For a total load of 1,000 MW, the economic scheduling of three units are:

$$P_{G_A} = 498.376 \text{ MW} \qquad (150 \text{ MW} < P_{G_A} < 600 \text{ MW})$$

$$P_{G_B} = 366.16 \text{ MW} \qquad (125 \text{ MW} < P_{G_B} < 500 \text{ MW})$$

$$\text{and } \quad P_{G_C} = 135.464 \text{ MW} \qquad (75 \text{ MW} < P_{G_C} < 300 \text{ MW})$$

Example 2.22: The fuel cost curve of two generators are given as:

$$C_A(P_{G_A}) = 800 + 45 P_{G_A} + 0.01 P_{G_A}$$

$$C_B(P_{G_B}) = 200 + 43 P_{G_B} + 0.003 P_{G_B}$$

and if the total load supplied is 700 MW, find the optimal dispatch with and without considering the generator limits where the limits have been expressed as:

$$50\ \text{MW} \leq P_{G_A} \leq 200\ \text{MW}$$

$$50\ \text{MW} \leq P_{G_B} \leq 600\ \text{MW}$$

Compare the system's increment at cost with and without generator limits considered.

Solution:

$$I_{C_A} = \frac{dC_A}{dP_{G_A}} = 45 + 0.02 P_{G_A}$$

$$I_{C_B} = \frac{dC_B}{dP_{G_B}} = 43 + 0.006 P_{G_B}$$

For economic operation, $I_{C_A} = I_{C_B} = \lambda$

Considering along with the given constraint equations:

$$\lambda = 45 + 0.02 P_{G_A}$$

$$\lambda = 43 + 0.006 P_{G_B}$$

$$P_{G_A} + P_{G_B} = 700\ \text{MW}$$

Solving these equations, $\lambda = 46.7$

$$P_{G_A} = 84.6\ \text{MW}$$

$$P_{G_B} = 615.4\ \text{MW}$$

In the above illustration, generator limits have not been included. If these limits are now included, it may be seen that Generator-B has violated the limit. Fixing it at the upper-most limits, let

$$P_{G_B} = 600\ \text{MW}$$

And obviously by so that $P_{G_A} = 100\ \text{MW}$ (since $P_{G_A} + P_{G_B} = 700\ \text{MW}$)

$$\therefore \quad \lambda_A = 45 + 0.02 \times 100 = 47$$

$$\lambda_B = 43 + 0.006 \times 600 = 46.6$$

Hence, it is observed that $\lambda_A \neq \lambda_B$, i.e., economic operation is not strictly maintained in this particular condition; incremental cost of Unit-A is now marginally more than that of Unit-B. However, in practice, this difference of λ_A and λ_B is not much; hence, the system operation is justified under this condition.

Example 2.23: The fuel cost curve of two generators are given as

$$C_1 = 625 + 35 P_{G_1} + 0.06 P_{G_1}^2$$

$$C_2 = 175 + 30 P_{G_2} + 0.005 P_{G_2}^2$$

if the total load supplied is 550 MW, find the optimal dispatch with and without considering the generator limits:

$$35 \text{ MW} \leq P_{G_1} \leq 175 \text{ MW}$$

$$33 \text{ MW} \leq P_{G_2} \leq 600 \text{ MW}$$

and also comment about the incremental cost of both cases.

Solution:

Given that total load $= P_{G_1} + P_{G_2} = 550 \text{ MW}$ $\qquad\qquad$ **(2.75)**

Cost of first unit, $C_1 = 625 + 35P_{G_1} + 0.06P_{G_1}^2$

The IFC of first unit, $\dfrac{dC_1}{dP_{G_1}} = 0.12P_{G_1} + 35 \text{ Rs./MWh}$

Cost of second unit, $C_2 = 175 + 30P_{G_2} + 0.005P_{G_2}^2$

The IFC of second unit, $\dfrac{dC_2}{dP_{G_2}} = 0.01P_{G_2} + 30 \text{ Rs./MWh}$

Case-I: Without considering generator limits:

For optimal dispatch of load, the necessary condition is

$$\frac{dC_1}{dP_{G_1}} = \frac{dC_2}{dP_{G_2}} = \lambda$$

$$0.12P_{G_1} + 35 = 0.01P_{G_2} + 30$$

$$0.12P_{G_1} + 0.01P_{G_2} = -5 \qquad\qquad \textbf{(2.76)}$$

Solving Equations (2.75) and (2.76), we get

Equation (2.75) $\times\, 0.01 \;\Rightarrow\; 0.01P_{G_1} + 0.01P_{G_2} = 5.5$

$$0.12P_{G_1} + 0.01P_{G_2} = -5$$

$$\rule{4cm}{0.4pt}$$

$$0.13P_{G_1} = 0.5$$

$$\text{or } \; P_{G_1} = 3.846 \text{ MW}$$

Substituting the P_{G_1} value in Equation (2.75), we get

$$P_{G_2} = 550 - 3.846 = 546.154 \text{ MW}$$

The above results are for the case without considering the generator limits. The IFCs are

$$\frac{dC_1}{dP_{G_1}} = 0.12P_{G_1} + 35 = 0.12(3.846) + 35 = 35.46 \text{ Rs./MWh}$$

and $\dfrac{dC_2}{dP_{G_2}} = 0.01P_{G_2} + 30 = 0.01(546.154) + 30 = 35.46 \text{ Rs./MWh}$

The IFC, $\lambda = 35.46 \text{ Rs./MWh}$

Case-II: Considering the generator limits:

$$35\,\text{MW} \le P_{G_1} \le 175\,\text{MW}$$

$$30\,\text{MW} \le P_{G_2} \le 600\,\text{MW}$$

From Case-I, the obtained power generations are

$$P_{G_1} = 3.846\,\text{MW}$$

$$P_{G_2} = 546.154\,\text{MW}$$

It is observed that the real-power generation of Unit-1 is violating the minimum generation limit. To achieve the optimum operation, fix up the generation of the first unit at its minimum generation, i.e., $P_{G_1} = 35\,\text{MW}$. Hence, for the load of $550\,\text{MW}$, $P_{G_1} = 35\,\text{MW}$ and $P_{G_2} = 550 - 35 = 515\,\text{MW}$.

Then, the IFCs are

$$\frac{dC_1}{dP_{G_1}} = 0.12 P_{G_1} + 35 = 0.12(35) + 35 = 39.2\,\text{Rs./MWh}$$

and $\dfrac{dC_2}{dP_{G_2}} = 0.01 P_{G_2} + 30 = 0.01\,(515) + 30 = 35.15\,\text{Rs./MWh}$

Hence, it is observed that $\lambda_1 \ne \lambda_2$, i.e., economic operation is not strictly maintained in this particular condition.

Comment on the results: When the generator limits are not considered, the economic operation of generating units is obtained at an IFC of 33.45 Rs./MWh. Their economic operation is not obtained when considering the generation limits, since the IFC of the first unit is somewhat marginally greater than that of the second unit.

KEY NOTES

- Economic operation of a power system is important in order to maintain the cost of electrical energy supplied to a consumer at a reasonable value.

- In analyzing the economic operation of a thermal unit, input–output modeling characteristics are of great significance.

- For operational planning, daily operation, and for economic scheduling, the data normally required are as follows:

For each generator

- Maximum and minimum output capacities.

- Fixed and incremental heat rate.

- Minimum shutdown time.

- Minimum stable output.

- Maximum run-up and run-down rates.

For each statlon

- Cost and calorific value of the fuel.

- Factors reflecting recent operational performance of the station.

- Minimum time between loading and unloading.

For the system

- Load cycle.

- Specified constraints imposed on transmission system capability.

- Spare capacity requirement.

- Transmission system parameters including maximum capacities and reliability factors.

- To analyze the power system network, there is a need of knowing the system variables. They are:

(i) Control variables.

(ii) Disturbance variables.

(iii) State variables.

- **Scheduling** is the process of allocation of generation among different generating units. **Economic scheduling** is the cost-effective mode of allocation of generation among the different units in such a way that the overall cost of generation should be minimum.

- **Input–output characteristics** establish the relationship between the energy input to the turbine and the energy output from the electrical generator.

- **Incremental fuel cost** is defined as the ratio of a small change in the input to the corresponding small change in the output.

- **Incremental efficiency** is defined as the reciprocal of incremental fuel rate.

- The input–output characteristics of hydro-power unit co-ordinates are water input or discharge (m³/s) versus output power (kW or MW).

Constraint Equations

The economic power system operation needs to satisfy the following types of constraints:

(1) Equality constraints.

(2) Inequality constraints.

(a) According to the nature:

(i) Hard-type constraints.

(ii) Soft-type constraints.

(b) According to the power system parameters:

(i) Output power constraints.

(ii) Voltage magnitude and phase-angle constraints.

(iii) Transformer tap position/settings constraints.

(iv) Transmission line constraints.

SHORT QUESTIONS AND ANSWERS

(1) Justify the production cost being considered as a function of real-power generation.

The production cost in the case of thermal and nuclear power stations is a function of fuel input. The real-power generation is a function of fuel input. Hence, the production cost would be a function of real-power generation.

(2) Give the expression for the objective function used for optimization of power system operation.

$$C = \sum_{i=1}^{n} C_i(P_{G_i})$$

(3) State the equality and inequality constraints on the optimization of product cost of a power station.

The equality constraint is the sum of real-power generation of all the various units that must always be equal to the total real-power demand on the system.

i.e., $\sum_{i=1}^{n} P_{G_i} = P_D$

The inequality constraint in each generating unit should not be operating above its rating or below some minimum generation.

i.e., $P_{G_i \text{(min)}} \leq P_{G_i} \leq P_{G_i \text{(max)}}$,

for $i = 1, 2, 3, \ldots, n$.

(4) What is an incremental fuel cost and what are its units?

Incremental fuel cost is the cost of the rate of increase of fuel input with the increase in power input. Its unit is Rs./MWh.

(5) How is the inequality constraint considered in the determination of optimum allocation?

If one or several generators reach their limit values, the balance real-power demand, which is equal to the difference between the total demand and the sum of the limit value, is optimally distributed among the remaining units by applying the equal incremental fuel cost rule.

(6) On what factors does the choice of a computation method depend on the determination of optimum distribution of load among the units?

The factors depend upon the following:

(i) Number of generating units.

(ii) The degree of polynomial representing the IC curve.

(iii) The presence of discontinuities in the IC curves.

(7) What does the production cost of a power plant correspond to?

The production cost of a power plant corresponds to the least of minimum or optimum production costs of various combinations of units, which can supply a given real-power demand on the station.

(8) To get the solution to an optimization problem, what will we define an objective's function?

Minimize the cost of production, min $C' = \min C(P_{G_n})$.

(9) Write the condition for optimality in allocating the total load demand among the various units.

The condition for optimality is the incremental fuel cost, $\dfrac{\partial C_i}{\partial P_{G_i}} = \lambda$.

(10) Write the separable objective function and why it is called so.

$$C = C_1(P_{G_1}) + C_2(P_{G_2}) + \cdots + C_n(P_{G_n})$$

i.e., $C = \sum_{i=1}^{n} C_i(P_{G_i})$

The above objective function consists of a summation of terms in which each term is a function of a separate independent variable. Hence, it is called separable objective function.

(11) Briefly discuss the optimization problem.

Minimize the overall cost of production, which is subjected to equality constraints and inequality constraints.

Equality constraint is: $\sum_{i=1}^{n} P_{G_i} - P_D = 0$

Inequality constraint is

$$P_{G_i(min)} \leq P_{G_i} \leq P_{G_i(max)}$$

(12) What is the reliable indicator of a country's or state's development?

It is the per capita consumption of electrical energy.

(13) State in words the condition for minimum fuel cost in a power system when losses are neglected.

The minimum fuel cost is obtained when the incremental fuel cost for all the stations is the same in the power system.

(14) What is the need of system variables and what are the variables?

To analyze the power system network, there is a need of knowing the system variables. They are:

- Control variables—P_G and Q_G.
- Disturbance variables—P_D and Q_D.
- State variables—V and δ.

(15) Define the control variables.

The real and reactive-power generations are called control variables since they are used to control the state of the system.

(16) Define the disturbance variables.

The real and reactive-power demands are called demand variables and they are beyond system control and are hence called uncontrolled or disturbance variables.

(17) Define the state variables.

The bus voltage magnitude V and its phase angle δ dispatch the state of the system. They are dependent variables that are being controlled by the control variables.

(18) What is the need of input–output characteristics of a steam unit?

It establishes the relationship between the energy input to the turbine and the energy output from the electrical generator.

(19) Define the incremental fuel or heat rate curve.

It is defined as the ratio of a small change in the input to the corresponding small change in the output.

$$\text{Incremental fuel rate} = \frac{\Delta \text{ input}}{\Delta \text{ output}} = \frac{\Delta F}{\Delta P_G}$$

(20) How do you get incremental cost curve?

The incremental cost curve is obtained by considering at various points, the change in cost of generation to the change in real-power generation, i.e., slope of the input–output curve.

(21) How you get the heat rate characteristic?

The heat rate characteristic is obtained from the plot of net heat rate in kCal/kWh versus power output in kW.

(22) Define the incremental efficiency.

It is defined as the reciprocal of incremental fuel rate and is given by

$$\text{Incremental efficiency} = \frac{\text{output}}{\text{input}} = \frac{dP_G}{dC}$$

(23) What are hard-type constraints? Give examples.

Hard-type constraints are definite and specific in nature. No flexibility will be taken place in violating these types of constraints.

E.g., The tapping range of an on-load tap-changing transformer.

(24) What are soft-type constraints? Give examples.

Soft-type constraints have some flexibility with them in violating these type of constraints.

E.g., Magnitudes of node voltages and the phase angle between them.

(25) What is the need of spare capacity constraints?

These constraints are required to meet:

- Errors in load prediction.
- The unexpected and fast changes in load demand.
- Unplanned loss of scheduled generation, i.e., the forced outages of one or more alternators on the system.

MULTIPLE-CHOICE QUESTIONS

(1) In a thermal-electric generating plant, the overall efficiency is improved when:

  (a) Boiler pressure is increased.

  (b) The difference between initial pressure and temperature and exhaust pressure and temperature are held at a maximum.

  (c) Load on the units is increased.

  (d) Its operating time is increased.

(2) When load on a thermal unit is increased, fuel input:

  (a) Increases.

  (b) Does not change.

  (c) Decreases.

  (d) None of these.

(3) Incremental heat rate curves, for thermal generating units, are used to determine the:

  (a) Fuel cost in rupees per hour.

  (b) Values at which the units should be loaded to result in minimum fuel costs.

  (c) Cost per unit of electrical output.

  (d) Heat produced per hour.

(4) When generating units are loaded to equal incremental costs, it results in:

  (a) Minimum fuel costs.

  (b) Fuel costs are at a maximum.

  (c) Fuel costs are not affected.

  (d) Maximum loading of generating units.

(5) One advantage of computer control of generating units is that:

  (a) Net output of the units is minimized.

  (b) All units under the control of the computer will be loaded to the same load.

  (c) Loading of the units will be frequently adjusted to maintain them at equal incremental costs.

  (d) Both (b) and (c).

(6) If the fuel cost of one unit, operating in parallel with other units, is increased and it is desired to maintain average fuel cost, the load on the unit will be:

  (a) Increased.

  (b) Held constant.

  (c) Decreased.

  (d) None of these.

(7) In a power system using both hydro- and thermal-generation, the proportion of hydro-generation can be increased by:

   (a) Increasing the price (gamma) of water.

   (b) Reducing the price of water.

   (c) Increasing the field currents of the hydro-generators.

   (d) None of these.

(8) Economic operation of power system is:

   (a) Unit commitment.

   (b) Load scheduling.

   (c) Controlling of voltage and its magnitude.

   (d) Both (a) and (b).

(9) Lagrangian multiplier method converts a non-linear constrained optimization problem into ————— non-linear optimization problem.

   (a) Gradient.

   (b) Linear.

   (c) Unconstrained.

   (d) All of these.

(10) Unit of heat rate curve is —————.

   (a) Million kCal/hr.

   (b) Rs.-hr.

   (c) Rs./MWh.

   (d) Rs./hr.

(11) Power balance equation is ————— constraint.

   (a) Equality.

   (b) Inequality.

   (c) Security constraints.

   (d) Branch transfer capacity constraint.

(12) Optimization problems with only objective function and without constraints is a ————— function.

   (a) Single-valued.

   (b) Multi-valued.

   (c) Both (a) and (b).

   (d) Either (a) or (b).

(13) Unit of λ is —————.

   (a) Rs./hr.

   (b) Rs./MW.

   (c) Rs./MWh.

   (d) MW/Rs.

(14) Which of the following has a negligible effect on the production cost?

   (i) Generation of real power.

   (ii) Real and reactive-power demands.

   (iii) System voltage and angle.

   (iv) Generation of reactive power.

     (a) (i) and (ii).

     (b) Except (iii).

     (c) (ii) and (iv).

     (d) All of these.

(15) An analytical method of getting the solution to optimization problem, the following graph is to be drawn:

   (a) Total real-power demand versus λ.

   (b) Total real-power generation versus total real-power demand.

   (c) Total real-power generation versus λ.

   (d) Total real-power generation versus fuel input.

(16) The control variables are:

   (a) P_D and Q_D.

   (b) P_G and Q_G.

   (c) V and δ.

   (d) Q and δ.

(17) P_D and Q_D are:

   (a) Control variables.

   (b) State variables.

   (c) Disturbance variables.

   (d) Constants.

(18) P_D and Q_D are:

   (a) Disturbance variables.

   (b) Demand variables.

   (c) Uncontrollable variables.

   (d) All of these.

(19) Cost curves can be obtained by:

   (a) Multiply the fuel input with cost of fuel.

   (b) Subtract the fuel input with cost of fuel.

   (c) Add the fuel input with cost of fuel.

   (d) None of these.

(20) Cost curves are expressed as:

   (a) Rs./million Cal.

   (b) Million Cal/hr × Rs./million Cal.

   (c) Rs./hr.

   (d) (b) and (c).

(21) The curve obtained by considering the change in cost of generation to change in real-power generation at various points is:

   (a) Fuel cost curve.

   (b) Input–output curve.

   (c) Incremental cost curve.

   (d) All of these.

(22) Incremental fuel cost, IC is given by:

   (i) Rs./MWh.

   (ii) Slope of the fuel cost curve.

   (iii) Tan $\beta = \Delta C/\Delta P_G$.

   (iv) $\Delta i/p/\Delta o/p$.

      (a) (i) and (ii).

      (b) (ii) and (iii).

      (c) All except (iv).

      (d) All of these.

(23) Incremental production cost of a given unit is made up of:

   (a) IC-incremental cost of labor, supplies, maintenance, etc.

   (b) IC+incremental cost of labor, supplies, maintenance, etc.

   (c) IC×incremental cost of labor, supplies, maintenance, etc.

   (d) IC% incremental cost of labor, supplies, maintenance, etc.

(24) The optimization problem is:

   (a) To allocate total load demand among various units such that the cost of generation is maintained constant.

   (b) To allocate total load demand among various units such that the cost of generation is minimized.

   (c) To allocate total load demand among various units such that the cost of generation is enormously increased.

   (d) To allocate total load demand among various units such that there is no effect with cost of generation.

(25) The method adopted to get an optimal solution to optimal scheduling problem depends on:

   (i) The mathematical equation representing I_c.

   (ii) No. of units.

   (iii) Need to represent any discontinuity in incremental cost curve.

   (iv) Change in location.

      (a) Only (i).

      (b) Only (ii).

      (c) All expect (iv).

      (d) All expect (ii).

(26) In a digital computer method of getting the solution to an optimization problem,

   (i) The number of terms included in expression for P_{G_i} depends on the degree of accuracy.

   (ii) α, β, γ coefficients are to be taken as output.

   (iii) α, β, γ coefficients are to be taken as input.

      (a) Both (i) and (ii).

      (b) Both (i) and (iii).

      (c) Only (i).

      (d) Only (iii).

(27) If the real-power inequality constraints are violated for any generator, then:

   (a) It is tied to the corresponding limit and the rest of the load is economically distributed among the remaining units.

   (b) It is tied to the corresponding limit and the total load is economically distributed among all the units.

   (c) It is not considered and the total load is economically distributed among all the units.

   (d) Any of the above methods.

(28) The method of getting the solution to an optimization problem with neglected transmission losses:

   (i) Does sense the changes in the loads.

   (ii) Does not sense the location of the changes in the load.

(iii) Does sense the changes in the load and the location of changes in the loads.

(iv) Does not sense both the location and the changes in the load.

    (a) (i) and (ii).

    (b) Either (iii) or (iv).

    (c) Only (iv).

    (d) Only (iii).

(29) To get an optimal solution to an optimization problem, we will define an objective function as:

    (a) $C^{*} = \sum_{i=1}^{n} C_{i}(P_{G_{i}})$.

    (b) $C^{*} = \sum_{i=1}^{n} C_{i}(P_{G_{i}} - \lambda)$.

    (c) $C^{*} = C - \lambda \left\{ \sum_{i=1}^{n} P_{G_{i}} - P_{D} \right\}$.

    (d) $C^{*} = \lambda - C \left\{ \sum_{i=1}^{n} P_{G_{i}} < P_{D} \right\}$.

(30) The condition for optimality is:

    (a) $\partial C_{1}/\partial P_{G_{1}} = \partial C_{2}/\partial P_{G_{2}} = \cdots = \partial C_{n}/\partial P_{G_{n}} = \lambda$.

    (b) $\partial C_{i}/\partial P_{G_{i}} - \lambda = 0$.

    (c) $\partial C_{i}/\partial P_{G_{i}} + \lambda = 0$.

    (d) Both (a) and (b).

(31) Which of the following is the real indicator of the state of development of a country?

    (a) Population.

    (b) Facilities.

    (c) Politics.

    (d) Per capita consumption of electricity.

(32) Equality and inequality constraints are

    (a) $\sum_{i=1}^{n} P_{G_{i}} - P_{D} = 0; \quad P_{G_{i}(min)} < P_{G_{i}} < P_{G_{i}(max)}$.

    (b) $\sum_{i=1}^{n} P_{G_{i}} + P_{D} = 0; \quad P_{G_{i}(min)} < P_{G_{i}} < P_{G_{i}(max)}$.

    (c) $\sum_{i=1}^{n} P_{G_{i}} - P_{D} = 0; \quad P_{G_{i}(min)} > P_{G_{i}} > P_{G_{i}(max)}$.

    (d) None of the above.

(33) In a mathematical determination, the optimization problem should be modified as:

    (a) Constrained optimization problem.

    (b) Normalized optimization problem.

    (c) Conditional optimization problem.

    (d) All the above.

REVIEW QUESTIONS

(1) Explain the important characteristics of a steam unit.

(2) Describe the need of economic dispatch.

(3) Explain why the production cost of electrical energy is treated as a function of real-power generation.

(4) Obtain the condition for optimum operation of a power system with 'n' plants.

(5) Bring out the difference between optimal operation of generators in thermal stations and optimal scheduling of hydro-thermal systems.

(6) Explain how the incremental production cost of a thermal power station can be determined.

(7) Explain the various factors to be considered in allocating generation to different power stations for optimum operation.

(8) Explain the significance of equality and inequality constraints in the economic allocation of generation among different plants in a system.

PROBLEMS

(1) Three power plants of a total capacity of 425 MW are scheduled for operation to supply a total system load of 300 MW. Find the optimum load scheduling if the plants have the following incremental cost characteristics and the generator constraints.

$$\frac{dC_1}{dP_{G_1}} = 30 + 0.15P_{G_1}, \quad 25 \le P_{G_1} \le 125$$

$$\frac{dC_2}{dP_{G_2}} = 40 + 0.20P_{G_2}, \quad 30 \le P_{G_2} \le 100$$

$$\frac{dC_3}{dP_{G_3}} = 15 + 0.18P_{G_3}, \quad 50 \le P_{G_3} \le 200$$

(2) A plant consists of two units. The incremental fuel characteristics for the two units are given as:

$$\frac{dC_1}{dP_{G_1}} = 0.15P_{G_1} + 30.0 \text{ Rs./MWh}$$

$$\frac{dC_2}{dP_{G_2}} = 0.25P_{G_2} + 20.0 \text{ Rs./MWh}$$

Find the optimal load sharing of two units when a total load of 300 MW is connected to the system. Also calculate the extra cost increased in Rs./hr if the total load is shared equally between them.

(3) The cost curves of the three plants are given as follows:

$$C_1 = 0.04P_{G_1}^2 + 20P_{G_1} + 230 \text{ Rs./hr}$$
$$C_2 = 0.06P_{G_2}^2 + 18P_{G_2} + 200 \text{Rs./hr}$$

$$C_3 = 0.15P_{G_3}^2 + 15P_{G_3} + 180 \text{ Rs./hr}$$

Determine the optimum sharing of a total load of 180 MW for which each plant would take up for minimum input cost of received power in Rs/MWh.

(4) The incremental fuel costs in rupees per MWh for a plant consisting of two units are:

$$\frac{dC_1}{dP_{G_1}} = 0.15P_{G_1} + 50.0$$

$$\frac{dC_2}{dP_{G_2}} = 0.2P_{G_2} + 40.0$$

Calculate the extra cost increased in Rs./hr, if a load of 210 MW is scheduled as $P_{G_1} = P_{G_2} = 105$ MW.

3

Economic Load Dispatch-II

OBJECTIVES

After reading this chapter, you should be able to:

- develop the mathematical model for economical load dispatch when losses are considered
- derive transmission loss expression
- study the optimal allocation of total load among the units
- develop a flowchart for the solution of optimization problem

3.1 INTRODUCTION

In case of an urban area where the load density is very high and the transmission distances are very small, the transmission loss could be neglected and the optimum strategy of generation could be based on the equal incremental production cost. If the energy is to be transported over relatively larger distances with low load density, the transmission losses, in some cases, may amount to about 20–30% of the total load; hence, it is essential to take these losses into account when formulating an economic load dispatch problem.

3.2 OPTIMAL GENERATION SCHEDULING PROBLEM: CONSIDERATION OF TRANSMISSION LOSSES

In a practical system, a large amount of power is being transmitted through the transmission network, which causes power losses in the network (P_L) as shown in Fig. 3.1.

In finding an optimal solution for economic scheduling problem (allocation of total load among the generating units), it is more realistic to consider the transmission line losses, which are about 5–15% of the total generation.

In general, the condition for optimality, when losses are considered, is different. Equal incremental fuel costs (IFCs) for all generating units will not give an optimal solution.

3.2.1 Mathematical modeling

Consider the objective function:

$$C = \sum_{i=1}^{n} C_i(P_{G_i})$$

(3.1)

Minimize Equation (3.1) subject to the following equality and inequality constraints:

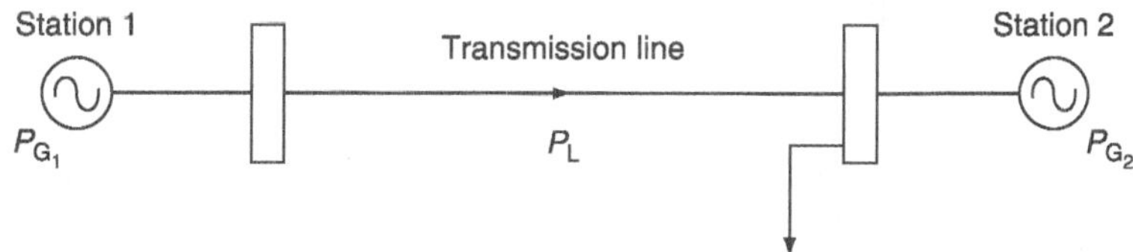

FIG. 3.1 *Transmission network*

(i) *Equality constraint*

The real-power balance equation, i.e., total real-power generations minus the total losses should be equal to the real-power demand:

$$\text{i.e.,} \quad \sum_{i=1}^{n} P_{G_i} - P_L = P_D$$

$$\text{(or)} \quad \sum_{i=1}^{n} P_{G_i} - P_D - P_L = 0 \tag{3.2}$$

where P_L is the total transmission losses (MW), P_D the total real-power demand (MW), and P_{G_i} the real-power generation at the ith unit (MW).

(ii) *Inequality constraints*

Always there will be upper and lower limits for real and reactive-power generation at each of the stations. The inequality constraints are represented:

(a) In terms of real-power generation as

$$P_{G_i\,(\min)} \leq P_{G_i} \leq P_{G_i\,(\max)} \tag{3.3}$$

(b) In terms of reactive-power generation as

$$Q_{G_i\,(\min)} \leq Q_{G_i} \leq Q_{G_i\,(\max)} \tag{3.4}$$

The reactive-power constraints are to be considered since the transmission line results in loss is a function of real and reactive-power generations and also the voltage at the station bus.

(c) In addition, the voltage at each of the stations should be maintained within certain limits:

$$\text{i.e.,} \quad V_{i(\min)} \leq V_i \leq V_{i(\max)} \tag{3.5}$$

The optimal solution should be obtained by minimizing the cost function satisfying constraint Equations (3.2)–(3.5).

3.3 TRANSMISSION LOSS EXPRESSION IN TERMS OF REAL-POWER GENERATION—DERIVATION

Transmission loss P_L is expressed without loss of accuracy as a function of real-power generations. The power loss is expressed using B-coefficients or loss coefficients.

The expression for transmission power loss is derived using Kron's method of reducing a system to an equivalent system with a single hypothetical load.

The expression is based on several assumptions as follows:

(i) All the lines in the system have the same $\dfrac{X}{R}$ ratio.

(ii) All the load currents have the same phase angle.

(iii) All the load currents maintain a constant ratio to the total current.

(iv) The magnitude and phase angle of bus voltages at each station remain constant.

(v) Power factor at each station bus remains constant.

We will derive an expression for the power loss of a system, having two generating stations, supplying an arbitrary number of loads through a transmission network as shown in Fig. 3.2(a).

To determine the current in any line, say k^{th} line, apply the superposition principle and determine the current passing through the line, I_k.

The current distribution factor of a transmission line w.r.t. a power source is the ratio of the current it would carry to the current that the source would carry when all other sources are rendered inactive, i.e., sources that are not supplying any current.

Let us assume that the entire load current is supplied by generating station-1 only as shown in Fig. 3.2(b).

Current in the kth line $= I_{k_1}$

Current distribution factor, $N_{k_1} = \dfrac{I_{k_1}}{I_{\text{L}}}$

If we assume that the entire load is supplied by the second generating station only as shown in Fig. 3.2(c):

The current flowing through the kth line $= I_{k_2}$

Current distribution factor, $N_{k_2} = \dfrac{I_{k_2}}{I_{\text{L}}}$

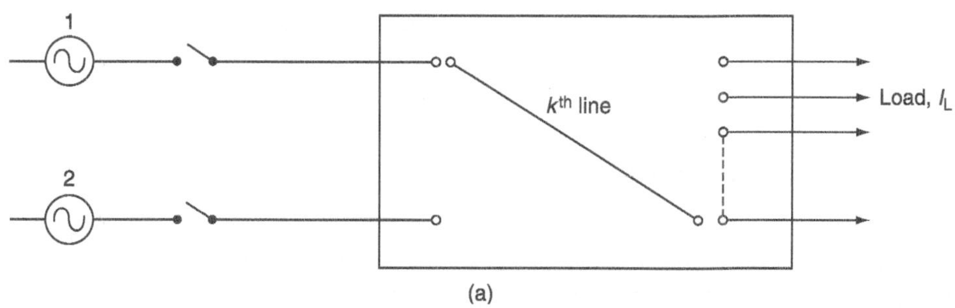

(a)

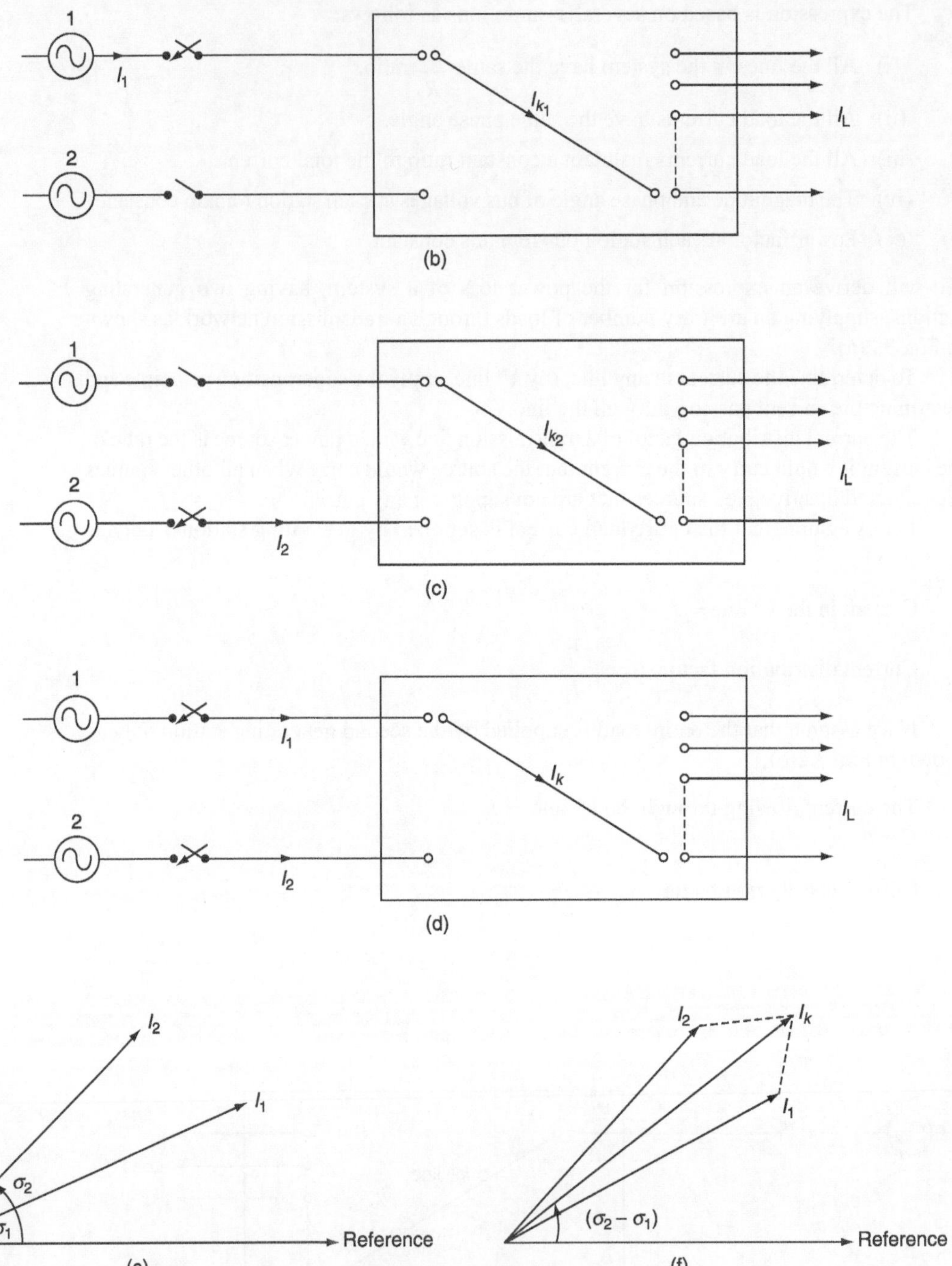

FIG. 3.2 (a) *Transmission network with two generating stations; (b) load supplied by generating station-1 only; (c) load supplied by generating station-2 only; (d) load supplied by two generating stations simultaneously; (e) source currents with respect to reference; (f) current in k^{th} line*

Because of assumptions (i) and (ii), the current distribution factors will be real numbers rather than complex numbers.

And also assuming that the total load is being supplied by both the stations as shown in Fig. 3.2(d):

The current in the kth line $= I_k$

$$\therefore I_k = I'_{K_1} + I'_{K_2}$$

From the relations,

$$N_{k_1} = \frac{I'_{K_1}}{I_1} \quad \text{and} \quad N_{k_2} = \frac{I'_{K_2}}{I_2}$$

$$\therefore I_k = N_{k_1} I_1 + N_{k_2} I_2 \tag{3.6}$$

Although the current distribution factors are real numbers, the various source currents supplying total load will not be in phase, i.e., I_1 and I_2 are not in phase.

Let the source currents I_1 and I_2 be expressed as $I_1 \angle \sigma_1$ and $I_2 \angle \sigma_2$ as shown in Fig. 3.2(e).

Then, $I_k = N_{k_1} I_1 \angle \sigma_1 + N_{k_2} I_2 \angle \sigma_2$ with a phase difference of $(\sigma_2 - \sigma_1)$ as shown in Fig. 3.2(f).

By adding I_{k_1} and I_{k_2} phasors, we have

$$|I_k|^2 = (N_{k_1} I_1)^2 + (N_{k_2} I_2)^2 + 2(N_{k_1} I_1)(N_{k_2} I_2) \cos(\sigma_2 - \sigma_1) \tag{3.7}$$

The station currents are related as

$$I_1 = \frac{P_{G_1}}{\sqrt{3} V_1 (\text{p.f.}_1)}$$

$$\text{and} \quad I_2 = \frac{P_{G_2}}{\sqrt{3} V_2 (\text{p.f.}_2)}$$

The power loss in the kth line can be calculated as $3 |I_k|^2 R_k$

i.e.,

$$\text{power loss} = 3|I_k|^2 R_k$$

$$= 3R_k \left\{ N_{k_1}^2 \left(\frac{P_{G_1}^2}{3V_1^2 (\text{p.f.}_1)^2} \right) + N_{k_2}^2 \frac{P_{G_2}^2}{3V_2^2 (\text{p.f.}_2)^2} \right.$$

$$\left. + 2(N_{k_1} N_{k_2}) \left(\frac{P_{G_1}}{\sqrt{3} V_1 (\text{p.f.}_1)} \frac{P_{G_2}}{\sqrt{3} V_2 (\text{p.f.}_2)} \right) \cos(\sigma_2 - \sigma_1) \right\}$$

$$= \left[\frac{P_{G_1}^2}{V_1^2 (\text{p.f.}_1)^2} N_k^2 R_k + \frac{P_{G_2}^2}{V_2^2 (\text{p.f.}_2)^2} N_{k_2}^2 R_k + \frac{2 P_{G_1} P_{G_2}}{V_1 V_2 (\text{p.f.}_1)(\text{p.f.}_2)} \cos((\sigma_2 - \sigma_1)(N_{k_1} N_{k_2} R_k)) \right]$$

If there are 'l' number of lines in the system, total power loss in the system can be calculated as

$$P_\mathrm{L} = \sum_{k=1}^{l} 3 I_{k_i}^2 R_K$$

i.e., $\quad P_\mathrm{L} = \dfrac{P_{\mathrm{G}_1}^2}{V_1^2(\mathrm{p.f_1})^2} \sum_{k=1}^{l} N_{k_1}^2 R_k + \dfrac{P_{\mathrm{G}_2}^2}{V_2^2(\mathrm{p.f_2})^2} \sum_{k=1}^{l} N_{k_2}^2 R_k$

$$+ \frac{2P_{\mathrm{G}_1} P_{\mathrm{G}_2} \cos(\sigma_2 + \sigma_1)}{V_1 V_2 (\mathrm{p.f_1})(\mathrm{p.f_2})} \sum_{k=1}^{l} N_{k_1} N_{k_2} R_k$$

This above expression can be written as

$$P_\mathrm{L} = B_{11} P_{\mathrm{G}_1}^2 + B_{22} P_{\mathrm{G}_2}^2 + 2 B_{12} P_{\mathrm{G}_1} P_{\mathrm{G}_2} \tag{3.8}$$

where

$$B_{11} = \frac{\displaystyle\sum_{k=1}^{l} N_{k_1}^2 R_k}{V_1^2(\mathrm{p.f_1})^2}; \quad B_{22} = \frac{\displaystyle\sum_{k=1}^{l} N_{k_2} R_k}{V_2^2(\mathrm{p.f_2})^2}$$

and $\quad B_{12} = \dfrac{\cos(\sigma_2 - \sigma_1)}{V_1 V_2 (\mathrm{p.f_1})(\mathrm{p.f_2})} \displaystyle\sum_{k=1}^{l} N_{k_1} N_{k_2} R_k$

Equation (3.8) expresses the total loss as a function of real-power generations, P_{G_1} and P_{G_2}.

The coefficients B_{11}, B_{22}, and B_{12} are called loss coefficients (or) B-coefficients and the unit is $(\mathrm{MW})^{-1}$ and is also considered to be a constant in view of the assumptions made.

The same procedure can be extended for systems having more number of stations. If the system has 'n' number of stations, supplying the total load through transmission lines, the transmission line loss is given by

$$P_\mathrm{L} = \sum_{p=1}^{n} \sum_{q=1}^{n} P_{\mathrm{G}_p} B_{p_q} P_{\mathrm{G}_q} \tag{3.9}$$

when $\quad n = 2$,

$$P_\mathrm{L} = \sum_{p=1}^{2} \sum_{q=1}^{2} P_{\mathrm{G}_p} B_{p_q} P_{\mathrm{G}_q}$$

$$= P_{\mathrm{G}_1} B_{11} P_{\mathrm{G}_1} + P_{\mathrm{G}_1} B_{12} P_{\mathrm{G}_2} + P_{\mathrm{G}_2} B_{21} P_{\mathrm{G}_1} + P_{\mathrm{G}_2} B_{22} P_{\mathrm{G}_2}$$

$$= B_{11} P_{\mathrm{G}_1}^2 + B_{22} P_{\mathrm{G}_2}^2 + 2 B_{12} P_{\mathrm{G}_1} P_{\mathrm{G}_2}$$

Similarly for $n = 3$,

$$P_\mathrm{L} = \sum_{p=1}^{3} \sum_{q=1}^{3} P_{\mathrm{G}_p} B_{p_q} P_{\mathrm{G}_q}$$

$$= B_{11} P_{\mathrm{G}_1}^2 + B_{22} P_{\mathrm{G}_2}^2 + B_{33} P_{\mathrm{G}_3}^2 + 2 B_{12} P_{\mathrm{G}_1} P_{\mathrm{G}_2} + 2 B_{23} P_{\mathrm{G}_2} P_{\mathrm{G}_3} + 2 B_{31} P_{\mathrm{G}_3} P_{\mathrm{G}_1} \tag{3.10}$$

Since the transmission lines are symmetrical, loss coefficients B_{pq} and B_{qp} are equal, i.e., $B_{pq} = B_{qp}$.

The B_{pq} coefficients are loss coefficients and can be represented in matrix form of an n-generator system as

$$B_{pq} = \begin{bmatrix} B_{11} & B_{12} & \cdots & B_{1n} \\ B_{21} & B_{22} & \cdots & B_{2n} \\ \cdots & \cdots & & \cdots \\ B_{n_1} & B_{n_2} & \cdots & B_{n_n} \end{bmatrix}_{n \times n}$$

The diagonal elements of these coefficients are all positive and strong (since generating stations are interconnected) as compared with the off-diagonal elements, which mostly are negative and are relatively weaker.

These coefficients are determined for a large system by an elaborate digital computer program starting from the assembly of the open-circuit impedance matrix of the transmission line, which is quite lengthy and time-consuming. Besides, the formulations of B-coefficients are based on several assumptions and do not take into account the actual conditions of the system; the solution for the plant generations cannot be expected to be the best for minimum cost of generation.

3.4 MATHEMATICAL DETERMINATION OF OPTIMUM ALLOCATION OF TOTAL LOAD WHEN TRANSMISSION LOSSES ARE TAKEN INTO CONSIDERATION

Consider a power station having 'n' number of units. Let us assume that each unit does not violate the inequality constraints and let the transmission losses be considered.

Assuming that the inequality constraint is satisfied, the objective function is redefined by augmenting Equation (3.1) with equality constraint (Equation (3.2)) using Lagrangian multiplier (λ) and is given by

$$C' = \sum_{i=1}^{n} C_i (P_{G_i}) - \lambda \left[\sum_{i=1}^{n} P_{G_i} - P_D - P_L (P_{G_i}) \right] \tag{3.11}$$

This augmented objective function is called constrained objective function.

In the above objective function, the real-power generations are the control variables and the condition for optimality becomes $\dfrac{\partial C'}{\partial P_{G_i}} = 0, \, i = 1, 2, \ldots, n$:

$$\text{i.e., } \frac{\partial C'}{\partial P_{G_i}} = \frac{\partial C_i}{\partial P_{G_i}} - \lambda \left[1 - 0 - \frac{\partial P_L}{\partial P_{G_i}} \right] = 0 \tag{3.12}$$

$$\text{(or) } \frac{\partial C_i}{\partial P_{G_i}} = \lambda \left[1 - 0 - \frac{\partial P_L}{\partial P_{G_i}} \right]$$

$$\therefore \frac{\partial C_i}{\partial P_{G_i}} = \lambda \left[1 - \frac{\partial P_L}{\partial P_{G_i}} \right] \tag{3.13}$$

$$\text{(or)} \quad \lambda = \frac{\partial C_i}{\partial P_{G_i}} \bigg/ \left[1 - \frac{\partial P_L}{\partial P_{G_i}} \right] \tag{3.14}$$

where $\dfrac{\partial P_L}{\partial P_{G_i}}$ represents the variation of total transmission loss with respect to real-power

generation of the ith station and is called incremental transmission loss (ITL) of the ith station. Equation (3.14) can be written as

$$\therefore \lambda = \frac{\dfrac{\partial C_i}{\partial P_{G_i}}}{[1 - (ITL)_i]}$$

$$\text{or} \quad \lambda = L_i \frac{\partial C_i}{\partial P_{G_i}} \tag{3.15}$$

where $L_i = \dfrac{1}{\left[1 - \dfrac{\partial P_L}{\partial P_{G_i}} \right]}$ and is called the penalty factor of the ith station. Equation (3.15)

can be utilized to obtain the optimal cost of operation.

The condition for optimality when the transmission losses are considered is that the IFC of each plant multiplied by its penalty factor must be the same for all the plants:

$$\text{i.e.,} \quad \frac{\partial C_1}{\partial P_{G_1}} L_1 = \frac{\partial C_2}{\partial P_{G_2}} L_2 = \cdots = \frac{\partial C_n}{\partial P_{G_n}} L_n = \lambda \tag{3.16}$$

Equation (3.12) is a set of n equations with $(n+1)$ unknowns. Here, the powers of n generators are unknown and λ is also unknown. These equations are known as **exact co-ordination equations** because they **co-ordinate the ITL with IFC.**

The following points should be kept in mind for the solution of economic load dispatch problems when transmission losses are included and co-ordinated:

- Although incremental production cost of a plant is always positive, ITL can either be positive or negative.

- The individual units will operate at different incremental production costs.

- The generation with the highest positive ITL will operate at the lowest incremental production cost.

For a small increase in received load by ΔP_D, the ith plant generation is only changed by ΔP_{G_i} and the generations of the remaining units are unaffected. Let ΔP_L be the change in transmission loss, the power balance equation becomes $\Delta P_{G_i} - \Delta P_L = \Delta P_D$.

Thus,

$$\frac{dC_i(P_{G_i})}{dP_{G_i}} \times L_i = \frac{dC_i(P_{G_i})}{dP_{G_i}} \times \frac{1}{1 - \dfrac{\Delta P_L}{\Delta P_{G_i}}} = \frac{dC_i(P_{G_i})}{dP_{G_i}} \times \frac{\Delta P_{G_i}}{\Delta P_D} = \frac{dC_i(P_{G_i})}{dP_D}$$

when $\dfrac{dC_i(P_{G_i})}{dP_D}$ is the incremental cost of the received power of the ith plant and the penalty factor $\dfrac{\Delta P_{G_i}}{\Delta P_D}$. This means that as ΔP_{G_i} increment has a larger proportion dissipated as loss, $\dfrac{\Delta P_{G_i}}{\Delta P_D}$ approaches unity and the penalty factor 'L_i' increases without bound. Thus, for a larger penalty factor 'L_i', unit 'i' should be operated at low incremental cost implying a low power output.

3.4.1 Determination of ITL formula

When a system consists of three generating units, i.e., $n = 3$, the transmission loss is

$$P_L = \sum_{p=1}^{3} \sum_{q=1}^{3} P_{G_p} B_{pq} P_{G_q}$$

$$= B_{11} P_{G_1}^2 + B_{22} P_{G_2}^2 + B_{33} P_{G_3}^2 + 2\,B_{12} P_{G_1} P_{G_2} + 2 B_{23} P_{G_2} P_{G_3} + 2\,B_{31} P_{G_3} P_{G_1} \quad (\because B_{pq} = B_{qp})$$

ITL of Generator-1 is obtained as

$$\frac{\partial P_L}{\partial P_{G_1}} = 2\,B_{11} P_{G_1} + 2 B_{12} P_{G_2} + 2 B_{13} P_{G_3}$$

$$= 2\,(B_{11} P_{G_1} + 2 B_{12} P_{G_2} + 2 B_{13} P_{G_3})$$

In general,

$$\frac{\partial P_L}{\partial P_{G_i}} = \sum_{j=1}^{n} 2 B_{ij} P_{G_j} \tag{3.17}$$

We know that the IFC of the ith unit is

$$\frac{\partial C_i}{\partial P_{G_i}} = (IC)_i = a_i P_{G_i} + b_i \tag{3.18}$$

Substitute Equations (3.17) and (3.18) in Equation (3.9); we get

$$\because \frac{\partial C'}{\partial P_{G_i}} = (a_i P_{G_i} + b_i) - \lambda\left[1 - 2\sum_{j=1}^{n} B_{ij} P_{G_j}\right] = 0$$

$$a_i P_{G_i} + b_i - \lambda + 2\lambda\left[\sum_{\substack{j=1 \\ j\neq i}}^{n} (B_{ij} P_{G_j} + B_{ii} P_{G_i})\right] = 0$$

$$[a_i + 2\,\lambda\,B_{ii}]P_{G_i} = \lambda - 2\lambda\sum_{\substack{j=1 \\ j\neq i}}^{n} (B_{ij} P_{G_j}) - b_i$$

Dividing the above equation by λ, we get

$$\left[\frac{a_i}{\lambda} + 2\,B_{ii}\right]P_{G_i} = 1 - 2\sum_{\substack{j=1 \\ j\neq i}}^{n} (B_{ij} P_{G_j}) - \frac{b_i}{\lambda}$$

$$\therefore P_{G_i} = \frac{1 - 2\displaystyle\sum_{\substack{j=1 \\ j\neq i}}^{n} (B_{ij} P_{G_j}) - \frac{b_i}{\lambda}}{\left(\dfrac{a_i}{\lambda} + 2B_{ii}\right)} \tag{3.19}$$

To solve this allocation problem, solve the co-ordination Equation (3.19) for a particular value of λ iteratively starting with an initial set of values of P_{G_i} (such as all P_{G_i} set to minimum values) and get the solution within a specified tolerance till all P_{G_i}'s converge, then check for power balance and if it is to be satisfied, then it is the optimal solution. If the power balance equation is not satisfied, modify the value of λ to a suitable value and solve the co-ordination equation.

3.4.2 Penalty factor

Consider Equation (3.12):

$$\frac{\partial C'}{\partial P_{G_i}} = \frac{\partial C_i}{\partial P_{G_i}} - \lambda\left[1 - 0 - \frac{\partial P_L}{\partial P_{G_i}}\right] = 0$$

$$\text{(or)} \quad \frac{\partial C_i}{\partial P_{G_i}} = \lambda\left[1 - 0 - \frac{\partial P_L}{\partial P_{G_i}}\right]$$

$$\therefore \frac{\partial C_i}{\partial P_{G_i}} = \lambda\left[1 - \frac{\partial P_L}{\partial P_{G_i}}\right]$$

$$\text{(or)} \quad \lambda = \frac{\partial C_i}{\partial P_{G_i}} \Big/ \left[1 - \frac{\partial P_L}{\partial P_{G_i}}\right]$$

The above expression can be written as

$$\therefore \lambda = \frac{\dfrac{\partial C_i}{\partial P_{G_i}}}{\left[1-(ITL)_i\right]} = \frac{\text{Incremental fuel cost}}{\left[1-\text{Incremental transmission loss}\right]}$$

or $\quad \lambda = L_i \dfrac{\partial C_i}{\partial P_{G_i}}$

where

$$L_i = \frac{1}{\left[1-\dfrac{\partial P_L}{\partial P_{G_i}}\right]}$$

is called the penalty factor of the ith station

$$L_i \equiv \frac{1}{\left[1-\dfrac{\partial P_L}{\partial P_{G_i}}\right]} = \frac{1}{\left[1-(ITL)_i\right]} \tag{3.20}$$

The penalty factor of any unit is defined as the ratio of a small change in power at that unit to the small change in received power when only that unit supplies this small change in received power.

3.5 FLOWCHART FOR THE SOLUTION OF AN OPTIMIZATION PROBLEM WHEN TRANSMISSION LOSSES ARE CONSIDERED

When transmission losses are taken into account, the solution of an optimization problem is represented by the following flowchart (Fig. 3.3).

Example 3.1: The fuel cost functions in Rs./hr for two thermal plants are given by

$$C_1 = 420 + 9.2P_1 + 0.004\,P_1^2$$
$$C_2 = 350 + 8.5P_2 + 0.0029\,P_2^2$$

where P_1, P_2 are in MW. Determine the optimal scheduling of generation if the total load is 640.82 MW. Estimate value of $\lambda = 12$ Rs./MWh. The transmission power loss is given by the expression

$$P_{L(pu)} = 0.0346\,P_{1(pu)}^2 + 0.00643\,P_{2(pu)}^2$$

Solution:

```
%MATLAB PROGRAM FOR ECONOMIC LOAD DISPATCH WITH LOSSES AND NO
GENERATOR
    %LIMITS (dispatch3.m)
    clc;
    clear;
```

```
          % uno    d      b        a
costdata=[1     420    9.2    0.004;
          2     350    8.5    0.0029];
ng=length(costdata(:,1));
for i=1:ng
  uno(i)=costdata(i,1);
  d(i)=costdata(i,2);
  b(i)=costdata(i,3);
  a(i)=costdata(i,4);
end
lambda=12;
pd=640.82;
delp=0.1;
dellambda=0;
lossdata=[0.0346 0.00643];
totgencost=0;
for i=1:ng
B(i)=lossdata(1,i);
end
iter=0;
disp('iter lambda pg1 pg2 ploss');
while (abs(delp)>=0.001)
  iter=iter+1;
  lambda=lambda+dellambda;
  pl=0;
  sum=0;
  delpla=0;
  for i=1:ng
        den=2*(a(i)+lambda*B(i)*0.01);
        p(i)=(lambda-b(i))/den;
        pl=pl+(B(i)*0.01*p(i)*p(i));
        sum=sum+p(i);
  end
  delp=pd+pl-sum;
  for i=1:ng
   den=2*(a(i)+lambda*B(i)*0.01)^2;
  delpla=delpla+(a(i)+B(i)*0.01*b(i))/den;
  end
  dellambda=delp/delpla;
  [iter;lambda;p(1);p(2);pl]'
end
for i=1:ng
  den=1-(B(i)*p(i)*2*0.01);
  l(i)=1/den;
end
totgencost=0;
for i=1:ng
totgencost=totgencost+(d(i)+b(i)*p(i)+a(i)*p(i)*p(i));
```

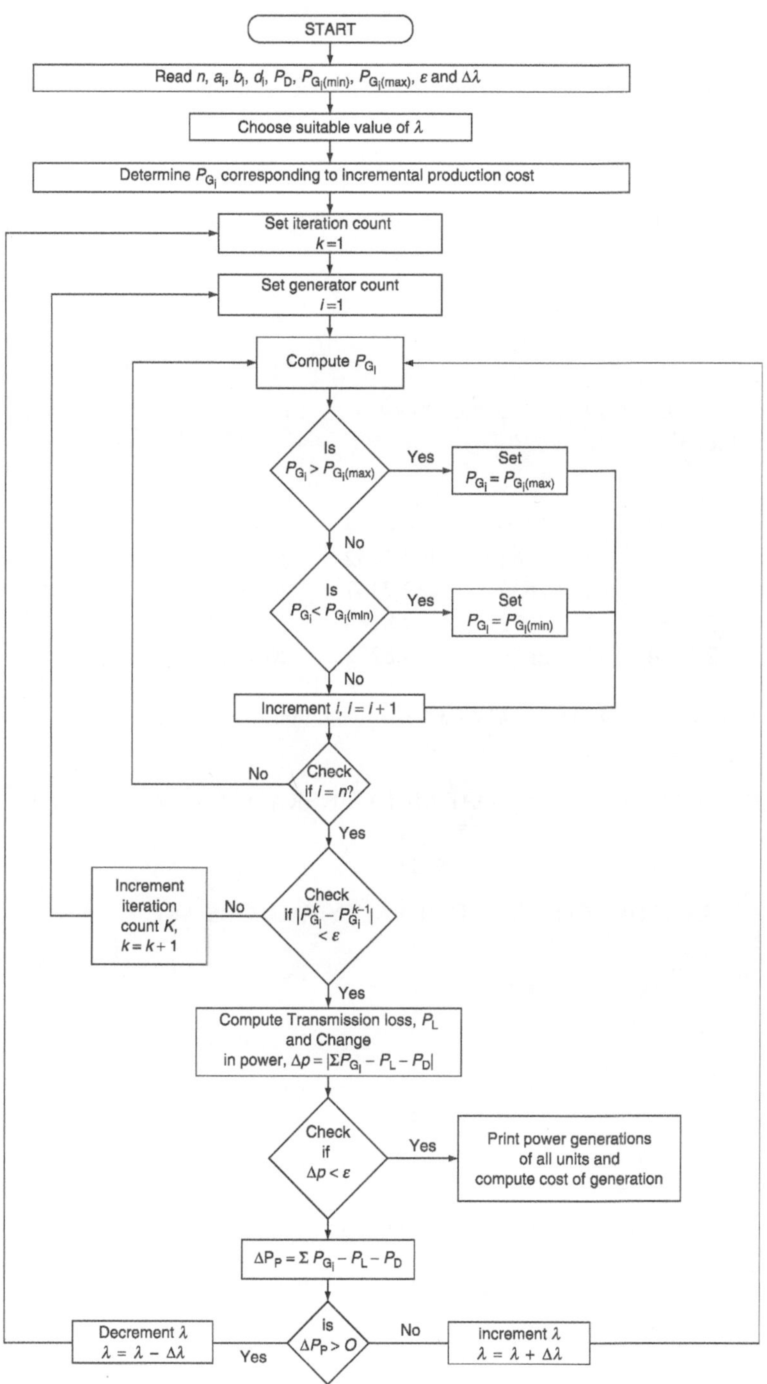

FIG. 3.3 *Flowchart*

```
ifc(i)=2*a(i)*p(i)+b(i);
end
disp('FINAL OUTPUT OF MATLAB PROGRAM dispatch3.m');
lambda
disp('GENERATING UNIT OPTIMALGENERATION(MW)');
[uno; p]'
disp('INCREMENTAL FUEL COST AND PENALTY FACTORS ARE');
disp('UNIT NO. IFC L');
[uno;ifc;l]'
disp('CHECK LAMBDA=IFC*L');
disp('UNIT NO. LAMBDA');
[uno;ifc.*l]'
disp('TOTAL GENERATION COST(Rs./h)');
totgencost
```

iter	lambda	pg1	pg2	ploss
1.0000	12.0000	171.7370	476.6314	24.8123
2.0000	12.0949	176.8464	488.7452	26.1805
3.0000	12.1027	177.2635	489.7367	26.2940
4.0000	12.1033	177.2972	489.8168	26.3032
5.0000	12.1034	177.2999	489.8232	26.3039

FINAL OUTPUT OF MATLAB PROGRAM dispatch3.m

lambda $= 12.1034$

GENERATING UNIT	OPTIMAL GENERATION (MW)
1.0000	177.2999
2.0000	489.8232

INCREMENTAL FUEL COSTS AND PENALTY FACTORS ARE:

UNIT NO.	IFC	L
1.0000	10.6184	1.1398
2.0000	11.3410	1.0672

CHECK LAMBDA $=$ IFC*L

UNIT NO.	LAMBDA
1.0000	12.1034
2.0000	12.1034

TOTAL GENERATION COST (Rs./hr) $= 7386.20$

Example 3.2: The fuel cost functions in Rs./hr for two thermal plants are given by

$$C_1 = 420 + 9.2P_1 + 0.004P_1^2, \quad 100 \leq P_2 \leq 200$$
$$C_2 = 350 + 8.5P_2 + 0.0029P_1^2, \quad 150 \leq P_3 \leq 500$$

where P_1, P_2, P_3 are in MW and plant outputs are subjected to the following limits. Determine the optimal scheduling of generation if the total load is 640.82 MW. Estimate the value of $\lambda = 12$ Rs./MWh.

$$P_{L(pu)} = 0.0346 \, P_{1(pu)}^2 + 0.00643 \, P_{2(pu)}^2$$

Solution:

```
%MATLAB  PROGRAM  FOR  ECONOMIC  LOAD  DISPATCH  WITH  LOSSES  AND
GENERATOR
    %LIMITS(dispatch4.m)
    clc;
    clear;

    %            uno      d       b       a        pmin      pmax
    costdata=[1        420     9.2     0.004      100      200;
              2        350     8.5     0.0029     150      500];

    ng=length(costdata(:,1));
    for i=1:ng
       uno(i)= costdata(i,1);
       d(i)=costdata(i,2);
       b(i)=costdata(i,3);
       a(i)=costdata(i,4);
       pmin(i)=costdata(i,5);
       pmax(i)=costdata(i,6);
    end
    lambda=12;
    pd=640.82;
    delp=0.1;
    dellambda=0;
    lossdata=[0.0346 0.00643];%NOT IN PU
    totgencost=0;
    for i=1:ng
    B(i)=lossdata(1,i);
    end
    while abs(delp)>=0.001
       lambda=lambda+dellambda;
       pl=0;
       sum=0;
       delpla=0;
       for i=1:ng
          den=2*(a(i)+lambda*B(i)*0.01);
          p(i)=(lambda-b(i))/den;
          pl=pl+(B(i)*0.01*p(i)*p(i));
          sum=sum+p(i);
       end
       delp=pd+pl-sum;
       for i=1:ng
          den=2*(a(i)+lambda*B(i)*0.01)^2;
       delpla=delpla+(a(i)+B(i)*0.01*b(i))/den;
       end
       dellambda=delp/delpla;
    end
    dellambda=0;
```

```
for i=1:ng
   pv(i)=0;
   pvfin(i)=0;
end
limvio=0;
for i=1:ng
   if p(i)<pmin(i)|p(i)>pmax(i)
       limvio=1;
       break;
   end
end
if limvio==0
   disp('GENERATION IS WITHIN THE LIMITS');
end
delp=0.1;
if limvio==1
       while (abs(delp)>=0.01)
   disp('GENERATION IS NOT WITHIN THE LIMITS');
   disp('VIOLATED GENERATOR NUMBER');
   i
       if p(i)<pmin(i)
          disp('GENERATION OF VIOLATED UNIT(MW)');
          p(i)
          disp('CORRESPONDING VOILATED LIMIT IS pmin');
       elseif p(i)>pmax(i)
          disp('GENERATION OF VIOLATED UNIT(MW)');
          p(i)
          disp('CORRESPONDING VIOLATED LIMIT IS pmax');
       end
   pl=0;
   sum=0;
   delpla=0;
   for i=1:ng
       pv(i)=0;
   end
 for i=1:ng
   if p(i)<pmin(i)|p(i)>pmax(i)
       pv(i)=1;
       pvfin(i)=1;
       if p(i)<pmin(i)
          p(i)=pmin(i);
       end
       if p(i)>pmax(i)
          p(i)=pmax(i);
       end
   end
  end
 end
```

```matlab
for i=1:ng
    if pvfin(i) ~=1
    den=2*(a(i)+lambda*B(i)*0.01)^2;
    delp1a=delp1a+(a(i)+B(i)*0.01*b(i))/den;
    end
    sum=sum+p(i);
end
delp=pd+p1-sum;
dellambda=delp/delp1a;
lambda=lambda+dellambda;
sum=0;
for i=1:ng
    if pvfin(i) ~=1
    den=2*(a(i)+lambda*B(i)*0.01);
    p(i)=(lambda-b(i))/den;
    end
    p1=p1+(B(i)*0.01*p(i)*p(i));
    sum=sum+p(i);
end
delp=pd+p1-sum;
end
end
for i=1:ng
    den=1-(B(i)*p(i)*2*0.01);
    l(i)=1/den;
end
for i=1:ng
    totgencost=totgencost+(d(i)+b(i)*p(i)+a(i)*p(i)*p(i));
    ifc(i)=2*a(i)*p(i)+b(i);
end
disp('FINAL OUTPUT OF MATLAB PROGRAM dispatch4.m');
lambda
disp('GENERATING UNIT OPTIMAL GENERATION(MW)');
[uno; p]'
disp('INCREMENTAL FUEL COSTS AND PENALTY FACTORS ARE');
disp('UNIT NO. IFC L');
[uno;ifc;l]'
disp('CHECK LAMBDA=IFC*L');
disp('UNIT NO. LAMBDA');
[uno;ifc.*l]'
disp('TOTAL GENERATION COST(Rs./hr)');
totgencost
```

Results:

GENERATION IS WITHIN THE LIMITS
FINAL OUTPUT OF MATLAB PROGRAM dispatch3.m

lambda = 12.1034

GENERATING UNIT	OPTIMAL GENERATION (MW)
1.0000	177.3001
2.0000	489.8236

INCREMENTAL FUEL COSTS AND PENALTY FACTORS ARE:

UNIT NO.	IFC	L
1.0000	10.6184	1.1399
2.0000	11.3410	1.0672

CHECK LAMBDA = IFC*L

UNIT NO.	LAMBDA
1.0000	12.1034
2.0000	12.1034

TOTAL GENERATION COST (Rs./hr) = 7386.20

Example 3.3 The IFC for two plants are

$$\frac{dC_1}{dP_{G_1}} = 0.075P_{G_1} + 18 \quad \text{Rs./MWh}$$

$$\frac{dC_2}{dP_{G_2}} = 0.08P_{G_2} + 16 \quad \text{Rs./MWh}$$

The loss coefficients are given as

$B_{11} = 0.0015/\text{MW}$, $B_{12} = -0.0004/\text{MW}$, and $B_{22} = 0.0032/\text{MW}$ for $\lambda = 25$ Rs./MWh. Find the real-power generations, total load demand, and the transmission power loss.

Solution:

$$P_{G_1} + P_{G_2} = P_D + P_L$$

And transmission loss, $P_L = \sum_{p=1}^{n} \sum_{p=1}^{n} P_{G_P} B_{pq} P_{G_q}$

For number of plants, $n = 2$, we have

$$P_L = \sum_{p=1}^{2} \sum_{q=1}^{2} P_{G_P} B_{pq} P_{G_q}$$

$$= P_{G_1} B_{11} P_{G_1} + P_{G_1} B_{12} P_{G_2} + P_{G_2} B_{21} P_{G_1} + P_{G_2} B_{22} P_{G_2}$$

$$= 0.0015 P_{G_1}^2 + 2(-0.0004) P_{G_1} P_{G_2} + 0.0032 P_{G_2}^2$$

$$= 0.0015 P_{G_1}^2 - 0.0008 P_{G_1} P_{G_2} + 0.0032 P_{G_2}^2$$

The ITL of Plant-1 is

$$(\text{ITL})_1 = \frac{\partial P_L}{\partial P_{G_1}} = 2(0.0015) P_{G_1} - 0.0008 P_{G_2}$$

$$= 2(0.0015) P_{G_1} - 0.0008 P_{G_2}$$

Penalty factor of Plant-1:

$$L_1 = \cfrac{1}{1 - \cfrac{\partial P_L}{\partial P_{G_1}}} = \cfrac{1}{1 - (0.003 P_{G_1} - 0.0008 P_{G_2})}$$

The ITL of Plant-2 is

$$(\text{ITL})_2 = \frac{\partial P_L}{\partial P_{G_2}} = -0.0008 P_{G_1} + 0.0064 P_{G_2}$$

and penalty factor of plant-2 is

$$L_2 = \cfrac{1}{1 - \cfrac{\partial P_L}{\partial P_{G_2}}} = \cfrac{1}{1 - (-0.0008 P_{G_1} + 0.0064 P_{G_2})}$$

Condition for optimum operation is

$$\frac{\partial C_1}{\partial P_{G_1}} L_1 = \frac{\partial C_2}{\partial P_{G_2}} L_2 = \lambda$$

$$\frac{\partial C_1}{\partial P_{G_1}} L_1 = \lambda$$

$$(0.075 P_{G_1} + 18) \frac{1}{1 - (0.003 P_{G_1} - 0.0008 P_{G_2})} = 25$$

$$0.075 P_{G_1} + 18 = 25[1 - (0.003 P_{G_1} - 0.0008 P_{G_2})]$$

or $0.15 P_{G_1} - 0.02 P_{G_2} = 7$ $\hspace{3cm}$ **(3.21)**

and $\dfrac{\partial C_2}{\partial P_{G_2}} L_2 = \lambda$

$$(0.08 P_{G_2} + 16) \frac{1}{1 - (-0.0008 P_{G_1} + 0.0064 P_{G_2})} = 25$$

$$(0.08 P_{G_2} + 16) = 25[1 - (-0.0008 P_{G_1} + 0.0064 P_{G_2})]$$

or $0.02 P_{G_1} - 0.24 P_{G_2} = -9$ $\hspace{3cm}$ **(3.22)**

Solving Equations (3.21) and (3.22),

Equation $(3.21) \times 0.24 \Rightarrow 0.036 P_{G_1} - 0.0048 P_{G_2} = 1.68$

Equation $(3.22) \times 0.02 \Rightarrow 0.0004 P_{G_1} - 0.0048 P_{G_2} = -0.18$

$$0.0356 P_{G_1} = 1.86$$

$$\therefore P_{G_1} = 52.247 \text{ MW}$$

Substituting the P_{G_1} value in Equation (3.21), we get

$$0.15(52.247) - 0.02P_{G_2} = 7$$

$$\therefore P_{G_2} = 41.852 \text{ MW}$$

Transmission loss, $P_L = 0.0015(52.247)^2 - 0.0008\,(52.247)(41.852) + 0.0032(41.852)^2$
$$= 7.95 \text{ MW}$$

Total load, $P_D = P_{G_1} + P_{G_2} - P_L$
$$= 52.247 + 41.852 - 7.95 = 86.149 \text{ MW}$$

Example 3.4: The cost curves of two plants are

$$C_1 = 0.05P_{G_1}^2 + 20P_{G_1} + 150\,\text{Rs./hr}$$
$$C_2 = (0.05P_{G_2}^2) + 15P_{G_1} + 180\,\text{Rs./hr}$$

The loss coefficient for the above system is given as $B_{11}=0.0015/\text{MW}$, $B_{12} = B_{21} = -0.0004/\text{MW}$, and $B_{22}=0.0032/\text{MW}$. Determine the economical generation scheduling corresponding to $\lambda = 25$ Rs./MWh and the corresponding system load that can be met with. If the total load connected to the system is 120 MW taking 4% change in the value of λ, what should be the value of λ in the next iteration?

Solution:

Given that the cost curves of two plants are

$$C_1 = 0.05P_{G_1}^2 + 20P_{G_1} + 150\,\text{Rs./hr}$$
$$C_2 = (0.05P_{G_2}^2) + 15P_{G_1} + 180\,\text{Rs./hr}$$

the incremental costs are

$$\frac{dC_1}{dP_{G_1}} = 0.1P_{G_1} + 20 \text{ Rs./MWh}$$

$$\frac{dC_2}{dP_{G_2}} = 0.1P_{G_2} + 15 \text{ Rs./MWh}$$

Transmission loss, $P_L = \sum_{p=1}^{n}\sum_{q=1}^{n} P_{G_p} B_{pq} P_{G_q}$

For two plants, $n=2$ and we have

$$
\begin{aligned}
P_L &= \sum_{p=1}^{2}\sum_{q=1}^{2} P_{G_p} B_{pq} P_{G_q} \\
&= P_{G_1} B_{11} P_{G_1} + P_{G_1} B_{12} P_{G_2} + P_{G_2} B_{21} P_{G_1} + P_{G_2} B_{22} P_{G_2} \\
&= 0.0015P_{G_1}^2 + 2(-0.0004)P_{G_1} P_{G_2} + 0.0032P_{G_2}^2 \\
&= 0.0015P_{G_1}^2 - 0.0008P_{G_1} P_{G_2} + 0.0032P_{G_2}^2
\end{aligned}
$$

The ITL of Plant-1 is

$$(\text{ITL})_1 = \frac{\partial P_L}{\partial P_{G_1}} = 2(0.0015)P_{G_1} - 0.0008P_{G_2}$$

$$= 0.003P_{G_1} - 0.0008P_{G_2}$$

The ITL of Plant-2 is

$$(ITL)_2 = \frac{\partial P_L}{\partial P_{G_2}} = -0.0008P_{G_1} + 0.0064P_{G_2}$$

The penalty factor of Plant-1 is

$$L_1 = \frac{1}{1 - \dfrac{\partial P_L}{\partial P_{G_1}}} = \frac{1}{1 - (0.003P_{G_1} - 0.0008P_{G_2})}$$

and the penalty factor of Plant-2 is

$$L_2 = \frac{1}{1 - \dfrac{\partial P_L}{\partial P_{G_2}}} = \frac{1}{1 - (-0.0008P_{G_1} + 0.0064P_{G_2})}$$

The condition for optimum operation is

$$\frac{\partial C_1}{\partial P_{G_1}} L_1 = \frac{\partial C_2}{\partial P_{G_2}} L_2 = \lambda$$

$$\frac{\partial C_1}{\partial P_{G_1}} L_1 = \lambda$$

$$(0.01P_{G_1} + 20) \frac{1}{1 - (0.003P_{G_1} - 0.0008P_{G_2})} = 30$$

$$0.01P_{G_1} + 20 = 30[1 - (0.003P_{G_1} - 0.0008P_{G_2})]$$

or $\quad 1.09P_{G_1} - 0.024P_{G_2} = 10$ $\qquad\qquad$ **(3.23)**

and $\quad \dfrac{\partial C_2}{\partial P_{G_2}} L_2 = \lambda$

$$(0.1P_{G_2} + 15) \frac{1}{1 - (-0.0008P_{G_1} + 0.0064P_{G_2})} = 30$$

$$(0.1P_{G_2} + 15) = 30[1 - (-0.0008P_{G_1} + 0.0064P_{G_2})]$$

or $\quad 0.024P_{G_1} - 0.292P_{G_2} = -15$ $\qquad\qquad$ **(3.24)**

Solving Equations (3.23) and (3.24), we get

Equation $(3.23) \times 0.024 \Rightarrow 0.02616P_{G_1} - 0.000576P_{G_2} = 0.24$

Equation $(3.24) \times 1.09 \Rightarrow 0.02616P_{G_1} - 0.31828P_{G_2} = -16.36$

$$0.3177P_{G_2} = 16.6$$

$$\therefore P_{G_2} = 52.25 \text{ MW}$$

Substituting the P_{G_2} value in Equation (3.23), we get

$$1.09P_{G_1} - 0.024(52.25) = 10$$

$$\therefore P_{G_1} = 10.325 \text{ MW}$$

Transmission loss, $P_L = 0.0015(10.325)^2 - 0.0008\,(10.325)(52.25) + 0.0032(52.25)^2$

$$= 8.465 \text{ MW}$$

The corresponding system load, $P_D = P_{G_1} + P_{G_2} - P_L$

$$= 10.325 + 52.25 - 8.465 = 54.11 \text{ MW}$$

For 4% change in value of λ, $\Delta\lambda = 4\%$ of $30 = 1.2$ Rs./MWh

New load connected to system, $P_D = 120$ MW

Change in load, $\Delta P_D = 120 - 54.11 = 65.89$ MW

Here, change in load, $\Delta P_D > 0$; hence, to get an optimum dispatch decrement λ by $\Delta\lambda$,

New value of $\lambda = \lambda' = \lambda - \Delta\lambda = 30 - 1.2 = 28.8$ Rs./MWh.

Example 3.5: A system consists of two power plants connected by a transmission line. The total load located at Plant-2 is as shown in Fig. 3.4. Data of evaluating loss coefficients consist of information that a power transfer of 100 MW from Station-1 to Station-2 results in a total loss of 8 MW. Find the required generation at each station and power received by the load when λ of the system in Rs. 100/MWh. The IFCs of the two plants are given by

$$\frac{dC_1}{dP_{G_1}} = 0.12P_{G_1} + 65 \text{ Rs./MWh}$$

$$\frac{dC_2}{dP_{G_2}} = 0.25P_{G_2} + 75 \text{ Rs./MWh}$$

Solution:
Total loss is

$$P_L = \sum_{p=1}^{n} \sum_{p=1}^{n} P_{G_p} B_{pq} P_{G_q}$$

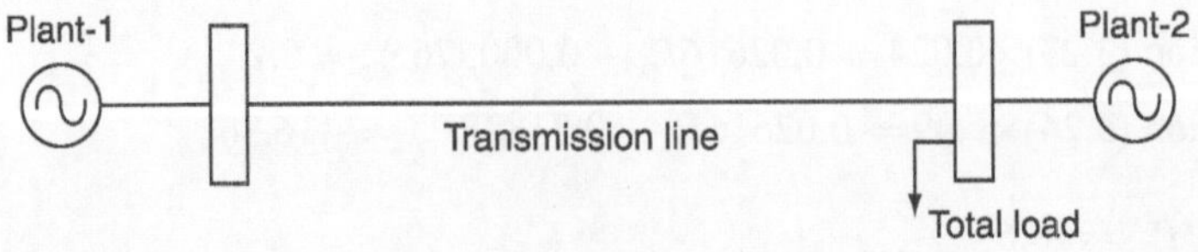

FIG. 3.4 *Illustration for Example 3.5*

Since $n=2$, we have

$$P_L = \sum_{p=1}^{2} \sum_{p=1}^{2} P_{G_p} B_{pq} P_{G_q}$$

$$= P_{G_1} B_{11} P_{G_1} + P_{G_1} B_{12} P_{G_2} + P_{G_2} B_{21} P_{G_1} + P_{G_2} B_{22} P_{G_2}$$

Since power transfer of 100 MW from Plant-1 to Plant-2 (i.e., $P_{G_1} = 100$ MW), $P_{G_2}, B_{21}, B_{22} = 0$

$$\therefore P_L = B_{11} P_{G_1}^2$$

Given: $P_L = 8$ MW

$$\therefore 8 = B_{11} (100)^2$$

$$\Rightarrow B_{11} = 8 \times 10^{-4}\ \text{MW}^{-1}$$

$$\therefore P_L = 8 \times 10^{-4} P_{G_1}^2$$

$$\frac{\partial P_L}{\partial P_{G_1}} = 8 \times 10^{-4} \times 2 P_{G_1} = 16 \times 10^{-4} P_{G_1}$$

$$\frac{\partial P_L}{\partial P_{G_2}} = 0$$

and the penalty factor of Plant-1 is

$$L_1 = \frac{1}{1 - \dfrac{\partial P_L}{\partial P_{G_1}}} = \frac{1}{1 - (16 \times 10^{-4} P_{G_1})}$$

And the penalty factor of Plant-2 is

$$L_2 = \frac{1}{1 - \dfrac{\partial P_L}{\partial P_{G_2}}} = \frac{1}{1} = 1$$

Now, the condition for optimality is

$$\frac{\partial C_1}{\partial P_{G_1}} L_1 = \frac{\partial C_2}{\partial P_{G_2}} L_2 = \lambda$$

For $\lambda = 100$ Rs./MWh

$$\therefore (0.12 P_{G_1} + 65) \left(\frac{1}{1 - 16 \times 10^{-4} P_{G_1}} \right) = (0.25 P_{G_2} + 75) \times 1 = 100$$

or $\quad 0.12 P_{G_1} + 65 = 100(1 - 16 \times 10^{-4} P_{G_1})$

or $\quad 0.12 P_{G_1} + 0.16 P_{G_1} = 100 - 65$

$$0.28 P_{G_1} = 35$$

$$\therefore P_{G_1} = \frac{35}{0.28} = 125 \text{ MW}$$

and $\quad 0.25 P_{G_2} + 75 = 100$

$$0.25 P_{G_2} = 25$$

$$P_{G_2} = \frac{25}{0.25}$$

$$\Rightarrow P_{G_2} = 100 \text{ MW}$$

Power received by the load $= (P_{G_1} + P_{G_2}) - \text{losses}$

$$= 125 + 100 - \{8 \times 10^{-4} \times P_{G_1}^2\}$$
$$= 125 + 100 - \{8 \times 10^{-4} \times 125^2\}$$
$$= 225 - 12.5$$
$$= 212.5 \text{ MW}$$

Example 3.6: For Example.3.5, with 212.5 MW received by the load, find the savings in Rs./hr obtained by co-ordinating the transmission losses rather than neglecting in determining the load division between the plants.

Solution:

By co-ordinating the losses to supply a load of 212.5 MW, the real-power generations at Plants 1 and 2 should be 125 and 100 MW, respectively.

When losses are neglected, total load $= 212.5$ MW is to be distributed between the two plants most economically. Condition for optimality,

$$\frac{dC_1}{dP_{G_1}} = \frac{dC_2}{dP_{G_2}}$$

$$0.12 P_{G_1} + 65 = 0.25 P_{G_2} + 75$$

$$0.12 P_{G_1} + (65 - 75) = 0.25 P_{G_2}$$

$$\Rightarrow P_{G_2} = \frac{0.12 P_{G_1} - 10}{0.25}$$

$$= 0.48 P_{G_1} - 40$$

$$P_D = P_{G_1} + P_{G_2} - P_L$$

Since the losses are not co-ordinated but neglected, we have

$$212.5 P_{G_1} + 0.49 P_{G_1} - 40 - 8 \times 10^{-4} P_{G_1}^2$$

or $\quad 8 \times 10^{-4} P_{G_1}^2 - 1.48 P_{G_1} \ 252.5 = 0$

By solving, we get

$$P_{G_1} = 1{,}659.8 \text{ MW} \quad \text{and} \quad 190.15 \text{ MW}$$

$$P_{G_1} = 1{,}659.8 \text{ MW} \Rightarrow \text{is not to be required to overcome that power demand } P_D$$

$$\therefore \ P_{G_1} = 190.15 \text{ MW is to be required}$$

$$\text{and} \quad P_{G_2} = 0.48 \times 190.15 - 40$$

$$= 51.27 \text{ MW}$$

$\therefore$ **Power generation:** with losses are being co-ordinated, $P_{G_1} = 125$ MW, $P_{G_2} = 100$ MW with losses are not being co-ordinated, $P_{G_1} = 190.15$ MW, $P_{G_2} = 51.27$ MW.

Increase in cost of Plant-1 when losses are co-ordinated:

$$= \int_{190.15}^{125} (0.12 P_{G_1} + 65) \, dP_{G_1}$$

$$= \left[0.12 \frac{P_{G_1}^2}{2} + 65 P_{G_1} \right]_{190.15}^{125}$$

$$= 0.06 \left[125^2 - 190.15^2 \right] + 65 [125 - 190.15]$$

$$= \text{Rs. } 5{,}466.67/\text{hr}$$

Increase in cost of Plant-2, because increase in generation:

$$= \int_{51.27}^{100} (0.25 P_{G_2} + 75) \, dP_{G_2}$$

$$= \left[0.25 \frac{P_{G_2}^2}{2} + 75 P_{G_2} \right]_{51.27}^{100}$$

$$= 0.125 [100^2 - 51.27^2] + 75(100 - 51.27) = 4{,}576.17$$

Savings in Rs./hr by co-ordinating the losses $= 5{,}466.67 - 4{,}576.17 = 890.50$ Rs./hr.

Example 3.7: On a system consisting of two generating plants, the incremental costs in Rs./MWh with P_{G_1} and P_{G_2} in MW are

$$\frac{dC_1}{dP_{G_1}} = 0.008 P_{G_1} + 8.0; \quad \frac{dC_2}{dP_{G_2}} = 0.012 P_{G_2} + 9.0$$

The system is operating on economic dispatch with $P_{G_1} = P_{G_2} = 500$ MW and $\dfrac{\partial P_L}{\partial P_{G_2}} = 0.2$. Find the penalty factor of Plant-1.

Solution:

Given that the system operates on economic dispatch with $P_{G_1} = P_{G_2} = 500$ MW, the condition for this optimal operation when considering the transmission loss is

$$\frac{dC_1}{dP_{G_1}} L_1 = \frac{dC_2}{dP_{G_2}} L_2 = \lambda$$

and also given that ITL of Plant-2,

$$(ITL)_2 = \frac{\partial P_L}{\partial P_{G_2}} = 0.2$$

The penalty factor of Plant-2,

$$L_2 = \frac{1}{1 - \dfrac{\partial P_L}{\partial P_{G_2}}}$$

$$= \frac{1}{1 - 0.2} = \frac{1}{0.8} = 1.25$$

$\therefore$ For optimal condition,

$$\frac{dC_1}{dP_{G_1}} L_1 = \frac{dC_2}{dP_{G_2}} L_2$$

$$(0.008 P_{G_1} + 80) L_1 = (0.012 P_{G_2} + 9.0) L_2$$

or $[(0.008 \times 500) + 80] L_1 = (0.012 \times 500 + 9.0) 1.25$

or $12 L_1 = 18.75$

or $L_1 = 1.5625$

$\therefore$ Penalty factor of Plant-1 $= 1.5625$.

Example 3.8: Determine the incremental cost of received power and the penalty factor of the plant shown in Fig. 3.5 if the incremental cost of production is

$$\frac{dC_1}{dP_{G_1}} = 0.1 P_{G_1} + 3.0 \text{ Rs./MWh}$$

Solution:

The penalty factor $= L_1 = \dfrac{1}{1 - \dfrac{\partial P_L}{\partial P_{G_1}}} = \dfrac{1}{1 - \dfrac{2}{10}} = \dfrac{10}{8}$

$\therefore$ Cost of received power $= \dfrac{dC_1}{dP_{G_1}} L_1$

$$= (0.1 P_{G_1} + 30) \times \frac{10}{8}$$

$$= (0.1 \times 10 + 30) \times \frac{10}{8}$$

$$= 4 \times \frac{10}{8} = 5$$

$$= 5.0 \text{ Rs./MWh}$$

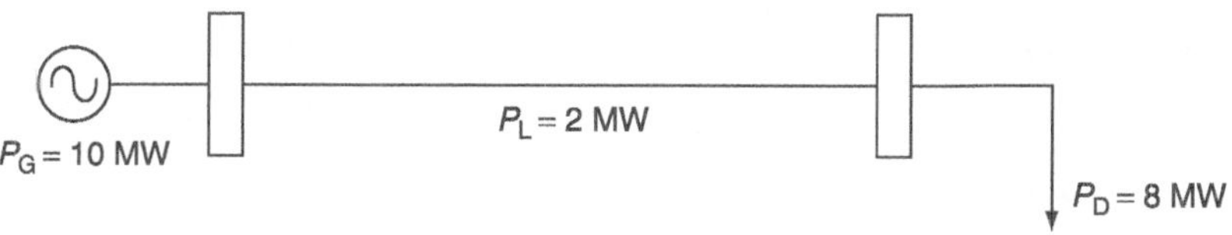

FIG. 3.5   *Illustration for Example 3.8*

Example 3.9: A 2-bus system consists of two power plants connected by a transmission line as shown in Fig. 3.6.

The cost-curve characteristics of the two plants are:

$$C_1 = 0.015P_{G_1}^{\,2} + 18P_{G_1} + 20 \text{ Rs./hr}$$

$$C_1 = 0.03P_{G_2}^{\,2} + 33P_{G_2} + 40 \text{ Rs./hr}$$

When a power of 120 MW is transmitted from Plant-1 to the load, a loss of 16.425 MW is incurred. Determine the optimal scheduling of plants and the load demand if the cost of received power is Rs. 26/MWh. Solve the problem using co-ordination equations and the penalty factor method approach.

Solution:

For two units, $P_L = P_{G_1}B_{11}P_{G_1} + 2P_{G_1}B_{12}P_{G_2} + P_{G_2}B_{21}P_{G_1}$.

Since the load is located at Bus-2 alone, the losses in the transmission line will not be affected by the generator of Plant-2.

i.e., $B_{12} = B_{21} = 0$ and $B_{22} = 0$

$$\therefore \; P_L = B_{11}P_{G_1}^2 \tag{3.25}$$

$$16.425 = B_{11} \times 120^2$$

$$B_{11} = 0.00114 \text{ MW}^{-1}$$

Using the co-ordination equation method:

The co-ordination equation for Plant-1 is

$$\frac{dC_1}{dP_{G_1}} + \lambda\frac{\partial P_L}{\partial P_{G_1}} = \lambda \tag{3.26}$$

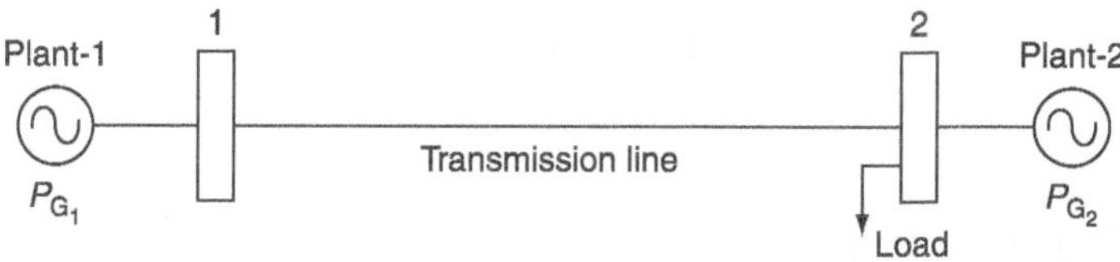

FIG. 3.6   *Illustration for Example 3.9*

$$P_{\mathrm{L}} = 0.00114 P_{\mathrm{G}_1}^2$$

$$\Rightarrow \frac{\partial P_{\mathrm{L}}}{\partial P_{\mathrm{G}_1}} = 0.00228 P_{\mathrm{G}_1} \qquad (3.27)$$

$$\frac{\mathrm{d}C_1}{\mathrm{d}P_{\mathrm{G}_1}} = 0.03 P_{\mathrm{G}_1} + 18 \qquad (3.28)$$

Substitute Equations (3.27) and (3.28) in Equation (3.26); then the equation for Plant-1 becomes

$$0.03 P_{\mathrm{G}_1} + 18 + \lambda(0.00228 P_{\mathrm{G}_1}) = \lambda$$
$$0.03 P_{\mathrm{G}_1} + 18 + 26(0.00228 P_{\mathrm{G}_1}) = 26$$
$$0.03 P_{\mathrm{G}_1} + 0.0593 P_{\mathrm{G}_1} + 18 = 26$$
$$0.0893 P_{\mathrm{G}_1} = 8$$

$$\therefore \ P_{\mathrm{G}_1} = \frac{8}{0.0893} = 89.6 \ \mathrm{MW}$$

The co-ordination equation for Plant-2 is

$$\frac{\mathrm{d}C_2}{\mathrm{d}P_{\mathrm{G}_2}} + \lambda \frac{\partial P_{\mathrm{L}}}{\partial P_{\mathrm{G}_2}} = \lambda \qquad (3.29)$$

$$\frac{\mathrm{d}C_2}{\mathrm{d}P_{\mathrm{G}_2}} + \lambda(0) = \lambda$$

$$\frac{\mathrm{d}C_2}{\mathrm{d}P_{\mathrm{G}_2}} = 0.06 P_{\mathrm{G}_2} + 22$$

$\therefore$ Equation (3.29) becomes

$$0.06 P_{\mathrm{G}_2} + 22 + 26(0) = 22$$

$$P_{\mathrm{G}_2} = \frac{26 - 22}{0.06} = 66.67 \ \mathrm{MW}$$

$\therefore$ The transmission loss, $P_{\mathrm{L}} = B_{11} P_{\mathrm{G}_1}^2$

$$= 0.00114 \times (89.6)^2$$

$$= 9.15 \ \mathrm{MW}$$

$\therefore$ The load, $P_{\mathrm{D}} = P_{\mathrm{G}_1} + P_{\mathrm{G}_2} - P_{\mathrm{L}}$

$$= 89.6 + 66.67 - 9.15 = 147.12 \ \mathrm{MW}$$

Penalty factor method:

The penalty factor of Plant-1 is

$$L_1 = \cfrac{1}{1-\cfrac{\partial P_L}{\partial P_{G_1}}} = \cfrac{1}{1-0.00228 P_{G_1}}$$

Now the condition for optimality is

$$\frac{dC_1}{dP_{G_1}} L_1 = \lambda$$

$$\frac{0.03 P_{G_1} + 18}{1 - 0.00228 P_{G_1}} = 26$$

$$\therefore P_{G_1} = 89.6 \text{ MW}$$

The penalty factor of Plant-2,

$$L_2 = 1 \quad \left(\text{since } \frac{\partial P_L}{\partial P_{G_2}} = 0 \right)$$

For optimality,

$$\frac{dC_2}{dP_{G_2}} L_2 = \lambda$$

$$(0.06 P_{G_2} + 22)1 = 26$$

$$\therefore P_{G_2} = 66.67 \text{ MW}$$

The transmission loss, $P_L = B_{11} P_{G_1}^2$

$$= 0.00114 \times (89.6)^2$$

$$= 9.15 \text{ MW}$$

$\therefore$ The load, $P_D = P_{G_1} + P_{G_2} - P_L$

$$= 89.6 + 66.67 - 9.15 = 147.12 \text{ MW}.$$

Example 3.10: Assume that the fuel input in British thermal unit (Btu) per hour for Units 1 and 2 are given by

$$C_1 = (8 P_{G_1} + 0.024 P_{G_1}^2 + 80)10^6$$
$$C_2 = (6 P_{G_1} + 0.04 P_{G_2}^2 + 120)10^6$$

The maximum and minimum loads on the units are 100 and 10 MW, respectively. Determine the minimum cost of generation when the following load is supplied as shown in Fig. 3.7. The cost of fuel is Rs. 2 per million Btu.

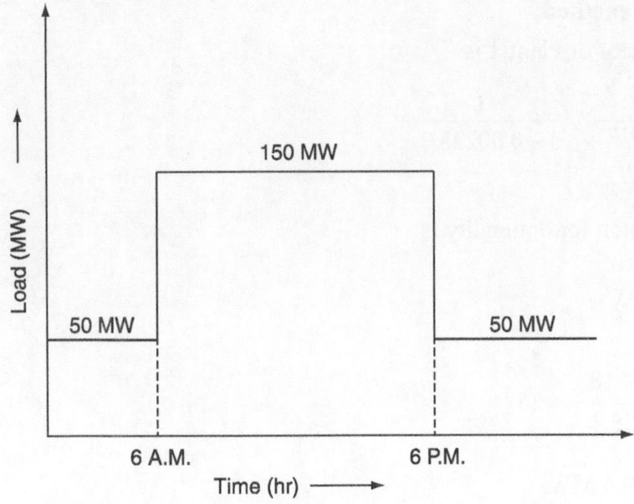

FIG. 3.7   *Illustration for Example 3.10*

Solution:

$$\frac{dC_1}{dP_{G_1}} = 0.048P_{G_1} + 8$$

$$\frac{dC_2}{dP_{G_2}} = 0.08P_{G_2} + 6$$

(i) When the load is 50 MW:

Condition for the economic scheduling is

$$\frac{dC_1}{dP_{G_1}} = \frac{dC_2}{dP_{G_2}}$$

$$0.048P_{G_1} + 8 = 0.08P_{G_2} + 6$$

$$0.048P_{G_1} - 0.08P_{G_2} = -2 \qquad\qquad\text{(3.30)}$$

$$\text{and}\quad P_{G_1} + P_{G_2} = 50 \qquad\qquad\text{(3.31)}$$

By solving Equations (3.30) and (3.31), we get

$$P_{G_1} = 15.625 \text{ MW}$$

$$P_{G_2} = 34.375 \text{ MW}$$

$$\therefore\quad C_1 = 210.868 \text{ million Btu/h}$$
$$C_2 = 373.5 \text{ million Btu/h}$$

(ii) When the load is 150 MW,

From Equation (3.30), $0.048P_{G_1} - 0.08P_{G_2} = -2$ **(3.32)**

$$P_{G_1} + P_{G_2} = 150 \qquad\qquad\text{(3.33)}$$

By solving Equations (3.32) and (3.33), we get

$$P_{G_1} = 78.126 \text{ MW}$$

$$P_{G_2} = 71.874 \text{ MW}$$

and $C_1 = 851.496$ million Btu/hr

$C_2 = 757.87$ million Btu/hr

$\therefore$ Total cost $=$ Rs. $(210.868 + 373.5 + 851.496 + 757.87) \times 2$

$$= \text{Rs. } 52,649.61/\text{hr.}$$

Example 3.11: Two power plants are connected together by a transmission line and load at Plant-2 as shown in Fig. 3.8. When 100 MW is transmitted from Plant-1, the transmission loss is 10 MW. The cost characteristics of two plants are

$$C_1 = 0.05P_{G_1}^2 + 13P_{G_1} \text{ Rs./hr}$$

$$C_2 = 0.06P_{G_2}^2 + 12P_{G_2} \text{ Rs./hr}$$

Find the optimum generation for $\lambda = 22$, $\lambda = 25$, and $\lambda = 30$.

Solution:

$$C_1 = 0.05P_{G_1}^2 + 13P_{G_1}$$

$$C_2 = 0.06P_{G_2}^2 + 12P_{G_2}$$

The IFC characteristics are

$$\frac{dC_1}{dP_{G_1}} = 0.1P_{G_1} + 13 \text{ Rs./MWh}$$

$$\frac{dC_2}{dP_{G_2}} = 0.12P_{G_2} + 12 \text{ Rs./MWh}$$

The transmission power loss, $P_L = \sum_{p=1}^{n} \sum_{q=1}^{n} P_{G_p} B_{pq}$

Here $n = 2$,

$$P_L = B_{11}P_{G_1}^2 + B_{22}P_{G_2}^2 + 2B_{12}P_{G_1}P_{G_2}$$

Since the load is connected at Bus-2 and the power transfer is from Plant-1 only, $B_{22} = 0$ and $B_{12} = 0$.

$$\therefore P_L = B_{11}P_{G_1}^2$$

$$10 = B_{11}(100)^2$$

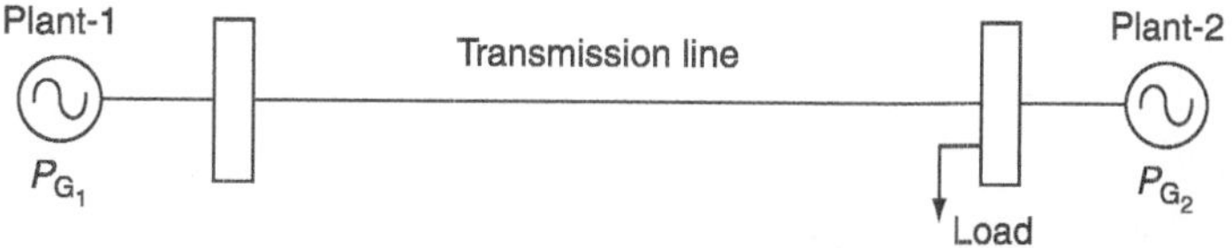

FIG. 3.8   *Single line diagram representing two power plants connected by a transmission line*

$$B_{11} = \frac{10}{(100)^2} = 0.001 \text{ MW}^{-1}$$

The ITL of Plant-1 is

$$(ITL)_1 = \frac{\partial P_L}{\partial P_{G_1}} = 2B_{11}P_{G_1} = 2(0.001)P_{G_1} = 0.002P_{G_1}$$

The penalty factor of Plant-1 is

$$L_1 = \frac{1}{1-(ITL)_1} = \frac{1}{1-0.002P_{G_1}}$$

$$(ITL)_2 = \frac{\partial P_L}{\partial P_{G_2}} = 0 \quad (\because \text{ the transmission power loss is not the function of } P_{G_2})$$

The penalty factor of Plant-2 is

$$L_2 = \frac{1}{1-(ITL)_2} = \frac{1}{1-0} = 1$$

For optimality, when transmission losses are considered, the condition is

$$\frac{dC_1}{dP_{G_1}}L_1 = \frac{dC_2}{dP_{G_2}}L_2 = \lambda$$

$$(0.1\,P_{G_1} + 13)\frac{1}{(1-0.002P_{G_1})} = (0.12P_{G_2} + 12)\times 1 = 22$$

$$\Rightarrow 0.1P_{G_1} + 13 = 22(1-0.002P_{G_1})$$

or $\quad 0.0144P_{G_1} = 9$

$$\therefore P_{G_1} = 62.5 \text{ MW},$$

and $\quad (0.12P_{G_2} + 12)1 = 22$

$$\therefore P_{G_2} = 83.33 \text{ MW}$$

Similarly, we have

For $\lambda = 25$,

$P_{G_1} = 80 \text{ MW}, \quad P_{G_2} = 108.33 \text{ MW}$

For $\lambda = 30$,

$P_{G_1} = 106.25 \text{ MW}, \quad P_{G_2} = 150 \text{ MW}$

Example 3.12: For Example 3.11, the data for the loss equations consist of the information that 200 MW transmitted from Plant-1 to the load results in a transmission loss of 20 MW. Find the optimum generation schedule considering transmission losses to supply a load of 204.41 MW. Also evaluate the amount of financial loss that may be incurred if at the time of scheduling transmission losses are not co-ordinated. Assume that the IFC characteristics of plants are given by

$$\frac{dC_1}{dP_{G_1}} = 0.025P_{G_1} + 14 \text{ Rs./MWh}$$

$$\frac{dC_2}{dP_{G_2}} = 0.05P_{G_2} + 16 \text{ Rs./MWh}$$

Solution:

$$P_L = B_{11}P_{G_1}^2 + 2B_{12}P_{G_1}P_{G_2} + B_{22}P_{G2}^2$$

$\therefore$ The load is at Plant-2, hence $B_2 = B_{22} = 0$

$\therefore P_L = B_{11}P_{G_1}^2$

And given that $P_{G_1} = 200$ MW, $P_L = 20$ MW

$$\Rightarrow 20 = B_{11}(200)^2$$

$$B_{11} = 0.0005 \text{MW}^{-1}$$

$$(\text{ITL})_1 = \frac{\partial P_L}{\partial P_{G_1}} = 2B_{11}P_{G_1} = 0.001P_{G_1}$$

The penalty factor of Plant-1 is

$$L_1 = \frac{1}{1 - (\text{ITL})_1} = \frac{1}{1 - 0.001P_{G_1}}$$

$$(\text{ILT})_2 = \frac{\partial P_L}{\partial P_{G_2}} = 0$$

The penalty factor of Plant-2 is

$$\Rightarrow L_2 = \frac{1}{1 - (\text{ITL})_2} = 1$$

For optimality,

$$\frac{dC_1}{dP_{G_1}}L_1 = \frac{dC_2}{dP_{G_2}}L_2$$

$$(0.025P_{G_1})\left(\frac{1}{1 - 0.001P_{G_1}}\right) = (0.05P_{G_2} + 16)1$$

$$\Rightarrow 0.025 P_{G_1} = (1 - 0.001 P_{G_1})(0.05 P_{G_2} + 16)$$

$$\Rightarrow 0.025 P_{G_1} = (0.05 P_{G_2} - 0.00005 P_{G_1} P_{G_2} + 16 - 0.016 P_{G_1})$$

$$0.041 P_{G_1} - 0.05 P_{G_2} + 0.00005 P_{G_1}^2 P_{G_2} = 16 \tag{3.34}$$

and $\quad P_{G_1} + P_{G_2} = P_D + P_L$

$$P_{G_1} + P_{G_2} = 204.41 + 0.0005 P_{G_1}^2$$

$$\Rightarrow P_{G_1} + P_{G_2} - 0.0005 P_{G_1}^2 = 204.41 \tag{3.35}$$

By solving Equations (3.34) and (3.35), we get

$$P_{G_1} = 133.3 \text{ MW} \quad \text{and} \quad P_{G_2} = 80 \text{ MW}$$

If the transmission losses are not co-ordinated, we have

$$\frac{dC_1}{dP_{G_1}} = \frac{dC_2}{dP_{G_2}}$$

$$0.025 P_{G_1} + 14 = 0.05 P_{G_2} + 16$$

$$\Rightarrow 0.025 P_{G_1} - 0.05 P_{G_2} = 2 \tag{3.36}$$

While in the system, a power balance equation always holds good.

$$P_{G_1} + P_{G_2} - P_D - P_L = 0$$

$$P_{G_1} + P_{G_2} - 204.41 - 0.0005 P_{G_1}^2 = 0 \tag{3.37}$$

By solving Equations (3.36) and (3.37), we get

$$P_{G_1} = 172.91 \text{ MW}$$

and $P_{G_2} = 46.45 \text{ MW}$ and $P_L = 0.0005 P_{G_1}^2 = 14.95 \text{ MW}$

From the results, it is clear that if the transmission losses are co-ordinated, the load on Plant-1 is increased from 133.3 to 172.91 MW.

Increase in fuel cost of Plant-1 is

$$\therefore \Delta C_1 = \int_{133.3}^{172.91} (0.025 P_{G_1} + 14)\, dP_{G_1} = 0.025 P_{G_1}^2 + 14 P_{G_1} \Big|_{133.3}^{172.91}$$

$$= 0.025(172.91)^2 + 14(172.91) - 0.025(133.3)^2 + 14(133.33)$$

$$= 706.15 \text{ Rs./hr}$$

The load on Plant-2 is decreased from 80 to 46.45 MW. The decrease in the fuel cost of Plant-2 is

$$\Delta C_2 = -\int_{80}^{46.45} (0.05 P_{G_2} + 16)\, dP_{G_2} = -\left[0.05 P_{G_2}^2 + 16 P_{G_2}\right]\Big|_{80}^{46.45}$$

$$= 642.70 \text{ Rs./hr}$$

The net financial loss $= \Delta C_1 - \Delta C_2 = 706.15 - 642.70$
$$= 63.45 \text{ Rs./hr}$$

Example 3.13: For the system shown in Fig. 3.9, with Bus-1 as the reference bus with a voltage of 1.0 $\angle 0°$ p.u., find the loss formula (B_{pq}) coefficients if the branch currents and impedances are:

$I_a = (1.00 - j0.15)$ p.u.; $Z_a = 0.02 + j0.15$ p.u.
$I_b = (0.50 - j0.05)$ p.u.; $Z_b = 0.03 + j0.15$ p.u.
$I_c = (0.20 - j0.05)$ p.u.; $Z_c = 0.02 + j0.25$ p.u.

If the base is 100 MVA, what will be the magnitudes of B_{pq} coefficients in reciprocal MW?

Solution:

The assumption in developing the expression for transmission loss is that all load currents maintain a constant ratio to the total current:

i.e., $$\frac{I_a}{I_a + I_b} = \frac{1.0 - j0.15}{1.5 - j0.25} = 0.6649$$

$$\frac{I_b}{I_a + I_b} = \frac{0.5 - j0.05}{1.5 - j0.25} = 0.3353$$

$\therefore$ The current distribution factors:

$$N_{a_1} = \frac{I_{a_1}}{I_a + I_b}$$

where I_{a_1} is the current in branch 'a' when Plant-1 is in operation; and I_{a_2} the current when Plant-2 is in operation.

$N_{a_1} = 0.6649;$ $N_{a_2} = 0.6649$

$N_{b_1} = 0.3353;$ $N_{b_2} = 0.3359$

$N_{c_1} = -0.3353;$ $N_{c_2} = 0.6649$

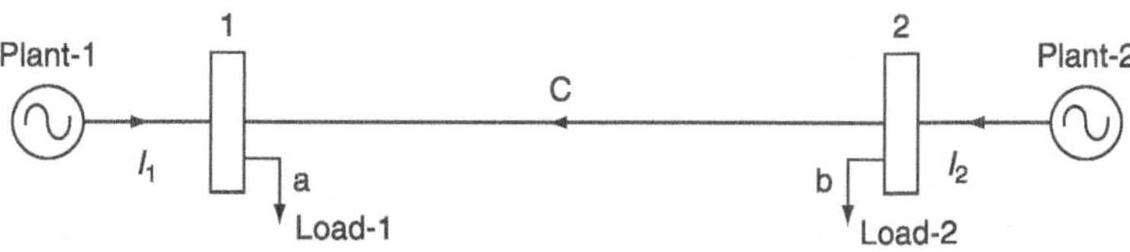

FIG. 3.9   *Illustration for Example 3.13*

Voltage at Bus-1 $= V_1 = 1.0 \; \angle 0° \text{ p.u.} = (1.0 + j\,0.0) \text{ p.u.}$
The bus voltage at Bus-2 is

$$V_2 = V_1 + I_c Z_c$$
$$= (1.00 + j\,0.0) + (0.20 - j\,0.05)(0.02 + j\,0.25)$$
$$= 1.0165 + j\,0.049 = 1.0176 \; \angle 2.76$$

The plant current are

$$I_{G_1} = I_a - I_c = (1.00 - j\,0.15) - (0.20 - j\,0.05) = 0.80 - j\,0.10 = 0.8062\angle - 7.11°$$

$$I_{G_2} = I_b + I_c = (0.50 - j\,0.10) + (0.20 - j\,0.05)$$

$$= (0.70 - j15) = 0.7519\angle - 12.09°$$

The plant currents are in the form of

$$I_1 = I_1 \; \angle \sigma_1$$

$$I_2 = I_2 \; \angle \sigma_2$$

$$\therefore \; \sigma_1 = -7.1° \quad \text{and} \quad \sigma_2 = -12.09°$$

$$\cos(\sigma_2 - \sigma_1) = \cos 4.99° = 0.996$$

The plant p.f.s are

$$\cos \phi_1 = \cos 7.1° = 0.9923$$

$$\cos \phi_2 = \cos (\text{angle between } V_2 \text{ and } I_{G_2})$$

$$= \cos(2.76° + 12.09°) = 0.9666$$

$$B_{11} = \frac{1}{|V_1|^2 (\cos \phi_1)^2}\left[N_{a_1}^2 R_a + N_{b_1}^2 R_b + N_{c_1}^2 R_c \right]$$

$$= \frac{1}{1 \times (0.9923)}\left[(0.6649)^2 \times 0.02 + (0.3353)^2 (0.03) + (-0.3353)^2 \times (0.02) \right]$$

$$= 0.01468 \text{ p.u.}$$

$$\text{or} \quad \frac{0.01468}{100} = 0.01468 \times 10^{-2} \text{ MW}^{-1}$$

$$B_{22} = \frac{1}{|V_2|^2 \cos \phi_2^2}\left[N_{a_2}^2 R_a + N_{b_2}^2 + R_b + N_c^2 R_c \right]$$

$$= \frac{1}{(0.0176)^2 (0.9666)^2}\left[(0.6649)^2 \times 0.02 + (0.3353)^2 \times 0.03 + (0.6649)^2 0.02 \right]$$

$$= 0.02175 \text{ p.u.}$$

or $\dfrac{0.02175}{100} = 0.02175 \times 10^{-2}$ MW^{-1}

$$B_{12} = \frac{\cos(\sigma_1 - \sigma_2)}{(|V_1||V_2|(\cos\phi_1)(\cos\phi_2)}\left[N_{a_1}N_{a_2}R_a + N_{b_1}N_{b_2}R_b + N_{c_1}N_{c_2}R_c\right]$$

$$= \frac{0.996}{1 \times 01.0176 \times 0.9923 \times 0.9666}[(0.6649 \times 0.6649 \times 0.02)$$

$$+ (0.3353 \times 0.3353 \times 0.03) + (-0.3353 \times 0.6649 \times 0.02)]$$

$$= 0.00791 \text{ p.u.}$$

or $\dfrac{0.00791}{100} = 0.00791 \times 10^{-2}$ MW^{-1}

Example 3.14: For Example 3.13, find the ITL at the operating conditions considered.

Solution:

Transmission power loss, $P_L = B_{11}P_{G_1}^2 + B_{22}P_{G_2}^2 + 2B_{12}P_{G_1}P_{G_2}$

$\because P_{G_1} = |V_1||I_{G_1}|\cos\phi_1$

$\qquad = 1 \times 0.8062 \times 0.9923 = 0.799$ p.u.

$P_{G_2} = V_2|I_{G_2}|\cos\phi_2$

$\qquad = (1.0176) \times 0.7159 \times 0.9666$

$\qquad = 0.7042$ p.u.

The ITL of Plant-1 is

$$(\text{ITL})_2 = \frac{\partial P_L}{\partial P_{G_1}} = 2B_{11}P_{G_1} + 2B_{12}P_{G_2}$$

$$= 2(0.01468)(0.7999) + 2(0.00791)(0.7042)$$

$$= 0.01127 \text{ p.u.}$$

$$(\text{ITL})_2 = \frac{\partial P_L}{\partial P_{G_2}} = 2B_{22}P_{G_2} + 2B_{12}P_{G_1}$$

$$= 2(0.02175)(0.7042) + 2(0.00791)(0.7999)$$

$$= 0.01285 \text{ p.u.}$$

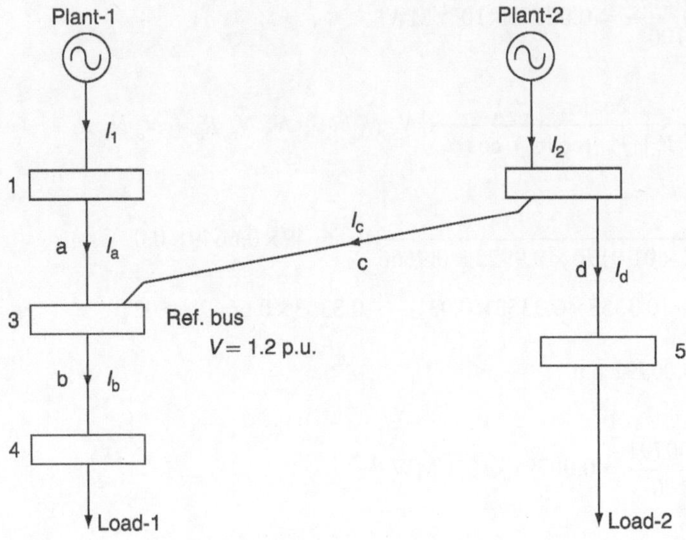

FIG. 3.10   *Illustration for Example 3.15*

Example 3.15: For the system shown in Fig. 3.10, with Bus-3 as the reference bus with a voltage of $1.2 \angle 0°$ p.u., find the loss formula (B_{pq}) coefficients of the system in p.u. and in actual units, if the branch currents and impedances are

$$I_a = 2.5 - j1.0 \text{ p.u.}; \quad Z_a = 0.02 + j\,0.08 \text{ p.u.}$$
$$I_b = 1.8 - j\,0.6 \text{ p.u.}; \quad Z_b = 0.03 + j\,0.09 \text{ p.u.}$$
$$I_c = 1.5 - j\,0.5 \text{ p.u.}; \quad Z_c = 0.013 + j\,0.05 \text{ p.u.}$$
$$I_d = 3.0 - j\,1.0 \text{ p.u.}; \quad Z_d = 0.015 + j\,0.06 \text{ p.u.}$$

Consider that the base has 100 MVA.

Solution:

The assumption in developing the expression for transmission loss is that all load currents maintain a constant ratio to the total current.

$$\frac{I_b}{I_b + I_d} = \frac{4 - j1.5}{7 - j2.5} = 0.575$$

$$\frac{I_d}{I_b + I_d} = \frac{3 - j1}{7 - j2.5} = 0.425$$

$\therefore$ Current distribution factors are

$$N_{a_1} = 1; \qquad N_{a_2} = 0$$
$$N_{b_1} = 0.575; \quad N_{b_2} = 0.425$$
$$N_{c_1} = -0.425; \quad N_{c_2} = 0.575$$
$$N_{d_1} = 0.425; \quad N_{d_2} = 0.575$$

The bus voltage at reference Bus '3' $= 1.2 + j\,0.0$ p.u.

The bus voltage at Plant-1, $V_1 = V_{\text{ref}} + I_a Z_a$

$$= 1.2 + j\,0.0 + (2.2 - j\,1)\,(0.02 + j\,0.08)$$

$$= 1.342\,\angle\,7.7^\circ \text{ p.u.}$$

The bus voltage at Plant-2, $V_2 = V_{\text{ref}} + I_c Z_c$

$$= 1.2 + (1.5 + j\,0.5)\,(0.013 - j\,0.05)$$

$$= 1.246\,\angle\,3.15^\circ \text{ p.u.}$$

The current phase angles at the plants $= (I_1 = I_a,\ I_2 = I_d + I_c)$

$$\sigma_1 = \tan^{-1}\left(\frac{-1}{2.5}\right) = -21.8^\circ$$

$$\sigma_2 = \tan^{-1}\left(\frac{-1.5}{4.5}\right) = -18.43^\circ$$

The plant power factors are

$$\cos \phi_1 = \cos\,(7.7^\circ + 21.8^\circ) = 0.87$$

$$\cos \phi_2 = \cos\,(3.5^\circ + 18.43^\circ) = 0.928$$

The loss coefficients are

$$B_{11} = \frac{1}{(V_1)^2 (\cos \phi_1)^2} = \sum_p N_{p_1}^2 R_p$$

$$= \frac{1}{|V_1|^2 (\cos \phi_1)^2}\left[N_{a_1}^2 R_a + N_{b_1}^2 R_b + N_{c_1}^2 R_c + N_{d_1}^2 R_d\right]$$

$$= \frac{1}{(1.342)^2 (0.87)^2}\left[1^2 \times 0.02 + (0.575)^2 \times 0.03 + (-0.425)^2 \times 0.013\right.$$

$$\left. + (0.425)^2 \times 0.015\right]$$
$$= 0.0257 \text{ p.u.}$$

$$B_{22} = \frac{1}{(V_2)^2 (\cos \phi_2)^2} \sum_p N_{p_2}^2 R_p$$

$$= \frac{1}{(1.246)^2 (0.928)^2}\left[0 \times 0.02 + (0.425)^2 \times 0.03 + (0.575)^2 \times 0.013\right.$$

$$\left. + (0.575)^2 \times 0.015\right]$$
$$= 0.00373 \text{ p.u.}$$

$$B_{12} = \frac{\cos(\sigma_2 - \sigma_1)}{|V_1||V_2|\cos\phi_1\cos\phi_2}\sum_p N_{p_1}N_{p_2}R_p$$

$$= \frac{1}{(1.324)(1.246)(0.87)(0.928)}[1\times0\times0.02 + (0.575)(0.425)(0.03)$$

$$+ (-0.425)(0.575)(0.013) + (0.425)(0.575)(0.015)]$$

$$= 0.00579 \text{ p.u.}$$

For obtaining the loss coefficient values in reciprocal megawatts, the loss coefficients in p.u. must be divided by the base value (i.e.,100 MVA):

$$B_{11} = \frac{0.0257}{100} = 0.0257\times10^{-2} \text{ MW}^{-1}$$

$$B_{22} = \frac{0.00373}{100} = 0.00373\times10^{-2} \text{ MW}^{-1}$$

$$B_{12} = \frac{0.00579}{100} = 0.00579\times10^{-2} \text{ MW}^{-1}$$

Example 3.16: A system consists of two generating plants with fuel costs of

$$C_1 = 0.05P_{G_1}^2 + 20P_{G_1} + 1.5$$

and $C_2 = 0.075P_{G_2}^2 + 22.5P_{G_2} + 1.6$

The system operates on economical dispatch with 100 MW of power generation by each plant. The ITL of Plant-2 is 0.2. Find the penalty factor of Plant-1.

Solution:

Given $C_1 = 0.05P_{G_1}^2 + 20P_{G_1} + 1.5$

$$C_2 = 0.075P_{G_2}^2 + 22.5P_{G_2} + 1.6$$

$$P_{G_1} = P_{G_2} = 100 \text{ MW}$$

and $\dfrac{\partial P_L}{\partial P_{G_2}} = \text{(ITL) of Plant-2} = 0.2$

The penalty factor of Plant-2,

$$L_2 = \frac{1}{1-(ITL)_2} = \frac{1}{1-\dfrac{\partial L}{\partial P_{G_2}}} = \frac{1}{1-0.2} = 1.25$$

Incremental fuel cost of Plant-1 $= \dfrac{dC_1}{dP_{G_1}} = 0.1P_{G_1} + 20$

Incremental fuel cost of Plant-2 $= \dfrac{dC_2}{dP_{G_2}} = 0.15P_{G_2} + 22.5$

For optimality, the condition is

$$\frac{dC_1}{dP_{G_1}}L_1 = \frac{dC_2}{dP_{G_2}}L_2 = \lambda$$

$$\Rightarrow \frac{dC_1}{dP_{G_1}}L_1 = \frac{dC_2}{dP_{G_2}}L_2$$

$$\Rightarrow (0.1P_{G_1} + 20)L_1 = (0.15P_{G_2} + 22.5)1.25$$

$$(0.1 \times 100 + 20)\, L_1 = (0.15 \times 100 + 22.5)\, 1.25$$

$$= (37.5)\,(1.25)$$

$$\therefore 30\, L_1 = 46.875$$

$$\text{or} \quad L_1 = 1.5625$$

i.e., the penalty factor of Plant-1 $= L_1 = 1.5625$.

Example 3.17: Two thermal plants are interconnected and supply power to a load as shown in Fig. 3.11.

The following are the incremental production costs of the plants:

$$\frac{dC_1}{dP_{G1}} = 20 + 10P_{G_1}\ \text{Rs./MWh}$$

$$\frac{dC_2}{dP_{G_2}} = 15 + 10P_{G_2}\ \text{Rs./MWh}$$

where P_{G_1} and P_{G_2} are expressed in p.u. in 100-MVA base. The transmission loss is given by

$$P_L = 0.1P_{G_1}^2 + 0.2P_{G_2}^2 + 0.1P_{G_1}P_{G_2}\ \text{p.u.}$$

If the incremental cost of received power is 50 Rs./MWh, find the optimal generation.

FIG. 3.11   *Illustration or Example 3.17*

Solution:

Given:
$$\frac{dC_1}{dP_{G_1}} = 20 + 10P_{G_1} \text{ Rs./MWh}$$

$$\frac{dC_2}{dP_{G_2}} = 15 + 10P_{G_2} \text{ Rs./MWh}$$

and
$$P_L = 0.1P_{G_1}^2 + 0.2P_{G_2}^2 + 0.1P_{G_1}P_{G_2} \text{ p.u.} \tag{3.38}$$

Generally,
$$P_L = B_{11}P_{G_1}^2 + B_{22}P_{G_2}^2 + 2B_{12}P_{G_1}P_{G_2} \tag{3.39}$$

Comparing the coefficients of Equations (3.38) and (3.39), we get

$$\therefore B_{11} = 0.1$$

$$B_{22} = 0.2$$

and
$$B_{12} = \frac{0.1}{2} = 0.05$$

Incremental cost of received power, $\lambda = 50$ Rs./MWh

The condition for optimum allocation of total load when transmission losses are considered is

$$\frac{dC_1}{dP_{G_1}}L_1 = \frac{dC_2}{dP_{G_2}}L_2 = \lambda$$

The ITL of Plant-1 is

$$(ITL)_1 = \frac{\partial P_L}{\partial P_{G_1}} = 2B_{11}P_{G_1} + 2B_{12}P_{G_2}$$

$$= 0.2P_{G_1} + 0.1P_{G_2}$$

The ITL of Plant-2 is

$$(ITL)_2 = \frac{\partial P_L}{\partial P_{G_2}} = 2B_{22}P_{G_2} + 2B_{12}P_{G_1}$$

$$= 0.4P_{G_2} + 0.1P_{G_1}$$

The penalty factor of Plant-1,

$$L_1 = \frac{1}{1 - (ITL)_1}$$

$$= \frac{1}{1 - \left(\dfrac{\partial P_L}{\partial P_{G_1}}\right)} = \frac{1}{1 - (0.2P_{G_1} + 0.1P_{G_2})}$$

The penalty factor of Plant-2,

$$L_2 = \frac{1}{1-(\text{ITL})_2} = \frac{1}{1-(0.4P_{G_2} + 0.1P_{G_1})}$$

$\therefore$ For optimum operation:

$$\frac{\partial C_1}{\partial P_{G_1}}L_1 = \frac{\partial C_2}{\partial P_{G_2}}L_2 = \lambda$$

$$(20+10P_{G_1})\frac{1}{1-(0.2P_{G_1}+0.1P_{G_2})} = (15+10P_{G_2})\frac{1}{1-(0.4P_{G_2}+0.1P_{G_1})} = 50$$

$$\Rightarrow (20+10P_{G_1}) = 50(1-0.2P_{G_1}-0.1P_{G_2})$$

or $\quad 20+10P_{G_1} = 50-10P_{G_1}-5P_{G_2}$

or $\quad 10P_{G_1}+10P_{G_2}+5P_{G_2} = 30$

$$20P_{G_1}+5P_{G_2} = 30 \tag{3.40}$$

and $\quad (15+10P_{G_2})\left(\dfrac{1}{1-0.4P_{G_2}-0.1P_{G_1}}\right) = 50$

or $\quad 15+10P_{G_2} = 50(1-0.4P_{G_2}-0.1P_{G_1})$

or $\quad 15+10P_{G_2} = 50-20P_{G_2}-5P_{G_1}$

$$5P_{G_1}+30P_{G_2} = 35$$

Solving Equations (3.40) and (3.41), we have

$$20P_{G_1}+5P_{G_2} = 30$$

Equations (3.41) $\times 4 \Rightarrow \qquad 20P_{G_1}+120P_{G_2} = 140$

$$\overline{\qquad\qquad 115P_{G_2} = 110 \qquad\qquad}$$

$\therefore P_{G_2} = 0.95652$ p.u.

Substituting the P_{G_2} value in Equation (3.40), we get

$$20P_{G_1}+5(0.95652) = 30$$

or $\quad 20P_{G_1}+4.7826 = 30$

$\therefore P_{G_1} = 1.26087$ p.u.

Substituting the P_{G_1} and P_{G_2} values in Equation (3.38), we have

$$P_L = 0.1 \,(1.26087)^2 + 0.2 \,(0.95652)^2 + 0.1 \,(1.26087)\,(0.95652)$$

$$= 0.158979 + 0.182986 + 0.12060$$

$$= 0.46256 \text{ p.u.}$$

$$= 0.46256 \times 100 = 46.256 \text{ MW}$$

P_{G_1} in MW = p.u. value × base MVA

$$P_{G_1} = 1.2608 \times 100$$

$$= 126.08 \text{ MW}$$

and $P_{G_2} = 95.652 \text{ MW}$

Example 3.18: A power system operates an economic load dispatch with a system λ of 60 Rs./MWh. If raising the output of Plant-2 by 100 kW (while the other output is kept constant) results in increased power losses of 12 kW for the system, what is the approximate additional cost per hour if the output of this plant is increased by 1 MW?

Solution:
For economic operation:

$$\frac{\partial C_1}{\partial P_{G_1}} L_1 = \frac{\partial C_2}{\partial P_{G_2}} L_2 = \lambda$$

If the Plant-2 output is increased by 1 MW, i.e., $\partial P_{G_2} = 1\,\text{MW}$, the additional cost, $\partial C_2 = ?$

$$\frac{\partial C_2}{\partial P_{G2}} L_2 = 60$$

Given:

$$\lambda = 60 \text{ Rs./MWh}, \ \partial P_{G_2} = 100 \text{ kW, and } \partial P_L = 12 \text{ kW:}$$

$$\therefore \frac{\partial P_L}{\partial P_{G2}} = \frac{12}{100} = 0.12$$

The penalty factor of Plant-2,

$$L_2 = L_2 = \frac{1}{1 - \left(\dfrac{\partial P_L}{\partial P_{G_2}}\right)} = \frac{1}{1 - 0.12} = 1.136$$

$$\therefore \frac{\partial C_2}{\partial P_{G_2}} = \frac{\lambda}{L_2} = \frac{60}{1.136} = 52.817$$

The fuel cost when the output is increased by 1 MW is

$$\partial C_2 = 52.817 \times dP_{G_2} = 52.817 \times 1 = 52.817 \text{ Rs./hr}$$

Example 3.19: A power system is supplied by only two plants, both of which operate on economical dispatch. At the bus of Plant-1, the incremental cost is 55 Rs./MWh and at Plant-2 is 50 Rs./MWh. Which plant has the higher penalty factor? What is the penalty factor of Plant-1 if the cost per hour of increasing the load on system by 1 MW is 75 Rs./hr?

Solution:

Given

$$\frac{\partial C_1}{\partial P_{G1}} = 55, \quad \frac{\partial C_2}{\partial P_{G_2}} = 50$$

The cost in Rs./hr to increase the total system load by 1 MW is called system λ:

$$\lambda = 75 \text{ Rs./MWh}$$

or

$$\partial P_{G_1} = 1 \text{ MW and Rs./hr} = 75 \text{ (given)}$$

$$\therefore \frac{\text{Rs.}}{\text{MWh}} = \frac{75}{1} = 75 = \frac{\partial C_1}{\partial P_{G_1}} = \lambda$$

For economical operation, both plants operating at common λ, i.e., $\lambda = 75$ Rs./MWh

$$\frac{\partial C_1}{\partial P_{G_1}} L_1 = \frac{\partial C_2}{\partial P_{G_2}} L_2 = \lambda$$

$$\therefore L_1 = \frac{\lambda}{\dfrac{\partial C_1}{\partial P_{G_1}}} = \frac{75}{55} = 1.364$$

$$\text{and} \quad L_2 = \frac{\lambda}{\dfrac{\partial C_2}{\partial P_{G_2}}} = \frac{75}{50} = 1.5$$

Therefore, L_2 is greater than L_1.

KEY NOTES

- When the energy is transported over relatively larger distances with low load density, the transmission losses in some cases may amount to about 20–30% of the total load. Hence, it becomes very essential to take these losses into account when formulating an economic dispatch problem.

- Consider the objective function:

  i.e., $C = \sum_{i=1}^{n} C_i \left(P_{G_i} \right)$

  Minimize the above function subject to the equality and inequality constraints.

Equality constraints

The real-power balance equation, i.e., total real-power generations minus the total losses should be equal to real-power demand:

i.e., $\sum_{i=1}^{n} P_{G_i} - P_L = P_D$

Inequality constraints

The inequality constraints are represented as:

(a) In terms of real-power generation as

$P_{G_i(min)} \leq P_{G_i} \leq P_{G_i(max)}$

(b) In terms of reactive-power generation as

$Q_{G_i(min)} \leq Q_{G_i} \leq Q_{G_i(max)}$

Always, there will be upper and lower limits for real and reactive-power generation at each of the stations.

(c) In addition, the voltage at each of the stations should be maintained within certain limits.

i.e., $V_{i(min)} \leq V_i \leq V_{i(max)}$

- **Current distribution factor** of a transmission line w.r.t a power source is the ratio of the current it would carry to the current that the source would carry when all other sources are rendered inactive i.e., the sources that do not supply any current.

- If the system has 'n' number of stations, supplying the total load through transmission lines, the transmission line loss is given by

$$P_L = \sum_{P=1}^{n} \sum_{q=1}^{n} P_{G_p} B_{pq} P_{G_q}$$

- The coefficients B_{11}, B_{12} and B_{22} are called loss coefficients or B-coefficients and are expressed in $(MW)^{-1}$.

- The transmission loss is expressed as a function of real-power generations.

- The incremental transmission loss is expressed as $\dfrac{\partial P_L}{\partial P_{G_i}}$.

- **The penalty factor** of any unit is defined as the ratio of a small change in power at that unit to the small change in received power when only that unit supplies this small change in received power and is expressed as

$$L_i = \frac{1}{1 - \dfrac{\partial P_L}{\partial P_{G_i}}}$$

- The condition for optimality when transmission losses are considered is

$$\frac{\partial C_1}{\partial P_{G_1}} L_1 = \frac{\partial C_2}{\partial P_{G_2}} L_2 = \cdots = \frac{\partial C_n}{\partial P_{G_n}} L_n = \lambda$$

SHORT QUESTIONS AND ANSWERS

(1) State in words the condition for minimum fuel cost in a power system when losses are considered.

The minimum fuel cost is obtained when the incremental fuel cost of each station multiplied by its penalty factor is the same for all the stations in the power system.

(2) Define the current distribution factor.

The current distribution factor of a transmission line with respect to a power source is the ratio of the current it would carry to the current that the source would carry when all other sources are rendered inactive, i.e., the sources that are not supplying any current.

(3) Write the expression for the total transmission loss in terms of real-power generations when $n=2$.

For $n=2$,

$$P_L = \sum_{p=1}^{2} \sum_{q=1}^{2} P_{G_p} B_{pq} P_{G_q}$$

$$= P_{G_1} B_{11} P_{G_1} + P_{G_1} B_{12} P_{G_2}$$
$$+ P_{G_2} B_{21} B_{21} P_{G_1} + P_{G_2} B_{22} P_{G_2}$$
$$= B_{11} P_{G_1}^{\ 2} + B_{22} P_{G_2}^{\ 2} + 2B_{12} P_{G1} P_{G2}$$

(4) In the study of an optimum allocation problem, what are the considerations that you will notice regarding equality and inequality constraints in the case of transmission loss consideration and why are reactive-power constraints taken?

Equality constraints,

$$\sum_{l=1}^{n} P_{G_l} - P_D - P_L = 0$$

Inequality constraints,

$$P_{G_i\,(min)} \le P_{G_i} \le P_{G_i\,(max)} \text{ and}$$

$$Q_{G_i\,(min)} \le Q_{G_i} \le Q_{G_i\,(max)}$$

$$V_{i\,(min)} \le V_i \le V_{i\,(max)}$$

Reactive-power constraints are to be taken since the transmission losses are functions of real and reactive-power generations and also the voltage at each bus.

(5) What are the assumptions considered in deriving the transmission loss expression?

The following assumptions are to be considered for deriving the transmission loss expression:

(a) All lines in the system have the same $\dfrac{X}{R}$ ratio.

(b) All the load currents have the same phase angle.

(c) All the load currents maintain a constant ratio to the total current.

(d) The magnitude and phase angle of bus voltages at each station remain constant.

(6) Write the transmission loss expression for the kth line, if there are two generating stations in terms of station voltages, real-power generations, and their power factors.

$$P_L = \frac{P_{G_1}^2}{(V_1)^2 (pf_1)^2} N_{K_1}^2 R_K + \frac{P_{G_2}^2}{(V_2)^2 (pf_2)^2} N_{K_2}^2 R_K$$

$$+ \frac{2P_{G_1} P_{G_2}}{V_1 V_2 (pf_1)(pf_2)} \cos(\sigma_2 - \sigma_1) N_{K_1} N_{K_2}$$

(7) A simple two-plant system has the IC's that are

$$dC_1/dP_{G_1} = 0.01 P_{G_1} + 2.0$$

$dC_2/dP_{G_2} = 0.01 P_{G_2} + 1.5$ and the total load on the system is distributed optimally between two stations as $P_{G_1} = 60$ MV and $P_{G_2} = 110$ MW, corresponding to $\lambda = 2.6$ and the loss coefficients of the system are given as

p	q	B_{pq}
1	1	0.0015
1	2	−0.0015
2	2	0.0025

Determine the transmission loss.

Transmission loss $= B_{11} P_{G_1}^2 + 2B_{12} P_{G_1} P_{G_2} + B_{22} P_{G_2}^2$

$$= (0.0015)(60)^2 + 2(-0.0015)$$
$$\times (60 \times 110) + (0.0025)$$
$$\times (110)^2 = 25.75 \text{ MW}$$

(8) What is your analysis by considering the optimization problem with and without transmission loss consideration?

To get the solution to optimization problem, i.e., to allocate the total load among various units:

When transmission losses are neglected, the condition is $\dfrac{\partial C_i}{\partial P_{G_i}} = \lambda$

i.e., the IC of all the units must be the same.

When transmission losses are considered, the condition is $\dfrac{\partial C_i}{\partial P_{G_i}} L_i = \lambda$

i.e., the product of IC of any unit and its penalty factor gives the optimum solution.

(9) Find the penalty factor of the plant shown in Fig. 3.12.

Here, $P_{G_1} = 59$ MW

$$P_D = 19 \text{ MW}$$

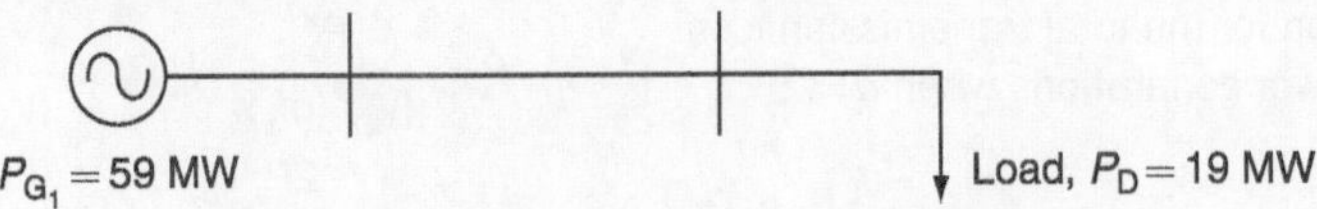

FIG. 3.12　*Illustration for Question number 9*

$$P_{G_1} = P_D + P_L$$
$$\Rightarrow P_L = P_{G_1} - P_D$$
$$= 59 - 19 = 40 \text{ MW}$$

$$\frac{\partial P_L}{\partial P_{G_1}} = \frac{40}{59}$$

Penalty factor, $L_1 = \dfrac{1}{1 - \dfrac{\partial P_L}{\partial P_{G_1}}} = \dfrac{1}{1 - \dfrac{40}{59}}$

$$= \frac{1}{19/59} = \frac{59}{19} = 3.105$$

(10) Write the expression for transmission loss in terms of B_{min} coefficients when there are three generating stations.

$$P_L = \sum_{m=1}^{3} \sum_{n=1}^{3} P_{G_m} B_{mn} P_{G_n}$$

$$P_L = B_{11} P_{G_1}^2 B_{22} P_{G_2}^2 + B_{33} P_{G_3}^2$$
$$+ 2B_{12} P_{G_1} P_{G_2} + 2B_{23} P_{G_2} P_{G}$$
$$+ 2B_{13} P_{G_1} P_{G_3}$$

(11) Write the condition for optimality when losses are taken into consideration.

$$L_i \frac{\partial C_i}{\partial P_{G_i}} = \lambda$$

i.e., $L_1 \dfrac{\partial C_1}{\partial P_{G_1}} = L_2 \dfrac{\partial C_2}{\partial P_{G_2}} = \cdots = L_n \dfrac{\partial C_n}{\partial P_{G_n}} = \lambda$

(12) Find the penalty factors of both the plants shown in Fig. 3.13.

Given: $P_{G_1} = 125$ MW

$P_{G_2} = 75$ MW,　$B_{11} = 0.0015$

Since the load is at Station-2, the transfer of power to the load is from only Station-1 and hence $B_{12} = B_{22} = B_{21} = 0$

$$P_L = B_{11} P_{G_1}^2 + 2B_{22} P_{G_1} P_{G_2} + B_{22} P_{G_2}^2$$

$$\therefore P_L = B_{11} P_{G_1}^2$$

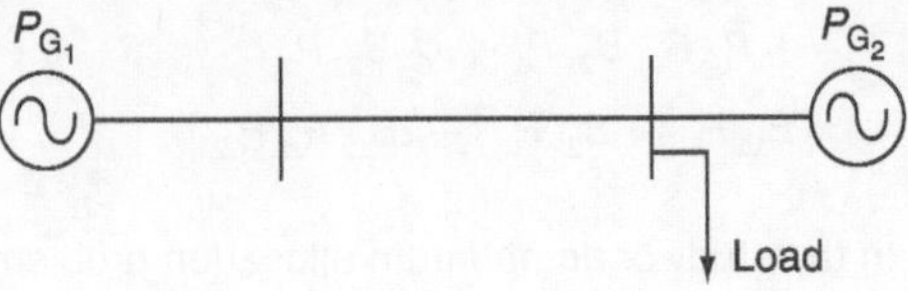

FIG. 3.13　*Illustration for Question number 12*

$$= (0.0015)(125)^2$$
$$= 0.0015 \times P_{G_1}^2$$

Penalty factor of Station-1,

$$L_1 = \frac{1}{1 - \dfrac{\partial P_L}{\partial P_{G_1}}}$$

$$= \frac{1}{1 - 0375}$$

$$= 1.6$$

$$\frac{\partial P_L}{\partial P_{G_1}} = 0.003 \, P_{G_1} = 0.002 \times 125 = 0.375$$

Penalty factor of Station-2,

$$L_2 = \frac{1}{1 - \dfrac{\partial P_L}{\partial P_{G_2}}}$$

$$= \frac{1}{1 - 0}$$

$$= 1.0.$$

(13) Define the optimization problem when transmission losses are considered.

$$\frac{\partial C_1}{\partial P_{G_1}} L_1 = \frac{\partial C_2}{\partial P_{G_2}} L_2 = \cdots = \frac{\partial C_n}{\partial P_{G_n}} L_n = \lambda$$

(14) What do you mean by ITL and penalty factor of the system? Write expressions for them.

ITL = Incremental transmission loss = $\dfrac{\partial P_L}{\partial P_{G_i}}$

It is defined as the ratio of the change in real-power loss to the change in real-power generation.

$$\text{Penalty factor} = L_i = \frac{1}{1 - \dfrac{\partial P_L}{\partial P_{G_i}}}.$$

(15) Why are the reactive-power constraints to be considered as inequality constraints in solving an optimization problem when transmission losses are considered?

The transmission loss is a function of real and reactive-power generations since reactive power is proportional to the square of the voltage.

(16) If the fuel cost in Rs./hr of a power station is related to the power generated in MW by $C_1 = 0.0002P_G{}^3 + 0.06P_G{}^2 + 300$, what is the incremental fuel cost at $P_G = 200$ MW?

$$\frac{\partial C_1}{\partial P_G} = 72 \, \text{Rs./hr.}$$

(17) What are the points that should be kept in mind for the solution of economic load dispatch problems when transmission losses are included and co-ordinated?

The following points should be kept in mind:

- Although the incremental production cost of a plant is always positive, ITL can be either positive or negative.

- The individual units will operate at different incremental production costs.

- The generation with highest positive ITL will operate at the lowest incremental production cost.

MULTIPLE-CHOICE QUESTIONS

(1) In the economic operation of a power system, the effect of increased penalty factor between a generating plant and system load center is to:

  (a) Decrease the load on the generating plant.

  (b) Increase the load on the plant.

  (c) Hold the plant load constant.

  (d) Decrease the load first and then increase.

(2) In a power system in which generating plants are remote from the load center, minimum fuel costs occur when:

  (a) Equal incremental costs are maintained at the generating station buses.

  (b) Equal incremental costs are referred to system load center.

  (c) Equal units are operated at the same load.

  (d) All the above.

(3) Unit of penalty factor is:

  (a) Rs.

  (b) MW^{-1}.

  (c) Rs./MWh.

  (d) No units.

(4) Unit commitment of more number of generating units is done using:

  (a) Gradient method.

  (b) Non-linear programming method.

  (c) Dynamic programming.

  (d) All the above.

(5) Economic dispatch is done first by ___________ and then by___________.

  (a) Unit commitment and then load scheduling.

  (b) Load scheduling and then unit commitment.

  (c) Either (a) or (b).

  (d) Unit commitment and load frequency control.

(6) Transmissiion losses are about:

  (a) 50% of the total generation.

  (b) 100% of the total generation.

  (c) 5–15% of the total generation.

  (d) None of these.

(7) In optimal scheduling of hydro-thermal units, the objective is:

  (a) Water discharge minimization.

  (b) Storage of water.

(c) Both (a) and (b).

(d) None of these.

(8) In optimal generation scheduling, the co-ordination equation for all 'i' values is:

(a) $IC_i = \lambda_i$.

(b) $IC_i = \lambda_i \, L_i$.

(c) $IC_i = \lambda_i / L_i$.

(d) $IC_i = \lambda_i + L_i$.

(9) Transmission loss by B-coefficients is PL = __________.

(a) $P^T B P$.

(b) $P^T B$.

(c) BP.

(d) All.

(10) The condition for optimality with consideration of transmission loss is:

(a) The incremental fuel costs in Rs./hr of all the units must be the same.

(b) The incremental fuel costs in Rs./hr of all the units must be the same.

(c) The incremental transmission losses in Rs./MWh of all the units must be the same.

(d) The incremental fuel cost of each multiplied by its penalty factor must be the same for all plants.

(11) Expression for transmission loss is derived using __________ method.

(a) Kron's.

(b) Penalty function.

(c) Kirchmayer's.

(d) Kuhn-Tucker.

(12) $\dfrac{\partial P_L}{\partial P_{G_i}} = $

(a) $\displaystyle\sum_{j=1}^{n} 2B_{ij} P_{G_j}$.

(b) $\displaystyle\sum_{j=1}^{n} 2B_{ji} P_{G_i}$.

(c) $\displaystyle\sum_{i=1}^{n} 2B_{ij} P_{G_j}$.

(d) $\displaystyle\sum_{i=1}^{n} 2B_{i} P_{ij}$.

(13) The equality constraint, when the transmission line losses are considered, is:

(a) $\displaystyle\sum_{i=1}^{n} P_{G_i} - P_L = 0$.

(b) $\displaystyle\sum_{i=1}^{n} P_{G_i} - P_D = P_L + P_G$.

(c) $\displaystyle\sum_{i=1}^{n} P_{G_i} - P_D = 0$.

(d) $\displaystyle\sum_{i=1}^{n} P_{G_i} - P_L = P_D$.

(14) Transmission loss is:

(a) A function of real-power generation.

(b) Independent of real-power generation.

(c) A function of reactive-power generation.

(d) A function of bus voltage magnitude and its angle.

(15) In Kron's method,

(a) Reduce the system to an equivalent system with a single hypothetical load.

(b) Reduce the system to an equivalent system without any load.

(c) Reduce the system to an equivalent system with a large number of loads.

(d) Enhance the system to an equivalent system with no power loss.

(16) The derivation of transmission line loss is not based on which assumption?

(a) All the load currents maintain a constant ratio.

(b) All the lines in the system have different $\dfrac{X}{R}$ ratios.

(c) All the load currents have same phase angle.

(d) The power factor at each station remains constant.

(17) The loss coefficient B_{12} is given by:

(a) $\dfrac{\displaystyle\sum_{k=1}^{l} N_{k_1}^2 R_k}{V_1^2 (\text{p.f}_1)^2}$.

(b) $\displaystyle\sum_{k=1}^{l} \dfrac{N_{k_2}^2 R_k}{V_2^2 (\text{p.f}_2)^2}$.

(c) $\dfrac{\cos(\sigma_2 - \sigma_1)}{V_1 V_2 (\text{p.f}_1)(\text{p.f}_2)} \sum\limits_{k=1}^{l} N_{k_1} N_{k_2} R_k.$

(d) $\dfrac{\cos(\sigma_2 - \sigma_1)}{V_1^2 (\text{p.f}_1)(\text{p.f}_2)} \sum\limits_{k=1}^{l} N_{k_1} N_{k_2} R_k.$

(18) Which of the following is correct?

(a) $\lambda = \dfrac{\partial C_i}{\partial P_{G_i}} \left(1 - \dfrac{\partial P_L}{\partial P_{G_i}} \right).$

(b) $\lambda = \dfrac{\partial C_i}{\partial P_{G_i}} \left(1 + \dfrac{\partial P_L}{\partial P_{G_i}} \right).$

(c) $\lambda = \dfrac{\partial P_L}{\partial P_{G_i}} \Big/ \left(1 - \dfrac{\partial C_i}{\partial P_{G_i}} \right).$

(d) $\lambda = \dfrac{\partial C_i}{\partial P_{G_i}} \Big/ \left(1 - \dfrac{\partial P_L}{\partial P_{G_i}} \right).$

(19) The penalty factor of the ith station is:

(a) $L_i = 1 - \dfrac{\partial P_L}{\partial P_{G_i}}.$

(b) $L_i = \dfrac{1}{1 - \dfrac{\partial P_L}{\partial P_{G_i}}}.$

(c) $L_i = \dfrac{\partial C_i}{\partial P_{G_i}} \Big| 1 - \dfrac{\partial P_L}{\partial P_{G_i}}.$

(d) $L_i = \dfrac{\partial P_L}{\partial P_{G_i}}.$

(20) $\dfrac{1}{\left(1 + \dfrac{\partial P_L}{\partial P_{G_i}} \right)}$ approximate penalty factor i^{th} plant is expressed as:

(a) $L_i = \dfrac{1}{\left(1 + \dfrac{\partial P_L}{\partial P_{G_i}} \right)}.$

(b) $L_i = \dfrac{1}{\left(1 - \dfrac{\partial P_L}{\partial P_{G_i}} \right)}.$

(c) $L_i = \dfrac{\partial C_i}{\partial P_{G_i}} \Big| - \dfrac{\partial P_L}{\partial P_{G_i}}.$

(d) $L_i = \dfrac{\partial P_L}{\partial P_{G_i}}.$

(21) The incremental transmission loss is:

(a) $1 - \dfrac{\partial P_L}{\partial P_{G_i}}.$

(b) $1 + \dfrac{\partial P_L}{\partial P_{G_i}}.$

(c) $\dfrac{\partial P_L}{\partial P_{G_i}}.$

(d) $\dfrac{1}{\dfrac{\partial P_L}{\partial P_{G_i}}}.$

(22) $\dfrac{\partial C_i}{\partial P_{G_i}} + \lambda \dfrac{\partial P_L}{\partial P_{G_i}} = \lambda$ is called the co-ordination equation because:

(a) It co-ordinates ITL with IC.

(b) It co-ordinates ITL with penalty factor.

(c) It co-ordinates real-power generation with reactive-power generation.

(d) It co-ordinates bus voltage magnitude with IC.

(23) The incremental cost of received power in Rs./MWh of the ith plant is:

(a) $\dfrac{L_i}{\partial C_i / \partial P_{G_i}}.$

(b) $L_i - \dfrac{\partial C_i}{\partial P_{G_i}}.$

(c) $L_i \dfrac{\partial C_i}{\partial P_{G_i}}.$

(d) $\dfrac{L_i}{1 - \dfrac{\partial C_i}{\partial P_{G_i}}}.$

(24) In solving optimization problem with transmission loss consideration, the condition for optimality is obtained as:

(a) The IC of all the plants must be the same.

(b) The IC of each plant multiplied with its penalty factor must be the same for all the plants.

(c) The IC of each plant divided by its penalty factor must be the same for all the plants.

(d) The IC of each plant subtracted from its penalty factor must be the same for all the plants.

(25) The matrix form of transmission loss expression is:

(a) $P_L = [P_G^T]_{1 \times k}[B]_{k \times k}[P_G]_{k \times 1}$.

(b) $P_L = [P_G^T]_{k \times 1}[B]_{k \times k}[P_G]_{1 \times k}$.

(c) $P_L = [P_G]_{1 \times k}[B]_{k \times k}[P_G^T]_{k \times 1}$.

(d) $P_L = [P_G]_{1 \times k}[B^T]_{k \times k}[P_G]_{k \times 1}$.

(26) The exact co-ordination equation of the ith plant is:

(a) $\dfrac{\partial C_i}{\partial P_{G_i}} = \lambda[1 - (\text{ITL})_i]$.

(b) $\dfrac{\partial C_i}{\partial P_{G_i}} = [(\text{ITL})_i - 1]$.

(c) $\dfrac{\partial C_i}{\partial P_{G_i}} = \dfrac{\lambda}{[1 - (\text{ITL})_i]}$.

(d) $\dfrac{\partial C_i}{\partial P_{G_i}} = \lambda\,[\text{ITL}_i]$.

(27) The optimization problem is solved by the computational method with the expression for P_{G_i}, which is given as:

(a) $P_{G_i} = \dfrac{1 + 2\sum\limits_{\substack{j=i \\ j \neq 1}}^{n} B_{ij}\, P_{G_i} - \dfrac{b_i}{\lambda}}{\dfrac{a_i}{\lambda}}$.

(b) $P_{G_i} = \dfrac{1 - 2\sum\limits_{\substack{j=i \\ j=1}}^{n} B_i\, P_{G_{ij}} - \dfrac{a_i}{\lambda}}{\dfrac{b_i}{\lambda} + 2B_{ii}}$.

(c) $P_{G_i} = \dfrac{1 - \sum\limits_{j=1}^{n} (B_{ij}\, a_{ij}) - \dfrac{B_{ii}}{\lambda}}{\dfrac{a_i}{\lambda} + 2B_{ij}}$.

(d) $P_{G_i} = \dfrac{1 - 2\sum\limits_{\substack{j=1 \\ j \neq i}}^{n} (B_{ij}\, P_{G_i}) - \dfrac{b_i}{\lambda}}{\dfrac{a_i}{\lambda} + 2B_{ii}}$.

(28) The penalty factor of the plant shown in Fig. 3.14 is:

(a) 5.

(b) 5.25.

(c) 1.25.

(d) 12.5.

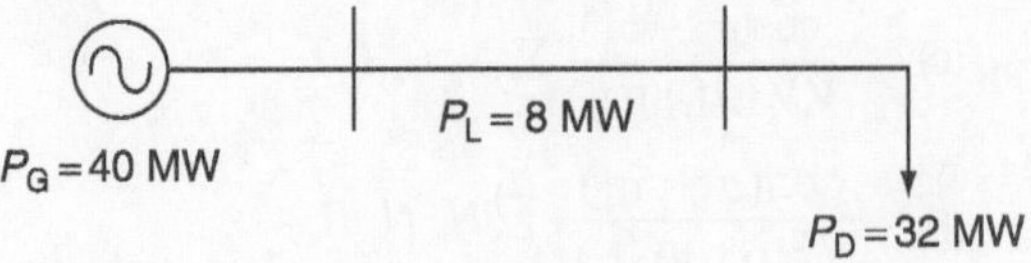

FIG. 3.14   *Illustration for Question number 28*

(29) The incremental cost of received power for the above plant if $\dfrac{dC_1}{dP_{G_1}} = 0.12P_{G_1} + 8.0$ Rs./MWh is:

(a) 1.25.

(b) 16.82.

(c) 16.00.

(d) 12.80.

(30) For Fig. 3.15, what is the penalty factor of the second plant if a power of 125 MW is transmitted from the first plant to load with an incurred loss of 15.625 MW?

(a) 24.

(b) 1.25.

(c) Zero.

(d) 1.

(31) To derive the transmission loss expression, which of the following assumptions are to be taken into consideration?

(i) All the lines in the system have the same R/X ratio.

(ii) P.f. at each station remains constant.

(iii) All the load currents maintain constant ratio to the total current.

(iv) All the load currents have their own different phase angles.

(a) (i) and (ii).

(b) (ii) and (iii).

(c) All except (iv).

(d) All of these.

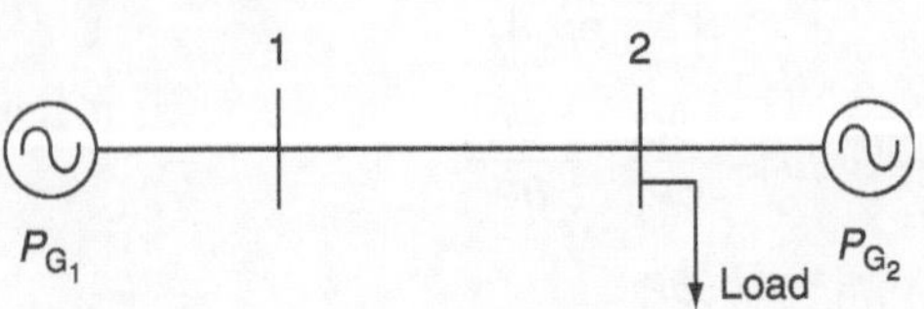

FIG. 3.15

(32) In deriving the expression for transmission power loss, which of the following principles are used?

  (i) Thevinin's theorem.

  (ii) Kron's method.

  (iii) Max. power-transfer theorem.

  (iv) Superposition theorem.

    (a) (i) only.

    (b) (ii) and (iii) only.

    (c) (ii) and (iv) only.

    (d) All except (i).

(33) The transmission loss is expressed as:

  (a) $\displaystyle P_L = \sum_{p=1}^{n} \sum_{q=1}^{n} P_{G_p}^2 B_{pq} P_{G_q}^2$.

  (b) $\displaystyle P_L = \sum_{p=1}^{n} \sum_{q=1}^{n} P_{G_p} B_{pq}^2 P_{G_q}$.

  (c) $\displaystyle P_L = \sum_{p=1}^{n} \sum_{q=1}^{n} (P_{G_p} B_{pq} P_{G_q})^2$.

  (d) $\displaystyle P_L = \sum_{p=1}^{n} \sum_{q=1}^{n} P_{G_p} B_{pq} P_{G_q}$.

(34) In finding the optimal solution, the objective function is redefined as constrained objective function and is given by:

  (a) $\displaystyle C = \sum_{i=1}^{n} C_i(P_{G_i}) - \lambda \left[\sum_{i=1}^{n} P_{G_i} - P_D - P_L(Q_{G_i}) \right]$.

  (b) $\displaystyle C = \sum_{i=1}^{n} C_i(P_{G_i}) - \lambda \left[\sum_{i=1}^{n} P_{G_i} - P_D(Q_{G_i}) - P_L \right]$.

  (c) $\displaystyle C = \sum_{i=1}^{n} C_i(P_{G_i}) - \lambda \left[\sum_{i=1}^{n} Q_{G_i} - Q_D - P_L(P_{G_i}) \right]$.

  (d) $\displaystyle C = \sum_{i=1}^{n} C_i(P_{G_i}) - \lambda \left[\sum_{i=1}^{n} P_{G_i} - P_D - P_L(Q_{G_i}) \right]$.

REVIEW QUESTIONS

(1) Derive the transmission loss formula and state the assumptions made in it.

(2) Obtain the condition for optimum operation of a power system with '*n*' plants when losses are considered.

(3) Briefly explain about the exact co-ordination equation and derive the penalty factor.

(4) What are *B*-coefficients? Derive them.

(5) Explain economic dispatch of thermal plants co-ordinating the system transmission line losses. Derive relevant equations and state the significance and role of penalty factor.

(6) Give a step-by-step procedure for computing economic allocation of power generation in a thermal system when transmission line losses are considered.

PROBLEMS

(1) A system consists of two generating plants with fuel costs of:

$$C_1 = 0.03 P_{G_1}^2 + 15 P_{G_1} + 1.0$$
$$\text{and} \quad C_2 = 0.04 P_{G_2}^2 + 21 P_{G_2} + 1.4.$$

The system operates on economical dispatch with 120 MW of power generation by each plant. The incremental transmission loss of Plant-2 is 0.15. Find the penalty factor of Plant-1.

(2) A system consists of two generating plants. The incremental costs in Rs./MWh with P_{G_1} and P_{G_2} in MW are:

$$\frac{dC_1}{dP_{G_1}} = 0.006 P_{G_1} + 7.0; \quad \frac{dC_2}{dP_{G_2}} = 0.01 P_{G_2} + 5.0.$$

The system operates on economic dispatch with $P_{G_1} = P_{G_2} = 400$ MW and $\dfrac{\partial P_L}{\partial P_{G_2}} = 0.3$. Find the penalty factor of Plant-1.

(3) The cost curves of two plants are as follows:

$$C_1 = 0.04P_{G_1}^{\,2} + 25P_{G_1} + 120$$

$$C_2 = 0.035P_{G_2}^{\,2} + 10P_{G_1} + 160$$

The loss coefficient for the above system is given as $B_{11} = 0.001/\text{MW}$, $B_{12} = B_{21} = -0.0002/\text{MW}$ and $B_{22} = 0.003/\text{MW}$. Determine the economical generation scheduling corresponding to $\lambda = 20$ Rs./MWh and corresponding system load that can be met with. If the total load connected to the system is 110 MW taking 3.5% change in the value of λ, what should be the value of λ in the next iteration?

4 Optimal Unit Commitment

OBJECTIVES

After reading this chapter, you should be able to:

- know the need of optimal unit commitment (UC)
- study the solution methods for UC
- solve the UC problem by dynamic programming (DP) approach
- prepare the UC table with reliability and start-up cost considerations

4.1 INTRODUCTION

The total load of the power system is not constant but varies throughout the day and reaches a different peak value from one day to another. It follows a particular hourly load cycle over a day. There will be different discrete load levels at each period as shown in Fig. 4.1.

Due to the above reason, it is not advisable to run all available units all the time, and it is necessary to decide in advance which generators are to startup, when to connect them to the network, the sequence in which the operating units should be shut down, and for how long. The computational procedure for making such decisions is called *unit commitment* (UC), and a unit when scheduled for connection to the system is said to be *committed*.

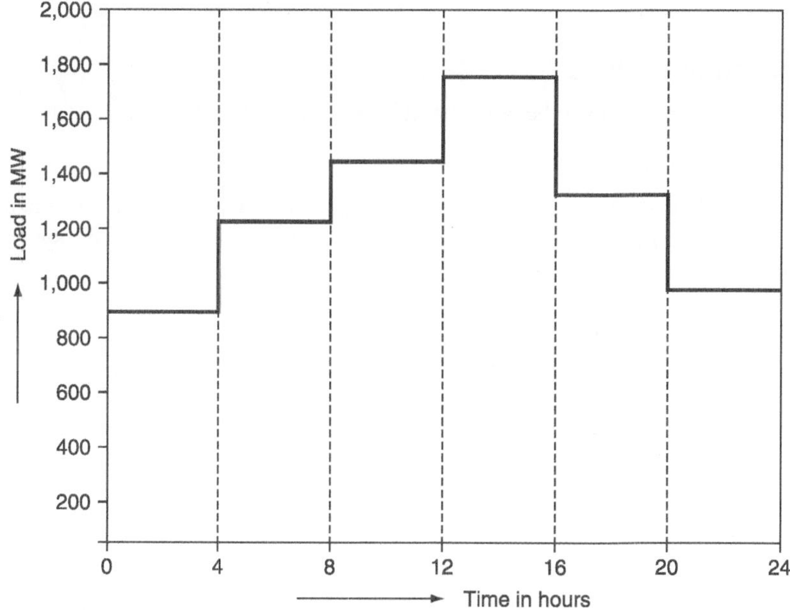

FIG. 4.1   *Discrete levels of system load of daily load cycle*

The problem of UC is nothing but to determine the units that should operate for a particular load. To 'commit' a generating unit is to 'turn it on', i.e., to bring it up to speed, synchronize it to the system, and connect it, so that it can deliver power to the network.

4.2 COMPARISON WITH ECONOMIC LOAD DISPATCH

Economic dispatch economically distributes the actual system load as it rises to the various units that are already on-line. However, the UC problem plans for the best set of units to be available to supply the predicted or forecast load of the system over a future time period.

4.3 NEED FOR UC

- The plant commitment and unit-ordering schedules extend the period of optimization from a few minutes to several hours.

- Weekly patterns can be developed from daily schedules. Likewise, monthly, seasonal, and annual schedules can be prepared by taking into consideration the repetitive nature of the load demand and seasonal variations.

- A great deal of money can be saved by turning off the units when they are not needed for the time. If the operation of the system is to be optimized, the UC schedules are required for economically committing units in plant to service with the time at which individual units should be taken out from or returned to service.

- This problem is of importance for scheduling thermal units in a thermal plant; as for other types of generation such as hydro, their aggregate costs (such as start-up costs, operating fuel costs, and shut-down costs) are negligible so that their on−off status is not important.

4.4 CONSTRAINTS IN UC

There are many constraints to be considered in solving the UC problem.

4.4.1 Spinning reserve

It is the term used to describe the total amount of generation available from all synchronized units on the system minus the present load and losses being supplied. Here, the synchronized units on the system may be named units spinning on the system.

$$\text{Spinning reserve} = \begin{bmatrix} \text{Total generation output of all} \\ \text{synchronized units at a} \\ \text{particular time} \end{bmatrix} - \begin{bmatrix} \text{Load at that time} + \\ \text{Losses at that time} \end{bmatrix}$$

Let $P_{G_{sp}}$ be the spinning reserve, $P_{G_{i_s}}$ the power generation of the ith synchronized unit, P_D the total load on the system, and P_L the total loss of the system:

$$\therefore P_{G_{sp}} = \sum_{i=1}^{n} P_{G_{i_s}} - (P_D + P_L)$$

The spinning reserve must be maintained so that the failure of one or more units does not cause too far a drop in system frequency. Simply, if one unit fails, there must be an ample reserve on the other units to make up for the loss in a specified time period.

The spinning reserve must be a given a percentage of forecasted peak load demand, or it must be capable of taking up the loss of the most heavily loaded unit in a given period of time.

It can also be calculated as a function of the probability of not having sufficient generation to meet the load.

The reserves must be properly allocated among fast-responding units and slow-responding units such that this allows the automatic generation control system to restore frequency and quickly interchange the time of outage of a generating unit.

- Beyond the spinning reserve, the UC problem may consider various classes of 'scheduled reserves' or off-line reserves. These include quick-start diesel or gas-turbine units as well as most hydro-units and pumped storage hydro-units that can be brought on-line, synchronized, and brought upto maximum capacity quickly. As such, these units can be counted in the overall reserve assessment as long as their time to come up to maximum capacity is taken into consideration.

- Reserves should be spread well around the entire power system to avoid transmission system limitations (often called 'bottling' of reserves) and to allow different parts of the system to run as 'islands', should they become electrically disconnected.

4.4.2 Thermal unit constraints

A thermal unit can undergo only gradual temperature changes and this translates into a time period (of some hours) required to bring the unit on the line. Due to such limitations in the operation of a thermal plant, the following constraints are to be considered.

(a) Minimum up-time: During the minimum up-time, once the unit is operating (up state), it should not be turned off immediately.

(b) Minimum down-time: The minimum down-time is the minimum time during which the unit is in 'down' state, i.e., once the unit is decommitted, there is a minimum time before it can be recommitted.

(c) Crew constraints: If a plant consists of two or more units, they cannot both be turned on at the same time since there are not enough crew members to attend to both units while starting up.

Start-up cost

In addition to the above constraints, because the temperature and the pressure of the thermal unit must be moved slowly, a certain amount of energy must be expended to bring the unit on-line and is brought into the UC problem as a start-up cost.

The start-up cost may vary from a maximum 'cold-start' value to a very small value if the unit was only turned off recently, and it is still relatively close to the operating temperature.

Two approaches to treating a thermal unit during its 'down' state:

- The first approach (cooling) allows the unit's boiler to cool down and then heat back up to a operating temperature in time for a scheduled turn-on.

- The second approach (banking) requires that sufficient energy be input to the boiler to just maintain the operating temperature.

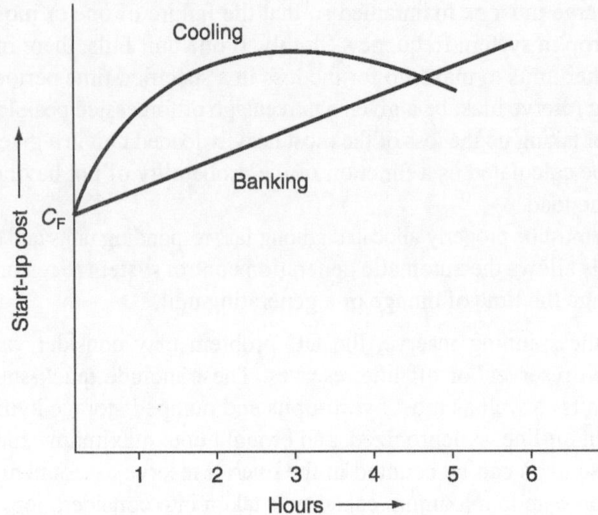

FIG. 4.2   *Time-dependent start-up costs*

The best approach can be chosen by comparing the costs for the above two approaches.

Let C_C be the cold-start cost (MBtu), C the fuel cost, C_F the fixed cost (includes crew expenses and maintainable expenses), α the thermal time constant for the unit, C_t the cost of maintaining unit at operating temperature (MBtu/hr), and t the time the unit was cooled (hr).

Start-up cost when cooling $= C_c \, (1 - e^{-t/\alpha}) \, C + C_F$;

Start-up cost when banking $= C_t \times t \times C + C_F$.

Upto a certain number of hours, the cost of banking $<$ cost of cooling is shown in Fig. 4.2.

The capacity limits of thermal units may change frequently due to maintenance or unscheduled outages of various equipments in the plant and this must also be taken into consideration in the UC problem.

The other constraints are as follows

4.4.3 Hydro-constraints

As pointed out already that the UC problem is of much importance for the scheduling of thermal units, it is not the meaning of UC that cannot be completely separated from the scheduling of a hydro-unit.

The hydro-thermal scheduling will be explained as separated from the UC problem. Operation of a system having both hydro and thermal plants is, however, far more complex as hydro-plants have negligible operation costs, but are required to operate under constraints of water available for hydro-generation in a given period of time.

The problem of minimizing the operating cost of a hydro-thermal system can be viewed as one of minimizing the fuel cost of thermal plants under the constraint of water availability for hydro-generation over a given period of operation.

4.4.4 Must run

It is necessary to give a must-run reorganization to some units of the plant during certain events of the year, by which we yield the voltage support on the transmission network or for such purpose as supply of steam for uses outside the steam plant itself.

4.4.5 Fuel constraints

A system in which some units have limited fuel or else have constraints that require them to burn a specified amount of fuel in a given time presents a most challenging UC problem.

4.5 COST FUNCTION FORMULATION

Let F_i be the cost of operation of the ith unit, P_{G_i} the output of the ith unit, and C_i the running cost of the ith unit. Then,

$$F_i = C_i\, P_{G_i}$$

C_i may vary depending on the loading condition.

Let C_{ij} be the variable cost coefficient for the ith unit when operating at the jth load for which the corresponding active power is $P_{G_{ij}}$.

Since the level of operation is a function of time, the cost efficiency may be described with yet another index to denote the time of operation, so that it becomes C_{ij_t} for the sub-interval 't' corresponding to a power output of $P_{G_{ij_t}}$.

If each unit is capable of operation at k discrete levels, then the running cost F_{i_t} of the ith unit in the time interval t is given by

$$F_{i_t} = \sum_{j=1}^{k} C_{i_{j_t}} P_{G_{ij_t}}$$

If there are n units available for operation in the time interval 't', then the total running cost of n units during the time interval 't' is

$$F_{i_{n_t}} = \sum_{i=1}^{n}\sum_{j=1}^{k} C_{ij_t} P_{G_{ij_t}}$$

For the entire time period of optimization, having T sub-intervals of time, the overall running cost for all the units may become

$$F_T = \sum_{t=1}^{T}\sum_{i=1}^{n}\sum_{j=1}^{k} C_{ij_t} P_{G_{ij_t}}$$

4.5.1 Start-up cost consideration

Suppose that for a plant to be brought into service, an additional expenditure C_{s_i} has to be incurred in addition to the running cost (i.e., start-up cost of the ith unit), the cost of starting 'x' number of units during any sub-interval t is given by

$$F_{SC} = \sum_{i=1}^{x} C_{s_i} \delta_{i_t}$$

where $\delta_{i_t} = 1$, if the ith unit is started in sub-interval 't' and otherwise $\delta_{it} = 0$.

4.5.2 Shut-down cost consideration

Similarly, if a plant is taken out of service during the scheduling period, it is necessary to consider the shut-down cost.

If 'y' number of units are be to shut down during the sub-interval 't', the shut-down cost may be represented as

$$F_{\text{Sd}} = \sum_{i=1}^{y} C_{\text{sd}_i} \sigma_{i_t}$$

where $\sigma_{i_t} = 1$, when the ith unit is thrown out of service in sub-interval 't'; otherwise $\sigma_{i_t} = 0$.

Over the complete scheduling period of T sub-intervals, the start-up cost is given by

$$F_{\text{SCT}} = \sum_{t=1}^{T} \sum_{i=1}^{x} C_{\text{st}} \delta_{i_t}$$

and the shut-down cost is

$$F_{\text{SdT}} = \sum_{t=1}^{T} \sum_{i=1}^{Y} C_{\text{sd}_t} \sigma_{i_t}$$

Now, the total expression for the cost function including the running cost, the start-up cost, and the shut-down cost is written in the form:

$$F_T = \sum_{t=1}^{T} \left\{ \sum_{i=1}^{N} \sum_{j=1}^{k} C_{ij_t} P_{G_{ij_t}} + \sum_{i=1}^{x} C_{\text{sc}_i} \delta_{i_t} + \sum_{i=1}^{y} C_{\text{sd}_i} \sigma_{i_t} \right\}$$

For each sub-interval of time t, the number of generating units to be committed to service, the generators to be shut down, and the quantized power loading levels that minimize the total cost have to be determined.

4.6 CONSTRAINTS FOR PLANT COMMITMENT SCHEDULES

As in the optimal point generation scheduling, the output of each generator must be within the minimum and maximum value of capacity:

$$\text{i.e., } P_{G_{i_{\min}}} \leq P_{G_{ij_t}} \leq P_{G_{i_{\max}}} \quad \text{for } i = 1, 2, \ldots, n; \quad j = 1, 2, \ldots, k; \quad t = 1, 2, \ldots, T$$

The optimum schedules of generation are prepared from the knowledge of the total available plant capacity, which must be in excess of the plant-generating capacity required in meeting the predicted load demand in satisfying the requirements for minimum running reserve capacity during the entire period of scheduling:

$$S_{\text{TAC}} \geq \sum_{i=1}^{n} P_{G_{i_{\max}}} \alpha_{i_t} + S_{r_{\min}}$$

where S_{TAC} is the total available capacity in any sub-interval 't', $S_{r_{\min}}$ the minimum running reserve capacity, $\alpha_{i_t} = 1$, if the ith unit is in operation during sub-interval 't'; otherwise $\alpha_{i_t} = 0$

In addition, for a predicted load demand P_{D}, the total generation output in sub-interval 't' must be in excess of the load demand by an amount not less than the minimum running reserve capacity $S_{r_{\min}}$.

$$\sum_{i=1}^{n}\sum_{j=1}^{k} P_{G_{ijt}} \geq (P_D + S_{r_{\min}}) \quad \text{(without considering the transmission losses)}$$

In case of consideration of transmission losses, the above equation becomes

$$\sum_{i=1}^{n}\sum_{j=1}^{k} P_{G_{ijt}} \geq (P_D + S_{r_{\min}} + P_L)$$

The generator start-up and shut-down logic indicators δ_{i_t} and σ_{i_t}, respectively, should be unity during the corresponding sub-intervals of operation

$$\text{or } (\alpha_{i_t} - \alpha_{i(t-1)}) = (\delta_{i_t} - \sigma_{i_t}) \quad \text{for} \quad i = 1, \ldots, n;\ t = 1, \ldots, T$$

4.7 UNIT COMMITMENT—SOLUTION METHODS

The most important techniques for the solution of a UC problem are:

1. Priority-list schemes.

2. Dynamic programming (DP) method.

3. Lagrange's relaxation (LR) method.

Now, we will explain the priority-list scheme and the DP method.

A simple shut-down rule or priority-list scheme could be obtained after an exhaustive enumeration of all unit combinations at each load level.

4.7.1 Enumeration scheme

A straightforward but highly time-consuming way of finding the most economical combination of units to meet a particular load demand is to try all possible combinations of units that can supply this load. This load is divided optimally among the units of each combination by the use of co-ordination equations so as to find the most economical operating cost of the combination. Then, the combination that has the least operating cost among all these is determined.

Some combinations will be infeasible if the sum of all maximum MW for the units committed is less than the load or if the sum of all minimum MW for the units committed is greater than the load.

Example 4.1: Let us consider a plant having three units. The cost characteristics and minimum and maximum limits of power generation (MW) of each unit are as follows:

Unit-1,
$$C_1 = 0.002842\,P_{G_1}^2 + 8.46\,P_{G_1} + 600.0\ \text{Rs./hr}, \quad 200 \leq P_{G_1} \leq 650$$

Unit-2,
$$C_2 = 0.002936\,P_{G_2}^2 + 8.32\,P_{G_2} + 420.0\ \text{Rs./hr}, \quad 150 \leq P_{G_2} \leq 450$$

Unit-3,
$$C_3 = 0.006449\,P_{G_3}^2 + 9.884\,P_{G_3} + 110.0\ \text{Rs./hr}, \quad 100 \leq P_{G_3} \leq 300$$

To supply a total load of 600 MW most economically, the combinations of units and their generation status are tabulated in Table 4.1.

Number of combinations $= 2^n = 2^3 = 8$

TABLE 4.1 *Combinations of the units and their status for the dispatch of a 600-MW load*

Combination	Status of units			$P_{G_{max}}$	$P_{G_{min}}$	P_{G_1}	P_{G_2}	P_{G_3}	C_1	C_2	C_3	Total generation cost $C_1 + C_2 + C_3$ (in Rs.)
	Unit-1	Unit-2	Unit-3									
1	Off	Off	Off	0	0	Infeasible	Infeasible	Infeasible	—	—	—	—
2	Off	Off	On	300	100	Infeasible	Infeasible	Infeasible	—	—	—	—
3	Off	On	Off	450	150	Infeasible	Infeasible	Infeasible	—	—	—	—
4	Off	On	On	750	250	0	450.00	150	—	4,758.54	1,882.805	6,641.345
5	On	Off	Off	650	200	600.00	0	0	6,698.4	—	—	6,698.400
6	On	Off	On	950	300	500.00	0	100	5,540.5	—	1,162.89	6,703.390
7	On	On	Off	1,100	350	292.77	307.23	0	3,320.43	3,253.28	—	6,573.71
8	On	On	On	1, 400	450	241.95	258.05	100	2,813.267	881.877	1,162.89	4,858.035

Note: The least expensive was not to supply the generation with all three units running or even any combination involving two units. Rather, the optimum commitment was to run only Unit-1, the most economic unit. By only running it, the load can be supplied by that unit operating closer to its best efficiency. If another unit is committed, both Unit-1 and the other unit will be loaded further from their best efficiency points such that the net cost is greater than Unit-1 alone.

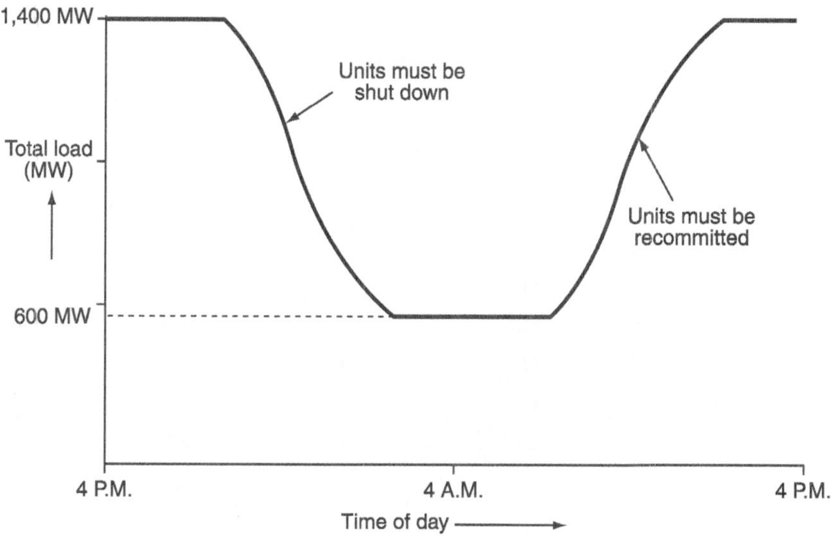

FIG. 4.3   *Simple peak–valley load pattern*

4.7.1.1 *UC operation of simple peak–valley load pattern: shut-down rule*

Let us assume that the load follows a simple 'peak–valley' pattern as shown in Fig. 4.3.

To optimize the system operation, some units must be shut down as the load decreases and is then recommitted (put into service) as it goes back up.

One approach called the 'shut-down rule' must be used to know which units to drop and when to drop them. A simple priority-list scheme is to be developed from the 'shut-down rule'.

Consider the example, with the load varying from a peak of 1,400 MW to a valley of 600 MW (Table 4.2). To obtain a 'shut-down rule', we simply use a brute-force technique wherein all combinations of units will be tried for each load level taken in steps of some MW (here 50 MW).

TABLE 4.2   *Shut-down rule derivation*

Load	Optimum generation		
	Unit-1	Unit-2	Unit-3
1,400	On	On	On
1,350	On	On	On
1,300	On	On	On

(Continued)

TABLE 4.2   *(Continued)*

Load	Optimum generation		
	Unit-1	Unit-2	Unit-3
1,250	On	On	On
1,200	On	On	On
1,150	On	On	On
1,100	On	On	Off
1,050	On	On	Off
1,000	On	On	Off
950	On	On	Off
900	On	On	Off
850	On	On	Off
800	On	On	Off
750	On	On	Off
700	On	Off	Off
650	On	Off	Off
600	On	Off	Off

From the above table, we can observe that for the load above 1,100 MW, running all the three units is economical; between 1,100 and 700 MW running the first and second units is economical. For below 700 MW, running of only Unit-1 is economical as shown in Fig. 4.4.

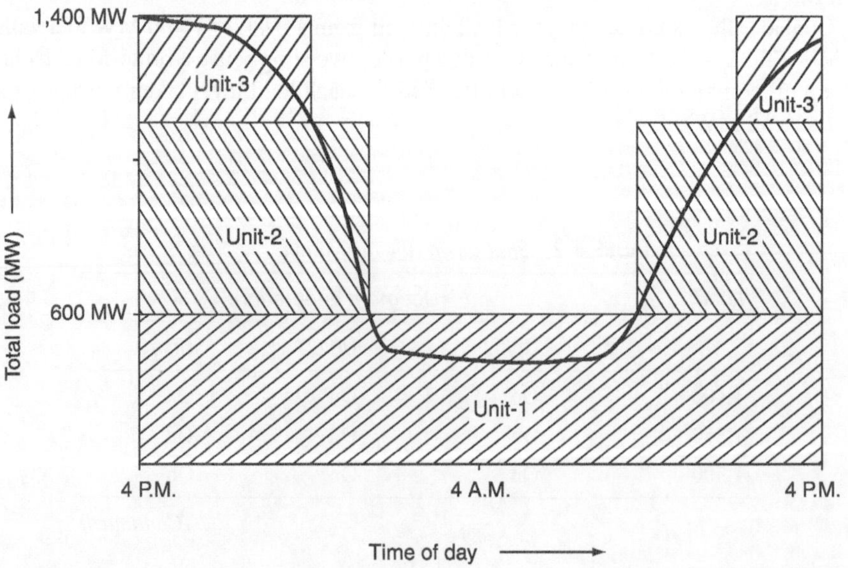

FIG. 4.4   *UC schedule using the shut-down rule*

TABLE 4.3 *Priority ordering of units*

Unit	Rs./MWh	$P_{G_{min}}$	$P_{G_{max}}$
2	9.834	150	450
1	9.838	200	650
3	11.1738	100	300

TABLE 4.4 *Priority list for supply of 1,400 MW*

Combination of units	For combination $P_{G_{min}}$	For combination $P_{G_{max}}$
2, 1, and 3	450	1,400
2 and 1	350	1,100
2	150	450

4.7.2 Priority-list method

A simple but sub-optimal approach to the problem is to impose priority ordering, wherein the most efficient unit is loaded first to be followed by the less efficient units in order as the load increases.

In this method, first we compute the full-load average production cost of each unit. Then, in the order of ascending costs, the units are arranged to commit the load demand.

For Example 4.1, we construct a priority list as follows:

First, the full-load average production cost will be calculated.

The full-load average production cost of Unit-1 = 9.79 Rs./MWh.

The full-load average production cost of Unit-2 = 9.48 Rs./MWh.

The full-load average production cost of Unit-3 = 11.188 Rs./MWh.

A priority order of these units based on the average production is as follows (Table 4.3):

By neglecting minimum up- or down-time, start-up costs, etc. the load demand can be met by the possible combinations as follows (Table 4.4):

4.7.2.1 Priority-list scheme versus shut-down sequence

In shut-down sequence, Unit-2 was shut down at 700 MW leaving Unit-1. With the priority-list scheme, both units would be held ON until the load had reached 450 MW and then Unit-1 would be dropped.

Many priority-list schemes are made according to a simple shut-down algorithm, such that they would have steps for shutting down a unit as follows:

1. During the dropping of load, at the end of each hour, determine whether the next unit on the priority list will have sufficient generation capacity to meet the load demand and to satisfy the requirement of the spinning reserve. If yes go to the next step and otherwise continue the operation with the unit as it is.

2. Determine the time in number of hours 'h' before the dropped unit (in Step 1) will be needed again for service.

3. If the number of hours (h) is less than minimum shut-down time for the unit, then keep the commitment of the unit as it is and go to Step 5; if not, go to the next step.

4. Now, calculate the first cost, which is the sum of hourly production costs for the next 'h' hours with the unit in 'up' state. Then, recalculate the same sum as second cost for the unit 'down' state and in the start-up cost for either cooling the unit or banking it, whichever is less expensive. If there are sufficient savings from shutting down the unit, it should be shut down, otherwise keep it on.

5. Repeat the above procedure for the next unit on the priority list and continue for the subsequent unit.

The various improvements to the priority-list schemes can be made by grouping of units to ensure that various constraints are met.

4.7.3 Dynamic programming

Dynamic programming is based on the principle of optimality explained by Bellman in 1957. It states that 'an optimal policy has the property, that, whatever the initial state and the initial decisions are, the remaining decisions must constitute an optimal policy with regard to the state resulting from the first decision'.

This method can be used to solve problems in which many sequential decisions are required to be taken in defining the optimum operation of a system, which consists of a distinct number of stages. However, it is suitable only when the decisions at the later stages do not affect the operation at the earlier stages.

4.7.3.1 *Solution of an optimal UC problem with DP method*

Dynamic programming has many advantages over the enumeration scheme, the main advantage being a reduction in the size of the problem.

The imposition of a priority list arranged in order of the full-load average cost rate would result in a correct dispatch and commitment only if

(i)   No-load costs are zero.

(ii)   Unit input–output characteristics are linear between zero output and full load.

(iii)   There are no other limitations.

(iv)   Start-up costs are a fixed amount.

In the DP approach, we assume that:

(i)   A state consists of an array of units with specified operating units and the rest are at off-line.

(ii)   The start-up cost of a unit is independent of the time if it has been off-line.

(iii)   There are no costs for shutting down a unit.

(iv)   There is a strict priority order and in each interval a specified minimum amount of capacity must be operating.

A feasible state is one at which the committed units can supply the required load and that meets the minimum amount of capacity in each period.

Practically, a UC table is to be made for the complete load cycle. The DP method is more efficient for preparing the UC table if the available load demand is assumed to increase in small but finite size steps. In DP it is not necessary to solve co-ordinate equations, while at the same time the unit combinations are to be tried.

Considerable computational saving can be achieved by using the branch and bound technique or a DP method for comparing the economics of combinations as certain combinations need not be tried at all.

The total number of units available, their individual cost characteristics, and the load cycle on the station are assumed to be known *a priori*. Further, it shall be assumed that the load on each unit or combination of units changes in suitably small but uniform steps of size Δ MW (say 1 MW).

Procedure for preparing the UC table using the DP approach:

Step 1: Start arbitrarily with consideration of any two units.

Step 2: Arrange the combined output of the two units in the form of discrete load levels.

Step 3: Determine the most economical combination of the two units for all the load levels. It is to be observed that at each load level, the economic operation may be to run either a unit or both units with a certain load sharing between the two units.

Step 4: Obtain the most economical cost curve in discrete form for the two units and that can be treated as the cost curve of a single equivalent unit.

Step 5: Add the third unit and repeat the procedure to find the cost curve of the three combined units. It may be noted that by this procedure, the operating combinations of the third and first and third and second units are not required to be worked out resulting in considerable saving in computation.

Step 6: Repeat the process till all available units are exhausted.

The main advantage of this DP method of approach is that having obtained the optimal way of loading 'K' units, it is quite easy to determine the optimal way of loading $(K+1)$ units.

Mathematical representation

Let a cost function $F_N(x)$ be the minimum cost in Rs./hr of generation of 'x' MW by N number of units, $f_N(y)$ the cost of generation of 'y' MW by the Nth unit, and $F_{N-1}(x-y)$ the minimum cost of generation of $(x-y)$ MW by remaining $(N-1)$ units.

The following recursive relation will result with the application of DP:

$$F_N(x) = \min_{y} \left\{ f_N(y) + F_{N-1}(x-y) \right\}$$

The most efficient economical combination of units can efficiently be determined by the use of the above relation. Here the most economical combination of units is such that it yields the minimum operating cost, for discrete load levels ranges from the minimum permissible load of the smallest unit to the sum of the capacities of all available units.

In this process, the total minimum operating cost and the load shared by each unit of the optimal combination are automatically determined for each load level.

Example 4.2: A power system network with a thermal power plant is operating by four generating units. Determine the most economical unit to be committed to a load demand of 8 MW. Also, prepare the UC table for the load changes in steps of 1 MW starting from the minimum to the maximum load. The minimum and maximum generating capacities and cost-curve parameters of the units listed in a tabular form are given in Table 4.5.

Solution:

We know that:

The cost function, $F_i = \dfrac{1}{2} a_i P_{G_i}^2 + b_i P_{G_i} + d_i$ **(4.1)**

Incremental fuel cost, $\dfrac{dC_i}{dP_{G_i}} = a_i P_{G_i} + b_i$ **(4.2)**

The total load $= P_D = 8$ MW (given)

By comparing the cost-curve parameters, we come to know that the cost characteristics of the first unit are the lowest. If only one single unit is to be committed, Unit-1 is to be employed.

Now, find out the cost of generation of power by the first unit starting from minimum to maximum generating capacity of that unit.

Let,

$f_1(1) =$ the main cost in Rs./hr for the generation of 1 MW by the first unit
$f_1(2) =$ the main cost in Rs./hr for the generation of 2 MW by the first unit
$f_1(3) =$ the main cost in Rs./hr for the generation of 3 MW by the first unit
$f_1(4) =$ the main cost in Rs./hr for the generation of 4 MW by the first unit
....
$f_1(8) =$ the main cost in Rs./hr for the generation of 8 MW by the first unit

$$f_i(P_{G_i}) = \frac{1}{2} a_i P_{G_i}^2 + b_i P_{G_i} + d_i$$

$$f_1(P_{G_i}) = \frac{1}{2}(0.74) P_{G_1}^2 + 22.9\, P_{G_1} = (0.37 P_{G_1} + 22.9) P_{G_1}$$

$$f_2(P_{G_2}) = \frac{1}{2}(1.56) P_{G_2}^2 + 25.9\, P_{G_2} = (0.78 P_{G_2} + 25.9) P_{G_2}$$

TABLE 4.5   *Capacities and cost-curve parameters of the units*

Unit number	Capacity (MW)		Cost-curve parameters		
	Min.	Max.	a	b	d
1	1.0	14.0	0.74	22.9	0
2	1.0	14.0	1.56	25.9	0
3	1.0	14.0	1.97	29.0	0
4	1.0	14.0	1.36	31.2	0

$$f_3(P_{G_3}) = \frac{1}{2}(1.97)P_{G_3}^2 + 29.0\,P_{G_3} = (0.985\,P_{G_3} + 29.0)P_{G_3}$$

$$f_4(P_{G_4}) = \frac{1}{2}(1.36)P_{G_4}^2 + 31.2\,P_{G_4} = (0.68\,P_{G_4} + 31.2)P_{G_4}$$

For the commitment of Unit-1 only

When only one unit is to be committed to meet a particular load demand, i.e., Unit-1 in this case due to its less cost parameters, then $F_1(x) = f_1(x)$.

where:

$F_1(x)$ is the minimum cost of generation of 'x' MW by only one unit

$f_1(x)$ is the minimum cost of generation of 'x' MW by Unit-1

$\therefore F_1(1) = f_1(1) = (0.37 \times 1 + 22.9)\,1 = 23.27$

$F_1(2) = f_1(2) = (0.37 \times 2 + 22.9)\,2 = 47.28$

$F_1(3) = f_1(3) = (0.37 \times 3 + 22.9)\,3 = 72.03$

Similarly,

$F_1(4) = f_1(4) = \ 97.52$

$F_1(5) = f_1(5) = 123.75$

$F_1(6) = f_1(6) = 150.72$

$F_1(7) = f_1(7) = 178.43$

$F_1(8) = f_1(8) = 206.88$

When Unit-1 is to be committed to meet a load demand of 8 MW, the cost of generation becomes 206.88 Rs./hr.

For the second unit

$f_2(1) = $ min. cost in Rs./hr for the generation of 1 MW by the second unit only

$$= (0.78 P_{G_2} + 25.9)P_{G_2}$$

$$= (0.78 \times 1 + 25.9)\,1 = 26.68$$

Similarly,

$f_2(2) = 54.92$

$f_2(3) = 84.72$

$f_2(4) = 116.08$

$f_2(5) = 149.0$

$f_2(6) = 183.48$

$f_2(7) = 219.52$

$f_2(8) = 257.12$

By observing $f_1(8)$ and $f_2(8)$, it is concluded that $f_1(8) < f_2(8)$, i.e., the cost of generation of 8 MW by Unit-1 is minimum than that by Unit-2.

For commitment of Unit-1 and Unit-2 combination

$F_2(8) = $ Minimum cost of generation of 8 MW by the simultaneous operation of two units i.e., Units-1 and 2.

$$= \min\begin{bmatrix} f_2(0)+F_1(8), & f_2(1)+F_1(7), & f_2(2)+F_1(6), \\ f_2(3)+F_1(5), & f_2(4)+F_1(4), & f_2(5)+F_1(3), \\ f_2(6)+F_1(2), & f_2(7)+F_1(1), & f_2(8)+F_1(0) \end{bmatrix}$$

$$= \min \begin{bmatrix} 206.88, & 205.11, & 205.64, \\ 208.47, & 213.6, & 221.03, \\ 230.76, & 242.79, & 257.12 \end{bmatrix}$$

$$\therefore F_2(8) = 205.11 \text{ Rs./hr}$$

In other words, the minimum cost of generation of 8 MW by the combination of Unit-1 and Unit-2 is 205.11 Rs./hr and for this optimal cost, Unit-1 supplies 7 MW and Unit-2 supplies 1 MW.

$$F_2(7) = \min \begin{bmatrix} f_2(0)+F_1(7), & f_2(1)+F_1(6), & f_2(2)+F_1(5), \\ f_2(3)+F_1(4), & f_2(4)+F_1(3), & f_2(5)+F_1(2), \\ f_2(6)+F_1(1), & f_2(7)+F_1(0) \end{bmatrix}$$

$$= \min \begin{bmatrix} 178.43, & 177.4, & 178.67 \\ 182.24, & 188.11, & 196.28, \\ 206.75, & 219.52 \end{bmatrix}$$

$$\therefore F_2(7) = 177.4 \text{ Rs./hr.}$$

i.e., the minimum cost of generation of 7 MW with the combination of Unit-1 (by 6-MW supply) and Unit-2 (by 1-MW supply) is 177.4 Rs./hr.

$$F_2(6) = \min \begin{bmatrix} f_2(0)+F_1(6), & f_2(1)+F_1(5), & f_2(2)+F_1(4), \\ f_2(3)+F_1(3), & f_2(4)+F_1(2), & f_2(5)+F_1(1), \\ f_2(6)+F_1(0) \end{bmatrix}$$

$$= \min \begin{bmatrix} 150.72, & 150.43, & 152.44 \\ 156.75, & 163.36, & 172.27, \\ 183.48 \end{bmatrix}$$

$$\therefore F_2(6) = 150.43 \text{ Rs./hr}$$

$$F_2(5) = \min \begin{bmatrix} f_2(0)+F_1(5), & f_2(1)+F_1(4), & f_2(2)+F_1(3), \\ f_2(3)+F_1(2), & f_2(4)+F_1(1), & f_2(5)+F_1(0) \end{bmatrix}$$

$$= \min \begin{bmatrix} 123.75, & 124.2, & 126.95 \\ 132.0, & 139.35 & 140.0 \end{bmatrix}$$

$$\therefore F_2(5) = 123.75 \text{ Rs./hr}$$

$$F_2(4) = \min \begin{bmatrix} f_2(0)+F_1(4), & f_2(1)+F_1(3), & f_2(2)+F_1(2), \\ f_2(3)+F_1(1), & f_2(4)+F_1(0) \end{bmatrix}$$

$$= \min \begin{bmatrix} 97.52, & 98.71, & 102.2, \\ 107.99, & 116.08 \end{bmatrix}$$

$$\therefore F_2(4) = 97.52 \text{ Rs./hr}$$

$$F_2(3) = \min \begin{bmatrix} f_2(0)+F_1(3), & f_2(1)+F_1(2), \\ f_2(2)+F_1(1), & f_2(3)+F_1(0) \end{bmatrix}$$

$$= \min \begin{bmatrix} 72.03, & 73.96, \\ 78.19, & 84.72 \end{bmatrix}$$

$$\therefore F_2(3) = 72.03 \text{ Rs./hr}$$

$$F_2(2) = \min [f_2(0) + F_1(2), \quad f_2(1) + F_1(1), \quad f_2(2) + F_1(0)]$$

$$= \min [47.28, \quad 49.95, \quad 54.92]$$

$$\therefore F_2(2) = 47.28 \text{ Rs./hr}$$

$$F_2(1) = \min [f_2(0) + F_1(1), \quad f_2(1) + F_1(0)]$$

$$= \min [23.27, \quad 26.68]$$

$$\therefore F_2(1) = 23.27 \text{ Rs./hr}$$

Now, the cost of generation by Unit-3 only is

$$f_3(P_{G_i}) = \frac{1}{2} a_i P_{G_i}^2 + b_i P_{G_i} + d_i$$

$$= (0.985) P_{G_3} + 29.0 P_{G_3}$$

$$
\begin{aligned}
&f_3(0) = 0; && f_3(5) = 169.625 \\
&f_3(1) = 29.985; && f_3(6) = 209.46 \\
&f_3(2) = 61.94; && f_3(7) = 251.265 \\
&f_3(3) = 95.865; && f_3(8) = 295.04 \\
&f_3(4) = 131.76;
\end{aligned}
$$

For commitment of Unit-1, Unit-2, and Unit-3 combination

$F_3(8) = $ The minimum cost of generation of 8 MW by the three units, i.e., Unit-1, Unit-2, and Unit-3

$$= \min \begin{bmatrix} f_3(0) + F_2(8), & f_3(1) + F_2(7), & f_3(2) + F_2(6), \\ f_3(3) + F_2(5), & f_3(4) + F_2(4), & f_3(5) + F_2(3), \\ f_3(6) + F_2(2), & f_3(7) + F_2(1), & f_3(8) + F_2(0) \end{bmatrix}$$

$$= \min \begin{bmatrix} 205.11, & 207.385, & 212.37 \\ 219.615, & 229.28, & 241.655 \\ 256.74, & 298.545, & 295.04 \end{bmatrix}$$

$$\therefore F_3(8) = 205.11 \text{ Rs./hr}$$

i.e., for the generation of 8 MW by three units, Unit-1 and Unit-2 will commit to meet the load of 8 MW with Unit-1 supplying 7 MW, Unit-2 supplying 1 MW, and Unit-3 is in an off-state condition.

$$F_3(7) = \min \begin{bmatrix} f_3(0) + F_2(7), & f_3(1) + F_2(6), & f_3(2) + F_2(5), \\ f_3(3) + F_2(4), & f_3(4) + F_2(3), & f_3(5) + F_2(2), \\ f_3(6) + F_2(1), & f_3(7) + F_2(0) \end{bmatrix}$$

$$= \min \begin{bmatrix} 177.4 & 180.415, & 185.69 \\ 193.385, & 203.79, & 216.905 \\ 232.73, & 251.265 \end{bmatrix}$$

$$\therefore F_3(7) = 177.4 \text{ Rs./hr}$$

$$F_3(6) = \min \begin{bmatrix} f_3(0) + F_2(6), & f_3(1) + F_2(5), & f_3(2) + F_2(4), \\ f_3(3) + F_2(3), & f_3(4) + F_2(2), & f_3(5) + F_2(1), \\ f_3(6) + F_2(0) \end{bmatrix}$$

$$= \min \begin{bmatrix} 150.43, & 153.735, & 159.46 \\ 167.895, & 179.04, & 192.89 \\ 209.46 \end{bmatrix}$$

$$\therefore F_3(6) = 150.43 \text{ Rs./hr}$$

$$F_3(5) = \min \begin{bmatrix} f_3(0) + F_2(5), & f_3(1) + F_2(4), & f_3(2) + F_2(3), \\ f_3(3) + F_2(2), & f_3(4) + F_2(1), & f_3(5) + F_2(0) \end{bmatrix}$$

$$= \min \begin{bmatrix} 123.75, & 127.505, & 133.97, \\ 143.145, & 155.03, & 169.625 \end{bmatrix}$$

$$\therefore F_3(5) = 123.75 \text{ Rs./hr}$$

$$F_3(4) = \min \begin{bmatrix} f_3(0) + F_2(4), & f_3(1) + F_2(3), & f_3(2) + F_2(2), \\ f_3(3) + F_2(1), & f_3(4) + F_2(0) \end{bmatrix}$$

$$= \min \begin{bmatrix} 97.52, & 102.015, & 109.22, \\ 119.135, & 131.76 \end{bmatrix}$$

$$\therefore F_3(4) = 97.52 \text{ Rs./hr}$$

$$F_3(3) = \min \begin{bmatrix} f_3(0) + F_2(3), & f_3(1) + F_2(2), & f_3(2) + F_2(1), \\ f_3(3) + F_2(0) \end{bmatrix}$$

$$= \min \begin{bmatrix} 72.03, & 77.265, & 85.21 \\ 95.865 \end{bmatrix}$$

$$\therefore F_3(3) = 72.03 \text{ Rs./hr}$$

$$F_3(2) = \min \left[f_3(0) + F_2(2), f_3(1) + F_2(1), f_3(2) + F_2(0) \right]$$

$$= \min \left[47.28, \quad 53.255, \quad 61.94 \right]$$

$$\therefore F_3(2) = 47.28 \text{ Rs./hr}$$

$$F_3(1) = \min \left[f_3(0) + F_2(1), f_3(1) + F_2(0) \right]$$

$$= \min \left[23.27, \quad 29.958 \right]$$

$$\therefore F_3(1) = 23.27 \text{ Rs./hr}$$

Cost of generation by the fourth unit

$$f_4(P_{G_i}) = \frac{1}{2}a_i P_{G_i}^2 + b_i P_{G_i} + d_i$$

$$= (0.68 P_{G_4} + 31.2) P_{G_4}$$

$$f_4(0) = 0$$
$$f_4(1) = 31.88 \text{ Rs./hr}$$
$$f_4(2) = 65.12 \text{ Rs./hr}$$
$$f_4(3) = 99.72 \text{ Rs./hr}$$
$$f_4(4) = 135.68 \text{ Rs./hr}$$
$$f_4(5) = 173.0 \text{ Rs./hr}$$
$$f_4(6) = 211.68 \text{ Rs./hr}$$
$$f_4(7) = 251.72 \text{ Rs./hr}$$
$$f_4(8) = 293.12 \text{ Rs./hr}$$

Minimum cost of generation by four units, i.e., Unit-1, Unit-2, Unit-3, and Unit-4

$F_4(8) = $ The minimum cost of generation of 8 MW by four units

$$= \min \begin{bmatrix} f_4(0)+F_3(8), & f_4(1)+F_3(7), & f_4(2)+F_3(6), \\ f_4(3)+F_3(5), & f_4(4)+F_3(4), & f_4(5)+F_3(3), \\ f_4(6)+F_3(2), & f_4(7)+F_3(1), & f_4(8)+F_3(0) \end{bmatrix}$$

$$= \min \begin{bmatrix} 205.11, & 209.28, & 215.55 \\ 223.47, & 233.2, & 245.03 \\ 258.96, & 274.99, & 293.12 \end{bmatrix}$$

$$\therefore F_4(8) = 205.11 \text{ Rs./hr}$$

i.e., for the generation of 8 MW by four units, Unit-1 and Unit-2 will commit to meet the load of 8 MW with Unit-1 supplying 7 MW, Unit-2 supplying 1 MW, and Unit-3 as well as Unit-4 are in an off-state condition:

$$F_4(7) = \min \begin{bmatrix} f_4(0)+F_3(7), & f_4(1)+F_3(6), & f_4(2)+F_3(5), \\ f_4(3)+F_3(4), & f_4(4)+F_3(3), & f_4(5)+F_3(2), \\ f_4(6)+F_3(1), & f_4(7)+F_3(0) \end{bmatrix}$$

$$= \min \begin{bmatrix} 177.4, & 182.31, & 188.87 \\ 197.24, & 207.71, & 220.28 \\ 234.94, & 251.72 \end{bmatrix}$$

$$\therefore F_4(7) = 177.4 \text{ Rs./hr}$$

$$F_4(6) = \min \begin{bmatrix} f_4(0)+F_3(6), & f_4(1)+F_3(5), & f_4(2)+F_3(4), \\ f_4(3)+F_3(43), & f_4(4)+F_3(2), & f_4(5)+F_3(1), \\ f_4(6)+F_3(0) \end{bmatrix}$$

$$= \min \begin{bmatrix} 150.43, & 155.63, & 162.64, \\ 171.75, & 182.96, & 196.27, \\ 211.68 \end{bmatrix}$$

$\therefore F_4(7) = 150.43$ Rs./hr

$$F_4(5) = \min \begin{bmatrix} f_4(0)+F_3(5), & f_4(1)+F_3(4), & f_4(2)+F_3(3), \\ f_4(3)+F_3(2), & f_4(4)+F_3(1), & f_4(5)+F_3(0) \end{bmatrix}$$

$$= \min \begin{bmatrix} 123.75, & 129.4, & 137.15, \\ 147.0, & 158.95, & 173.0 \end{bmatrix}$$

$\therefore F_4(5) = 123.75$ Rs./hr

$$F_4(4) = \min \begin{bmatrix} f_4(0)+F_3(4), & f_4(1)+F_3(3), & f_4(2)+F_3(2), \\ f_4(3)+F_3(21), & f_4(4)+F_3(0) \end{bmatrix}$$

$$= \min \begin{bmatrix} 97.52, & 103.91, & 112.4 \\ 122.99, & 135.68 \end{bmatrix}$$

$\therefore F_4(4) = 97.52$ Rs./hr

$$F_4(3) = \min \begin{bmatrix} f_4(0)+F_3(3), & f_4(1)+F_3(2), & f_4(2)+F_3(1), \\ f_4(3)+F_3(0) \end{bmatrix}$$

$$= \min \begin{bmatrix} 72.03, & 79.16, & 88.39 \\ 99.72] \end{bmatrix}$$

$\therefore F_4(3) = 72.03$ Rs./hr

$$F_4(2) = \min \begin{bmatrix} f_4(0)+F_3(2), & f_4(1)+F_3(1) \\ f_4(3)+F_3(0) \end{bmatrix}$$

$$= \min \begin{bmatrix} 47.28, & 55.154 \\ 65.12 \end{bmatrix}$$

$\therefore F_4(2) = 47.28$ Rs./hr

$$F_4(1) = \min [f_4(0)+F_3(1), \quad f_4(1)+F_3(0)]$$

$$= \min [23.27 \quad 31.88]$$

$\therefore F_4(1) = 23.27$ Rs./hr

From the above criteria, it is observed that for the generation of 8 MW, the commitment of units is as follows:

$f_1(8) = F_1(8) =$ the minimum cost of generation of 8 MW in Rs./hr by Unit-1 only
$= 206.88$ Rs./hr

$F_2(8) =$ the minimum cost of generation of 8 MW by two units with Unit-1 supplying
7 MW and Unit-2 supplying 1 MW
$= 205.11$ Rs./hr

$F_3(8) =$ the minimum cost of generation of 8 MW by three units with Unit-1 supplying 7 MW, Unit-2 supplying 1 MW, and Unit-3 is in an off-state condition.
$= 205.11$ Rs./hr

$F_4(8) =$ minimum cost of generation of 8 MW by four units with Unit-1 supplying 7 MW, Unit-2 supplying 1MW, and Unit-3 and Unit-4 are in an off-state condition.
$= 205.11$ Rs./hr

By examining the costs $F_1(8)$, $F_2(8)$, $F_3(8)$, and $F_4(8)$, we have concluded that for meeting the load demand of 8 MW, the optimal combination of units to be committed is Unit-1 with 7 MW and Unit-2 with 1 MW, respectively, at an operating cost of 205.11 Rs./hr

For preparing the UC table, the ordering of units is not a criterion. For any order, we get the same solution that is independent of numbering units.

To get a higher accuracy, the step size of the load is to be reduced, which results in a considerable increase in time of computation and required storage capacity.

Status 1 of any unit indicates unit running or unit committing and Status 0 of any unit indicates that the unit is not running.

The UC table is prepared once and for all for a given set of units (Table 4.6). As the load cycle on the station changes, it would only mean changes in starting and stopping of units without changing the basic UC table.

The UC table is used in giving the information of which units are to be committed to supply a particular load demand. The exact load sharing between the units committed is to be obtained by solving the co-ordination equations as below.

Total load, $P_D = 8$ MW

$$C_1 = \frac{1}{2} a_1 P_{G_1}^2 + b_1 P_{G_1}$$

$$C_2 = \frac{1}{2} a_2 P_{G2}^2 + b_1 P_{G_1}$$

$$\frac{dC_1}{dP_{G_1}} = a_1 P_{G_1} + b_1 = 0.74 P_{G_1} + 22.9 \tag{4.3}$$

$$\frac{dC_2}{dP_{G_2}} = a_2 P_{G_2} + b_2 = 1.56 P_{G_2} + 25.9 \tag{4.4}$$

$$P_{G_1} + P_{G_2} = 8 \text{ MW (given)} \tag{4.5}$$

$$\Rightarrow P_{G_2} = 8 - P_{G_1}$$

TABLE 4.6 *The UC table for the above-considered system*

Load range	Unit			
	1	2	3	4
1–5	1	0	0	0
6–13	1	1	0	0
14–18	1	1	1	0
19–56	1	1	1	1

For an optimal load sharing,

$$\frac{dC_i}{dP_{G_i}} = \lambda$$

i.e., $\dfrac{dC_1}{dP_{G_1}} = \dfrac{dC_2}{dP_{G_2}}$

$$0.74P_{G_1} + 22.9 = 1.56P_{G_2} + 25.9$$
$$= 1.56(8 - P_{G_1}) + 25.9$$
$$= 12.48 - 1.56P_{G_1} + 25.9$$
$$\Rightarrow 2.3P_{G_1} = 15.48$$
$$P_{G_1} = \frac{15.48}{2.3} = 6.73 \text{ MW}$$

i.e., load shared by the first unit, $P_{G_1} = 6.73$ MW

and $P_{G_2} = 8 - P_{G_1} = 8 - 6.73 = 1.27$ MW

i.e., load shared by the second unit, $P_{G_2} = 1.27$ MW

Lagrangian multiplier, $\lambda = 0.74P_{G_1} + 22.9 = 1.56P_{G_2} + 25.9$

$$= 27.88 \text{ Rs./MWh}$$

$f_1(6.73) = $ cost of generation of 6.73 MW by the first unit

$$= \frac{1}{2}a_1 P_{G_1}^2 + b_1 P_{G_1} + d_1$$

$$= \left[\frac{1}{2} \times 0.74 \times (6.73)^2\right] + (22.9)(6.73) + 0.0$$

$$= 170.87 \text{ Rs./hr}$$

$f_2(1.27) = $ cost of generation of 1.27 MW by the second unit

$$= \frac{1}{2}a_2 P_{G_2}^2 + b_2 P_{G_2} + d_2$$

$$= \frac{1}{2} \times 1.56 \times (1.27)^2 + (25.9)(1.27)$$

$$= 34.15 \text{ Rs./hr}$$

The total minimum operating cost with an optimal combination of Unit-1 and Unit-2 is

$$f_1 + f_2 = 205.11 \text{ Rs./hr}$$

To prepare the UC table, the load is to vary in steps of 1 MW starting from a minimum generating capacity to a maximum generating capacity of a station in suitable steps.

4.8 CONSIDERATION OF RELIABILITY IN OPTIMAL UC PROBLEM

In addition to the economy of power generation, the reliability or continuity of power supply is also another important consideration. Any supply undertaking has assured all its consumers to provide reliable and quality of service in terms of the specified range of voltage and frequency.

The aspect of reliability in addition to economy is to be properly co-ordinated in preparing the UC table for a given system.

The optimal UC table is to be modified to include the reliability considerations.

Sometimes, there is an occurrence of the failure of generators or their derating conditions due to small and minor defects. Under that contingency of forced outage, in order to meet the load demand, 'static reserve capacity' is always maintained at a generating station such that the total installed capacity exceeds the yearly peak demand by a certain margin. This is a planning problem.

In arriving at the economic UC decision at any particular period, the constraint taken into consideration was merely a fact that the total generating capacity on-line was at least equal to the total load demand. If there was any margin between the capacity of units committed and the load demand, it was incidental. Under actual operation, one or more number of units had failed randomly; it may not be possible to meet the load demand for a certain period of time. To start the spare (standby) thermal unit and to bring it on the line to take up the load will involve long periods of time usually from 2 to 8 hr and also some starting cost. In case of a hydro-generating unit, it could be brought on-line in a few minutes to take up the load.

Hence, to ensure continuity of supply to meet random failures, the total generating capacity on-line must have a definite margin over the load requirements at any point of time. This margin is called the *spinning reserve*, which ensures continually by meeting the demand upto a certain extent of probable loss of generating capacity. While rules of thumb have been used, based on past experience to determine the system spinning reserve at any time, a recent better approach called Patton's analytical approach is the most powerful approach to solve this problem.

Consider the following points in the aspect of reliability consideration in the UC problem:

1. The probability of outage of any unit that increases with its operating time and a unit, which is to provide a spinning reserve at any particular time, has to be started several hours later. Hence, the security of supply problem has to be treated in totality over a period of one day.

2. The loads are never known with complete certainty.

3. The spinning reserve has to be facilitated at suitable generating stations of the system and not necessarily at each generating station.

A unit's useful life span undergoes alternate periods of operation and repair as shown in Fig. 4.5.

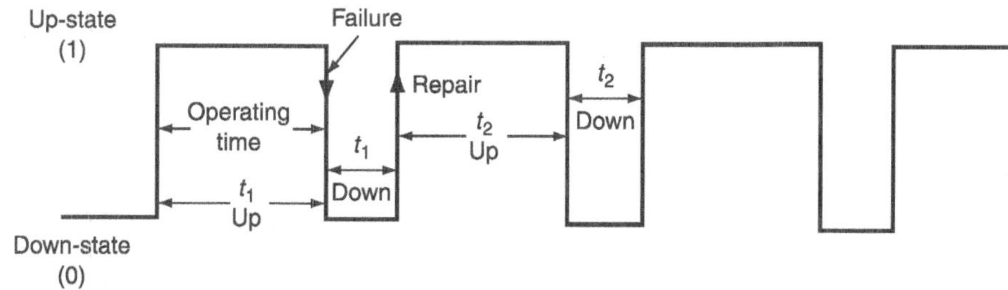

FIG. 4.5 *Random outage phenomena of a generating unit excluding the scheduled outages*

A unit operating time is also called unit 'up-time' (t_{up}) and its repair time as its 'down-time' (t_{down}).

The lengths of individual operating and repair periods are a random phenomenon with much longer periods of operation compared to repair periods.

This random phenomenon with a longer operating period of a unit is described by using the following parameters.

Mean time to failure (mean 'up' time):

$$\overline{T}_{up} = \frac{\sum_i t_{i(operation)}}{\text{No. of cycles}}$$

$$\Rightarrow \overline{T}_{up} = \frac{\sum_i t_{i(up)}}{\text{No. of cycles}} \tag{4.6}$$

Mean time to repair (mean 'down' time):

$$\overline{T}_{down} = \frac{\sum_i t_{i(down)}}{\text{No. of cycles}} \tag{4.7}$$

$$\therefore \text{Mean cycle time} = \overline{T}_{up} + \overline{T}_{down}$$

The rate of failure and the rate of repair can be defined by inversing Equations (4.6) and (4.7) as

Rate of failure $(\lambda) = \dfrac{1}{T_{up}}$ failures/year

Rate of repair $(\mu) = \dfrac{1}{\overline{T}_{down}}$ repairs/year

The failure and repair rates are to be estimated from the past data of units or other similar units elsewhere.

The rates of failure are affected by relative maintenance and the rates of repair are affected by the size, composition, and skill of repair teams.

By making use of the ratio definition of generating units, the probability of a unit being in an 'up' state and 'down' state can be expressed as

Probability of the unit in the 'up' state is

$$P_{up} = \frac{\overline{T}_{up}}{\overline{T}_{up} + \overline{T}_{down}}$$

$$= \frac{\dfrac{1}{\lambda}}{\dfrac{1}{\lambda} + \dfrac{1}{\mu}} = \frac{\mu}{\lambda + \mu} \tag{4.8}$$

The probability of the unit in the 'down' state is

$$P_{down} = \frac{\overline{T}_{down}}{\overline{T}_{up} + \overline{T}_{down}}$$

$$= \frac{\dfrac{1}{\mu}}{\dfrac{1}{\lambda} + \dfrac{1}{\mu}} = \frac{\lambda}{\lambda + \mu} \tag{4.9}$$

Obviously, $P_{up} + P_{down} = 1$ $\hspace{2cm}$ **(4.10)**

P_{up} and P_{down} are also known as availability and unavailability of the unit.

In any system with k number of units, the probability of the system state changes, i.e., when k units are present in a system, the system state changes due to random outages.

The random outage (failure) of a unit can be considered as an event independent of the state of the other unit.

Let a particular system state 'i', in which x_i units are in the 'down' state and y_i units are in the 'up' state:

i.e., $\quad x_i + y_i = k$

The probability of the system being in state 'i' is expressed as

$$P_1 = \prod_{j \in y_i} P_{j(up)} \prod_{j \in x_i} P_{j(down)} \hspace{2cm} \text{(4.11)}$$

Π indicates probability multiplication of the system state.

4.8.1 Patton's security function

Some intolerable or undesirable condition of system operation is termed as a 'breach of system security'.

In an optimal UC problem, the only breach of security considered is the insufficient generating capacity of the system at a particular instant of time.

The probability that the available generating capacity at a particular time is less than the total load demand on the system at that time is complicatively estimated by one function known as *Patton's security function*.

Patton's security function is defined as

$$S = \sum_i P_i r_i \hspace{2cm} \text{(4.12)}$$

where P_i is the probability of the system being in the ith state and r_i is the probability that the system state i causes a breach of system security.

In considering all possible system states to determine the security function, from the practical point of view, this sum is to be taken over the states in which not more than two units are on forced outage, i.e., states with more than two units out may be neglected as the probability of their occurrence will be too small.

$r_i = 1$, if the available generating capacity (sum of capacities of units committed) is less than the total load demand, i.e., $\sum P_{G_i} < P_D$. Otherwise $r_i = 0$.

The security function S gives a quantitative estimation of system insecurity.

4.9 OPTIMAL UC WITH SECURITY CONSTRAINT

From a purely economical point of view, a UC table is prepared from which we know which units are committed for a given load on the system.

For each period, we will estimate the security function

$$S = \sum_i P_i r_i$$

For any system, we will define maximum tolerable insecurity level (MTIL). This is a management decision and the value is based on past experience.

Whenever the security function exceeds MTIL ($S > $ MTIL), it is necessary to modify the UC table to include the aspects of security. It is normally achieved by committing the next most economical unit to supply the load. With the new unit being committed, we will estimate the security function and check whether it is $S <$ MTIL.

The procedure of committing the next most economical unit is continued upto $S <$ MTIL. If $S =$ MTIL, the system does not have proper reliability. Adding units goes upto one step only because for another, it is not necessary to add the next units more than one unit since there is a presence of spinning reserve.

4.9.1 Illustration of security constraint with Example 4.2

Reconsider Example 4.2 and the daily load curve for the above system as given in Fig. 4.6.

The economically optimal UC for this load curve is obtained by the use of the UC Table 4.6 (which was previously prepared) (Table 4.7).

Considering period E, in which the minimum load is 5 MW and Unit-1 is being committed to meet the load. We will check for this period whether the system is secure or not.

Assume the rate of repair, $\mu = 99$ repairs/year

And rate of failure, $\lambda = 1$ failure/year for all four units

And also assume that MTIL $= 0.005$

We have to estimate the security function S for this period E:

$$S = \sum_i P_i r_i$$

Value of r_i depending on whether there is a breach of security or not.

There are two possible states for Unit-1:

    operating state (or) 'up' state

                (or)

    forced outage state (or) 'down' state

The probability of Unit-1 being in the 'up' state,

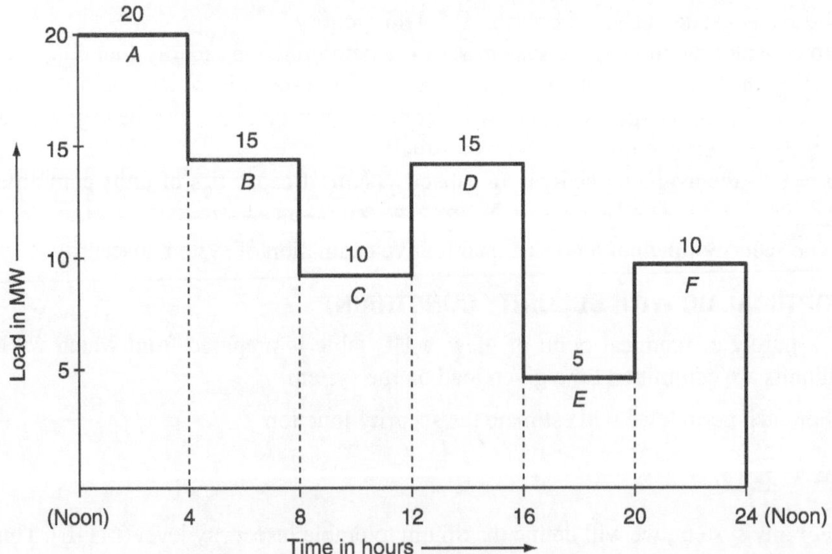

FIG. 4.6   *Daily load curve*

TABLE 4.7   *Economically optimal UC table for load curve shown in Fig. 4.4*

Period	Units committed			
	1	2	3	4
A	1	1	1	1
B	1	1	1	0
C	1	1	0	0
D	1	1	1	0
E	1	0	0	0
F	1	1	0	0

$$P_{1_{up}} = \frac{\mu}{\lambda + \mu} = \frac{99}{1 + 99} = 0.99$$

$r_1 = 0$, since the generation of Unit-1 (max. capacity) is greater than the load (i.e., 14 MW $>$ 5 MW).

There is no breach of security when the Unit-1 is in the 'up' state.

The probability of Unit-1 being in the 'down' state:

$$P_{1_{down}} = \frac{\lambda}{\lambda + \mu} = \frac{1}{1 + 99} = 0.01$$

$r_2 = 1$, since Unit-1 is in the down state $(P_{G_1} = 0)$, the load demand of 5 MW cannot be met.

There is a breach of security when Unit-1 is in the 'down' state. Now, find the value of the security function.

$$S = \sum_{i=1}^{2} P_i r_i$$

where i represents the state of Unit-1.

If n is the number of units, number of states $= 2^n$

For $n = 1$, states $= 2^1 = 2$ (i.e., up and down states)

$\therefore S = P_1 r_1 + P_2 r_2$

$\quad = P_{1_{up}} r_1 + P_{1_{down}} r_2$

$\quad = 0.99 \times 0 + 0.01 \times 1$

$\quad = 0.01$

It is observed that $0.01 > 0.005$,   i.e., $S > \text{MTIL}$

Since in this case, $S > \text{MTIL}$ represents system insecurity. Therefore, it is necessary to commit the next most economical unit, i.e., Unit-2, to improve the security. When both Units-1 and 2 are operating, estimate the security function as follows:

Here, number of units, $n = 2$

$\therefore$ Number of states $= 2^n = 2^2 = 4$

$r_i = 0$, represents no breach of security and

$r_i = 1$, represents breach of security

TABLE 4.8 *Representation of breach of security for the possible combinations*

Units	States			
1	up	down	up	down
2	down	up	up	down
ri	0	0	0	1

When taking either   up   down   up   combinations of states,
                        down   up   up
there is no breach of security, since $r_i = 0$

For the combination   down

                      down,

there is a breach of security (Table 4.8).

$$\therefore S = \sum_i P_i r_i$$

$$= P_{down} \times P_{down} \times 1$$

$$= 0.01 \times 0.01 \times 1$$

$$= 0.001$$

It is observed that $0.001 < 0.005$

Therefore, the combination of Unit-1 and Unit-2 does meet the MTIL of 0.005.

For all other periods of a load cycle, check whether the security function is less than MTIL. It is also found that for all other periods except E, the security function is less than MTIL. Now, we will obtain the optimal and security-constrained UC table for Example 4.2 (Table 4.9).

4.10 START-UP CONSIDERATION

From the optimal and secured UC table given in Table 4.9, depending on the load in a particular period, it is observed that some units are to be decommitted and restarted in the next period. Whenever a unit is to be restarted, it involves some cost as well as some time before the unit is put on-line. For thermal units, it is necessary to build up certain temperature

TABLE 4.9 *Optimal and secure UC table for Example 4.2*

Period	Unit committed			
	1	*2*	*3*	*4*
A	1	1	1	1
B	1	1	1	0
C	1	1	0	0
D	1	1	1	0
E	1	1*	0	0
F	1	1	0	0

* Unit is committed from the point of view of security considerations.

and pressure gradually before the unit can supply any load demand. The cost involved in restarting any unit after the decommitting period is known as **START-UP cost**.

Depending on the condition of the unit, the start-up costs will be different. If the unit is to be started from a cold condition and brought upto normal temperature and pressure, the start-up costs will be maximum since some energy is required to build up the required pressure and temperature of the steam. Sometimes, the unit may be switched off and the temperature of steam may not be in a cold condition. This particular condition is called the **banking condition**.

From the **UC table given in Table 4.7,** it is observed that during Period B, Unit-3 is operating and during Period C, it is decommitted. It is restarted during Period D.

In Period C, check whether it is economical to run only two units or allow all the three units (Units-1, 2, and 3) to continue to run such that the start-up costs are eliminated.

Let us assume that the start-up cost of each unit=Rs. 500.

Case A: Unit-3 is not in operation in Period C, i.e., only two Units-1 and 2 are operating.

For Period B or D, total load $= 15$ MW

This is to be shared by three units, i.e., $P_{G_1} + P_{G_2} + P_{G_3} = 15$

$$\frac{dC_1}{dP_{G_1}} = 0.74P_{G_1} + 22.9 \tag{4.13}$$

$$\frac{dC_2}{dP_{G_2}} = 1.56P_{G_2} + 25.9 \tag{4.14}$$

$$\frac{dC_3}{dP_{G_3}} = 1.97P_{G_3} + 29.0 \tag{4.15}$$

Subtracting Equation (4.14) from Equation (4.13), we get

$$0.74P_{G_1} - 1.56P_{G_2} = 3 \tag{4.16}$$

Subtracting Equation (4.15) from Equation (4.13), we get

$$0.74P_{G_1} - 1.97P_{G_3} = 6.1 \tag{4.17}$$

or $0.74P_{G_1} - 1.97(15 - P_{G_1} - P_{G_2}) = 6.1$

or $2.71P_{G_1} + 1.97P_{G_2} = 35.65 \tag{4.18}$

By solving Equations (4.13) and (4.16), we have

$P_{G_1} = 10.8$ MW, $P_{G_2} = 3.2$ MW

$P_{G_3} = 15 - P_{G_1} - P_{G_2} = 15 - 108 - 3.2 = 1$ MW

$C_1 = (0.37P_{G_1} + 22.9)P_{G_1} = 290.48$ Rs./hr

$C_2 = (0.78P_{G_2} + 25.9)P_{G_2} = 90.87$ Rs./hr

$C_3 = (0.985P_{G_3} + 29)P_{G_3} = 29.98$ Rs./hr

For Period B, the operating time is 4 hr.

$$\therefore \text{Total cost, } C = [C_1 + C_2 + C_3]\, t$$
$$= [290.48 + 90.87 + 29.98] \times 4$$
$$= \text{Rs. } 1,645.34$$

Total operating cost during Period B is Rs. 1,645.34.

In Period C, 10 MW of load is to be shared by Units-1 and 2

$$\text{i.e., } P_{G_1} + P_{G_2} = 10 \text{ MW} \tag{4.19}$$

By solving Equations (4.16) and (4.19), we get

$$P_{G_1} = 8.086 \text{ MW} \quad \text{and} \quad P_{G_2} = 1.913 \text{ MW}$$

Total operating cost for Period C

$$= \left[(0.37 P_{G_1} + 22.9) P_{G_1} + (0.78 P_{G_2} + 25.9) P_{G_2} \right] \times 4$$
$$= \text{Rs. } 1,047.05.$$

For Period D, the total operating cost is the same as that of Period B = Rs. 1,645.34. Therefore, the total operating cost for Periods B, C, and D is

$$= \text{Rs. } [1,645.34 + 1,047.05 + 1,645.34]$$
$$= \text{Rs. } 4,337.73.$$

In Period D, Unit-3 is restarted to commit the load, hence the start-up cost of Unit-3 is added to the total operating cost for periods B, C, and D:

Start-up cost for Unit-3 = Rs. 500 (given)

$\therefore$ Total cost of operating of units during period B, C, and D is

$$= 4,337.73 + 500$$
$$= \text{Rs. } 4,837.73$$

Case B : Unit-3 is allowed to run in Period C.

Hence, 10-MW load is to be shared by units 1, 2, and 3.

$$\text{i.e., } P_{G_1} + P_{G_2} + P_{G_3} = 10 \tag{4.20}$$

Substituting P_{G_3} from Equation (4.20) in Equation (4.17), we get

$$0.74 P_{G_1} - 1.97(10 - P_{G_1} - P_{G_2}) = 6.1 \tag{4.21}$$

$$\text{or} \qquad 2.71 P_{G_1} + 1.97 P_{G_2} = 25.8 \tag{4.22}$$

By solving Equations (4.16) and (4.21), we get

$$P_{G_1} = 8.1 \text{ MW}, \quad P_{G_2} = 1.9 \text{ MW}, \quad \text{and} \quad P_{G_3} = 0 \text{ MW}$$

From the above powers, it is observed that P_{G_3} violates the minimum generation capacity (i.e., $0 < 1$).

Hence, set the generation capacity of Unit-3 at minimum capacity, i.e., $P_{G_3} = 1$ MW.

Then the remaining 9 MW is optimally shared by Unit-1 and Unit-2 as

$$P_{G_1} = 7.4 \text{ MW}, \quad P_{G_2} = 1.6 \text{ MW}, \quad \text{and} \quad P_{G_3} = 1 \text{ MW}$$

The operating cost at Period C

$$= \left[(0.37P_{G_1} + 22.9)P_{G_1} + (0.78P_{G_2} + 25.9)P_{G_2} + (0.985P_{G_3} + 29)P_{G_3} \right] \times 4 \text{ hr}$$

$$= \text{Rs. } 1,048.57$$

Total cost for Periods B, C, and D $= \text{Rs. } 1,645.34 + \text{Rs. } 1,048.57 + \text{Rs. } 1,645.34$

$$= \text{Rs. } 4,339.25.$$

Rs. $4,339.25 <$ Rs. $4,837.73$

i.e., $\left[\begin{array}{l}\text{The operating cost when}\\ \text{Unit-3 is committed in Period C}\\ \text{i.e., all the three units are continuously}\\ \text{run for Periods B, C, and D}\end{array}\right] < \left[\begin{array}{l}\text{The operating cost when Unit-3}\\ \text{is decommitted in Period C and}\\ \text{restarted in Period D}\end{array}\right]$

$\therefore$ It is concluded that to run Unit-3 in Period C is the economical way.

Now, the optimal UC table is modified as

Period	Unit committed			
	1	2	3	4
A	1	1	1	1
B	1	1	1	0
C	1	1	1**	0
D	1	1	1	0
E	1	1*	0	0
F	1	1	0	0

* Unit is committed from the point of security consideration.
** Unit is committed from the point of start-up considerations.

Hence, it is economical to allow all the three units to continue to run in Periods B, C, and D, i.e., in Period C continuation of Unit-3 is economical.

Example 4.3: A power system network with a thermal power plant is operating by four generating units. Determine the most economical units to be committed to a load demand of 10 MW. Also prepare the UC table for the load changes in steps of 1 MW starting from the minimum to the maximum load. The minimum and maximum generating capacities and cost-curve parameters of the units listed in a tabular form are as given in Table 4.10.

Solution:
We know:

The cost function, $F_i = \dfrac{1}{2}a_i P_{G_i}^{\,2} + b_i P_{G_i} + d_i$ (4.23)

Incremental fuel cost, $\dfrac{dC_i}{dP_{G_i}} = a_i P_{G_i} + b_i$ (4.24)

The total load $= P_D = 10$ MW (given)

By comparing the cost-curve parameters, we come to know that the cost characteristics of the first unit are the lowest. If only one single unit is to be committed, Unit-1 is to be employed.

TABLE 4.10   *Capacities and cost-curve parameters of the units*

Unit number	Capacity (MW)		Cost-curve parameters		
	Min	Max	a	b	d
1	1.0	15.0	0.68	22.8	823
2	1.0	15.0	1.53	25.9	120
3	1.0	15.0	1.98	29.0	480
4	1.0	15.0	2.23	30.0	500

Now, find the cost-of-generation of power by the first unit starting from the minimum to the maximum generating capacity of that unit.

Let

$f_1(1)$ = the main cost in Rs./hr for the generation of 1 MW by the first unit

$f_1(2)$ = the main cost in Rs./hr for the generation of 2 MW by the first unit

$f_1(3)$ = the main cost in Rs./hr for the generation of 3 MW by the first unit

$f_1(4)$ = the main cost in Rs./hr for the generation of 4 MW by the first unit

....

$f_1(10)$ = the main cost in Rs./hr for the generation of 10 MW by the first unit

$$f_i(P_{G_i}) = \frac{1}{2} a_i P_{G_i}^2 + b_i P_{G_i} + d_i$$

$$f_1(P_{G_1}) = \frac{1}{2}(0.68)P_{G_1}^2 + 22.8P_{G_1} + 823 = (0.34P_{G_1} + 22.8)P_{G_1} + 823$$

$$f_2(P_{G_2}) = \frac{1}{2}(1.53)P_{G_2}^2 + 25.9P_{G_2} + 120 = (0.765P_{G_2} + 25.9)P_{G_2} + 120$$

$$f_3(P_{G_3}) = \frac{1}{2}(1.98)P_{G_3}^2 + 29.0P_{G_3} + 480 = (0.99P_{G_3} + 29.0)P_{G_3} + 480$$

$$f_4(P_{G_4}) = \frac{1}{2}(2.23)P_{G_4}^2 + 30P_{G_4} + 500 = (0.68P_{G_4} + 31.2)P_{G_4} + 500$$

For the commitment of Unit-1 only

When only one unit is to be committed to meet a particular load demand, i.e., Unit-1, in this case, due to its low cost parameters, then $F_1(x) = f_1(x)$.

where

$F_1(x)$ is the minimum cost of generation of 'x' MW by only one unit

$f_1(x)$ is the minimum cost of generation of 'x' MW by Unit-1

$\therefore F_1(1) = f_1(1) = (0.34 \times 1 + 22.8)\,1 + 823 = 846.14$

$F_1(2) = f_1(2) = (0.34 \times 2 + 22.8)\,2 + 823 = 869.96$

$F_1(3) = f_1(3) = (0.34 \times 3 + 22.8)\,3 + 823 = 894.46$

Similarly,

$$F_1(4)=f_1(4)=916.64$$
$$F_1(5)=f_1(5)=945.50$$
$$F_1(6)=f_1(6)=972.04$$
$$F_1(7)=f_1(7)=996.26$$
$$F_1(8)=f_1(8)=1,027.16$$
$$F_1(9)=f_1(9)=1,055.74$$
$$F_1(10)=f_1(10)=1,085.00$$

When Unit-1 is to be committed to meet a load demand of 10 MW, the cost of generation becomes 1,085 Rs./hr.

For the second unit:

$f_2(1)=$ minimum cost in Rs./hr for the generation of 1 MW by the second unit only

$$= (0.765P_{G_2} + 25.9)P_{G_2} + 120$$

$$= (0.765 \times 1 + 25.9)1 + 120 = 146.665$$

Similarly,

$$f_2(2)=174.860$$
$$f_2(3)=204.585$$
$$f_2(4)=235.840$$
$$f_2(5)=268.625$$
$$f_2(6)=302.940$$
$$f_2(7)=338.785$$
$$f_2(8)=376.160$$
$$f_2(9)=415.065$$
$$f_2(10)=455.500$$

By observing $f_1(10)$ and $f_2(10)$, it is concluded that $f_1(10)<f_2(10)$, i.e., the cost of generation of 10 MW by Unit-1 is minimum than that by Unit-2.

For commitment of Unit-1 and Unit-2 combination

$F_2(10)=$ Minimum cost of generation of 10 MW by the simultaneous operation of two units, i.e., Units-1 and 2

$$= \min \begin{bmatrix} f_2(0)+F_1(10), & f_2(1)+F_1(9), & f_2(2)+F_1(8), \\ f_2(3)+F_1(7), & f_2(4)+F_1(6), & f_2(5)+F_1(5), \\ f_2(6)+F_1(4), & f_2(7)+F_1(8), & f_2(8)+F_1(2) \\ f_2(9)+F_1(1), & f_2(10)+F_1(0) \end{bmatrix}$$

$$= \min \begin{bmatrix} 1,085, & 1,202.405, & 1,202.02, \\ 1,203.845, & 1,207.88, & 1,214.125, \\ 1,219.58, & 1,233.245, & 1,261.205 \\ 1,246.12 & 455.5 \end{bmatrix}$$

$$\therefore F_2(10) = 455.5 \text{ Rs./hr}$$

In other words, the minimum cost of generation of 10 MW by the combination of Unit-1 and Unit-2 is 455.5 Rs./hr and for this optimal cost, Unit-1 supplies 0 MW and Unit-2 supplies 10 MW.

$$F_2(9) = \min \begin{bmatrix} f_2(0)+F_1(9), & f_2(1)+F_1(8), & f_2(2)+F_1(7), \\ f_2(3)+F_1(6), & f_2(4)+F_1(5), & f_2(5)+F_1(4), \\ f_2(6)+F_1(3), & f_2(7)+F_1(2), & f_2(8)+F_1(1), \\ f_2(9)+F_1(0) \end{bmatrix}$$

$$= \min \begin{bmatrix} 1{,}055.74, & 1{,}173.825, & 1{,}174.12 \\ 1{,}176.625, & 1{,}181.34, & 1{,}185.265, \\ 1{,}197.4 \,, & 1{,}208.745 & 1{,}222.3 \\ 415.065 \end{bmatrix}$$

$$\therefore F_2(9) = 415.065 \text{ Rs./hr}$$

i.e., the minimum cost of generation of 9 MW with the combination of Unit-1 (by 0-MW supply) and Unit-2 (by 9-MW supply) is 415.065 Rs./hr.

Similarly,
$$F_2(8) = \min \begin{bmatrix} 1{,}027.16, & 1{,}145.925, & 1{,}146.9 \\ 1{,}150.08, & 1{,}152.48, & 1{,}163.085, \\ 1{,}172.9 & 1{,}184.9 & 376.16 \end{bmatrix}$$

$$\therefore F_2(8) = 376.16 \text{ Rs./hr}$$

$$F_2(7) = \min \begin{bmatrix} 999.26, & 1{,}118.705, & 1{,}120.36 \\ 1{,}121.225, & 1{,}130.3 & 1{,}138.585 \\ 1{,}149.08 & 338.785 \end{bmatrix}$$

$$\therefore F_2(7) = 338.785 \text{ Rs./hr}$$

$$F_2(6) = \min \begin{bmatrix} 972.04, & 1{,}092.165, & 1{,}091.5, \\ 1{,}099.045, & 1{,}105.8 & 1{,}114.765 \\ 338.785 \end{bmatrix}$$

$$\therefore F_2(6) = 338.785 \text{ Rs./hr}$$

$$F_2(5) = \min \begin{bmatrix} 945.5, & 1{,}063.305 & 1{,}069.32, \\ 1{,}074.545, & 1{,}081.98 & 268.625 \end{bmatrix}$$

$$\therefore F_2(5) = 268.625 \text{ Rs./hr}$$

$$F_2(4) = \min \begin{bmatrix} 916.64, & 1{,}041.125, & 1{,}044.82 \\ 1{,}050.725, & 235.84 \end{bmatrix}$$

$$\therefore F_2(4) = 235.84 \text{ Rs./hr}$$

$$F_2(3) = \min [894.46, \quad 1{,}016.625 \quad 1{,}0221 \quad 204.585]$$

$$\therefore F_2(3) = 204.585 \text{ Rs./hr}$$

$$F_2(2) = \min [869.96, \quad 992.805 \quad 174.86]$$

$$\therefore F_2(2) = 174.86 \text{ Rs./hr}$$

$$F_2(1) = \min\,[846.14 \quad 146.665]$$

$$\therefore F_2(1) = 146.665 \text{ Rs./hr}$$

Now, the cost of generation by Unit-3 only:

$$f_3(P_{G_i}) = \frac{1}{2} a_i P_{G_i}^2 + b_i P_{G_i} + d_i$$

$$= (0.99 P_{G_3} + 29.0) P_{G_3} + 480$$

$$f_3(0) = 0; \qquad f_3(5) = 649.75; \quad f_3(9) = 821.19$$

$$f_3(1) = 509.99; \quad f_3(6) = 689.64; \quad f_3(10) = 869.00$$

$$f_3(2) = 541.96; \quad f_3(7) = 731.51$$

$$f_3(3) = 575.91; \quad f_3(8) = 775.36$$

$$f_3(4) = 611.84;$$

For commitment of Unit-1, Unit-2, and Unit-3 combination:

$F_3(10) =$ The minimum cost of generation of 10 MW by the three units i.e., Unit-1, Unit-2, and Unit-3

$$= \min \begin{bmatrix} f_3(0)+F_2(10), & f_3(1)+F_2(9), & f_3(2)+F_2(8), \\ f_3(3)+F_2(7), & f_3(4)+F_2(6), & f_3(5)+F_2(5), \\ f_3(6)+F_2(4), & f_3(7)+F_2(3), & f_3(8)+F_2(2) \\ f_3(9)+F_2(1), & f_3(10)+F_2(0) \end{bmatrix}$$

$$= \min \begin{bmatrix} 455.5, & 925.055, & 918.12 \\ 914.695, & 914.78, & 918.375 \\ 925.48, & 936.095, & 950.22 \\ 967.855, & 869 \end{bmatrix}$$

$$\therefore F_3(10) = 455.5 \text{ Rs./hr}$$

i.e., for the generation of 10 MW by three Units, Unit-2 alone will commit to meet the load of 10 MW and Units-1 and 3 are in an off-state condition:

$$F_3(9) = \min \begin{bmatrix} f_3(0)+F_2(9), & f_3(1)+F_2(8), & f_3(2)+F_2(7), \\ f_3(3)+F_2(6), & f_3(4)+F_2(5), & f_3(5)+F_2(4), \\ f_3(6)+F_2(3), & f_3(7)+F_2(2), & f_3(8)+F_2(1) \\ f_3(9)+F_2(0) \end{bmatrix}$$

$$= \min \begin{bmatrix} 415.065 & 886.15, & 880.745 \\ 878.85, & 880.465, & 885.59 \\ 894.225, & 906.37 & 922.025 \\ 821.19 \end{bmatrix}$$

$$\therefore F_3(9) = 415.065 \text{ Rs./hr}$$

$$F_3(8) = \min \begin{bmatrix} 376.16, & 848.775, & 844.9 \\ 844.535, & 847.68, & 854.335 \\ 864.5 & 878.175 & 775.36 \end{bmatrix}$$

$\therefore F_3(8) = 376.16$ Rs./hr

$$F_3(7) = \min \begin{bmatrix} 338.785, & 812.93, & 810.585, \\ 811.75, & 816.425, & 824.61 \\ 836.305, & 731.51 & \end{bmatrix}$$

$\therefore F_3(7) = 338.785$ Rs./hr

$$F_3(6) = \min \begin{bmatrix} 302.94, & 778.615, & 777.8, \\ 780.495, & 786.7 & 796.415 \\ 689.64 & & \end{bmatrix}$$

$\therefore F_3(6) = 302.94$ Rs./hr

$$F_3(5) = \min \begin{bmatrix} 268.625, & 745.83, & 746.545 \\ 750.77 & 758.505 & 649.75 \end{bmatrix}$$

$\therefore F_3(5) = 268.625$ Rs./hr

$$F_3(4) = \min \begin{bmatrix} 235.84, & 714.575, & 716.82 \\ 722.575 & 611.84 & \end{bmatrix}$$

$\therefore F_3(4) = 235.84$ Rs./hr

$F_3(3) = \min [204.585, \quad 684.85, \quad 688.625, \quad 575.91]$

$\therefore F_3(3) = 204.585$ Rs./hr

$F_3(2) = \min [174.86, \quad 565.655, \quad 541.96]$

$\therefore F_3(2) = 174.86$ Rs./hr

$F_3(1) = \min [509.99, \quad 146.665]$

$\therefore F_3(1) = 146.665$ Rs./hr

Cost of generation by the fourth unit

$$f_4(P_{G_i}) = \frac{1}{2} a_i P_{G_i}^2 + b_i P_{G_i} + d_i$$

$$= (1.115 P_{G_4} + 30) P_{G_4} + 500$$

$f_4(0) = 0$

$f_4(1) = 531.115$ Rs./hr

$f_4(2) = 564.46$ Rs./hr

$f_4(3) = 600.035$ Rs./hr

$f_4(4) = 637.84$ Rs./hr

$$f_4(5) = 677.875 \text{ Rs./hr}$$

$$f_4(6) = 720.14 \text{ Rs./hr}$$

$$f_4(7) = 764.635 \text{ Rs./hr}$$

$$f_4(8) = 811.36 \text{ Rs./hr}$$

$$f_4(9) = 860.315 \text{ Rs./hr}$$

$$f_4(10) = 911.5 \text{ Rs./hr}$$

Minimum cost of generation by four units, i.e., Unit-1, Unit-2, Unit-3, and Unit-4:

$F_4(10) = $ The minimum cost of generation of 10 MW by four units

$$= \min \begin{vmatrix} f_4(0)+F_3(10), & f_4(1)+F_3(9), & f_4(2)+F_3(8), \\ f_4(3)+F_3(7), & f_4(4)+F_3(6), & f_4(5)+F_3(5), \\ f_4(6)+F_3(4), & f_4(7)+F_3(3), & f_4(8)+F_3(2) \\ f_4(9)+F_3(1), & f_4(10)+F_3(0) \end{vmatrix}$$

$$= \min \begin{vmatrix} 455.5, & 946.18, & 940.62 \\ 938.82, & 940.78, & 946.5 \\ 955.98, & 969.22, & 986.22 \\ 1,006.98 & 911.5 \end{vmatrix}$$

$$\therefore F_4(10) = 455.5 \text{ Rs./hr}$$

i.e., for the generation of 10 MW by four units, Unit-2 will commit to meet the load of 10 MW, and Unit-1, Unit-3, and Unit 4 are in an off-state condition:

$$F_4(9) = \min \begin{vmatrix} f_4(0)+F_3(9), & f_4(1)+F_3(8), & f_4(2)+F_3(7), \\ f_4(3)+F_3(6), & f_4(4)+F_3(5), & f_4(5)+F_3(4), \\ f_4(6)+F_3(3), & f_4(7)+F_3(2), & f_4(8)+F_3(1), \\ f_4(9)+F_3(0) \end{vmatrix}$$

$$= \min \begin{vmatrix} 415.065, & 907.275, & 885.245 \\ 902.975, & 906.465, & 913.715 \\ 955.98, & 969.22, & 986.22 \\ 924.725, & 936.495, & 958.025 \\ 860.315 \end{vmatrix}$$

$$\therefore F_4(9) = 415.065 \text{ Rs./hr}$$

$$F_4(8) = \min \begin{vmatrix} 376.16, & 869.9, & 867.4, \\ 868.66, & 873.68, & 882.36, \\ 895, & 911.3, & 811.36 \end{vmatrix}$$

$$\therefore F_4(8) = 376.16 \text{ Rs./hr}$$

$$F_4(7) = \min \begin{vmatrix} 338.785, & 834.055, & 833.085, \\ 835.875, & 842.425, & 852.735, \\ 866.805, & 764.635 \end{vmatrix}$$

$$\therefore F_4(7) = 338.785 \text{ Rs./hr}$$

$$F_4(6) = \min \begin{bmatrix} 302.94, & 799.74, & 800.3 & 804.62, \\ 812.7, & 824.54, & 720.14 & \end{bmatrix}$$

$\therefore F_4(6) = 302.94$ Rs./hr

$$F_4(5) = \min \begin{bmatrix} 268.625, & 766.955, & 769.045 \\ 774.895, & 784.505, & 677.875 \end{bmatrix}$$

$\therefore F_4(3) = 72.03$ Rs./hr

$$F_4(4) = \min \begin{bmatrix} 235.84, & 735.7, & 739.32 \\ 746.7, & 637.84 & \end{bmatrix}$$

$\therefore F_4(4) = 235.84$ Rs./hr

$F_4(3) = \min [204.585, \quad 705.975, \quad 711.125, \quad 600.035]$

$\therefore F_4(3) = 204.585$ Rs./hr

$F_4(2) = \min [46.96, \quad 554.255, \quad 564.46]$

$\therefore F_4(2) = 46.96$ Rs./hr

$F_4(1) = \min [23.14, \quad 531.115]$

$\therefore F_4(1) = 23.14$ Rs./hr

From the above criteria, it is observed that for the generation of 10 MW, the commitment of units is as follows:

$f_1(10) = F_1(10) =$ the minimum cost of generation of 10 MW in Rs./hr by Unit-1 only
$\qquad = 1085$ Rs./hr

$F_2(10) =$ the minimum cost of generation of 10 MW by two units with Unit-1 supplying 0 MW and Unit-2 supplying 10 MW
$\qquad = 455.5$ Rs./hr

$F_3(10) =$ the minimum cost of generation of 10 MW by three units with Unit-2 supplying 10 MW, Unit-1 and Unit-3 is in an off-state condition
$\qquad = 455.5$ Rs./hr

$F_4(10) =$ the minimum cost of generation of 10 MW by four units with Unit-2 supplying 10 MW, and Unit-1, Unit-3 and Unit-4 are in an off-state condition
$\qquad = 455.5$ Rs./hr

By examining the costs $F_1(10)$, $F_2(10)$, $F_3(10)$, and $F_4(10)$, we have concluded that for meeting the load demand of 10 MW, the optimal combination of units to be committed is Unit-1, Unit-3, and Unit-4 in an off-state condition and Unit-2 supplying a 10-MW load at an operating cost of 455.5 Rs./hr.

For preparing the UC table, the ordering of units is not a criterion. For any order, we get the same solution that is independent of numbering units.

TABLE 4.11   *The UC table for the above-considered system*

Load range	Unit			
	1	2	3	4
1–10	0	1	0	0

To get a higher accuracy, the step size of the load is to be reduced, which results in considerable increase in time of computation and required storage capacity.

Status 1 of any unit indicates unit running or unit committing and Status 0 of any unit indicates unit not running.

The UC table is prepared once and for all for a given set of units (Table 4.11). As the load cycle on the station changes, it would only mean changes in starting and stopping of units without changing the basic UC table.

KEY NOTES

- Unit commitment is a problem of determining the units that should operate for a particular load.

- To 'commit' a generating unit is to 'turn it on'.

- The constraints considered for unit commitment are:

  (i) Spinning reserve.

  (ii) Thermal unit constraints.

  (iii) Hydro-constraints.

  (iv) Must-run constraints.

  (v) Fuel constraints.

- The solution methods to a UC problem are:

  (i) Priority-list scheme.

  (ii) Dynamic programming method (DP).

  (iii) Lagrange's relaxation method (LR).

- In the priority ordering method, the most efficient unit is loaded first to be followed by the less efficient units in order as the load increases.

- The main advantage of the DP method is resolution in the dimensionality of problems, i.e., having obtained the optimal way of loading K number of units, it is quite easy to determine the optimal way of loading $(K+1)$ number of units.

MULTIPLE-CHOICE QUESTIONS

(1) Due to the load variation, it is not advisable to:

   (a) Run all available units at all the times.

   (b) Run only one unit at each discrete load level.

   (c) Both (a) and (b).

   (d) None of these.

(2) A unit when scheduled for connection to the system is said to be:

   (a) Loaded.

   (b) Disconnected.

   (c) Committed.

   (d) None of these.

(3) To determine the units that should operate for a particular load is the problem of:

   (a) Unit commitment.

   (b) Optimal load scheduling.

   (c) Either (a) or (b).

   (d) None of these.

(4) To commit a generating unit is:

   (a) To bring it upto speed.

   (b) To synchronize it to the system.

   (c) To connect it so that it can deliver power to the network.

   (d) All of these.

(5) Economic dispatch problem is applicable to various units, Which of the following is suitable?

   (a) The units are already on-line.

   (b) To supply the predicted or forecast load of the system over a future time period.

   (c) Both (a) and (b).

   (d) None of these.

(6) Unit commitment problem plans for the best set of units to be available. Which of the following is suitable?

    (a) The units are already on-line.

    (b) To supply the predicted or forecast load of the system over a future time period.

    (c) Both (a) and (b).

    (d) None of these.

(7) Spinning reserve is defined as:

    (a) $\begin{bmatrix} \text{Total generation output} \\ \text{of all synchronized units} \\ \text{at a particular time} \end{bmatrix} - \begin{bmatrix} \text{load at} + \text{losses at} \\ \text{that time} \quad \text{that time} \end{bmatrix}.$

    (b) $\begin{bmatrix} \text{Total generation output} \\ \text{of all synchronized units} \\ \text{at a particular time} \end{bmatrix} + \begin{bmatrix} \text{load at} + \text{losses at} \\ \text{that time} \quad \text{that time} \end{bmatrix}.$

    (c) $\begin{bmatrix} \text{Total generation output} \\ \text{of all synchronized units} \\ \text{at a particular time} \end{bmatrix} - \begin{bmatrix} \text{load at} - \text{losses at} \\ \text{that time} \quad \text{that time} \end{bmatrix}.$

    (d) None of these.

(8) Spinning reserve must be:

    (a) Maintained so that the failure of one or more units does not cause too far a drop in system frequency.

    (b) Capable of taking up the loss of most heavily loaded unit in a given period of time.

    (c) Calculated as a function of the probability of not having sufficient generation to meet the load.

    (d) All of these.

(9) Because of temperature and pressure of thermal unit that must be moved slowly, a certain amount of energy must be expended to bring the unit on-line and is brought into the UC problem as a:

    (a) Running cost.

    (b) Fixed cost.

    (c) Fuel cost.

    (d) Start-up cost.

(10) Unit commitment problem is of much importance for:

    (a) Scheduling of thermal units.

    (b) Scheduling of hydro-units.

    (c) Scheduling of both thermal and hydro-units.

    (d) None of these.

(11) Thermal unit constraints considered in a UC problem are:

    (a) Minimum up and minimum down times.

    (b) Crew constraints.

    (c) Start-up costs.

    (d) All of these.

(12) The start-up cost may vary from a maximum cold-start value to a very small value if the thermal unit:

    (a) Was only turned off recently.

    (b) Is still relatively close to the operating temperature.

    (c) Is still operating at normal temperature.

    (d) Both (a) and (b).

(13) Unit commitment problem is:

    (a) Of much importance for scheduling of thermal units.

    (b) Cannot be completely separated from the scheduling of hydro-units.

    (c) Used for hydro-thermal scheduling.

    (d) Both (a) and (b).

(14) The constraints considered in a UC problem are:

    (a) Thermal unit and hydro-unit constraints.

    (b) Spinning reserve.

    (c) Must-run and fuel constraints.

    (d) All the above.

(15) The method used for obtaining the solution to a UC problem is:

    (a) Priority-list scheme.

    (b) Dynamic programming method.

    (c) Lagrange's relaxation method.

    (d) All the above.

(16) A straightforward but highly time-consuming way of finding the most economical combination of units to meet a particular load demand is:

    (a) Enumeration scheme.

    (b) Priority-list scheme.

(c) DP method.

(d) All of these.

(17) Which is correct regarding the shut-down rule?

(a) To know which units to drop and when.

(b) From which a simple priority-list scheme is developed.

(c) Both (a) and (b).

(d) To know which units to start from shut-down condition.

(18) In the priority-list method of solving an optimal UC problem:

(a) Most efficient unit is loaded first to be followed by the less efficient unit in order as load increases.

(b) Less efficient unit is loaded first to be followed by the most efficient unit in order as load increases.

(c) Most efficient unit is loaded first to be followed by the less efficient unit in order as load decreases.

(d) Either (a) or (b).

(19) In the priority-list method, the units are arranged to commit the load demand in the order of:

(a) Ascending costs of units.

(b) Descending costs of units.

(c) Either (a) or (b).

(d) Independent of costs of units.

(20) The chief advantage of the DP method over the enumerate scheme is:

(a) Reduction in time of computation.

(b) Reduction in the dimensionality of the problem.

(c) Reduction in the number of units.

(d) All of these.

(21) In the DP method, the cost function $F_N(x)$ represents:

(a) Minimum cost in Rs/hr of generation of N MW by x number of units.

(b) Minimum cost in Rs/hr of generation of x MW by N number of units.

(c) Minimum cost in Rs/hr of generation of N MW by the xth unit.

(d) Minimum cost in Rs/hr of generation of x MW by the Nth unit.

(22) In the DP method, the cost function $F_N(y)$ represents:

(a) Cost of generation of N MW by y number of units.

(b) Cost of generation of y MW by N number of units.

(c) Cost of generation of N MW by the yth unit.

(d) Cost of generation of y MW by the Nth unit.

(23) The recursive relation results with the application of the DP method of solving the UC problem is:

(a) $f_N(x) = \overset{min}{y} \left\{ F_N(y) + F_{N-1}(x-y) \right\}$.

(b) $F_N(x) = \overset{min}{y} \left\{ f_N(y) + F_{N-1}(x-y) \right\}$.

(c) $F_N(x) = \overset{min}{y} \left\{ f_N(x) + f_{N-1}(x-y) \right\}$.

(d) $F_N(x) = \overset{min}{y} \left\{ F_N(x) + f_{N+1}(x-y) \right\}$.

(24) For preparing the UC table, which of the following is not a criterion?

(a) Ordering of units.

(b) Ordering of costs of units.

(c) Ordering of range of load.

(d) All of these.

(25) In a UC table, unit running or unit committing is indicated by:

(a) Status 0.

(b) Status 1.

(c) Status +.

(d) Status >.

(26) In a UC table, the status of the unit not running is indicated by:

(a) Status 0.

(b) Status 0.

(c) Status +.

(d) Status −.

(27) The unscheduled or maintenance outages of various equipments of a thermal plant must be taken into account in:

(a) Optimal scheduling problem.

(b) UC problem.

(c) Load frequency controlling problem.

(d) All of these.

(28) Unit up-time is nothing but:

   (a) A unit operating time.

   (b) A unit repair time.

   (c) A unit total lifetime.

   (d) A unit designing time.

(29) Unit down-time is nothing but:

   (a) A unit operating time.

   (b) A unit repair time.

   (c) A unit total lifetime.

   (d) A unit designing time.

(30) In reliability aspects of a UC problem, the lengths of an individual operating and repair periods of a unit considered at a random phenomenon with:

   (a) Much longer periods of operation compared to repair periods.

   (b) Much longer periods of repair compared to operation periods.

   (c) Equal periods of operation and repair.

   (d) Either (a) or (b).

(31) Mean up-time of a unit ($\overline{T}_{up}$) is:

   (a) Mean time to failure.

   (b) Mean time to repair.

   (c) Mean of failure and repair times.

   (d) Mean of total time.

(32) Mean down-time of a unit ($\overline{T}_{down}$) is:

   (a) Mean time to failure.

   (b) Mean time to repair.

   (c) Mean of failure and repair times.

   (d) Mean of total time.

(33) Mean cycle time of a unit is:

   (a) $\overline{T}_{up} / \overline{T}_{down}$.

   (b) $\overline{T}_{up} - \overline{T}_{down}$.

   (c) $\overline{T}_{up} + \overline{T}_{down}$.

   (d) $\overline{T}_{up} \times \overline{T}_{down}$.

(34) Rate of failure of a unit is expressed as:

   (a) $\lambda = \dfrac{1}{\overline{T}_{up}}$.

   (b) $\lambda = \dfrac{1}{\overline{T}_{down}}$.

   (c) $\mu = \dfrac{1}{\overline{T}_{down}}$.

   (d) $\mu = \dfrac{1}{\overline{T}_{up}}$.

(35) Rate of repair of a unit is expressed as:

   (a) $\lambda = \dfrac{1}{\overline{T}_{up}}$.

   (b) $\lambda = \dfrac{1}{\overline{T}_{down}}$.

   (c) $\mu = \dfrac{1}{\overline{T}_{down}}$.

   (d) $\mu = \dfrac{1}{\overline{T}_{up}}$.

(36) The rate of failure of a unit affected by:

   (a) Relative maintenance.

   (b) Size, composition of repair team.

   (c) Skill of repair team.

   (d) All of these.

(37) The rate of repair of a unit is affected by:

   (a) Relative maintenance.

   (b) Size, composition of repair team.

   (c) Skill of repair team.

   (d) Both (b) and (c).

(38) P_{up} and P_{down} of any unit represent:

   (a) Unavailability and availability of a unit.

   (b) Availability and unavailability of a unit.

   (c) Either (a) or (b).

   (d) Both (a) and (b).

(39) $P_{up} + P_{down} =$

   (a) Zero.

   (b) 1.

   (c) -1.

   (d) Infinite.

(40) A breach of system security considered in optimal UC problem is:

   (a) Sufficient generating capacity of the system at a particular instant of time.

   (b) Insufficient generating capacity of the system at a particular instant of time.

   (c) Insufficient generating capacity of the system at all times.

   (d) Either (a) or (b).

(41) Use of Patton's security function in the UC problem is the estimation of the probability that the available generating capacity at a particular time is:

   (a) Less than the total load demand.

   (b) More than the total load demand.

   (c) Equal to the total load demand.

   (d) Independent of the total load demand.

(42) Patton's security function S gives a quantitative estimation of:

   (a) System security.

   (b) System insecurity.

   (c) System stability.

   (d) System variables.

(43) It is necessary to modify the UC table to include security aspects by committing the next most economical unit to supply the load when:

   (a) $S < $ MTIL.

   (b) $S > $ MTIL.

   (c) $S = $ MTIL.

   (d) $S = $ MTIL/2.

(44) The procedure of committing a most economical unit, to include security aspect in the UC table, is continued upto:

   (a) $S < $ MTIL.

   (b) $S > $ MTIL.

   (c) $S = $ MTIL.

   (d) $S = $ MTIL/2.

(45) System insecurity is represented by:

   (a) $S < $ MTIL.

   (b) $S > $ MTIL.

   (c) $S = $ MTIL.

   (d) $S = $ MITL/2.

(46) If the unit is to be started from a cold condition and brought upto normal temperature and pressure, the start-up cost will be:

   (a) Minimum.

   (b) Maximum.

   (c) Having no effect.

   (d) None of these.

(47) When any unit is in the UP state, there is:

   (a) Breach of security.

   (b) No breach of security.

   (c) Stability.

   (d) All of these.

(48) If $\sum P_{G_i} > P_D$, the probability that the system state 'i' causes a breach of system security becomes:

   (a) $r_i = 1$.

   (b) $r_i = 0$.

   (c) $r_i = -1$.

   (d) $r_i = \infty$.

(49) If $\sum P_{G_i} < P_D$, the probability that the system state 'i' causes a breach of system security becomes:

   (a) $r_i = 1$.

   (b) $r_i = 0$.

   (c) $r_i = -1$.

   (d) $r_i = \infty$.

SHORT QUESTIONS AND ANSWERS

(1) What is a UC problem?

It is not advisable to run all available units at all times due to the variation of load. It is necessary to decide in advance:

   (i) Which generators to start up.

  (ii) When to connect them to the network.

 (iii) The sequence in which the operating units should be shut down and for how long.

The computational procedure for making the above such decisions is called the problem of UC.

(2) What do you mean by commitment of a unit?

To commit a generating unit is to turn it ON, i.e., to bring it upto speed, synchronize it to the system, and connect it, so that it can deliver power to the network.

(3) Why is the UC problem important for scheduling thermal units?

As for other types of generation such as hydro, the aggregate costs such as start-up costs, operating fuel costs, and shut-down costs are negligible so that their ON–OFF status is not important.

(4) Compare the UC problem with economic load dispatch.

Economic load dispatch economically distributes the actual system load as it rises to the various units already on-line. But the UC problem plans for the best set of units to be available to supply the predicted or forecast load of the system over future time periods.

(5) What are the different constraints that can be placed on the UC problem?

- Spinning reserve.
- Thermal unit constraints.
- Hydro-constraints.
- Must-run constraints.
- Fuel constraints.

(6) What are the thermal unit constraints considered in the UC problem?

The thermal unit constraints considered in the UC problem are:

- Minimum up-time.
- Minimum down-time.
- Crew constraints.
- Start-up cost.

(7) Why must the spinning reserve be maintained?

Spinning reserve must be maintained so that failure of one or more units does not cause too far a drop in system frequency, i.e., if one unit fails, there must be ample reserve on the other units to make up for the loss in a specified time period.

(8) Why are thermal unit constraints considered in a UC table?

A thermal unit can undergo only gradual temperature changes and this translates into a time period of some hours required to bring the unit on the line. Due to such limitations in the operation of a thermal plant, the thermal unit constraints are to be considered in the UC problem.

(9) What is a start-up cost and what is its significance?

Because of temperature and pressure of a thermal unit that must be moved slowly, a certain amount of energy must be moved slowly, a certain amount of energy must be expended to bring the unit on-line, and it is brought into the UC problem as a start-up cost.

The start-up cost may vary from a maximum cold-start value to a very small value if the unit was only turned off recently and is still relatively close to the operating temperature.

(10) Write the expressions of a start-up cost when cooling and when banking.

Start-up cost when cooling $= C_c \, (1-e^{-t/\alpha}) \, C + C_F$

Start-up cost when banking $= C_t \times t \times C + C_F$

where C_c is the cold-start cost (MBtu), C is the fuel cost, C_F is the fixed cost (includes crew expenses and maintainable expenses), α is the thermal time constant for the unit, C_t is the cost of maintaining a unit at operating temperature (MBtu/hr), and t is the time the unit was cooled (hr).

(11) What are the techniques used for getting the solution to the UC problem?

- Priority-list scheme.
- Dynamic programming (DP) method.
- Lagrange's relaxation (LR) method.

(12) What are the steps of an enumeration scheme of finding the most economical combination of units to meet a load demand?

- To try all possible combinations of units that can supply the load.
- To divide this load optimally among the units of each combination by the use of co-ordination equations, so as to find the most economical operating cost of the combination.
- Then to determine the combination that has the least operating cost among all these.

(13) What is a shut-down rule of the UC operation?

If the operation of the system is to be optimized, units must be shut down as the load goes down and then recommitted as it goes back up. To know which units to drop and when, one approach called the shut-down rule must be used from which a simple priority-list scheme is developed.

(14) What is a priority-list method of solving a UC problem?

In this method, first the full-load average production cost of each unit, which is simply the net heat rate at full load multiplied by the fuel cost, is computed. Then, in the order of ascending costs, the units are arranged to commit the load demand.

(15) In a priority-list method, which unit is loaded first and to be followed by which units?

The most efficient unit is loaded first, to be followed by the less efficient units in the order as load increases.

(16) What is the chief advantage of the DP method over other methods in solving the UC problem?

Resolution in the dimensionality of problems, i.e., having obtained the optimal way of loading K number of units, it is quite easy to determine the optimal way of loading $(K+1)$ number of units.

(17) What is the thermal constraint minimum up-time?

Minimum up-time is the time during which if the unit is running, it should not be turned off immediately.

(18) What is minimum down-time?

If the unit is stopped, there is a certain minimum time required to start it and put it on the line.

(19) What is spinning reserve?

To ensure the continuity of supply to meet random failures, the total generating capacity on-line must have a definite margin over the load requirements at any point of time. This margin is called spinning reserve, which ensures continuation by meeting the demand upto a certain extent of probable loss of generating capacity.

(20) What do you mean by a breach of system security?

Some intolerable or undesirable conditions of system operation is termed as a breach of system security.

(21) In an optimal UC problem, what is considered as a breach of security?

Insufficient generating capacity of the system at a particular instant of time.

(22) What is Patton's security function? Give its expression.

Patton's security function estimates the probability that the available generating capacity at a particular time is less than the total load demand on the system at that time.

It is expressed as
$$S = \sum_i P_i r_i$$

where P_i is the probability of the system being in the ith state and r_i the probability that the system state 'i' causes a breach of system security.

(23) How the optimal UC table is modified with consideration of security constraints?

Whenever the security function exceeds MTIL ($S >$ MTIL), the UC table is modified by committing the next most economic unit to supply the loads. With the new unit being committed, the security function is then estimated and checked whether it is $S <$ MTIL or not.

(24) What is the significance of must-run constraints considered in preparing the UC table?

Some units are given a must-run recognization during certain times of the year for the reason of voltage support on the transmission network or for such purposes as supply of steam for uses outside the steam plant itself.

REVIEW QUESTIONS

(1) Using the DP method, how do you find the most economical combination of the units to meet a particular load demand?

(2) Explain the different constraints considered in solving a UC problem.

(3) Compare an optimal UC problem with an economical load dispatch problem.

(4) Explain the need of an optimal UC problem.

(5) Describe the reliability consideration in an optimal UC problem.

(6) Describe the start-up cost consideration in an optimal UC problem.

PROBLEMS

(1) A power system network with a thermal power plant is operating by four generating units. Determine the most economical unit to be committed to a load demand of 10 MW. Prepare the unit commitment table for the load changes in steps of 1 MW starting from minimum load to maximum load. The minimum and maximum generating capacities and cost-curve parameters of units listed in a tabular form are given in the following table.

Unit number	Capacity (MW)		Cost-curve parameters		
	Min.	Max.	a	b	d
1	1.0	15.0	0.68	22.8	0
2	1.0	15.0	1.53	25.9	0
3	1.0	15.0	1.98	29.0	0
4	1.0	15.0	2.23	30.0	0

(2) Prepare the unit commitment table with the application of DP approach for the system having four thermal generating units, which have the following characteristic parameters. Also obtain the most economical station operating cost for the complete range of station capacity.

Generating unit parameters

Unit number	Capacity (MW)		Cost-curve parameters	
	Min.	Max.	a (Rs./ MW²)	b (Rs./ MW)
1	1.0	16.0	0.98	28.8
2	1.0	16.0	1.83	30.0
3	1.0	16.0	2.24	32.8
4	1.0	16.0	2.75	38.0

(3) For the power plant of Problem 2, and for the daily load cycle given in the figure, prepare the reliability constrained optimal unit commitment table. Also include the start-up consideration from the point of view of overall economy with the start-up cost of any unit being Rs. 75.

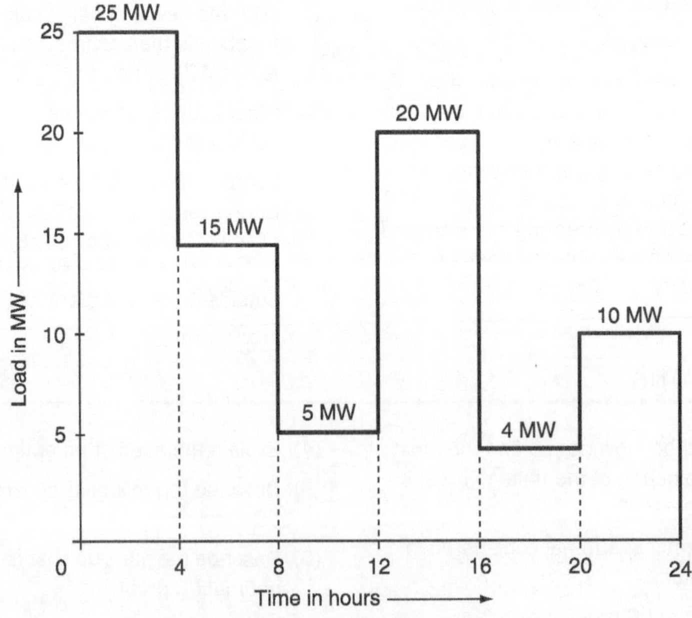

Daily load curve

5 Optimal Power-Flow Problem—Solution Technique

OBJECTIVES

After reading this chapter, you should be able to:

- know the optimal power flow problem concept
- study the major steps for optimal power flow solution techniques
- formulate the mathematical modeling for optimal power flow problem with and without inequality constraints
- develop algorithms for optimal power flow problems with and without inequality constraints

5.1 INTRODUCTION

The problem of optimizing the performance of a power system network is formulated as a general optimization problem. It is required to state from which aspect the performance of the power system network is optimized.

In optimization problem, the objective function becomes 'to minimize the overall cost of generation in economic scheduling and unit commitment problem':

- It is based on allocating the total load on a station among various units in an optimal way with cases being taken into consideration in a load-scheduling problem.

- It is based on allocating the total load on the system among the various generating stations.

The optimal power flow problem:

- refers to the load flow that gives maximum system security by minimizing the overloads,

- aims at minimum operating cost and minimum losses,

- should be based on operational constraints, and

- is a static optimization problem with the cost function as a scalar objective function.

The solution technique for an optimal power flow problem was first proposed by Dommel and Tinney and has following three major steps:

(i) It is based on the load flow solution by the Newton–Raphson (N–R) method.

(ii) A first-order gradient method consists of an algorithm that adjusts the gradient for minimizing the objective function.

(iii) Use of penalty functions to account for inequality constraints on dependent variables.

The optimization problem of minimizing the instantaneous operating costs in terms of real and reactive-power flows is studied in this unit.

The optimal power flow problem without considering constraints, i.e., unconstrained optimal power flow problem, is first studied and then the optimal power flow problem with inequality constraints is studied. The inequality constraints are introduced first on control variables and then on dependent variables.

5.2 OPTIMAL POWER-FLOW PROBLEM WITHOUT INEQUALITY CONSTRAINTS

The primary objective of the optimal power flow solution is to minimize the overall cost of generation. This is represented by an objective function (or) cost function as:

$$\min\ C = \sum_{i=1}^{n} C_i\left(P_{G_i}\right) \tag{5.1}$$

subject to power (load) flow constraints:

$$P_i - \sum_{j=1}^{m} V_i\, Y_{ij} V_j\, \cos\left(\theta_{ij} + \delta_j - \delta_i\right) = 0 \tag{5.2}$$

$$Q_i + \sum_{j=1}^{m} V_i\, Y_{ij} V_j\, \sin\left(\theta_{ij} + \delta_j - \delta_i\right) = 0 \quad \text{for each } P\text{--}Q \text{ bus} \tag{5.3}$$

$$\text{and}\quad P_i - \sum_{j=1}^{m} V_i\, Y_{ij} V_j\, \cos\left(\theta_{ij} + \delta_j - \delta_i\right) = 0 \quad \text{for each } P\text{--}V \text{ bus} \tag{5.4}$$

where $V_i = |V_i| \angle \delta_i$ is the voltage at bus 'i', $V_j = |V_j| \angle \delta_j$ the voltage at bus 'j', $Y_{ij} = |Y_{ij}| \angle \theta_{ij}$ the mutual admittance between the ith and j^{th} buses, P_i the specified real-power at bus i, Q_i the specified reactive power at bus i, the net real power injected into the system at the ith bus, $P_i = P_{G_i} - P_{D_i}$, and the net reactive power injected into the system at the ith bus, $Q_i = Q_{G_i} - Q_{D_i}$.

For load buses (or) P–Q buses, P and Q are specified and hence Equations (5.2) and (5.3) form the equality constraints.

For P–V buses, P and $|V|$ are specified as the function of some vectors and are represented as:

$$f(x, y) = 0$$

where x is a vector of dependent variables and is represented as

$$x = \begin{bmatrix} \left.\begin{matrix} |V_i| \\ \delta_i \end{matrix}\right\} & \text{for all } P\text{--}Q \ \text{ buses (or) load buses} \\ \delta_i & \text{for all } P\text{--}V \ \text{ buses} \end{bmatrix} \tag{5.5}$$

and y is a vector of independent variables and is represented as

$$y = \begin{cases} \begin{bmatrix} P_i \\ Q_i \end{bmatrix} & \text{for } P{-}Q \text{ buses} \\ \begin{bmatrix} P_i \\ |V_i| \end{bmatrix} & \text{for } P{-}V \text{ buses} \\ \begin{bmatrix} |V_s| \\ |\delta_s| \end{bmatrix} & \text{for slack buses} \end{cases} \tag{5.6}$$

Out of these independent variables (Equation (5.6)), certain variables are chosen as control variables, which are to be varied to yield an optimal value of the objective function. The remaining independent variables are called *fixed or disturbance or uncontrollable parameters*.

Let 'u' be the vector of control variables and 'p' the vector of fixed or disturbance variables.

Hence, the vector of independent variables can be represented as the combination of vector of control variables 'u' and vector of fixed or disturbance or uncontrollable variables 'p' and is expressed as

$$y = \begin{cases} u \\ p \end{cases} \tag{5.7}$$

The choice of 'u' and 'p' depends on what aspect of power system is to be optimized. The control parameters may be:

(i)   voltage magnitude at $P{-}V$ buses,

(ii) P_{G_i} at generator buses with controllable power,

(iii)   slack bus voltage and regulating transformer tap setting as additional control variables, and

(iv)   in the case of buses with reactive-power control, Q_{G_i} is taken as a control variable.

Now, the optimal power flow problem can be stated as

$$\min C = C(x, u) \tag{5.8}$$

subject to equality constraints:

$$f(x, u, p) = 0 \tag{5.9}$$

Define the corresponding Lagrangian function $\mathcal{L}$ by augmenting the equality constraint to the objective function through a Lagrangian multiplier λ as

$$\min \cdot \mathcal{L} = C(x, u) + \lambda^{\mathrm{T}}[f(x, u, p)] \tag{5.10}$$

where λ is a vector of the Lagrangian multiplier of suitable dimension and is the same as that of equality constraint, $f(x, u, p)$.

The necessary conditions for an optimal solution are as follows:

$$\frac{\partial \mathcal{L}}{\partial x} = \frac{\partial C}{\partial x} + \left[\frac{\partial f}{\partial x}\right]^{T} \lambda = 0 \tag{5.11}$$

$$\frac{\partial \mathcal{L}}{\partial u} = \frac{\partial C}{\partial u} + \left[\frac{\partial f}{\partial u}\right]^{T} \lambda = 0 \tag{5.12}$$

$$\frac{\partial \mathcal{L}}{\partial \lambda} = f(x, u, p) = 0 \tag{5.13}$$

Equation (5.13) is the same as equality constraints, and the expressions for $\dfrac{\partial C}{\partial x}$ and $\dfrac{\partial f}{\partial u}$ are not very involved.

Consider the general load flow problem of the N–R method by considering a set of 'n' non-linear algebraic equations,

$$f_i(x_1, x_2 \ldots x_n) = 0 \quad \text{for } i = 1, 2 \ldots n \tag{5.14}$$

Let $\quad x_1^0, x_2^0 \ldots x_n^0$ be the initial values.

and let $\Delta x_1^0, \Delta x_2^0 \ldots \Delta x_n^0$ be the corrections, which on being added to the initial guess, give an actual solution:

$$\therefore f_i(x_1^0 + \Delta x_1^0, \, x_2^0 + \Delta x_2^0, \ldots, x_n^0 + \Delta x_n^0) = 0 \quad \text{for } i = 1, 2 \ldots n \tag{5.15}$$

By expanding these equations according to Taylor's series around the initial guess, we get

$$f_i(x_1^0, x_2^0 \ldots x_n^0) + \left[\left(\frac{\partial f_i}{\partial x_1}\right)^0 \Delta x_1^0 + \left(\frac{\partial f_i}{\partial x_2}\right)^0 \Delta x_2^0 + \ldots \left(\frac{\partial f_i}{\partial x_2}\right)^0 \Delta x_n^0\right]$$

$$+ \text{ higher order terms} = 0 \tag{5.16}$$

where $\left[\dfrac{\partial f_i}{\partial x_1}\right]^0, \left[\dfrac{\partial f_i}{\partial x_2}\right]^0, \ldots, \left[\dfrac{\partial f_i}{\partial x_n}\right]^0$ are the derivatives of f_i with respect to $x_1, x_2 \ldots x_n$ evaluated at $(x_1^0, x_2^0 \ldots x_n^0)$.

Neglecting higher order terms, we can write Equation (5.16) in a matrix form as

$$\begin{bmatrix} f_1^0 \\ f_2^0 \\ \cdot \\ \cdot \\ \cdot \\ f_n^0 \end{bmatrix} + \begin{bmatrix} \left(\dfrac{\partial f_i}{\partial x_1}\right)^0 & \left(\dfrac{\partial f_1}{\partial x_2}\right)^0 & \cdots & \left(\dfrac{\partial f_1}{\partial \chi_n}\right)^0 \\ \left(\dfrac{\partial f_2}{\partial x_1}\right)^0 & \left(\dfrac{\partial f_2}{\partial x_2}\right)^0 & \cdots & \left(\dfrac{\partial f_2}{\partial x_n}\right)^0 \\ & & & \\ \left(\dfrac{\partial f_n}{\partial x_1}\right)^0 & \left(\dfrac{\partial f_n}{\partial x_2}\right)^0 & \cdots & \left(\dfrac{\partial f_n}{\partial x_n}\right)^0 \end{bmatrix} \begin{bmatrix} \Delta x_1^0 \\ \Delta x_2^0 \\ \cdot \\ \cdot \\ \cdot \\ \Delta x_n^0 \end{bmatrix} \cong \begin{bmatrix} 0 \\ 0 \\ \cdot \\ \cdot \\ \cdot \\ 0 \end{bmatrix} \tag{5.17}$$

or in a vector matrix form as

$$f^0 + J^0 \Delta x^0 \cong 0 \tag{5.18}$$

where J^0 is known as the Jacobian matrix and is obtained by differentiating the function vector 'f' with respect to 'x' and evaluating it at x^0.

By comparing Equations (5.11), (5.12), and (5.13) with Equations (5.17) and (5.18), it is observed that $\left[\dfrac{\partial f}{\partial x}\right]$ is the Jacobian matrix and the partial derivatives of equality constraints with respect to dependent variables are obtained as Jacobian elements in a load flow solution.

The equality constraints are basically the power flow equations, i.e., real and reactive-power flow equations:

$$\left.\begin{array}{l} P_i \\ Q_i \end{array}\right\} \text{equality constraints for } P\text{–}Q \text{ bus}$$

P_i is the equality constraint for P–V bus

'x' is the dependent variable like $|V_i|$, δ_i

Then, $\dfrac{\partial f}{\partial x}$ may be expressed as partial derivatives of $\dfrac{\partial f}{\partial |V_m|}, \dfrac{\partial P_i}{\partial \delta_m}, \dfrac{\partial Q_i}{\partial |V_m|}$, and $\dfrac{\partial Q_i}{\partial \delta_m}$.

The Equations (5.11), (5.12), and (5.13) are non-linear algebraic equations and can be solved iteratively by employing a simple technique that is a 'gradient method' and is also called the *steepest descent method*.

The basic technique employed in the steepest descent method is to adjust the control parameters 'u' so as to move from one feasible solution point to a new feasible solution point in the direction of the steepest descent (or negative gradient). Here, the starting point of feasible solution is one where a set of values 'x' (i.e., dependent variables) satisfies Equation (5.13) for given 'u' and 'p'. The new feasible solution point refers to a location where the lower objective function is achieved.

These moves are to be repeated in the direction of negative gradient till minimum value is reached. Hence, this method of obtaining a solution to non-linear algebraic is also called the *negative gradient method*.

5.2.1 Algorithm for computational procedure

The algorithm for obtaining an optimal solution by the steepest descent method is given below:

Step 1: Make an initial guess for control variables (u^0).

Step 2: Find the feasible load flow solution by the N–R method. The N–R method is an iterative method and the solution does not satisfy the constraint equation (5.13). Hence, to satisfy Equation (5.12), 'x' is improved as follows:

$$x^{r+1} = x^r + \Delta x$$

Δx is obtained by solving the set of linear equations of the Jacobian matrix of Equation (5.18) as given below:

$$f(x^r + \Delta x, y) = f(x^r, y) + \frac{\partial f}{\partial x}(x^r, y)\,\Delta x = 0$$

$$\Rightarrow \frac{\partial f}{\partial x}(x^{r},\, y)\,\Delta x = -f(x^{r},\, y)$$

$$\Rightarrow \Delta x = -\left[\frac{\partial f}{\partial x}(x^{r}, y)\right]^{-1} f(x^{r},\, y)$$

$$\Delta x \Rightarrow -\left[J^{r}\right]^{-1} f(x^{r},\, y)$$

The final results of Step-2 provide a feasible solution of 'x' and the Jacobian matrix.

Step 3: Solve Equation (5.11) for λ and it is obtained as

$$\lambda = -\left\{\left[\frac{\partial f}{\partial x}\right]^{\mathrm{T}}\right\}^{-1} \frac{\partial C}{\partial x} \tag{5.19}$$

Step 4: Substitute λ from Equation (5.19) into Equation (5.12) and calculate the gradient:

$$\frac{\partial}{\partial u}\mathcal{L} = \nabla\mathcal{L} = \frac{\partial C}{\partial u} + \left[\frac{\partial f}{\partial u}\right]^{\mathrm{T}}\lambda \tag{5.20}$$

For computing the gradient, the Jacobian matrix, $J = \dfrac{\partial f}{\partial x}$, is already known from **Step 2.**

Step 5: If the gradient $\nabla\mathcal{L}$ is nearly zero within the specified tolerance, the optimal solution is obtained. Otherwise,

Step 6: Find a new set of control variables as

$$u_{\mathrm{new}} = u_{\mathrm{old}} + \Delta u \tag{5.21}$$

where $\Delta u = -\alpha\nabla\mathcal{L}.$ $\tag{5.22}$

Here, Δu is a step in the negative direction of the gradient.

The parameter α is a positive scalar, which controls the step i's (size of steps), and the choice of α is very important.

Too small a value of α guarantees the convergence but slows down its rate. Too high a value of it causes oscillations around the optimal solution. Several methods are suggested for determining the best value of α for a given problem and for an optimum choice of step size.

α is a problem-dependent constant. Experience and proper judgment are necessary in choosing a value of it. **Steps** 1, 2, and 5 are repeated for a new value of 'u' till an optimal solution is reached.

5.3 OPTIMAL POWER-FLOW PROBLEM WITH INEQUALITY CONSTRAINTS

5.3.1 Inequality constraints on control variables

In Section 5.2, the unconstrained optimal power flow problem and the computational procedure for obtaining the optimal solution are discussed. Now, in this section, the inequality constraints are introduced on control variables, and then the method of obtaining a solution to the optimal power flow problem is discussed.

The permissible values of control variables, in fact, are always constrained, such that

$$u_{\min} \leq u \leq u_{\max} \tag{5.23}$$

For example, if the real power or reactive-power generation are taken as control variables, then inequality constraints become

$$P_{G_i(\min)} \leq P_{G_i} \leq P_{G_i(\max)}$$

$$Q_{G_i(\min)} \leq Q_{G_i} \leq Q_{G_i(\max)} \tag{5.24}$$

In finding the optimal power flow solution, **Step 6** of the algorithm of Section 5.2.1 gives the change in control variable as

$$\Delta u = -\alpha \nabla \mathcal{L}$$

where $\nabla \mathcal{L} = \dfrac{\partial \mathcal{L}}{\partial u}$ and the new control variable, $u_{\text{new}} = u_{\text{old}} + \Delta u$.

This new value of control variable must be checked whether it violates the inequality constraints on the control variable or not:

i.e., $\quad u_{i_{(\min)}} \leq u_{i_{(\text{new})}} \leq u_{i_{(\max)}}$

If the correction Δu causes to exceed one of the limits, 'u_i' is set equal to the corresponding limit, i.e., the new value of u_i is determined as

$$\left. \begin{array}{l} \text{if} \quad u_{i(\text{new})} > u_{i(\max)}, \quad \text{set } u_{i(\text{new})} = u_{i(\max)} \\ \text{if} \quad u_{i(\text{new})} < u_{i(\min)}, \quad \text{set } u_{i(\text{new})} = u_{i(\min)} \end{array} \right\} \tag{5.25}$$

otherwise set $\quad u_{i(\text{new})} = u_{i(\text{old})} + \Delta u_i$

After a control variable reaches any of the limits, its component in the gradient should continue to be computed in later iteration, as the variable may come within limits at some later stages.

The optimality condition under inequality constraints can be rewritten as Kuhn–Tucker conditions given below:

$$\frac{\partial \mathcal{L}}{\partial u_i} = 0 \quad \text{if } u_{i(\min)} \leq u_i \leq u_{i(\max)}$$

$$\frac{\partial \mathcal{L}}{\partial u_i} \leq 0 \quad \text{if } u_i = u_{i(\max)}$$

$$\frac{\partial \mathcal{L}}{\partial u_i} \geq 0 \quad \text{if } u_i = u_{i(\min)} \tag{5.26}$$

Therefore, now, in Step 5 of the algorithm of Section 5.2.1, the gradient vector has to satisfy the optimality condition given by Equation (5.26).

5.3.2 Inequality constraints on dependent variables—penalty function method

In this section, the optimal solution to an optimal power flow problem will be obtained with the introduction of inequality constraints on dependent variables and penalties for their violation.

The inequality constraints on dependent variables specified in terms of upper and lower limits are

$$x_{(min)} \leq x \leq x_{(max)} \tag{5.27}$$

where x is a vector of dependent variables.

For example, if the bus voltage magnitude $|V_i|$ is taken as a dependent variable, the inequality constraint becomes

$$|V|_{min} \leq |V| \leq |V|_{max} \quad \text{on a } P\text{--}Q \text{ bus} \tag{5.28}$$

The above-mentioned inequality constraints can be handled conveniently by a method known as the *penalty function method*. In this method, the objective function is augmented by penalties for the violations of inequality constraints. Due to this augmented objective function, the solution lies sufficiently close to the constraint limits when the violations of these limits have taken place. The penalty function method, in this case, is valid since these constraints are seldom rigid limits in the strict sense but are, in fact, soft limits (e.g., $|V| \leq 1.0$ on a P–Q bus really means $|V|$ should not exceed 1.0 too much and $|V| = 1.01$ may still be permissible).

When inequality constraints are violated, the objective function can be modified by augmenting penalties as

$$C' = C(x, u) + \sum_j \omega_j \tag{5.29}$$

where ω_j is the penalty introduced for each of the violated inequality constraints. A suitable penalty function is defined as

$$\omega_j = \gamma_j \left(x_j - x_{j(max)} \right)^2 \quad \text{when } x_j > x_{j(max)}$$
$$= \gamma_j \left[x_{j(min)} - x_j \right]^2 \quad \text{when } x_j < x_{j(min)} \tag{5.30}$$

where γ_j is called a penalty factor since it controls the degree of penalty and is a real positive number.

A plot of the penalty function, which is proposed, is shown in Fig. 5.1. The plot clearly indicates how the rigid limits are replaced by soft limits. The necessary optimality conditions:

$$\frac{\partial \mathcal{L}}{\partial x} = \frac{\partial C}{\partial x} + \left[\frac{\partial f}{\partial x} \right]^{\mathrm{T}} \lambda = 0$$

$$\frac{\partial \mathcal{L}}{\partial u} = \frac{\partial C}{\partial u} + \left[\frac{\partial f}{\partial u} \right]^{\mathrm{T}} \lambda = 0$$

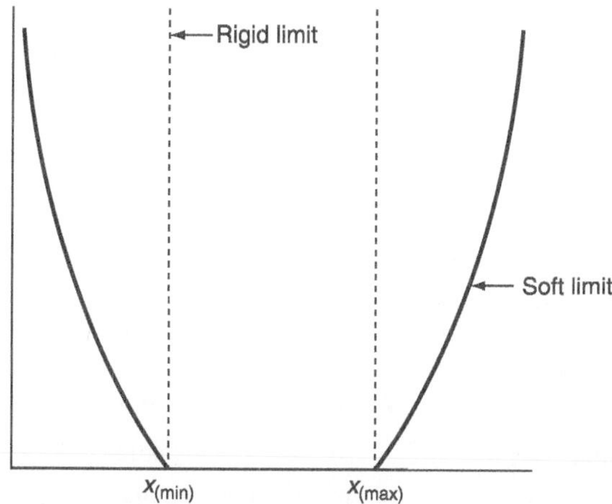

FIG. 5.1 *Penalty function*

$$\frac{\partial \mathcal{L}}{\partial \lambda} = \mathbf{f}(x, u, p) = 0$$

would now be modified as given below, while the condition of Equation (5.13), i.e., load flow equations, remains unchanged:

$$\frac{\partial \mathcal{L}}{\partial x} = \frac{\partial C}{\partial x} + \sum_j \frac{\partial \omega_j}{\partial x} + \left[\frac{\partial f}{\partial x}\right]^{\mathrm{T}} \lambda = 0 \tag{5.31}$$

$$\frac{\partial \mathcal{L}}{\partial u} = \frac{\partial C}{\partial u} + \sum_j \frac{\partial \omega_j}{\partial u} + \left[\frac{\partial f}{\partial u}\right]^{\mathrm{T}} \lambda = 0 \tag{5.32}$$

In the above equations, the vector $\dfrac{\partial \omega_j}{\partial x}$ can be calculated from the penalty function ω_j.

The vector $\dfrac{\partial \omega_j}{\partial x}$ can be obtained from Equation (5.30) and would contain one non-zero term corresponding to dependent variable x_j.

The vector $\dfrac{\partial \omega_j}{\partial u} = 0$, since the penalty functions on dependent variables are independent of control variables.

If we choose a higher value for γ_j, the penalty function ω_j can be made steeper so that the solution lies closer to the rigid limits, but the convergence becomes poorer. In normal practice, it is required to start with a lower value of γ_j and then increase it during the optimization process if the solution violates constraints above a certain tolerance limit.

It is concluded that the solution to optimal power flow problem can be achieved by superimposing the N–R method of load flow on the optimal power flow problem with respect to relevant inequality constraints. These solutions are often required for system planning and operation.

KEY NOTES

- The optimal power flow problem:

  (i) refers to load flow, which gives maximum system security by minimizing the overloads,

  (ii) aims at minimum operating cost and minimum losses,

  (iii) should be based on operational constraints, and

  (iv) is a static optimization problem with cost function as the scalar objective function.

- The solution technique for optimal power flow problem proposed by Dommel and Tinney has the following three major steps:

  (i) it is based on the load flow solution by the N–R method.

  (ii) a first-order gradient method consists of an algorithm that adjusts the gradient for minimizing the objective function.

  (iii) use of penalty functions to account for inequality constraints on dependent variables.

- For P–V buses, P and $|V|$ are specified as functions of some vectors and are represented as $f(x, y) = 0$

where x is a vector of dependent variables and y is a vector of independent variables.

- The vector of independent variables can be represented as the combination of vector of control variables 'u' and vector of fixed or disturbance or uncontrollable variables 'p' and is expressed as

$$y = \begin{Bmatrix} u \\ p \end{Bmatrix}$$

- The control parameters are:

  (i) voltage magnitude at P–V buses,

  (ii) P_{G_i} at generator buses with controllable power,

  (iii) slack bus voltage and regulating transformer tap setting as additional control variables, and

  (iv) in the case of buses with reactive-power control, Q_{G_i} is taken as control variable.

- Optimal power flow problem can be stated as

  $\min C = C(x, u)$

  subject to equality constraints:

  $f(x, u, p) = 0$

SHORT QUESTIONS AND ANSWERS

(1) The solution technique proposed by Dommel and Tinney for the optimal power flow problem is based on three major steps. What are they?

- Load flow solution by N–R method.

- A first-order gradient method.

- Use of penalty functions to account for inequality constraints on dependent variables.

(2) Write the expressions of power flow equality constraints in terms of optimal power flow problem.

$$P_i - \sum_{j=1}^{m} v_i y_{ij} v_j \cos(\theta_{ij} + \delta_j - \delta_i).$$

$$Q_i + \sum_{j=1}^{m} v_i y_{ij} v_j \sin(\theta_{ij} + \delta_j - \delta_i).$$

(3) Write the necessary conditions for obtaining an optimal solution to the optimal power flow problem without inequality constraints.

$$\frac{\partial \mathcal{L}}{\partial x} = \frac{\partial C}{\partial x} + \left[\frac{\partial f}{\partial x}\right]^T \lambda = 0.$$

$$\frac{\partial \mathcal{L}}{\partial u} = \frac{\partial C}{\partial u} + \left[\frac{\partial f}{\partial u}\right]^T \lambda = 0.$$

$$\frac{\partial \mathcal{L}}{\partial \lambda} = f(x, u, p) = 0.$$

(4) What is an optimal power flow problem?

- A general optimization problem refers to load flow, which gives maximum system security by minimizing the overloads.

- Optimal power flow problem is a static optimization problem with cost function as a scalar objective function.

(5) Which parameters are obtained as Jacobian elements in an optimal power flow problem?

Partial derivatives of equality constraints with respect to dependent variables. i.e., $\dfrac{\partial f}{\partial x}$.

(6) What is the basic technique employed in the steepest descent method?

To adjust the control variables so as to move from one feasible solution point to a new feasible solution point where the lower objective function is achieved.

(7) Why the steepest descent method is called the negative gradient method?

The moves from one feasible solution point to a new feasible point are to be repeated in the direction of negative gradient till a minimum value is reached.

(8) What is the effect of too small a value of α and too high value of α on the convergence of a solution?

The too small value of α guarantees the convergence but slows down the rate of convergence, whereas too high a value of it causes an oscillatory solution around the optimal solution.

(9) When will the Kuhn–Tucker conditions become optimality conditions?

While introducing inequality constraints on control variables.

(10) When will the penalty function method be adopted in solving optimal power-flow problem?

While introducing inequality constraints on dependent variables.

(11) What is the effect of choosing a higher value for γ_j the penalty factor?

The penalty function ω_j can be made steeper so that the solution lies closer to the rigid limits, but convergence becomes poorer.

MULTIPLE-CHOICE QUESTIONS

(1) The solution technique for an optimal power flow problem proposed by Dommel and Tinney has the steps based on:

  (i) Load flow solution by the N–R method.

  (ii) A first-order gradient method.

  (iii) Use of penalty functions to account for inequality constraints on dependent variables.

  (iv) Non-linearities present in the operation methods

    (a)   (i) and (iii)   (b)   (ii) and (iii)
    (c)   All except (iii)   (d)   All except (iv).

(2) According to the Dommel and Tinney technique, __________ method is employed for obtaining the optimal solution.

(a) Divergence method.

(b) Kuhn–Tucker method.

(c) First-order gradient method.

(d) Lagrangian multiplier method.

(3) In an optimal power flow solution, the objective function $\min C = \sum_i C_i (P_{G_i})$ subject to the equality constraints:

(a) $P_i - \sum_{j=1}^{m} v_i y_{ij} v_j \sin(\theta_{ij} + \delta_j - \delta_i) = 0.$

$Q_i + \sum_{j=1}^{m} v_i y_{ij} v_j \cos(\theta_{ij} + \delta_j - \delta_i) = 0.$

(b) $P_i - \sum_{j=1}^{m} v_i y_{ij} v_j \cos(\theta_{ij} + \delta_j - \delta_i) = 0.$

$Q_i + \sum_{j=1}^{m} v_i y_{ij} v_j \sin(\theta_{ij} + \delta_j - \delta_i) = 0.$

(c) $P_i + \sum_{j=1}^{m} v_i y_{ij} v_j \cos(\theta_{ij} + \delta_j - \delta_i) = 0.$

$Q_i - \sum_{j=1}^{m} v_i y_{ij} v_j \sin(\theta_{ij} + \delta_j - \delta_i) = 0.$

(d) $P_i + \sum_{j=1}^{m} v_i y_{ij} v_j \sin(\theta_{ij} + \delta_j - \delta_i) = 0.$

$Q_i - \sum_{j=1}^{m} v_i y_{ij} v_j \cos(\theta_{ij} + \delta_j - \delta_i) = 0$

(4) The equality constraints of an optimal power flow problem are specified as function $f(x, y) = 0$, where x is:

(a) Vector of dependent variables.

(b) Vector of independent variables.

(c) Vector of control variables.

(d) Vector of uncontrolled variables.

(5) The equality constraints of an optimal power flow problem are specified as function $f(x, y) = 0$, where y is:

(a) Vector of dependent variables.

(b) Vector of independent variables.

(c) Vector of control variables.

(d) Vector of uncontrolled variables.

(6) The independent variables are:

  (a) Control variables.

  (b) Disturbance variables.

  (c) Both (a) and (b).

  (d) None of these.

(7) The control parameter in an optimal power flow problem is:

  (a) Voltage magnitude at the P–V bus.

  (b) P_{G_i} and Q_{G_i} at the generator bus.

  (c) Slack bus voltage.

  (d) All of these.

(8) If x is the vector of dependent variables, y is the vector of independent variables, u is the vector of control variables, and p is the vector of disturbance variables, then among the following which is correct?

  (a) $x = \begin{Bmatrix} u \\ p \end{Bmatrix}$.

  (b) $y = \begin{Bmatrix} u \\ p \end{Bmatrix}$.

  (c) $x = \begin{Bmatrix} x \\ y \\ p \end{Bmatrix}$.

  (d) None of these.

(9) In an optimal power flow solution, the equality constraints are basically:

  (a) Voltage equations.

  (b) Power flow equations.

  (c) Current flow equations.

  (d) Both (a) and (c).

(10) Which of the following is obtained as Jacobian elements in a load-flow solution?

  (a) Partial derivatives of equality constraints with respect to dependent variables, $\dfrac{\partial f}{\partial x}$.

  (b) Partial derivatives of equality constraints with respect to independent variables, $\dfrac{\partial f}{\partial y}$.

  (c) Partial derivatives of equality constraints with respect to control variables, $\dfrac{\partial f}{\partial u}$.

  (d) Partial derivatives of equality constraints with respect to controlled variables, $\dfrac{\partial f}{\partial p}$.

(11) In an optimal power flow problem, the basic technique is to adjust the control variable u so as to move from one feasible solution point to a new solution point with a lower value of objective function. This technique is:

  (a) Steepest descent method.

  (b) Negative gradient method.

  (c) Either (a) or (b).

  (d) None of these.

(12) The new set of control variables is $u_{new} = u_{old} + \Delta u$. The change in control variable Δu is expressed as

  (a) $\Delta u = -\alpha \nabla \mathcal{L}$.

  (b) $\Delta u = -\alpha \mathcal{L}$.

  (c) $\Delta u = -\nabla \alpha$.

  (d) None of these.

(13) α is a parameter and too small a value of α results in the following:

  (a) Guarantees the convergence.

  (b) Slows down the rate of convergence.

  (c) Increases the rate of convergence.

    (i) Only (a).

    (ii) (b) Only.

    (iii) (a) and (c).

    (iv) (a) and (b).

(14) ___________ value of α causes an oscillatory solution around the optimal solution.

  (a) Too high.

  (b) Too low.

  (c) In between too high and too low.

  (d) None of these.

(15) The Kuhn–Tucker condition $\dfrac{\partial}{\partial u_i} \mathcal{L} = 0$, if

  (a) $u_{i(min)} \le u_i \le u_{i(max)}$.

  (b) $u_i = u_{i(min)}$.

  (c) $u_i = u_{i(max)}$.

  (d) None of these.

(16) $\dfrac{\partial \mathcal{L}}{\partial u_i} < 0$, if

  (a) $u_{i(min)} \le u_i \le u_{i(max)}$.

  (b) $u_i = u_{i(min)}$.

(c) $u_i = u_{i(max)}$.

(d) None of these.

(17) $\dfrac{\partial \mathcal{L}}{\partial u_i} > 0$, if

(a) $u_{i(min)} \leq u_i \leq u_{i(max)}$.

(b) $u_i = u_{i(min)}$.

(c) $u_i = u_{i(max)}$.

(d) None of these.

(18) The inequality constraints on dependent variables are conveniently handled by_____________ method.

(a) Penalty function.

(b) Kuhn–Tucker.

(c) Newton–Raphson.

(d) None of these.

(19) The inequality constraint limits are usually not very _____________ limit (soft/rigid) but are in fact _____________ limits (soft/rigid).

(20) In the above, the penalty introduced (ω_j) for each violation of _____________ constraint.

(a) Equality.

(b) Inequality.

(c) Either (a) or (b).

(d) None of these.

(21) For the optimal power flow problem, the equality constraints are specified as function, $f(x, y) = 0$, where:

(a) x is a vector of a dependent variable.

   y is a vector of an independent variable.

(b) x is a vector of an independent variable.

   y is a vector of a dependent variable.

(c) x is a vector of a dependent and an independent variable.

   y is a vector of a constant.

(d) x is a vector of a control variable.

   y is a vector of an uncontrolled variable.

(22) To obtain the optimal solution to an optimal power flow problem, a simple technique that can be employed is

(a) A positive gradient method.

(b) Negative gradient method.

(c) Fast decoupled method.

(d) Priority ordering.

(23) The penalty introduced for each violated inequality constraint is ω_j. For a higher value of ω_j,

(a) The penalty function can be made steeper.

(b) The solution lies closer to the rigid limits.

(c) Rate of convergence becomes poorer.

(d) Rate of convergence becomes higher.

   (i) (a) and (b).

   (ii) (a) and (c).

   (iii) All except (d).

   (iv) All of these.

(24) In an optimal power flow solution, the equality constraints are specified as a function of:

(a) Vector of dependent variables.

(b) Vector of independent variables.

(c) Vector of constants.

(d) Both (a) and (b).

(25) Control variable 'u' and disturbance variable 'p' come under:

(a) Dependent variables.

(b) Independent variables.

(c) Both (a) and (b).

(d) None of these.

(26) The optimal power flow problem with inequality constraints on dependent variables can be solved conveniently by

(a) Negative gradient method.

(b) Cost function method.

(c) Penalty function method.

(d) Steepest descent method.

(27) Penalty functions on dependent variables are _____________ of the control variables.

(a) Dependent.

(b) Independent.

(c) Dependent in one case and independent on another case.

(d) None of these.

(28) The optimal power flow problem:

 (a) Refers to the load flow that gives maximum system security by minimizing the overloads.

 (b) Aims at minimum operating cost and minimum losses.

 (c) Should be based on operational constraints.

 (d) All of these.

(29) The optimal power flow problem is:

 (a) A static optimization problem with the cost function as a scalar objective function.

 (b) A dynamic optimization problem with the cost function as a scalar objective function.

 (c) Fully static and partially dynamic optimization problem with the cost function as an objective function.

 (d) None of these.

(30) For a higher value of the penalty factor,

 (a) The penalty function can be made steeper.

 (b) The solution lies closer to the rigid limit.

 (c) Convergence becomes poorer.

 (d) All of these.

REVIEW QUESTIONS

(1) Discuss optimal power flow problems without inequality constraints.

(2) Obtain an optimal power flow solution with inequality constraints on control variables.

(3) Explain the penalty function method of obtaining an optimal power flow solution with inequality constraints on dependent variables.

(4) Develop an algorithm for obtaining the optimal power flow solution without inequality constraints by the steepest descent method.

6 Hydro-Thermal Scheduling

OBJECTIVES

After reading this chapter, you should be able to:

- know the importance of hydro-thermal co-ordination
- develop the mathematical modeling of long-term hydro-thermal co-ordination
- study the Kirchmayer's method for short-term hydro-thermal co-ordination
- study the advantages of hydro-thermal plants combination

6.1 INTRODUCTION

No state or country is endowed with plenty of water sources or abundant coal and nuclear fuel. For minimum environmental pollution, thermal generation should be minimum. Hence, a mix of hydro and thermal-power generation is necessary. The states that have a large hydro-potential can supply excess hydro-power during periods of high water run-off to other states and can receive thermal power during periods of low water run-off from other states. The states, which have a low hydro-potential and large coal reserves, can use the small hydro-power for meeting peak load requirements. This makes the thermal stations to operate at high load factors and to have reduced installed capacity with the result economy. In states, which have adequate hydro as well as thermal-power generation capacities, power co-ordination to obtain a most economical operating state is essential. Maximum advantage of cheap hydro-power should be taken so that the coal reserves can be conserved and environmental pollution can be minimized. The whole or a part of the base load can be supplied by the run-off river hydro-plants, and the peak or the remaining load is then met by a proper mix of reservoir-type hydro-plants and thermal plants. Determination of this by a proper mix is the determination of the most economical operating state of a hydro-thermal system. The hydro-thermal co-ordination is classified into long-term co-ordination and short-term co-ordination.

6.2 HYDRO-THERMAL CO-ORDINATION

Initially, there were mostly thermal power plants to generate electrical power. There is a need for the development of hydro-power plants due to the following reasons.

(i) Due to the increment of power in the load demand from all sides such as industrial, agricultural, commercial, and domestic.

(ii) Due to the high cost of fuel (coal).

(iii) Due to the limited range of fuel.

The hydro-plants can be started easily and can be assigned a load in very short time. However, in the case of thermal plants, it requires several hours to make the boilers, super heater, and turbine system ready to take the load. For this reason, the hydro-plants can handle fast-changing loads effectively. The thermal plants in contrast are slow in response. Hence, due to this, the thermal plants are more suitable to operate as base load plants, leaving hydro-plants to operate as peak load plants.

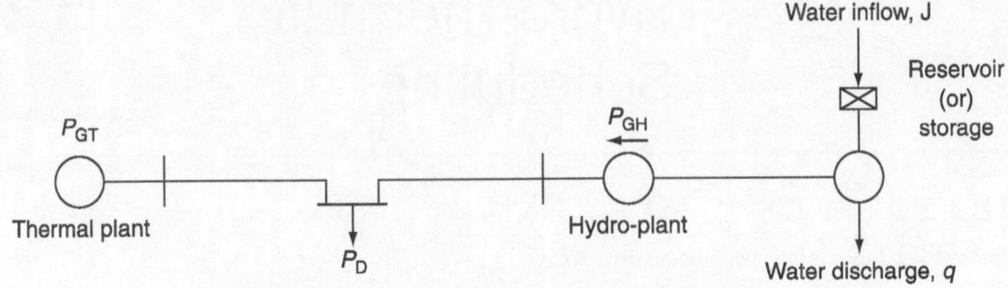

FIG. 6.1   *Fundamental hydro-thermal system*

The maximum advantage of cheap hydro-power should be taken so that the coal reserves can be conserved and environmental pollution can be minimized. In a hydro-thermal system, the whole or a part of the base load can be supplied by the run-off river hydro-plants and the peak or the remaining load is then met by a proper co-ordination of reservoir-type hydro-plants and thermal plants.

The operating cost of thermal plants is very high and at the same time its capital cost is low when compared with a hydro-electric plant. The operating cost of a hydro-electric plant is low and its capital cost is high such that it has become economical as well as convenient to run both thermal as well as hydro-plants in the same grid.

In the case of thermal plants, the optimal scheduling problem can be completely solved at any desired instant without referring to the operation at other times. It is a *static optimization problem*.

The operation of a system having both hydro and thermal plants is more complex as hydro-plants have a negligible operating cost but are required to run under the constraint of availability of water for hydro-generation during a given period of time. This problem is the **'dynamic optimization problem'** where the time factor is to be considered.

The optimal scheduling problem in a hydro-thermal system can be stated as to minimize the fuel cost of thermal plants under the constraint of water availability for hydro-generation over a given period of operation.

Consider a simple hydro-thermal system, shown in Fig. 6.1, which consists of one hydro and one thermal plant supplying power to load connected at the center in between the plants and is referred to as the fundamental system.

To solve the optimization problem in this system, consider the real power generations of two plants $P_{G_{\text{Thermal}}}$ and $P_{G_{\text{Hydro}}}$ as control variables. The transmission power loss is expressed in terms of the B coefficient as

$$P_L = \sum_{p=1}^{n} \sum_{q=1}^{n} P_{GP} B_{PQ} P_{GQ}$$

6.3 SCHEDULING OF HYDRO-UNITS IN A HYDRO-THERMAL SYSTEM

(i)   In case of hydro-units without thermal units in the system, the problem is simple. The economic scheduling consists of scheduling water release to satisfy the hydraulic constraints and to satisfy the electrical demand.

(ii) Where hydro-thermal systems are predominantly hydro, scheduling may be done by scheduling the system to produce minimum cost for the thermal systems.

(iii) In systems where there is a close balance between hydro and thermal generation and in systems where the hydro-capacity is only a fraction of the total capacity, it is generally desired to schedule generation such that thermal generating costs are minimized.

6.4 CO-ORDINATION OF RUN-OFF RIVER PLANT AND STEAM PLANT

A run-off river hydro-plant operates as the water is available in needed quantities. These plants are provided with a small pondage or reservoir, which makes it possible to meet the hourly variation of load.

The ratio of run-off during the rainy season to the run-off during the dry season may be as large as 100. As such the run-off river plants have very little from capacity. The usefulness of these run-off river plants can be considerably increased if such a plant is properly co-ordinated with a thermal plant. When such co-ordination exists, the hydro-plant may carry the base load upto its installed capacity during the period of high stream flows and the thermal plant may carry the peak load. During the period of lean flow, the thermal plant supplies the base load and the hydro-plant supplies the peak load. Thus, the load met by a thermal plant can be adjusted to conform to the available river flow. This type of co-ordination of a run-off river hydro-plant with a thermal plant results in a greater utilization factor of the river flow and a saving in the amount of fuel consumed in the thermal plant.

6.5 LONG-TERM CO-ORDINATION

Typical long-term co-ordination may be extended from one week to one year or several years. The co-ordination of the operation of reservoir hydro-power plants and steam plants involves the best utilization of available water in terms of the scheduling of water released. In other words, since the operating costs of hydro-plants are very low, hydro-power can be generated at very little incremental cost. In a combined operational system, the generation of thermal power should be displaced by available hydro-power so that maximum decrement production costs will be realized at the steam plant. The long-term scheduling problem involves the long-term forecasting of water availability and the scheduling of reservoir water releases for an interval of time that depends on the reservoir capacities and the chronological load curve of the system. Based on these factors during different times of the year, the hydro and steam plants can be operated as base load plants and peak load plants and vice versa.

For the long-term drawdown schedule, a basic best policy selection must be made. The best policy is that should the water be used under the assumption that it will be replaced at a rate based on the statistically expected rate or should the water be released using a worst-case prediction?

Long-term scheduling is made based on an optimizing policy in view of statistically treated unknowns such as load, hydraulic inflows, and unit availability (i.e., steam and hydro-plants).

The useful techniques employed for this type of scheduling problems include:

(i) the simulation of an entire long-term operational time period for a given set of operating conditions by using the dynamic programming method,

(ii) composite hydraulic simulation models, and

(iii) statistical production cost models.

For the long-term scheduling of a hydro-thermal system, there should be required generation to meet the requirements of load demand and both hydro and thermal generations should be so scheduled so as to maintain the minimum fuel costs. This requires that the available water should be put to an optimum use.

6.6 SHORT-TERM CO-ORDINATION

The economic system operation of thermal units depends only on the conditions that exist from instant to instant. However, the economic scheduling of combined hydro-thermal systems depends on the conditions existing over the entire operating period.

This type of hydro-thermal scheduling is required for one day or one week, which involves the hour-by-hour scheduling of all available generations on a system to get the minimum production cost for the given time. Such types of scheduling problems, the load, hydraulic inflows, and unit availabilities are assumed to be known.

Here also, the problem is how to supply load, as per the load cycle during the period of operation so that generation by thermal plants will be minimum. This condition will be satisfied when the value of hydro-power generation rather than its amount is a maximum over a certain period. The basic problem is that determining the degree to which the minimized economy of operating the hydro-units at other than the maximum efficiency loading may be tolerated for an increased economy with an increased load or vice versa to result in the lowest total thermal power production costs over the specified operating period.

The factors on which the economic operation of a combined hydro-thermal system depends are as follows:

- Load cycle.

- Incremental fuel costs of thermal power stations.

- Expected water inflow in hydro-power stations.

- Water head that is a function of water storage in hydro-power stations.

- Hydro-power generation.

- Incremental transmission loss (ITL).

The following are the few important methods for short-term hydro-thermal co-ordination:

1. Constant hydro-generation method.

2. Constant thermal generation method.

3. Maximum hydro-efficiency method.

4. Kirchmayer's method.

6.6.1 Constant hydro-generation method

In this method, a scheduled amount of water at a constant head is used such that the hydro-power generation is kept constant throughout the operating period.

6.6.2 Constant thermal generation method

Thermal power generation is kept constant throughout the operating period in such a way that the hydro-power plants use a specified and scheduled amount of water and operate on varying power generation schedules during the operating period.

6.6.3 Maximum hydro-efficiency method

In this method, during peak load periods, the hydro-power plants are operated at their maximum efficiency; during off-peak load periods they operate at an efficiency nearer to their maximum efficiency with the use of a specified amount of water for hydro-power generation.

Kirchmayer's method is explained in Section 6.8.

6.7 GENERAL MATHEMATICAL FORMULATION OF LONG-TERM HYDRO-THERMAL SCHEDULING

To mathematically formulate the optimal scheduling problem in a hydro-thermal system, the following assumptions are to be made for a certain period of operation T (a day, a week, or a year):

(i) The storage of a hydro-reservoir at the beginning and at the end of period of operation T are specified.

(ii) After accounting for the irrigation purpose, water inflow to the reservoir and load demand on the system are known deterministically as functions of time with certainties.

The optimization problem here is to determine the water discharge rate $q(t)$ so as to minimize the cost of thermal generation.

Objective function is

$$\min C_T = \int_0^t C'\left(P_{GT}(t)\right)\mathrm{d}t \tag{6.1}$$

Subject to the following constraints:

(i) The real power balance equation

$$P_{GT}(t) + P_{GH}(t) = P_L(t) + P_D(t)$$

i.e., $\quad P_{GT}(t) + P_{GH}(t) - P_L(t) - P_D(t) = 0 \quad$ for $t \in (0,T)$ $\tag{6.2}$

where $\quad P_{GT}(t)$ is the real power thermal generation at time 't',

$\quad P_{GH}(t)$ the real power hydro generation at time 't',

$\quad P_L(t)$ real power loss at time 't', and

$\quad P_D(t)$ the real power demand at time 't'.

(ii) Water availability equation:

$$X'(t) - X'(0) - \int_0^T J(t)\,\mathrm{d}t + \int_0^T q(t)\,\mathrm{d}t = 0 \tag{6.3}$$

where $\quad X'(t)$ is the water storage at time 't',

$\quad X'(0)$ the water storage at the beginning of operation time, T,

$\quad X'(T)$ the water storage at the end of operation time, T,

$\quad J(t)$ the water inflow rate, and

$\quad q(t)$ the water discharge rate.

(iii) Real power hydro-generation

The real power hydro-generation $P_{GH}(t)$ is a function of water storage $X'(t)$ and water discharge rate $q(t)$

i.e., $P_{GH}(t) = f(X'(t), q(t))$ (6.4)

6.7.1 Solution of problem-discretization principle

By the discretization principle, the above problem can be conveniently solved. The optimization interval T is sub-divided into N equal sub-intervals of Δt time length and over each sub-interval, it is assumed that all the variables remain fixed in value.

The same problem can be reformulated as

$$\min_{q^k\,(k=1,2,\dots,N)} \Delta t \sum_{K=1}^{N} C'(P_{GT}^K) = \min_{q^k(k=1,2,\dots,N)} \sum_{K=1}^{N} C(P_{GT}^K)$$ (6.5)

subject to the following constraints:

(i) Power balance equation

$$P_{GT}^K + P_{GH}^K - P_L^K - P_D^K = 0$$ (6.6)

where P_{GT}^K is the thermal generation in Kth interval,

P_{GH}^K the hydro generation in Kth interval,

P_L^K the transmission power loss in Kth interval and is expressed as
$$P_L^K = B_{TT}(P_{GT}^K)^2 + 2B_{TH}\,P_{GT}^K\,P_{GH}^K + B_{HH}(P_{GH}^K)^2, \text{and}$$

P_D^K is the load demand in the Kth interval.

(ii) Water availability equation:

$$X'^K - X'^{(K-1)} - j^K \Delta t + q^K \Delta t = 0$$ (6.7)

where X'^K is the water storage at the end of interval K, j^K the water inflow rate in interval K, and q^K the water discharge rate in interval K.

Dividing Equation (6.7) by Δt, it becomes

$$X^K - X^{K-1} - j^K + q^K = 0 \quad \text{for } K = 1, 2 \dots N$$ (6.8)

where $X^K = \dfrac{X'^K}{\Delta t}$ is the water storage in discharge units.

x^0 and x^N are specified as water storage rates at the beginning and at the end of the optimization interval, respectively.

(iii) The real power hydro-generation in any sub-interval can be written as

$$P_{GH}^K = h_o\{1 + 0.5\,e(X^K + X^{K-1})\}\,(q^K - \rho)$$ (6.9)

where $h_o = 9.81 \times 10^{-3} \, h'_o$;

h'_o is the basic water head which is corresponding to dead storage,

e the water head correction factor to account for the variation in head with storage, and

ρ the non-effective discharge (due to the need of which a hydro generation can run at no-load condition).

Equation (6.9) can be obtained as follows:

$$P_{GH}^K = 9.81 \times 10^{-3} \, h_{av}^K \, (q^K - \rho) \, \text{MW}$$

where $(q^K - \rho)$ is the effective discharge in m³/s and h_{av}^K is the average head in the Kth interval and is given as

$$h_{av}^K = h'_0 + \frac{\Delta t \, (X^K + X^{K-1})}{2A}$$

where A is the area of cross-section of the reservoir at the given storage

$$h_{av}^K = h'_o \, (1 + 0.5 \, e(X^K + X^{K-1}))$$

where $e = \dfrac{\Delta t}{A \, h'_o}$, which is tabulated for various storage values

$$\therefore P_{GH}^K = h_o \, \{1 + 0.5 e(X^K + X^{K-1})\}(q^K - \rho)$$

where $h_o = 9.81 \times 10^{-3} \, h'_o$.

The optimization problem is mathematically stated for any sub-interval 'K' by the objective function given by Equation (6.5), which is subjected to equation constraints given by Equations (6.6), (6.8), and (6.9).

In the above optimization problem, it is convenient to choose water discharges in all sub-intervals except one sub-interval as independent variables and hydro-generations, thermal generations, water storages in all sub-intervals and except water discharge as dependent variables; i.e., independent variables are represented by q^K, for $K = 2, 3, \ldots$, N and for $K \neq 1$. Dependent variables are represented by P_{GT}^K, P_{GH}^K, X^K, and q^1, for $K = 1, 2, \ldots, N$. [Since the water discharge in one sub-interval is a dependent variable.]

Equation (6.8) can be written for all values of $K = 1, 2, \ldots, N$:

i.e., $\quad X^1 - X^o - j^1 + q^1 = 0 \quad$ for $K = 1$

$$X^2 - X^1 - j^2 + q^2 = 0 \quad \text{for } K = 2$$

$$X^N - X^{(N-1)} - j^N + q^N = 0 \quad \text{for } k = N^{th} \text{ interval}$$

By adding the above set of equations, we get

$$X^N - X^o - (j^1 + j^2 \ldots + j^N) + (q^1 + q^2 \ldots + q^N) = 0$$

or $\quad X^N - X^o - \displaystyle\sum_{K=1}^{N} j^K + \sum_{K=1}^{N} q^K = 0 \qquad\qquad$ (6.10)

Equation (6.10) is known as the *water availability equation*.

For $K = 2, 3, \ldots, N$, there are $(N-1)$ number of water discharges (q's), which can be specified as independent variables and the remaining one, i.e., q^1, is specified as a dependent variable and it can be determined from Equation (6.10) as

$$q^1 = X^\circ - X^N + \sum_{K=1}^{N} j^K - \sum_{K=2}^{N} q^K \tag{6.11}$$

6.7.2 Solution technique

For obtaining a solution to the optimization problem in a hydro-thermal system, a non-linear programming technique in conjunction with the first-order gradient method is used.

Define the Lagrangian function $\mathcal{L}$ by augmenting the objective function (cost function) given by Equation (6.5) with equality constraints given by Equations (6.6), (6.8), and (6.9) through Lagrangian multipliers.

$$\mathcal{L} = \sum_{k=1}^{N} (C(P_{GT}^K) - \lambda_1^K \, (P_{GH}^K + P_{GT}^K - P_L^K - P_D^K) + \lambda_2^K \, (X^K - X^{K-1} - j^K + q^K)$$

$$+ \lambda_3^K (P_{GH}^K - h_0 \{1 - 0.5 \, e(X^K + X^{K-1})\} \, (q^K - \rho)] \tag{6.12}$$

where λ_1^K, λ_2^K, and λ_3^K are the Lagrangian multipliers that are dual variables. These are obtained by taking the partial derivatives of the Lagrangian function with respect to the dependent variables and equating them to zero.

$$\frac{\partial \mathcal{L}}{\partial P_{GT}^K} = \frac{\partial C(P_{GT}^K)}{\partial P_{GT}^K} - \lambda_1^K \left(1 - \frac{\partial P_L^K}{\partial P_{GT}^K}\right) = 0 \tag{6.13}$$

$$\frac{\partial \mathcal{L}}{\partial P_{GH}^K} = -\lambda_1^K \left(1 - \frac{\partial P_L^K}{\partial P_{GT}^K}\right) + \lambda_3^K = 0 \tag{6.14}$$

$$\left(\frac{\partial \mathcal{L}}{\partial x^K}\right)_{\substack{K \neq 0 \\ \neq N}} = \lambda_2^K - \lambda_2^{K+1} - \lambda_3^K \, h_0 \, 0.5 e(q^K - \rho) - \lambda_3^{K+1} \, h_0 \, 0.5 e(q^{K+1} - \rho) = 0 \tag{6.15}$$

Substituting Equation (6.8) in Equation (6.12) and differentiating the resultant equation with respect to q^1, we get

$$\frac{\partial \mathcal{L}}{\partial q^1} = \lambda_2^1 - \lambda_3^1 \, h_0 \, \{1 + 0.5 e(2X^\circ + j^1 - 2q^1 + \rho)\} = 0 \tag{6.16}$$

From the above equations, for any sub-interval, the Lagrangian multipliers can be obtained as follows:

(i) λ_1^K can be obtained from Equation (6.13),

(ii) λ_3^K can be obtained from Equation (6.14), and

(iii) λ_2' can be obtained from Equation (6.16) and remaining $\lambda_{2\,(K \neq 1)}^K$ can be obtained from Equation (6.15).

The partial derivatives of the Lagrangian function with respect to independent variables give the gradient vector:

$$\left[\frac{\partial \mathcal{L}}{\partial q^K}\right]_{K \neq 1} = \lambda_2^K - \lambda_3^K h_0 \{1 + 0.5 e(2X^{K-1} + j^K - 2q^K + \rho)\} \tag{6.17}$$

For optimality, the gradient vector should be zero $\left[\text{i.e., } \dfrac{\partial}{\partial q^K}\mathcal{L}=0\right]$, if there are no inequality constraints on the independent variables, i.e., on control variables (water discharges).

If not we have to find out the new values of control variables that will optimize the objective function, this can be achieved by moving in the negative direction of the gradient vector to a point, where the value of objective function is nearer to the optimal value.

It is an iterative process and this process is repeated till all the components of the gradient vector are closer to zero within a specified tolerance.

6.7.3 Algorithm

Step 1: Assume an initial set of independent variables, q^K for all sub-intervals except the first sub-interval $\{$i.e., $q^2, q^3 \ldots q^N\}$.

Step 2: Obtain the values of dependent variables x^K, P_{GH}^K, P_{GT}^K, and q^1 using Equations (6.8), (6.9), (6.6), and (6.11), respectively.

Step 3: Obtain the Lagrangian multipliers λ_1^K, λ_3^K, λ_2, and λ_2^K using Equations (6.13), (6.14), (6.16), and (6.15), respectively.

Step 4: Obtain the gradient vector $\dfrac{\partial \mathcal{L}}{\partial q^K}_{(K\neq 1)}$ and check whether all its elements are close to zero within a specified tolerance, if so the optimal value is reached; if not, go to the next step.

Step 5: Obtain new values of control variables using the first-order gradient method,

$$\text{i.e.,} \quad q_{(new)}^K = q_{(old)}^K - \alpha\left(\dfrac{\partial \mathcal{L}}{\partial q^K}\right)_{K\neq 1} \tag{6.18}$$

where α is a positive scalar, which defines the step length, and having a value depends on the problem on hand, then go to Step 2 and repeat the process.

The inequality constraints of the problem on dependent and independent variables can be handled in the case of an optimal power flow solution. Inequality constraints on independent variables check the **Kuhn–Tucker condition** (*given in optimal power flow, Chapter V*). The inequality constraints on dependent variables can be handled by augmenting the objective function through a penalty function.

The above-mentioned solution method can be directly extended to a system having multihydro and multithermal plants.

Drawback: It requires large memory since the independent variables, dependent variables, and gradients need to be stored simultaneously.

A modified technique known as decomposition overcomes the above drawback. In the decomposition technique, optimization is carried out over each sub-interval and a complete cycle of iteration is repeated, if the water availability equation does not check at the end of the cycle.

Example 6.1: A typical hydro-thermal system is shown in Fig. 6.2. For a typical day, the load on the system varies in steps of eight hours each as 9, 12, and 8 MW, respectively. There is no water inflow into the reservoir of the hydro-plant. The initial water storage in the reservoir is 120 m³/s and the final water storage should be 75 m³/s, i.e., the total water available for hydro-generation during the day is 30 m³/s.

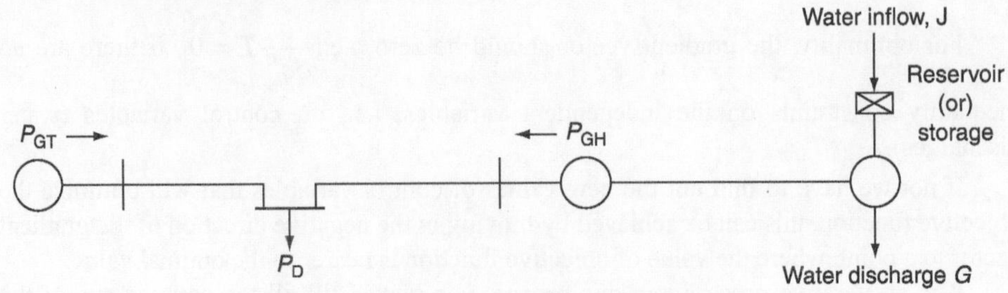

FIG. 6.2   *Fundamental hydro-thermal system*

Basic head is 30 m. Water head correction factor e is given to be 0.004. Assume for simplicity that the reservoir is rectangular so that e does not change with water storage. Let the non-effective water discharge be assumed as 3 m³/s. The fuel cost-curve characteristics of the thermal plant is $C_T = 0.2\,P_{GT}{}^{2} + 50\,P_{GT} + 130$ Rs./hr. Find the optimum generation schedule by assuming the transmission losses neglected.

Solution:

Given:

Fuel cost of the thermal plant, $C_T = 0.2P_{GT}^2 + 50P_{GT} + 130$ Rs./hr

Incremental fuel cost, $\dfrac{\partial C}{\partial P_{GT}} = 0.4P_{GT} + 50$ Rs./MWh

Total time of operation, $T = 24$ hr

No. of sub-intervals, $N = 3$

Duration of each sub-interval, $\Delta t = 8$ hr

Initial water storage in reservoir, $x'(0) = 120$ m³/s

Final water storage, $x'(3) = 75$ m³/s

Basic water head, $h'_o = 30$ m

Water-head correction factor, $e = 0.04$

Non-effective water discharge, $\rho = 3$ m³/s

Since there are three sub-intervals, $(N\text{-}1)$, the number of water discharges of the corresponding sub-intervals can be specified as independent variables and the remaining one is specified as a dependent variable, i.e., the water discharges q^2 and q^3 are considered as independent variables and dependent variable q^1.

Let us assume the initial values to be

$q^2 = 15$ m³/s

$q^3 = 15$ m³/s

for the problem formulation P_{GH}, P_{GT}, x, and q^1 are treated as independent variables.

The dependent variable q^1 (water discharge in the first sub-interval) can be obtained by Equation (6.11).

$$U^1 = x^0 - x^N + \sum_{k=1}^{N} j^K - \sum_{k=2}^{N} q^K$$

$$\Rightarrow q^1 = 120 - 75 - (15 + 20) = 10 \text{ m}^3/\text{s}$$

We have the water availability equation,

$$x^K - x^{k-1} - j^K + q^K = 0 \quad \text{for } K = 1, 2, \ldots N$$

From the above equation, we have

$$x^1 = x^0 + j^1 - q^1 = 120 - 10 = 110 \text{ m}^3/\text{s}$$

$$x^2 = x^1 + j^2 - q^2 = 110 - 15 = 95 \text{ m}^3/\text{s}$$

We know the real power hydro-generation at any interval K by Equation (6.9):

$$P_{GH}^K = h_o \left\{1 + 0.5e\left(x^K + x^{K-1}\right)\right\}\left(q^K - e\right)$$

$$= 9.81 \times 10^{-3} h_0' \left\{1 + 0.5e\left(x^K + x^{K-1}\right)\right\}\left(q^K - \rho\right)$$

$$P_{GH}^1 = 9.81 \times 10^{-3} \times 30 \left\{1 + 0.5 \times 0.004\left(x^1 + x^0\right)q^1 - \rho\right\}$$

$$= 9.81 \times 10^{-3} \times 30 \ \{1 + 0.5 \times 0.004 \ (110 + 120)\} \ (10 - 3)$$

$$= 3.0077 \text{ MW}$$

$$P_{GH}^2 = 9.81 \times 10^{-3} \times 30 \left\{1 + 0.5 \times 0.004\left(x^2 + x^1\right)q^2 - \rho\right\}$$

$$= 9.81 \times 10^{-3} \times 30 \ \{1 + 0.5 \times 0.004 \ (95 + 110)\} \ (15 - 3)$$

$$= 4.9795 \text{ MW}$$

$$P_{GH}^3 = 9.81 \times 10^{-3} \times 30 \left\{1 + 0.5 \times 0.004\left(x^3 + x^2\right)q^2 - \rho\right\}$$

$$= 9.81 \times 10^{-3} \times 30 \ \{1 + 0.5 \times 0.004 \ (75 + 95)\} \ (20 - 3)$$

$$= 6.7041 \text{ MW}$$

The thermal power generations during the sub-intervals are

$$P_{GT}^1 = P_D^1 - P_{GH}^1 = 9 - 3.0077 = 5.9923 \text{ MW}$$

$$P_{GT}^2 = P_D^2 - P_{GH}^2 = 12 - 4.9795 = 7.0205 \text{ MW}$$

$$P_{GT}^3 = P_D^3 - P_{GH}^3 = 8 - 6.7041 = 1.2959 \text{ MW}$$

λ_1^K can be obtained from Equation (6.13):

$$\text{i.e.,} \quad \frac{\partial \mathcal{L}}{\partial P_{GT}^K} = \frac{\partial C\left(P_{GT}^K\right)}{\partial P_{GT}^K} - \lambda_1^K \left(1 - \frac{\partial P_L^K}{\partial P_{GT}^K}\right) = 0$$

By neglecting transmission losses, we have

$$\lambda_1^K = \frac{\partial C\left(P_{\mathrm{GT}}^K\right)}{\partial P_{\mathrm{GT}}^K} = 0.4 P_{\mathrm{GT}}^K + 50 \text{ Rs./MWh}$$

$$\Rightarrow \lambda_1^1 = 0.4 P_{\mathrm{GT}}^1 + 50 = 0.4 \times 5.9923 + 50 = 52.3969 \text{ Rs./MWh}$$

$$\lambda_1^2 = 0.4 P_{\mathrm{GT}}^2 + 50 = 0.4 \times 7.0205 + 50 = 52.8082 \text{ Rs./MWh}$$

$$\lambda_1^3 = 0.4 P_{\mathrm{GT}}^3 + 50 = 0.4 \times 1.2959 + 50 = 50.5183 \text{ Rs./MWh}$$

From Equation (6.14),

$$\frac{\partial \mathcal{L}}{\partial P_{\mathrm{G}}^K} = \lambda_3^K - \lambda_1^K \left(1 - \frac{\partial P_{\mathrm{L}}^K}{\partial P_{\mathrm{G}}^K} \right) = 0$$

By neglecting transmission losses, we have

$$\Rightarrow \lambda_3^K = \lambda_1^K$$

$$\therefore \lambda_3^1 = \lambda_1^1 = 52.3969 \text{ Rs./MWh}$$

$$\lambda_3^2 = \lambda_1^2 = 52.8082 \text{ Rs./MWh}$$

$$\lambda_3^3 = \lambda_1^3 = 50.5183 \text{ Rs./MWh}$$

From Equation (6.16), we have

$$\frac{\partial \mathcal{L}}{\partial q^1} = \lambda_2^1 - \lambda_3^1 h_{\mathrm{o}} \left\{ 1 + 0.5 e \left(2\lambda^{\circ} + j^1 - 2q^1 + \rho \right) \right\} = 0$$

$$\Rightarrow \lambda_2^1 = \lambda_3^1 h_{\mathrm{o}} \left\{ 1 + 0.5 e \left(2x^{\circ} + j^1 - 2q^1 + \rho \right) \right\}$$

$$= 52.3969 \times 9.81 \times 10^{-3} \times 30 \; \{1 + 0.5 \times 0.004 \, (2 \times 120 - 2 \times 10 + 3)$$

$$\text{(since } j = 0)$$

$$= 22.2979 \text{ Rs./MWh}$$

From Equation (6.15), we have

$$\left(\frac{\partial \mathcal{L}}{\partial x^k} \right)_{\substack{K \neq 0 \\ \neq N}} = \lambda_2^K - \lambda_2^{K+1} - \lambda_3^K h_{\mathrm{o}} \; 0.5 e \left(q^k - \rho \right) - \lambda_3^{K+1} h_{\mathrm{o}} \; 0.5 e \left(q^{K+1} - \rho \right) = 0$$

For $K = 1$,

$$\frac{\partial \mathcal{L}}{\partial x^1} = \lambda_2^1 - \lambda_2^2 - \lambda_3^1 0.5 h_{\mathrm{o}} e \left(q^1 - \rho \right) - \lambda_3^2 0.5 h_{\mathrm{o}} e \left(q^2 - \rho \right) = 0$$

$$\therefore \lambda_2^2 = \lambda_2^1 - \lambda_3^1 0.5 h_{\mathrm{o}} e \left(q^1 - \rho \right) - \lambda_3^2 0.5 h_{\mathrm{o}} e \left(q^2 - \rho \right)$$

$$= 22.2979 - \{52.3969 \times 0.5 \times 9.81 \times 10^{-3} \times 30 \times 0.004(10 - 3)\}$$
$$- \{52.8082 \times 0.5 \times 9.81 \times 10^{-3} \times 30 \times 0.004(15 - 3)\}$$

$$= 22.2979 - 0.5889$$

$$= 21.709 \text{ Rs./MWh}$$

and for $K = 2$

$$\therefore \lambda_2^3 = \lambda_2^2 - \lambda_3^2 \, 0.5 h_o e \left(q^2 - \rho \right) - \lambda_3^3 0.5 h_o e \left(q^3 - \rho \right)$$

$$= 21.709 - \{ 52.8082 \times 0.5 \times 9.81 \times 10^{-3} \times 30 \times 0.004 (15 - 3) \}$$
$$- \{ 50.5183 \times 0.5 \times 9.81 \times 10^{-3} \times 30 \times 0.004 (20 - 3) \}$$

$$= 21.709 - 0.8784$$

$$= 20.8305 \text{ Rs./MWh}$$

i.e., $\lambda_2^1 = 22.2979$ Rs./MWh

$\lambda_2^2 = 21.709$ Rs./MWh

$\lambda_2^3 = 20.8305$ Rs./MWh

From Equation (6.17), the gradient vector is

$$\left(\frac{\partial \mathcal{L}}{\partial q^k} \right)_{K \neq 1} = \lambda_2^K - \lambda_3^K h_o \left\{ 1 + 0.5 \times 0.005 \left(2x^1 - 2q^2 + \rho \right) \right\} \quad (\text{since } j = 0 \,)$$

$$= 21.709 - 52.8082 \times 9.8 \times 10^{-3} \times 30 \, \{ 1 + 0.5 \times 0.004$$
$$(2 \times 110 - 2 \times 15 + 3) \}$$

$$= 0.1685$$

$$\left(\frac{\partial \mathcal{L}}{\partial q^3} \right) = \lambda_2^3 - \lambda_3^3 h_o \left\{ 1 + 0.5 \times e \left(2x^2 - 2q^3 + \rho \right) \right\}$$

$$= 20.8305 - 50.5183 \times 9.81 \times 10^{-3} \times 30 \{ 1 + 0.5 \times 0.004$$
$$(2 \times 95 - 2 \times 20 + 3) \}$$

$$= 1.4134$$

If the tolerance value for the gradient vector is 0.1, since for the above iteration, the gradient vector is not zero (≤ 0.1), i.e., the optimality is not satisfied here. Then, for the second iteration, obtain the new values of control variables (q_{new}^K, for $K \neq 1$) by using the first-order gradient method as follows:

$$q_{\text{new}}^K = q_{\text{old}}^K - \alpha \left(\frac{\partial}{\partial q^K} \right)_{K \neq 1} \mathcal{L} \quad (\because \, \alpha \text{ is a positive scalar})$$

$$\therefore q_{\text{new}}^2 = \left(q^2 \right)^1 = q_{\text{old}}^2 - \alpha \left(\frac{\partial}{\partial q^2} \right) \mathcal{L}$$

Let us consider $\alpha = 0.5$,

$$\therefore q_{\text{new}}^2 = \left(q^2 \right)^1 = 15 - 0.5 (0.1685) = 14.9157 \text{ m}^3/\text{s}$$

Similarly, $q_{\text{new}}^3 = \left(q^3 \right)^1 = 15 - 0.5 (1.4134) = 19.2933 \text{ m}^3/\text{s}$

and from Equation (6.11),

$$q^1 = x^o - x^N + \sum_{K=1}^{N} j^K - \sum_{K=2}^{N} q^K$$

$$\Rightarrow q^1 = x^o - x^3 - (q^2 + q^3) \quad (\text{since } j^K = 0)$$

$$q^1 = 120 - 75 - (14.9157 + 19.2933)$$

$$= 10.791 \text{ m}^3/\text{s}$$

To obtain the optimal generation schedule in hydro-thermal co-ordination, the procedure is repeated for the next iteration and checked for a gradient vector. If the gradient vector becomes zero within a specified tolerance, then that will be the optimum generation schedule, otherwise the iterations are to be carried out.

6.8 SOLUTION OF SHORT-TERM HYDRO-THERMAL SCHEDULING PROBLEMS—KIRCHMAYER'S METHOD

In this method, the co-ordination equations are derived in terms of penalty factors of both plants for obtaining the optimum scheduling of a hydro-thermal system and hence it is also known as the *penalty factor method* of solution of short-term hydro-thermal scheduling problems.

Let P_{GT_i} be the power generation of ith thermal plant in MW,

P_{GH_j} be the power generation of jth hydro-plant in MW,

$\dfrac{dC_i}{dP_{GT_i}}$ be the incremental fuel cost of ith thermal plant in Rs./MWh,

w_j be the quantity of water used for power generation at j^{th} hydro-plant in m^3/s,

$\dfrac{dw_j}{dP_{GH_j}}$ be the incremental water rate of jth hydro-plant in m^3/s/MW,

$\dfrac{\partial P_L}{\partial P_{GT_i}}$ be the incremental transmission loss of ith thermal plant,

$\dfrac{\partial P_L}{\partial P_{GH_j}}$ be the incremental transmission loss of jth hydel plant,

λ be the Lagrangian multiplier,

γ_j be the constant which converts the incremental water rate of hydel plant j into an incremental cost,

n be the total number of plants,

α be the number of thermal plants,

$n - \alpha$ be the number of hydro-plants, and

T be the time interval during which the plant operation is considered.

Here, the objective is to find the generation of individual plants, both thermal as well as hydel that the generation cost (cost of fuel in thermal) is optimum and at the same time total demand (P_D) and losses (P_L) are continuously met.

As it is a short-range problem, there will not be any appreciable change in the level of water in the reservoirs during the interval (i.e., the effects of rainfall and evaporation are neglected) and hence the head of water in the reservoir will be assumed to be constant.

Let K_j be the specified quantity of water, which must be utilized within the interval T at each hydro-station j.

Problem formulation

The objective function is to minimize the cost of generation:

i.e., $\min \sum_{i=1}^{\alpha} \int_{0}^{T} C_i \, dt$ (6.19)

subject to the equality constraints

$$\sum_{i=1}^{\alpha} P_{GT_i} + \sum_{j=\alpha+1}^{n} P_{GH_j} = P_D + P_L \qquad (6.20)$$

and $\int_{0}^{T} w_j \, dt = K_j \quad \text{for } j = \alpha+1, \alpha+2, \ldots, n$ (6.21)

where w_j is the turbine discharge in the jth plant in m³/s and K_j the amount of water in m³ utilized during the time period T in the jth hydro-plant.

The coefficient γ must be selected so as to use the specified amount of water during the operating period.

Now, the objective function becomes

$$\min C = \sum_{i=1}^{\alpha} \int_{0}^{T} C_i \, dt + \sum_{j=\alpha+1}^{n} \gamma_j K_j$$

Substituting K_j from Equation (6.21) in the above equation, we get

$$\min C = \sum_{i=1}^{\alpha} \int_{0}^{T} C_i \, dt + \sum_{j=\alpha+1}^{n} \gamma_j \int_{0}^{T} w_j \, dt \qquad (6.22)$$

For a particular load demand P_D, Equation (6.20) results as

$$\sum_{i=1}^{\alpha} \Delta P_{GT_i} + \sum_{j=\alpha+1}^{n} \Delta P_{GH_j} - \sum_{i=1}^{\alpha} \frac{\partial P_L}{\partial P_{GT_i}} \Delta P_{GT_i} - \sum_{j=\alpha+1}^{n} \frac{\partial P_L}{\partial P_{GH_j}} \Delta P_{GH_j} = 0 \qquad (6.23)$$

For a particular hydro-plant x, Equation (6.23) can be rewritten as

$$\Delta P_{GH_x} - \frac{\partial P_L}{\partial P_{GH_x}} \Delta P_{GH_x} = -\sum_{i=1}^{\alpha} \Delta P_{GT_i} - \sum_{\substack{j=\alpha+1 \\ j \neq x}}^{n} \Delta P_{GH_j}$$

$$+ \sum_{i=1}^{\alpha} \frac{\partial P_L}{\partial P_{GT_i}} \Delta P_{GT_i} - \sum_{\substack{j=\alpha+1 \\ j \neq x}}^{n} \frac{\partial P_L}{\partial P_{GH_j}} \Delta P_{GH_j} = 0$$

By rearranging the above equation, we get

$$\left(1 - \frac{\partial P_L}{\partial P_{GH_x}}\right) \Delta P_{GH_x} = -\sum_{i=1}^{\alpha} \left(1 - \frac{\partial P_L}{\partial P_{GT_i}}\right) \Delta P_{GT_i} - \sum_{j=\alpha+1}^{n} \left(1 - \frac{\partial P_L}{\partial P_{GH_j}}\right) \Delta P_{GH_j} \qquad (6.24)$$

From Equation (6.22), the condition for minimization is

$$\Delta\left[\sum_{i=1}^{\alpha}\int_0^T C_i \, dt + \sum_{j=\alpha+1}^{n}\gamma_j\int_0^T w_j \, dt\right]=0 \tag{6.25}$$

The above equation can be written as

$$\sum_{i=1}^{\alpha}\frac{dC_i}{dP_{GT_i}}\Delta P_{GT_i} + \sum_{j=\alpha+1}^{n}\gamma_j\frac{dw_j}{dP_{GH_j}}\Delta P_{GH_j}=0 \tag{6.26}$$

For hydro-plant x,

$$\gamma_x\frac{dw_x}{dP_{GH_x}}\Delta P_{GH_x} = -\sum_{i=1}^{\alpha}\frac{dC_i}{dP_{GT_i}}\Delta P_{GT_i} - \sum_{\substack{j=\alpha+1 \\ j\neq x}}^{n}\gamma_j\frac{dw_j}{dP_{GH_j}}\Delta P_{GH_j}$$

Multiplying the above equation by $\left(1-\dfrac{\partial P_L}{\partial P_{GH_x}}\right)$,

$$\left(1-\frac{\partial P_L}{\partial P_{GH_x}}\right)\gamma_x\frac{dw_x}{dP_{GH_x}}\Delta P_{GH_x} = \left(1-\frac{\partial P_L}{\partial P_{GH_x}}\right)$$

$$\left[-\sum_{i=1}^{\alpha}\frac{dC_i}{dP_{GT_i}}\Delta P_{GT_i} - \sum_{\substack{j=\alpha+1 \\ j\neq x}}^{n}\gamma_j\frac{dw_j}{dP_{GH_j}}\Delta P_{GH_j}\right] \tag{6.27}$$

Substitute for $\left(1-\dfrac{\partial P_L}{\partial P_{GH_x}}\right)\Delta P_{GH_x}$ from Equation (6.24) in Equation (6.27), we get

$$\gamma_x\frac{dw_x}{dP_{GH_x}}\left[-\sum_{i=1}^{\alpha}\left(1-\frac{\partial P_L}{\partial P_{GT_i}}\right)\Delta P_{GT_i} - \sum_{\substack{j=\alpha+1 \\ j\neq x}}^{n}\left(1-\frac{\partial P_L}{\partial P_{GH_j}}\right)\Delta P_{GH_j}\right]$$

$$=\left(1-\frac{\partial P_L}{\partial P_{GH_x}}\right)\left[-\sum_{i=1}^{\alpha}\frac{dC_i}{dP_{GT_i}}\Delta P_{GT_i} - \sum_{\substack{j=\alpha+1 \\ j\neq x}}^{n}\gamma_j\frac{dw_j}{dP_{GH_j}}\Delta P_{GH_j}\right]$$

Rewriting the above equation as

$$\sum_{i=1}^{\alpha}\left[\frac{\partial C_i}{\partial P_{GT_i}}\left(1-\frac{\partial P_L}{\partial P_{GH_x}}\right)-\gamma_x\frac{dw_x}{dP_{GH_x}}\left(1-\frac{\partial P_L}{\partial P_{GT_i}}\right)\right]\Delta P_{GT_i}$$

$$+\sum_{\substack{j=\alpha+1 \\ j\neq x}}^{n}\gamma_j\frac{dw_j}{dP_{GH_j}}\left(1-\frac{\partial P_L}{\partial P_{GH_x}}\right)-\gamma_x\frac{dw_x}{dP_{GH_x}}\left(1-\frac{\partial P_L}{\partial P_{GH_j}}\right)\Delta P_{GH_j}=0 \tag{6.28}$$

$\therefore \Delta P_{GT_i} \neq 0$ and $\Delta P_{GH_j} \neq 0$, Equation (6.28) becomes

$$\frac{dC_i}{dP_{GT_i}}\left(1 - \frac{\partial P_L}{\partial P_{GH_x}}\right) - \gamma_x \frac{dw_x}{dP_{GH_x}}\left(1 - \frac{\partial P_L}{\partial P_{GT_i}}\right) = 0 \quad \text{for } i = 1.2, \ldots, \alpha \tag{6.29}$$

and

$$\gamma_j \frac{dw_j}{dP_{GH_j}}\left(1 - \frac{\partial P_L}{\partial P_{GH_x}}\right) - \gamma_x \frac{dw_x}{dP_{GH_x}}\left(1 - \frac{\partial P_L}{\partial P_{GH_j}}\right) = 0 \quad \text{for } j = \alpha+1, \alpha+2 \ldots n \tag{6.30}$$

Equations (6.29) and (6.30) can be written in the form:

$$\frac{dC_i}{dP_{GT_i}} \frac{1}{\left(1 - \dfrac{\partial P_L}{\partial P_{GT_i}}\right)} = \gamma_x \frac{dw_x}{dP_{GH_x}} \frac{1}{\left(1 - \dfrac{\partial P_L}{\partial P_{GH_x}}\right)} \tag{6.31}$$

and

$$\gamma_j \frac{dw_j}{dP_{GH_j}} \frac{1}{\left(1 - \dfrac{\partial P_L}{\partial P_{GH_j}}\right)} = \gamma_x \frac{dw_x}{dP_{GH_x}} \frac{1}{\left(1 - \dfrac{dP_L}{dP_{GH_x}}\right)} \tag{6.32}$$

From Equations (6.31) and (6.32), we have

$$\Rightarrow \frac{dC_i}{dP_{GT_i}} \frac{1}{\left(1 - \dfrac{\partial P_L}{\partial P_{GT_i}}\right)} = \gamma_j \frac{dw_j}{dP_{GH_j}} \frac{1}{\left(1 - \dfrac{\partial P_L}{\partial P_{GH_j}}\right)} = \lambda \tag{6.33}$$

$$\frac{dC_i}{dP_{GT_i}} \frac{1}{\left(1 - \dfrac{\partial P_L}{\partial P_{GH_j}}\right)} = (I_c)_i \, L_i = \lambda \quad \text{for } i = 1, 2, \ldots, \alpha \tag{6.34}$$

$$\gamma_j \frac{dw_j}{dP_{GH_j}} \frac{1}{\left(1 - \dfrac{\partial P_L}{\partial P_{GH_j}}\right)} = \gamma_j (I_w)_j \, L_j = \lambda \quad \text{for } j = \alpha+1, \alpha+2, \ldots, N \tag{6.35}$$

where $(I_c)_i$ is the incremental fuel cost of the ith thermal plant and $(I_w)_j$ the incremental water rate of the jth hydro-plant.

Equations (6.34) and (6.35) may be expressed approximately as

$$\therefore \frac{dC_i}{dP_{GT_i}}\left(1 + \frac{\partial P_L}{\partial P_{GT_j}}\right) = \lambda \quad \text{for } i = 1, 2, \ldots, \alpha \tag{6.36}$$

$$\gamma_j \frac{dw_j}{dP_{\text{GH}_j}} \left(1 + \frac{\partial P_{\text{L}}}{\partial P_{\text{GH}_j}}\right) = \lambda \quad \text{for } j = \alpha+1, \alpha+2 \ldots n \tag{6.37}$$

where $\left(1 + \dfrac{\partial P_{\text{L}}}{\partial P_{\text{GT}_i}}\right)$ and $\left(1 + \dfrac{\partial P_{\text{L}}}{\partial P_{\text{GH}_j}}\right)$ are the approximate penalty factors of the ith thermal plant and the jth hydro-plant, respectively.

Equations (6.34) and (6.35) are the co-ordinate equations, which are used to obtain the optimal scheduling of the hydro-thermal system when considering the transmission losses.

In the above equations, the transmission loss P_{L} is expressed as

$$P_{\text{L}} = \sum_{i=1}^{\alpha}\sum_{k=1}^{\alpha} P_{\text{GT}_i} B_{ik} P_{\text{GT}_k} + \sum_{g=\alpha+1}^{n}\sum_{j=\alpha+1}^{n} P_{\text{GH}_g} B_{gj} P_{\text{GH}_j} + 2\sum_{i=1}^{\alpha}\sum_{j=\alpha+1}^{n} P_{\text{GT}_i} B_{ij} P_{\text{GH}_j} \tag{6.38}$$

The power generation of a hydro-plant P_{GH_j} is directly proportional to its head and discharge rate w_j.

When neglecting the transmission losses, the co-ordination equations become

$$\frac{\partial C_i}{\partial P_{\text{GT}_i}} = \lambda$$

$$\gamma_j \frac{\partial w_j}{\partial P_{\text{GH}_j}} = \lambda$$

Example 6.2: A two-plant system having a steam plant near the load center and a hydro-plant at a remote location is shown in Fig. 6.3. The load is 500 MW for 16 hr a day and 350 MW, for 8 hr a day.

The characteristics of the units are

$$C_1 = 120 + 45P_{\text{GT}} + 0.075P_{\text{GT}}^2$$

$$w_2 = 0.6P_{\text{GH}} + 0.00283P_{\text{GH}}^2 \ \text{m}^3/\text{s}$$

Loss coefficient, $B_{22} = 0.001 \ \text{MW}^{-1}$

Find the generation schedule, daily water used by the hydro-plant, and daily operating cost of the thermal plant for $\gamma_j = 85.5$ Rs./m³-hr.

Solution:

Given: $C_1 = 120 + 45P_{\text{GT}} + 0.075P_{\text{GT}}^2$

Co-ordination equation for thermal unit is

$$\frac{dC_1}{dP_{\text{GT}}} = 45 + 0.15P_{\text{GT}} \ \text{Rs./MWh} = \lambda$$

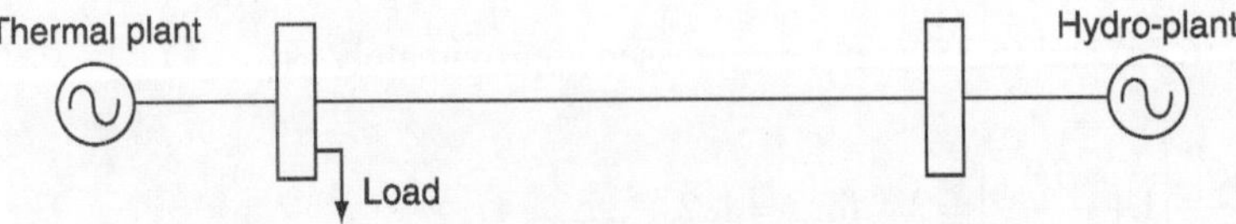

FIG. 6.3   *A typical two-plant hydro-thermal system*

For the hydro-unit, the co-ordination equation is

$$\gamma_j \frac{dw}{dP_{GH}} \left[\frac{1}{1 - \dfrac{\partial P_L}{\partial P_{GH_j}}} \right] = \lambda$$

$$\Rightarrow 85.5 \left(0.6 + 5.66 \times 10^{-3} P_{GH}\right) \left[\frac{1}{1 - 0.002 P_{GH}} \right] = \lambda$$

$$L_T = \frac{1}{1 - \dfrac{\partial P_L}{\partial P_{GT}}} = 1$$

Since the load is nearer to the thermal plant, the transmission loss is only due to the hydro-plant and therefore $B_{TT} = B_{TH} = B_{HT} = 0$:

$$\therefore P_L = B_{HH} P_{GH}^2 = 0.001 P_{GH}^2$$

$$\frac{\partial P_L}{\partial P_{GH}} = 0.002 P_{GH}$$

Power balance equation, $P_{GT} + P_{GH} = P_D + P_L$ and the condition for optimal scheduling is

$$\frac{dC_1}{dP_{GT}} L_T = \gamma_j \frac{dw}{dP_{GH}} \left[\frac{1}{1 - \dfrac{\partial P_L}{\partial P_{GH}}} \right] = \lambda$$

When $P_D = 500$ MW

$$0.15 P_{GT} + 45 = 85.5 \left(0.6 + 5.66 \times 10^{-3} P_{GH}\right) \left[\frac{1}{1 - 0.002 P_{GH}} \right]$$

$$(0.15 P_{GT} + 45)(1 - 0.002 P_{GH}) = 85.5 \left(0.6 + 5.66 \times 10^{-3} P_{GH}\right)$$

$$0.15 P_{GT} + 45 - 3 \times 10^{-4} P_{GT} P_{GH} - 0.09 P_{GH} = 51.3 + 0.48393 P_{GH}$$

$$0.57393 P_{GH} - 0.15 P_{GT} + 3 \times 10^{-4} P_{GT} P_{GH} + 6.3 = 0 \tag{6.39}$$

and

$$P_{GT} + P_{GH} = 400 + 0.001 P_{GH}^2$$

$$P_{GT} = 400 + 0.001 P_{GH}^2 - P_{GH} \tag{6.40}$$

Substituting Equation (6.40) in Equation (6.39), we get

$$0.57393 P_{GH} - 0.15 \left(400 + 0.001 P_{GH}^2 - P_{GH}\right) + 3 \times 10^{-4}$$

$$P_{GH} \left(400 + 0.001 P_{GH}^2 - P_{GH}\right) + 6.3 = 0$$

By solving the above equation, we get

$$P_{GH} = 81.876 \text{ MW}$$

By substituting the P_{GH} value in Equation (6.40), we get

$$P_{GT} = 424.8 \text{ MW}$$

$$P_L = 6.70367 \text{ MW}$$

When $P_D = 350$ MW

Equation (6.40) can be modified as

$$P_{GT} = 350 + 0.001P_{GH}^2 - P_{GH} \qquad\qquad (6.41)$$

Substituting Equation (6.41) in Equation (6.39), we get

$$0.57393P_{GH} - 0.15\left(350 + 0.001P_{GH}^2 - P_{GH}\right) + 3 \times 10^{-4}$$

$$P_{GH}\left(350 + 0.001P_{GH}^2 - P_{GH}\right) + 6.3 = 0$$

By solving the above equation, we get

$$P_{GH} = 58.5851 \text{ MW}$$

By substituting the P_{GH} value in Equation (6.41), we get

$$P_{GT} = 294.847 \text{ MW}$$

$$P_L = 3.43221 \text{ MW}$$

Daily water used by the hydro-plant

$$w = 0.6P_{GH} + 0.00283P_{GH}^2 \text{ m}^3/\text{s}$$

$=$ Daily water quantity used for a 500 MW load for 16 hr $+$ daily water quantity used for a 350 MW load for 8 hr

$= \{[0.6 \times 81.876 + 0.00283 \times (81.876)^2] \times 14 + [0.6 \times 58.586 + 0.00283 \times (58.586)^2] \times 8\} \times 3600$

$= 5.21449 \times 10^6 \text{ m}^3$

Daily operating cost of the thermal plant is:

$$C_1 = \left(120 + 45P_{GT} + 0.075P_{GT}^2\right)$$

$=$ Operating cost of the thermal plant for meeting the 500 MW load for 16 hr $+$ operating cost of the thermal plant for meeting the 350 MW load for 8 hr

$= [120 + 45 \times 424.8 + 0.075(424.8)^2] \times 16 + [120 + 45 \times 294.85 + 0.075(424.8)^2] \times 8$

$= $ Rs. 6,83,589.96 per day

Example 6.3: A two-plant system that has a hydro-plant near the load center and a steam plant at a remote location is shown in Fig. 6.4. The load is 400 MW for 14 hr a day and 200 MW, for 10 hr a day.

The characteristics of the units are

$$C_1 = 150 + 60P_{GT} + 0.1\,P_{GT}^2 \text{ Rs/hr}$$

$$w_2 = 0.8P_{GH} + 0.000333P_{GH}^2 \text{ m}^3/\text{s}$$

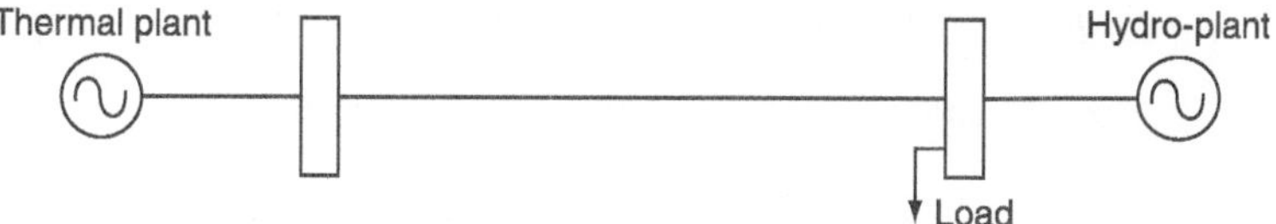

FIG. 6.4 *A typical two-plant hydro-thermal system*

Loss coefficient, $B_{22} = 0.001$ MW^{-1}

Find the generation schedule, daily water used by the hydro-plant, and the daily operating cost of a thermal plant for $\gamma_j = 77.5$ Rs./m^{3}hr.

Solution:

Equations for thermal and hydro-plants are

$$\frac{dC_1}{dP_{GT}} = 0.2P_{GT} + 60 = \lambda$$

$$\gamma_j \frac{dw}{dP_{GH}} \left| \frac{1}{1 - \dfrac{\partial P_L}{\partial P_{GH}}} \right| = \lambda$$

$$77.5(0.8 + 0.000666P_{GH}) = \lambda$$

Since the load is nearer to the hydro-plant, the transmission loss is only due to the thermal plant and therefore $B_{HH} = B_{TH} = B_{HT} = 0$:

$$\therefore P_L = B_{HH}P_{GT}^2 = 0.001P_{GT}^2$$

$$\frac{\partial P_L}{\partial P_{GH}} = 0.002P_{GH}$$

When $P_D = 400$ MW

The power balance equation is

$$P_{GT} + P_{GH} = P_D + P_L$$

$$= 400 + 0.001\,P_{GT}^2$$

$$P_{GH} = 400 + 0.001P_{GT}^2 - P_{GT} \tag{6.42}$$

The condition for optimal scheduling problem is

$$\frac{dC_1}{dP_{GT}} \frac{1}{\left[1 - \dfrac{\partial P_L}{\partial P_{GT}} \right]} = \gamma_j \frac{dw}{dP_{GH}} L_H = \lambda$$

$$(0.2P_{GT} + 60)\frac{1}{(1 - 0.002\,P_{GT})} = 77.5(0.8 + 6.6 \times 10^{-4}P_{GH})$$

$$0.2P_{GT} + 60 = 77.5\,(0.8 + 6.6 \times 10^{-4}\,P_{GH})\,(1 - 0.002P_{GT})$$

$$0.2P_{GT} + 60 = 62 + 0.051615P_{GH} - 0.124P_{GT} - 1.032 \times 10^{-4}P_{GH}P_{GT}$$

$$0.2P_{GT} + 0.124P_{GT} + 1.032 \times 10^{-4}P_{GH}P_{GT} - 0.0516P_{GH} - 2 = 0 \qquad (6.43)$$

Substituting P_{GH} from Equation (6.42) in Equation (6.43), we get

$$0.2P_{GT} + 0.124P_{GT} + 1.032 \times 10^{-4}P_{GT}(400 + 0.001\,P_{GT}^2 - P_{GT})$$
$$- 0.0516(400 + 0.001P_{GT}^2 - P_{GT}) - 2 = 0$$

By solving the above equation, we get

$$P_{GT} = 55.4\ \text{MW}$$

By substituting the P_{GT} value in Equation (6.42), we get

$$P_{GH} = 347.66\ \text{MW}$$
$$P_L = 3.069\ \text{MW}$$

When $P_D = 200$ MW

From Equation (6.42), the power balance equation becomes

$$P_{GH} = 200 + 0.001\,P_{GT}^2 - P_{GT} \qquad (6.44)$$

Substituting P_{GH} from Equation (6.44) in Equation (6.43), we get

$$0.2P_{GT} + 0.124\,P_{GT} + 1.032 \times 10^{-4}P_{GT}(200 + 0.001P_{GT}^2 - P_{GT})$$
$$- 0.0516(200 + 0.001P_{GT}^2 - P_{GT}) - 2 = 0$$

By solving the above equation, we get

$$P_{GT} = 31.575\ \text{MW}$$

By substituting the P_{GH} value in Equation (6.44), we get

$$P_{GH} = 169.421\ \text{MW}$$

$$P_L = 0.9969\ \text{MW}$$

Daily operating cost of the thermal plant

$$C_1 = 150 + 60P_{GT} + 0.1P_{GT}^2$$

$\qquad$ = Daily operating cost of the thermal plant for meeting a 400 MW load for 14 hr
$\qquad$ + daily operating cost of the thermal plant for meeting a 200 MW load for 10 hr

$\qquad = [150 + 60 \times 55.4 + 0.1 \times (55.4)^2] \times 14 + [150 + 60 \times 31.575 + 0.1 \times (31.575)^2] \times 10$

$\qquad = $ Rs. 74,374.80

Daily operating cost of the hydro-plant

$$w = 0.8P_{GH} + 0.000333P_{GH}^2\ \text{m}^3/\text{s}$$

$\qquad$ = Daily water quantity used for the 400 MW load for 14 hr + daily water quantity
$\qquad$ used for the 200 MW load for 10 hr

$\qquad = \{[0.8 \times 347.66 + 0.000333 \times (347.66)^2] \times 14 + [0.8 \times 169.421 + 0.000333 \times (169.421)^2]$
$\qquad \times 10 \times 3600$

$\qquad = 21.2696 \times 10^6\ \text{m}^3$

Example 6.4: A two-plant system that has a thermal station near the load center and a hydro-power station at a remote location is shown in Fig. 6.5.

The characteristics of both stations are

$$C_1 = (26 + 0.045P_{GT})P_{GT} \text{ Rs./hr}$$

$$w_2 = (7 + 0.004P_{GH})P_{GH} \text{ m}^3/\text{s}$$

$$\text{and} \quad \gamma_2 = \text{Rs. } 4 \times 10^{-4}/\text{m}^3$$

The transmission loss coefficient, $B_{22} = 0.0025$ MW^{-1}.

Determine the power generation at each station and the power received by the load when $\lambda = 65$ Rs./MWh.

Solution:

Here, $n = 2$

Transmission loss, $\quad P_L = \displaystyle\sum_{p}^{n}\sum_{q=1}^{n} P_{GP} B_{Pq} P_{Gq}$

$$= B_{11}P_{G1}^2 + B_{22}P_{G2}^2 + 2B_{12}P_{G1}P_{G2}$$

Since the load is near the thermal station, the power flow is from the hydro-station only; therefore, $B_{12} = B_{11} = 0$:

$$\therefore P_L = B_{22}P_{GH}^2 = 0.0025P_{GH}^2 \quad (\text{since } B_{22} = B_{HH})$$

$$\Rightarrow \frac{\partial P_L}{\partial P_{GT}} = 0 \quad \text{and} \quad \frac{\partial P_L}{\partial P_{GH}} = 2B_{HH}P_{GH} = 0.005P_{GH}$$

For the thermal power station, the co-ordination equation is

$$\frac{dC_1}{dP_{GT}}\frac{1}{\left(1 - \dfrac{\partial P_L}{\partial P_{GT}}\right)} = \lambda$$

$$\frac{\partial C_1}{\partial P_{GT}}\left(\frac{1}{1.0}\right) = \lambda$$

$$\Rightarrow \frac{dC_1}{dP_{GT}} = \lambda$$

$$(26 + 0.09P_{GT}) = 65$$

$$\therefore P_{GT} = 433.33 \text{ MW}$$

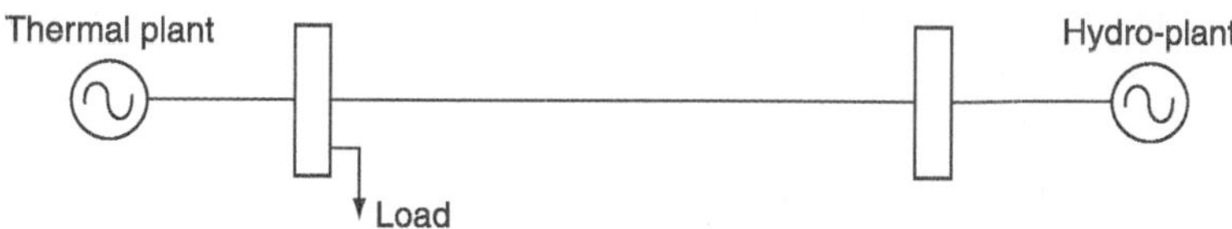

FIG. 6.5   *Two-plant system*

For a hydro-power station, the co-ordination equation is

$$\gamma_j \frac{dw_2}{dP_{GH}}\left|\frac{1}{1-\dfrac{\partial P_L}{\partial P_{GH}}}\right| = \lambda$$

$$4\times 10^{-4}\left(7+8\times 10^{-3}\,P_{GH}\right)\left(\frac{1}{1-5\times 10^{-3}P_{GH}}\right) = 65$$

By solving the above equation, we get

$P_{GH} = 199.99$ MW

Transmission loss, $P_L = B_{22}P_{GH}^2 = 0.0025(199.99)^2 = 99.993$ MW

Therefore, the power received by the load, $P_D = P_{GT} + P_{GH} - P_L = 433.33 + 622.38 - 193.68 = 533.327$ MW.

Example 6.5: For the system of Example 6.4, if the load is 750 MW for 14 hr a day and 500 MW for 10 hr on the same day, find the generation schedule, daily water used by the hydro-plant, and the daily operating cost of thermal power.

Solution:
When load, $P_D = 750$ MW

The power balance equation, $P_{GT} + P_{GH} = P_D + P_L$

$$= 750 + 0.0025\,P_{GH}^2$$

$$P_{GT} = 750 + 0.0025\,P_{GH}^2 - P_{GH} \tag{6.45}$$

The condition for optimality is

$$\frac{dC_1}{dP_{GT}}L_T = \gamma_j \frac{dw}{dP_{GH}}\left[\frac{1}{1-\dfrac{\partial P_L}{\partial P_{GH}}}\right] = \lambda$$

$$(26+0.09P_{GT}) = 4\times 10^{-4}\left(7+8\times 10^{-3}\,P_{GH}\right)\left(\frac{1}{1-5\times 10^{-3}P_{GH}}\right)$$

$$(26+0.09\,P_{GT})\,(1-5\times 10^{-3}P_{GH}) = 28\times 10^{-4} + 32\times 10^{-7}\,P_{GH} \tag{6.46}$$

Substituting P_{GT} from Equation (6.45) in Equation (6.46), we get

$$[26+0.09\,(750+0.0025P_{GH}^2 - P_{GH})](1-5\times 10^{-3}P_{GH}) = 28\times 10^{-4} + 32\times 10^{-7}P_{GH}$$

$$-1.125\times 10^{-6}\,P_{GH}^3 + 6.75\times 10^{-4}\,P_{GH}^2 - 0.5574P_{GH} + 25.9922 = 0$$

By solving the above equation, we get

$P_{GH} = 200$ MW

Substituting the P_{GH} value in Equation (6.45), we get

$$P_{GT} = 650 \text{ MW} \quad \text{and} \quad P_L = 100 \text{ MW}$$

When load, $P_D = 400$ MW

Equation (6.45) can be modified as

$$P_{GT} = 400 + 0.0025\,P_{GH}^2 - P_{GH} \tag{6.47}$$

Substituting the above equation in Equation (6.46), we get

$$[26 + 0.09(400 + 0.0025P_{GH}^2 - P_{GH})](1 - 5 \times 10^{-3} P_{GH}) = 28 \times 10^{-4} + 32 \times 10^{-7} P_{GH}$$

$$-1.125 \times 10^{-6} P_{GH}^3 + 6.75 \times 10^{-4} P_{GH}^2 - 0.3999 P_{GH} + 61.9972 = 0$$

By solving the above equation, we get

$$P_{GH} = 200 \text{ MW}$$

Substituting the P_{GH} value in Equation (6.47), we get

$$P_{GT} = 300 \text{ MW} \quad \text{and} \quad P_L = 100 \text{ MW}$$

Daily operating cost of the hydro-plant

$$w_2 = (7 + 0.004P_{GH})\,P_{GH} \text{ m}^3/\text{s}$$

= Daily water quantity used for a 750 MW load for 14 hr + daily water quantity used for a 400 MW load for 10 hr

$$= \{[7 \times 200 + 0.004 \times (200)^2] \times 14 + [7 \times 200 + 0.004 \times (200)^2] \times 10\} \times 3600$$

$$= 134.784 \times 10^6 \text{ m}^3$$

Daily operating cost of the thermal plant

$$C_1 = (26 + 0.045P_{GT})\,P_{GT} \text{ Rs./hr}$$

= Daily operating cost of a thermal plant for meeting a 750 MW load for 14 hr + daily operating cost of thermal plant for meeting a 400 MW load for 10 hr

$$= [26 \times 650 + 0.045 \times (650)^2] \times 14 + [26 \times 300 + 0.045 \times (300)^2] \times 10$$

$$= \text{Rs. } 6,21,275$$

Example 6.6: A load is feeded by two plants, one is thermal and the other is a hydro-plant. The load is located near the thermal power plant as shown in Fig. 6.6. The characteristics of the two plants are as follows:

$$C_T = 0.04P_{GT}^{\,2} + 30P_{GT} + 20 \text{ Rs./hr}$$

$$w_H = 0.0012\,P_{GH}^{\,2} + 7.5P_{GH} \text{ m}^3/\text{s}$$

$$\gamma_H = 2.5 \times 10^{-5} \text{ Rs./m}^3$$

FIG. 6.6   *Two-plant system*

The transmission loss co-efficient is $B_{22} = 0.0015$ MW^{-1}. Determine the power generation of both thermal and hydro-plants, the load connected when $\lambda = 45$ Rs./MWh.

Solution:

Given:

$$C_T = 0.04P_{GT}^2 + 30P_{GT} + 20 \text{ Rs./hr}$$

$$\frac{dC_T}{dP_{GT}} = 0.08P_{GT} + 30$$

and $\quad w_H = 0.0012P_{GH}^2 + 7.5P_{GH} \text{ m}^3/\text{s}$

$$\frac{dw_H}{dP_{GH}} = (0.0024P_{GH} + 7.5)\text{m}^3/\text{s}$$

$$= (0.0024P_{GH} + 7.5) \times 3{,}600 \text{ m}^3/\text{hr}$$

$$= 8.64P_{GH} + 27{,}000 \text{ m}^3/\text{h}$$

$$\gamma_H = 2.5 \times 10^{-5}\,\text{Rs./m}^3$$

Transmission loss, $\quad P_L = \sum_{p=1}^{2}\sum_{q=1}^{2} P_{GP}\, B_{pq}\, P_{Gq}$

$$= B_{11}P_{G1}^2 + B_{22}P_{G2}^2 + 2B_{12}P_{G1}P_{G2}$$

The load is located near the thermal plants; hence, the power flow to the load is only from the hydro-plant:

i.e., $B_{11} = B_{12} = 0$

$\therefore P_L = B_{22}P_{G2}^2 = B_{22}\,P_{GH}^2 = 0.0015\,P_{GH}^2$

The incremental transmission loss of the thermal plant is

$$(\text{ITL})_T = \frac{\partial P_L}{\partial P_{GT}} = 0$$

Penalty factor of the thermal plant, $\quad L_T \equiv \dfrac{1}{\left[1 - \dfrac{\partial P_L}{\partial P_{GT}}\right]} = 1$

The incremental transmission loss of the hydro-plant is

$$(\text{ITL})_H = \frac{\partial P_L}{\partial P_{GH}} = 2 \times B_{22}P_{GH} = 0.003P_{GH}$$

Penalty factor of the hydro-plant, $\quad L_H \equiv \dfrac{1}{\left[1 - \dfrac{\partial P_L}{\partial P_{GH}}\right]} = \dfrac{1}{1 - 0.003P_{GH}}$

The condition for hydro-thermal co-ordination is

$$\frac{dC_T}{dP_{GT}}L_T = \gamma_H \frac{dw_H}{dp_{GH}}L_H = \lambda$$

$$(0.08P_{GT}+30)\,1=2.5\times10^{-5}(8.64P_{GH}+27{,}000)\times\dfrac{1}{1-0.003P_{GH}}=45$$

$$0.08P_{GT}+30=(0.000216P_{GH}+0.675)\dfrac{1}{1-0.003P_{GH}}=45$$

$$0.08P_{GT}+30=45$$

$$\therefore P_{GT}=187.5\text{ MW}$$

and $(0.000216P_{GH}+0.675)\dfrac{1}{1-0.003P_{GH}}=45$

or $(0.000216P_{GH}+0.675)=(1-0.003P_{GH})45$

$$0.135216P_{GH}=44.325$$

$$\therefore P_{GH}=327.809\text{ MW}$$

Transmission loss, $P_L=B_{22}\,P_{GH}{}^2=0.0015(327.809)^2=161.188$ MW

The load connected, $P_D=P_{GT}+P_{GH}-P_L=187.5+327.809-161.188=354.121$ MW

Example 6.7: For Example 6.6, determine the daily water used by the hydro-plant and the daily operating cost of the thermal plant with the load connected for totally 24 hr.

Solution:

From Example 6.6,

$$\text{The load connected, }P_D=354.121\text{ MW}$$
$$\text{Generation of the thermal plant, }P_{GT}=187.5\text{ MW}$$
$$\text{Generation of the hydro-plant, }P_{GH}=327.809\text{ MW}$$

The daily water used is

$$w_H=0.0012\,P_{GH}{}^2+7.5\,P_{GH}\ \text{m}^3/\text{s}$$
$$=[0.0012\,P_{GH}{}^2+7.5\,P_{GH}]\times3{,}600\ \text{m}^3/\text{hr}$$
$$=[0.0012\,P_{GH}{}^2+7.5\,P_{GH}]\times3{,}600\times24\ \text{m}^3/\text{day}$$

Substituting the value of $P_{GH}=327.809$ MW in the above equation, we have

$$w_H=[0.0012(327.809)^2+7.5\times327.809]\times3{,}600\times24$$
$$=223.56\times10^6\,\text{m}^3$$

Daily operating cost of the thermal plant $=(0.04\,P_{GT}{}^2+30\,P_{GT}+20)$ Rs./h
$$=\text{Rs. }[0.04(187.5)^2+30\,(187.5)+20]\times24$$
$$=\text{Rs.}1{,}69{,}230$$

Example 6.8: In a two-plant operation system, the hydro-plant operates for 8 hr during each day and the steam plant operates throughout the day. The characteristics of the steam and hydro-plants are

$$C_T=0.025P_{GT}{}^2+14P_{GT}+12\ \text{Rs./hr}$$
$$w_H=0.002P_{GH}{}^2+28P_{GH}\ \text{m}^3/\text{s}$$

When both plants are running, the power flow from the steam plant to the load is 190 MW and the total quantity of water used for the hydro-plant operation during 8 hr is $220\times10^6\,\text{m}^3$.

Determine the generation of a hydro-plant and cost of water used. Neglect the transmission losses.

Solution:

The cost of the thermal plant is

$$C_T = (0.025\,P_{GT}^{\,2} + 14P_{GT} + 12)\ \text{Rs./hr}$$

The incremental fuel cost of the thermal plant is

$$\frac{dC_T}{dP_{GT}} = 0.05P_{GT} + 14$$

and for the hydro-plant, $w_H = (0.002P_{GH}^{\,2} + 28P_{GH})\ \text{m}^3/\text{s}$

The incremental water flow is

$$\frac{dw_H}{dP_{GH}} = (0.004P_{GH} + 28)\ \text{m}^3/\text{s}$$

For hydro-thermal scheduling, the optimal condition is

$$\frac{dC_T}{dP_{GT}}L_T = \gamma_H \frac{dw_H}{dP_{GH}}L_H = \frac{dC_T}{dP_{GT}} = 0.05P_{GT} + 14 = \lambda \tag{6.48}$$

(since losses are neglected, $L_T = 1$)

Power flow to the load from the thermal plant, $P_{GT} = 190$ MW (given). By substituting the value of $P_{GT} = 190$ MW in the above equation, we get

$$\lambda = 0.05(190) + 14 = 23.5\ \text{Rs./MWh}$$

The total quantity of water used during a one-hour operation is

$$w_H = 0.0012\,P_{GH}^{\,2} + 7.5P_{GH}\ \text{m}^3/\text{s}$$

$$= [0.0012\,P_{GH}^{\,2} + 7.5P_{GH}] \times 3,600\ \text{m}^3/\text{hr}$$

For an 8-hr operation, the quantity of water used is

$$w_H = [0.0012P_{GH}^{\,2} + 7.5P_{GH}] \times 3,600 \times 8 = 220 \times 10^6\ \text{m}^3$$

$$[0.0012P_{GH}^{\,2} + 7.5P_{GH}] = \frac{220 \times 10^6}{3,600 \times 8} = 7,638.888$$

or $0.0012P_{GH}^{\,2} + 7.5P_{GH} - 7,638.888 = 0$

$$P_{GH} = \frac{-28 \pm \sqrt{(28)^2 - 4(0.002)(-7,638.888)}}{2 \times (0.002)}$$

$$= \frac{-28 \pm 29.07}{0.004} = 267.698\ \text{MW}$$

Let the cost of water be γ_{H} Rs./hr/m³/s.

From Equation (6.48)

$$\gamma_{\mathrm{H}} \frac{dw_{\mathrm{H}}}{dP_{\mathrm{GH}}} \times 1 = \lambda \qquad \text{(since losses are neglected, } L_{\mathrm{H}} = 1\text{)}$$

$$\gamma_{\mathrm{H}} \left[0.004 P_{\mathrm{GH}} + 28\right] = \lambda$$

$$\text{or} \quad \gamma_{\mathrm{H}} \left[0.004(267.698) + 28\right] = 23.5$$

$$\therefore \gamma_{\mathrm{H}} = 0.8083 \text{ Rs./hr/m}^3\text{/s}$$

Example 6.9: A two-plant system with no transmission loss shown in Fig. 6.7(a) is to supply a load shown in Fig. 6.7(b).

The data of the system are as follows:

$$C_1 = (32 + 0.03 P_{\mathrm{GT}}) P_{\mathrm{GT}}$$

$$w_2 = (8 + 0.004 P_{\mathrm{GH}}) P_{\mathrm{GH}} \text{ m}^3\text{/s}$$

The maximum capacity of the hydro-plant and the steam plant are 450 and 250 MW, respectively. Determine the generating schedule of the system so that 150.3478 million m³ water is used during the 24-hr period.

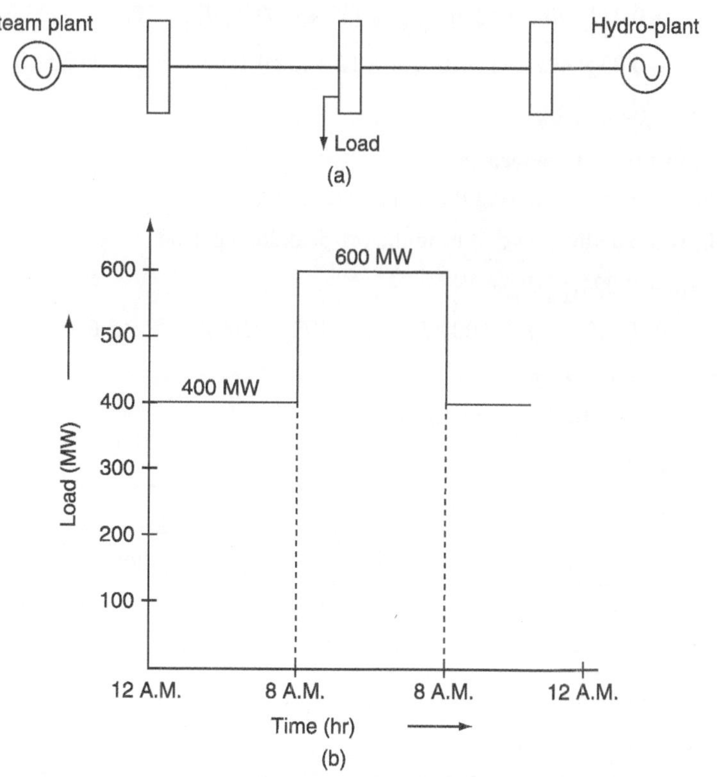

FIG. 6.7   *(a) Two-plant system; (b) daily load curve*

Solution:

(i) Constant hydro-generation

If P_{GH} is the hydro-power in MW generated in 24 hr, then we have

$$(8 + 0.004P_{GH})P_{GH} \times 24 \times 60 \times 60 = 150.3478 \times 10^6$$

$$8P_{GH} + 0.004P_{GH}^2 = 1{,}740.136$$

$$0.004P_{GH}^2 + 8P_{GH} - 1{,}740.136 = 0$$

By solving the above equation, we get

$$P_{GH} = 197.929 \text{ MW}$$

During the peak load of 600 MW

Hydro-generation, $P_{GH} = 197.929$ MW

Thermal generation, $P_{GT} = 600 - P_{GH} = 600 - 197.929 = 402.071$ MW

During off-peak load of 400 MW

Hydro-generation, $P_{GH} = 197.929$ MW

Thermal generation, $P_{GT} = 400 - P_{GH} = 400 - 197.929 = 402.071$ MW

The running cost of a steam plant for 24 hr is

$$C_1 = (32 + 0.03P_{GT})P_{GT} \times 12/\text{at }_{600 \text{ MW}} + (32 + 0.03P_{GT})P_{GT} \times 12/\text{at }_{400 \text{ MW}}$$

$$= (32 + 0.03 \times 402.071)\,402.071 \times 12 + (32 + 0.03 \times 202.071)\,202.071 \times 12$$

$$= \text{Rs. } 3{,}04{,}888.288$$

(ii) Constant thermal generation

If P_{GH} is the hydro-power during the peak load period

$(P_{GH} - 200)$ is the hydro-power during the off-peak load period

Given $w_2 = (8 + 0.004P_{GH})P_{GH}$ m^3/s

$$\{(8 + 0.004P_{GH})P_{GH} + [8 + 0.004(P_{GH} - 200)](P_{GH} - 200)\} \times 12 \times 3{,}600 = 150.3478 \times 10^6$$

After simplification, we get

$$8 \times 10^{-3}\,P_{GH}^2 + 14.4P_{GH} - 4{,}920.273 = 0$$

$$\therefore P_{GH} = 93.74793 \text{ MW}$$

The generation scheduling is given as follows:

	Hydro	Thermal $(P_D - P_{GH})$
Peak (600)	293.75 MW	306.25 MW
Off-peak (400)	93.75 MW	306.25 MW

The steam plant operating cost for 24 hr is

$$C_1 = (32 + 0.03P_{GT})P_{GT}$$

$$= (32 + 0.03 \times 306.25)\,306.25 \times 12 + (32 + 0.03 \times 306.25)\,306.25 \times 12$$

$$= \text{Rs. } 3{,}02{,}728.125$$

(iii) Equal incremental plant costs

Let P'_{GT} and P'_{GH} be the steam generation and hydro-generation during peak loads, P''_{GT} and P''_{GH} the steam generation and hydro-generation during off-peak loads, respectively.

For peak load conditions:

$$\frac{dC_1}{dP'_{GT}} = 32 + 0.03P'_{GT} = \lambda' \tag{6.49}$$

$$\gamma_2 \frac{dw_2}{dP'_{GH}} = \lambda'$$

$$\gamma_2(8 + 0.004P'_{GH}) = \lambda' \tag{6.50}$$

The value of λ' should be so chosen as to make

$$P'_{GT} + P'_{GH} = 600 \tag{6.51}$$

For off-peak periods:

$$\frac{dC_1}{dP''_{GT}} = 32 + 0.03P''_{GT} = \lambda'' \tag{6.52}$$

$$\gamma_2 \frac{dw_2}{dP''_{GH}} = \lambda''$$

$$\gamma_2(8 + 0.004P'_{GH}) = \lambda'' \tag{6.53}$$

The value of λ'' should be chosen so as to make

$$P''_{GT} + P''_{GH} = 400 \tag{6.54}$$

For the whole operating period, γ_2 should be chosen so as to use the same value of water, i.e., 150.3478 million m³ during the 24-hr period.

$$\left\{ \left(8 + 0.004P'_{GH}\right)P'_{GH} + (8 + 0.004P''_{GH})P''_{GH} \right\} \times 12 \times 3{,}600 = 150.3478 \times 10^6 \tag{6.55}$$

All the above equations can be solved by a hit-and-trail or an iterative method:

$$P'_{GT} = 276.362 \text{ MW}, \quad P'_{GH} = 323.638 \text{ MW}$$

$$\lambda' = 48.58172 \text{ Rs./MWh}$$

$$-(8 + 0.004 \times 323.638) \times 323.638 + 3{,}480.273 = (8 + 0004P''_{GH})P''_{GH}$$

$$8P''_{GH} = +0.004P''^2_{GH} = 472.2$$

By solving the above equation, we get

$$P''_{GH} + 57.38 \text{ MW} P''_{GT} = 342.62 \text{ MW}$$

$$\lambda'' = 52.5572 \text{ Rs./MWh}$$

$$\gamma_2 = 6.2131 \text{ Rs./hr/m}^3/\text{s}$$

The thermal operating cost

$$C_1 = (32 + 0.03\ P_{GT})P_{GT}$$
$$= (32 + 0.03 \times 276.362) \times 276.362 \times 12 + (32 + 0.03 \times 342.62) \times 342.62 \times 12$$
$$= \text{Rs. } 3,07,444.279$$

(iv) Maximum hydro-efficiency method

Let it be assumed that the maximum efficiency of a hydro-unit occurs at 275 MW. Therefore, the hydro-power plant supply is 275 MW during the peak load. The amount of water used during peak load hours:

$$w_2 = (8 + 0.004P_{GH})P_{GH}\ \text{m}^3/\text{s}$$
$$= (8 + 0.004 \times 275) \times 275 \times 12 \times 3,600 = 108.108 \times 10^6$$

Water available for off-peak hydro-generation:

$$= \text{total water available} - \text{water available at peak load}$$
$$= 150.34 \times 10^6 - 108.108 \times 10^6 = 42.2398 \times 10^6\ \text{m}^3$$

The real power generation of a hydro-plant P_{GH} during off-peak hours is found by using

$$(8 + 0.004P_{GH})\,P_{GH} \times 12 \times 3600 = 4,22,39,800$$
$$0.004P_{GH}^2 + 8P_{GH} - 977.77 = 0$$
$$P_{GH(\text{Off-peak load})} = 115.5461\ \text{MW}$$

The generation scheduling is given as follows:

	Hydro	Thermal $(P_D - P_{GH})$
Peak (600)	273 MW	325 MW
Off-peak (400)	115.546 MW	284.4539 MW

The daily operating cost of a thermal plant

$$C_1 = (32 + 0.03P_{GT})P_{GT}$$
$$= (32 + 0.03 \times 325)\,325 \times 12 + (32 + 0.03 \times 284.4539)\,284.4539 \times 12$$
$$= \text{Rs. } 12,43,023.55$$

Example 6.10: A thermal station and a hydro-station supply an area jointly. The hydro-station is run 16 hr daily and the thermal station is run through 24 hr. The incremental fuel cost characteristics of the thermal plant are

$$C_T = 6 + 12\,P_{GT} + 0.04P_{GT}^2\ \text{Rs./hr}$$

If the load on the thermal station, when both plants are in operation, is 350 MW, the incremental water rate of a hydro-power plant $\dfrac{d\omega}{dP_{GH}} = 28 + 0.03P_{GH}\ \text{m}^3/\text{MW-s}$. The total quantity of water utilized during a 16-hr operation of the hydro-plant is 450 million m³. Find the generation of the hydro-plant and cost of water use. Assume that the total load on the hydro-plant is constant for the 16-hr period.

Solution:

Given: $C_T = 6 + 12P_{GT} + 0.04P_{GT}^2$

$$\frac{dC_1}{dP_{GT}} = 12 + 0.08P_{GT} = \lambda$$

$P_{GT} = 350$ MW (given)

$\therefore\ 12 + 0.08 \times 350 = \lambda$

$\lambda = 40$ Rs./MWh

The total quantity of water used during 16 hr of operation of a hydro-plant is

$$(28 + 0.03P_{GH})P_{GH} \times 16 \times 3{,}600 = 450 \times 10^6$$
$$0.03P_{GH}^2 + 28P_{GH} = 7{,}812.5$$
$$0.03P_{GH}^2 + 28P_{GH} - 7{,}812.5 = 0$$

By solving the above equation, we get

$P_{GH} = 224.849$ MW

If the cost of water used is γ, then we have

$$\gamma(28 + 0.03P_{GH}) = \lambda$$
$$\gamma(28 + 0.03 \times 224.849) = 40$$
$$\therefore\ \gamma = 1.15122 \text{ Rs./hr/m}^3\text{/s}$$

6.9 ADVANTAGES OF OPERATION OF HYDRO-THERMAL COMBINATIONS

The following advantages are obtained by operation combination of hydro-thermal power plants.

6.9.1 Flexibility

The power system reliability and security can be obtained by the combined operation of hydro and thermal units. It provides the reserve capacity to meet the random phenomena of forced outage of units and unexpected load implied on a system.

Thermal plants require an appreciable time for starting and for being put into service. Hydro-plants can be started and put into operation very quickly with lower operating costs. Hence, it is required to operate hydro-plants economically as base-load plants as well as peak load plants. Hydro-plants are most preferable to operate as peak load plants such that their operation improves the flexibility of the system operation and makes the thermal plant operation easier.

6.9.2 Greater economy

The run-off river hydro-plants would generally meet the entire or part of the base loads, and thermal plants should be set up to increase the firm capacity of the system. The remaining power demand can be met by a combination of reservoir-type hydro-plants, thermal plants, and nuclear plants. In every power system, a certain ratio of hydro-power to total power demand will result in a minimum overall cost of supply.

6.9.3 Security of supply

Water availability must depend on the season. It is high during the rainy season and may be reduced due to the occurrence of draught during longer plants. Problems arise in the thermal power plant operation due to transportation of coal, unavailability of labor, etc. It is found that the forced outages of hydro-plants are few compared to those in thermal plants.

The above facts suggested the operation of hydro-thermal systems to maintain the reliability and security of supply to the consumers.

6.9.4 Better energy conservation

During heavy run-off periods, the generation of hydro-power is more, which results in the conservation of fossil fuels. During draught periods, more steam power has to be generated such that the availability of water needs the minimum needs like drinking and agricultural events.

6.9.5 Reserve capacity maintenance

For the operation of a power system, it is necessary that every system has some certain reserve capacity to meet the forced outages and unexpected load demands. By the combined operation of hydro and thermal plants, the reserve capacity maintenance is reduced.

Example 6.11: MATLAB program on hydro-thermal scheduling without inflow and without loss. Find the optimum generation for a hydro-thermal system for a typical day, wherein load varies in three steps of 8 hr each as 15, 25, and 8 MW, respectively. There is no water inflow into the reservoir of the hydro-plant. The initial water storage in the reservoir is 180 m³/s and the final water storage should be 100 m³/s. The basic head is 35 m and the water-head correction factor e is 0.005. Assume for simplicity that the reservoir is rectangular so that ρ does not change with water storage. Let the non-effective water discharge be assumed as 4 m³/s. The incremental fuel cost (IFC) of the thermal power plant is $\dfrac{dC}{dP_{GT}} = 2.0P_{GT} + 25.0$ Rs./hr. Further transmission losses may be neglected.

PROGRAM IS UNDER THE FILE NAME hydrothermal.m

```
clc;
clear;
load =[15;25;8];                        %load in MW
time =[8;8;8];                          %time of each load
ni=length(load(:,1);
x(1)=180;
x(ni+1)=100;
h0=35;
e=0.005;
ro=4;
ifc=[2.0 25.0];
delt=time(1);
alpha=0.5;
```

```matlab
%assuming in initial discharges
q(2)=20;
q(3)=22;
maxgrad=1;
iter=0;
    while(maxgrad>0.1)
            iter=iter+1;
            sumq=0;
    for I=2:ni
            sumq=sumq+q(i);
    end
q(1)=x(1)-x(ni+1)-sumq;
    for I=2:ni
       x(i)=x(i-1)-q(i-1); %water storage before ith interval
    end
    for i=1:ni
       pd(i)=load(i,1);
       pgh(i)= 9.81*0.001*h0*(1+0.5*e*(x(i)+x(i+1)))*(q(i)-ro);
       if pgh(i)>pd(i)
          pgh(i)=pd(i);
          pgt(i)=0;
       else
    pgt(i)=pd(i)-pgh(i);
    end
       lambda1(i)=ifc(1,1)*pgt(i)+ifc(1,2);
       lambda3(i)=lambda1(i);%since losses are neglected
       if i==1
       lambda2(i)=lambda3(i)*9.81*0.001*h0*(1+0.5*e*(2*x(i)-
2*q(i)+ro));
       else
       lambda2(i)=lambda2(i-1)-0.5*9.81*0.001*h0*e*(lambda3
(i-1)*(q(i-1)-ro)+lambda3(i)*(q(i)-ro));
       end
       if i~=1
       grad(i)=lambda2(i)-lambda3(i)*9.81*0.001*h0*(1+0.5*e*(2*x
(i)-2*q(i)+ro));
       end
    end
    maxgrad=(max(abs(grad)));
    for i=2:ni
       q(i)=q(i)-alpha*(grad(i));
    end
    end
    pgh
    pgt
    iter
```

RESULTS:

pgh = 12.4706	21.8178	5.4672
pgt = 2.5294	3.1822	2.5328
netPG = 15	25	8
iter = 15		

Example 6.12: MATLAB program on hydro-thermal scheduling with inflow and without losses. Find the optimum generation for a hydro-thermal system for a typical day, wherein load varies in three steps of 8 hr each as 15, 25, and 8 MW, respectively. There is water inflow into the reservoir of the hydro-plant in three intervals of 2, 4, and 3 m^3/s. The initial water storage in the reservoir is 180 m^3/s and the final water storage should be 100 m^3/s. The basic head is 35 m and the water-head correction factor e is 0.005. Assume for simplicity that the reservoir is rectangular so that ρ does not change with water storage. Let the non-effective water discharge be assumed as 4 m^3/s. The IFC of the thermal power plant is $\dfrac{dC}{dP_{GT}} = 2.0P_{GT} + 25.0$ Rs./hr. Further transmission losses may be neglected.

PROGRAM IS UNDER THE FILE NAME hydrothermalinflow.m

```
clc;
clear;
load=[15;25;8];                                    %load in MW
time=[8;8;8];                                      %time of each load
ni=length(load(:,1));
x(1)=180;
x(ni+1)=100;
h0=35;
e=0.005;
ro=4;
ifc=[2.0 25.0];
delt=time(1);
alpha=0.5;
J=[2;4;3];
%assuming initial discharges
q(2)=20;
q(3)=22;
maxgrad=1;
iter=0;
    while(maxgrad>0.1)
        iter=iter+1;
        sumq=0;
        sumj=0;
for i=2:ni
        sumq=sumq+q(i);
end
for i=1:ni
        sumj=sumj+J(i,1);
end
```

```
q(1)=x(1)-x(ni+1)-sumq+sumj;
for i=2:ni
    x(i)=x(i-1)-q(i-1)+J(i,1); %water storage before ith
    interval
end
for i=1:ni
    pd(i)=load(i,1);
    pgh(i)=9.81*0.001*h0*(1+0.5*e*(x(i)+x(i+1)))*(q(i)-ro);
    if pgh(i)>pd(i)
        pgh(i)=pd(i);
        pgt(i)=0;
    else
    pgt(i)=pd(i)-pgh(i);
    end
    lambda1(i)=ifc(1,1)*pgt(i)+ifc(1,2);
    lambda3(i)=lambda1(i); %since losses are neglected
    if i=1
    lambda2(i)=lambda3(i)*9.81*0.001*h0*(1+0.5*e*(2*x(i)-
2*q(i)+J(i,1)+ro));
    else
   lambda2(i)=lambda2(i-1)-0.5*9.81*0.001*h0*e*(lambda3
(i-1)*(q(i-1)-ro)+lambda3(i)*(q(i)-ro));
    end
    if i~=1
   grad(i)=lambda2(i)-
   lambda3(i)*9.81*0.001*h0*(1+0.5*e*(2*x(i)-
   2*q(i)+J(i,1)+ro));
    end
end
maxgrad=max(abs(grad));
for i=2:ni
    q(i)=q(i)-alpha*(grad(i));
end
    end
pgh
pgt
iter
```

RESULTS:

```
pgh=14.0553      23.6463      7.3583
pgt=0.9447        1.3537      0.6417
netPG=15         25           8
iter=15
```

Example 6.13: MATLAB program on hydro-thermal scheduling without inflow and with losses. Find the optimum generation for a hydro-thermal system for a typical day, wherein load varies in three steps of 8 hr each as 15, 25, and 8 MW, respectively. There is no water inflow into the reservoir of the hydro-plant. The initial water storage in the reservoir is 180 m³/s and the final water storage should be 100 m³/s. The basic head is 35 m and the water-head correction factor e is 0.005. Assume for simplicity that the reservoir is rectangular so that ρ does not change with water storage. Let the non-effective water discharge be assumed as 4 m³/s. The IFC of the thermal power plant is

$\dfrac{dC}{dP_{GT}} = 2.0 P_{GT} + 25.0$ Rs./hr. Further transmission losses are considered and are taken

as follows: $\left(1 - \dfrac{\partial P_L^K}{\partial P_{GT}^K}\right) = 0.85$.

PROGRAM IS UNDER THE FILE NAME hydrothermalloss.m

```
clc;
clear;
load=[15;25;8]; %load in MW
time=[8;8;8]; %time of each load
ni=length(load(:,1));
x(1)=180;
x(ni+1)=100;
h0=35;
e=0.005;
ro=4;
ifc=[2.0 25.0];
delt=time(1);
alpha=0.5;
dpl=0.5;

%assuming initial discharges

q(2)=20;
q(3)=22;
maxgrad=1;
iter==0;
    while(maxgrad>0.1)
        iter=iter+1;
        sumq=0;
for i=2:ni
    sumq=sumq+q(i);
end
    q(1)=x(1)-x(ni+1)-sumq;
    for i=2:ni
    x(i)=x(i-1)-q(i-1);              %water storage before ith interval
end
for i=1:ni
    pd(i)=load(i,1);
    pgh(i)=9.81*0.001*h0*(1+0.5*e*(x(i)+x(i+1)))*(q(i)-ro);
    if pgh(i)>pd(i)
        pgh(i)=pd(i);
        pgt(i)=0;
```

```
        else
        pgt(i)=pd(i)-pgh(i);
        end
        lambda1(i)=(ifc(1,1)*pgt(i)+ifc(1,2))/(1-dp1);
        lambda3(i)=lambda1(i)*(1-dp1);
        if i==1
        lambda2(i)=lambda3(i)*9.81*0.001*h0*(1+0.5*e*(2*x(i)-
2*q(i)+ro));
        else
    lambda2(i)=lambda2(i-1)-0.5*9.81*0.001*h0*e*(lambda3
(i-1)*(q(i-1)-ro)+lambda3(i)*(q(i)-ro));
        end
        if i~=1
    grad(i)=lambda2(i)-
    lambda3(i)*9.81*0.001*h0*(1+0.5*e*(2*x(i)-2*q(i)+ro));
        end
    end
    maxgrad=(max(abs(grad)));
    for i=2:ni
        q(i)=q(i)-alpha*(grad(i));
    end
    end
    pgh
    pgt
    iter
    grad
```

RESULTS:

```
pgh=12.4706      21.8178       5.4672
pgt=2.5294       3.1822        2.5328
grad=0          -0.0765        0.0826
netPG=15         25            8
iter=15
```

Example 6.14: MATLAB program on hydro-thermal scheduling with inflow and with losses. Find the optimum generation for a hydro-thermal system for a typical day, wherein load varies in three steps of 8 hr each as 15, 25 and 8 MW, respectively. There is water inflow into the reservoir of the hydro-plant in three intervals of 2, 4, 3 m³/s. The initial water storage in the reservoir is 180 m³/s and the final water storage should be 100 m³/s. The basic head is 35 m and the water-head correction factor e is 0.005. Assume for simplicity that the reservoir is rectangular so that ρ does not change with water storage. Let the non-effective water discharge be assumed as 4 m³/s. The IFC of the thermal power plant is $\dfrac{dC}{dP_{GT}} = 2.0P_{GT} + 25.0$ Rs./hr. Further transmission losses are considered and are

taken as follows: $\left(1 - \dfrac{\partial P_L^K}{\partial P_{GT}^K}\right) = 0.85$.

PROGRAM IS UNDER THE FILE NAME hydrothermalinflowloss.m

```
clc;
clear;
load=[15;25;8]; %load in MW
time=[8;8;8]; %time of each load
ni=length(load(:,1));
x(1)=180;
x(ni+1)=100;
h0=35;
e=0.005;
ro=4;
ifc=[2.0 25.0];
delt=time(1);
alpha=0.5;
J=[2;4;3];
dp1=0.15;
%assuming initial discharges
q(2)=20;
q(3)=22;
maxgrad=1;
iter=0;
   while(maxgrad>0.1)
        iter=iter+1;
        sumq=0;
        sumj=0;
   for i=2:ni
        sumq=sumq+q(i);
   end
for i=1:ni
        sumj=sumj+J(i,1);
end
q(1)=x(1)-x(ni+1)-sumq+sumj;
for i=2:ni
        x(i)=x(i-1)-q(i-1)+J(i,1);   %water storage before ith
                                     interval
end
for i=1:ni
   pd(i)=load(i,1);
   pgh(i)=9.81*0.001*h0*(1+0.5*e*(x(i)+x(i+1)))*(q(i)-ro);
   if pgh(i)>pd(i)
        pgh(i)=pd(i);
        pgt(i)=0;
   else
   pgt(i)=pd(i)-pgh(i);
   end
   lambda1(i)=(ifc(1,1)*pgt(i)+ifc(1,2))/(1-dp1);
   lambda3(i)=lambda1(i)*(1-dp1);
   if i=1
```

```
    lambda2(i)=lambda3(i)*9.81*0.001*h0*(1+0.5*e*(2*x(i)-
2*q(i)+J(i,1)+ro));
        else
    lambda2(i)=lambda2(i-1)-0.5*9.81*0.001*h0*e*(lambda3
(i-1)*(q(i-1)-ro)+lambda3(i)*(q(i)-ro));
        end
        if i~=1
    grad(i)=lambda2(i)-
    lambda3(i)*9.81*0.001*h0*(1+0.5*e*(2*x(i)-
    2*q(i)+J(i,1)+ro));
        end
    end
    maxgrad=max(abs(grad));
    for i=2:ni
        q(i)=q(i)-alpha*(grad(i));
    end
    end
    pgh
    pgt
    iter
```

RESULTS:

```
pgh = 14.0553      23.6463      7.3583
pgt = 0.9447        1.3537      0.6417
netPG = 15         25          8
iter = 15
```

KEY NOTES

- The optimal scheduling problem in the case of thermal plants can be completely solved at any desired instant without referring to the operation at other times. It is a *static optimization problem.*

- The optimal scheduling problem in the hydro-thermal system is a *dynamic optimization problem* where the time factor is to be considered.

- The optimal scheduling problem in a hydro-thermal system can be stated as minimizing the fuel cost of thermal plants under the constraint of water availability for hydro-generation over a given period of operation.

- The methods of hydro-thermal co-ordination are:
  - Constant hydro-generation method.
  - Constant thermal generation method.
  - Maximum hydro-efficiency method.
  - Kirchmayer's method.

- **Constant hydro-generation method**—A scheduled amount of water at a constant head is used such that the hydro-power generation is kept constant throughout the operating period.

- **Constant thermal generation method**—Thermal power generation is kept constant throughout the operating period in such a way that the hydro-power plants use a specified and scheduled amount of water and operate on varying power generation schedule during the operating period.

- **Maximum hydro-efficiency method**—During peak-load periods, the hydro-power plants are operated at their maximum efficiency; during off-peak load periods, they operate at an efficiency nearer to their maximum efficiency with the use of a specified amount of water for hydro-power generation.

- **Kirchmayer's method**—The co-ordination equations are derived in terms of penalty factors of both plants for obtaining the optimum scheduling of the hydro-thermal system and hence it is also known as the *penalty factor method* of solution of short-term hydro-thermal scheduling problems.

- Long-term hydro-thermal scheduling problems can be solved by the discretization principle.

- In the long-term hydro-thermal scheduling problem, it is convenient to choose water discharges in all sub-intervals except one sub-interval as independent variables and hydro-generations, thermal generations, water storages in all sub-intervals, and excepted water discharge as dependent variables,

  i.e., Independent variables are represented by q^K, for $K = 2, 3,...,N$

  $\neq 1$

- Dependent variables are represented by

  $P_{GH}^K, P_{GT}^K, X^{K,}$ and q^1, for $K = 1, 2, \cdots, N$. [Since the water discharge in one sub-interval is a dependent variable.]

- For optimality of long-term hydro-thermal scheduling, the gradient vector should be zero, i.e.,

  $$\frac{\partial}{\partial q^K} \mathcal{I} = 0.$$

SHORT QUESTIONS AND ANSWERS

(1) Why is the optimal scheduling problem in the case of thermal plants referred to as a static optimization problem?

Optimal scheduling problem can be completely solved at any desired instant without referring to the operation at other times.

(2) The optimization problem in the case of a hydro-thermal system is referred to as a dynamic problem. Why is it so?

The operation of the system having hydro and thermal plants have negligible operation costs but is required under the constraint of water availability for hydro-generation over a given period of time.

(3) What is the statement of optimization problem of hydro-thermal system?

Minimize the fuel cost of thermal plants under the constraint of water availability for hydro-generation over a given period of time.

(4) In the optimal scheduling problem of a hydro-thermal system, which variables are considered as control variables?

Thermal and hydro-power generations (P_{GT} and P_{GH}).

(5) Fast-changing loads can be effectively met by which type of plants?

Hydro-plants.

(6) Generally, which type of plants are more suitable to operate as base-load and peak load plants?

Thermal plants are suited for base-load plants and hydro-plants are suited for peak load plants.

(7) Whole or part of the base load can be supplied by which type of hydro-plants?

Run-off river type.

(8) The peak load or remaining base load is met by which type of plants?

A proper co-ordination of reservoir-type hydro-plants and thermal plants.

(9) In the optimal scheduling problem of a hydro-thermal system, what parameters are assumed to be known as the function of time with certainty?

Water inflow to the reservoir and load demand.

(10) What is the mathematical statement of the optimization problem in the hydro-thermal system?

Determine the water discharge rate $q(t)$ so as to minimize the cost of thermal generation.

(11) Write the objective function expression of hydro-thermal scheduling problem.

$$\min C_T = \int_0^T C' \left(P_{GT}(t) \right) dt$$

(12) Write the constraint equations of the hydro-thermal scheduling problem.

$$P_{GT}(t) + P_{GH}(t) - P_L(t) - P_D(t) = 0$$

for $t \in (0,T)$—Real power balance equation

$$X'(T) - X'(0) - \int_0^T J(t)\, dt + \int_0^T q(t)\, dt = 0$$

—Water availability equation

$$P_{GH}(t) = (X'(t), q(t))$$

(13) By which principle can the optimal scheduling problem of a hydro-thermal system be solved?

Discretization principle.

(14) Write the expression for real power hydro-generation in any sub-interval 'K'?

$$P_{GH}^K = h_0\left\{1 + 0.5e\left(X^k + X^{k-1}\right)\right\}\left(q^k - \rho\right)$$

(15) Define the terms of the above real power hydro-generation.

$$P_{GH}^K = h_0\left\{1 + 0.5e\left(X^k + X^{k-1}\right)\right\}\left(q^k - \rho\right)$$

where $h_0 = 9.81 \times 10^{-3}\, h_0^1$, h_0^1 is the basic water head that corresponds to dead storage, e the water-head correction factor to account for the variation in head with water storage, X^k the water storage at interval k, q^k the water discharge at interval k, and ρ the non-effective discharge.

(16) In the optimal scheduling problem of a hydro-thermal system, which variables are used to choose as independent variables?

Water discharges in all sub-intervals except one sub-interval:

i.e., $q_{k \neq 1}^k$, for q^2, $q^3 \cdots q^N$

where $k = 2, 3 \ldots N$ (k is sub-interval).

(17) Which parameters are used as dependent variables?

Thermal, hydro-generations, water storages at all sub-intervals, and water discharge at excepted sub-intervals are used as dependent variables,

i.e., P_{GT}^k, P_{GH}^k, X^k, and q^1

(18) In solving the optimal scheduling problem of a hydro-thermal system, for 'N' sub-intervals (i.e., $k = 1, 2, \ldots, N$), N-1 number of water discharges q's can be specified as independent variables except one sub-interval. Write the expression for water discharge in the excepted sub-interval, which is taken as a dependant variable.

$$q' = X^0 - X^N + \sum_{k=1}^{N} J^k - \sum_{k=2}^{N} q^k$$

(19) Which technique is used to obtain the solution to the optimization problem of the hydro-thermal system?

A non-linear programming technique in conjunction with a first-order negative gradient method is used to obtain the solution to the optimization problem.

(20) Write the expression for a Lagrangian function obtained by augmenting the objective function with constraint equations in the case of a hydro-thermal scheduling problem.

$$\mathcal{I} = \sum_{k=1}^{N}\begin{bmatrix}C(P_{GT}^k) - \lambda_1^k(P_{GT}^k + P_{GH}^k - P_L^k - P_D^k) + \\ \lambda_2^k(X^k - X^{k-1} - J^k + q^k) + \\ \lambda_3^k(P_{GH}^k - h_0\left\{1 + 0.5e(X^k + X^{k-1})\right\}\end{bmatrix}(q^k - \rho)$$

(21) What is the gradient vector?

The partial derivatives of the Lagrangian function with respect to independent variables are

i.e., $\left[\dfrac{\partial \mathcal{I}}{\partial q^k}\right]_{\substack{k \neq 1 \\ k=2\cdots N}} = \lambda_2^k - \lambda_3^k h_0\left\{\begin{bmatrix}1 + 0.5e \\ \left(2X^{k-1} + \right. \\ \left. J^k - 2q^k + \rho\right)\end{bmatrix}\right\}$

(22) What is the condition for optimality in the case of a hydro-thermal scheduling problem?

The gradient vector should be zero:

i.e., $\left[\dfrac{\partial \mathcal{I}}{\partial q^k}\right]_{\substack{k \neq 1 \\ k=2\ldots N}} = 0$

(23) The condition for optimality in a hydro-thermal scheduling problem is that the gradient vector should be zero. If this condition is violated, how will we obtain the optimal solution?

Find the new values of control variables, which will optimize the objective function. This can be achieved by moving in the negative direction of the gradient vector to a point where the value of the objective function is nearer to an optimal value.

(24) For a system with a multihydro and a multithermal plant, the non-linear programming technique in conjunction with the first-order gradient method is also directly applied. However, what is the drawback?

It requires large memory since the independent and dependant variables, and gradients need to be stored simultaneously.

(25) By which method can the drawback of the non-linear programming technique be overcome when applied to a multihydro and multithermal plant system and what is its procedure?

By the method of decomposition technique. In this technique, the optimization is carried out over each sub-interval and a complete cycle of iteration is replaced, if the water availability equation does not check at the end of the cycle.

(26) For short-range scheduling of a hydro-thermal plant, which method is useful?

Kirchmayer's method or the penalty factor method is useful for short-range scheduling.

(27) What is Kirchmayer's method of obtaining the optimum scheduling of a hydro-thermal system?

In Kirchmayer's method or the penalty factor method, the co-ordination equations are derived in terms of penalty factors of both hydro and thermal plants.

(28) What is the condition for optimality in a hydro-thermal scheduling problem when considering transmission losses?

$$\frac{\partial C_i}{\partial P_{GT_i}} \left(1 \Big/ 1 - \frac{\partial P_L}{\partial P_{GT_i}} \right)_{\text{for } i=1,2,\dots,\alpha} = \gamma_j$$

$$\frac{\partial w_j}{\partial P_{GH_j}} \left(1 \Big/ 1 - \frac{\partial P_L}{\partial P_{GH_j}} \right)_{\text{for } j=\alpha+1,\alpha+2,\dots,n} = \lambda$$

where i represents the thermal plant and j represents the hydro-plant.

(29) What is the meaning of the terms $\dfrac{\partial C_i}{\partial P_{GT_i}}$ and $\dfrac{\partial w_j}{\partial P_{GH_j}}$?

$\dfrac{\partial C_i}{\partial P_{GT_i}}$ is the incremental cost of the ith thermal plant and

$\dfrac{\partial \omega_j}{\partial P_{GH_j}}$ is the incremental water rate of the jth hydro-plant.

(30) What is short-term hydro-thermal co-ordination?

Short-term hydro-thermal co-ordination is done for a fixed quantity of water to be used in a certain period (i.e., 24 hr).

(31) What are the scheduling methods for short-term hydro-thermal co-ordination?

- Constant hydro-generation method.
- Constant steam generation method.
- Maximum hydro-efficiency method.
- Equal incremental production costs and solution of co-ordination equations (Kirchmayer's method).

(32) What is the significance of the co-efficient γ_j?

γ_j represents the incremental water rates into incremental costs which must be so selected as to use the desired amount of water during the operating period.

(33) Write the condition for optimality in the problem of a short range hydro-thermal system according to Kirchmayer's method when neglecting transmission losses.

$$\frac{\partial c_i}{\partial P_{GT_i}} = \gamma_j \frac{\partial \omega_j}{\partial P_{GH_j}} = \lambda \ (\text{Co-ordination equation})$$

(34) What is the significance of terms

$$\left(\frac{1}{1 - \dfrac{\partial P_L}{\partial P_{GT_i}}} \right) \quad \text{and} \quad \left(\frac{1}{1 - \dfrac{\partial P_L}{\partial P_{GH_j}}} \right) \ ?$$

$$\left(\frac{1}{1 - \dfrac{\partial P_L}{\partial P_{GT_i}}} \right) \quad \text{is the penalty factor of the } i^{\text{th}} \text{ thermal}$$

plant and

$$\left(\frac{1}{1 - \dfrac{\partial P_L}{\partial P_{GH_i}}} \right) \quad \text{is the penalty factor of the } j^{\text{th}} \text{ hydro-plant}$$

These terms are very much useful in getting the optimality in a hydro-thermal scheduling problem, which is solved by Kirchmayer's method.

(35) Write the condition for optimality in an optimal scheduling problem of a short range hydro-thermal system with approximate penalty factors.

$$\frac{\partial c_i}{\partial P_{GT_i}} \left(1 + \frac{\partial P_L}{\partial P_{GT_i}} \right) = \gamma_j$$

$$\text{for } i = 1,2\dots\alpha$$

$$\frac{\partial \omega_j}{\partial P_{GHj}} \left(1 + \frac{\partial P_L}{\partial P_{GH_j}} \right) = \lambda$$

$$\text{for } j = \alpha+1, \alpha+2 \dots n$$

MULTIPLE-CHOICE QUESTIONS

(1) When compared to a hydro-electric plant, the operating cost of the thermal plant is very __________ and its capital cost is __________ .

  (a) High, low.

  (b) High, high.

  (c) Low, low.

  (d) Low, high.

(2) When compared to a thermal plant, the operating cost and capital cost of a hydro-electric plant are:

  (a) High and low.

  (b) Low and high.

  (c) Both high.

  (d) Both low.

(3) The optimal scheduling problem in the case of thermal plants is:

  (a) Static optimization problem.

  (b) Dynamic optimization problem.

  (c) Static as well as dynamic optimization problem.

  (d) Either static or dynamic optimization problem.

(4) The operation of the system having hydro and thermal plants is more complex. In this case, the optimal scheduling problem is:

  (a) Static optimization problem.

  (b) Dynamic optimization problem.

  (c) Static as well as dynamic optimization problem.

  (d) Either static or dynamic optimization problem.

(5) The optimal scheduling problem in the case of a thermal plant can be completely solved at any desired instant:

  (a) With reference to the operation at other times.

  (b) Without reference to the operation at other times.

  (c) Case (a) or case (b) that depends on the size of the plant.

  (d) None of these.

(6) The time factor is considered in solving the optimization problem of __________ .

  (a) Hydro plants.

  (b) Thermal plants.

  (c) Hydro-thermal plants.

  (d) None of these.

(7) The objective function to the optimization problem in a hydro-thermal system becomes:

  (a) Minimize the fuel cost of thermal plants.

  (b) Minimize the time of operation.

  (c) Maximize the water availability for hydro-generation.

  (d) All of these.

(8) The optimal scheduling problem of a hydro-thermal system is solved under the constraint of:

  (a) Fuel cost of thermal plants for thermal generation.

  (b) Time of operation of the entire system.

  (c) Water availability for hydro-generation over a given period.

  (d) Availability of coal for thermal generation over a given period.

(9) To solve the optimization problem in a hydro-thermal system, which of the following variables are considered as control variables?

  (a) $P_{D\text{ thermal}}$ and $P_{G\text{ hydro}}$.

  (b) $Q_{D\text{ thermal}}$ and $Q_{D\text{ hydro}}$.

  (c) $P_{G\text{ thermal}}$ and $P_{D\text{ hydro}}$.

  (d) $P_{G\text{ thermal}}$ and $P_{G\text{ hydro}}$.

(10) In which system is the generation scheduled generally such that the operating costs of thermal generation are minimized?

  (a) Systems where there is a close balance between hydro and thermal generation.

  (b) Systems where the hydro-capacity is only a fraction of the total capacity.

  (c) Both (a) and (b).

  (d) None of these.

(11) Thermal plants are more suitable to operate as __________ plants leaving hydro-plants to operate as __________ plants.

(a) Base load, base load.

(b) Peak load, peak load.

(c) Peak load, base load.

(d) Base load, peak load.

(12) In hydro-thermal systems, the whole are part of the base load that can be supplied by:

(a) Run-off river-type hydro-plants.

(b) Reservoir-type hydro-plants.

(c) Thermal plants.

(d) Reservoir-type hydro-plants and thermal plants with proper co-ordination.

(13) In a hydro-thermal system, the peak load can be met by:

(a) Run-off river-type hydro-plants.

(b) Reservoir-type hydro-plants.

(c) Thermal plants.

(d) Reservoir-type hydro-plants and thermal plants with proper co-ordination.

(14) For an optimal scheduling problem, it is assumed, which parameter is known deterministically as a function of time?

(a) Water inflow to the reservoir.

(b) Power generation.

(c) Load demand.

(d) Both (a) and (c).

(15) In a hydro-thermal system, the optimization problem is stated as determining _____________ so as to minimize the cost of thermal generation.

(a) Load demand (P_D).

(b) Water storage (X).

(c) Water discharge rate ($q(t)$).

(d) Water inflow rate ($J(t)$).

(16) Which of the following equations is considered as a constraint to the optimization problem of a hydro-thermal system?

(a) Real power balance equation.

(b) Water availability equation.

(c) Real power hydro-generation as a function of water storage.

(d) All of these.

(17) The water availability equation is:

(a) $X'(T) - X'(0) - \int\limits_0^T J(t)\,dt + \int\limits_0^T q(t)\,dt = 0$.

(b) $P_{GH}(t) + P_{GH}(t) - P_L(t) - P_D(t) = 0,\ t \in (0,T)$.

(c) $P_{GH}(t) = f(X'(t),\ q(t))$.

(d) None of these.

(18) In the optimization problem of a hydro-thermal system, the constraint real power hydro-generation is a function of:

(i) Water inflow rate ($J(t)$).

(ii) Water storage (X).

(iii) Water discharge rate ($q(t)$)

(a) (i) and (ii).

(b) (ii) and (iii).

(c) (i) and (iii).

(d) None of these.

(19) The optimization scheduling problem of a hydro-thermal system can be conveniently solved by _____________ principle.

(a) Dependence.

(b) Discretization.

(c) Dividing.

(d) None of these.

(20) In the discretization principle, the real power hydro-generation at any sub-interval 'k' can be expressed as:

(a) $P_{GH}^k = h_0\left\{1 + 0.5e\left(X^{k-1} + X^k\right)\right\}\left(q^k - \rho\right)$.

(b) $P_{GH}^k = h_0\left\{1 - 0.5e\left(X^{k-1} - X^k\right)\right\}\left(q^k - \rho\right)$.

(c) $P_{GH}^k = h_0\left\{1 + 0.5e\left(X^k + X^{k-1}\right)\right\}\left(q^k - \rho\right)$.

(d) $P_{GH}^k = h_0\left\{1 + 0.5e\left(X^{k+1} + X^k\right)\right\}\left(q^k + \rho\right)$.

(21) In the optimization problem of a hydro-thermal system, which of the following are closed as independent variables?

(a) Water storages in all sub-intervals except one sub-interval.

(b) Water inflows in all sub-intervals except one sub-interval.

(c) Water discharges in all sub-intervals except one sub-interval.

(d) Hydro and thermal generations, water storages at all sub-intervals, and water discharge at one sub-interval.

(22) In the optimal scheduling problem of a hydro-thermal system, which of the following are closed as dependent variables?

(a) Water storages in all sub-intervals except one sub-interval.

(b) Water inflows in all sub-intervals except one sub-interval.

(c) Water discharges in all sub-intervals except one sub-interval.

(d) Hydro and thermal generations, water storages at all sub-intervals, and water discharge at one sub-interval.

(23) To obtain the solution to the optimization problem of a hydro-thermal system, which of the following technique is used?

(a) Non-linear programming technique in conjunction with the first-order gradient method.

(b) Linear programming technique in conjunction with the first-order gradient method.

(c) Non-linear programming technique in conjunction with the multiple-order gradient method.

(d) Linear programming technique in conjunction with the multiple-order gradient method.

(24) In a hydro-thermal system for optimality, the condition is:

(a) Gradient vector should be zero.

(b) Gradient vector should be positive.

(c) Gradient vector should be negative.

(d) None of these.

(25) For multihydro and multithermal plants, the optimization problem can be solved by a modified technique, which is known as:

(a) Discretization technique.

(b) Decomposition technique.

(c) Decoupled technique.

(d) None of these.

(26) In Kirchmayer's method of solution of optimization problem in a hydro-thermal system, the co-ordination equations are derived in terms of __________ of both plants.

(a) Power generations.

(b) Power demands.

(c) Penalty factors.

(d) All of these.

(27) γ_j is used as a constant, in an optimization problem of a hydro-thermal system, which converts:

(a) Fuel cost of a thermal plant into an incremental fuel cost.

(b) Incremental water rate of a hydro-plant into an incremental cost.

(c) Incremental water inflow rate into an incremental discharge rate.

(d) None of these.

(28) The power generation of a hydro-plant P_{GH} is directly proportional to:

(a) Plant head.

(b) Water discharge ω_j.

(c) Both (a) and (b).

(d) None of these.

(29) The main advantages of the operation of a hydro-thermal system are:

(i) Greater economy.

(ii) Security of supply and flexibility.

(iii) Better energy conservation.

(iv) Reduction in reserve capacity maintenance.

Regarding the above statement, which is correct?

(a) (i) and (ii).

(b) (ii) and (iii).

(c) all except (iii).

(d) All of these.

(30) The co-ordination equations used to obtain the optimal scheduling of a hydro-thermal system when considering transmission losses are:

(a) $$\frac{dC_I}{dP_{GT_I}}\left(\frac{1}{1-\dfrac{\partial P_L}{\partial P_{GT_I}}}\right)=\lambda=\frac{d\omega_J}{dP_{GH_J}}\left(\frac{1}{1-\dfrac{\partial P_L}{\partial P_{GH_J}}}\right).$$

(b) $$\frac{dC_I}{dP_{GH_I}}\left(\frac{1}{1-\dfrac{\partial P_L}{\partial P_{GH_I}}}\right)=\lambda=\gamma_J\frac{d\omega_J}{dP_{GT_J}}\left(\frac{1}{1-\dfrac{\partial P_L}{\partial P_{GT_J}}}\right).$$

(c) $$\frac{dC_I}{dP_{GT_I}}\left(\frac{1}{1-\dfrac{\partial P_L}{\partial P_{GT_I}}}\right)=\lambda=\gamma_J\frac{d\omega_J}{dP_{GH_J}}\left(\frac{1}{1-\dfrac{\partial P_L}{\partial P_{GH_J}}}\right).$$

(d) None of these.

(31) As far as possible, hydro-plants are used for base-load operation since:

(a) Their capital cost is high.

(b) Their operation is easy.

(c) Their capital cost is low.

(d) Their efficiency is low.

(32) A thermal plant gives minimum cost per unit of energy generated when used as a __________ plant.

(a) Peak load.

(b) Base-load plant.

(c) Simultaneously as base-load and peak load plant.

(d) None of these.

(33) In the combined operation of steam plant and run-off river plants, the sites of hydro and steam plants can be found with the help of __________ .

(a) Demand curve.

(b) Input–output curve.

(c) Load curve.

(d) Chronological load curve.

(34) Long-term hydro-thermal co-ordination can be done by:

(a) Plotting the basic rule curve.

(b) Plotting no spill-rule curve.

(d) Plotting the full reservoir storage curve.

(d) All of these.

(35) __________ hydro-thermal co-ordination is done for the available water and is to be used in a given period (24 hr).

(a) Long-term.

(b) Short-term.

(c) Both (a) and (b).

(d) None of these.

(36) Hydro-thermal co-ordination is necessary only in countries with:

(a) Ample coal resources.

(b) Ample water resources.

(c) Both (a) and (b).

(d) None of these.

(37) In short-term hydro-thermal co-ordination,

(a) No spill-rule curve is used.

(b) Spill-rule curve is used.

(c) Here no rule curve is used due to constraints.

(d) None of these.

(38) The units of incremental water rate are:

(a) m^3/s-MW.

(b) m^3-s/MW.

(c) m-s/MW.

(d) m^2-s/MW.

(39) Hydro-generation is a function of:

(a) Water head.

(b) Water discharge.

(c) Water inflow.

(d) Both (a) and (b).

(40) In the long-term hydro-thermal co-ordination,

(a) Basic rule curve is plotted.

(b) No spill curve.

(c) No full reservoir storage curve.

(d) All of these.

(41) In the combined operation of a steam and a run-off river plant, the sizes of hydro and steam plants can be obtained with the help of:

(a) Load curve.

(b) Demand curve.

(c) Chronological load curve.

(d) None of these.

REVIEW QUESTIONS

(1) Explain the hydro-thermal co-ordination and its importance.

(2) Describe the types of hydro-thermal co-ordination.

(3) What are the factors on which economic operation of a combined hydro-thermal system depends?

(4) What are the important methods of hydro-thermal co-ordination? Explain them in brief.

(5) Explain the mathematical formulation of long-term hydro-thermal scheduling.

(6) Explain the solution method of long-term hydro-thermal scheduling by discretization principle.

(7) Develop an algorithm for the solution of long-term hydro-thermal scheduling problem.

(8) Derive the condition for optimality of short-term hydro-thermal scheduling problem.

(9) What are the advantages of hydro-thermal plants combinations?

PROBLEMS

(1) The system shown in Fig. 6.8(a) is to supply a load shown in Fig. 6.8(b). The data of the system are as follows:

$$C_T = (16 + 0.01P_{GT})P_{GT}\,\text{Rs./hr}$$

$$w_2 = (4 + 0.0035P_{GH})P_{GH}\,\text{m}^3/\text{s}$$

The maximum capacity of a hydro-plant and a steam plant are 400 and 270 MW, respectively. Determine the generating schedule of the system so that 130.426 million m³ water is used during the 24-hr period.

(2) A thermal station and a hydro-station supply an area jointly. The hydro-station is run 12 hr daily and the thermal station is run throughout 24 hr. The incremental fuel cost characteristic of the thermal plant is

$$C_T = 3 + 5P_{GT} + 0.02P_{GT}^2\,\text{Rs./hr}$$

If the load on the thermal station, when both plants are in operation, is 250 MW, the incremental water rate of the hydro-power plant is

$$\frac{d\omega}{dP_{GH}} = 24 + 0.04P_{GH}\,\text{m}^3\,/\,\text{MW-s}$$

The total quantity of water utilized during the 12-hr operation of a hydro-plant is 450 million m³. Find the generation of the hydro-plant and the cost of water used. Assume that the total load on the hydro-plant is constant for the 12-hr period.

(3) A two-plant system that has a thermal station near the load center and a hydro-power station at a remote location is shown in Fig. 6.9.

The characteristics of both stations are:

$$C_T = (20 + 0.03P_{GT})P_{GT}\,\text{Rs./hr}$$

$$w_2 = (8 + 0.002P_{GH})P_{GH}\,\text{m}^3/\text{s}$$

and $\gamma_2 = \text{Rs. } 5 \times 10{-}4/\text{m}^3$

The transmission loss co-efficient, $B_{22} = 0.0005$.

(i) if the load is 700 MW for 15-hr a day and 500 MW for 9 hr on the same day, find the generation schedule, daily water used by hydro-plant, and the daily operating cost of the thermal power.

(ii) Determine the power generation at each station and the power received by the load when $\lambda = 50$ Rs./MWh.

(4) A two-plant system that has a hydro-station near the load center and a thermal power station at a remote location is shown in Fig. 6.10.

The characteristics of both stations are

$$C_T = (20 + 0.03P_{GT})P_{GT}\,\text{Rs./hr}$$

$$w_2 = (8 + 0.002P_{GH})P_{GH}\,\text{m}^3/\text{s}$$

and $\gamma_2 = \text{Rs. } 5.5/\text{m}^3$

The transmission loss co-efficient, $B_{22} = 0.0005$.

(i) If the load is 700 MW for 15 hr a day and 500 MW for 9 hr on the same day, find the generation schedule, daily water used by the hydro-plant, and the daily operating cost of thermal power.

(ii) Determine the power generation at each station and the power received by the load when $\lambda = 50$ Rs. /MWh.

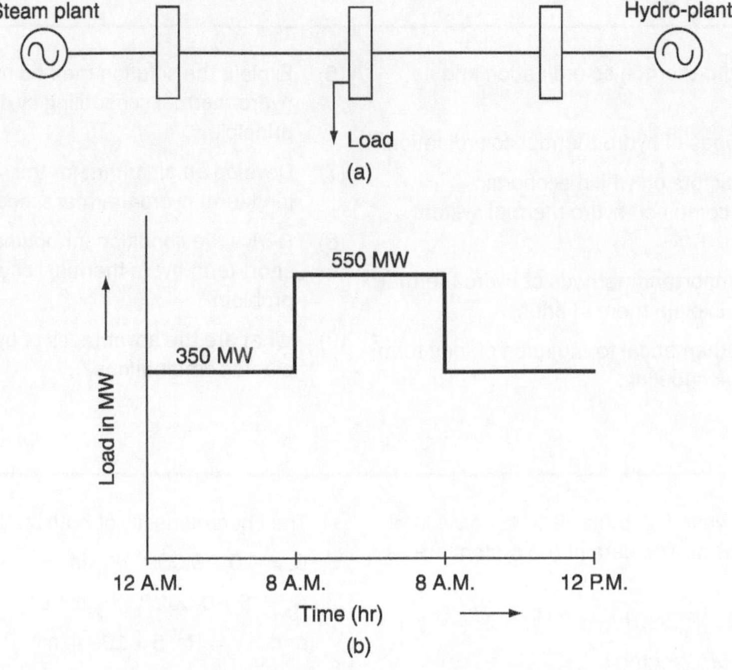

FIG. 6.8   *Illustration for Problem 1; (a) two-plant system; (b) daily load curve*

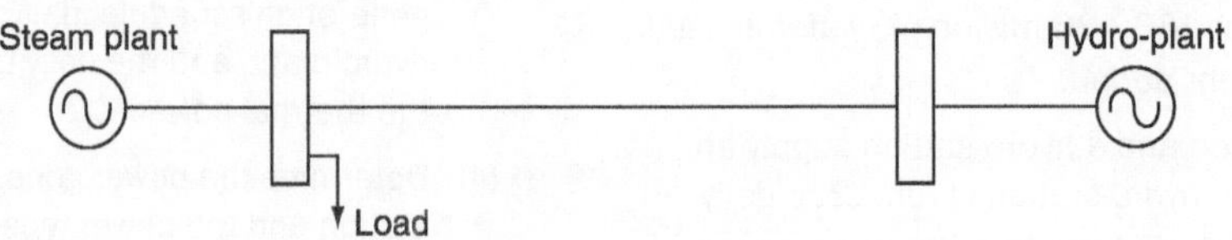

FIG. 6.9   *Two-plant system*

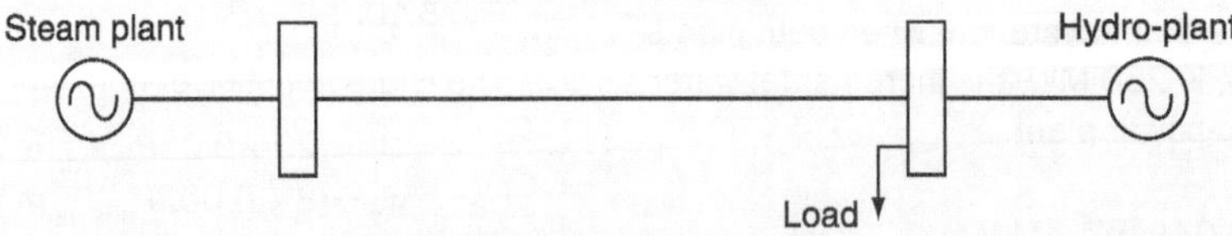

FIG. 6.10   *Two-plant system*

7 Load Frequency Control-I

OBJECTIVES

After reading this chapter, you should be able to:

- study the governing characteristics of a generator
- study the load frequency control (LFC)
- develop the mathematical models for different components of a power system
- observe the steady state and dynamic analysis of a single-area power system with and without integral control

7.1 INTRODUCTION

In a power system, both active and reactive power demands are never steady and they continually change with the rising or falling trend. Steam input to turbo-generators or water input to hydro-generators must, therefore, be continuously regulated to match the active power demand, failing which the machine speed will vary with consequent change in frequency and it may be highly undesirable. The maximum permissible change in frequency is $\pm 2\%$. Also, the excitation of the generators must be continuously regulated to match the reactive power demand with reactive power generation; otherwise, the voltages at various system buses may go beyond the prescribed limits. In modern large interconnected systems, manual regulation is not feasible and therefore automatic generation and voltage regulation equipment is installed on each generator. The controllers are set for a particular operating condition and they take care of small changes in load demand without exceeding the limits of frequency and voltage. As the change in load demand becomes large, the controllers must be reset either manually or automatically.

7.2 NECESSITY OF MAINTAINING FREQUENCY CONSTANT

Constant frequency is to be maintained for the following functions:

- All the AC motors should require constant frequency supply so as to maintain speed constant.

- In continuous process industry, it affects the operation of the process itself.

- For synchronous operation of various units in the power system network, it is necessary to maintain frequency constant.

- Frequency affects the amount of power transmitted through interconnecting lines.

- Electrical clocks will lose or gain time if they are driven by synchronous motors, and the accuracy of the clocks depends on frequency and also the integral of this frequency error is loss or gain of time by electric clocks.

7.3 LOAD FREQUENCY CONTROL

Load frequency control (LFC) is the basic control mechanism in the power system operation. Whenever there is a variation in load demand on a generating unit, there is momentarily an occurrence of unbalance between real-power input and output. This difference is being supplied by the stored energy of the rotating parts of the unit.

The kinetic energy of any unit is given by

$$\text{KE} = \frac{1}{2}\, I\omega^2$$

where I is the moment of inertia of the rotating part and ω the angular speed of the rotating part.

If KE reduces, ω decreases; then the speed falls, hence the frequency reduces. The change in frequency Δf is sensed and through a speed-governor system, it is fed back to control the position of the inlet valve of the prime mover, which is connected to the generating unit. It changes the input to the prime mover suitably and tries to bring back the balance between the real-power input and output. Hence, it can be stated that the frequency variation is dependent on the real-power balance of the system.

The LFC also controls the real-power transfer through the interconnecting transmission lines by sensing the change in power flow through the tie lines.

7.4 GOVERNOR CHARACTERISTICS OF A SINGLE GENERATOR

Prime movers driving the generators are fitted with governors, which are regarded as primary control elements in the LFC system. Governors sense the change in a speed control mechanism to adjust the opening of steam valves in the case of steam turbines and the opening of water gates in the case of water turbines. The characteristics of a typical governor of a steam turbine are shown in Fig. 7.1, which is linearized by dotted lines for studying the system behavior.

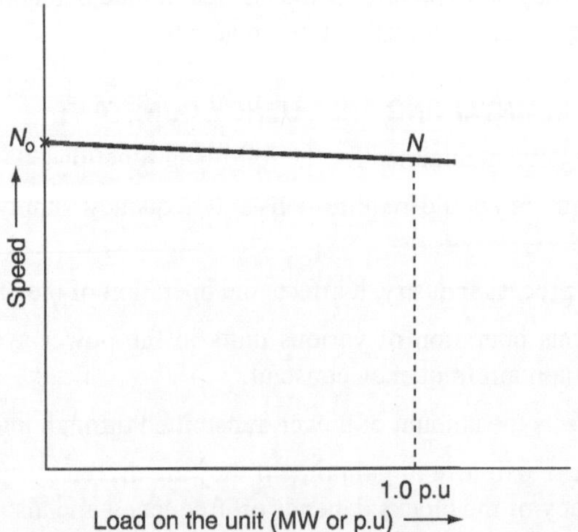

FIG. 7.1   *Characteristics of a typical governor of a steam turbine*

The amount of speed drop as the load on the turbine is increased from no load to its full-load value is $(N_0 - N)$, where N_0 is the speed at no load and N is the speed at rated load.

The steady-state speed regulation in per unit is given by

$$R = \frac{N_0 - N}{N}$$

The value of R varies from 2% to 6% for any generating unit. Since the frequency and speed are directly related, the speed regulation can also be expressed as the ratio of the change in frequency from no load to its full load to the rated frequency of the unit:

i.e., $\quad R = \dfrac{f_0 - f}{f}$

If there is a 4% speed regulation of a unit, then for a rated frequency of 50 Hz, there will be a drop of 2 Hz in frequency.

If the generation is increased by ΔP_G due to a static frequency drop of Δf, then the speed regulation can be defined as the ratio of the change in frequency to the corresponding change in real-power generation:

i.e., $\quad R = -\dfrac{\Delta f}{\Delta P_G}$

The unit of R is taken as Hz per MW. In practice, power is measured in per unit and hence R is in Hz/p.u. MW.

In Fig. 7.2, the turbine is operating with 99% of no-load speed at 25% of full-load power and if the load is increased to 50%, the speed drops to 98%. Let 'A' be the initial operating point of the turbine at 50% load and if the load is dropped to 25%, the speed increases to 99%. In order to keep the speed at 25% of the load same as at 'A', the governor setting has to be changed by changing the spring tension in the fly-ball type of governor.

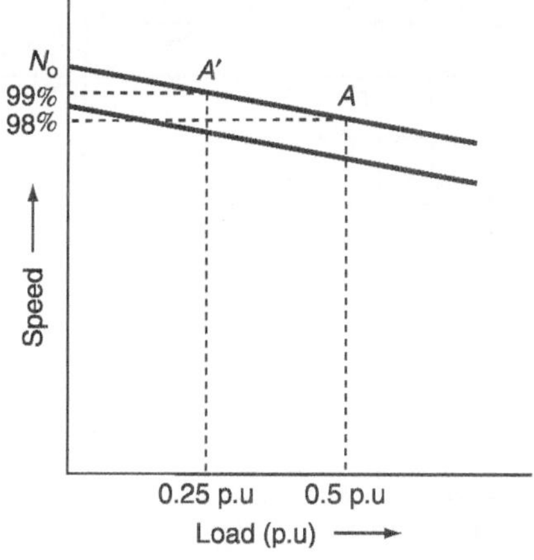

FIG. 7.2 *Speed-governor setting*

This will result in speed characteristics indicated by the dotted line parallel to the first one and below it, passing through the point A', which is the point of intersection of the new speed line and 25% load line. Hence, the turbine can be adjusted to carry any given load at any desired speed.

This type of shifting the speed or frequency characteristic parallel to itself is known as supplementary control. It is adopted in on-line control to ensure proper load division among the running units and to maintain the system frequency. There is another method of changing the slope of the governor characteristics. This is achieved by changing the ratio of the lever L (refer the speed control mechanism) of the governor and thereby adjusting the parameter R to ensure proper co-ordination with the other units of the system. This adjustment can be made during the off-line condition only.

7.5 ADJUSTMENT OF GOVERNOR CHARACTERISTIC OF PARALLEL OPERATING UNITS

When two generators are running in parallel, the governor characteristic of the first unit (Line 1) is shown towards the right, while that of the second unit (Line 2) is shown towards the left of the frequency axis as shown in Fig. 7.3.

The characteristics are obviously different and hence corresponding to the rated frequency f_r, the two units carry loads P_1 and P_2 so that the system load $P_D = P_1 + P_2$. If the system load is now increased to P'_D, the system frequency will drop down to f', since the units can only increase their output by decreasing the speed.

To restore the system frequency, the characteristic of one of the units say of Unit 1 needs to be shifted upwards as indicated by the dotted characteristic, so that it can carry the increased load. The share of Unit 1 will be P'_1 and that of Unit 2 will be P_2 so that the increased total load, $P'_D = P'_1 + P_2$.

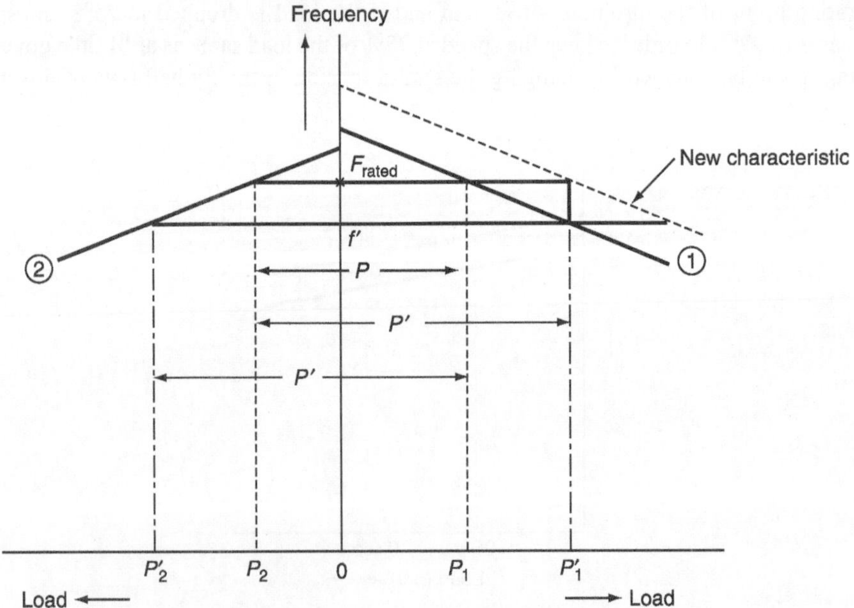

FIG. 7.3   *Sharing of load by two units (parallel) with a speed-governor characteristics setting*

7.6 LFC: (*P–f* CONTROL)

The LFC, also known as generation control or $P\text{–}f$ control, deals with the control of loading of the generating units for the system at normal frequency. The load in a power system is never constant and the system frequency remains at its nominal value only when there is a match between the active power generation and the active power demand. During the period of load change, the deviation from the nominal frequency, which may be called frequency error (Δf), is an index of mismatch and can be used to send the appropriate command to change the generation by adjusting the LFC system. It is basically controlling the opening of the inlet valves of the prime movers according to the loading condition of the system. In the case of a multi-area system, the LFC system also maintains the specified power interchanges between the participating areas. In a smaller system, this control is done manually, but in large systems automatic control devices are used in the loop of the LFC system.

The LFC system, however, does not consider the reactive power flow in the system even though the reactive power flow is also affected to some extent during the fluctuating load condition. But since there is no counterpart of the reactive power in the mechanical side of the system, it does not come within the loop of the LFC system.

7.7 *Q–V* CONTROL

In this control, the terminal voltage of the generator is sensed and converted into proportionate DC signal and then compared to DC reference voltage. The error in between a DC signal and a DC reference voltage, i.e., $\Delta \left| V_i \right|$ is taken as an input to the $Q\text{–}V$ controller. A control output ΔQ_{ci} is applied to the exciter.

7.8 GENERATOR CONTROLLERS (*P–f* AND *Q–V* CONTROLLERS)

The active power P is mainly dependent on the internal angle δ and is independent of the bus voltage magnitude $\left| V \right|$. The bus voltage is dependent on machine excitation and hence on reactive power Q and is independent of the machine angle δ. Change in the machine angle δ is caused by a momentary change in the generator speed and hence the frequency. Therefore, the load frequency and excitation voltage controls are non-interactive for small changes and can be modeled and analyzed independently.

Figure 7.4 gives the schematic diagram of load frequency ($P\text{–}f$) and excitation voltage ($Q\text{–}V$) regulators of a turbo-generator. The objective of the MW frequency or the $P\text{–}f$ control mechanism is to exert control of frequency and simultaneously exchange of the real-power flows via interconnecting lines. In this control, a frequency sensor senses the change in frequency and gives the signal Δf_i. The $P\text{–}f$ controller senses the change in frequency signal (Δf_i) and the increments in tie-line real powers (ΔP_{tie}), which will indirectly provide information about the incremental state error ($\Delta \delta_i$). These sensor signals (Δf_i and ΔP_{tie}) are amplified, mixed, and transformed into a real-power control signal ΔP_{ci}. The valve control mechanism takes ΔP_{ci} as the input signal and provides the output signal, which will change the position of the inlet valve of the prime mover. As a result, there will be a change in the prime mover output and hence a change in real-power generation ΔP_{Gi}. This entire $P\text{–}f$ control can be yielded by automatic load frequency control (ALFC) loop.

The objective of the MVAr-voltage or $Q\text{–}V$ control mechanism is to exert control of the voltage state $\left| V_i \right|$. A voltage sensor senses the terminal voltage and converts it into an equivalent proportionate DC voltage. This proportionate DC voltage is compared with a

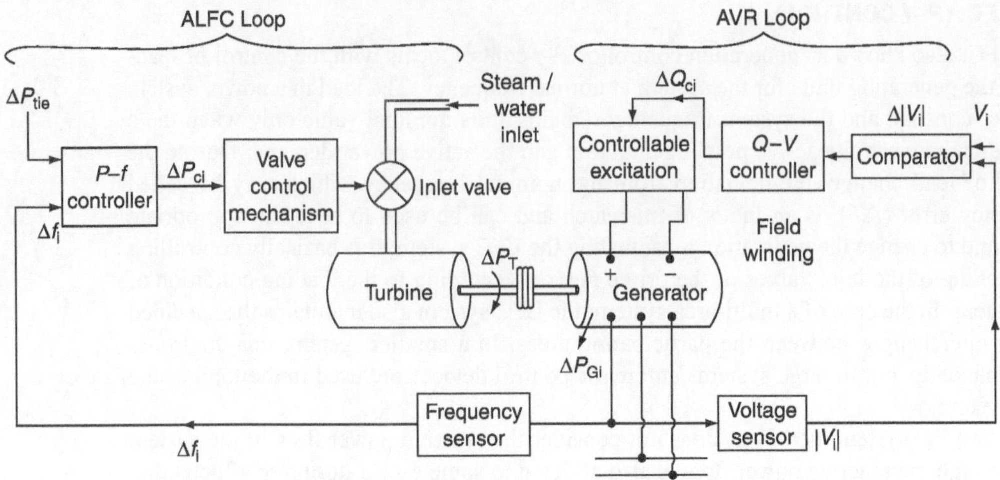

FIG. 7.4   *Schematic diagram of P–f controller and Q–V controller*

reference voltage V_{iref} by means of a comparator. The output obtained from the comparator is error signal $\Delta|V_i|$ and is given as input to Q–V controller, which transforms it to a reactive power signal command ΔQ_{ci} and is fed to a controllable excitation source. This results in a change in the rotor field current, which in turn modifies the generator terminal voltage. This entire Q–V control can be yielded by an automatic voltage regulator (AVR) loop.

In addition to voltage regulators at generator buses, equipment is used to control voltage magnitude at other selected buses. Tap-changing transformers, switched capacitor banks, and static VAr systems can be automatically regulated for rapid voltage control.

7.9 *P–f* CONTROL VERSUS *Q–V* CONTROL

Any static change in the real bus power ΔP_i will affect only the bus voltage phase angles (δ_i) (since $P \propto \delta$), but will leave the bus voltage magnitudes almost unaffected.

Static change in the reactive bus power ΔQ_i affects essentially only the bus voltage magnitudes (since $Q \propto V^2$), but leave the bus voltage phase angles almost unchanged.

Static change in reactive bus power at a particular bus affects the magnitude of that bus voltage most strongly, but in less degree the magnitudes of the bus voltages at remote buses.

7.10 DYNAMIC INTERACTION BETWEEN *P–f* AND *Q–V* LOOPS

In a static sense, for small deviations, there is a little interaction between P–f and Q–V loops. During dynamic perturbations, we encounter considerable coupling between two control loops for two following reasons:

- As the voltage magnitude fluctuates at a bus, the real load of that bus will likewise change as a result of the voltage load characteristic $\dfrac{\partial P_{di}}{\partial V_i}$.

- As the voltage magnitude fluctuates at a bus, the power transmitted over the lines connected to that bus will change. In other words, the change in Q–V loop will affect the generated emf, which also affects the magnitude of real power.

A dynamic perturbation in the Q–V loop will thus affect the real-power balance in the system. In general, the Q–V loop is much faster than the P–f loop due to the mechanical inertial constants in the P–f loop. If it can be assumed that the transients in the Q–V loop are essentially over before the P–f loop reacts, then the coupling between the two loops can be ignored.

7.11 SPEED-GOVERNING SYSTEM

The speed governor is the main primary tool for the LFC, whether the machine is used alone to feed a smaller system or whether it is a part of the most elaborate arrangement. A schematic arrangement of the main features of a speed-governing system of the kind used on steam turbines to control the output of the generator to maintain constant frequency is as shown in Fig. 7.5.

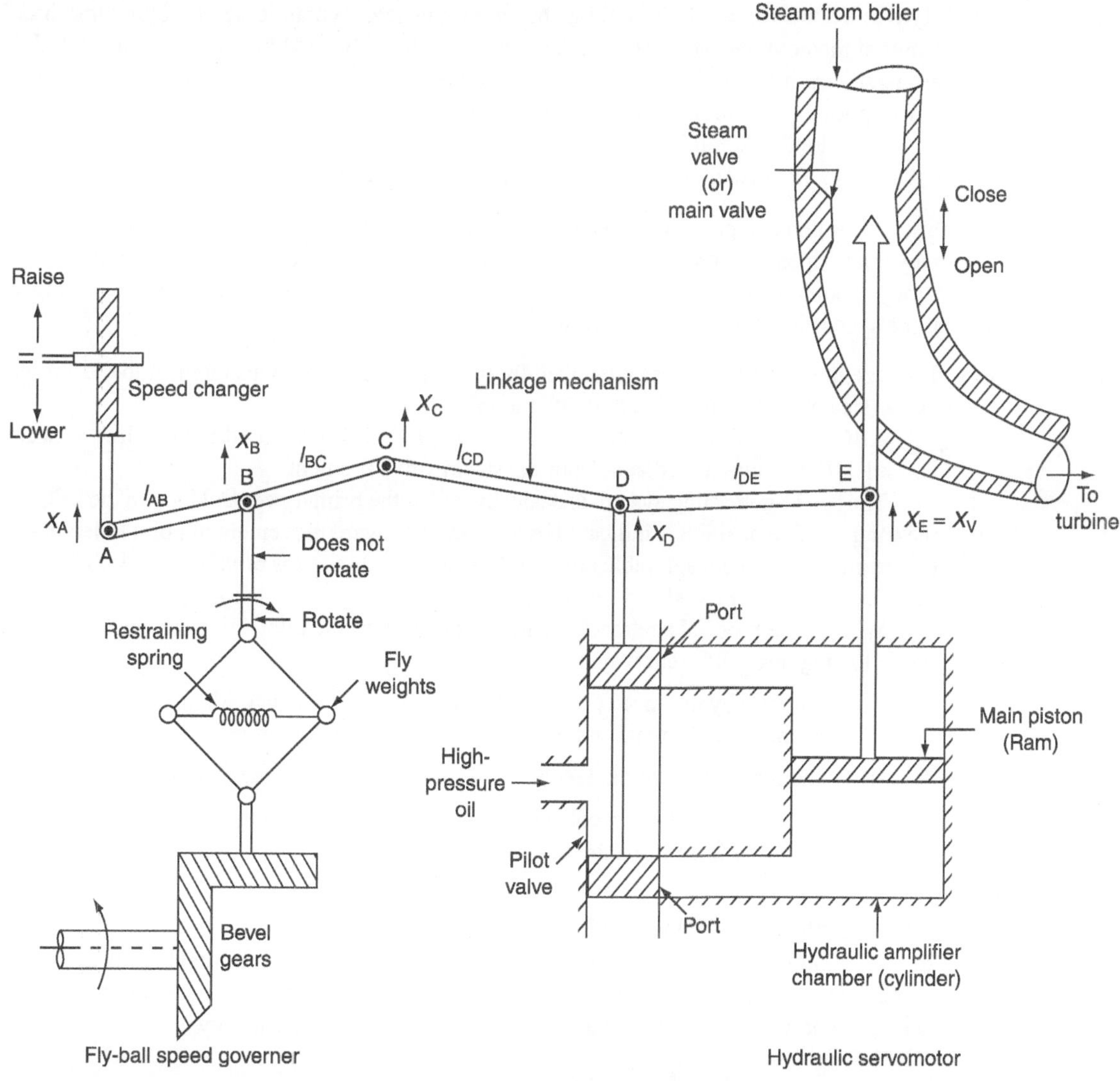

FIG. 7.5 *Speed-governor system*

Its main parts or components are as follows:

Fly-ball speed governor: It is a purely mechanical, speed-sensitive device coupled directly to and builds directly on the prime movers to adjust the control valve opening via linkage mechanism. It senses a speed deviation or a power change command and converts it into appropriate valve action. Hence, this is treated as the heart of the system, which controls the change in speed (frequency). As the speed increases, the fly balls move outwards and the point B on linkage mechanism moves upwards. The reverse will happen if the speed decreases.

The horizontal rotating shaft on the lower left may be viewed as an extension of the shaft of a turbine-generator set and has a fixed axis as shown in Fig. 7.5. The vertical shaft, above the fly-ball mechanism, also rotates between fixed bearings. Although its axis is fixed, it can move up and down, transferring its vertical motion to the pilot point B.

Hydraulic amplifier: It is nothing but a single-state hydraulic servomotor interposed between the governor and valve. It consists of a pilot valve and the main piston. With this arrangement, hydraulic amplification is obtained by converting the movement of low-power pilot valve into movement of high-power level main piston.

In hydraulic amplification, a large mechanical force is necessary so that the steam valve could be opened or closed against high-pressure inlet steam.

Speed changer: It provides a steady-state power output setting for the turbines. Its upward movement opens the upper pilot valve so that more steam is admitted to the turbine under steady conditions. This gives rise to higher steady-state power output. The reverse will happen if the speed changer moves downward.

Linkage mechanism: These are linked for transforming the fly-balls moment to the turbine valve (steam valve) through a hydraulic amplifier.

ABC is a rigid link pivoted at point B and CDE is another rigid pivoted link at point D. Link DE provides a feedback from the steam valve moment.

The speed-governing system is basically called the primary control loop in the LFC. If the control valve position is indicated by x_{E}, a small upward movement of point E decreases the steam flow by a considerable amount. It is measured in terms of valve power ΔP_{v}. This flow decrement gets translated into decrement in turbine power output ΔP_{T}.

With the help of linkage mechanism, the position of the pilot valve can be changed in the following three different ways:

(i) Directly by the speed changer: A small upward moment of linkage point A corresponds to a decrease in the steady-state power output or reference power ΔP_{ref}.

(ii) Indirectly through the feedback due to the position changes in the main system.

(iii) Indirectly through feedback due to the position changes in linkage point E resulting from a change in speed.

7.11.1 Speed-governing system model

In this section, we develop the mathematical model based on small deviations around a nominal steady state. Let us assume that the steam is operating under steady state and is delivering power P_{G}^{0} from the generator at nominal speed or frequency f°.

Under this condition, the prime mover valve has a constant setting x_E^0, the pilot valve is closed, and the linkage mechanism is stationary. Now, we will increase the turbine power by ΔP_C with the help of the speed changer. For this, *the movement of linkage point A* moves downward by a small distance Δx_A and is given by

$$\Delta x_A = K \, \Delta P_C \tag{7.1}$$

With the movement Δx_A, the link point C move upwards by an amount Δx_C and so does the link point D by an amount Δx_D upwards. Due to the movement of link point D, the pilot valve moves upwards, then the high-pressure oil is admitted into the cylinder of the hydraulic amplifier and flows on to the top of the main piston. Due to this, the piston moves downward by an amount Δx_E and results in the opening of the steam valve. Due to the opening of the steam valve, the flow of steam from the boiler increases and the turbine power output increases, which leads to an increase in power generation by ΔP_G. The increased power output causes an accelerating power in the system and there is a slight increase in frequency say by Δf if the system is connected to a finite size (i.e., not connected to infinite bus).

Now with the increased speed, the fly balls of the governor move downwards, thus causing the link point B to move slightly downwards by a small distance Δx_B proportional to Δf. Due to the downward movement of link point B, the link point C also moves downwards by an amount Δx_C, which is also proportional to Δf.

It should be noted that all the downward movements are assumed to be positive in directions as indicated in Fig. 7.5. Now model the above events mathematically.

The net *movement of link point C* contributes two factors as follows:

(i) Δx_A contribution: The lowering of speed changer by an amount Δx_A results in the upward moment of link point C proportional to Δx_A:

i.e., $\Delta x'_C = \Delta x_A \, l_{AB} = -\Delta x_C \, l_{BC}$

or $\Delta x'_C = -\left[\dfrac{l_{BC}}{l_{AB}}\right] \Delta x_A$

Substituting Δx_A from Equation (7.1) in the above equation, we get

$$\Delta x'_C = -\left[\frac{l_{BC}}{l_{AB}}\right] K \Delta P_C$$

$$= -K_1 \Delta P_C \tag{7.2}$$

where $K_1 = K \left[\dfrac{l_{BC}}{l_{AB}}\right]$

(ii) Δf contribution: Increase in frequency Δf causes an outward moment of fly balls and in turn causes the downward movement of point B by an amount Δx_B, which is proportional to $K'_2 \Delta f$, i.e., movement of point 'C' with point 'A' remaining fixed at Δx_A is

$$\left(\frac{l_{BC} + l_{AB}}{l_{AB}}\right) K'_2 \Delta f = K_2 \Delta f$$

$$\therefore \Delta x''_{C} = K_2 \Delta f \tag{7.3}$$

Therefore, the net movement of link point C can be expressed as

$$\Delta x_{C} = \Delta x'_{C} + \Delta x''_{C} \tag{7.4}$$

Substituting the values of $\Delta x'_{C}$ and $\Delta x''_{C}$ from Equations (7.2) and (7.3) in Equation (7.4), we get

$$\Delta x_{C} = -K_1 \Delta P_{C} + K_2 \Delta f \tag{7.5}$$

The constants K_1 and K_2 depend upon the length of linkage arms AB and BC and also depend upon the proportional constants of the speed changer and the speed governor.
The *movement of link point D,* Δx_{D} is the amount by which the pilot valve opens and it is contributed by the movement of point C, Δx_{C}, and movement of point E, Δx_{E}.
Therefore, the net movement of point D can be expressed as

$$\Delta x_{D} = \Delta x'_{D} + \Delta x''_{D} \tag{7.6}$$

where $\Delta x'_{D} \, (l_{CD} + l_{DE}) = \Delta x_{C}(l_{DE})$

$$\Delta x'_{D} = \frac{l_{DE}}{(l_{CD} + l_{DE})} \Delta x_{C}$$

$$= K_3 \Delta x_{C} \tag{7.7}$$

and $\Delta x''_{D} \, (l_{CD} + l_{DE}) = \Delta x_{E} \, (l_{CD})$

$$\Delta x''_{D} = \frac{l_{CD}}{(l_{CD} + l_{DE})} \Delta x_{E}$$

$$= K_4 \Delta x_{E} \tag{7.8}$$

Substituting the values of $\Delta x'_{D}$ and $\Delta x''_{D}$ from Equations (7.7) and (7.8) in Equation (7.6), we get

$$\Delta x_{D} = K_3 \Delta x_{C} + K_4 \, \Delta x_{E} \tag{7.9}$$

The movement Δx_{D}, results in the opening of the pilot valve, which leads to the admission of high-pressure oil into the hydraulic amplifier cylinder; then the downward movement of the main piston takes place and thus the steam valve opens by an amount Δx_{E}.

Two assumptions are made to represent the mathematical model of the *movement of point E*:

(i)   The main piston and steam valve have some inertial forces, which are negligible when compared to the external forces exerted on the piston due to high-pressure oil.

(ii) Because of the first assumption, the amount of oil admitted into the cylinder is proportional to the port opening Δx_{D}, i.e., the volume of oil admitted into the cylinder is proportional to the time integral of Δx_{D}.

The movement Δx_E is obtained as

$$\Delta x_E = \frac{\text{volume of oil admitted}}{\text{area of cross-section of the piston}} = \frac{1}{A} \int_0^t (-\Delta x_D) \, dt$$

where A is the area of cross-section of the piston:

$$\therefore \Delta x_E = K_5 \int_0^t (-\Delta x_D) \, dt \tag{7.10}$$

where $K_5 = \dfrac{1}{A}$

The constant K_5 depends upon the fluid pressure and the geometry of the orifice and cylinder of the hydraulic amplifier.

In Equation (7.10), the negative sign represents the movements of link points D and E in the opposite directions. For example, the small downward movement of Δx_D causes the movement Δx_E in the positive direction (i.e., upwards).

Taking the Laplace transform of Equations (7.5), (7.9), and (7.10), we get

$$\Delta x_C(s) = -K_1 \Delta P_C(s) + K_2 \Delta F(s) \tag{7.11}$$

$$\Delta x_D(s) = -K_3 \Delta x_C(s) + K_4 \Delta x_E(s) \tag{7.12}$$

$$\Delta x_E(s) = -K_5 \frac{1}{s} \Delta x_D(s) \tag{7.13}$$

Eliminating $\Delta x_C(s)$ and $\Delta x_D(s)$ in the above equations and substituting $\Delta x_D(s)$ from Equation (7.12) in Equation (7.13), we get

$$\Delta x_E(s) = -K_5 \frac{1}{s} \left[K_3 \Delta x_C(s) + K_4 \Delta x_E(s) \right]$$

Substituting $\Delta x_C(s)$ from Equation (7.11) in the above equation, we get

$$\Delta x_E(s) = -K_5 \frac{1}{s} \left[K_3 \left(K_1 \Delta P_C(s) + K_2 \Delta F(s) \right) + K_4 \Delta x_E(s) \right]$$

$$\Delta x_E(s) \left(1 + \frac{K_4 K_5}{s} \right) = \frac{1}{s} \left(K_1 K_3 K_5 \Delta P_C(s) - K_2 K_3 K_5 \Delta F(s) \right)$$

$$\Delta x_E(s) \left(K_4 + \frac{s}{K_5} \right) = K_1 K_3 \Delta P_C(s) - K_2 K_3 \Delta F(s)$$

$$\Delta x_E(s) = \frac{K_1 K_3 \Delta P_C(s) - K_2 K_3 \Delta F(s)}{\left[K_4 + \dfrac{s}{K_5} \right]} \tag{7.14}$$

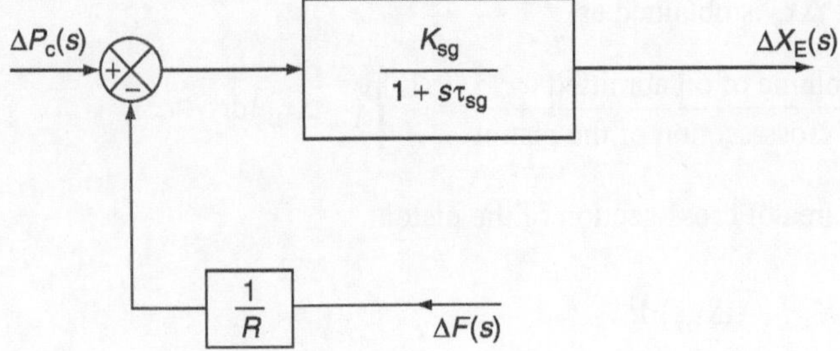

FIG. 7.6   *Block diagram model of a speed-governor system*

Equation (7.14) can be modified as

$$\Delta x_{\mathrm{E}}\left(s\right) = \left[\Delta P_{\mathrm{C}}\left(s\right) - \frac{1}{R}\Delta F\left(s\right)\right]\left[\frac{K_{\mathrm{sg}}}{1+s\tau_{\mathrm{sg}}}\right] \tag{7.15}$$

where $R \triangleq \dfrac{K_1}{K_2}$ is the speed regulation of the governor it is also termed as regulation constant or setting, $K_{\mathrm{sg}} \triangleq \dfrac{K_1 K_3}{K_4}$ the gain of the speed governor, and $\tau_{\mathrm{sg}} \triangleq \dfrac{1}{K_4 K_5}$ the time constant of the speed governor. Normally, $\tau_{\mathrm{sg}} \leq 100$ ms.

Equation (7.15) can be represented in a block diagram model as shown in Fig. 7.6, which is the linearized model of the speed-governor mechanism.

From the block diagram, $\left[\Delta P_{\mathrm{C}}\left(s\right) - \dfrac{1}{R}\Delta F\left(s\right)\right]$ is the net input to the speed-governor system and $\Delta x_{\mathrm{E}}(s)$ is the output of the speed governor.

7.12 TURBINE MODEL

We are interested not in the turbine valve position but in the generator power increment ΔP_{G}. The change in valve position Δx_{E} causes an incremental increase in turbine power ΔP_{T} and due to electromechanical interactions within the generator, it will result in an increased generator power ΔP_{G}, i.e., $\Delta P_{\mathrm{T}} = \Delta P_{\mathrm{G}}$, since the generator incremental loss is neglected. This overall mechanism is relatively complicated particularly if the generator voltage simultaneously undergoes wild swing due to major network disturbances.

At present, we can assume that the voltage level is constant and the torque variations are small. Then an incremental analysis will give a relatively simple dynamic relationship between Δx_{E} and ΔP_{G}. Such an analysis reveals considerable differences, not only between steam turbines and hydro-turbines, but also between various types (reheat and non-reheat) of steam turbines. Therefore, the transfer function, relates the change in the generated power output with respect to the change in the valve position, varies with the type of the prime mover.

7.12.1 Non-reheat-type steam turbines

Figure 7.7 (a) shows a single-stage non-reheat type steam turbine.

In this model, the turbine can be characterized by a single gain constant K_t and a single time constant τ_t as

$$G_{\mathrm{T}}\left(s\right) = \frac{\Delta P_{\mathrm{T}}(s)}{\Delta x_{\mathrm{E}}(s)} = \frac{K_t}{1+s\tau_t} \tag{7.16}$$

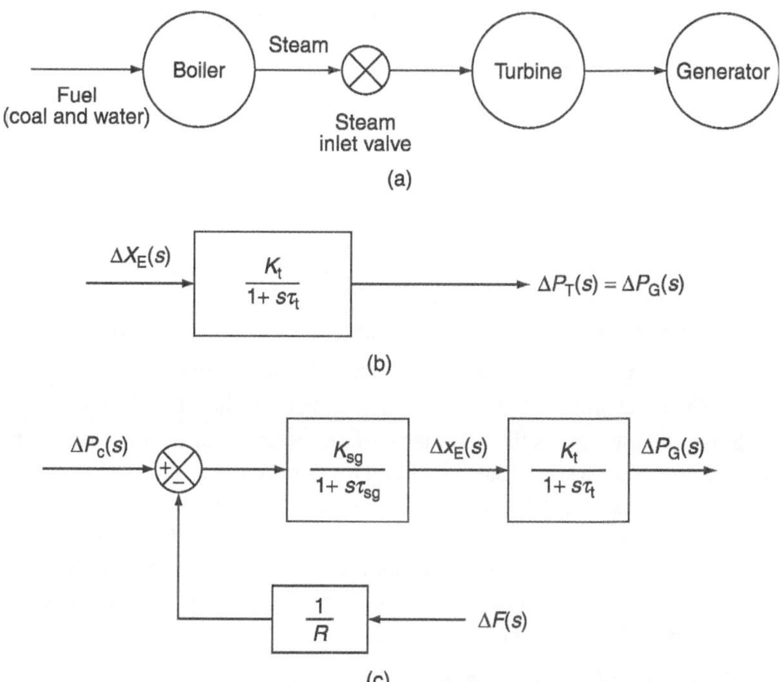

FIG. 7.7   *(a) Single-stage non-reheat-type steam turbine; (b) block diagram representation of a non-reheat-type steam turbine; (c) transfer function representation of speed control mechanism of a generator with a non-reheat-type steam turbine*

Typically, the time constant τ_t lies in the range of 0.2 to 2.

On opening the steam valve, the steam flow will not reach the turbine cylinder instantaneously. The time delay experienced in this is in the order of 2 s in the steam pipe.

From Equation (7.16), we have

$$\Delta P_G(s) = \frac{K_t}{1+s\tau_t}\,\Delta x_E(s) \quad (\text{since } \Delta P_T = \Delta P_G) \tag{7.17}$$

We can represent Equation (7.17) by a block diagram as shown in Fig. 7.7(b).

Figure 7.7(c) shows the linearized model of a non-reheat-type turbine controller including the speed-governor mechanism.

From Fig. 7.7(c), the combined transfer function of the turbine and the speed-governor mechanism will be $\dfrac{K_{sg}K_t}{(1+s\tau_{sg})(1+s\tau_t)}$.

Therefore, $\quad \Delta P_G(s) = \dfrac{K_{sg}K_t}{(1+s\tau_{sg})(1+s\tau_t)}\left[\Delta P_c(s) - \dfrac{1}{R}\Delta F(s)\right]$

In general, it is obtained that the turbine response is low with the response time of several seconds.

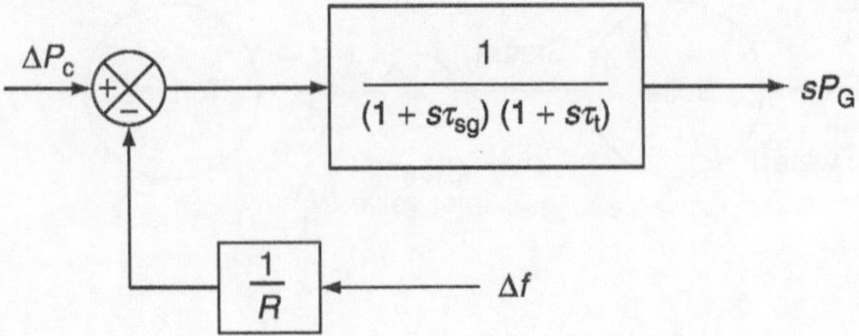

FIG. 7.8   *Block diagram of a simplified turbine governor*

7.12.2 Incremental or small signal for a turbine-governor system

Let the command incremental signal be ΔP_{C}. Then in the steady state, we get $\Delta P_{\text{G}} = K_{\text{sg}} K_{\text{t}} \Delta P_{\text{C}}$. Let $K_{\text{sg}} K_{\text{t}} = 1$; the block diagram of Fig. 7.7(c) is reduced to that shown in Fig. 7.8.

This block diagram gives the derivation of an incremental or small signal model. The model is adopted for large signal use by adding a saturation-type non-linear element, which introduces the obvious fact that the steam valve must operate between certain limits. The valve can neither be more open than fully open nor more closed than fully closed.

This model of Fig. 7.8 may also be modified to account for reheat cycles in the turbine and more accurate representation of fluid dynamics in the steam inlet pipes or in the hydraulic turbines in the penstock.

7.12.3 Reheat type of steam turbines

Modern generating units have reheat-type steam turbines as prime movers for higher thermal efficiency.

Figure 7.9 shows a two-stage reheat-type steam turbine.

In such turbines, steam at high pressure and low temperature is withdrawn from the turbine at an intermediate stage. It is returned to the boiler for resuperheating and then reintroduced into the turbine at low pressure and high temperature. This increases the overall thermal efficiency. Mostly, two factors influence the dynamic response of a reheat-type steam turbine:

(i)   Entrained steam between the inlet steam valve and the first state of turbine.

(ii) The storage action in the reheater, which causes the output of the low-pressure stage to lag behind that of the high-pressure stage.

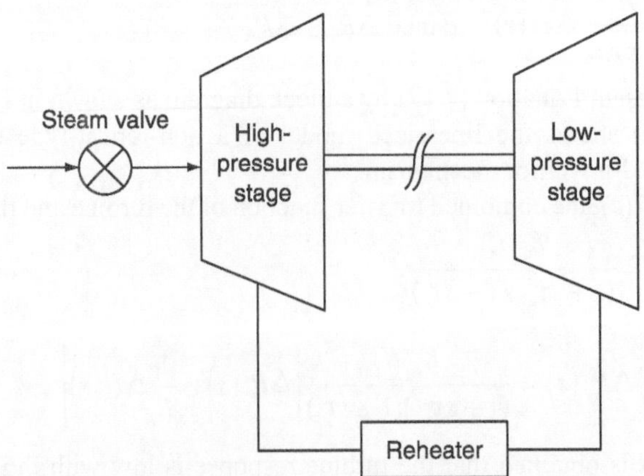

FIG. 7.9   *A two-stage reheat type of a steam turbine*

Thus, in this case, the turbine transfer function is characterized by two time constants. It involves an additional time lag τ_r associated with the reheater in addition to the turbine time constant τ_t. Hence, the turbine transfer function will be of a second order and is given by

$$G_T\left(s\right) = -\frac{\Delta P_G(s)}{\Delta x_E(s)} = \left[\frac{K_t}{1+s\tau_t}\right]\frac{\left(1+sK_r\tau_r\right)}{\left(1+s\tau_r\right)} \tag{7.18}$$

The time constant τ_r has a value in the range of 10 s and approximates the time delay for charging the reheat section of the boiler. K_r is a reheat coefficient and is equal to the proportion of torque developed in the high-pressure section of the turbine:

$K_r = (1 - \text{fraction of the steam reheated})$

When there is no reheat $K_r = 1$ and the transfer function reduces to a single time constant given in Equation (7.16).

The transfer functions as given by Equations (7.16) and (7.18) give good representation within the first 20 s following the incremental disturbance. They do not account for the slower boiler dynamics. To get an easy analyzation, it can be assumed that the prime mover or turbine is modeled by a single equivalent time constant τ_t as given in Equation (7.16).

7.13 GENERATOR–LOAD MODEL

The generator–load model gives the relation between the change in frequency (Δf) as a result of the change in generation (ΔP_G) when the load changes by a small amount (ΔP_D).

When neglecting the change in generator loss, $\Delta P_G = \Delta P_T$ (change in turbine power output), net-surplus power at the bus bar $= (\Delta P_G - \Delta P_D)$. This surplus power can be absorbed by the system in two different ways:

(i) By increasing the stored kinetic energy of the generator rotor at a rate $\dfrac{\mathrm{d}W_{KE}}{\mathrm{d}t}$.

Let W^0_{KE} be the stored KE before the disturbance at normal speed and frequency f^0, and W_{KE} be the KE when the frequency is $(f^0 + \Delta f)$.

Since the stored KE is proportional to the square of speed and the frequency is proportional to the speed,

$$\frac{W_{KE}}{W^0_{KE}} = \left(\frac{f^0 + \Delta f}{f^0}\right)^2$$

$$W_{KE} = W^0_{KE}\left(\frac{f^0 + \Delta f}{f^0}\right)^2 \tag{7.19}$$

$$W_{KE} = W^0_{KE}\left(1 + \frac{2\Delta f}{f^0} + \text{Higher-order terms}\right) \tag{7.20}$$

Neglecting higher-order terms, since $\dfrac{\Delta f}{f^0}$ is small:

$$\therefore W_{KE} \approx W_{KE}^0 \left(1 + \frac{2\Delta f}{f^0}\right)$$

Differentiating the above expression with respect to 't', we get

$$\therefore \frac{dW_{KE}}{dt} = \frac{2W_{KE}^0}{f^0}\frac{d}{dt}(\Delta f) \tag{7.21}$$

Let H be the inertia constant of a generator (MW-s/MVA) and P_r the rating of the turbo-generator (MVA):

$$W_{KE}^0 = H \times P_r \text{ (MW-s or M-J)} \tag{7.22}$$

Hence, Equation (7.21) becomes

$$\frac{dW_{KE}}{dt} = \frac{2HP_r}{f^0}\frac{d}{dt}(\Delta f) \tag{7.23}$$

(ii) The load on the motors increases with increase in speed. The load on the system being mostly motor load, hence some portion of the surplus power is observed by the motor loads. The rate of change of load with respect to frequency can be regarded as nearly constant for small changes in frequency.

$$\text{i.e.,} \quad \left(\frac{\partial P_D}{\partial f}\right)\Delta f = B\Delta f \tag{7.24}$$

where the constant B is the area parameter in MW/Hz and can be determined empirically. B is positive for a predominantly motor load.

Now, the surplus power can be expressed as

$$\Delta P_G - \Delta P_D = \frac{dW_{KE}}{dt} + \left(\frac{\partial P_D}{\partial f}\right)\Delta f$$

From Equations (7.23) and (7.24), the above equation can be modified as

$$\Delta P_G - \Delta P_D = \frac{2HP_r}{f^0}\frac{d}{dt}(\Delta f) + B\Delta f \tag{7.25}$$

Dividing throughout by P_r of Equation (7.25), we get

$$\Delta P_G\,(\text{p.u}) - \Delta P_D\,(\text{p.u}) = \frac{2H}{f^0}\frac{d}{dt}(\Delta f) + B\,(\text{p.u})\Delta f$$

Taking Laplace transform on both sides, we get

$$\Delta P_G\,(s) - \Delta P_D\,(s) = \frac{2H}{f^0}s\Delta F\,(s) + B\Delta F\,(s)$$

$$= \left(\frac{2H}{f^0} s + B \right) \Delta F(s)$$

$$\therefore \Delta F(s) = \frac{\Delta P_G(s) - \Delta P_D(s)}{\dfrac{2H}{f^0} s + B} = \frac{1}{B} \left[\frac{\Delta P_G(s) - P_D(s)}{\left(1 + \dfrac{2H}{Bf^0} s \right)} \right]$$

$$= \left(\frac{K_{ps}}{1 + \tau_{ps}} \right) \left[\Delta P_G(s) - \Delta P_D(s) \right] \qquad (7.26)$$

where $\tau_{ps} = \dfrac{2H}{Bf^0}$ is the power system time constant (normally 20 s) and $K_{ps} = \dfrac{1}{B}$ the power system gain.

Equation (7.26) can be represented in a block diagram model as given in Fig. 7.10.

The overall block diagram of an isolated power system is obtained by combining individual block diagrams of a speed-governor system, a turbine system, and a generator–load model and is as shown in Fig. 7.11.

This representation being a third-order system, the characteristic equation for the system will be of the third order.

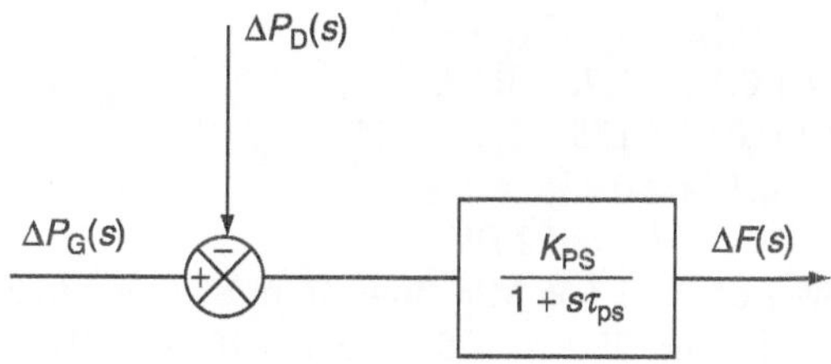

FIG. 7.10   *Block diagram representation of a generator–load model*

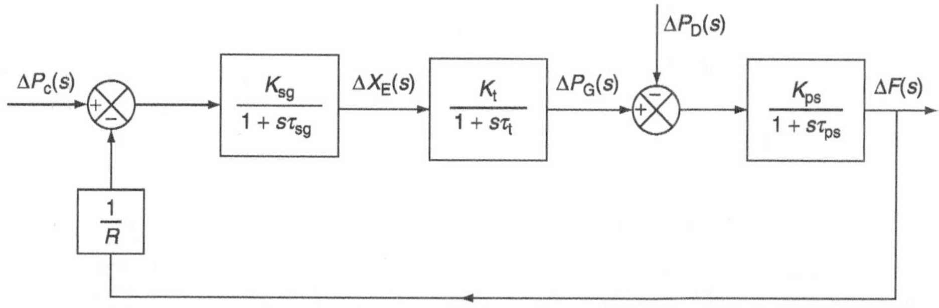

FIG. 7.11   *Complete block diagram representation of an isolated power system*

Example 7.1: Two generating stations 1 and 2 have full-load capacities of 200 and 100 MW, respectively, at a frequency of 50 Hz. The two stations are interconnected by an induction motor and synchronous generator set with a full-load capacity of 25 MW as shown in Fig. 7.12. The speed regulation of Station-1, Station-2, and induction motor and synchronous generator set are 4%, 3.5%, and 2.5%, respectively. The loads on respective bus bars are 750 and 50 MW, respectively. Find the load taken by the motor-generator set.

Solution:

Let a power of A MW flow from Station-1 to Station-2:

$\therefore$ Total load on Station-1 $= (75 + A)$ MW

Total load on Station-2 $= (50 - A)$

$$\% \text{ drop in speed at Station-1} = \frac{4}{200}\,(750 + A)$$

$$\% \text{ drop in speed at Station-2} = \frac{3.5}{100}\,(50 - A)$$

The reduction in frequency will result due to the power flow from Station-1 through the interconnector of M-G set.

$$\therefore \% \text{ drop in speed at M-G set} = \frac{2.5}{25}(A) = \frac{2.5A}{25}$$

(reduction in frequency at Station-1 + reduction in frequency at M-G set)
$$= (\text{reduction in frequency at Station-2})$$

$$\therefore \frac{4}{200}(75 + A) + \frac{2.5A}{25} = \frac{3.5}{100}(50 - A)$$

$$0.02\,(75 + A) + 0.1A = 0.035\,(50 - A)$$
$$1.5 + 0.02A + 0.1A = 1.75 - 0.03A$$
$$0.02A + 0.1A + 0.03A = 175 - 1.5 = 0.25$$
$$0.15A = 0.25$$
$$A = 1.666 \text{ MW}$$

i.e., a power of $A = 1.666$ MW flows from Station-1 to Station-2.

$\therefore$ Total load at Station-1 $= 75 + A = 75 + 1.666 = 76.666$ MW

Total load at Station-2 $= 50 - A = 50 - 1.666 = 48.334$ MW

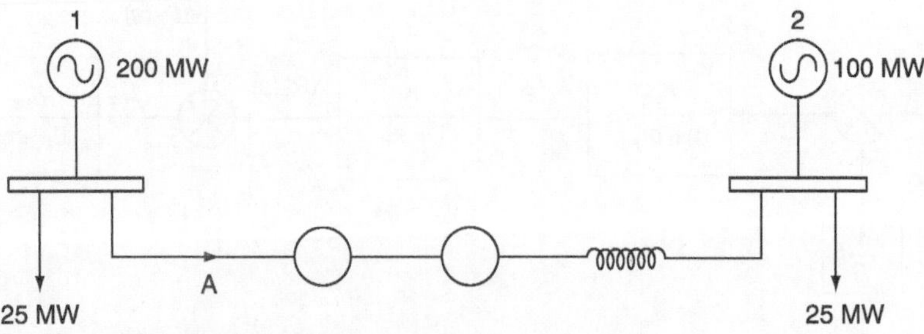

FIG. 7.12   *Illustration for Example 7.1*

Example 7.2: A 125 MVA turbo-alternator operator on full load operates at 50 Hz. A load of 50 MW is suddenly reduced on the machine. The steam valves to the turbine commence to close after 0.5 s due to the time lag in the governor system. Assuming the inertia to be constant, $H = 6$ kW-s per kVA of generator capacity, calculate the change in frequency that occurs in this time.

Solution:

By definition, $H = \dfrac{\text{stored energy}}{\text{capacity of the machine}}$

$\therefore$ Energy stored at no load $= 6 \times 125 \times 1{,}000 = 750$ MJ

Excessive energy input to rotating parts in 0.5 s $= 50 \times 0.5 \times 1{,}000 = 25$ MJ

As a result of this, there is an increase in the speed of the motor and hence an increase in frequency:

$$W_{KE} = W_{KE}^0 \left(\frac{f^0 + \Delta f}{f^0} \right)^2$$

$$\therefore f_{new} = \sqrt{\frac{750 + 25}{750}} \times 50 \text{ Hz} = 50.83 \text{ Hz}$$

7.14 CONTROL AREA CONCEPT

In real practice, the system of a single generator that feeds a large and complex area has rarely occurred. Several generators connected in parallel, located also at different locations, will meet the load demand of such a geographically large area. All the generators may have the same response characteristics to the changes in load demand.

It is possible to divide a very large power system into sub-areas in which all the generators are tightly coupled such that they swing in unison with change in load or due to a speed-changer setting. Such an area, where all the generators are running coherently is termed as a control area. In this area, frequency may be same in steady state and dynamic conditions. For developing a suitable control strategy, a control area can be reduced to a single generator, a speed governor, and a load system.

7.15 INCREMENTAL POWER BALANCE OF CONTROL AREA

In this section, we shall develop a dynamic model in terms of incremental power and frequency dynamics of a control area 'i' connected via tie lines as shown in Fig. 7.13.

Now assume that control area 'i' experiences a real load change ΔP_{Di} (MW). Due to the actions of the turbine controllers, its output increases by ΔP_{Gi} (MW). The net-surplus power in the area $(\Delta P_{Gi} - \Delta P_{Di})$ will be absorbed by the system in three ways:

- By increasing the area kinetic energy $W_{KE,\,i}$ at the rate $\dfrac{dW_{KEi}}{dt}$.

- By an increased load consumption. All typical loads (because of the dominance of motor loads) experience an increase, $B = \left(\dfrac{\partial P_D}{\partial f} \right)$ MW/Hz, with speed or frequency.

- By increasing the flow of power via tie lines with the total amount $\Delta P_{tie,\,i}$ MW, which is defined positive for outflow from the area.

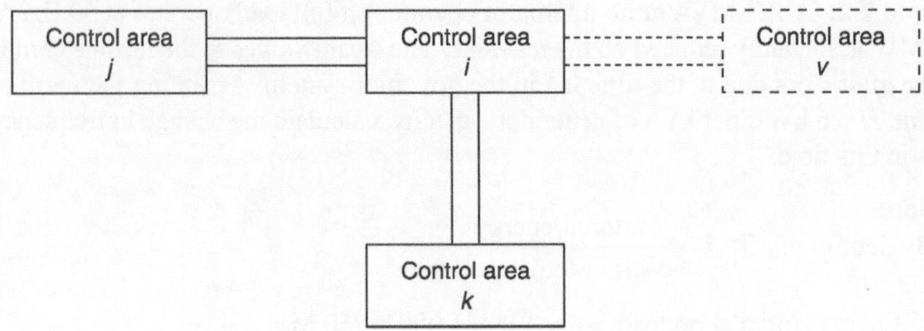

FIG. 7.13   *Interconnected control area*

Hence, the net-surplus power can be expressed as

$$\Delta P_{Gi} - \Delta P_{Di} = \frac{\mathrm{d}W_{KEi}}{\mathrm{d}t} + B_i \Delta f_i + \Delta P_{\mathrm{tie},\,i} \tag{7.27}$$

ΔP_{tie} is the difference between scheduled real power and actual real power through interconnected lines and it is taken as the input to the LFC system.

7.16 SINGLE AREA IDENTIFICATION

The first two terms on the right-hand side of Equation (7.27) represent a generator–load model (with the subscript 'i' absent). If the third term is absent, it means that there is no interchange of power between area 'i' and any other area. Thus, it becomes a single-area case. A single area is a coherent area in which all the generators swing in unison to the changes in load or speed-changer settings and in which the frequency is assumed to be constant throughout both in static and dynamic conditions. This single control area can be represented by an isolated power system consisting of a turbine, its speed governor, generator, and load.

7.16.1 Block diagram representation of a single area

The block diagram of an isolated power system, which in essence is a single-area system, is the same as the block diagram given in Fig. 7.11.

7.17 SINGLE AREA—STEADY-STATE ANALYSIS

The block diagram of an LFC of an isolated power system of a third-order model is represented in Fig. 7.11.

There are two incremental inputs to the system and they are:

(i)   The change in the speed-changer position, ΔP_{C} (reference power input).

(ii) The change in the load demand, ΔP_{D}.

In this section, we will analyze the response of a single-area system to steady-state changes by three ways:

(i)   Constant speed-changer position with variable load demand (uncontrolled case).

(ii)   Constant load demand with variable speed-changer position (controlled case).

(iii) Variable speed-changer position as well as load demand.

7.17.1 Speed-changer position is constant (uncontrolled case)

With the model given in Fig. 7.11 and with $\Delta P_C = 0$, the response of an uncontrolled single area LFC can be obtained as follows.

Let us consider a simple case wherein the speed changer has a fixed setting, which means $\Delta P_c = 0$ and the load demand alone changes. Such an operation is known as free governor operation or uncontrolled case since the speed changer is not manipulated (or controlled to achieve better frequency constancy).

For a sudden step change of load demand,

$$\Delta P_D(s) = \frac{\Delta P_D}{s}$$

For such an operation, the steady-state change of frequency Δf is to be estimated from the block diagram of Fig. 7.14 as

$$\Delta F(s)\Big|_{\Delta PC(s)=0} = -\left[\frac{\dfrac{K_{ps}}{(1+s\tau_{ps})}}{1+\dfrac{(K_{sg}K_tK_{ps}/R)}{(1+s\tau_{sg})\,(1+s\tau_t)} \times \dfrac{K_{ps}}{(1+s\tau_{ps})}}\right] \times \frac{\Delta P_D(s)}{s}$$

$$= -\left[\frac{K_{ps}}{(1+s\tau_{ps}) + \dfrac{(K_{sg}K_tK_{ps}/R)}{(1+s\tau_{sg})\,(1+s\tau_t)}}\right] \times \frac{\Delta P_D(s)}{s} \qquad (7.28)$$

Applying the final value theorem, we have

$$\Delta f\Big|_{\substack{\text{steady state} \\ \Delta P_c = 0}} = \underset{s \to 0}{\text{Lt}}\; s\Delta F(s)\Big|_{\Delta P_C = 0}$$

$$= -\left(\frac{K_{ps}}{1 + \dfrac{K_{ps}K_tK_{sg}}{R}}\right)\Delta P_D \qquad (7.29)$$

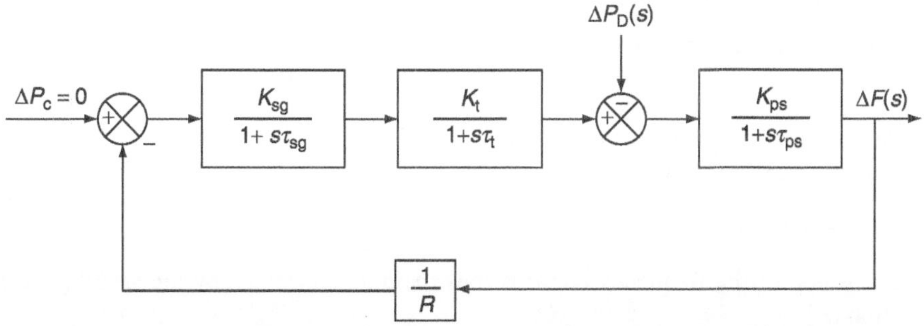

FIG. 7.14   *Block diagram representation of an isolated power system setting $\Delta P_c = 0$*

The gain K_t is fixed for the turbine and K_{ps} is fixed for the power system. The gain K_{sg} of the speed governor is easily adjustable by changing the lengths of various links of the linkage mechanism. K_{sg} is so adjusted such that $K_{sg}K_t \approx 1$.

Therefore Equation (7.29) can be simplified as:

$$\Delta f = -\left(\frac{K_{ps}}{1+\dfrac{K_{ps}}{R}}\right)\Delta P_D$$

Also we know from the dynamics of the generator–load model, $K_{ps} = \dfrac{1}{B}$

where $B = \dfrac{\partial P_D}{\partial f}$ MW/Hz

$$= \frac{\dfrac{\partial P_D}{\partial f}}{P_r} \quad \text{in p.u.MW/unit change in frequency}$$

$$\Delta f = -\left(\frac{\dfrac{1}{B}}{1+\dfrac{1}{BR}}\right)\Delta P_D$$

$$\therefore \Delta f = -\left(\frac{1}{B+\dfrac{1}{R}}\right)\Delta P_D = -\frac{1}{\beta}\Delta P_D \tag{7.30}$$

where the factor $\beta = \left(B + \dfrac{1}{R}\right)$ and is known as the area frequency response characteristic (AFRC) or area frequency regulation characteristic.

Equation (7.30) gives the steady-state response of frequency to the changes in load demand. The speed regulation is usually so adjusted that changes in frequency are small (of the order of 5%) from no load to full load. Figure 7.15 gives the linear relationship between frequency and load for a free governor operation, with speed changes set to give a scheduled frequency of 100% at full load.

The droop or the slope of the relationship is $-\left(\dfrac{1}{B+1/R}\right)$.

Power system parameter B is generally much less than $\left(\text{i.e., } B \ll \dfrac{1}{R}\right)$, so that B can be neglected in Equation (7.30), which results in

$$\Delta f = -R(\Delta P_D) \tag{7.31a}$$

The droop of the frequency curve is thus mainly determined by the speed-governor regulation (R).

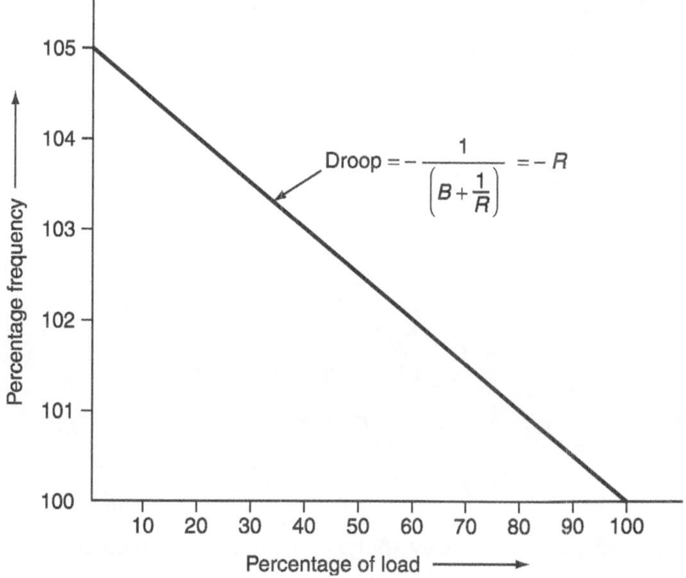

FIG. 7.15   *Steady-state load frequency characteristics of a speed-governing system*

The increase in load demand (ΔP_D) is met under steady-state conditions partly by the increased generation (ΔP_G) due to the opening of the steam valve and partly by the decreased load demand due to droop in frequency.

The increase in generation is expressed as

$$\Delta P_G = -\frac{1}{R}\,\Delta f$$

Substituting Δf from Equation (7.30), we get

$$\therefore \Delta P_G = -\frac{1}{R}\left(-\left(\frac{1}{B + \dfrac{1}{R}}\right)\Delta P_D\right) = \left[\frac{1}{R} \times \frac{R}{BR + 1}\right]\Delta P_D$$

$$= \left(\frac{1}{BR + 1}\right)\Delta P_D \tag{7.31b}$$

And a decrease in the system load is expressed as

$$B\Delta f = B\left(-\left(\frac{1}{B + \dfrac{1}{R}}\right)\Delta P_D\right) = B\left(\frac{R}{BR + 1}\right)\Delta P_D = \left(\frac{BR}{BR + 1}\right)\Delta P_D \tag{7.31c}$$

From Equations (7.31(b)) and (7.31(c)), it is observed that contribution of the decrease in the system load is much less than the increase in generation.

7.17.2 Load demand is constant (controlled case)

Consider a step change in a speed-changer position with the load demand remaining fixed:

$$\text{i.e.,} \quad \Delta P_C(s) = \frac{\Delta P_c}{s} \quad \text{and} \quad \Delta P_D = 0$$

The steady-state change in frequency can be obtained from the block diagram of Fig. 7.16:

$$\Delta F(s)\Big|_{\Delta P_D(s)=0} = \frac{\dfrac{K_{sg}K_t K_{ps}}{(1+s\tau_{sg})(1+s\tau_t)(1+s\tau_{ps})}}{1+\dfrac{K_{sg}K_t K_{ps}}{(1+s\tau_{sg})(1+s\tau_t)(1+s\tau_{ps})}\times\dfrac{1}{R}}\ \frac{\Delta P_c}{s}$$

$$= \frac{K_{sg}K_t K_{ps}}{(1+s\tau_{sg})(1+s\tau_t)(1+s\tau_{ps})+ K_{sg}K_t K_{ps}\times\dfrac{1}{R}}\ \frac{\Delta P_c}{s}$$

The steady-state value is obtained by applying the final-value theorem:

$$\Delta f\Big|_{\Delta P_D(s)=0}^{\text{steady state}},\ \underset{s\to 0}{\text{Lt}}\ s\Delta F(s) = \frac{K_{sg}K_t K_p}{1+\dfrac{K_{sg}K_t K_p}{R}}\Delta P_C$$

$$= \frac{K_p}{1+\dfrac{K_p}{R}}\Delta P_C \quad (\text{since } K_{sg}K_t \approx 1)$$

$$= \left(\frac{\dfrac{1}{B}}{1+\dfrac{1}{BR}}\right)\Delta P_C$$

$$\Delta f = \left(\frac{1}{B+\dfrac{1}{R}}\right)\Delta P_c \tag{7.32}$$

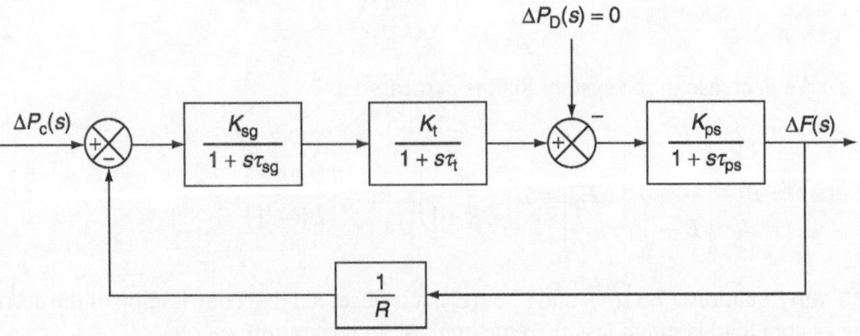

FIG. 7.16   *Block diagram representation of an isolated power system setting $\Delta P_D = 0$*

7.17.3 Speed changer and load demand are variables

By superposition, if the speed-changer setting is changed by ΔP_c while the load demand also changes by ΔP_D, the steady-state change in frequency is obtained from Equations (7.30) and (7.32) as

$$\Delta f = \left(\frac{1}{B+\dfrac{1}{R}}\right)(\Delta P_C - \Delta P_D) = -\left(\frac{1}{B+\dfrac{1}{R}}\right)\Delta P_D + \left(\frac{1}{B+\dfrac{1}{R}}\right)\Delta P_C$$

From the above equation, we can observe that the change in load demand causes the changes in frequency, which can be compensated by changing the position of the speed changer.

If $\Delta P_C = \Delta P_D$, then Δf will become zero.

7.18 STATIC LOAD FREQUENCY CURVES

The block diagram representation of a turbine-speed-governor model is shown in Fig. 7.17(a) and their static load frequency curves are shown in Fig. 7.17(b).

The curve relates power generation P_G and frequency f with control parameter P_C.

From the block diagram shown in Fig. 7.17(a), we get the static algebraic relation

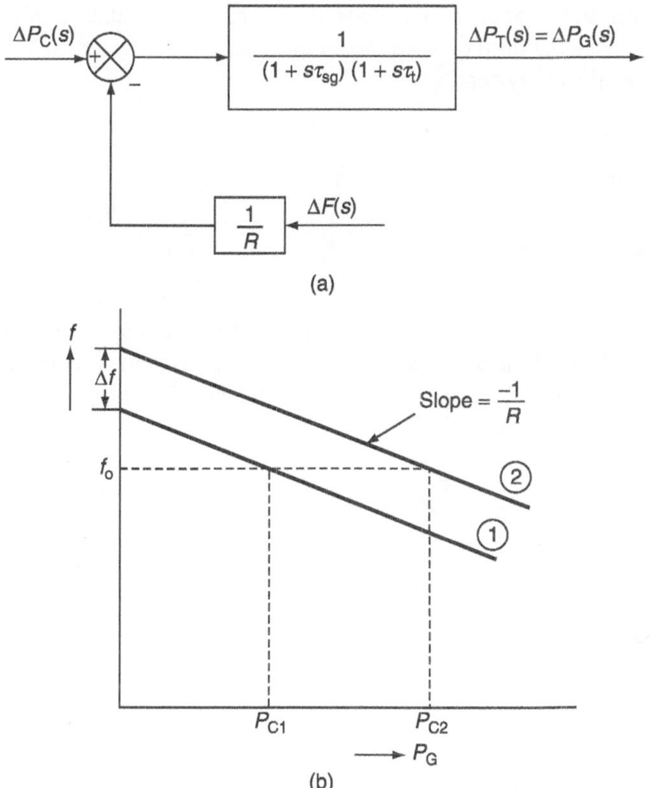

FIG. 7.17 *(a) Block diagram of a turbine-speed-governor model; (b) static load frequency curves for the turbine governor*

$$\Delta P_{\text{G}} = \Delta P_{\text{C}} - \frac{1}{R}\Delta f$$

from which the local shape of the speed-power curves may be inferred.

Figure 7.17(b) gives the two static load-frequency curves. Adjust power generation, P_{G}, by using a speeder meter (speed changer) upto $P_{\text{G}} = P_{\text{C1}}$, where P_{C1} is the desired command power at synchronous speed $\omega^0(f^0)$. With free governor operation (i.e., $\Delta P_{\text{C}} = 0$), the fixed speed-changer position P_{C1} predicts the straight-line relationship. This straight line (1) has a slope of $-R$.

To get more generation at the same synchronous speed of $\omega^0(f^0)$, adjust P_{C1} to P_{C2} with a speeder meter. This results in the load frequency curve (2). The speed regulation R refers to the variation in frequency with power generation. Better the regulation results, less the droops speed-power (load) characteristics of LFC.

7.19 DYNAMIC ANALYSIS

The meaning of dynamic response is how the frequency changes as a function of time immediately after disturbance before it reaches the new steady-state condition. The analyzation of dynamic response requires the solution of dynamic equation of the system for a given disturbance. The solution involves the solution of different equations representing the dynamic behavior of the system.

The inverse Laplace transform of $\Delta F(s)$ gives the variation of frequency with respect to time for a given step change in load demand. Comparing the relative values of time constants, we can reduce the third ordered model to a first ordered system.

For a practical LFC system,

$$\tau_{\text{sg}} < \tau_{\text{t}} << \tau_{Ps}$$

Typical values are:

$$\tau_{\text{sg}} = 0.4 \text{ s}$$
$$\tau_{\text{t}} = 0.5 \text{ s}$$
$$\tau_{Ps} = 20 \text{ s}$$

If τ_{sg} and τ_{t} are considered negligible compared to τ_{Ps} and by adjusting $K_{\text{sg}} K_{\text{t}} = 1$, the block diagram of LFC of the power system of an isolated system is reduced to a first-order system as shown in Fig. 7.18 with $\Delta P_{\text{c}} = 0$ for an uncontrolled case.

From Fig. 7.18, the change in frequency is given by

$$\Delta F(s)\Big|_{\Delta P_{\text{c}}(s)=0} = \left[\frac{\dfrac{K_{\text{ps}}}{1+s\tau_{\text{ps}}}}{1 + \dfrac{K_{\text{ps}}}{1+s\tau_{\text{ps}}} \times \dfrac{1}{R}}\right]\frac{-\Delta P_{\text{D}}}{s}$$

$$= \left[\frac{-K_{\text{ps}}}{(1+s\tau_{\text{ps}}) + \dfrac{K_{\text{ps}}}{R}}\right]\frac{\Delta P_{\text{D}}}{s}$$

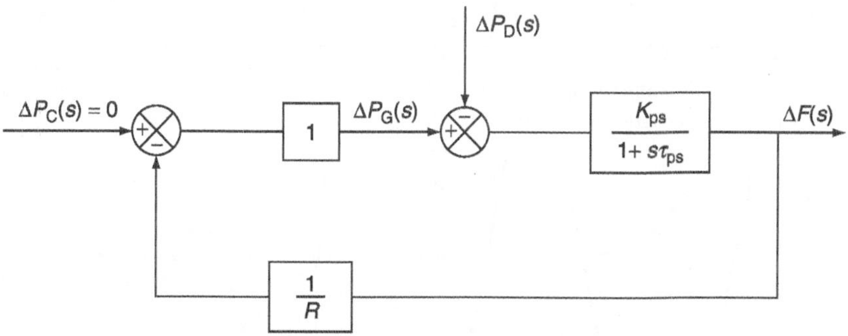

FIG. 7.18   *First-order approximate block diagram of LFC of an isolated area*

$$= \left[\frac{-K_{ps}}{s\tau_{ps} + \dfrac{K_{ps} + R}{R}} \right] \frac{\Delta P_D}{s}$$

$$= \left[\frac{-\dfrac{K_{ps}}{\tau_{ps}}}{s + \dfrac{K_{ps} + R}{R\tau_{ps}}} \right] \frac{\Delta P_D}{s}$$

$$= -\frac{K_{ps} \times \Delta P_D}{\tau_{ps}} \left(\frac{1}{s\left(s + \dfrac{K_{ps} + R}{R\tau_{ps}}\right)} \right)$$

$$= -\frac{K_{ps} \times \Delta P_D}{\tau_{ps}} \times \frac{R\tau_{ps}}{(K_{ps} + R)} \left[\frac{1}{s} - \frac{1}{\left(s + \dfrac{K_{ps} + R}{R\tau_{ps}}\right)} \right]$$

$$\therefore \; \Delta f(t) = L^{-1} \, \Delta F(s)$$

$$= \frac{-RK_{ps}}{R + K_{ps}} \left(1 - e^{-\dfrac{t}{\tau_{ps}}\left(\dfrac{R + K_{ps}}{R}\right)} \right) \Delta P_D$$

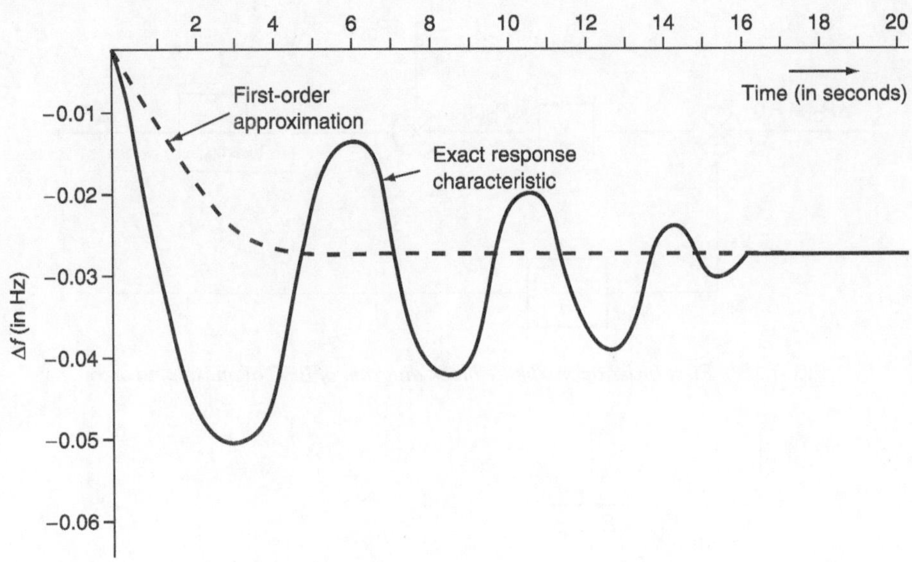

FIG. 7.19   *Dynamic response of frequency change (Δf) for a step-load change*

The plot of change in frequency versus time for a first-order approximation and exact response are shown in Fig. 7.19:

$$\Delta P_{\mathrm{D}} = 0.01 \text{ p.u}, \quad K_{\mathrm{ps}} = 100, \quad R = 3, \quad \tau_{\mathrm{sg}} = 0.4 \text{ s}, \quad \tau_{\mathrm{t}} = 0.5 \text{ s}, \quad \text{and} \quad \tau_{\mathrm{Ps}} = 20 \text{ s}$$

Example 7.3: An isolated control area consists of a 200-MW generator with an inertia constant of $H = 5$ kW-s/kVA having the following parameters (Fig. 7.20(a)):

Power system gain constant, $K_{\mathrm{ps}} = 100$
Power system time constant, $\tau_{\mathrm{ps}} = 20$ s
Speed regulation, $R = 3$
Normal frequency, $f^0 = 50$ Hz

Obtain the frequency error and plot the graph of deviation of frequency when a step-load disturbance of (i) 0.5%, (ii) 1%, and (iii) 2% is applied (Fig. 7.20(b)).

From Fig. 7.20(b), the steady-state change in frequency $\Delta f_{\mathrm{ss}} = -0.0145$ Hz.

Similarly (ii) for a step-load change of 1%, the steady-state change in frequency $\Delta f_{\mathrm{ss}} = -0.029$ Hz; (iii) for a step-load change of 2%, steady-state change in frequency $\Delta f_{\mathrm{ss}} = -0.0583$ Hz

Example 7.4: For Example 7.3, show the effect of governor action and turbine dynamics (Fig. 7.21(a)), if they are not to be neglected and given that $\tau_{\mathrm{sg}} = 0.4$ s and $\tau_{\mathrm{t}} = 0.5$ s for a step-load change of (i) 0.5% and (ii) 1% (Fig. 7.21(b)).

From Fig. 7.21(b), the steady-state change in frequency $\Delta f_{\mathrm{ss}} = -0.0235$ Hz.

Similarly (ii) for a step-load change of 1%, the steady-state change in frequency $\Delta f_{\mathrm{ss}} = -0.047$ Hz.

Example 7.5: Obtain the resultant frequency plot when combining Examples 7.3 and 7.4 for a step-load disturbance of 0.5% (Figs. 7.22(a) and (b)).

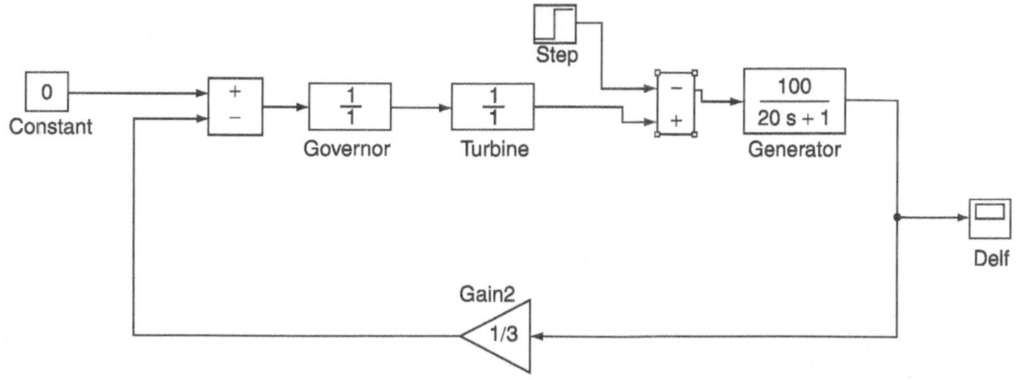

FIG. 7.20 *(a) Simulation block diagram of single area without a speed-governor system; (b) response of the change in frequency for Fig. 7.20(a) for a step-load change of 0.5%*

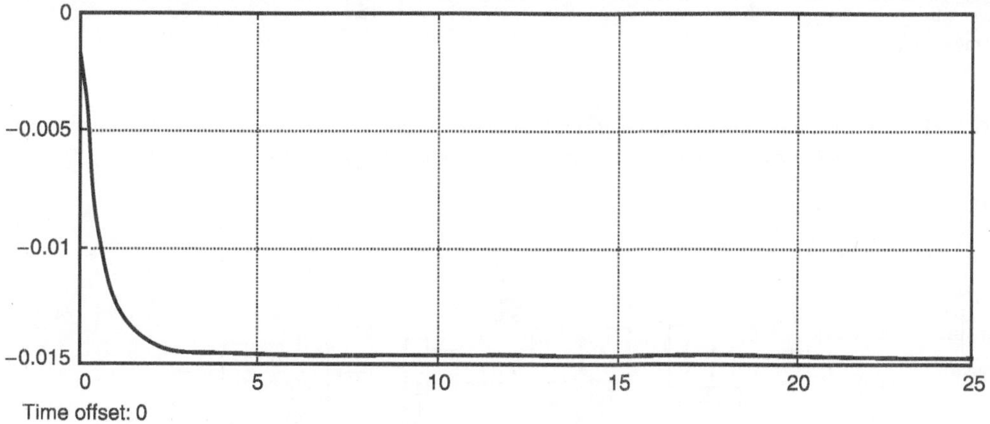

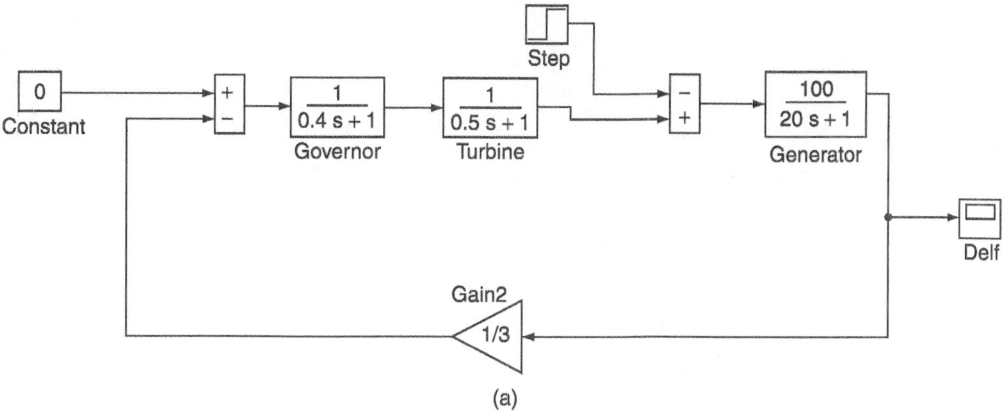

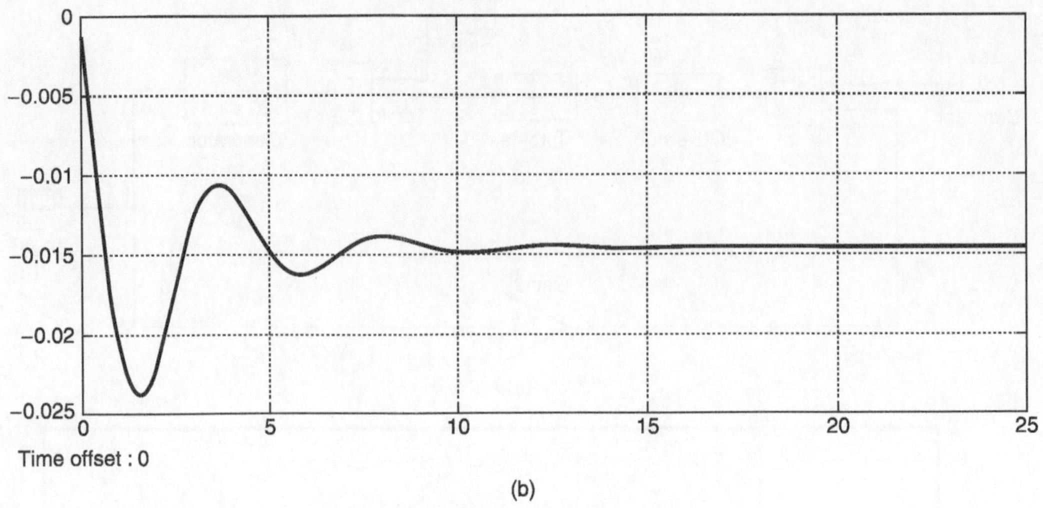

FIG. 7.21 *(a) Simulation block diagram of single area with a speed-governor system; (b) response of the change in frequency for Fig. 7.21(a) for a step-load change of 0.5%*

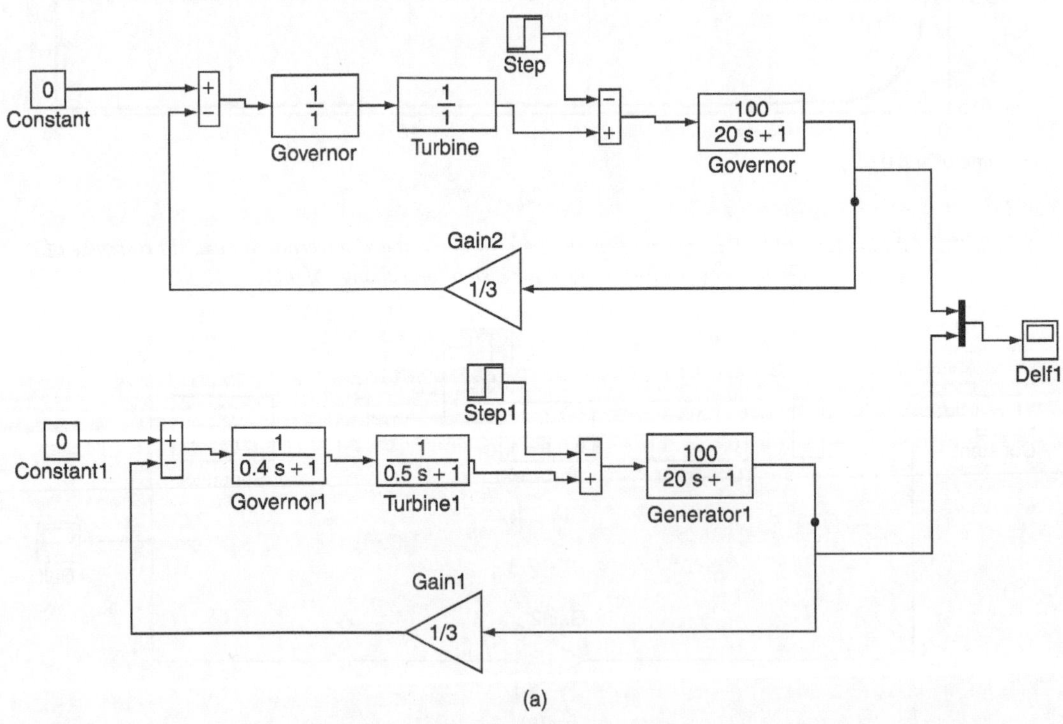

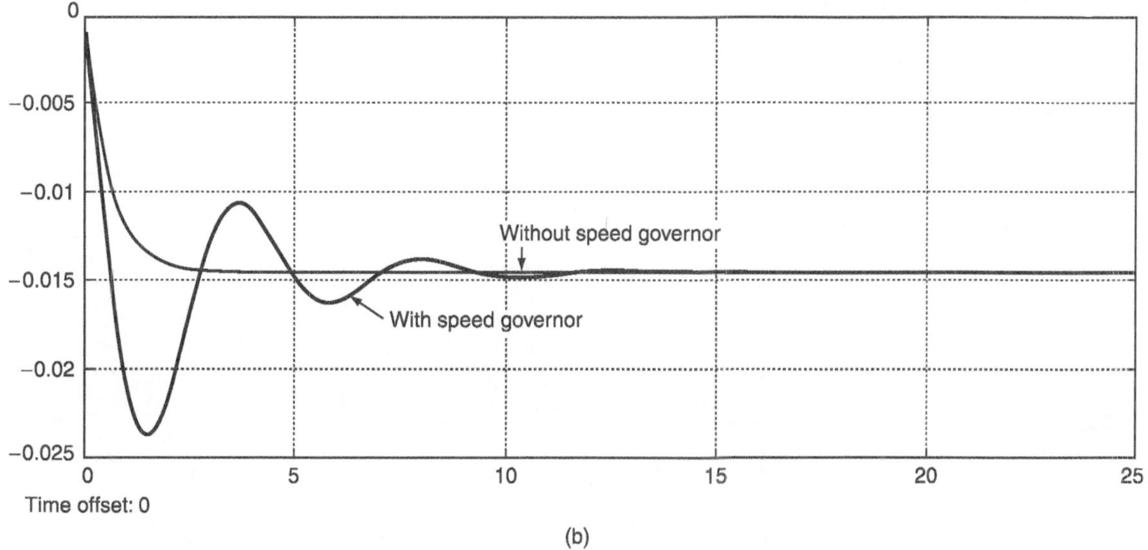

(b)

FIG. 7.22   *(a) Simulation block diagram of a single area without and with a speed-governor system; (b) response of the change in frequency for Fig. 7.22(a) for a step-load change of 0.5%*

Example 7.6: An isolated control area consists of a 200-MW generator with an inertia constant of $H = 5$ kw-s/kVA having the following parameters (Fig. 7.23(a)):

Power system gain constant, $K_{ps} = 100$
Power system time constant, $\tau_{ps} = 20$s
Speed regulation, $R = 3$
Normal frequency, $f^0 = 50$ Hz
Governor time constant, $\tau_{sg} = 0.4$ s
Turbine time constant, $\tau_t = 0.5$ s

Obtain the frequency error and plot the graph of deviation of frequency when a step change of 1% in the speed-changer position is applied (Fig. 7.23(b)).

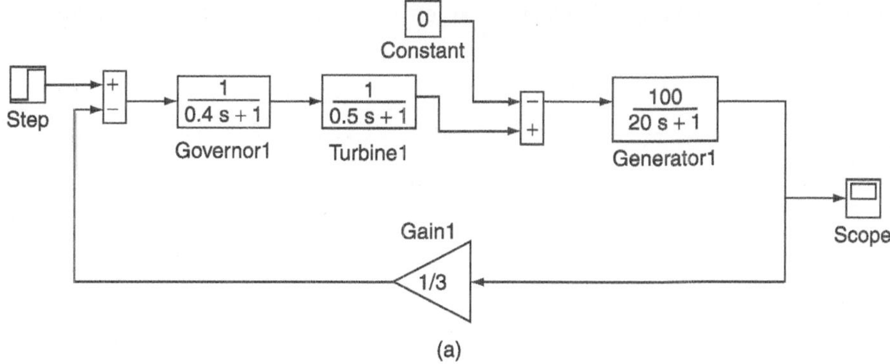

(a)

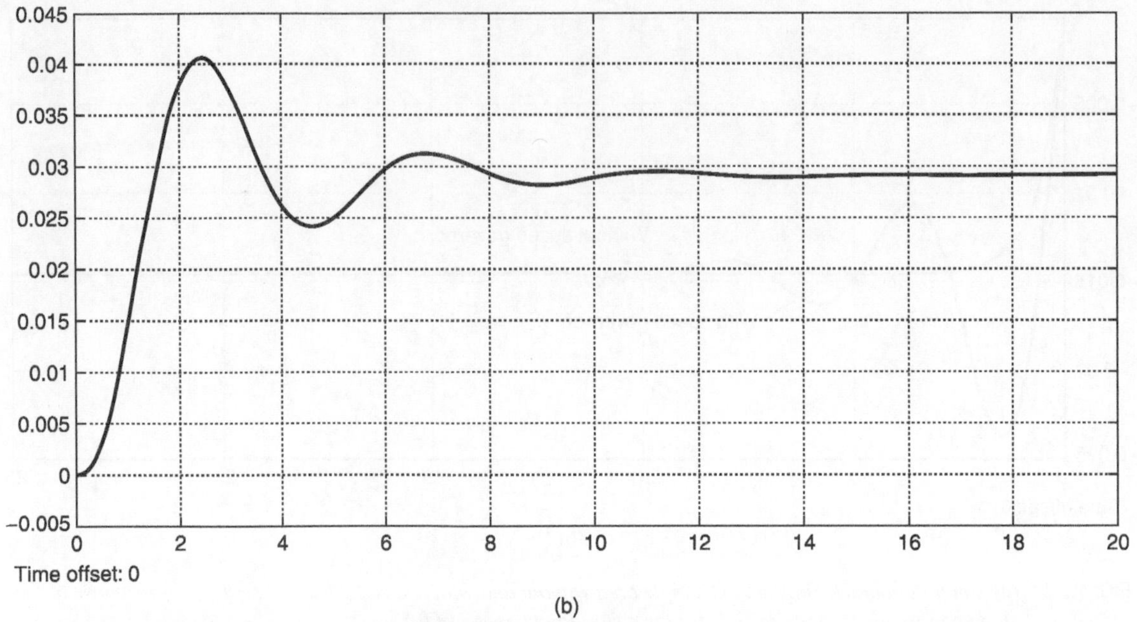

FIG. 7.23   *(a) Simulation block diagram of a simulated system with a step change in the speed-changer position; (b) frequency response for Example 7.6*

Example 7.7: An isolated control area consists of a 200-MW generator with an inertia constant of $H = 5$ kW-s/kVA having the following parameters (Fig. 7.24(a)):

Power system gain constant, $K_{ps} = 100$
Power system time constant, $\tau_{ps} = 20$ s
Speed regulation, $R = 3$
Normal frequency, $f^0 = 50$ Hz
Governor time constant, $\tau_{sg} = 0.4$ s
Turbine time constant, $\tau_t = 0.5$ s

Obtain the frequency error and plot the graph of deviation of frequency when a step change of 1% in both the speed-changer position and the load is applied (Fig. 7.24(b)).

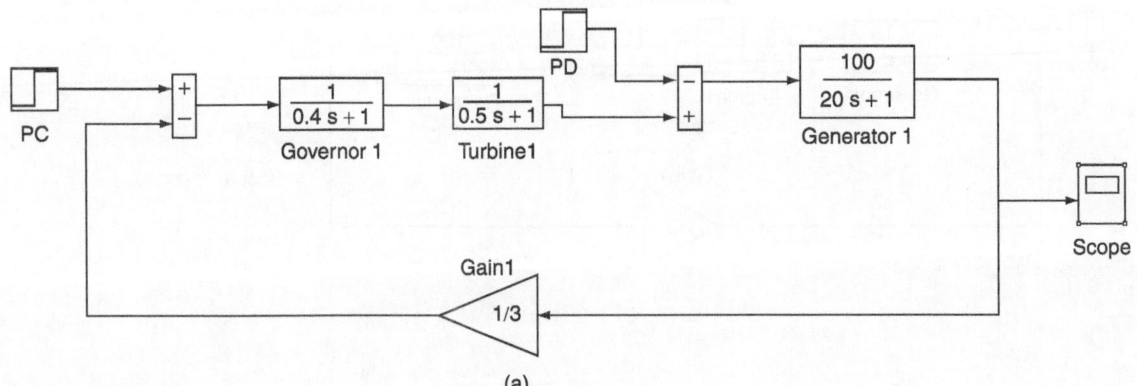

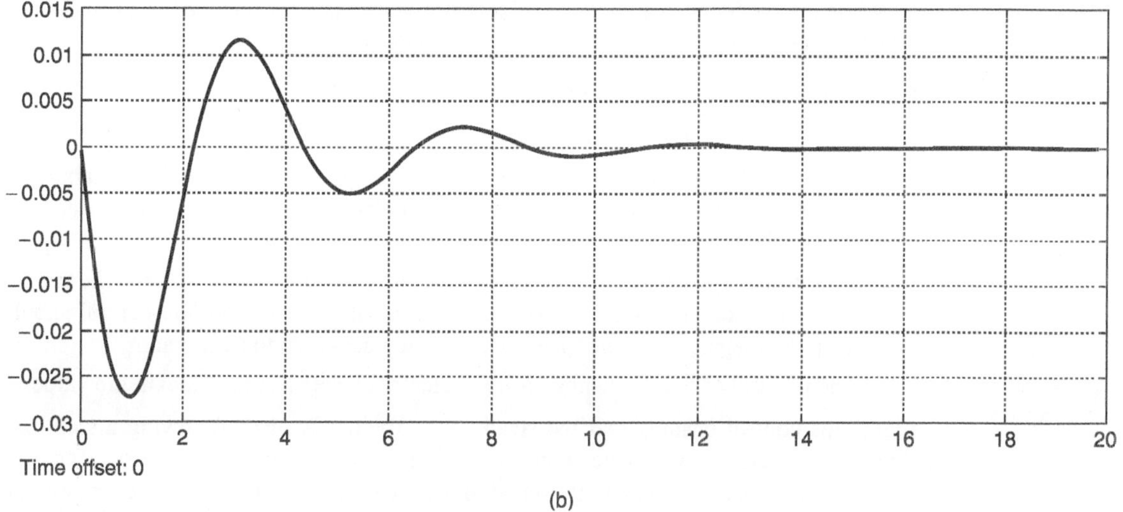

Time offset: 0

(b)

FIG. 7.24   *(a) Simulation block diagram of a single-area system with P_C and P_D; (b) frequency response of Example 7.5*

Example 7.8:   Find the static frequency drop if the load is suddenly increased by 25 MW on a system having the following data:

$$\text{Rated capacity } P_r = 500 \text{ MW}$$
$$\text{Operating Load } P_D = 250 \text{ MW}$$
$$\text{Inertia constant } H = 5 \text{ s}$$
$$\text{Governor regulation } R = 2 \text{ Hz p.u. MW}$$
$$\text{Frequency } f = 50 \text{ Hz}$$

Also find the additional generation.

Solution:

Assuming the frequency characteristic to be linear, we have

$$B = \partial P_D / \partial f = \frac{250}{50} \text{ MW/Hz}$$

$\partial P_D / \partial f$ expressed in p.u., $B = \dfrac{250}{50 \times 500} = 0.01 \text{ p.u. MW/Hz}$

ΔP_D in p.u. $= \dfrac{25}{500} = 0.05$

Area frequency response characteristic (AFRC)

$$\beta = B + \frac{1}{R} = 0.01 + \frac{1}{2} = 0.06$$

The static frequency drop $\Delta f = \dfrac{-\Delta P_D}{\text{AFRC}} = \dfrac{-0.05}{0.06} = 0.098 \text{ Hz}.$

Hence, the system frequency drops to $(50 - 0.098) = 49.902$ Hz.

The amount of additional generation $\Delta P_G = \dfrac{-\Delta f}{R} = \dfrac{0.098}{2} = 4.9 \times 10^{-2} \text{ p.u. MW}$

$$= 4.9 \times 10^{-2} \times 500 = 24.5 \text{ MW}$$

While the sudden increase in load is 25 MW, the increase in generation is 24.5 MW and 0.5 MW is the loss of load due to the drop in frequency.

7.20 REQUIREMENTS OF THE CONTROL STRATEGY

The following are the basic requirements needed for the control strategy:

- The system frequency control is obtained through a closed loop. Since stability is the major problem associated with a closed-loop control, maintenance of the stability will be the main objective.

- The frequency deviation due to a step-load change should return to zero. The control that offers above is called 'isochronous control'. In addition, the control should keep the magnitude of the transient frequency deviation to a minimum.

- The integral of the frequency error should not exceed a certain maximum value.

Isochronous control ensures that the steady-state frequency error following a step-load change will be zero. However, no control can eliminate transient frequency error. The time error of synchronous clocks is proportional to the integral of this transient frequency error. Therefore, it is necessary to put a limit on the value of this integral.

- The total load should be divided among the individual generators of the control area for optimum economy.

The first three requirements are satisfied when the addition of the integral-control to the system takes place.

7.20.1 Integral control

The integral control is composed of a frequency sensor and an integrator. The frequency sensor measures the frequency error Δf and this error signal is fed into the integrator. The input to the integrator is called the 'Area Control Error' (i.e., ACE $= \Delta f$).

The ACE is the change in area frequency, which when used in an integral-control loop, forces the steady-state frequency error to zero.

The integrator produces a real-power command signal ΔP_C and is given by

$$\Delta P_C = -K_I \int \Delta f \, dt \tag{7.33}$$
$$= -K_I \int (ACE) \, dt$$

The signal ΔP_C is fed to the speed-changer causing it to move. Here, K_I is called the integral gain constant, which controls the rate of integration. The frequency sensor and the integrator are connected in the system as a closed control loop as shown in the block diagram in Fig. 7.25.

Figure 7.25 consists of Fig. 7.11 augmented by additional loops showing the generation of ACE and its use in changing the area command powers; R is the speed-regulation feedback parameter. $\Delta P_G(s)$, $\Delta P_D(s)$, and $\Delta F(s)$ are the incremental changes in the generation, system load, and frequency, respectively. The block diagram of Fig. 7.25 is the single-area power system (isolated power system) with integral control for small incremental changes.

The negative sign in the integral controller is for producing a negative or decrease command for a positive frequency error. The gain constant K_I is positive and controls the rate of integration, and thus the speed of the response of the control loop. The integrator is an electronic integrator of the same type as used in analog computers.

In view of hardware, we can understand the presence of the integrator by considering the ACE voltages distributed to the speed changers (speeder motors) of individual generator

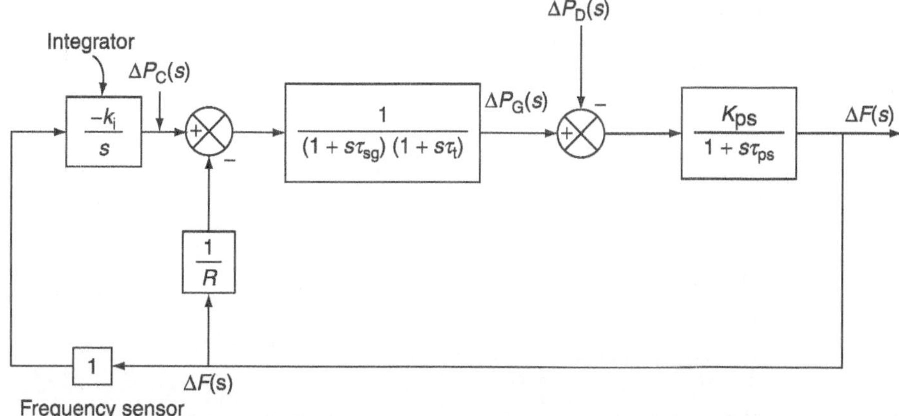

FIG. 7.25   *Proportional plus integral control of LFC of a single-area system*

units that participate in supplementary control within a given area. These motors turn at a rate of θ proportional to the ACE voltage and continue to turn until they are driven to zero.

The integral control will give rise to zero steady-state frequency error ($\Delta f_{\text{steady state}} = 0$) due to a step-load change. As long as the error remains, the integrator output will increase, causing the speed changer to move. The integrator output and thus the speed-changer position attain a constant value only when the frequency error has been reduced to zero. This is proved through a simplified mathematical analysis as follows.

7.21 ANALYSIS OF THE INTEGRAL CONTROL

The following assumptions are made in order to obtain a simple analysis. These assumptions do not distort the essential features. Also, the errors introduced on account of these assumptions affect only the transient and not the steady-state response.

Assumptions

- The time constant of the speed-governor mechanism τ_{sg} and that of the turbine τ_t are both neglected, i.e., $\tau_{sg} = \tau_t = 0$.
- The speed changer is an electromechanical device and hence its response is not instantaneous. However, it is assumed to be instantaneous in the present analysis.
- All non-linearities in the equipment, such as dead zone, etc., are neglected.
- The generator can change its generation ΔP_G as fast as it is commanded by the speed changer.
- The ACE is a continuous signal.

The Laplace transformation of Equation (7.33) gives

$$\Delta P_C(s) = -\frac{K_I}{s}\Delta F(s) \tag{7.34}$$

and, for a step change of load demand ΔP_D,

$$\Delta P_D(s) = \frac{\Delta P_D}{s} \tag{7.35}$$

From the block diagram of Fig. 7.25, we have

$$\left[\frac{-1}{R}\Delta F(s) - \frac{K_I}{s}\Delta F(s)\right]\frac{1}{(1+s\tau_{sg})(1+s\tau_t)} = \Delta P_G(s)$$

$$-\left[\frac{1}{R} + \frac{K_I}{s}\right]\frac{1}{(1+s\tau_{sg})(1+s\tau_t)}\Delta F(s) = \Delta P_G(s) \tag{7.36}$$

and

$$\Delta F(s) = \left(\frac{K_{ps}}{1+s\tau_{ps}}\right)[\Delta P_G(s) - \Delta P_D(s)]$$

$$= \frac{-K_{ps}}{1+s\tau_{ps}}\Delta P_D(s) + \frac{K_{ps}}{1+s\tau_{ps}}\Delta P_G(s)$$

Substituting for $\Delta P_G(s)$ from Equation (7.36) in the above equation, we get

$$\Delta F(s) = \frac{-K_{ps}}{1+s\tau_{ps}}\Delta P_D(s) + \frac{K_{ps}}{1+s\tau_{ps}}\left[\frac{1}{R} + \frac{K_I}{s}\right]\left[\frac{-1}{(1+s\tau_{sg})(1+s\tau_t)}\right]\Delta F(s)$$

$$\left[1 + \frac{K_{ps}}{1+s\tau_{ps}}\left[\frac{1}{R} + \frac{K_I}{s}\right]\left[\frac{1}{(1+s\tau_{sg})(1+s\tau_t)}\right]\right]\Delta F(s) = \frac{-K_{ps}}{1+s\tau_{ps}}\Delta P_D(s)$$

$$\Delta F(s) = \frac{\dfrac{-K_{ps}}{1+s\tau_{ps}}}{\left[1 + \dfrac{K_{ps}}{1+s\tau_{ps}}\left[\dfrac{1}{R} + \dfrac{K_I}{s}\right]\left[\dfrac{1}{(1+s\tau_{sg})(1+s\tau_t)}\right]\right]}\Delta P_D(s) \tag{7.37}$$

$$= \frac{-sK_{ps}R(1+s\tau_{sg})(1+s\tau_t)}{\left[s(1+s\tau_{ps})(1+s\tau_{sg})(1+s\tau_t)R + K_{ps}(s+RK_I)\right]} \times \frac{\Delta P_D}{s} \tag{7.38}$$

Equation (7.38) becomes a fourth-order system.

The steady-state value of $\Delta f(t)$ can be obtained by applying the final-value theorem, viz.,

$$\Delta f(t)\Big|_{\text{steady-state}} = \lim_{s\to 0}\left[s\Delta F(s)\right] = 0$$

Hence, the static- or steady-state frequency error will be zero with integral control.

The nature of transient variation of $\Delta f(t)$ can be found by taking the inverse Laplace transform of Equation (7.39). According to assumption (i), Equation (7.39) simplifies to

$$\Delta F(s) = \frac{-K_{ps}}{\left[1 + s\tau_{ps} + K_{ps}\left[\dfrac{1}{R} + \dfrac{K_I}{s}\right]\right]}\Delta P_D(s)$$

$$\Delta F(s) = \frac{-K_{ps}}{\tau_{ps}}\left(\frac{s\dfrac{\Delta P_D}{s}}{s^2 + \left(1 + \dfrac{K_{ps}}{R}\right)\dfrac{s}{\tau_{ps}} + \dfrac{K_{ps}K_I}{\tau_{ps}}}\right) \tag{7.39}$$

$$\left(\text{since}\quad \Delta P_D(s)=\frac{\Delta P_D}{s}\right)$$

The nature of $\Delta f(t)$ depends on the roots of the characteristic equation of Equation (7.39)

$$s^2+\left(1+\frac{K_{ps}}{R}\right)\frac{s}{\tau_{ps}}+\frac{K_{ps}K_I}{\tau_{ps}}=0 \tag{7.40}$$

The above equation can be rewritten as

$$\left[s+\frac{\left(1+\dfrac{K_{ps}}{R}\right)}{2\tau_{ps}}\right]^2+\frac{K_IK_{ps}}{\tau_{ps}}-\left(\frac{\left(1+\dfrac{K_{ps}}{R}\right)}{2\tau_{ps}}\right)^2=0$$

$$\text{or}\quad (s+a)^2+\omega^2=0 \tag{7.41}$$

where $\quad \alpha=\dfrac{\left(1+\dfrac{K_{ps}}{R}\right)}{2\tau_{ps}}$ is a positive real number

$$\text{and}\quad \omega=\left[\frac{K_1K_{Ps}}{\tau_{Ps}}-\left(\frac{\left(1+\dfrac{K_{ps}}{R}\right)}{2\tau_{ps}}\right)^2\right]^{1/2}$$

The nature of the roots of Equation (7.41) depends on whether $\omega^2=0$, $\omega^2>0$, or $\omega^2<0$.

Case (i): $\omega^2=0$

The characteristic equation has a repeated root (viz., α repeated twice). Hence, the expression for $\Delta f(t)$ contains terms of the type

$$e^{-\alpha t}\quad \text{and}\quad t\,e^{-\alpha t}$$

Consequently, the response [viz., $\Delta f(t)$] is a critically damped one. For this critical case,

$$\omega^2=\frac{K_1K_{ps}}{\tau_{ps}}-\left(\frac{\left(1+\dfrac{K_{ps}}{R}\right)}{2\tau_{ps}}\right)^2=0$$

Solving the above for K_I, we get

$$K_I=\frac{1}{4\tau_{ps}\,K_{ps}}\left(1+\frac{K_{ps}}{R}\right)^2=K_{Icrit} \tag{7.42}$$

Case (ii): $\omega^2>0$

Now, $(s+\alpha)^2=-\omega^2$, where ω^2 is a positive real number.

$$(s+\alpha)=+j\omega$$

$$\text{(or)}\quad s=(-\alpha\pm j\omega)$$

The time response $\Delta f(t)$ will therefore consist of damped oscillatory terms of the type

$$\mathrm{e}^{-\alpha t}\sin \omega t \quad \text{and} \quad \mathrm{e}^{-\alpha t}\cos \omega t.$$

This case is called a supercritical case. In this case, $K_\mathrm{I} > K_{\mathrm{I_{crit}}}$.

Case (iii): $\omega^2 < 0$

Then, ω^2 is a negative real number.

So, $(s + \alpha)^2 = -\omega^2$ is a positive real number

$$= \gamma^2 \text{ (say)}$$

$$\therefore (s + \alpha) = +\gamma \quad [\text{since } \gamma < \alpha]$$

$$\text{or} \quad s = (-\alpha + \gamma) \quad \text{or} \quad (-\alpha - \gamma)$$

$$= \beta_1 \text{ or} -\beta_2 \text{ (say)}$$

Accordingly, in this case, the time response $\Delta f(t)$ will comprise terms of the type

$$\mathrm{e}^{-\beta_1 t} \quad \text{and} \quad \mathrm{e}^{-\beta_2 t}$$

Hence, the response will be damped and non-oscillatory. The control, in this case, is called the subcritical integral control. In this case, $K_\mathrm{I} < K_{\mathrm{Icrit}}$.

In all the three cases described above, $\Delta f(t)$ will approach zero. This was proved earlier using the final-value theorem. It can be observed that the transient frequency error does remain finite. This is a proof that the control is both stable and isochronous. Thus, the first two control requirements stated earlier (Section 7.20) are fulfilled with this integral control. This control is also called 'proportional plus integral control'. The proportional control is provided by the closed loop of gain constant of $1/R$.

The actual simulated time responses of a single-area control system with and without integral control are as shown in Fig. 7.26.

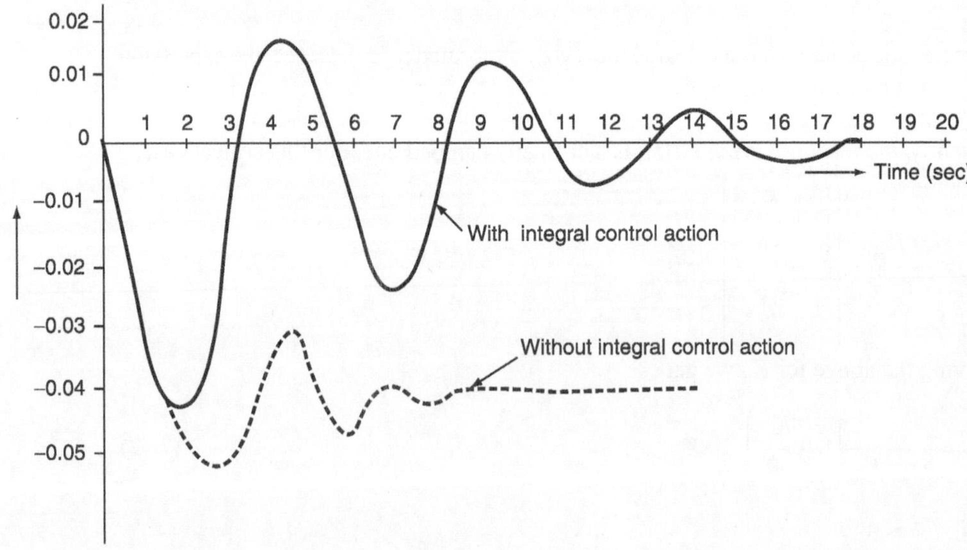

FIG.7.26   *Dynamic response of LFC of a single-area system with and without integral control action*

7.22 ROLE OF INTEGRAL CONTROLLER GAIN (K_I) SETTING

The role played by the gain setting of an integral controller in the control of frequency error is described below.

With subcritical in gain settings (i.e., $K_I < K_{Icrit}$), a sluggish, non-oscillatory response is obtained. The slowness of the response makes the integral of $\Delta f(t)$, and hence the time error, relatively large. However, with this setting, the generator need not 'chase' rapid load fluctuations, which ultimately cause equipment wear.

7.23 CONTROL OF GENERATOR UNIT POWER OUTPUT

The collective performance of all generators in the system is studied by assuming the equivalent generator having an inertia constant of H_{eq} to be equal to the sum of the inertia constants of all the generating units. Similarly, the effects of the system loads are lumped into a single damping constant B.

For a system having 'n' generators and a composite load-damping constant of B the steady-state frequency deviation following a load change ΔP_D is

$$\Delta f_{(s)} = \frac{-\Delta P_D}{\left(\dfrac{1}{R_1} + \dfrac{1}{R_2} + \cdots + \dfrac{1}{R_n}\right) + B} = \frac{-\Delta P_D}{\left(\dfrac{1}{R_{eq}} + B\right)}$$

The composite frequency response characteristic of the system is

$$\frac{-\Delta P_D}{\Delta f_{(s)}} = \frac{1}{R_{eq}} + B$$

It is normally expressed in MW/Hz. Sometimes, it is referred to as the stiffness of the system.

Example 7.9: An isolated control area consists of a 200-MW generator with an inertia constant of $H = 5$ kW-s/kVA having the following parameters (Fig. 7.27(a)):

Power system gain constant, $K_{ps} = 104$

Power system time constant, $\tau_{ps} = 22$ s

Speed regulation, $R = 3$

Normal frequency, $f^0 = 50$ Hz

Governor time constant, $\tau_{sg} = 0.3$ s

Turbine time constant, $\tau_t = 0.4$ s

Obtain the frequency error and plot the graph of deviation of frequency when a step-load change of 0.48 p.u. with an integral controller action of $k_i = 0.1$ is applied (Fig. 7.27(b)).

Example 7.10: An isolated control area consists of a 200-MW generator with an inertia constant of $H = 5$ kW-s/kVA having the following parameters (Fig. 7.28(a)):

Power system gain constant, $K_{ps} = 100$

Power system time constant, $\tau_{ps} = 20$ s

Speed regulation, $R = 2.5$

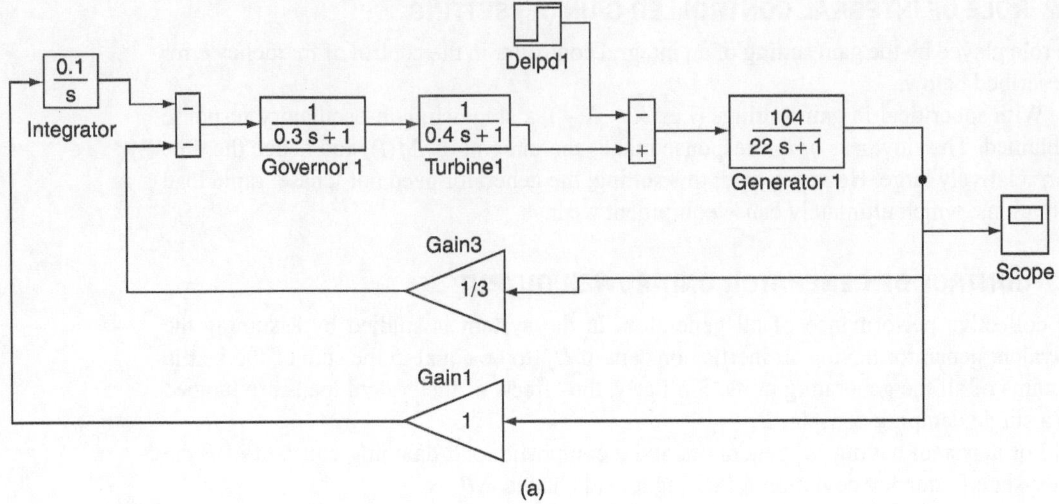

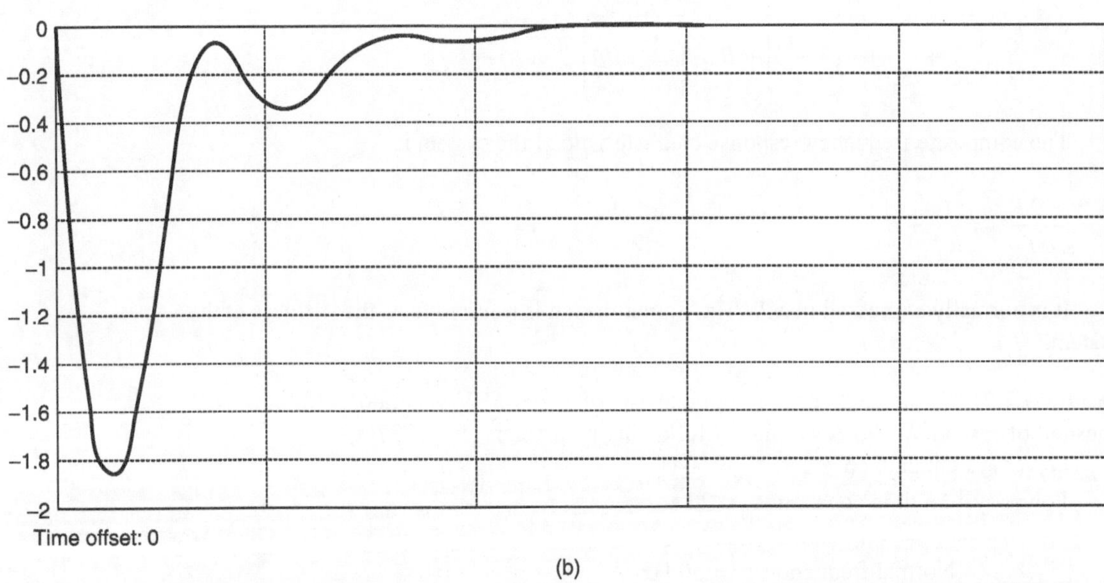

FIG. 7.27   *(a) Simulation block diagram for a single-area system with integral control action; (b) frequency response characteristics of Example 7.9*

$$\text{Normal frequency, } f^0 = 50 \text{ Hz}$$
$$\text{Governor time constant, } \tau_{sg} = 0.3 \text{ s}$$
$$\text{Turbine time constant, } \tau_t = 0.4 \text{ s}$$
$$\text{Integrator gain constant, } k_i = 0.15$$

Obtain the frequency error and plot the graph of deviation of frequency when a step-load disturbance of 2% with and without the integral control action is applied (Fig. 7.28(b)).

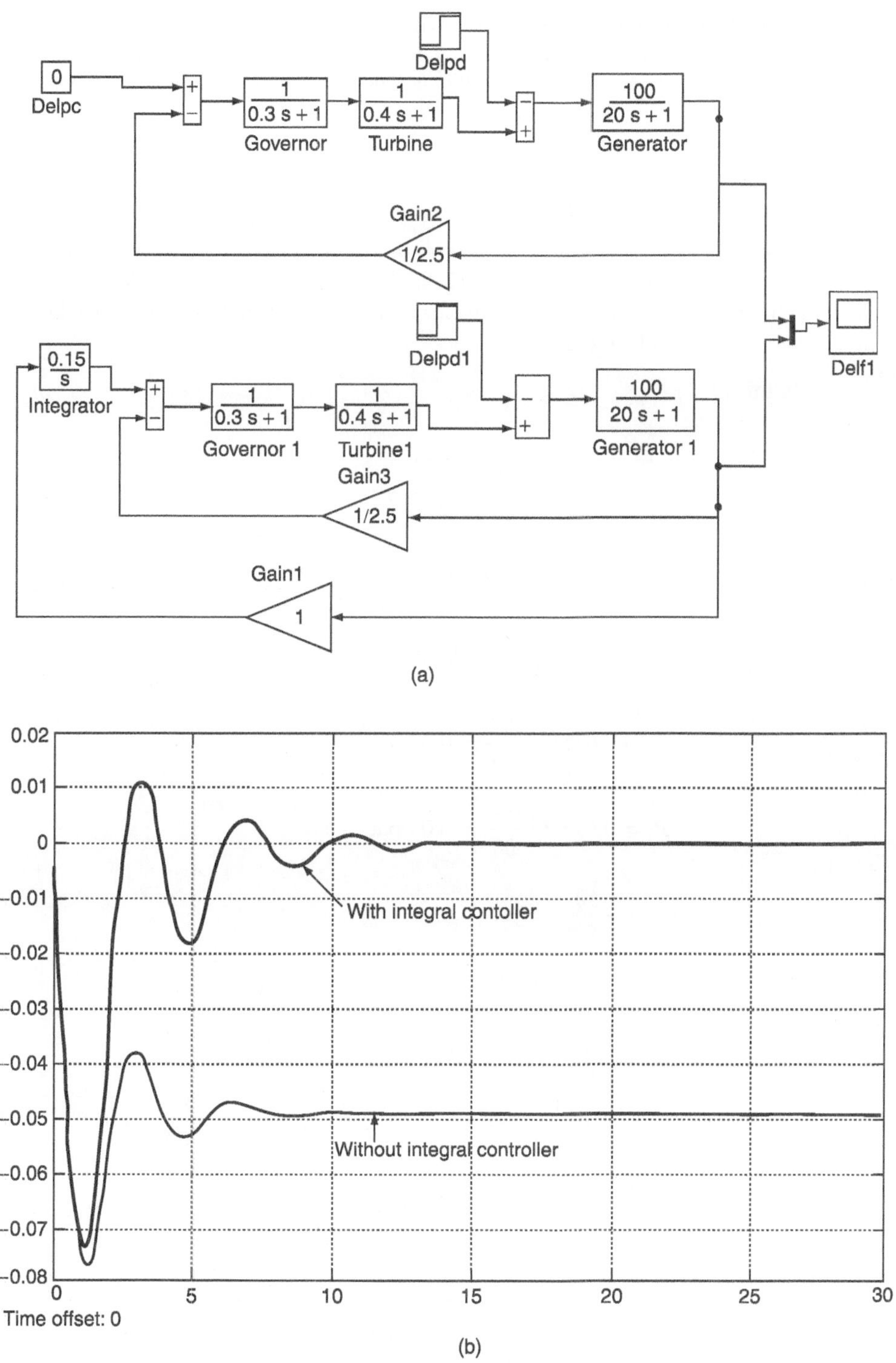

FIG. 7.28   *(a) Simulation diagram of a single-area system without and with integral control action; (b) frequency response characteristics of Example 7.10*

Example 7.11: Given a single area with three generating units as shown in Fig. 7.29:

Unit	Rating (MVA)	Speed droop R (per unit on unit base)
1	100	0.010
2	500	0.015
3	500	0.015

The units are loaded as $P_1 = 80$ MW; $P_2 = 300$ MW; $P_3 = 400$ MW. Assume $B = 0$; what is the new generation on each unit for a 50-MW load increase? Repeat with $B = 1.0$ p.u. (i.e., 1.0 p.u. on load base).

Solution:

(a) $$\Delta f = \frac{-\Delta P}{\sum_{i=1}^{3} \dfrac{1}{R_i} + B} = \frac{-\Delta P}{\dfrac{1}{R_1} + \dfrac{1}{R_2} + \dfrac{1}{R_3} + B}$$

with $B = 0$; at a common base of 1,000 MVA

$$R_1 = 0.01 \times \frac{1,000}{100} = 0.1 \text{ p.u.}$$

$$R_2 = 0.015 \times \frac{1,000}{500} = 0.03 \text{ p.u.}$$

$$R_3 = 0.015 \times \frac{1,000}{500} = 0.03 \text{ p.u.}$$

$$\Delta P = \frac{50}{1,000} = 0.05 \text{ p.u.}$$

$$\Delta f = \frac{-0.05}{\dfrac{1}{0.1} + \dfrac{1}{0.03} + \dfrac{1}{0.03}} = -652.17 \times 10^{-6} \text{ p.u.}$$

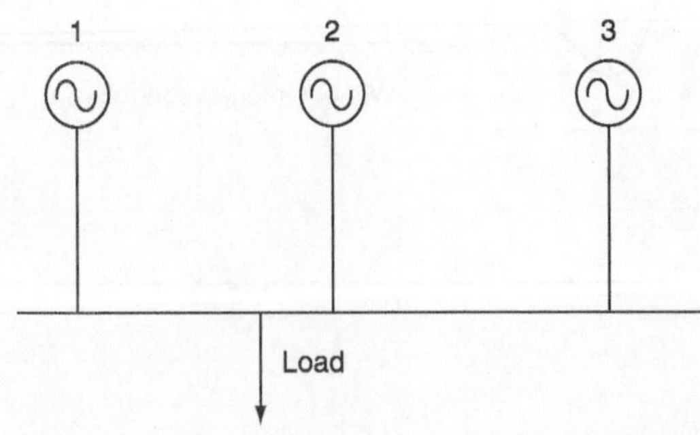

FIG. 7.29   *A single area with three generating units*

$$f = f^0 + \Delta f$$

$$= 50 - 652.17 \times 10^{-6}\,(50) = 49.96 \text{ Hz}$$

Changes in unit generation:

$$\Delta P_1 = \frac{-\Delta f}{R_1} = 0.00652 \text{ p.u.} = 6.52 \text{ MW}$$

$$\Delta P_2 = \frac{-\Delta f}{R_2} = 21.739 \times 10^{-3} \text{ p.u.} = 21.74 \text{ MW}$$

$$\Delta P_3 = \frac{-\Delta f}{R_3} = 21.739 \times 10^{-3} \text{ p.u.} = 21.74 \text{ MW}$$

Total = 50 MW

New generation:

$$P_1' = P_1 + \Delta P_1 = 80 + 6.52 = 86.52 \text{ MW}$$

$$P_2' = P_2 + \Delta P_2 = 300 + 21.74 = 321.74 \text{ MW}$$

$$P_3' = P_3 + \Delta P_3 = 400 + 21.74 = 421.74 \text{ MW}$$

(b) with $B = 1$ p.u. (on load base)

$$\Delta f = \frac{-0.05}{\dfrac{1}{0.1} + \dfrac{1}{0.03} + \dfrac{1}{0.03} + 1} = 643.78 \times 10^{-6}$$

$$\therefore f = f^\circ + \Delta f = 50 - 643.78 \times 10^{-6}\,(50) = 49.9614 \text{ Hz}$$

Changes in unit generation:

$$\Delta P_1 = \frac{-\Delta f}{R_1} = \frac{643.78 \times 10^{-6}}{0.1} = 6.44 \text{ MW}$$

$$\Delta P_2 = \frac{-\Delta f}{R_2} = \frac{643.78 \times 10^{-6}}{0.03} = 21.459 \text{ MW}$$

$$\Delta P_3 = \frac{-\Delta f}{R_3} = \frac{643.78 \times 10^{-6}}{0.03} = 21.459 \text{ MW}$$

New generation:

$$P_1' = P_1 + \Delta P_1 = 80 + 6.44 = 86.44 \text{ MW}$$

$$P_2' = P_2 + \Delta P_2 = 300 + 21.459 = 321.459 \text{ MW}$$

$$P_3' = P_3 + \Delta P_3 = 400 + 21.459 = 421.459 \text{ MW}$$

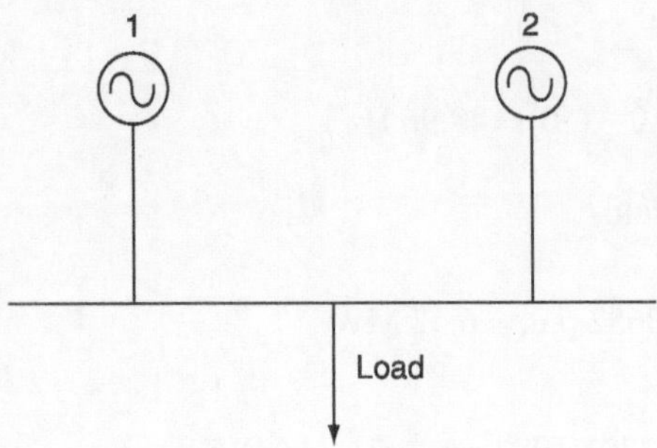

FIG. 7.30   *Single area with two generating units*

Example 7.12: Given a single area with two generating units as shown in Fig. 7.30:

Unit	Rating (MVA)	Speed droop R (per unit on unit base)
1	400	0.04
2	800	0.05

The units share a load of $P_1 = 200$ MW; $P_2 = 500$ MW. The units are operating in parallel, sharing 700 MW at 1.0 (50 Hz) frequency. The load is increased by 130 MW.

With $B = 0$, find the steady-state frequency deviation and the new generation on each unit.

With $B = 0.804$, find the steady-state frequency deviation and the new generation on each unit.

Solution:

(a) $\Delta f = \dfrac{-\Delta P}{\dfrac{1}{R_1} + \dfrac{1}{R_2} + B}$

At a common base of 1,000 MVA:

$$R_1 = 0.04 \times \frac{1,000}{400} = 0.1$$

$$R_2 = 0.05 \times \frac{1,000}{800} = 0.0625$$

$$\Delta f = \frac{-130/1,000}{\dfrac{1}{0.1} + \dfrac{1}{0.0625}} = -5 \times 10^{-3} \text{ p.u.}$$

$$f = f^0 + \Delta f = 50 - 5 \times 10^{-3}(50) = 49.75 \text{ Hz}$$

Change in unit generation:

$$\Delta P_1 = \frac{-\Delta f}{R_1} = \frac{5 \times 10^{-3}}{0.1} = 50 \text{ MW}$$

$$\Delta P_2 = \frac{-\Delta f}{R_2} = \frac{5 \times 10^{-3}}{0.0625} = 80 \text{ MW}$$

$$\textbf{Total} = \textbf{130 MW}$$

New generation:

$$P_1' = P_1 + \Delta P_1 = 200 + 50 = 250 \text{ MW}$$

$$P_2' = P_2 + \Delta P_2 = 500 + 80 = 580 \text{ MW}$$

(b) With $B = 0.804$ (on load base)

$$\Delta f = \frac{-130/1{,}000}{\dfrac{1}{0.1} + \dfrac{1}{0.0625} + 0.804} = -4.85 \times 10^{-3} \text{ p.u.}$$

$$f = f^0 + \Delta f = 50 - 4.85 \times 10^{-3}\,(50) = 49.457 \text{ Hz}$$

Change in unit generation:

$$\Delta P_1 = \frac{-\Delta f}{R_1} = \frac{4.85 \times 10^{-3}}{0.1} = 48.5 \text{ MW}$$

$$\Delta P_2 = \frac{-\Delta f}{R_2} = \frac{4.85 \times 10^{-3}}{0.0625} = 77.6 \text{ MW}$$

New generation:

$$P_1' = P_1 + \Delta P_1 = 200 + 48.5 = 248.5 \text{ MW}$$

$$P_2' = P_2 + \Delta P_2 = 500 + 77.6 = 577.6 \text{ MW}$$

Example 7.13: A 500-MW generator has a speed regulation of 4%. If the frequency drops by 0.12 Hz with an unchanged reference, determine the increase in turbine power. And also find by how much the reference power setting should be changed if the turbine power remains unchanged.

Solution:

Case 1:

Speed regulation, $R = 4\%$ of 50 Hz

$$= \frac{4}{100} \times 50 = 2 \text{ Hz}$$

$$R = \frac{2 \text{ Hz}}{500 \text{ MW}} = 0.004 \text{ Hz/MW}$$

Given a drop in frequency, $\Delta f = -0.12$ Hz

Increase in turbine power, $\Delta P = -\dfrac{1}{R}\Delta f$

$$= -\frac{1}{0.004} \times (-0.12)$$

$$= 30 \text{ MW}$$

$\therefore$ Turbine power increase, $\Delta P = 30 \text{ MW}$

Case 2:

If the turbine power remains unchanged, the reference power setting at the point of the block diagram must be changed such that the signal to the increase in generation is blocked:

$$\text{i.e.,} \quad \Delta P_{ref} - \frac{1}{R}\Delta f = 0$$

$$\Delta P_{ref} = \frac{1}{R}\Delta f = 30 \text{ MW}$$

Example 7.14: Two generating units having the capacities 600 and 900 MW and are operating at a 50 Hz supply. The system load increases by 150 MW when both the generating units are operating at about half of their capacity, which results in the frequency falling by 0.5 Hz. If the generating units are to share the increased load in proportion to their ratings, what should be the individual speed regulations? What should the regulations be if expressed in p.u. Hz/p.u. MW?

Solution:

$$\text{Rated capacity of Unit-1} = 600 \text{ MW}$$
$$\text{Rated capacity of Unit-2} = 900 \text{ MW}$$
$$\text{System frequency, } f = 50 \text{ Hz}$$
$$\text{System load increment, } \Delta P = 150 \text{ MW}$$
$$\text{Falling in frequency, } \Delta f = 0.5 \text{ Hz}$$

We know that $\Delta P = -\dfrac{1}{R}\Delta f$

$$\therefore \Delta P_1 = \frac{1}{-R_1}\Delta f \tag{7.43}$$

$$\Delta P_2 = -\frac{1}{R_2}\Delta f \tag{7.44}$$

If the load is shared in proportional to their ratings,

$$\Delta P_1 = 150 \times \frac{600}{1,500} = 60 \text{ MW}$$

$$\Delta P_2 = 150 \times \frac{900}{1,500} = 90 \text{ MW}$$

$$\therefore \text{ From Equation (7.43), } R_1 = \frac{-\Delta f}{\Delta P_1} = \frac{-0.5}{60} = -0.00833 \text{ Hz/p.u. MW}$$

$$R_2 = -\frac{\Delta f}{\Delta P_2} = -\frac{0.5}{90} = 0.0055 \text{ Hz/MW}$$

$$R_1 = -\frac{0.00833}{50} \times \frac{600}{1} = 0.099 \simeq 0.1 \text{ p.u. Hz/p.u. MW}$$

$$R_2 = -\frac{0.0055}{50} \times \frac{900}{1} = 0.099 \simeq 0.1 \text{ p.u. Hz/p.u. MW}$$

It is observed that the speed regulations in p.u. Hz/p.u. MW are attaining the same value, even when they are based on their individual ratings and they have different regulations.

Example 7.15: A single-area system has the following data:

$$\text{Speed regulation, } R = 4 \text{ Hz/p.u. MW}$$
$$\text{Damping coefficient, } B = 0.1 \text{ p.u. MW/Hz}$$
$$\text{Power system time constant, } T_p = 10 \text{ s}$$
$$\text{Power system gain, } K_p = 75 \text{ Hz/p.u. MW}$$

When a 2% load change occurs, determine the AFRC and the static frequency error. What is the value of the steady-state frequency error if the governor is blocked?

Solution:

$$\text{AFRC} = \beta = B + \frac{1}{R}$$

$$= 0.1 + \frac{1}{4}$$

$$= 0.35 \text{ MW/Hz}$$

$$\text{Static frequency error } = \Delta f_{ss} = \frac{-\Delta P_D}{\left(B + \dfrac{1}{R}\right)} = \frac{-\Delta P_D}{\beta}$$

$$= -\frac{2}{100} \times \frac{1}{0.35}$$

$$= -\,0.0571 \text{ Hz}$$

If the governor is blocked, the feedback loop will not be present; therefore, R will become infinite:

$$\text{i.e.,} \quad R \to \infty, \quad \therefore \frac{1}{R} = 0$$

$$\therefore \beta = B + \frac{1}{R} = B = 0.1 \text{ p.u. MW/Hz}$$

$$\text{Static frequency error} = \Delta f_{ss} = \frac{-\Delta P_D}{B + \dfrac{1}{R}} = \frac{-\Delta P_D}{B}$$

$$= \frac{-2}{100} \times \frac{1}{0.1}$$

$$= -\,0.2 \text{ Hz}$$

i.e., frequency falls by 0.0571 Hz.

$$\therefore \text{ New frequency, } f' = 50 - 0.0571$$
$$= 49.94 \text{ Hz}$$

Observation:

With speed-governor action:

Frequency falls by 0.0571 Hz

$\therefore$ New frequency, $f' = 50 - 0.0571$

$\qquad = 49.94$ Hz

Without speed-governor action:

Frequency falls by 0.2 Hz

$\therefore$ New frequency, $f' = 50 - 0.2 = 49.8$ Hz

From the above results, it is noted that the speed-governor action is necessary for obtaining a reduction in the steady-state frequency error.

Example 7.16: A 200-MVA synchronous generator is operated at 3,000 rpm, 50 Hz. A load of 40 MW is suddenly applied to the machine and the station valve to the turbine opens only after 0.4 s due to the time lag in the generator action. Calculate the frequency to which the generated voltage drops before the steam flow commences to increase so as to meet the new load. Given that the valve of H of the generator is 5.5 kW-s per kVA of the generator energy.

Solution:

Given:

Rating of the generator $= 200$ MVA

Load applied on the m/c $= 40$ MW

Time taken by the valve to open $= 0.4$ s

$H = 5.5$ kW-s/kVA

$\quad = 11 \times 10^5$ s

Energy stored at no-load $= 5.5 \times 200 \times 1{,}000 = 1{,}100$ MW-s $= 1{,}100$ MJ

Before the steam valve opens, the energy lost by the rotor $= 40 \times 0.4 = 16$ MJ.

The energy lost by the rotor results in a reduction in the speed of the rotor and hence the reduction in frequency.

We know

$$\omega = \omega^0 \left(\frac{f^0 + \Delta f}{f^0} \right)^2$$

$\therefore$ Frequency at the end of 0.4 s $= f_{\text{new}} = \sqrt{\dfrac{1{,}100 - 16}{1{,}100}} \times 50$

$$= \sqrt{0.09854} \times 50 = 0.9927 \times 50$$

$$= 49.635 \text{ Hz}$$

Example 7.17: Two generators of rating 100 and 200 MW are operated with a droop characteristic of 6% from no load to full load. Determine the load shared by each generator, if a load of 270 MW is connected across the parallel combination of those generators.

Solution:

The two generators are operating with parallel connection; the % drop in frequency from two generators due to different loads must be same.

Let power supplied by (100 MW) Generator-1 $= x$

Percentage drop in frequency $= 6\%$

$\therefore$ Percentage drop in the speed of Generator-1 $= \dfrac{6x}{100}$

Total load across the parallel connection $= 270$ MW

Power supplied by (200 MW) Generator-2 $= (270 - x)$

$\therefore$ Percentage drop in the speed of Generator-2 $= \dfrac{6(270 - x)}{200}$

Percent drop in frequency (or speed) of both machines must be the same:

$$\therefore \frac{6x}{100} = \frac{6(270 - x)}{200}$$

By solving the above equation, we get

$x = 90$ MW

$\therefore$ Load shared by Generator-1 (100 MW unit) $= 90$ MW

Load shared by Generator-2 (200 MW unit) $= 270 - x$

$= 270 - 90 = 180$ MW

KEY NOTES

- **Necessity of maintaining frequency constant**

  (i) All the AC motors should be given a constant frequency supply so as to maintain the speed constant.

  (ii) In continuous process industry, it affects the operation of the process itself.

  (iii) For synchronous operation of various units in the power system network, it is necessary to maintain the frequency constant.

  (iv) Frequency affects the amount of power transmitted through interconnecting lines.

- **Load frequency control (LFC)** is the basic control mechanism in the power system operation whenever there is a variation in load demand on a generating unit momentarily if there is an occurrence of unbalance between real-power input and output. This difference is being supplied by the stored energy of the rotating parts of the unit.

- Prime movers driving the generators are fitted with governors, which are regarded as primary control elements in the LFC system. Governors sense the change in a speed control mechanism to adjust the opening of steam valves in the case of steam turbines and the opening of water gates in the case of water turbines.

- The steady-state speed regulation in per unit is given by

$$R = \frac{N_0 - N}{N}$$

The value of R varies from 2% to 6% for any generating unit.

- The speed governor is the main primary tool for the LFC, whether the machine is used alone to feed a smaller system or whether it is a part of the most elaborate arrangement.

  Its main parts are fly-ball speed governor, hydraulic amplifier, speed changer, and linkage mechanism.

- **Control area** is possible to divide a very large power system into sub-areas in which all the generators are tightly coupled such that they swing in unison with change in load or due to a speed-changer setting. Such an area, where all the generators are running coherently is termed as a control area.

- **A single area** is a coherent area in which all the generators swing in unison to the changes in load or speed-changer settings and in which the frequency is assumed to be constant throughout both in static and dynamic conditions.

- **Dynamic response** is how the frequency changes as a function of time immediately after disturbance before it reaches the new steady-state condition. The canalization of dynamic response requires the solution of a dynamic equation of the system for a given disturbance.

- **Integral control consists** of a frequency sensor and an integrator. The frequency sensor measures the frequency error Δf and this error signal is fed into the integrator. The input to the integrator is called the **'area control error (ACE)'**.

- The ACE is the change in area frequency, which when used in an integral-control loop forces the steady-state frequency error to zero.

SHORT QUESTIONS AND ANSWERS

(1) What is the effect of speed of a generator on its frequency?

The effect of speed of a generator on its frequency is

$$f = \frac{pN}{120} \text{ Hz}$$

where p is the number of poles and N the speed in rpm.

(2) Why should the system frequency be maintained constant?

Constant frequency is to be maintained for the following functions:

- All the AC motors should be given constant frequency supply so as to maintain the speed constant.

- In continuous process industry, it affects the operation of the process itself.

- For synchronous operation of various units in the power system network, it is necessary to maintain the frequency constant.

(3) What is the nature of the generator–load frequency characteristic?

The nature of the generator is drooping straight-line characteristics.

(4) How do load frequency characteristics change during on-line control?

By shifting the load frequency characteristics as a whole up or down varying the inlet valve opening of the prime mover.

(5) How do load frequency characteristics change during off-line control?

By changing the slope of the load characteristics by varying the lever ratio of the speed governor.

(6) State why P–f and Q–V control loops can be treated as non-interactive?

The active power P is mainly dependent on the internal angle δ and is independent of bus voltage magnitude $|V|$. The bus voltage is dependent on machine excitation and hence on reactive power Q and is independent of the machine angle δ. The change in the machine angle δ is caused by a momentary change in the generator speed and hence the frequency. Therefore, the load frequency and excitation voltage controls are non-interactive for small changes and can be modeled and analyzed independently.

(7) What will be the order of the system for non-reheat steam turbine and reheat turbine?

The order of the system for non-reheat and reheat steam turbine are first order and second order, respectively.

(8) What are the transfer functions of non-reheat steam turbine and reheat turbine? What will be the value of their time constants?

The transfer function of non-reheat type of steam turbine is

$$G_{T(s)} = \frac{K_t}{1 + s\tau_t}, \quad \tau_1 = 0.2 \text{ to } 2 \text{ s}$$

The transfer function of reheat type of steam turbine is

$$G_T(s) = \frac{\Delta P_G(s)}{\Delta X_E(s)} = \left[\frac{K_t}{(1 + s\tau_t)}\right] \frac{(1 + sK_r\tau_r)}{(1 + s\tau_r)}$$

The time constant τ_r has a value in the range of 10 s.

(9) Under what condition will the model developed for a turbine be valid?

The condition for the turbine is the first 20 s following the incremental disturbance.

(10) Explain the control area concept.

It is possible to divide a very large power system into sub-areas in which all the generators are tightly coupled such that they swing in unison with change in load or due to a speed-changer setting. Such an area, where all the generators are running coherently, is termed the control area. In this area, frequency may be same in steady-state and dynamic conditions. For developing a suitable control strategy, a control area can be reduced to a single generator, a speed governor, and a load system.

(11) What is meant by single-area power system?

A single area is a coherent area in which all the generators swing in unison to the changes in load or speed-changer settings and in which the frequency is assumed to be constant throughout both in static and dynamic conditions. This single control area can be represented by an isolated power system consisting of a turbine, its speed governor, generator, and load.

(12) What is meant by dynamic response in LFC?

The meaning of dynamic response is how the frequency changes as a function of time immediately after disturbance before it reaches the new steady-state condition.

(13) What is meant by uncontrolled case?

For uncontrolled case, $\Delta P_c = 0$; i.e., constant speed-changer position with variable load.

(14) What is the need of a fly-ball speed governor?

This is the heart of the system, which controls the change in speed (frequency).

(15) What is the need of a speed changer?

It provides a steady-state power output setting for the turbines. Its upward movement opens the upper pilot valve so that more steam is admitted to the turbine under steady conditions. This gives rise to higher steady-state power output. The reverse happens for downward movement of the speed changer.

(16) What is meant by area control error?

The area control error (ACE) is the change in area frequency, which when used in an integral-control loop forces the steady-state frequency error to zero.

(17) What is the nature of the steady-state response of the uncontrolled LFC of a single area?

The nature of the steady-state response of a single area is the linear relationship between frequency and load for free governor operation.

(18) How and why do you approximate the system for the dynamic response of the uncontrolled LFC of a single area?

The characteristic equation of the LFC of an isolated power system is third order, dynamic response that can be obtained only for a specific numerical case.

However, the characteristic equation can be approximated as first order by examining the relative magnitudes of the different time constants involved.

(19) What are the basic requirements of a closed-loop control system employed for obtaining the frequency constant?

The basic requirements are as follows:

(i) Good stability;

(ii) Frequency error, accompanying a step-load change, returns to zero;

(iii) The magnitude of the transient frequency deviation should be minimum;

(iv) The integral of the frequency error should not exceed a certain maximum value.

(20) What are the basic components of an integral controller

It consists of a frequency sensor and an integrator.

(21) Why should the integrator of the frequency error not exceed a certain maximum value?

The frequency error should not exceed a maximum value so as to limit the error of synchronous clocks.

(22) What are the assumptions made in the simplified analysis of the integral control?

- The time constant of the speed-governing mechanism τ_{sg} and that of the turbine are both neglected, i.e., it is assumed that $\tau_{sg} = \tau_t = 0$.

- The speed changer is an electromechanical device and hence its response is not instantaneous. However, it is assumed to be instantaneous in the present analysis.

- All non-linearities in the equipment, such as dead zone, etc., are neglected.

- The generator can change its generation ΔP_G as fast as it is commanded by the speed-changer.

- The ACE is a continuous signal.

(23) State briefly how the time response of the frequency error depends upon the gain setting of the integral control.

If K_I is less than its critical value, then the response will be damped non-oscillatory. $\Delta f(t)$ reduces to zero in a longer time. Hence, the response is sluggish. This is an overdamped case. This is the subcritical case of integral control.

If K_I is greater than its critical value, the time response would be damped oscillatory. $\Delta f(t)$ approaches zero faster. This is an underdamped case. This is the supercritical case of the control.

If K_I equals its critical value, no oscillations would be present in the time response and $\Delta f(t)$ approaches zero in less time than in the subcritical case. The integral of the frequency error would be the least in this case.

MULTIPLE-CHOICE QUESTIONS

(1) If the load on an isolated generator is increased without increasing the power input to the prime mover:

   (a) The generator will slow down.

   (b) The generator will speed up.

   (c) The generator voltage will increase.

   (d) The generator field.

(2) Governors of controlling the speed of electric-generating units normally provide:

   (a) A flat-speed load characteristic.

   (b) An increase in speed with an increasing load.

   (c) A decrease in speed with an increasing load.

   (d) None

(3) When two identical AC-generating units are operated in parallel on governor control, and one machine has a 5% governor droop and the other a 10% droop, the machine with the greater governor droop will:

   (a) Tend to take the greater portion of the load changes.

   (b) Share the load equally with the other machine.

   (c) Tend to take the lesser portion of the load changes.

   (d) None.

(4) On LFC installations, error signals are developed proportional to the frequency error. If the frequency declines, the error signal will act to:

   (a) Increase the prime mover input to the generators.

   (b) Reduce the prime mover input to the generators.

   (c) Increase generator voltages.

   (d) None.

(5) If KE reduces

   (a) w decreases.

   (b) Speed falls.

   (c) Frequency reduces.

   (d) All.

(6) The changing of slope of a speed governer characteristic is achevied by changing the ratio of lever L of governer and can be made during

   (a) On-line condition only.

   (b) Off-line condition only.

   (c) Both (a) and (b).

   (d) Either (a) or (b).

(7) Unit of R is ____________ .

   (a) Hz/MVAr.

   (b) Hz/MVA.

   (c) Hz/MW.

   (d) Hz-s.

(8) Unit of B is ____________ .

   (a) MVAr/Hz.

   (b) MVA/Hz.

   (c) MW/Hz.

   (d) MW-s.

(9) Unit of H of a synchronous machine is:

   (a) MJ/MW.

   (b) MJ/MVA.

   (c) MJ/s.

   (d) MW-s.

(10) KE and frequency of a synchronous machine are related as:

   (a) $KE = f$.

   (b) $KE = 1/f$.

(c) $KE = f^2$.

(d) None of these.

(11) Input signals to an ALFC loop is ___________.

(a) ΔP_{ref}

(b) ΔP_{D}

(c) Both (a) and (b).

(d) None of these.

(12) Two main control loops in generating stations are:

(a) ALFC.

(b) AVR.

(c) Both (a) and (b).

(d) None of these.

(13) The speed regulation can be expressed as

(a) Ratio of change in frequency from no-load to full load to the rated frequency of the unit.

(b) Ratio of change in frequency to the corresponding change in real-power generation.

(c) (a) and (b).

(d) None of these.

(14) In an ALFC loop, Δf can be reduced using ___________ controller.

(a) Differential.

(b) Integral.

(c) Proportional.

(d) All of these.

(15) Time constant of a power syste when compared to a speed governor is:

(a) Less.

(b) More.

(c) Same.

(d) None of these.

(16) Δf is of the order of ___________ Hz.

(a) 0 to 0.05.

(b) -0.05 to 0.

(c) Both (a) and (b).

(d) None of these.

(17) In a power system ___________ are continuously changing.

(a) Active and reactive power generation.

(b) Active and reactive power demands.

(c) Voltage and its angle.

(d) All of these.

(18) In a normal state, the frequency and voltage are kept at specified values that carefully maintain a balance between:

(a) Real-power demand and real-power generation.

(b) Reactive power demand and reactive power generation.

(c) Both.

(d) None of these.

(19) Real-power balance will control the variations in ___________ .

(a) Voltage.

(b) Frequency.

(c) Both.

(d) None of these.

(20) The excitations of the generators must be continuously regulated:

(a) To match the reactive power generations with reactive power demand.

(b) To control the variations in voltage.

(c) Both.

(d) None of these.

(21) ___________ is the basic control mechanism in the power system.

(a) LFC.

(b) Voltage.

(c) Both.

(d) None of these

(22) Setting of speed-load characteristic parallel to itself is known as ___________ and its adapted as on-line control.

(a) Primary control.

(b) Supplementary control.

(c) Basic.

(d) All of these.

(23) The basic function of LFC is:

(a) To maintain frequency for variations in real-power demand.

(b) To maintain voltage for variations in reactive power demand.

   (c) To maintain both voltage and frequency for variations in real-power demand.

   (d) To maintain both voltage and frequency for variations in real-power demand.

(24) The degree of unbalance between real-power generation and real-power demand is indicated by the index:

   (a) Speed regulation R.

   (b) Change in voltage,

   (c) Frequency error.

   (d) None.

(25) The LFC system ___________ in the system.

   (a) Does consider the reactive power flow.

   (b) Does not consider the reactive power flow.

   (c) Does not consider the real-power flow.

(26) ___________ controls the excitation voltage and modifies the excitation.

   (a) Change in real-power, ΔP_a.

   (b) Change in frequency Δ.

   (c) Change in tie-line power, ΔP_{tie}.

   (d) Change in reactive power ΔQ_{ci}.

(27) The p–f controller is employed to:

   (a) Control the frequency.

   (b) Monitor the active power flows in interconnection.

   (c) Control the voltage.

     (i) Only (a).

     (ii) Only (b).

     (iii) (b) and (c).

     (iv) (a) and (b).

(28) Which of the following is correct regarding p–f controller?

   (a) It senses the frequency error.

   (b) It changes the tie-line powers.

   (c) Provides the information about incremental error in power angle $\Delta\delta$.

     (i) (a) and (b).

     (ii) (b) and (c).

     (iii) (a) and (c).

     (iv) All of these.

(29) The control signal that will change the position of the inlet valve of the prime mover is:

   (a) ΔP_{ci}.

   (b) ΔP_{gi}.

   (c) ΔP_{di}.

   (d) None of these.

(30) The objective of Q–V controller is to transform the:

   (a) Terminal voltage error signal into a reactive power control signal, ΔQ_{ci}.

   (b) Terminal voltage error signal into a real-power control signal, ΔP_{ci}.

   (c) Frequency error signal into a real-power control signal, ΔP_{ci}.

   (d) None of these.

(31) The active power P is:

   (a) Mainly dependent on the internal torque angle, δ.

   (b) Almost independent of the voltage magnitude.

   (c) totally dependent on both the torque angle and the voltage.

   (d) Mainly dependent on voltage and independent of torque angle, δ.

     (i) (a) and (d).

     (ii) (b) and (c).

     (iii) (a) and (b).

     (iv) Only (d).

(32) The bus voltage V is:

   (a) Dependent on the internal torque angle, δ.

   (b) Almost independent of active power, P.

   (c) Dependent on machine excitation and hence on reactive power.

   (d) Almost independent of internal torque angle, δ.

     (i) (a) and (d).

     (ii) (b) and (c).

     (iii) (a) and (b).

     (iv) (c) and (d).

(33) Usually p–f controller and Q–V controller for ___________ change, can be considered as ___________ type.

   (a) Dynamic, non-interacting.

(b) Static, interacting.

(c) Static, non-interacting.

(d) None of these.

(34) AVR loop is ___________ control mechanism.

(a) Slow.

(b) Faster.

(c) Slow in some cases and faster in some other cases.

d) None of these.

(35) ALFC loop is ___________ control mechanism.

(a) Slow.

(b) Faster.

(c) Slow as well as fast.

(d) None of these.

(36) Which of the following indicates the large-signal analysis of power system dynamics?

(a) Large and sudden variations in the system variables due to sudden disturbances.

(b) Mathematical model is a set of non-linear differential equations.

(c) Mathematical model is a set of linear differential equations.

(d) Small and gradual variations of system variables.

(i) (a) and (b).

(ii) (b) and (c).

(iii) (c) and (d).

(iv) None of these.

(37) Laplace transform methods are employed to determine the response of the system in ___________ analysis.

(a) Large signal.

(b) Small signal.

(c) Both.

(d) None of these.

(38) A signal area system is one in which:

(a) It is not connected to any other system.

(b) Total demand on the system should be fully met by its own local generation.

(c) All generators swing together.

(d) All of these.

(39) In a signal area system, all generators working remain in synchronism maintaining their relative power angles; such a group of generators is called ___________ .

(a) Swing group.

(b) Synchro group.

(c) Coherent group.

(d) None of these.

(40) The heart of the speed governor system, which controls the change in speed is:

(a) Linkage mechanism.

(b) Fly-ball speed governor.

(c) Speed changer.

(d) Hydraulic amplifier.

(41) In a hydraulic amplifier:

(a) High-power-level pilot valve moment is converted into low-power-level main piston movement.

(b) Low-power pilot valve moment is converted into low-power-level piston movement.

(c) Low-power-level pilot valve moment is converted into high-power-level piston movement.

(d) Low-power-level pilot valve moment is converted into high-power-level pilot valve moment.

(42) Linkage mechanism provides:

(a) The moment of control valve in propositional to the inlet steam.

(b) The feedback from the control valve moment.

(c) Both (a) and (b).

(d) None of these.

(43) The primary control loop in generator control is:

(a) Linkage mechanism.

(b) Fly-ball speed governor.

(c) Speed changer.

(d) Hydraulic amplifier.

(44) The position of the pilot valve can be affected through linkage mechanism in ___________ way.

(a) Directly by the speed changer.

(b) Indirectly through feedback due to position changes of the main system.

(c) Indirectly through feedback due to position changes of the linkage point E resulting from a change in speed.

(d) All of these.

(45) For non-reheat type of steam turbine, the mathematical model is:

(a) $\dfrac{\Delta P_G(s)}{\Delta X_E(s)} = \dfrac{K_t}{1+s\tau_t}$

(b) $\dfrac{\Delta P_G(s)}{\Delta X_E(s)} = 1 - \dfrac{K_t}{1+s\tau_t}$

(c) $\dfrac{\Delta P_G(s)}{\Delta X_E(s)} = \left(\dfrac{k_E}{1+s\tau_t}\right)\left(\dfrac{1+s\tau_r k_r}{1+s\tau_t}\right)$

(d) None of these.

(46) In reheat type of steam turbine,

(a) Steam at high pressure with low temperature is transformed into steam at low pressure with higher temperature.

(b) Steam at low pressure with higher temperature is transformed into steam at high pressure with low temperature.

(c) Steam at low pressure with low temperature is transformed into steam at high pressure with higher temperature.

(d) None of these.

(47) Transfer function of reheat type of steam turbine is of ____________ order.

(a) First.

(b) Second.

(c) Third.

(d) None of these.

(48) Transfer function of non-reheat type of steam turbine is of ____________ order.

(a) First.

(b) Second.

(c) Third.

(d) None of these.

(49) The surplus power ($\Delta P_G - \Delta P_D$) can be absorbed by a system:

(a) By increasing the stored $K \in$ of the system at the rate $\dfrac{d}{dt}(w_{KE})$.

(b) By motor loads.

(c) There is no absorption of surplus power by the system.

(d) Both (a) and (b).

(50) The block diagram of the LFC of an isolated power system is of ____________ model.

(a) First.

(b) Second.

(c) Third.

(d) Fourth.

REVIEW QUESTIONS

(1) Develop the block diagram of the LFC of a single-area system.

(2) Compare the steady state and dynamic operations of an isolated system.

(3) Draw the schematic diagram of a speed-governing system and explain its components on the dynamic response of an uncontrolled system with necessary equations. Hence, obtain the transfer function of a speed-governing system.

(4) How do the governor characteristics of the prime mover affect the control of system frequency and system load?

(5) Explain why it is necessary to maintain the frequency of the system constant.

(6) What do you mean by LFC?

(7) Draw a neat sketch of a typical turbine speed-governing system and derive its block diagram representation.

(8) For a single-area system, show that the static error in frequency can be reduced to zero using frequency control and comment on the dynamic response of an uncontrolled system with necessary equations.

(9) Explain the p–f and Q–V control loops of power system.

(10) What is meant by control area and ACE?

(11) Explain clearly about proportional plus integral LFC with a block diagram.

(12) Discuss the adverse effects of change in the voltage and the frequency of a power system. Mention the acceptable ranges of these changes.

PROBLEMS

(1) A 250-MVA synchronous generator is operating at 1,500 rpm, 50 Hz. A load of 50 MW is suddenly applied to the machine and the station valve to the turbine opens only after 0.35 s due to the time lag in the generator action. Calculate the frequency at which the generated voltage drops before the steam flow commences to increase to meet the new load. Given that the valve of H of the generator is 3.5 kW-s per kVA of the generator energy.

(2) Two generating stations A and B have full-load capacities of 250 and 100 MW, respectively. The interconnector connecting the two stations has an induction motor/synchronous generator (Plant C) of full-load capacity 30 MW; percentage changes of speeds of A, B, and C are 4, 3, and 2, respectively. The loads on bus bars A and B are MW and 50 MW, respectively. Determine the load taken by Plant C and indicate the direction of the power flow.

(3) A 750-MW generator has a speed regulation of 3.5%. If the frequency drops by 0.1 Hz with an unchanged reference, determine the increase in turbine power. And also find by how much the reference power setting should be changed if the turbine power remains unchanged.

8 Load Frequency Control-II

OBJECTIVES

After reading this chapter, you should be able to:

- develop the block diagram models for a two-area power system
- observe the steady state and dynamic analysis of a two-area power system with and without integral control
- develop the dynamic-state variable model for single-area, two-area, and three-area power system networks

8.1 INTRODUCTION

An extended power system can be divided into a number of load frequency control (LFC) areas, which are interconnected by tie lines. Such an operation is called a *pool operation*. A power pool is an interconnection of the power systems of individual utilities. Each power system operates independently within its own jurisdiction, but there are contractual agreements regarding internal system exchanges of power through the tie lines and other agreements dealing with operating procedures to maintain system frequency. There are also agreements relating to operational procedures to be followed in the event of major faults or emergencies. The basic principle of a pool operation in the normal steady state provides:

(i) Maintaining of scheduled interchanges of tie-line power: The interconnected areas share their reserve power to handle anticipated load peaks and unanticipated generator outages.

(ii) Absorption of own load change by each area: The interconnected areas can tolerate larger load changes with smaller frequency deviations than the isolated power system areas.

For analyzing the dynamics of the LFC of an n-area power system, primarily consider two-area systems.

Two control areas 1 and 2 are connected by a single tie line as shown in Fig. 8.1.

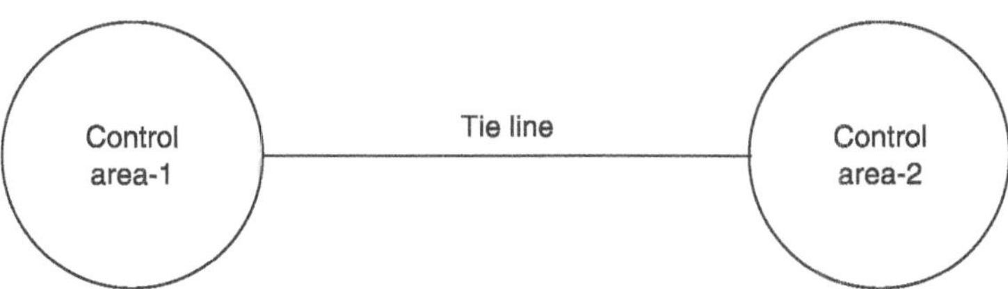

FIG. 8.1   *Two control areas interconnected through a single tie line*

Here, the control objective is to regulate the frequency of each area and to simultaneously regulate the power flow through the tie line according to an interarea power agreement.

In the case of an isolated control area, the zero steady-state error in frequency (i.e., $\Delta f_{\text{steady state}} = 0$) can be obtained by using a proportional plus integral controller, whereas in two-control area case, proportional plus integral controller will be installed to give zero steady-state error in a tie-line power flow (i.e., $\Delta P_{\text{TL}} = 0$) in addition to zero steady-state error in frequency.

For the sake of convenience, each control area can be represented by an equivalent turbine, generator, and governor system.

In the case of a single control area, the incremental power ($\Delta P_{\text{G}} - \Delta P_{\text{D}}$) was considered by the rate of increase of stored KE and increase in area load caused by the increase in frequency.

In a two-area case, the tie-line power must be accounted for the incremental power balance equation of each area, since there is power flow in or out of the area through the tie line.

Power flow out of Control area-1 can be expressed as

$$P_{\text{TL}_1} = \frac{|E_1||E_2|}{X_{\text{TL}}}\sin(\delta_1 - \delta_2) \tag{8.1}$$

where $|E_1|$ and $|E_2|$ are voltage magnitudes of Area-1 and Area-2, respectively, δ_1 and δ_2 are the power angles of equivalent machines of their respective areas, and X_{TL} is the tie-line reactance.

If there is change in load demands of two areas, there will be incremental changes in power angles ($\Delta\delta_1$ and $\Delta\delta_2$). Then, the change in the tie-line power is

$$P_{\text{TL}_1} + \Delta P_{\text{TL}_1} = \frac{|E_1||E_2|}{X_{\text{TL}}}\sin[(\delta_1 - \delta_2) + (\Delta\delta_1 - \Delta\delta_2)]$$

$$= \frac{|E_1||E_2|}{X_{\text{TL}}}[\sin(\delta_1 - \delta_2)\cos(\Delta\delta_1 - \Delta\delta_2) + \cos(\delta_1 - \delta_2)\sin(\Delta\delta_1 - \Delta\delta_2)]$$

$$= \frac{|E_1||E_2|}{X_{\text{TL}}}[\sin(\delta_1 - \delta_2) + \cos(\delta_1 - \delta_2)(\Delta\delta_1 - \Delta\delta_2)]$$

$$[\text{since } (\Delta\delta_1 - \Delta\delta_2) \approx 0]$$

$$= \frac{|E_1||E_2|}{X_{\text{TL}}}\sin(\delta_1 - \delta_2) + \frac{|E_1||E_2|}{X_{\text{TL}}}[\cos(\delta_1 - \delta_2)(\Delta\delta_1 - \Delta\delta_2)]$$

Therefore, change in incremental tie-line power can be expressed as

$$\Delta P_{\text{TL}_1} = \frac{|E_1||E_2|}{X_{\text{TL}}}[\cos(\delta_1 - \delta_2)(\Delta\delta_1 - \Delta\delta_2)]$$

$$\Delta P_{\text{TL}_1(\text{p.u.})} = T_{12}(\Delta\delta_1 - \Delta\delta_2) \tag{8.2}$$

where $\quad T_{12} = \frac{|E_1||E_2|}{X_{\text{TL}}P_1}\cos(\delta_1 - \delta_2) \tag{8.3}$

T_{12} is known as the synchronizing coefficient or the stiffness coefficient of the tie-line. Equation (8.3) can be written as

$$T_{12} = \frac{P_{\max_{12}}}{P_1} \cos(\delta_1 - \delta_2)$$

where $P_{\max_{12}} = \dfrac{|E_1||E_2|}{X_{TL}} =$ Static transmission capacity of the tie line.

Consider the change in frequency as

$$\Delta\omega = \frac{d}{dt}(\Delta\delta)$$

$$2\pi\Delta f = \frac{d}{dt}(\Delta\delta)$$

$$\Delta f = \frac{1}{2\pi} \times \frac{d}{dt}(\Delta\delta) \ \text{Hz}$$

In other words,

$$\frac{d}{dt}(\Delta\delta) = 2\pi\Delta f$$

$$\int \frac{d}{dt}(\Delta\delta) = \int 2\pi\Delta f$$

$$\Delta\delta = 2\pi \int \Delta f \ dt \ \text{radians}$$

Hence, the changes in power angles for Areas-1 and 2 are

$$\Delta\delta_1 = 2\pi \int \Delta f_1 dt$$

and $\quad \Delta\delta_2 = 2\pi \int \Delta f_2 dt$

Since the incremental power angles are related in terms of integrals of incremental frequencies, Equation (8.2) can be modified as

$$\Delta P_{TL_1} = 2\pi T_{12}\left(\int \Delta f_1 dt - \int \Delta f_2 dt\right) \tag{8.4}$$

Δf_1 and Δf_2 are the incremental frequency changes of Areas-1 and 2, respectively. Similarly, the incremental tie-line power out of Area-2 is

$$\Delta P_{TL_2} = 2\pi T_{21}\left(\int \Delta f_2 dt - \int \Delta f_1 dt\right) \tag{8.5}$$

where $\quad T_{21} = \dfrac{|E_1||E_2|}{X_{TL}P_2}\cos(\delta_2 - \delta_1) \tag{8.6}$

Dividing Equation (8.6) by Equation (8.3), we get

$$\frac{T_{21}}{T_{12}} = \frac{P_1}{P_2} = a_{12}$$

Therefore, $T_{21} = a_{12}T_{12}$

and hence $\Delta P_{TL_2} = a_{12}\Delta P_{TL_1}$ (8.7)

From Equation (7.25) **(LFC-1)**, surplus power in p.u. is

$$\Delta P_G - \Delta P_D = \frac{2H}{f^0}\frac{d}{dt}(\Delta f) + B\Delta f \quad \text{(for a single-area case)}$$

For a two-area case, the surplus power can be expressed in p.u. as

$$\Delta P_{G_1} - \Delta P_{D_1} = \frac{2H_1}{f^0}\frac{d}{dt}(\Delta f_1) + B_1\Delta f_1 + \Delta P_{TL_1} \qquad (8.8)$$

Taking Laplace transform on both sides of Equation (8.8), we get

$$\Delta P_{G_1}(s) - \Delta P_{D_1}(s) = \frac{2H_1}{f^0}s(\Delta F_1(s)) + B_1\Delta F_1(s) + \Delta P_{TL_1}(s)$$

Rearranging the above equation as follows, we get

$$\Delta P_{G_1}(s) - \Delta P_{D_1}(s) = \Delta F_1(s)\left(\frac{2H_1}{f^0}s + B_1\right) + \Delta P_{TL_1}(s)$$

$$\Delta F_1(s) = \left[\Delta P_{G_1}(s) - \Delta P_{D_1}(s) - \Delta P_{TL_1}(s)\right]\left[\frac{\dfrac{1}{B_1}}{1 + \left(\dfrac{2H_1}{B_1 f^0}\right)s}\right]$$

$$\Delta F_1(s) = \left[\Delta P_{G_1}(s) - \Delta P_{D_1}(s) - \Delta P_{TL_1}(s)\right]\frac{K_{ps_1}}{1 + s\tau_{ps_1}} \qquad (8.9)$$

where $K_{ps_1} = 1/B_1$

$$\tau_{ps_1} = \frac{2H_1}{B_1 f^0}$$

By comparing Equation (8.9) with single-area Equation (7.26), the only additional term is the appearance of signal $\Delta P_{TL_1}(s)$

Equation (8.9), can be represented in a block diagram model as shown in Fig. 8.2. Taking Laplace transformation on both sides of Equation (8.4), we get

$$\Delta P_{TL_1}(s) = 2\pi T_{12}\left(\frac{\Delta F_1(s)}{s} - \frac{\Delta F_2(s)}{s}\right)$$

$$= \frac{2\pi T_{12}}{s}(\Delta F_1(s) - \Delta F_2(s)) \qquad (8.10)$$

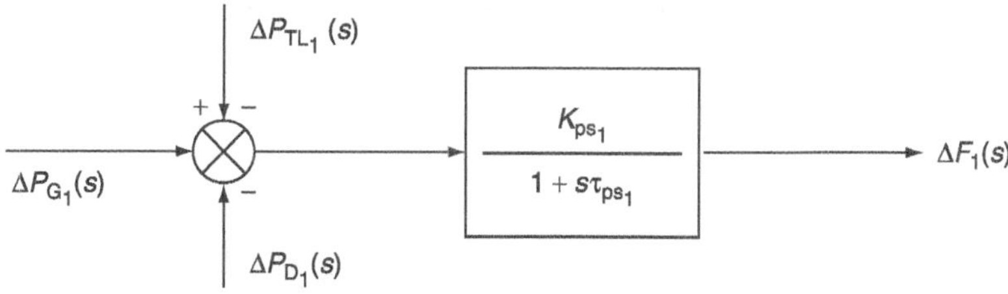

FIG. 8.2   *Block diagram representation of Equation (8.9) (for Control area-1)*

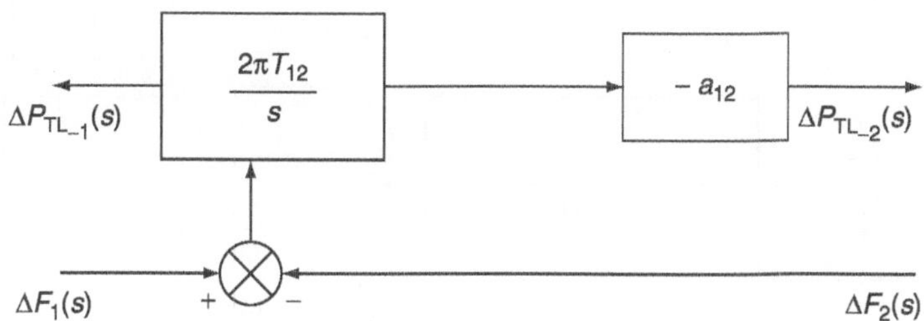

FIG. 8.3   *Block diagram representation of Equations (8.10) and (8.11)*

For Control area-2, we have

$$\Delta P_{TL_2}(s) = 2\pi T_{21}\left(\frac{\Delta F_2(s)}{s} - \frac{\Delta F_1(s)}{s}\right)$$

$$= -2\pi a_{12}T_{12}\left(\frac{\Delta F_1(s)}{s} - \frac{\Delta F_2(s)}{s}\right)$$

$$= -\frac{2\pi a_{12}T_{12}}{s}(\Delta F_1(s) - \Delta F_2(s)) \tag{8.11}$$

The block diagram representation of Equations (8.10) and (8.11) is shown in Fig. 8.3.

8.2 COMPOSITE BLOCK DIAGRAM OF A TWO-AREA CASE

By the combination of basic block diagrams of Control area-1 and Control area-2 and with the use of Figs. 8.2 and 8.3, the composite block diagram of a two-area system can be modeled as shown in Fig. 8.4.

8.3 RESPONSE OF A TWO-AREA SYSTEM—UNCONTROLLED CASE

For an uncontrolled case, $\Delta P_{c_1} = \Delta P_{c_2} = 0$, i.e., the speed-changer positions are fixed.

8.3.1 Static response

In this section, the changes or deviations, which result in the frequency and tie-line power under steady-state conditions following sudden step changes in the loads in the two areas, are determined.

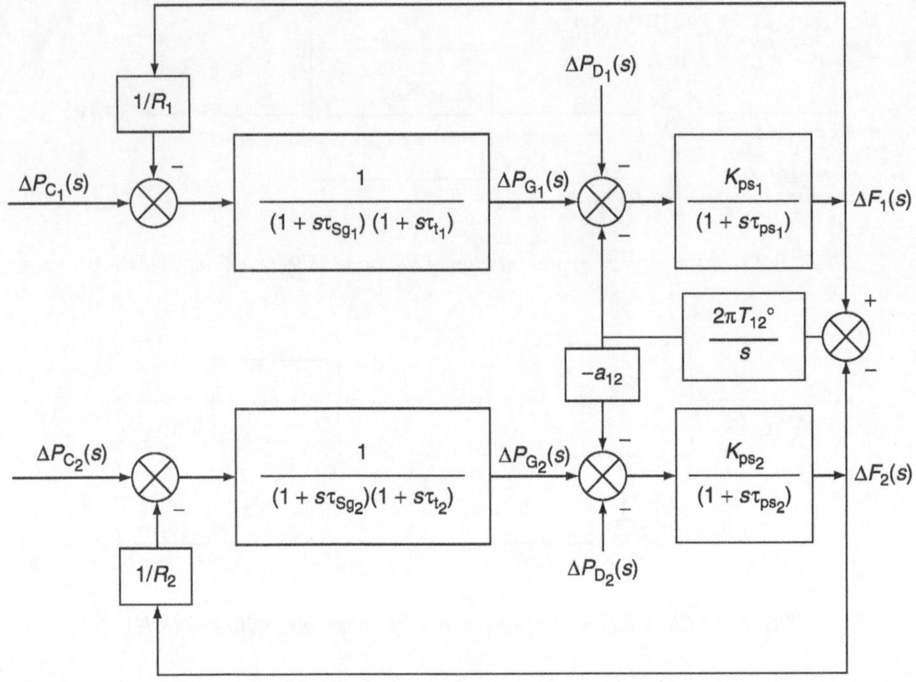

FIG. 8.4 *Block diagram representation of a two-area system with an LFC*

Let ΔP_{D_1}, ΔP_{D_2} be sudden (incremental) step changes in the loads of Control area-1 and Control area-2, simultaneously.

ΔP_{G_1}, ΔP_{G_2} are the incremental changes in the generation in Area-1 and Area-2 as a result of the load changes.

Δf is the static change in frequency. This will be the same for both the areas and ΔP_{TL_1} is the static change in the tie-line power transmitted from Area-1 to Area-2. Since only the static changes are being determined, the incremental changes in generation can be determined by the static loop gains. So, we have

$$\Delta P_{G_1} = -\frac{\Delta f}{R_1} \tag{8.12}$$

and $\quad \Delta P_{G_2} = -\frac{\Delta f}{R_2} \quad$ for static changes $\tag{8.13}$

For the two areas, the dynamics are described by:

$$(\Delta P_{G_1} - \Delta P_{D_1}) = \frac{2H_1}{f^0} \frac{d}{dt}(\Delta f_1) + B_1\,\Delta f_1 + \Delta P_{TL_1} \tag{8.14}$$

and $\quad (\Delta P_{G_2} - \Delta P_{D_2}) = \frac{2H_2}{f^0} \frac{d}{dt}(\Delta f_2) + B_2\,\Delta f_2 + \Delta P_{TL_2} \tag{8.15}$

Under steady-state conditions, we have

$$\frac{\mathrm{d}}{\mathrm{d}t}(\Delta f) = 0 \tag{8.16}$$

After substituting Equations (8.12), (8.13), and (8.16) in Equations (8.14) and (8.15), we get

$$\frac{-\Delta f}{R_1} - \Delta P_{D_1} = B_1\,\Delta f + \Delta P_{TL_1} \tag{8.17}$$

and

$$\frac{-\Delta f}{R_2} - \Delta P_{D_2} = B_2\Delta f - a_{12}\,\Delta P_{TL_1} \tag{8.18}$$

Since $\Delta P_{TL_2} = -a_{12}\Delta P_{TL_1}$ and $\Delta f_1 = \Delta f_2 = \Delta f$, from Equation (8.17), we have

$$\Delta P_{TL_1} = -\left(\frac{1}{R_1} + B_1\right)\Delta f - \Delta P_{D_1} \tag{8.18(a)}$$

Substituting ΔP_{TL_1} from Equation (8.18(a)) in Equation (8.18), we get

$$\frac{-\Delta f}{R_2} - \Delta P_{D_2} = B_2\,\Delta f - a_{12}\left[-\left(\frac{1}{R_1} + B_1\right)\Delta f - \Delta P_{D_1}\right]$$

$$\frac{-\Delta f}{R_2} - B_2\Delta f = \Delta P_{D_2} + a_{12}\left(\frac{1}{R_1} + B_1\right)\Delta f + a_{12}\Delta P_{D_1}$$

$$-\left(\frac{1}{R_2} + B_2\right)\Delta f - a_{12}\left(\frac{1}{R_1} + B_1\right)\Delta f = \Delta P_{D_2} + a_{12}\Delta P_{D_1}$$

$$-\left\{\left(\frac{1}{R_2} + B_2\right) + a_{12}\left(\frac{1}{R_1} + B_1\right)\right\}\Delta f = \Delta P_{D_2} + a_{12}\,\Delta P_{D_1}$$

$$\Delta f = \frac{\Delta P_{D_1} + a_{12}\Delta P_{D_2}}{\left(B_2 + \dfrac{1}{R_2}\right) + a_{12}\left(B_1 + \dfrac{1}{R_1}\right)} \tag{8.18(b)}$$

Substituting Δf from Equation (8.18(b)) in Equation (8.18 (a)), we get

$$\Delta P_{TL_1} = -\left(\frac{1}{R_1} + B_1\right)\left[\frac{\Delta P_{D_1} + a_{12}\Delta P_{D_2}}{\left(B_2 + \dfrac{1}{R_2}\right) + a_{12}\left(B_1 + \dfrac{1}{R_1}\right)}\right] - \Delta P_{D_1}$$

$$= \frac{\left(B_1 + \dfrac{1}{R_1}\right)\Delta P_{D_2} - \left(B_2 + \dfrac{1}{R_2}\right)\Delta P_{D_1}}{\left(B_2 + \dfrac{1}{R_2}\right) + a_{12}\left(B_1 + \dfrac{1}{R_1}\right)} \tag{8.18(c)}$$

Equations (8.18(b)) and (8.18(c)) are modified as

$$\text{Tie-line frequency,} \quad \Delta f = \frac{\Delta P_{D_1} + a_{12} \Delta P_{D_2}}{\beta_2 + a_{12} \beta_1} \tag{8.19}$$

$$\text{Tie-line power,} \quad \Delta P_{TL_1} = \frac{\beta_1 \Delta P_{D_2} - \beta_2 \Delta P_{D_1}}{\beta_2 + a_{12} \beta_1} \tag{8.20}$$

where $\beta_1 = \left(B_1 + \dfrac{1}{R_1} \right)$

$$\beta_{12} = \left(B_2 + \frac{1}{R_2} \right)$$

Equations (8.19) and (8.20) give the values of the static changes in frequency and tie-line power, respectively, as a result of sudden step-load changes in the two areas. It can be observed that the frequency and tie-line power deviations do not reduce to zero in an uncontrolled case.

Consider two identical areas,

$$B_1 = B_2 = B, \quad \beta_1 = \beta_2 = \beta, \quad R_1 = R_2 = R \quad \text{and} \quad a_{12} = +1$$

Hence, from Equations (8.19) and (8.20), we have

$$\Delta f = \frac{\left(\Delta P_{D_2} + \Delta P_{D_1} \right)}{2\beta} \text{ Hz} \tag{8.21}$$

$$\text{and} \quad \Delta P_{TL_1 (p.u.)} = \frac{\Delta P_{D_2} - \Delta P_{D_1}}{2} = -\Delta P_{TL_2} \text{ (p.u.) MW} \tag{8.22}$$

If a sudden load change occurs only in Area-2 (i.e., $\Delta P_{D_1} = 0$), then we have

$$\Delta f = \frac{\Delta P_{D_2}}{2\beta} \text{ Hz} \tag{8.23}$$

$$\text{and} \quad \Delta P_{TL_1} = \frac{\Delta P_{D_2}}{2} \text{ p.u.} \tag{8.24}$$

Equations (8.23) and (8.24) illustrate the advantages of pool operation (i.e., grid operation) as follows:

- Equations (8.19) represents the change in frequency according to the change in load in either of a two-area system interconnected by a tie line. When considering that those two areas are identical, Equation (8.19) becomes Equation (8.21). Hence, it is concluded that if a load disturbance occurs in only one of the areas (i.e., $\Delta P_{D_1} = 0$ or $\Delta P_{D_2} = 0$), the change in frequency (Δf) is only half of the steady-state error, which would have occurred with no interconnection (i.e., an isolated case). Thus, with several systems interconnected, the steady-state frequency error would be reduced.

- Half of the added load (in Area-2) is supplied by Area-1 through the tie line.

The above two advantages represent the necessity of interconnecting the systems.

8.3.2 Dynamic response

To describe the dynamic response of the two-area system as shown in Fig. 8.4, a system of seventh-order differential equations is required. The solution of these equations would be tedious. However, some important characteristics can be brought out by an analysis rendered simple by the following assumptions. A power system of two identical control areas is considered for the analysis:

(i) $\tau_{gt} = \tau_t = 0$ for both the areas.

(ii) The damping constants of two areas are neglected,

i.e., $B_1 = B_2 = 0$

By virtue of the second assumption, Equations (8.14) and (8.15) become

$$(\Delta P_{G_1} - \Delta P_{D_1}) = \frac{2H_1}{f^0}\frac{d}{dt}(\Delta f_1) + \Delta P_{TL_1} \tag{8.25}$$

$$(\Delta P_{G_2} - \Delta P_{D_2}) = \frac{2H_2}{f^0}\frac{d}{dt}(\Delta f_2) + \Delta P_{TL_2} \tag{8.26}$$

Taking Laplace transformation on both sides of Equations (8.25) and (8.26) and by rearrangement, we get

$$\Delta F_1(s) = \frac{f^0}{2H_1 s}\left[\Delta P_{G_1}(s) - \Delta P_{D_1}(s) - \Delta P_{TL_1}(s)\right] \tag{8.27}$$

$$\Delta F_2(s) = \frac{f^0}{2H_2 s}\left[\Delta P_{G_2}(s) - \Delta P_{D_2}(s) - \Delta P_{TL_2}(s)\right] \tag{8.28}$$

From the block diagram of Fig. 8.4, the following equations can be obtained:

$$\Delta P_{G_1}(s) = -\frac{\Delta F_1(s)}{R}$$

$$\Delta P_{G_2}(s) = -\frac{\Delta F_2(s)}{R} \tag{8.29}$$

$$\Delta P_{TL_1}(s) = \frac{2\pi T_{12}^0}{s}\left[\Delta F_1(s) - \Delta F_2(s)\right]$$

$$\Delta P_{TL_2}(s) = -\Delta P_{TL_1}(s) \tag{8.30}$$

$\left(a_{12} = -\dfrac{P_1}{P_2} = -1\right.$, since two control areas are identical$\left.\right)$

By solving Equations (8.27)–(8.30), we get

$$\Delta P_{TL_1}(s) = \frac{\Delta P_{D_2}(s) - \Delta P_{D_1}(s)}{\left(s^2 + \left(\dfrac{f^0}{2RH}\right)s + \dfrac{2\pi f^0 T_{12}}{H}\right)}\frac{\pi f^0 T_{12}}{H} \tag{8.31}$$

From the above equation, the following observations can be made:

(i) The denominator is of the form:

$$s^2 + 2\alpha s + \omega^2 = (s + \alpha)^2 + (\omega^2 - \alpha^2) \tag{8.32}$$

where $\quad \alpha = \dfrac{f^0}{4RH} \quad$ and $\quad \omega = \sqrt{\dfrac{2\pi f^0 T_{12}}{H}}$

and α and ω^2 are both real and positive. Hence, it can be concluded from the roots of characteristic equation that the time response is stable and damped.

The three conditions are:

If $\alpha = \omega_n$, system is critically damped

$\alpha > \omega_n$, system becomes overdamped

$$\alpha < \omega_n, \text{ then } s_{12} = -\alpha \pm j\sqrt{\omega_n^2 - \alpha^2}$$

$$= -\alpha \pm j\omega_n \sqrt{1 - \left(\frac{\alpha}{\omega_n}\right)^2}$$

$$= -\alpha \pm j\omega_0$$

where $\alpha = $ damping factor or decrement of attenuation

$$\omega_0 = \text{damped angular frequency} = \sqrt{\frac{2\pi f^0 T_{12}^0}{H} - \left(B + \frac{1}{R}\right)^2 \frac{f^{02}}{16H^2}}$$

Since parameter α also depends on B, but $B \leq \dfrac{1}{R}$ in practice, therefore, the effect of coefficient B is neglected on damping.

(ii) After a disturbance, the change in tie-line power oscillates at the damped angular frequency.

(iii) The damping of the tie-line power variation is strongly dependent upon the parameter α, which is equal to $\dfrac{f^0}{4RH}$. Since f^0 and H are essentially constant, the damping is a function of the R parameters. If the R value is low, damping becomes strong and vice versa.

The transient change in the tie-line power will be of undamped oscillations of frequency, $\omega_0 = \omega$.

If $R = \infty$, i.e., if the speed governor is not present ($\alpha = 0$), the variation in frequency deviation and the tie-line power would be as shown in Fig. 8. 5.

It can be seen that the steady-state frequency deviation is the same for both the areas and does not vanish. The tie-line power deviation also does not become zero.

Although the above approximate analysis has confirmed stability, it has been found through more accurate analyses that with certain parameter combinations, the system becomes unstable.

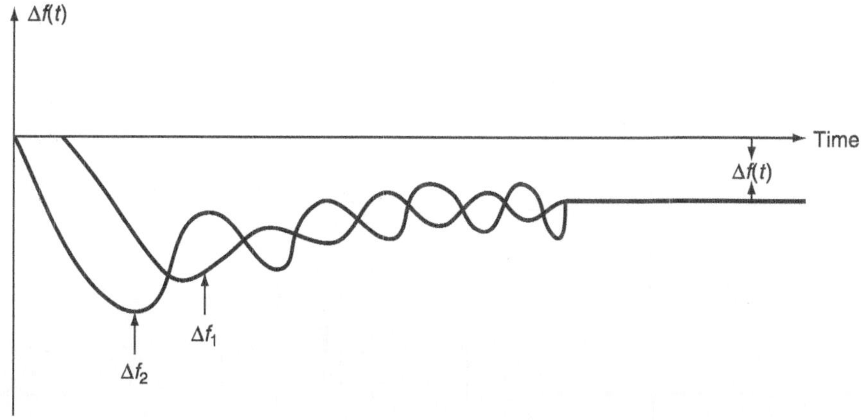

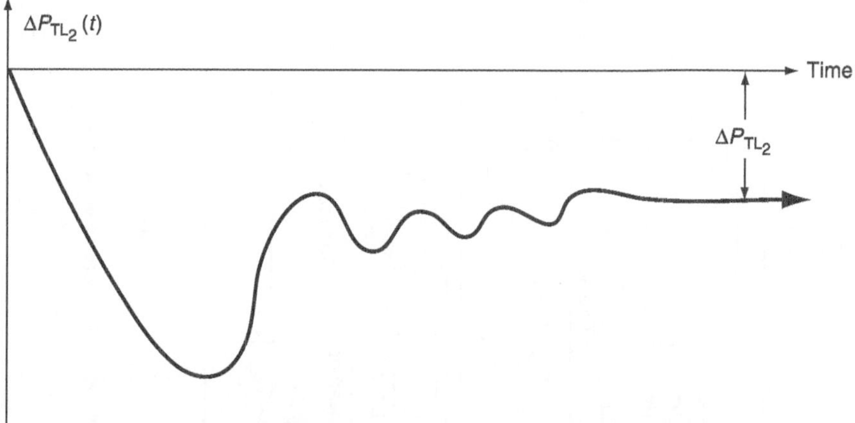

FIG. 8.5   *Frequency deviation and tie-line power change following a step-load change in Area-2 (two areas are identical)*

Example 8.1: A two-identical area power system has the following parameters (Fig. 8.6(a)):

Power system gain constant, $K_{ps} = 105$
Power system time constant, $\tau_{ps} = 22$ s
Speed regulation, $R = 2.5$
Normal frequency, $f^0 = 50$ Hz
Governor time constant, $\tau_{sg} = 0.3$ s
Turbine time constant, $\tau_t = 0.5$ s
Integration time constant, $k_i = 0.15$
Bias parameter, $b = 0.326$
$2\pi T_{12} = 0.08$

Plot the change in the tie-line power and change in frequency of Control-area 1 if there exists a step-load change of 2% in Area-1 (Fig. 8.6(b)).

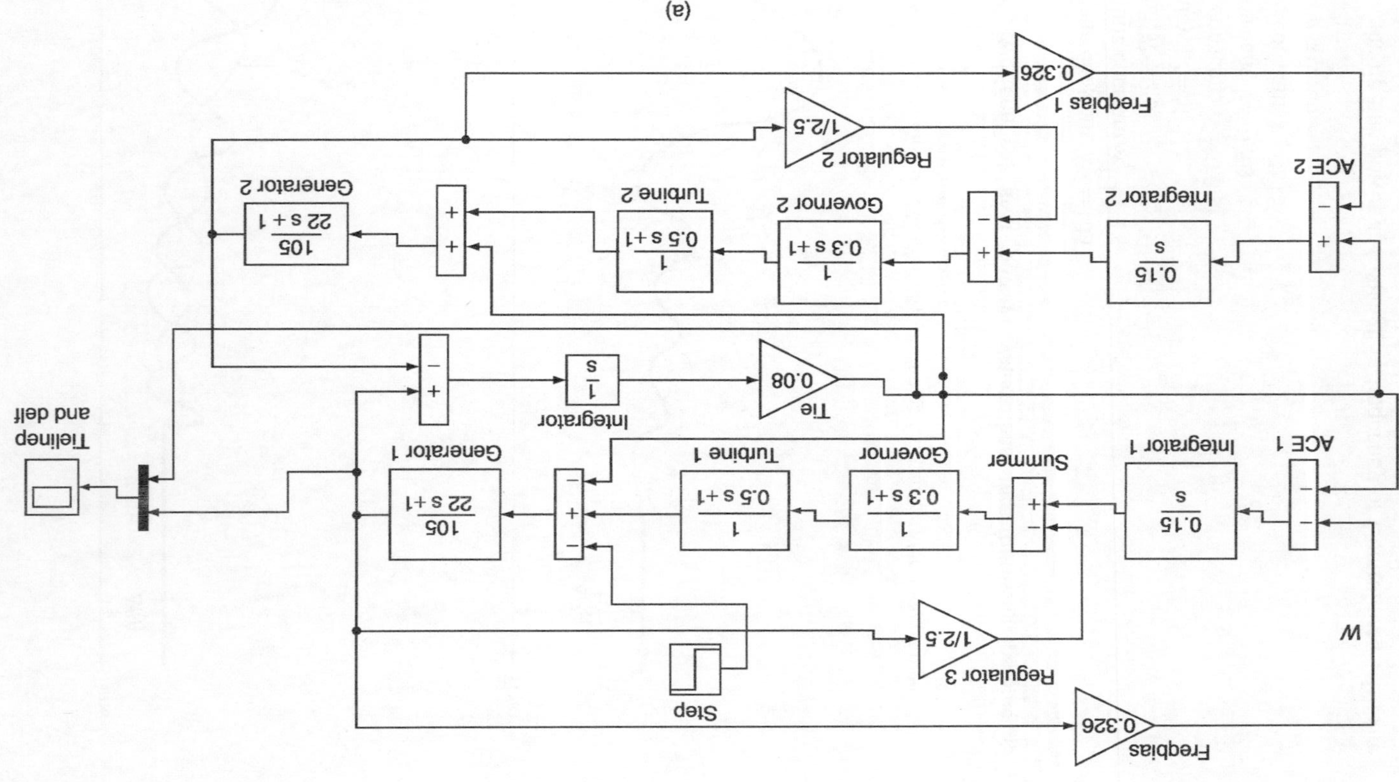

(a)

FIG. 8.6   (*Continued on next page*)

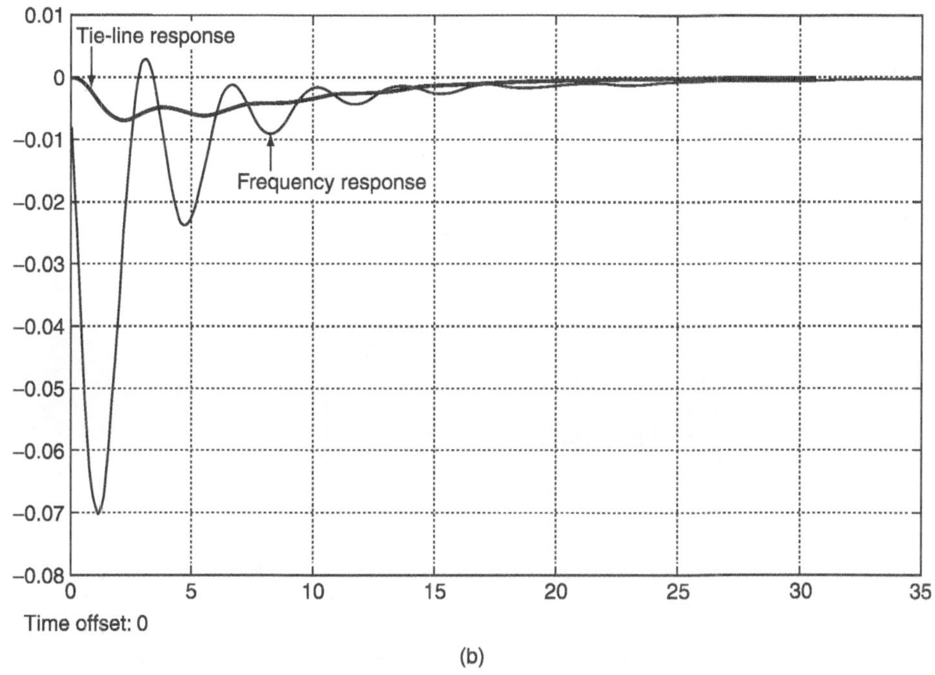

(b)

FIG. 8.6   *(a) Simulation block diagram for a two-identical area system of Example 8.1;*
(b) frequency and tie-line response for Example 8.1

Example 8.2: A two-area power system has the following parameters (Fig. 8.7(a)):

For Area-1:

Power system gain constant, $K_{ps} = 120$
Power system time constant, $\tau_{ps} = 20$ s
Speed regulation, $R = 2.5$
Normal frequency, $f^0 = 50$ Hz
Governor time constant, $\tau_{sg} = 0.2$ s
Turbine time constant, $\tau_t = 0.4$ s
Integration time constant, $k_i = 0.1$
Bias parameter, $b = 0.425$

For Area-2:

Power system gain constant, $K_{ps} = 100$
Power system time constant, $\tau_{ps} = 22$ s
Speed regulation, $R = 3$
Normal frequency, $f^0 = 50$ Hz
Governor time constant, $\tau_{sg} = 0.3$ s
Turbine time constant, $\tau_t = 0.5$ s
Integration time constant, $k_i = 0.15$
Bias parameter, $b = 0.326$
$2\pi T_{12} = 0.08$

Plot the change in the tie-line power and change in frequency of Control-area 1 if there
exists a step-load change of 2% in Area-1 (Fig. 8.7(b)).

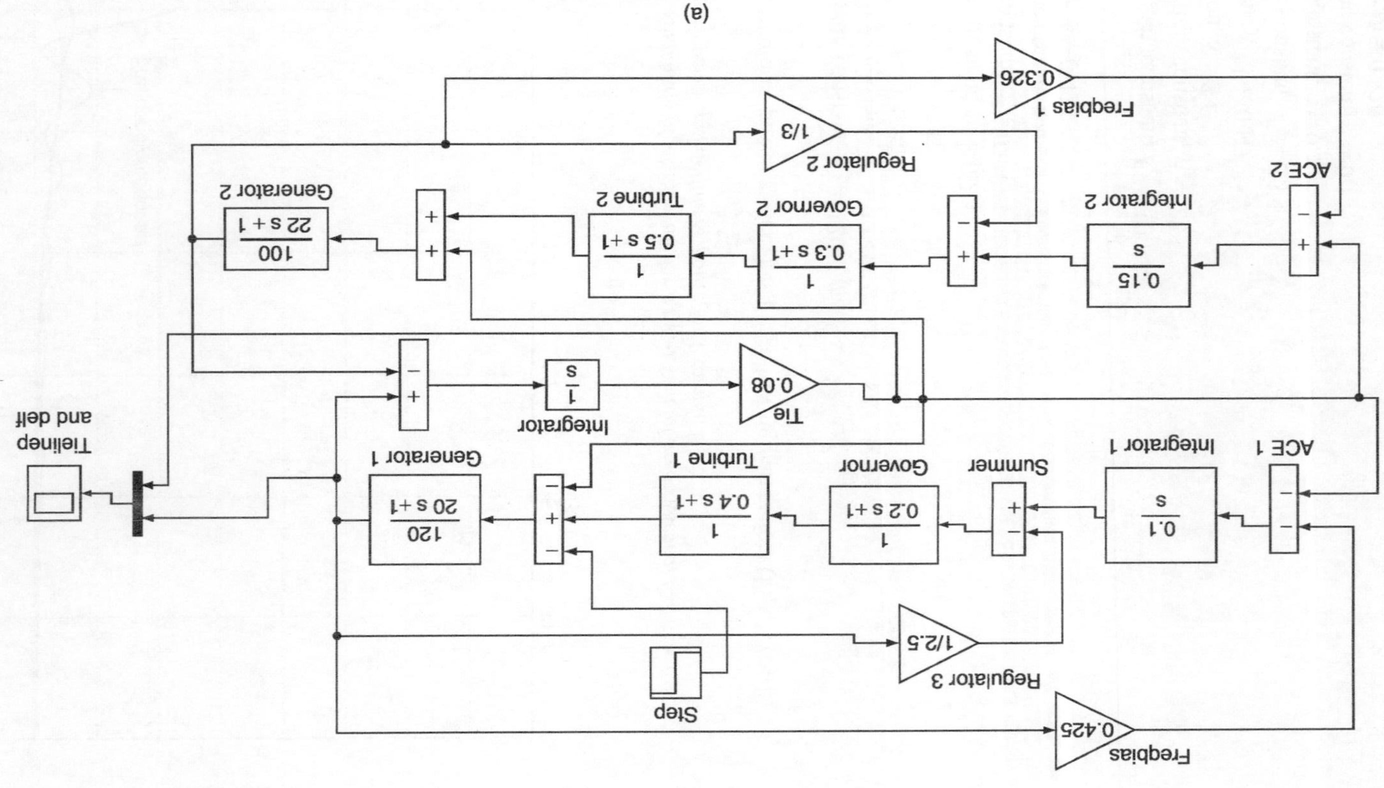

(a)

FIG. 8.7   *(Continued on next page)*

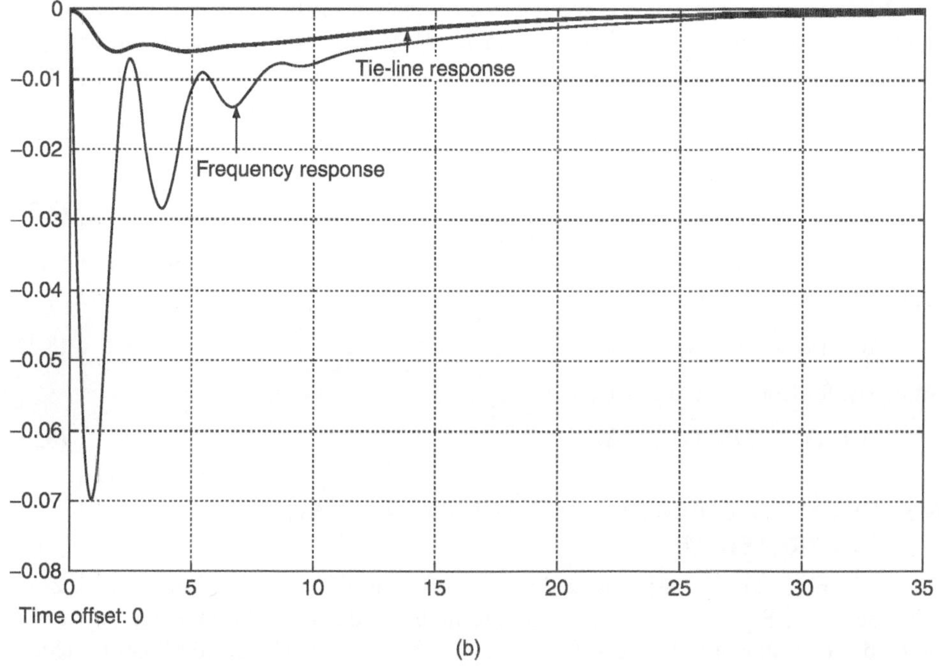

FIG. 8.7   *(a) Simulation block diagram of Example 8.2; (b) Frequency and tie-line power response of Example 8.2*

Example 8.3: Determine the frequency of oscillations of the tie-line power deviation for a two-identical-area system given the following data:

$R = 3.0$ Hz/p.u.; $H = 5$ s; $f^0 = 60$ Hz

The tie-line has a capacity of 0.1 p.u. and is operating at a power angle of 45°.

Solution:

The synchronizing-power coefficient of the line is given by

$$T_{12}^0 = P_{\mathrm{m}} \cos \delta_{12} = 0.1 \times \cos 45° = 0.0707 \text{ p.u.}$$

Hence, the frequency of oscillations is given by

$$\omega_0 = \sqrt{\frac{2\pi\, f^0 T_{12}^0}{H} - \left(\frac{f^0}{4RH}\right)^2}$$

$$= \sqrt{\frac{2\pi \times 60 \times 0.0707}{5} - \left(\frac{60}{4 \times 3 \times 5}\right)^2} = 2.1 \text{ rad/s}$$

$$\therefore f_0 = \frac{2.1}{2\pi} = 0.33 \text{ Hz}$$

8.4 AREA CONTROL ERROR—TWO-AREA CASE

In a single-area case, ACE is the change in frequency. The steady-state error in frequency will become zero (i.e., $\Delta f_{\mathrm{ss}} = 0$) when ACE is used in the integral-control loop.

In a two-area case, ACE is the linear combination of the change in frequency and change in tie-line power. In this case to make the steady-state tie-line power zero (i.e., $\Delta P_{TL} = 0$), another integral-control loop for each area must be introduced in addition to the integral frequency loop to integrate the incremental tie-line power signal and feed it back to the speed-changer.

Thus, for Control area-1, we have

$$\mathrm{ACE}_1 = \Delta P_{TL_1} + b_1 \Delta f_1 \tag{8.33}$$

where $b_1 = \text{constant} = \text{area}$ frequency bias. Taking Laplace transform on both sides of Equation (8.33), we get

$$\mathrm{ACE}_1(s) = \Delta P_{TL_1}(s) + b_1 \Delta F_1(s) \tag{8.34}$$

Similarly, for Control area-2, we have

$$\mathrm{ACE}_2(s) = \Delta P_{TL_2}(s) + b_2 \Delta F_2(s) \tag{8.35}$$

8.5 COMPOSITE BLOCK DIAGRAM OF A TWO-AREA SYSTEM (CONTROLLED CASE)

By the combination of basic block diagrams of Control area-1 and Control area-2 and with the use of Figs. 8.2 and 8.3, the composite block diagram of a two-area system can be modeled as shown in Fig. 8.4. Figure 8.8 can be obtained by the addition of integrals of ACE_1 and ACE_2 to the block diagram shown in Fig. 8.4. It represents the composite block diagram of a two-area system with integral-control loops. Here, the control signals $\Delta P_{c_1}(s)$ and $\Delta P_{c_2}(s)$ are generated by the integrals of ACE_1 and ACE_2. These control errors are obtained through the signals representing the changes in the tie-line power and local frequency bias.

8.5.1 Tie-line bias control

The speed-changer command signals will be obtained from the block diagram shown in Fig. 8.6 as

$$\Delta P_{c_1} = -K_{I_1} \int (\Delta P_{TL_1} + b_1 \Delta f_1)\, \mathrm{d}t \tag{8.36}$$

$$\text{and} \quad \Delta P_{c_2} = -K_{I_2} \int (\Delta P_{TL_2} + b_2 \Delta f_2)\, \mathrm{d}t \tag{8.37}$$

The constants K_{I_1} and K_{I_2} are the gains of the integrators. The first terms on the right-hand side of Equations (8.36) and (8.37) constitute and are known as tie-line bias controls. It is observed that for decreases in both frequency and tie-line power, the speed-changer position decreases and hence the power generation should decrease, i.e., if the ACE is negative, then the area should increase its generation.

So, the right-hand side terms of Equations (8.36) and (8.37) are assigned a negative sign.

8.5.2 Steady-state response

That the control strategy, described in the previous section, eliminates the steady-state frequency and tie-line power deviations that follow a step-load change, can be proved as follows:

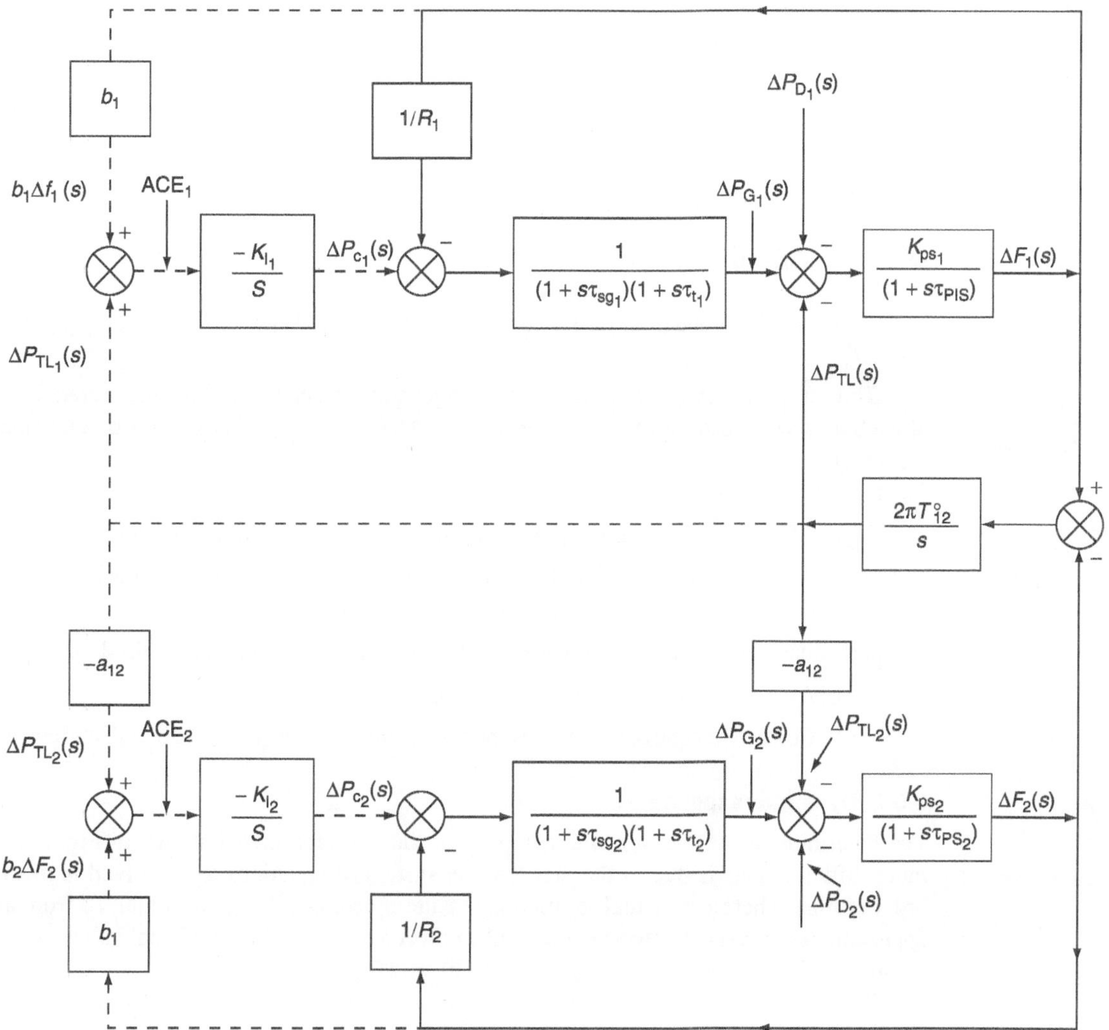

FIG. 8.8   *Two-area system with integral control*

Let the step changes in loads ΔP_{D_1} and ΔP_{D_2} simultaneously occur in Control area-1 and Control area-2, respectively, or in either area. A new static equilibrium state, i.e., steady-state condition is reached such that the output signal of all integrating blocks will become constant. In this case, the speed-changer command signals ΔP_{c_1} and ΔP_{c_2} have reached constant values. This obviously requires that both the integrands (input signals) in Equations (8.36) and (8.37) be zero.

Input of integrating block $\left(\dfrac{-K_{I_1}}{s}\right)$ is

$$\Delta P_{TL_1(\mathrm{ss})} + b_1 \Delta f_{1(\mathrm{ss})} = 0 \qquad\qquad (8.38)$$

Input of integrating block $\left(\dfrac{-K_{I_2}}{s}\right)$ is

$$\Delta P_{TL_2(ss)} + b_2 \Delta f_{2(ss)} = 0 \tag{8.39}$$

and input of integrating block $\left(\dfrac{-2\pi T_{12}}{s}\right)$ is

$$\Delta f_1 - \Delta f_2 = 0 \tag{8.40}$$

Equations (8.38) and (8.39) are simultaneously satisfied only for $\Delta P_{TL_1(ss)} = \Delta P_{TL_2(ss)} = 0$ and $\Delta f_{1(ss)} = \Delta f_{2(ss)} = 0$.

Thus, under a steady-state condition, change in tie-line power and change in frequency of each area will become zero. To achieve this, ACEs in the feedback loops of each area are integrated.

The requirements for integral control action are:

(i) ACE must be equal to zero at least one time in all 10-minute periods.

(ii) Average deviation of ACE from zero must be within specified limits based on a percentage of system generation for all 10-minute periods.

The performance criteria also apply to disturbance conditions, and it is required that:

(i) ACE must return to zero within 10-minute periods.

(ii) Corrective control action must be forthcoming within 1 minute of a disturbance.

8.5.3 Dynamic response

The determination of the dynamic response of the two-area model shown in Fig. 8.6 is more difficult. This is due to the fact that the system of equations to be solved is of the order of nine. Therefore, actual solution is not attempted. But the results obtained from an approximate analysis of a two-identical-area power system for three different values of the parameter 'b', are presented in Figs. 8.9(a), (b), and (c).

The graphs of Fig. 8.9(a) correspond to the case of $b = 0$. It can be seen that the tie-line power deviation reduces to zero while the frequency does not.

The graphs of Fig. 8.9(b) correspond to the other extreme case of $b = \infty$. Now, the frequency error vanishes. But, the tie-line power does not vanish.

The graphs of Fig. 8.9(c) show an intermediate case wherein both the frequency and the tie-line power errors decrease to zero. This is the desired case.

Therefore, it can be concluded that the stability is not always guaranteed. Hence, there is a need for proper parameter selection and adjustment of their values.

8.6 OPTIMUM PARAMETER ADJUSTMENT

The graphs given in Fig. 8.9(c) stress the need for proper parameter settings. The choice of b and K_I constants affects the transient response to load changes. The frequency bias b should be high enough such that each area adequately contributes to frequency control. It is proved that choosing $b = \beta$ gives satisfactory performance of the interconnected system.

The integrator gain K_I should not be too high, otherwise, instability may result. Also the time interval at which LFC signals are dispatched, two or more seconds, should be low enough so that LFC does not attempt to follow random or spurious load changes.

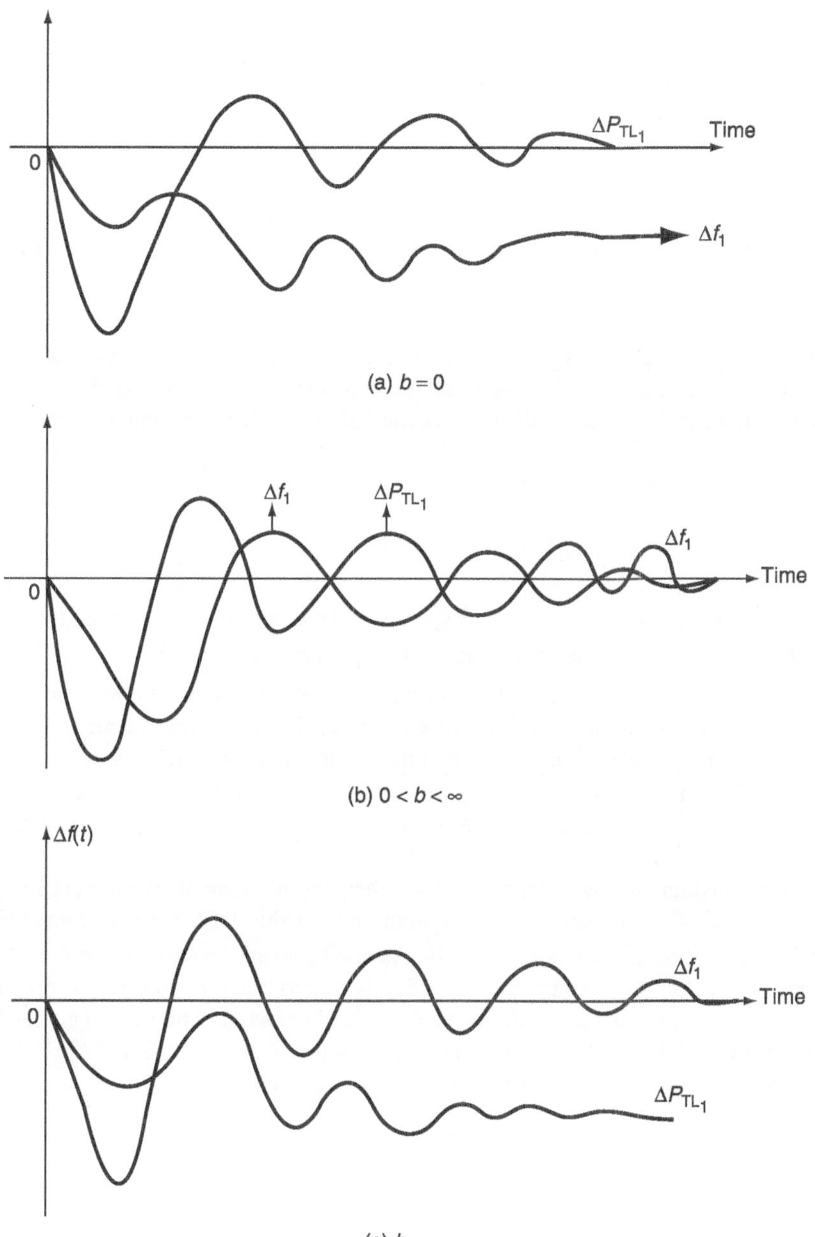

FIG. 8.9   *Approximate dynamic response of two-identical-area power systems with three different values of b parameters*

First, a set of parameters, which ensure stability of the control, is selected. For example, b_1 and b_2 cannot both be zero, i.e., one of them should be chosen for the control strategy. Later, the values of these parameters are adjusted so that a best or an optimum response is obtained. In other words, the values of parameters, which give rise to an optimum response, are to be determined.

The procedure is as follows:

The popular error criterion, known as the integral of the squared errors (ISE), is chosen for the control parameters Δf_1, Δf_2, and ΔP_{TL_1}. For a two-area system, the ISE criterion function C would be

$$C = \int_0^\infty \left[\alpha_1 (\Delta P_{TL_1})^2 + \alpha_2 (\Delta f_1)^2 + \alpha_3 (\Delta f_2)^2 \right] dt \qquad (8.41)$$

where α_1, α_2, and α_3 are the weight factors, which provide appropriate importance, i.e., weightage to the errors ΔP_{TL_1}, Δf_1, and Δf_2. There is no need to choose ΔP_{TL_2}, since $\Delta P_{TL_2} = -a_{12}\Delta P_{TL_1}$.

Since Δf_1 and Δf_2 behave in a similar manner, we need to consider only one of them. So, let us consider Δf_1 only. This is a parameter selection. Then, $\alpha_3 = 0$. Also, let $\alpha_2 = \alpha$. Since, we are interested only in the relative magnitudes of C for parameter setting, we can set $\alpha_1 = 1$.

With these, Equation (7.41) reduces to

$$C = \int_0^\infty \left[(\Delta P_{TL_1})^2 + \alpha (\Delta f_1)^2 \right] dt \qquad (8.42)$$

For a two-area system, ΔP_{TL_1} and Δf_1 would be functions of the integrator gain constants K_{I_1} and K_{I_2} as well as the frequency bias parameters b_1 and b_2.

The procedure for obtaining the optimum parameter values would be as follows:

First, a convenient and suitable value is chosen for the weight factor 'α'. Then, for different assumed values of K_{I_1}, K_{I_2}, b_1, and b_2, the values of ΔP_{TL_1} and Δf_1 are determined at different instants of time. With these values and α, the value of C is computed using Equation (8.42). The set of values of K_{I_1}, K_{I_2}, b_1, and b_2 for which C is a minimum is the optimum one.

If we consider the two identical areas, then the number of parameters reduces to two, viz., $K_{I_1} = K_{I_2} = K_I$ and $b_1 = b_2 = b$. In this case, values of C for different values of K_I and b can be plotted as shown in Fig. 8.10. As can be seen, the variation of C with K_I for different fixed values of b is plotted to get a family of curves called constant-b contours. For illustration, only three curves are shown in Fig. 8.10. In practice, a number of curves have to be determined and drawn. It can be seen that C is minimum for $b = 0.2$ and $K_I = 1.0$.

In this case, the optimum control strategy would, therefore, be

$$P_{C_1} = -\int (\Delta P_{TL_1} + 0.2\Delta f_1)\, dt$$

$$\text{and} \quad P_{C_2} = -\int (\Delta P_{TL_2} + 0.2\Delta f_1)\, dt \qquad (8.43)$$

In practice, the frequency and tie-line power deviations are measured at fixed intervals of time in a sample-data fashion. The sampling rate (the rate at which the frequency deviation and tie-line power deviation samples are measured) should be sufficiently high to avoid errors due to sampling.

Note: LFC provides enough control during normal changes in load and frequency, i.e., changes that are not too large. During emergencies, when large imbalances between generation and load occur, LFC is bypassed and other emergency controls are applied, which is beyond the scope of this book.

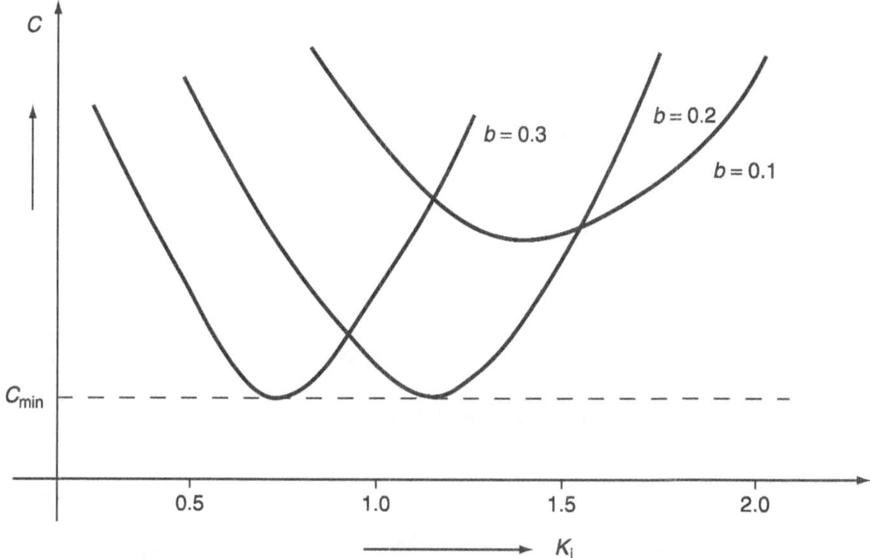

FIG. 8.10   *Constant b-contours of the ISE criterion function C*

8.7 LOAD FREQUENCY AND ECONOMIC DISPATCH CONTROLS

Economic load dispatch and LFC play a vital role in modern power system. In LFC, zero steady-state frequency error and a fast, dynamic response were achieved by integral controller action. But this control is independent of economic dispatch, i.e., there is no control over the economic loadings of various generating units of the control area.

Some control over loading of individual units can be exercised by adjusting the gain factors (K_I) of the integral signal of the ACE as fed to the individual units. But this is not a satisfactory solution.

A suitable and satisfactory solution is obtained by using independent controls of load frequency and economic dispatch. The load frequency controller provides a fast-acting control and regulates the system around an operating point, whereas the economic dispatch controller provides a slow-acting control, which adjusts the speed-changer settings every minute in accordance with a command signal generated by the central economic dispatch computer.

EDC—economic dispatch controller
CEDC—central economic dispatch computer

The speed-changer setting is changed in accordance with the economic dispatch error signal, (i.e., $P_{G(desired)} - P_{G(actual)}$) conveniently modified by the signal $\int ACE\ dt$ at that instant of time. The central economic dispatch computer (CEDC) provides the signal $P_{G(desired)}$, and this signal is transmitted to the local economic dispatch controller (EDC). The system they operate with economic dispatch error is only for very short periods of time before it is readily used (Fig. 8.11).

This tertiary control can be implemented by using EDC and EDC works on the cost characteristics of various generating units in the area. The speed-changer settings are once again operated in accordance with an economic dispatch computer program.

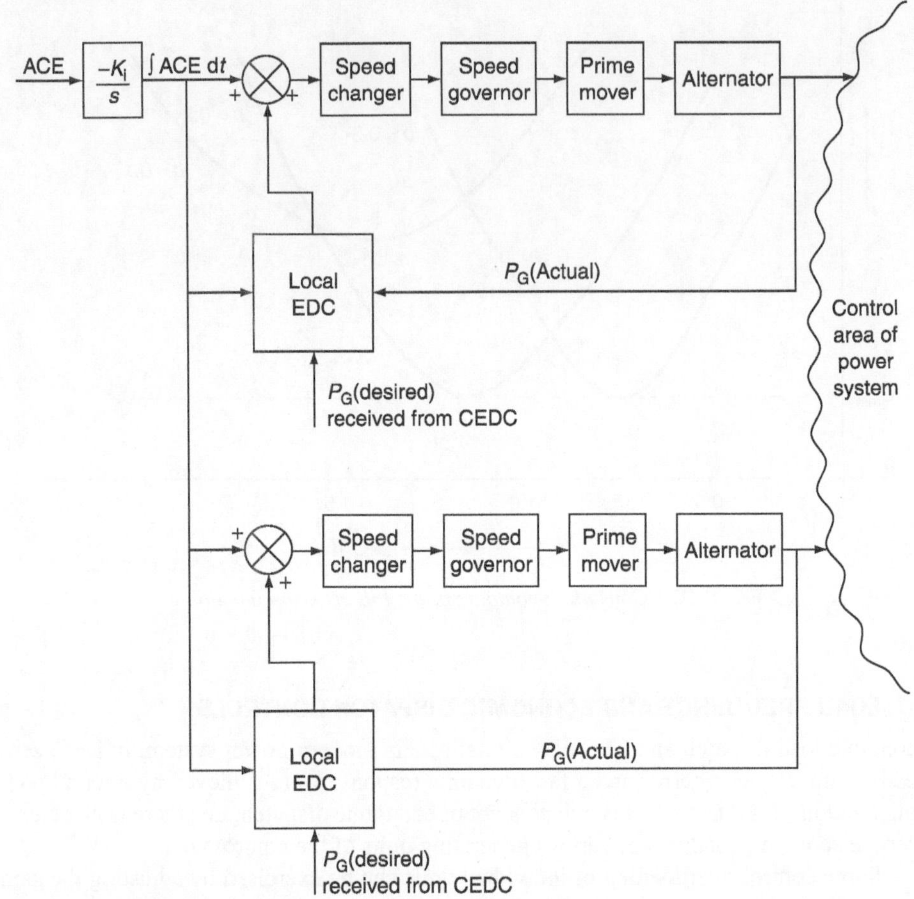

FIG. 8.11   *Load frequency and economic dispatch control of the control area of a power system*

The CEDCs are provided at a central control center. The variable part of the load is carried by units that are controlled from the central control center. Medium-sized fossil fuel units and hydro-units are used for control. During peak load hours, lesser efficient units, such as gas-turbine units or diesel units, are employed in addition; generators operating at partial output (with spinning reserve) and standby generators provide a reserve margin.

The central control center monitors information including area frequency, outputs of generating units, and tie-line power flows to interconnected areas. This information is used by ALFC in order to maintain area frequency at its scheduled value and net tie-line power flow out of the area at its shedding value. Raise and lower reference power signals are dispatched to the turbine governors of controlled units.

Economic dispatch is co-ordinated with LFC such that the reference power signals dispatched to controlled units move the units toward their economic loading and satisfy LFC objectives.

8.8 DESIGN OF AUTOMATIC GENERATION CONTROL USING THE KALMAN METHOD

A modern gigawatt generator with its multistage reheat turbine, including its automatic load frequency control (ALFC) and automatic voltage regulator (AVR) controllers, is characterized by an impressive complexity. When all its non-negligibility dynamics are taken into account, including cross-coupling between control channels, the overall dynamic model may be of the twentieth order.

The dimensionality barrier can be overcome by means of computer-aided optimal control design methods originated by Kalman. A computer-oriented technique called optimum linear regulator (OLR) design has proven to be particularly useful in this regard.

The OLR design results in a controller that minimizes both transient variable excursions and control efforts. In terms of power system, this means optimally damped oscillation with minimum wear and tear of control valves.

OLR can be designed using the following steps:

  (i) Casting the system dynamic model in state-variable form and introducing appropriate control forces.

  (ii) Choosing an integral-squared-error control index, the minimization of which is the control goal.

  (iii) Finding the structure of the optimal controller that will minimize the chosen control index.

8.9 DYNAMIC-STATE-VARIABLE MODEL

The LFC methods discussed so far are not entirely satisfactory. In order to have more satisfactory control methods, optimal control theory has to be used. For this purpose, the power system model must be in a state-variable model.

8.9.1 Model of single-area dynamic system in a state-variable form

From the block diagram of an uncontrolled single-area system shown in Fig. 8.12, we get the following 's-domain' equations:

$$\Delta X_{\mathrm{E}}(s) = \left(\frac{K_{sg}}{1 + s\tau_{sg}} \right) \left[\Delta P_{\mathrm{C}}(s) - \frac{1}{R} \Delta F(s) \right]$$

$$\Delta P_{\mathrm{G}}(s) = \left(\frac{K_{t}}{1 + s\tau_{t}} \right) \Delta X_{\mathrm{E}}(s)$$

$$\Delta F(s) = \left(\frac{K_{ps}}{1 + s\tau_{ps}} \right) \left(\Delta P_{\mathrm{G}}(s) - \Delta P_{\mathrm{D}}(s) \right)$$

In time domain, the above equations can be expressed as

$$\Delta X_{\mathrm{E}} + \tau_{sg} \frac{\mathrm{d}}{\mathrm{d}t}(\Delta X_{\mathrm{E}}) = \Delta P_{\mathrm{C}} - \frac{1}{R} \Delta f$$

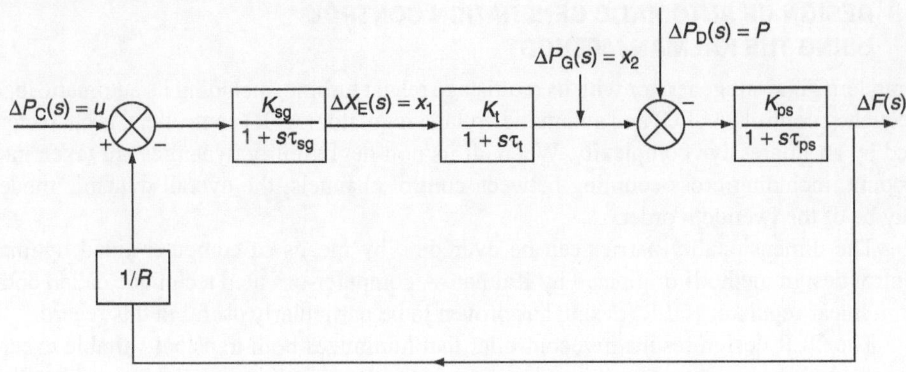

FIG. 8.12   *State-space model of a single-area system*

$$\Delta P_G + \tau_t \frac{d}{dt} \Delta P_G = \Delta X_E$$

$$\Delta f + \tau_{ps} \frac{d}{dt}(\Delta f) = K_{ps}\Delta P_G - K_{ps}\Delta P_D$$

Let us choose the state variables $\begin{bmatrix} x_1 \\ x_2 \\ x_3 \end{bmatrix} = \begin{bmatrix} \Delta X_E \\ \Delta P_G \\ \Delta f \end{bmatrix}$, input, $u = \Delta P_C$, and disturbance, $d = \Delta P_D$

The above equations are written in a state-variable form:

$$\begin{bmatrix} \dot{X}_1 \\ \dot{X}_2 \\ \dot{X}_3 \end{bmatrix} = \begin{bmatrix} \dfrac{-1}{\tau_{sg}} & 0 & \dfrac{-1}{R\tau_{sg}} \\ \dfrac{1}{\tau_t} & \dfrac{-1}{\tau_t} & 0 \\ 0 & \dfrac{K_{ps}}{\tau_{ps}} & \dfrac{-1}{\tau_{ps}} \end{bmatrix} \begin{bmatrix} x_1 \\ x_2 \\ x_3 \end{bmatrix} + \begin{bmatrix} \dfrac{1}{\tau_{sg}} \\ 0 \\ 0 \end{bmatrix} u + \begin{bmatrix} 0 \\ 0 \\ \dfrac{-K_{ps}}{\tau_{ps}} \end{bmatrix} p$$

The state equation would then be written in a more general form as

$$\dot{X} = Ax + Bu + Jp$$

where x is the n-dimensional state vector, u the m-dimensional control-force vector $= [u] = [\Delta P_c]$, and p the disturbance force vector $= [p] = [\Delta P_D]$.

8.9.2 Optimum control index (*I*)

The optimum linear regulator design is based on the integral-squared-error index of the form.

$$I = \int_{0}^{\infty} (q_1 x_1^2 + \cdots + q_n x_n^2 + r_1 u_1^2 + \cdots + r_m u_m^2)\, dt$$

where q's and r's are positive penalty factors. Let us consider index I for a single-area system as

$$I = \int_{0}^{\infty} [q_1(\Delta P_v)^2 + q_2(\Delta P_r)^2 + q_3(\Delta f)^2 + ru^2]\, dt$$

Here, the idea of optimum control is to minimize the index (I). Consider the term $q_3 (\Delta f)^2$, squaring of frequency error will contribute to 'I' independent of its sign. If Δf is doubled, its contribution to 'I' will quadruple. The integral causes Δf to add to 'I' during its entire duration. The penalty factors q_i distribute the penalty weight among the state-variable errors. If the error in a particular variable is of little significance, we simply set its penalty factor to zero.

Similarly, in the case of control-force increments, the penalty factors r_i distribute the penalties among the m control force. (Here, none of r's are set to zero.) If they are set to zero, then the control force will assume an infinite magnitude without affecting 'I'. An infinite control force could do its correcting job in zero time. This would obviously be a very unrealistic regulator.

If all q's and r's constitute the diagonal elements of the two penalty matrices, then we have

$$Q \approx \begin{bmatrix} q_1 & .. & .. & 0 \\ . & q_2 & . & . \\ . & . & q_3 & . \\ 0 & . & . & q_n \end{bmatrix} \quad \text{and} \quad R \approx \begin{bmatrix} r_1 & .. & .. & 0 \\ . & r_2 & . & . \\ . & . & r_3 & . \\ 0 & . & . & r_m \end{bmatrix}$$

Index-I in a compact form is

$$I = \int_{0}^{\infty} (x^{\mathrm{T}} Q x + u^{\mathrm{T}} R u)\, dt$$

8.9.3 Optimum control problem and strategy

The task that an OLR or an optimum controller must perform is the fulfillment of the optimum control problem, which can be stated as follows: consider a system that is initially under steady-state condition. If it is disturbed by a set of step type of disturbances, it goes through a transient state first. It is required that, after the expiry of the transient period, it should return to the original or new prescribed steady-state condition. The problem is to determine the set of control forces $\bar{u} = \bar{u}_{\text{opt}}$, which will not only take the system to the original or new prescribed steady state, but will also do so by simultaneously minimizing the chosen control criterion function or optimum control index (I).

The value of $\bar{u} = \bar{u}_{\text{opt}}$, which fulfills the above optimum control requirement, is the desired optimum control strategy. An optimum controller or OLR is the one that carries out the above strategy.

8.9.4 Dynamic equations of a two-area system

From the block diagram of uncontrolled two-area systems shown in Fig. 8.4, get the following 's-domain' equations:

$$\Delta F_1(s) = \frac{K_{ps_1}}{(1+s\tau_{ps_1})}\left[\Delta P_{G_1}(s) - \Delta P_{D_1}(s) - \Delta P_{TL_1}(s)\right]$$

$$\Delta F_2(s) = \frac{K_{ps_2}}{(1+s\tau_{ps_2})}\left[\Delta P_{G_2}(s) - \Delta P_{D_2}(s) - a_{12}\,\Delta P_{TL_1}(s)\right]$$

$$\Delta X_{E_1}(s) = \frac{1}{(1+s\tau_{sg_1})}\left[\Delta P_{C_1}(s) - F_1(s)/R_1\right]$$

$$\Delta X_{E_2}(s) = \frac{1}{(1+s\tau_{sg_2})}\left[\Delta P_{C_2}(s) - F_2(s)/R_2\right]$$

$$\Delta P_{G_1}(s) = \frac{1}{(1+s\tau_{t_1})}\left[\Delta X_{E_1}(s)\right]$$

$$\Delta P_{G_2}(s) = \frac{1}{(1+s\tau_{t_2})}\left[\Delta X_{E_2}(s)\right]$$

$$\Delta P_{TL_1}(s) = \frac{2\pi T_{12}^0}{s}\left[\Delta F_1(s) - F_2(s)\right]$$

where $X_{E_1}(s)$ and $X_{E_2}(s)$ are the Laplace transforms of the movements of the main positions in the speed-governing mechanisms of the two areas.

By taking inverse Laplace transform for the above equations, we get a set of seven differential equations. These are the time-domain equations, which describe the small-disturbance dynamic behavior of the power system.

Consider the first equation,

$$(1+s\tau_{ps_1})\Delta F_1(s) = K_{ps_1}\left[\Delta P_{G_1}(s) - \Delta P_{D_1}(s) - P_{TL_1}(s)\right]$$

or $\quad s\tau_{ps_1}\Delta E_1(s) = -\Delta F_1(s) + K_{ps_1}\left[\Delta P_{G_1}(s) - \Delta P_{D_1}(s) - P_{TL_1}(s)\right]$

or $\quad s\Delta F_1(s) = \dfrac{1}{\tau_{ps_1}}\left[-\Delta F_1(s) + K_{ps_1}\left\{\Delta P_{G_1}(s) - \Delta P_{D_1}(s) - \Delta P_{TL_1}(s)\right\}\right]$

Taking the inverse Laplace transform of the above equation, we get

$$\frac{d}{dt}\left[\Delta f_1\right] = \frac{1}{\tau_{ps_1}}\left[-\Delta f_1 + K_{ps_1}\Delta P_{G_1} - K_{ps_1}\Delta P_{G_1} - K_{ps_1}\Delta P_{D_1} - K_{ps_1}\Delta P_{TL_1}\right]$$

In a similar way, the remaining equations can be rearranged and an inverse Laplace transform is found. Then, the entire set of differential equations is

$$\frac{d}{dt}(\Delta f_1) = \frac{1}{\tau_{ps_1}}\left[-\Delta f_1 + K_{ps_1}\Delta P_{G_1} - K_{ps_1}\Delta P_{D_1} - K_{ps_1}\Delta P_{TL_1}\right]$$

$$\frac{\mathrm{d}}{\mathrm{d}t}(\Delta f_2) = \frac{1}{\tau_{\mathrm{ps}_2}}\left[-\Delta f_2 + K_{\mathrm{ps}_2}\Delta P_{\mathrm{G}_2} - K_{\mathrm{ps}_2}\Delta P_{\mathrm{D}_2} - K_{\mathrm{ps}_2}\Delta P_{\mathrm{TL}_1}a_{12}\right]$$

$$\frac{\mathrm{d}}{\mathrm{d}t}(\Delta X_{\mathrm{E}_1}) = \frac{1}{\tau_{\mathrm{sg}_1}}\left[-\Delta X_{\mathrm{E}_1} + \Delta P_{\mathrm{C}_1} - \Delta f_1/R_1\right]$$

$$\frac{\mathrm{d}}{\mathrm{d}t}(\Delta X_{\mathrm{E}_2}) = \frac{1}{\tau_{\mathrm{sg}_2}}\left[-\Delta X_{\mathrm{E}_2} + \Delta P_{\mathrm{C}_2} - \Delta f_2/R_2\right]$$

$$\frac{\mathrm{d}}{\mathrm{d}t}(\Delta P_{\mathrm{G}_1}) = \frac{1}{\tau_{t_1}}\left[-\Delta P_{\mathrm{G}_1} + \Delta X_{\mathrm{E}_1}\right]$$

$$\frac{\mathrm{d}}{\mathrm{d}t}(\Delta P_{\mathrm{G}_2}) = -\frac{1}{\tau_{t_2}}\left[-\Delta P_{\mathrm{G}_2} + \Delta X_{\mathrm{E}_2}\right]$$

$$\text{and}\quad \frac{\mathrm{d}}{\mathrm{d}t}(\Delta P_{\mathrm{TL}_1}) = 2\pi T_{12}^0\left[\Delta f_1 - \Delta f_2\right]$$

8.9.4.1 *State variables and state-variable model*

The state variables are a minimum number of those variables, which contain sufficient information about the past history with which all future states of the system can be determined for known control inputs. For the two-area system under consideration, the state variables would be Δf_1, Δf_2, ΔX_{E_1}, ΔX_{E_2}, ΔP_{sg_1}, ΔP_{sg_2}, and ΔP_{TL_1}; seven in number. Denoting the above variables by $x_1, x_2, x_3, x_4, x_5, x_6$, and x_7 and arranging them in a column vector as

$$\bar{X} = \begin{bmatrix} x_1 \\ x_2 \\ x_3 \\ x_4 \\ x_5 \\ x_6 \\ x_7 \end{bmatrix} \equiv \begin{bmatrix} \Delta f_1 \\ \Delta f_2 \\ \Delta X_{\mathrm{E}_1} \\ \Delta X_{\mathrm{E}_2} \\ \Delta P_{\mathrm{sg}_1} \\ \Delta P_{\mathrm{sg}_2} \\ \Delta P_{\mathrm{TL}_1} \end{bmatrix}_{7\times1}$$

where $\bar{X}$ is called a state vector.

The control variables ΔP_{C_1} and ΔP_{C_2} are denoted by the symbols u_1 and u_2, respectively, as

$$\bar{u} = \begin{bmatrix} u_1 \\ u_2 \end{bmatrix} \equiv \begin{bmatrix} \Delta P_{\mathrm{C}_1} \\ \Delta P_{\mathrm{C}_2} \end{bmatrix}_{2\times1}$$

where $\bar{u}$ is called the control vector or the control-force vector.

The disturbance variables ΔP_{D_1} and ΔP_{D_2}, since they create perturbations in the system, are denoted by p_1 and p_2, respectively, as

$$\bar{p} = \begin{bmatrix} p_1 \\ p_2 \end{bmatrix} = \begin{bmatrix} \Delta P_{D_1} \\ \Delta P_{D_2} \end{bmatrix}_{2 \times 1}$$

where $\bar{p}$ is called the disturbance vector.

The above state equations can be written in a matrix form as

$$
\begin{bmatrix} \dot{x}_1 \\ \dot{x}_2 \\ \dot{x}_3 \\ \dot{x}_4 \\ \dot{x}_5 \\ \dot{x}_6 \\ \dot{x}_7 \end{bmatrix}
=
\begin{bmatrix}
-\dfrac{1}{\tau_{ps_1}} & 0 & 0 & 0 & \dfrac{K_{ps_1}}{\tau_{ps_1}} & 0 & -\dfrac{K_{ps_1}}{\tau_{ps_1}} \\[2ex]
0 & -1/\tau_{ps_2} & 0 & 0 & 0 & \dfrac{K_{ps_2}}{\tau_{ps_2}} & -\dfrac{K_{ps_2}}{\tau_{ps_2}} \\[2ex]
-\dfrac{1}{R_1 \tau_{sg_1}} & 0 & -1/\tau_{sg_1} & 0 & 0 & 0 & 0 \\[2ex]
0 & -1/R_2 \tau_{sg_2} & 0 & -1/\tau_{sg_2} & 0 & 0 & 0 \\[2ex]
0 & 0 & -1/\tau_{t_1} & 0 & -1/\tau_{t_1} & 0 & 0 \\[2ex]
0 & 0 & 0 & 1/\tau_{t_2} & 0 & -1/\tau_{t_2} & 0 \\[2ex]
2\pi T_{12}^0 & -2\pi T_{12}^0 & 0 & 0 & 0 & 0 & 0
\end{bmatrix}
\begin{bmatrix} x_1 \\ x_2 \\ x_3 \\ x_4 \\ x_5 \\ x_6 \\ x_7 \end{bmatrix}
$$

$$
+
\begin{bmatrix}
0 & 0 \\
0 & 0 \\
1/\tau_{sg_1} & 0 \\
0 & 1/\tau_{sg_2} \\
0 & 0 \\
0 & 0 \\
0 & 0
\end{bmatrix}
\begin{bmatrix} u_1 \\ u_2 \end{bmatrix}
+
\begin{bmatrix}
-\dfrac{K_{ps_1}}{\tau_{ps_1}} & 0 \\[2ex]
0 & -K_{ps_2}/\tau_{ps_2} \\[1ex]
0 & 0 \\
0 & 0 \\
0 & 0 \\
0 & 0 \\
0 & 0
\end{bmatrix}
\begin{bmatrix} p_1 \\ p_2 \end{bmatrix}
\tag{8.44}
$$

where $\dot{x}_i = \dfrac{d_{x_i}}{dt}$; $i = 1, 2, 3, \ldots, 7$.

The above matrix equation can be written in the vector form as

$$\dot{\bar{x}} = [A]\bar{X} + [B]\bar{u} + [J]\bar{p} \tag{8.45}$$

where $[A]$ is called the system matrix, $[B]$ the input distribution matrix, and $[J]$ the disturbance distribution matrix.

In the present case, their dimensions are (7×7), (7×2), and (7×2), respectively. Equation (8.45) is a shorthand form of Equation (8.44), and Equation (8.44) constitutes the dynamic 'state-variable model' of the considered two-area system.

The differential equations can be put in the above form only if they are linear. If the differential equations are non-linear, then they can be expressed in the more general form as

$$\dot{\bar{X}} = f(\bar{x}, \bar{u}, \bar{p}) \tag{8.46}$$

8.9.5 State-variable model for a three-area power system

The block diagram representation of this model is shown in Fig. 8.13.

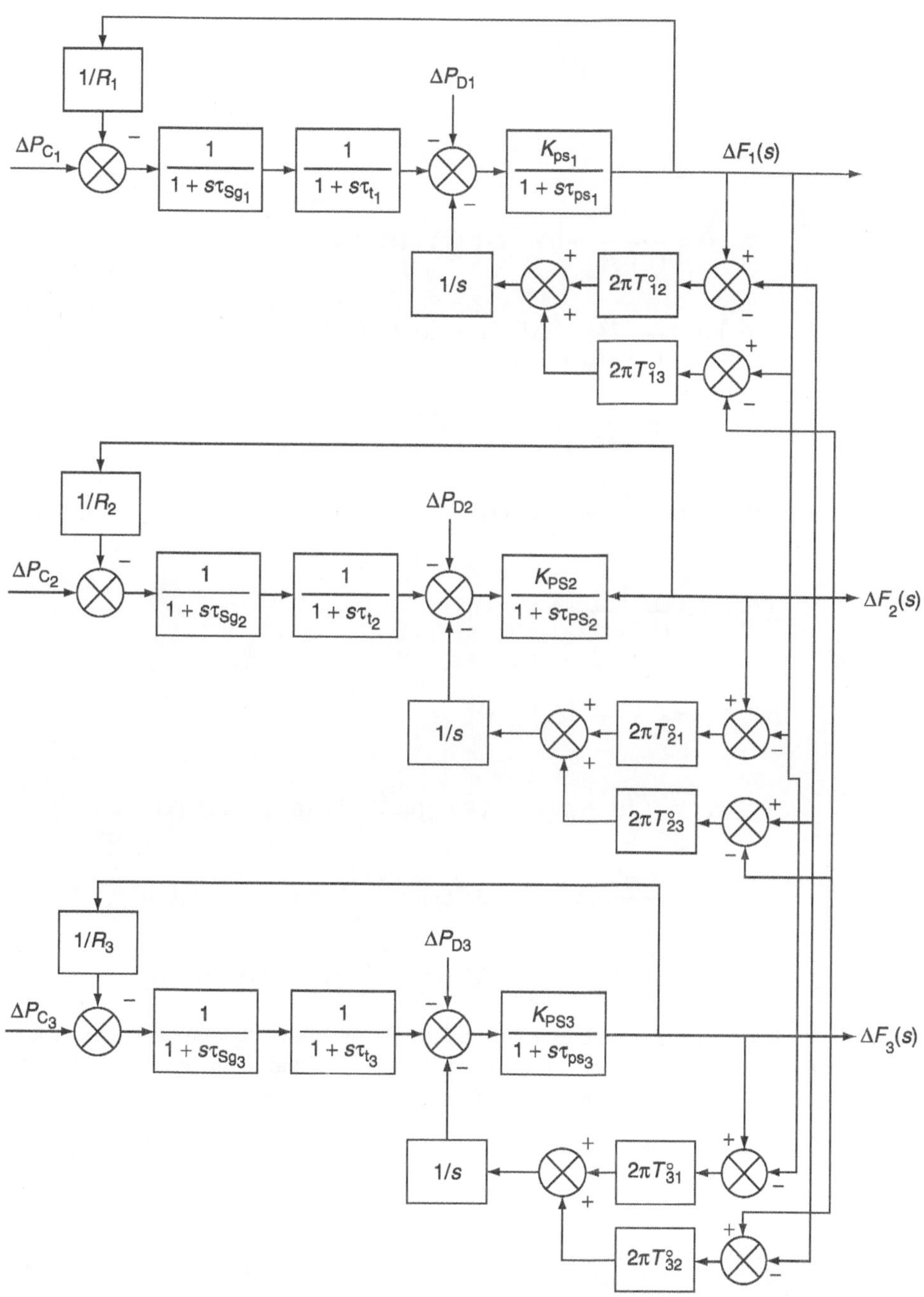

FIG. 8.13 *Three-area model*

From the block diagram, the following equations are written as

$$\Delta F_1(s) = \frac{Kp_{s_1}}{(1+s\tau_{ps_1})}[\Delta P_{G_1}(s) - \Delta P_{D_1}(s) - \Delta P_{TL_1}(s)]$$

$$\Delta F_2(s) = \frac{Kp_{s_2}}{(1+s\tau_{ps_2})}[\Delta P_{G_2}(s) - \Delta P_{D_2}(s) - \Delta P_{TL_2}(s)]$$

$$\Delta F_3(s) = \frac{Kp_{s_3}}{(1+s\tau_{ps_3})}[\Delta P_{G_3}(s) - \Delta P_{D_3}(s) - \Delta P_{TL_3}(s)]$$

$$\Delta X_{E_1}(s) = \frac{1}{(1+s\tau_{sg_1})}\left[\Delta P_{C_1}(s) - \Delta F_1(s)/R_1\right]$$

$$\Delta X_{E_2}(s) = \frac{1}{(1+s\tau_{sg_2})}\left[\Delta P_{C_2}(s) - \Delta F_2(s)/R_2\right]$$

$$\Delta X_{E_3}(s) = \frac{1}{(1+s\tau_{sg_3})}\left[\Delta P_{C_3}(s) - \Delta F_3(s)/R_3\right]$$

$$\Delta P_{G_1}(s) = \frac{1}{(1+s\tau_{t_1})}\left[\Delta X_{E_1}(s)\right]$$

$$\Delta P_{G_2}(s) = \frac{1}{(1+s\tau_{t_2})}\left[\Delta X_{E_2}(s)\right]$$

$$\Delta P_{G_3}(s) = \frac{1}{(1+s\tau_{t_3})}\left[\Delta X_{E_3}(s)\right]$$

$$\Delta P_{TL_1}(s) = \frac{2\pi T_{12}^0}{s}\left[\Delta F_1(s) - \Delta F_2(s)\right] + \frac{2\pi T_{13}^0}{s}\left[\Delta F_1(s) - \Delta F_3(s)\right]$$

$$\Delta P_{TL_2}(s) = \frac{2\pi T_{21}^0}{s}\left[\Delta F_2(s) - \Delta F_1(s)\right] + \frac{2\pi T_{23}^0}{s}\left[\Delta F_2(s) - \Delta F_3(s)\right]$$

$$\Delta P_{TL_3}(s) = \frac{2\pi T_{31}^0}{s}\left[\Delta F_3(s) - \Delta F_1(s)\right] + \frac{2\pi T_{32}^0}{s}\left[\Delta F_3(s) - \Delta F_2(s)\right]$$

Taking the inverse Laplace transform for the above equations, which we get in a similar way, the remaining equations can be rearranged and the inverse Laplace transform can be found. Then, the entire set of differential equations is

$$\frac{d}{dt}(\Delta f_1) = \frac{1}{\tau_{ps_1}}\left[-\Delta f_1 + K_{ps_1}\Delta P_{G_1} - K_{ps_1}\Delta P_{D_1} - K_{ps_1}\Delta P_{TL_1}\right]$$

$$\frac{d}{dt}(\Delta f_2) = \frac{1}{\tau_{ps_2}}\left[-\Delta f_2 + K_{ps_2}\Delta P_{G_2} - K_{ps_2}\Delta P_{D_2} - K_{ps_2}\Delta P_{TL_2}\right]$$

$$\frac{d}{dt}(\Delta f_3) = \frac{1}{\tau_{ps_3}}\left[-\Delta f_3 + K_{ps_3}\Delta P_{G_3} - K_{ps_3}\Delta P_{D_3} - K_{ps_3}\Delta P_{TL_3}\right]$$

$$\frac{\mathrm{d}}{\mathrm{d}t}(\Delta X_{E_1}) = \frac{1}{\tau_{sg_1}}\left[-\Delta X_{E_1} + \Delta P_{C_1} - \Delta f_1/R_1\right]$$

$$\frac{\mathrm{d}}{\mathrm{d}t}(\Delta X_{E_2}) = \frac{1}{\tau_{sg_2}}\left[-\Delta X_{E_2} + \Delta P_{C_2} - \Delta f_2/R_2\right]$$

$$\frac{\mathrm{d}}{\mathrm{d}t}(\Delta X_{E_3}) = \frac{1}{\tau_{sg_3}}\left[-\Delta X_{E_3} + \Delta P_{C_3} - \Delta f_3/R_3\right]$$

$$\frac{\mathrm{d}}{\mathrm{d}t}(\Delta P_{G_1}) = \frac{1}{\tau_{t_1}}\left[-\Delta P_{G_1} + \Delta X_{E_1}\right]$$

$$\frac{\mathrm{d}}{\mathrm{d}t}(\Delta P_{G_2}) = -\frac{1}{\tau_{t_2}}\left[-\Delta P_{G_2} + \Delta X_{E_2}\right]$$

$$\frac{\mathrm{d}}{\mathrm{d}t}(\Delta P_{G_3}) = -\frac{1}{\tau_{t_3}}\left[-\Delta P_{G_3} + \Delta X_{E_3}\right]$$

and $\quad \dfrac{\mathrm{d}}{\mathrm{d}t}(\Delta P_{TL_1}) = 2\pi T_{12}^0\left[\Delta f_1 - \Delta f_2\right] + 2\pi T_{13}^0\left[\Delta f_1 - \Delta f_3\right]$

and $\quad \dfrac{\mathrm{d}}{\mathrm{d}t}(\Delta P_{TL_2}) = 2\pi T_{21}^0\left[\Delta f_2 - \Delta f_1\right] + 2\pi T_{23}^0\left[\Delta f_2 - \Delta f_3\right]$

and $\quad \dfrac{\mathrm{d}}{\mathrm{d}t}(\Delta P_{TL_3}) = 2\pi T_{31}^0\left[\Delta f_3 - \Delta f_1\right] + 2\pi T_{32}^0\left[\Delta f_3 - \Delta f_2\right]$

The above equations are written in a vector form as shown below:

$$\dot{X} = [A]X + [B]\bar{u} + [J]\bar{p}$$

where $\dot{x} = \dfrac{\mathrm{d}x}{\mathrm{d}t}$

$$X = \begin{bmatrix} x_1 \\ x_2 \\ x_3 \\ x_4 \\ x_5 \\ x_6 \\ x_7 \\ x_8 \\ x_9 \\ x_{10} \\ x_{11} \\ x_{12} \end{bmatrix} \equiv \begin{bmatrix} \Delta f_1 \\ \Delta f_2 \\ \Delta f_3 \\ \Delta X_{E_1} \\ \Delta X_{E_2} \\ \Delta X_{E_3} \\ \Delta P_{sg_1} \\ \Delta P_{sg_2} \\ \Delta P_{sg_3} \\ \Delta P_{TL_1} \\ \Delta P_{TL_2} \\ \Delta P_{TL_3} \end{bmatrix}_{12\times 1}$$

where X is called a state vector.

The control variables ΔP_{C_1}, ΔP_{C_2}, and ΔP_{C_3} are denoted by the symbols u_1, u_2, and u_3, respectively, as

$$U = \begin{bmatrix} u_1 \\ u_2 \\ u_3 \end{bmatrix} \equiv \begin{bmatrix} \Delta P_{C_1} \\ \Delta P_{C_2} \\ \Delta P_{C_3} \end{bmatrix}_{3 \times 1}$$

where u is called the control vector or the control-force vector.

The disturbance variables ΔP_{D_1}, ΔP_{D_2}, and ΔP_{D_3}, since they create perturbations in the system, are denoted by p_1, p_2, and p_3, respectively, as

$$\bar{p} = \begin{bmatrix} p_1 \\ p_2 \\ p_3 \end{bmatrix} = \begin{bmatrix} \Delta P_{D_1} \\ \Delta P_{D_2} \\ \Delta P_{D_3} \end{bmatrix}_{3 \times 1}$$

where $\bar{p}$ is called the disturbance vector.

$$A = \begin{bmatrix}
\dfrac{-1}{\tau_{ps_1}} & 0 & 0 & 0 & 0 & 0 & \dfrac{K_{ps_1}}{\tau_{ps_1}} & 0 & 0 & -\dfrac{K_{ps_1}}{\tau_{ps_1}} & 0 & 0 \\[12pt]
0 & \dfrac{-1}{\tau_{ps_2}} & 0 & 0 & 0 & 0 & 0 & \dfrac{K_{ps_1}}{\tau_{ps_1}} & 0 & 0 & -\dfrac{K_{ps_1}}{\tau_{ps_1}} & 0 \\[12pt]
0 & 0 & \dfrac{-1}{\tau_{ps_2}} & 0 & 0 & 0 & 0 & 0 & \dfrac{K_{ps_1}}{\tau_{ps_1}} & 0 & 0 & -\dfrac{K_{ps_1}}{\tau_{ps_1}} \\[12pt]
\dfrac{-1}{R_1 \tau_{gs_1}} & 0 & 0 & 0 & 0 & 0 & \dfrac{-1}{\tau_{gs_1}} & 0 & 0 & 0 & 0 & 0 \\[12pt]
0 & \dfrac{-1}{R_2 \tau_{gs_2}} & 0 & 0 & 0 & 0 & 0 & \dfrac{-1}{\tau_{gs_2}} & 0 & 0 & 0 & 0 \\[12pt]
0 & 0 & \dfrac{-1}{R_2 \tau_{gs_3}} & 0 & 0 & 0 & 0 & 0 & \dfrac{-1}{\tau_{gs_3}} & 0 & 0 & 0 \\[12pt]
0 & 0 & 0 & \dfrac{-1}{\tau_{t_1}} & 0 & 0 & \dfrac{1}{\tau_{t_1}} & 0 & 0 & 0 & 0 & 0 \\[12pt]
0 & 0 & 0 & 0 & \dfrac{-1}{\tau_{t_2}} & 0 & 0 & \dfrac{1}{\tau_{t_2}} & 0 & 0 & 0 & 0 \\[12pt]
0 & 0 & 0 & 0 & 0 & \dfrac{-1}{\tau_{t_3}} & 0 & \dfrac{1}{\tau_{t_3}} & 0 & 0 & 0 & 0 \\[12pt]
m & p & q & 0 & 0 & 0 & 0 & 0 & 0 & 0 & 0 & 0 \\[6pt]
r & n & s & 0 & 0 & 0 & 0 & 0 & 0 & 0 & 0 & 0 \\[6pt]
t & u & o & 0 & 0 & 0 & 0 & 0 & 0 & 0 & 0 & 0
\end{bmatrix}$$

where $m = 2\pi(T_{12}^0 + T_{13}^0)$

$\quad\quad\quad n = 2\pi(T_{21}^0 + T_{23}^0)$

$$o = 2\pi(T_{31}^0 + T_{32}^0)$$

$$p = -2\pi T_{12}^0$$

$$q = -2\pi T_{13}^0$$

$$r = -2\pi T_{21}^0$$

$$s = -2\pi T_{23}^0$$

$$t = -2\pi T_{31}^0$$

$$\bar{u} = -2\pi T_{32}^0$$

$$[B] = \begin{bmatrix} 0 & 0 & 0 \\ 0 & 0 & 0 \\ 0 & 0 & 0 \\ 1/\tau_{sg_1} & 0 & 0 \\ 0 & 1/\tau_{sg_2} & 0 \\ 0 & 0 & 1/\tau_{sg_3} \\ 0 & 0 & 0 \\ 0 & 0 & 0 \\ 0 & 0 & 0 \\ 0 & 0 & 0 \\ 0 & 0 & 0 \\ 0 & 0 & 0 \end{bmatrix}$$

$$[J] = \begin{bmatrix} -\dfrac{K_{ps_1}}{\tau_{ps_1}} & 0 & 0 \\ 0 & -\dfrac{K_{ps_2}}{\tau_{ps_2}} & 0 \\ 0 & 0 & -\dfrac{K_{ps_3}}{\tau_{ps_3}} \\ 0 & 0 & 0 \\ 0 & 0 & 0 \\ 0 & 0 & 0 \\ 0 & 0 & 0 \\ 0 & 0 & 0 \\ 0 & 0 & 0 \\ 0 & 0 & 0 \\ 0 & 0 & 0 \\ 0 & 0 & 0 \end{bmatrix}$$

8.9.6 Advantages of state-variable model

The state-variable modeling of a power system offers the following advantages:

(i)   Modern control theory is based upon this standard form.

(ii) By arranging system parameters into matrices $[A]$, $[B]$, and $[J]$, a very organized methodology of solving system equations, either analytically or by computer, is developed. This is important for large systems where a lack of organization easily results in errors.

Example 8.4: Two interconnected Area-1 and Area-2 have the capacity of 2,000 and 500 MW, respectively. The incremental regulation and damping torque coefficient for each area on its own base are 0.2 p.u. and 0.8 p.u., respectively. Find the steady-state change in system frequency from a nominal frequency of 50 Hz and the change in steady-state tie-line power following a 750 MW change in the load of Area-1.

Solution:

Rated capacity of Area-1 $= P_{1(\text{rated})} = 2,000$ MW

Rated capacity of Area-2 $= P_{2(\text{rated})} = 500$ MW

Speed regulation, $R = 0.2$ p.u.

Nominal frequency, $f = 50$ Hz

Change in load power of Area-1, $\Delta P_1 = 75$ MW

Speed regulation, $R = 0.2 = 0.2$ p.u. $\times 50 = 10$ Hz/p.u. MW

Damping torque coefficient, $B = 0.8$ p.u. MW/p.u. Hz

$$= \frac{0.8}{50} = 0.016 \text{ p.u. MW/Hz}$$

Change in load of Area-1, $\Delta P_{D_1} = 75$ MW

$$\text{p.u. change in load of Area-1} = \frac{\Delta P_{D_1}}{P_{1(\text{rated})}} = \frac{75}{2,000} = 0.0375 \text{ p.u. MW}$$

$$\text{p.u. change in load of Area-2} = \frac{\Delta P_{D_2}}{P_{2(\text{rated})}} = 0$$

$$\frac{P_{1(\text{rated})}}{P_{2(\text{rated})}} = a_{12} = \frac{2,000}{500} = 4$$

Steady-state change in system frequency, $\Delta f_{ss} = \dfrac{-\Delta P_{D_2} + a_{12}\Delta P_{D_1}}{\beta_2 + a_{12}\beta_1}$

$$\Delta f_{ss} = -\left(\frac{\Delta P_{D_2} + a_{12}\Delta P_{D_1}}{\beta(1 + a_{12})}\right) \quad (\because \beta_1 = \beta_2 = \beta)$$

$$= \left(\frac{-a_{12}\Delta P_{D_1}}{\beta(1 + a_{12})}\right) \text{Hz} \quad \left[\because \Delta P_{D_2} = 0\right]$$

where $\beta = B + \dfrac{1}{R} = 0.016 + \dfrac{1}{10} = 0.116$ p.u. MW/Hz

$$\therefore \Delta f_{ss} = -\left[\frac{-4 \times 0.0375}{0.116(1+4)}\right] = +0.2586 \text{ Hz}$$

Steady-state change in tie-line power following load change in Area-1:

$$\Delta f_{\text{tie-1(ss)}} = \frac{(\beta_1 \Delta P_{D_2} - \beta_2 \Delta P_{D_1})}{\beta_2 + a_{12}\beta_1} = \frac{\beta(\Delta P_{D_2} - \Delta P_{D_1})}{\beta(1 + a_{12})}$$

$$= \frac{-\Delta P_{D_1} \beta}{\beta(1 + a_{12})} = \frac{-\Delta P_{D_1}}{(1 + a_{12})}$$

$$\Rightarrow \Delta P_{\text{tie-1(ss)}} = \frac{-\Delta P_{D_1}}{(1 + a_{12})} = \frac{-0.0375}{1+4} = -0.0075 \text{ p.u. MW}$$

$$= -0.0075 \times P_{1(\text{rated})} = -0.0075 \times 2000$$

$$= -15 \text{ MW}$$

Example 8.5: Solve Example 8.4, without governor control action.

Solution:

Without the governor control action, $R = 0$

$$\beta = B + \frac{1}{R} = 0.016 + \frac{1}{0}$$

$$\because R = 0, \qquad \beta = B = 0.016 \qquad [\because \beta_1 = \beta_2 = \beta]$$

$$\Delta f_{ss} = -\left(\frac{\Delta P_{D_2} + a_{12}\,\Delta P_{D_1}}{(\beta_2 + a_{12}\,\beta_1)}\right) = -\frac{\Delta P_{D_2} + a_{12}\,\Delta P_{D_1}}{\beta(1 + a_{12})}$$

$$\because \Delta f_{ss} = -\frac{a_{12}\,\Delta P_{D_1}}{\beta(1 + a_{12})} \qquad\qquad [\because \Delta P_{D_2} = 0]$$

$$= -\left[\frac{4 \times 0.0375}{0.016(1+4)}\right] = -1.875 \text{ Hz}$$

Steady-state change in tie-line power following load change in Area-1:

$$\Delta P_{\text{tie-1(steady state)}} = \frac{-\beta\,\Delta P_{D_1}}{\beta(1 + a_{12})} = \frac{-\Delta P_{D_1}}{1 + a_{12}}$$

$$= \frac{0.0375}{1+4} = -0.0075 \text{ p.u. MW}$$

$$= -0.0075 \times 2000$$

$$= -15 \text{ MW}$$

It is observed from the result that the power flow through the tie line is the same in both the cases of with governor action and without governor action, since it does not depend on speed regulation R.

Example 8.6: Find the nature of dynamic response if the two areas of the above problem are of uncontrolled type, following a disturbance in either area in the form of a step change in electric load. The inertia constant of the system is given as $H=3$ s and assume that the tie line has a capacity of 0.09 p.u. and is operating at a power angle of 30° before the step change in load.

Solution:

Given:

Speed regulation, $R=0.2$ p.u. $=0.2 \times 50 = 10$ Hz/p.u. MW

Damping coefficient, $B=0.8$ p.u. MW/p.u. Hz

$$= \frac{0.8}{50} = 0.016 \text{ p.u. MW/Hz}$$

Inertia constant, $H=3$ s

Nominal frequency, $f^0=50$ Hz

Tie-line capacity, 0.1 p.u. $= \dfrac{P_{\text{tie(max)}}}{P_{\text{rated}}}$

From the theory of dynamic response, we know that

$$\Delta P_{\text{tie}-1(s)} = \frac{\Delta P_{D_2}(s) - \Delta P_{D_1}(s)}{s^2 + 2\alpha s + \omega_n^2}$$

$$\alpha = \frac{f^0}{4H}\left(B + \frac{1}{R}\right) = \frac{50}{4 \times 3}\left[0.016 + \frac{1}{10}\right] = 0.4833$$

$$T_{12}^0 = \frac{P_{\text{tie(max)}}}{P_{\text{rated}}}\cos(\delta_1^0 - \delta_2^0) = 0.1 \times \cos(30°)$$

$$= 0.866 \times 0.1$$

$$= 0.0866$$

$$\therefore \omega_n^2 = \frac{2\pi T_{12}^0 f^0}{H} = \frac{2\pi \times 0.0866 \times 50}{3} = 9.068$$

$$\Rightarrow \omega_n = 3.0114 \text{ rad/s}$$

It is observed that the damped oscillation type of dynamic response has resulted since $\alpha < \omega_n$:

$$\therefore \text{ Damped angular frequency} = \omega_d = \sqrt{\omega_n^2 - \alpha^2}$$

$$= \sqrt{9.068 - 0.2335}$$

$$= 2.9723 \text{ rad/s}$$

$$\therefore \text{ Damped frequency} = f_d = \frac{\omega_d}{2\pi} = \frac{2.9723}{2\pi} = 0.473 \text{ Hz}$$

Example 8.7: Two control areas have the following characteristics:

Area-1: Speed regulation $= 0.02$ p.u.
Damping coefficient $= 0.8$ p.u.
Rated MVA $= 1{,}500$

Area-2: Speed regulation $= 0.025$ p.u.

Damping coefficient $= 0.9$ p.u.

Rated MVA $= 500$

Determine the steady-state frequency change and the changed frequency following a load change of 120 MW, which occurs in Area-1. Also find the tie-line power flow change.

Solution:

Given $R_1 = 0.1$ p.u.; $R_2 = 0.098$ p.u.

$B_1 = 0.8$ p.u.; $B_2 = 0.9$ p.u.

$P_{1\,\text{rated}} = 1{,}500$ MVA; $P_{2\,\text{rated}} = 1{,}500$ MVA

Change in load of Area-1,

$$\Delta P_{D_1} = 120 \text{ MW}, \quad \Delta P_{D_2} = 0$$

p.u. change in load of Area-1 $= \dfrac{120}{1{,}500} = 0.08$ p.u.

$$\Delta f_{ss} = \frac{\Delta P_{D_2} + a_{12}\,\Delta P_{D_1}}{\beta_2 + a_{12}\,\beta_1}$$

$$\beta_1 = B_1 + \frac{1}{R_1} = 0.8 + \frac{1}{0.02} = 50.8$$

$$\beta_2 = B_2 + \frac{1}{R_2} = 0.9 + \frac{1}{0.025} = 40.9$$

$$a_{12} = \frac{P_{1(\text{rated})}}{P_{2(\text{rated})}} = \frac{1{,}500}{500} = 3$$

$\therefore$ Steady-state frequency change, $\Delta f_{ss} = \dfrac{a_{12}\,\Delta P_{D_1}}{\beta_2 + a_{12}\,\beta_1}$

$$= \frac{-3 \times 0.08}{40.9 + 3(50.8)}$$

$$= -\frac{3 \times 0.08}{193.3} = -0.0012415 \text{ p.u. Hz}$$

i.e., Steady-state change in frequency, $\Delta f_{ss} = 0.0012415 \times 50$

$$= 0.062 \text{ Hz}$$

$\therefore$ New value of frequency, $f = f^0 - \Delta f_{ss} = 50 - 0.062$

$$= 49.937 \text{ Hz}$$

Steady-state change in tie-line power $= \Delta P_{(tie\text{-}1)ss} = \dfrac{\beta_1 \, \Delta P_{D_2} - \beta_2 \, \Delta P_{D_1}}{\beta_2 + a_{12} \, \beta_1} \text{ p.u. MW}$

$$= \dfrac{50.8(0) - 40.9(0.08)}{40.9 + 3(50.8)}$$

$$= -0.0169 \text{ p.u. MW}$$

$$\therefore \Delta P_{(tie\text{-}1)} = 0.0169 \times P_{1(rated)} = -0.0169 \times 1500 = 25.35 \text{ MW}$$

Example 8.8: In Example 8.6, if the disturbance also occurs in Area-2, which results in a change in load by 75 MW, determine the frequency and tie-line power changes.

Solution:

Change in load of Area-1, $\Delta P_{D_1} = 120$ MW

$$\text{p.u. change in load of Area-1} = \dfrac{\Delta P_{D_1}}{P_{1rated}} = \dfrac{120}{1,500} = 0.08 \text{ p.u. MW}$$

Change in load of Area-2, $\Delta P_{D_2} = 75$ MW

$$\text{p.u. change in load of Area-2} = \dfrac{\Delta P_{D_2}}{P_{2rated}} = \dfrac{75}{500} = 0.15 \text{ p.u. MW}$$

Steady-state frequency change, $\Delta f_{ss} = \dfrac{\Delta P_{D_2} + a_{12} \, \Delta P_{D_1}}{\beta_2 + a_{12} \, \beta_1}$

$$= \dfrac{0.15 + 3(0.08)}{193.3} = 0.002 \text{ p.u Hz}$$

$\therefore$ Steady-state frequency change $= 0.002 \times 50 = 0.1$ Hz

$\therefore$ New value of frequency $= f^0 - \Delta f_{ss} = 50 - 0.1 = 49.899$ Hz

Steady-state change in tie-line power, $\Delta P_{tie(ss)} = \dfrac{\beta_1 \, \Delta P_{D_2} - \beta_2 \, \Delta P_{D_1}}{\beta_2 + a_{12} \, \beta_1} \text{ p.u. MW}$

$$= \dfrac{(50.8)(0.15) - (40.9)(0.08)}{40.9 + 3(50.8)}$$

$$= 7.62 - 3.273$$

$$= \dfrac{4.348}{133.3} = 0.02249 \text{ p.u. MW}$$

Example 8.9: Two areas of a power system network are interconnected by a tie line, whose capacity is 250 MW, operating at a power angle of 45°. If each area has a capacity of 2,000 MW and the equal speed-regulation coefficient of 3 Hz/p.u. MW, determine the frequency of oscillation of the power for a step change in load. Assume that both areas have the same inertia constants of $H = 4$ s. If a step-load change of 100 MW occurs in one of the areas, determine the change in tie-line power.

Solution:

Given:

Tie-line capacity, $P_{tie(max)} = 250$ MW

Power angle of two areas, $(\delta_1^0 - \delta_2^0) = 45°$

Capacity of each area, $P_{rated} = 2,000$ MW

Speed-regulation coefficient $= R_1 = R_2 = R = 3$Hz/p.u. MW

Inertia constant, $H = 4$ s

$$\alpha = \frac{f^0}{4H}\left(B + \frac{1}{R}\right) = \frac{50}{4 \times 4}\left(0 + \frac{1}{3}\right)$$

$$= \frac{50}{4 \times 4 \times 3} = \frac{50}{48} = 1.04$$

$$T_{12}^0 = \frac{P_{tie_{12}(max)}}{P_{rated}}\cos(\delta_1^0 - \delta_2^0)$$

$$= \frac{250}{2,000}\cos(45°) = 0.125\cos(45°) = 0.0883$$

$$\omega_n^2 = \frac{2\pi T_{12}^0}{H} = \frac{2\pi \times 0.0883 \times 50}{4} = 6.935$$

$$\therefore \omega_n = 2.6334 \text{ rad/s}$$

Since, $\alpha < \omega_n$, the dynamic response will be of a damped oscillation type.

Damped angular frequency, $\omega_d = \sqrt{\omega_n^2 - \alpha^2}$

$$= \sqrt{6.935 - 1.0816}$$

$$= 2.4193 \text{ rad/s}$$

$\therefore$ Frequency of oscillation, $f_d = \frac{\omega_d}{2\pi} = \frac{2.4193}{2 \times \pi} = 0.385$ Hz

If a step-load change of 100 MW occurs in any one of the areas, the total load change will be shared equally by both areas since the two areas are equal, i.e., a power of $\frac{100}{2} = 50$ MW will flow from the other area into the area where a load change occurs.

Example 8.10: Two power stations A and B of capacities 75 and 200 MW, respectively, are operating in parallel and are interconnected by a short transmission line. The generators of stations A and B have speed regulations of 4% and 2%, respectively. Calculate the output of each station and the load on the interconnection if

(a) the load on each station is 100 MW,

(b) the loads on respective bus bars are 50 and 150 MW, and

(c) the load is 130 MW at Station A bus bar only.

Solution:

Given:

Capacity of Station-A $= 75$ MW

Capacity of Station-B $= 200$ MW

Speed regulation of Station-A generator, $R_A = 4\%$

Speed regulation of Station-B generator, $R_B = 2\%$

(a) If the load on each station $= 100$ MW

$$\text{i.e., } P_1 + P_2 = 100 + 100 = 200 \text{ MW} \tag{8.47}$$

$$\text{Speed regulation} = \frac{N_0 - N}{N_0} = \frac{f_0 - f}{f_0}$$

$$\frac{P_1}{75} = \frac{1 - f}{0.04}$$

$$\Rightarrow (1 - f) = \frac{0.04}{75} \times P_1 \tag{8.48}$$

$$\frac{P_2}{200} = \frac{1 - f}{0.02}$$

$$\therefore (1 - f) = 0.0001 P_2 \tag{8.49}$$

From Equations (8.48) and (8.49), we have

$$0.000533 P_1 = 0.0001 P_2$$

$$5.33 P_1 = P_2 \tag{8.50}$$

$$P_1 + P_2 = 200$$

Substituting Equation (8.50) in Equation (8.47), we get

$$P_1 + 5.33\, P_1 = 200$$

$$6.33 P_1 = 200$$

$$\text{or} \quad P_1 = \frac{200}{6.33} = 31.60 \text{ MW}$$

$$\therefore P_2 = 200 - 31.59 = 168.40 \text{ MW}$$

The power generations and tie-line power are indicated in Fig. 8.14(a).

(b) If the load on respective bus bars are 50 and 150 MW, then we have

$$\text{i.e., } \quad P_1 + P_2 = 50 + 150 = 200 \text{ MW}$$

$$5.33\, P_1 = P_2$$

$$P_1 + 5.33\, P_1 = 200$$

$$\Rightarrow 6.33 P_1 = 200$$

$$P_1 = 31.6 \text{ MW}$$

$$\therefore P_2 = 200 - 31.60 = 168.4 \text{ MW}$$

The power generations and tie-line power are indicated in Fig. 8.14(b).

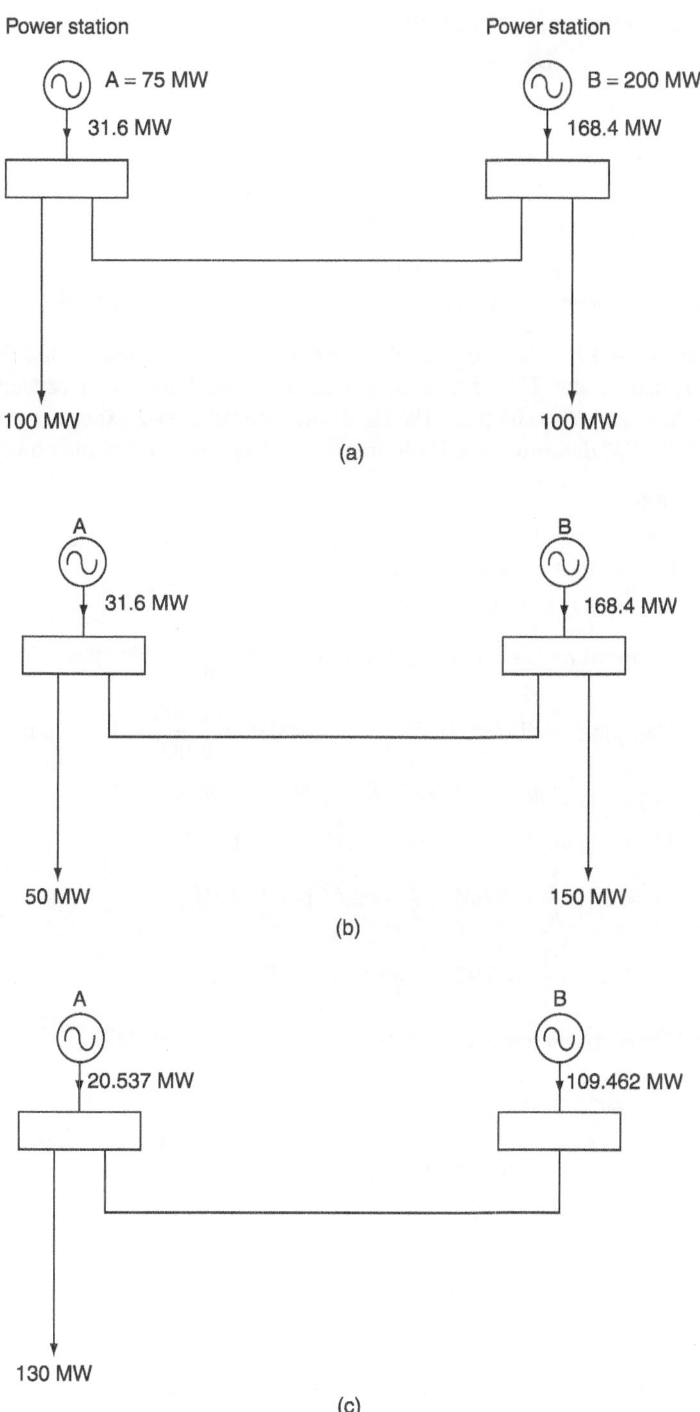

FIG. 8.14   *(a) Illustration for Example 8.10; (b) illustration for Example 8.10; (c) illustration for Example 8.10*

(c) If the load is 130 MW at A only, then we have

$$P_1 + P_2 = 130$$

$$5.33P_1 = P_2$$

$$\therefore P_1 = 5.33\,P_1 = 130$$

$$6.33P_1 = 130$$

$$\Rightarrow \quad P_1 = 20.537 \text{ MW}$$

$$\therefore P_2 = 130 - P_1 = 109.462 \text{ MW}$$

The power generations and tie-line power are indicated in Fig. 8.14(c).

Example 8.11: The two control areas of capacity 2,000 and 8,000 MW are interconnected through a tie line. The parameters of each area based on its own capacity base are $R = 1$ Hz/p.u. MW and $B = 0.02$ p.u. MW/Hz. If the Control area-2 experiences an increment in load of 180 MW, determine the static frequency drop and the tie-line power.

Solution:

Capacity of Area-1 = 2,000 MW

Capacity of Area-2 = 8,000 MW

Taking 8,000 MW as base,

$$\therefore \text{ Speed regulation of Area- 1, } R_1 = 1 \times \frac{8,000}{2,000} = 4 \text{ Hz / p.u. MW}$$

$$\text{Damping coefficient of Area-1, } B_1 = 0.02 \times \frac{2,000}{8,000} = 0.005 \text{ p.u. MW/Hz}$$

Speed regulation of Area-2, $R_2 = 1$ Hz/p.u. MW

Damping coefficient of Area-2, $B_2 = 0.02$ p.u. MW/Hz

$$\beta_1 = B_1 + \frac{1}{R_1} = 0.005 + \frac{1}{4} = 0.255 \text{ p.u. MW/Hz}$$

$$\beta_2 = B_2 + \frac{1}{R_2} = 0.02 + \frac{1}{1} = 1.02 \text{ p.u. MW/Hz}$$

Given an increment of Area-2 in load, $\Delta P_{D_2} = 180 \text{ MW} = \dfrac{180}{8,000} = 0.025 \text{ p.u. MW}$

$$\Delta P_{D_1} = 0$$

$\therefore$ Static change in frequency, $\Delta f_{ss} = -\dfrac{\Delta P_{D_2} + a_{12}\,\Delta P_{D_1}}{\beta_2 + a_{12}\beta_1}$

$$= -\frac{0.0225}{0.255 + 1.02}$$

$$= -0.017647 \text{ p.u. Hz}$$

Static change in tie-line power, $\Delta P_{tie(ss)} = \dfrac{\beta_1\,\Delta P_{D_2} - \beta_2\,\Delta P_{D_1}}{\beta_2 + a_{12}\,\beta_1}$

$$= \frac{0.255 \times 0.0225}{0.255 + 1.02} = 0.0051 \text{ p.u. MW}$$

Note: Here, a_{12} value determination is not required since values of R_1, B_1, and β_1 are obtained according to the base values.

Alternate method:

Find $a_{12} \dfrac{P_{1\text{rated}}}{P_{2\text{rated}}} = \dfrac{2,000}{8,000} = 0.25$. Then, obtain the $\Delta f_{(ss)}$ and $\Delta P_{\text{tie(ss)}}$ values.

Here, there is no need to obtain, R_1, B_1, R_2, and B_2 separately.

Example 8.12: Two generating stations A and B having capacities 500 and 800 MW, respectively, are interconnected by a short line. The percentage speed regulations from no-load to full load of the two stations are 2 and 3, respectively. Find the power generation at each station and power transfer through the line if the load on the bus of each station is 200 MW.

Solution:

Given data:

Capacity of Station-A $= 500$ MW

Capacity of Station-B $= 800$ MW

Percentage speed regulation of Station-A $= 2\% = 0.02$

Percentage speed regulation of Station-B $= 3\% = 0.03$

Load on bus of each station $= P_{DA} = P_{DB} = 200$ MW

Total load, $P_D = 400$ MW

Speed regulation of Station-A:

$$R_A = (0.02)\frac{50}{500} = 0.002 \text{ Hz/MW}$$

Speed regulation of Station-B:

$$R_B = (0.03) \times \frac{50}{800} = 0.001875 \text{ Hz/MW}$$

Let P_{GA} be the power generation of Station-A and P_{GB} the power generation of Station-B:

$$P_{GB} = \text{Total load} - P_{GA} = (400 - P_{GA})$$
$$\Rightarrow 0.002 P_{GA} = 0.001875\,(400 - P_{GA})$$
$$= 0.75 = 193.55 \text{ MW}$$
$$(0.002 + 0.001875)\,P_{GA} = 0.75$$
$$\Rightarrow \quad P_{GA} = 193.55 \text{ MW}$$
$$P_{GB} = 206.45 \text{ MW}$$
$$P_{GA} = 193.55 \text{ MW}$$
$$\therefore \quad P_{GB} = 206.45 \text{ MW}$$

The power transfer through the line from Station-B to Station-A

$$= P_{GB} - (\text{load at bus bar of B})$$

$$= 206.45 - 200$$

$$= 6.45 \text{ MW}$$

Example 8.13: Two control areas of 1,000 and 2,000 MW capacities are interconnected by a tie line. The speed regulations of the two areas, respectively, are 4 Hz/p.u. MW and 2.5 Hz/p.u. MW. Consider a 2% change in load occurs for 2% change in frequency in each area. Find steady-state change in frequency and tie-line power of 10 MW change in load occurs in both areas.

Solution:

Capacity of Area-1 = 1,000 MW

Capacity of Area-2 = 2,000 MW

Speed regulation of Area-1, $R_1 = 4$ Hz/p.u. MW (on 1,000-MW base)

Speed regulation of Area-2, $R_2 = 2$ Hz/p.u. MW

Let us choose 2,000 MW as base, 2% change in load for 2% change in frequency

Damping coefficient of Area-1, $B_1 = \dfrac{0.02 \times 1,000}{0.02 \times 50} = 20$ MW/Hz

$$= \dfrac{20}{2,000} = 0.01 \text{ p.u. MW/Hz}$$

Similarly, damping coefficient of Area-2 on 2,000-MW base

$$B_2 = \dfrac{0.02}{0.02} \times \dfrac{2,000}{50} = 40 \text{ MW/Hz}$$

$$= \dfrac{40}{2,000} = 0.02 \text{ p.u. MW/Hz}$$

Speed regulation of Area - 1 on 2,000-MW base $= R_1$

$$= 4 \times \dfrac{2,000}{1,000} = 8 \text{ Hz/p.u. MW}$$

Speed regulation of Area-2, $R_2 = 2$ Hz/p.u. MW

$$\therefore \quad \beta_1 = B_1 + \dfrac{1}{R_1} = 0.01 + \dfrac{1}{8} = 0.135 \text{ p.u. MW/Hz}$$

$$\beta_2 = B_2 + \dfrac{1}{R_2} = 0.02 + \dfrac{1}{2} = 0.52 \text{ p.u. MW/Hz}$$

If a 10-MW change in load occurs in Area-1, then we have

$$\Delta P_{D_1} = \dfrac{10}{2,000} = 0.005; \quad \Delta P_{D_2} = 0$$

Steady-state change in frequency,

Steady-state change in frequency, $\Delta f(ss) = -\dfrac{\Delta P_{D_2} + a_{12}\Delta P_{D_1}}{\beta_2 + a_{12}\,\beta_1} = -0.007633 \text{ p.u. Hz}$

or $\Delta f_{(ss)} = -0.007633 \times 50 = 0.38$ Hz

Steady-state change in tie-line power:

$$\Delta P_{\text{tie(ss)}} = \dfrac{\beta_1 \Delta P_{D_2} - \beta_2 \Delta P_{D_1}}{\beta_2 + a_{12}\,\beta_1}$$

$$= \frac{0.135 \times 0 - 0.52 \times 0.005}{0.52 + 0.135}$$

$$= -0.003964 \text{ p.u. MW}$$

$$= -0.003964 \times 2{,}000$$

$$= -7.938 \text{ MW}$$

i.e., the power transfer of 7.938 MW is from Area-2 to Area-1.

If a 10-MW change in load occurs in Area-2, then we have

$$\Delta P_{D_2} = \frac{10}{2{,}000} = 0.005; \quad \Delta P_{D_1} = 0$$

$\therefore$ Steady-state change in frequency, $\Delta f_{ss} = -\left(\dfrac{0.005}{0.52 + 0.135} \right)$

$$= -0.007633 \text{ p.u. Hz}$$

$$= -0.38 \text{ Hz}$$

Steady-state change in tie-line power: $\Delta P_{\text{tie(ss)}} = \left(\dfrac{0.135 \times 0.005 - 0.52 \times 0}{0.52 + 0.135} \right)$

$$= 0.00103 \text{ p.u. MW}$$

$$= 0.00103 \times 2000$$

$$= 2.061 \text{ MW}$$

i.e., A power of 2.061 MW is transferred from Area-1 to Area-2.

Example 8.14: Two similar areas of equal capacity of 5,000 MW, speed regulation $R = 3$ Hz/p.u. MW, and $H = 5$ s are connected by a tie line with a capacity of 500 MW, and are operating at a power angle of 45°. For the above system, the frequency is 50 Hz; find:

(a) The frequency of oscillation of the system.

(b) The steady-state change in the tie-line power if a step change of 100 MW load occurs in Area-2.

(c) The frequency of oscillation of the system in the speed-governor loop is open.

Solution:

Given:

Capacity of each control area $= P_{1(\text{rated})} \; P_{2(\text{rated})} = 500 \text{ MW}$

Speed regulation, $R = 2$ Hz/p.u. MW

Inertia constant, $H = 5$ s

Power angle $= 45°$

Supply frequency, $f^0 = 50$ Hz

(a) Stiffness coefficient, $T_{12} = \dfrac{P_{\max(\text{tie})}}{P_{1(rated)}} \cos (\delta_1^0 - \delta_2^0)$

$$= \frac{500}{5{,}000} \cos 45° = 0.0707$$

$$\omega_n^2 = \frac{2\pi T_{12} f^0}{H} = \frac{2\pi \times 0.0707 \times 50}{5} = 4.4422$$

$$\omega_n = 2.1076 \text{ rad/s}$$

$$\alpha = \frac{f^0}{4H}\left(B + \frac{1}{R}\right) = \frac{50}{4 \times 5}\left[0 + \frac{1}{4}\right] = \frac{50}{4 \times 5 \times 4} = \frac{50}{80} = 0.625$$

Since $\alpha < \omega_n$, damped oscillations will be present.

$\therefore$ Damped angular frequency, $\omega_d = \sqrt{\omega_n^2 - \alpha^2}$

$$= \sqrt{4.4422 - (0.625)^2}$$

$$= 2.0128 \text{ rad/s}$$

$$f_d = \frac{\omega_d}{2\pi} = 0.32 \text{ Hz}$$

(b) Since the two areas are similar, each area will supply half of the increased load:

$$\therefore \beta_1 = \beta_2$$

$\Delta P_{\text{tie}} = 50$ MW from Area-1 to Area-2.

If the speed-governor loop is open, then $R \to \infty \Rightarrow \dfrac{1}{R} = 0$

$$\therefore \alpha = \frac{f^0}{4DH} = 0$$

Damped angular frequency, $\omega_d = \sqrt{\omega_n^2 - \alpha^2}$

$$= \omega_d = \omega_n = 2.1076 \text{ rad/s}$$

$$f_d = \frac{\omega d}{2\pi} = \frac{2.1076}{2 \times \pi} = 0.335 \text{ Hz}$$

KEY NOTES

- An extended power system can be divided into a number of LFC areas, which are interconnected by tie lines. Such an operation is called a **pool operation.**

  The basic principle of a pool operation in the normal steady state provides:

  (i) Maintaining of scheduled interchanges of tie-line power.

  (ii) Absorption of own load change by each area.

- The advantages of a pool operation are as follows:

  (i) Half of the added load (in Area-2) is supplied by Area-1 through the tie line.

  (ii) The frequency drop would be only half of that which would occur if the areas were operating without interconnection.

- The speed-changer command signals will be:

$$\Delta P_{c_1} = -K_{l_1} \int (\Delta P_{TL_1} + b_1 \Delta f_1)\, dt$$

and

$$\Delta P_{c_2} = -K_{I_2} \int (\Delta P_{TL_2} + b_2 \Delta f_2)\, dt$$

The constants K_{I_1} and K_{I_2} are the gains of the integrators. The first terms on the right-hand side of the above equations constitute what is known as a tie-line bias control.

- The load frequency controller provides a fast-acting control and regulates the system around an operating point, whereas the EDC provides a slow-acting control, which adjusts the speed-changer settings every minute in accordance with a command signal generated by the CEDC.

SHORT QUESTIONS AND ANSWERS

(1) What are the advantages of a pool operation?

The advantages of a pool operation (i.e., grid operation) are:

(i) Half of the added load (in Area-2) is supplied by Area-1 through the tie line.

(ii) The frequency drop would be only half of that which would occur if the areas were operating without interconnection.

(2) Without speed-changer position control, can the static frequency deviation be zero?

No, the static frequency deviation cannot be zero.

(3) State the additional requirement of the control strategy as compared to the single-area control.

The tie-line power deviation due to a step-load change should decrease to zero.

(4) Write down the expressions for the ACEs.

The ACE of Areas-1 and 2 are:

$$ACE_1(s) = \Delta P_{TL_1}(s) + b_1 \Delta F_1(s).$$

$$ACE_2(s) = \Delta P_{TL_2}(s) + b_2 \Delta F_2(s).$$

(5) What is the criterion used for obtaining optimum values for the control parameters?

Integral of the sum of the squared error criterion is the required criterion.

(6) Give the error criterion function for the two-area system.

$$C = \int_0^\infty \left[\alpha_1 (\Delta P_{TL_1})^2 + \alpha_2 (\Delta f_1)^2 + \alpha_3 (\Delta f_2)^2 \right] dt.$$

(7) What is the order of differential equation to describe the dynamic response of a two-area system in an uncontrolled case?

It is required for a system of seventh-order differential equations to describe the dynamic response of a two-area system. The solution of these equations would be tedious.

(8) What is the difference of ACE in single-area and two-area power systems?

In a single-area case, ACE is the change in frequency. The steady-state error in frequency will become zero (i.e., $\Delta f_{ss} = 0$) when ACE is used in an integral-control loop.

In a two-area case, ACE is the linear combination of the change in frequency and change in tie-line power. In this case to make the steady-state tie-line power zero (i.e., $\Delta P_{TL} = 0$), another integral-control loop for each area must be introduced in addition to the integral frequency loop to integrate the incremental tie-line power signal and feed it back to the speed-changer.

(9) What is the main difference of load frequency and economic dispatch controls?

The load frequency controller provides a fast-acting control and regulates the system around an operating point, whereas the EDC provides a slow-acting control, which adjusts the speed-changer settings every minute in accordance with a command signal generated by the CEDC.

(10) What are the steps required for designing an optimum linear regulator?

An optimum linear regulator can be designed using the following steps:

(i) Casting the system dynamic model in a state-variable form and introducing appropriate control forces.

(ii) Choosing an integral-squared-error control index, the minimization of which is the control goal.

(iii) Finding the structure of the optimal controller that will minimize the chosen control index.

MULTIPLE-CHOICE QUESTIONS

(1) Changes in load division between AC generators operation in parallel are accomplished by:

(a) Adjusting the generator voltage regulators.

(b) Changing energy input to the prime movers of the generators.

(c) Lowering the system frequency.

(d) Increasing the system frequency.

(2) When the energy input to the prime mover of a synchronous AC generator operating in parallel with other AC generators is increased, the rotor of the generator will:

(a) Increase in average speed.

(b) Retard with respect to the stator-revolving field.

(c) Advance with respect to the stator-revolving field.

(d) None of these.

(3) When two or more systems operate on an interconnected basis, each system:

(a) Can depend on the other system for its reserve requirements.

(b) Should provide for its own reserve capacity requirements.

(c) Should operate in a 'flat frequency' mode.

(4) When an interconnected power system operates with a tie-line bias, they will respond to:

(a) Frequency changes only.

(b) Both frequency and tie-line load changes.

(c) Tie-line load changes only.

(5) In a two-area case, ACE is:

(a) Change in frequency.

(b) Change in tie-line power.

(c) Linear combination of both (a) and (b).

(d) None of the above.

(6) An extended power system can be divided into a number of LFC areas, which are interconnected by tie lines. Such an operator is called

(a) Pool operation.

(b) Bank operation.

(c) (a) and (b).

(d) None.

(7) For the static response of a two-area system,

(a) $\Delta P_{ref_1} = \Delta_{ref_2}$.

(b) $\Delta P_{ref_1} = 0$.

(c) $\Delta P_{ref_2} = 0$.

(d) Both (b) and (c).

(8) Area of frequency response characteristic 'β' is:

(a) $1/R$.

(b) B.

(c) $B + 1/R$.

(d) $B - 1/R$.

(9) The tie-line power equation is $\Delta P_{12} = \underline{\hspace{2cm}}$.

(a) $T(\Delta\delta_1 + \Delta\delta_2)$.

(b) $T/(\Delta\delta_1 + \Delta\delta_2)$.

(c) $T/(\Delta\delta_1 - \Delta\delta_2)$.

(d) $T(\Delta\delta1 - \Delta\delta2)$.

(10) The unit of synchronizing coefficients 'T' is:

(a) MW-s.

(b) MW/s.

(c) MW-rad.

(d) MW/rad.

(11) For a two-area system, Δf is related to increased step load M_1 and M_2 with area frequency response characteristics β_1 and β_2 is:

(a) $M_1 + M_2/\beta_1 + \beta_2$.

(b) $(M_1 + M_2)(\beta_1 + \beta_2)$.

(c) $-(M_1 + M_2)/(\beta_1 + \beta_2)$.

(d) None of these.

(12) Tie-line power flow for the above question (11) is $\Delta P_{12} = \underline{\hspace{2cm}}$.

(a) $(\beta_1 M_2 + \beta_2 M_1)/\beta_1 + \beta_2$.

(b) $(\beta_1 M_2 - \beta_2 M_1)/\beta_1 + \beta_2$.

(c) $(\beta_1 M_1 - \beta_2 M_2)/\beta_1 + \beta_2$.

(d) None of these.

(13) Advantage of a pool operation is:

(a) Added load can be shared by two areas.

(b) Frequency drop reduces.

(c) Both (A) and (B).

(d) None of these.

(14) Damping of frequency oscillations for a two-area system is more with:

(a) Low-R.

(b) High-R.

(c) $R = \alpha$.

(d) None of these.

(15) ACE equation for a general power system with tie-line bias control is:

(a) $\Delta P_{ij} + B_i\, \Delta f_i$.

(b) $\Delta P_{ij} - B_i\, \Delta f_i$.

(c) $\Delta P_{ij} / B_i\, \Delta f_i$.

(d) None of these.

(16) For a two-area system Δf, ΔP_L, R_1, R_2, and D are related as $\Delta f =$ ___________.

(a) $\Delta P_L / R_1 + R_2$.

(b) $-\Delta P_L / (1/R_1 + R_2 + B)$.

(c) $-\Delta P_L / (B + R_1 + 1/R_2)$.

(d) None of these.

(17) If the two areas are identical, then we have:

(a) $\Delta f_1 = 1/\Delta f_2$.

(b) $\Delta f_1\, \Delta f_2 = 2$.

(c) $\Delta f_1 = \Delta f_2$.

(d) None of these.

(18) Tie-line between two areas usually will be a ___________ line.

(a) HVDC.

(b) HVAC.

(c) Normal AC.

(d) None of these.

(19) Dynamic response of a two-area system can be represented by a ___________ order transfer function.

(a) Third.

(b) Second.

(c) First.

(d) Zero.

(20) Control of ALFC loop of a multi-area system is achieved by using ___________ mathematical technique.

(a) Root locus.

(b) Bode plots.

(c) State variable.

(d) Nyquist plots.

REVIEW QUESTIONS

(1) Obtain the mathematical modeling of the line power in an interconnected system and its block diagram.

(2) Obtain the block diagram of a two-area system.

(3) Explain how the control scheme results in zero tie-line power deviations and zero-frequency deviations under steady-state conditions, following a step-load change in one of the areas of a two-area system.

(4) Deduce the expression for static-error frequency and tie-line power in an identical two-area system.

(5) Explain about the optimal two-area LFC.

(6) What is meant by tie-line bias control?

(7) Derive the expression for incremental tie-line power of an area in an uncontrolled two-area system under dynamic state for a step-load change in either area.

(8) Draw the block diagram for a two-area LFC with integral controller blocks and explain each block.

(9) What are the differences between uncontrolled, controlled, and tie-line bias LFC of a two-area system.

(10) Explain the method involved in optimum parameter adjustment for a two-area system.

(11) Explain the combined operation of an LFC and an ELDC system.

PROBLEMS

(1) Two interconnected areas 1 and 2 have the capacity of 250 and 600 MW, respectively. The incremental regulation and damping torque coefficient for each area on its own base are 0.3 and 0.07 p.u. respectively. Find the steady-state change in system frequency from a nominal frequency of 50 Hz and the change in steady-state tie-line power following a 850 MW change in the load of Area-1.

(2) Two control areas of 1,500 and 2,500 MW capacities are interconnected by a tie line. The speed regulations of the two areas, respectively, are 3 and 1.5 Hz/p.u. MW. Consider that a 2% change in load occurs for a 2% change in frequency in each area. Find the steady-state change in the frequency and the tie-line power of 20 MW change in load occurring in both areas.

(3) Find the nature of dynamic response if the two areas of the above problem are of uncontrolled type, following a disturbance in either area in the form of a step change in an electric load. The inertia constant of the system is given as $H = 2$ s and assume that the tie line has a capacity of 0.08 p.u. and is operating at a power angle of 35° before the step change in load.

9 Reactive Power Compensation

OBJECTIVES

After reading this chapter, you should be able to:

- know the need of reactive power compensation
- discuss the objectives of load compensation
- discuss the operation of uncompensated and compensated transmission lines
- discuss the concept of sub-synchronous resonance (SSR)
- study the voltage-stability analysis

9.1 INTRODUCTION

In an ideal AC-power system, the voltage and the frequency at every supply point would remain constant, free from harmonics and the power factor (p.f.) would remain unity. For the optimum performance at a particular supply voltage, each load could be designed such that there is no interference between different loads as a result of variations in the current taken by each one.

Most electrical power systems in the world are interconnected to achieve reduced operating cost and improved reliability with lesser pollution. In a power system, the power generation and load must balance at all times. To some extent, it is self-regulating. If an unbalance between power generation and load occurs, then it results in a variation in the voltage and the frequency. If voltage is propped up with reactive power support, then the load increases with a consequent drop in frequency, which may result in system collapse. Alternatively, if there is an inadequate reactive power, the system's voltage may collapse.

Here, the quality of supply means maintaining constant-voltage magnitude and frequency under all loading conditions. It is also desirable to maintain the three-phase currents and voltages as balanced as possible so that underheating of various rotating machines due to unbalancing could be avoided.

In a three-phase system, the degree to which the phase currents and voltages are balanced must also be taken into consideration to maintain the quality of supply.

To achieve the above-mentioned requirements from the supply point of view as well as the loads, which can deteriorate the quality of supply, we need load compensation.

Load compensation is the control of reactive power to improve the quality of supply in an AC-power system by installing the compensating equipment near the load.

9.2 OBJECTIVES OF LOAD COMPENSATION

The objectives of load compensation are:

(i) p.f. Correction.

(ii) Voltage regulation improvement.

(iii) Balancing of load.

9.2.1 p.f. Correction

Generally, load compensation is a local problem. Most of the industrial loads absorb the reactive power since they have lagging p.f.'s. The load current tends to be larger than it is required to supply the real power alone. So, p.f. correction of load is achieved by generating reactive power as close as possible to the load, which requires it to generate it at a distance and transmit it to the load, as this results not only in a large conductor size but also in increased losses. It is desirable to operate the system near unity p.f. economically.

9.2.2 Voltage regulation Improvement

All loads vary their demand for reactive power, although they differ widely in their range and rate of variation. The voltage variation is due to the imbalance in the generation and consumption of reactive power in the system. If the generated reactive power is more than that being consumed, voltage levels go up and vice versa. However, if both are equal, the voltage profile becomes flat. The variation in demand for reactive power causes variation (or regulation) in the voltage at the supply point, which can interfere with an efficient operation of all plants connected to that point. So, different consumers connected to that point get affected. To avoid this, the supply utility places bounds to maintain supply voltages within defined limits. These limits may vary from typically $\pm\,6\%$ averaged over a period of a few minutes or hours.

To improve voltage regulation, we should strengthen the power system by increasing the size and number of generating units as well as by making the network more densely interconnected. This approach would be uneconomic and would introduce problems such as high fault levels, etc. In practice, it is much more economic to design the power system according to the maximum demand for active power and to manage the reactive power by means of compensators locally.

9.2.3 Load balancing

Most power systems are three-phased and are designed for balanced operation since their unbalanced operation gives rise to wrong phase-sequence components of currents (negative and zero-sequence components). Such components produce undesirable results such as additional losses in motors, generators, oscillating torque in AC machines, increased ripples in rectifiers, saturation of transformers, excessive natural current, and so on. These undesirable effects are caused mainly due to the harmonics produced under an unbalanced operation. To suppress these harmonics, certain types of equipment including compensators are provided, which yield the balanced operation of the power system.

9.3 IDEAL COMPENSATOR

An ideal compensator is a device that can be connected at or near a supply point and in parallel with the load. The main functions of an ideal compensator are instantaneous p.f. correction to unity, elimination or reduction of the voltage regulation, and phase balance of the load currents and voltages. In performing these interdependent functions, it will consume zero power.

The *characteristics* of an ideal compensator are to:

- provide a controllable and variable amount of reactive power without any delay according to the requirements of the load,

- maintain a constant-voltage characteristic at its terminals, and

- should operate independently in the three phases.

9.4 SPECIFICATIONS OF LOAD COMPENSATION

The specifications of load compensation are:

- Maximum and continuous reactive power requirement in terms of absorbing as well as generation.

- Overload rating and duration.

- Rated voltage and limits of voltage between which the reactive power rating must not be exceeded.

- Frequency and its variation.

- Accuracy of voltage regulation requirement.

- Special control requirement.

- Maximum harmonic distortion with compensation in series.

- Emergency procedure and precautions.

- Response time of the compensator for a specified disturbance.

- Reliability and redundancy of components.

9.5 THEORY OF LOAD COMPENSATION

In this section, relationships between the supply system, the load, and the compensator were to be developed. The supply system, the load, and the compensator can be modeled in different ways. Here, the supply system is modeled as a Thevenin's equivalent circuit with reactive power requirements. The compensator is modeled as a variable impedance/as a variable source (or sink) of reactive current/power. According to requirements, the selection of model used for each component can be varied.

The assumption made in developing the relationships between supply system, the load, and the compensator is that the load and system characteristics are static/constant (or) changing slowly so that phasor representation can be used.

9.5.1 p.f. Correction

Consider a single-phase load with admittance $Y_L = G_L + jB_L$ with a source voltage as shown in Fig. 9.1(a).
The load current I_L is given by

$$I_L = V_s (G_L + jB_L) = V_s G_L + jV_s B_L = I_a + jI_r$$

where I_a is the active component of the load current
I_r the reactive component of the load current.

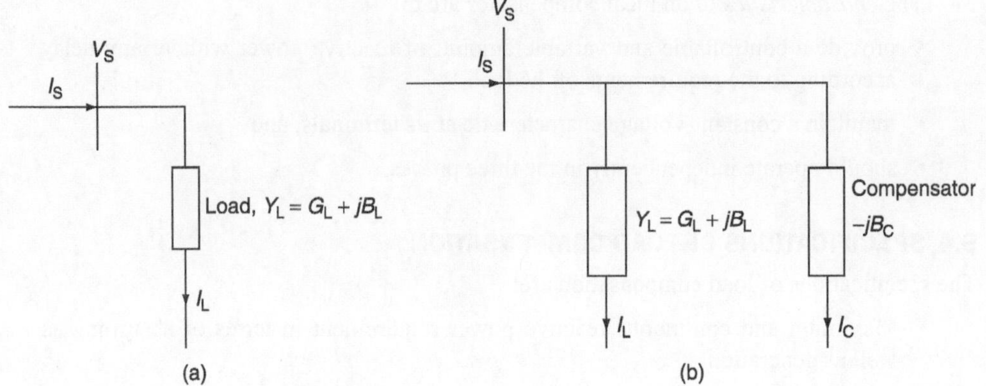

FIG. 9.1   *Representation of single-phase load; (a) without compensation; (b) with compensation*

Apparent power of the load, $S_L = V_s I_L^*$

$$= V_s^2 G_L - jV_s^2 B_L$$

$$= P_L + jQ_L$$

where P_L is the active power of the load
Q_L the reactive power of the load.

For inductive loads, B_L is negative and Q_L is positive by convention.

The current supplied to the load is larger than when it is necessary to supply the active power alone by the factor

$$\frac{I_L}{I_a} = \frac{1}{\cos \phi_L} \quad (\because I_a = I_L \cos \phi_L)$$

The objective of the p.f. correction is to compensate for the reactive power, i.e., locally providing a compensator having a purely reactive admittance jB_C in parallel with the load as shown in Fig. 9.1(b). The current supplied from the source with the compensator is

$$I_s = I_L + I_C$$
$$= V_s (G_L + jB_L) - V_s (jB_C)$$
$$= V_s G_L = I_a \quad (\because B_L = B_C)$$

which makes the p.f. to unity, since I_a is in phase with the source voltage V_s.

The current of the compensator, $I_c = V_s Y_c = -jV_s B_C$

The apparent power of the compensator, $S_C = V_s I_C^*$

$$S_c = jV_s^2 B_C = -jQ_C \quad (\because S_C = P_C - jQ_C, \text{ for pure compensation } P_C = 0)$$

We know that

$$Q_L = P_L \tan \phi_L$$

For a fully compensated system, i.e., $Q_L = Q_C$

$$\therefore Q_C = S_L \sin \phi_L$$

$$= S_L \sqrt{1 - \cos^2 \phi_L} \quad (\because \sin^2 Q_L + \cos^2 Q_L = 1)$$

The degree of compensation is decided by an economic trade-off between the capital cost of the compensator and the savings obtained by the reactive power compensation of the supply system over a period of time.

9.5.2 Voltage regulation

It is defined as the proportional change in supply voltage magnitude associated with a defined change in load current, i.e., from no-load to full load. It is caused by the voltage drop in the supply impedance carrying the load current.

When the supply system is balanced, it can be represented as single-phase model as shown in Fig. 9.2(a). The voltage regulation is given by $\dfrac{|V_s| - |V_L|}{|V_L|}$, where $|V_L|$ is the load voltage.

9.5.2.1 *Without compensator*

From the phasor diagram of an uncompensated system, shown in Fig. 9.2(b), the change in voltages is given by

$$\Delta V = V_s - V_L = Z_s I_L \tag{9.1}$$

where $Z_s = R_s + jX_s$ and the load current, $I_L = \dfrac{P_L - jQ_L}{V_L}.$ \tag{9.2}

Substituting Z_s and I_L in Equation (9.1), we get

$$\therefore \Delta V = (R_s + jX_s)\left(\frac{P_L - jQ_L}{V_L}\right)$$

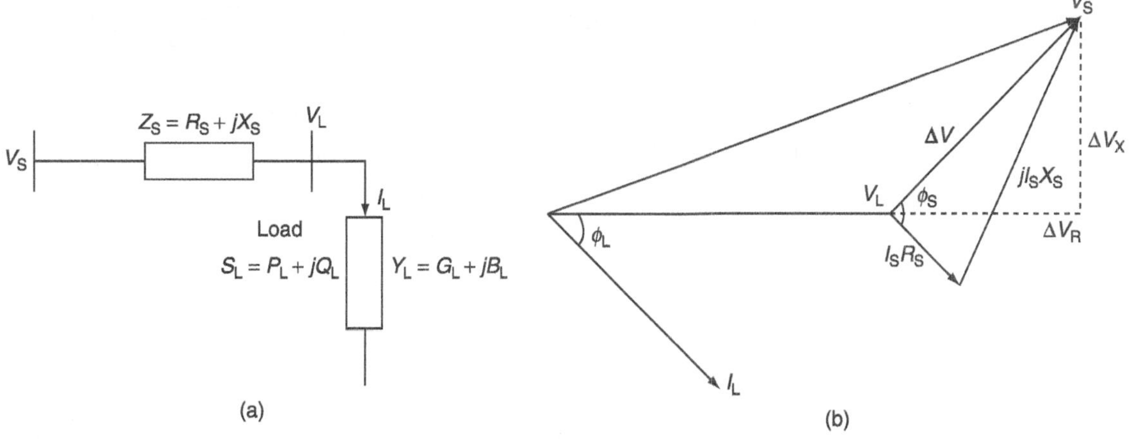

FIG. 9.2   *(a) Circuit model of an uncompensated load and supply system; (b) phasor diagram for an uncompensated system*

$$= \frac{R_s P_L + X_s Q_L}{V_L} + j \frac{X_s P_L - R_s Q_L}{V_L} \tag{9.3}$$

$$\Delta V = \Delta V_R + j \Delta V_X$$

From Equation (9.3), it is observed that the change in voltage depends on both real and reactive powers of the load considering the line parameters to be constant.

9.5.2.2 *With compensator*

In this case, a purely reactive compensator is connected across the load as shown in Fig. 9.3(a) to make the voltage regulation zero, i.e., the supply voltage ($|V_s|$) equals the load voltage ($|V_L|$). The corresponding phasor diagram is shown in Fig. 9.3(b).

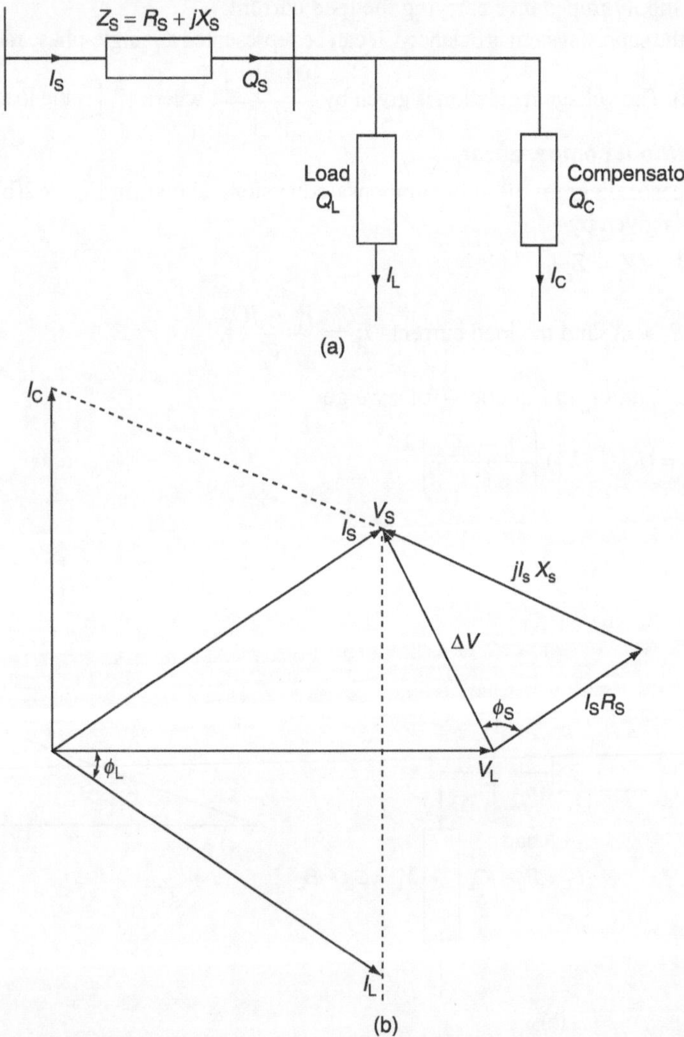

FIG. 9.3   *(a) Circuit model of a compensated load and supply system; (b) phasor diagram for a compensated system*

The supply reactive power with a compensator is

$$Q_s = Q_C + Q_L$$

Q_C is adjusted in such a way that $\Delta V = 0$

i.e., $|V_s| = |V_L|$

From Equations (9.1) and (9.3), we get

$$|V_s^2| = \left[|V_L| + \frac{R_s P_L + X_s Q_s}{|V_L|}\right]^2 + \left[\frac{X_s P_L - R_s Q_s}{V_L}\right]^2 \tag{9.4}$$

Simplifying and rearranging Equation (9.4),

$$|V_s^2| = |V_L^2| + \left[\frac{R_s P_L + X_s Q_s}{|V_L^2|}\right]^2 + 2(R_s P_L + X_s Q_s) + \frac{X_s^2 P_L^2 + R_s^2 Q_s^2 - 2X_s P_L R_s Q_s}{|V_L^2|}$$

$$= |V_L^2| + \frac{R_s^2 P_L^2 + X_s^2 Q_s^2 + 2R_s P_L Q_s X_s}{|V_L^2|} + 2(R_s P_L + X_s Q_s) + \frac{X_s^2 P_L^2 + R_s^2 Q_s^2 - 2X_s P_L R_s Q_s}{|V_L^2|}$$

$$|V_s^2 V_L^2| = |V_L^4| + R_s^2 P_L^2 + X_s^2 Q_s^2 + 2(R_s P_L + X_s Q_s)|V_L^2| + X_s^2 P_L^2 + R_s^2 Q_s^2$$

$$= Q_s^2 \left(R_s^2 + X_s^2\right) + Q_s \left(2|V_L^2| X_s\right) + |V_L^4| + P_L^2 \left(R_s^2 + X_s^2\right) + 2R_s P_L |V_L^2|$$

$$\therefore Q_s^2 \left(R_s^2 + X_s^2\right) + Q_s \left(2|V_L^2| X_s\right) + |V_L^4| + P_L^2 \left(R_s^2 + X_s^2\right) + 2R_s P_L |V_L^2| - |V_s^2 V_L^2| = 0$$

The above equation can be represented in a compact form as

$$a Q_s^2 + b Q_s + c = 0$$

where $\quad a = R_s^2 + X_s^2$

$$b = 2|V_L^2| X_s$$

$$c = (V_L^2 + R_s P_L)^2 + X_s^2 P_L^2 - |V_s^2 V_L^2|$$

$$Q_s = \frac{-b \pm \sqrt{b^2 - 4ac}}{2a}$$

The value of Q_C is found using the above equation by using the compensator reactive power balance equation $|V_s| = |V_L|$ and $Q_C = Q_s - Q_L$.

Here, the compensator can perform as an ideal voltage regulator, i.e., the magnitude of the voltage is being controlled, its phase varies continuously with the load current, whereas the compensator acting as a p.f. corrector reduces the reactive power supplied by the system to zero i.e., $Q_s = 0 = Q_L + Q_C$.

Equation (9.3) can be reduced to

$$\Delta V = \frac{R_s P_L + jX_s P_L}{V_L} = (R_s + jX_s)\frac{P_L}{V_L} \tag{9.5}$$

So, ΔV is independent of the load reactive power. From this, we conclude that a pure reactive compensator cannot maintain both constant voltage and unity p.f. simultaneously.

9.6 LOAD BALANCING AND P.F. IMPROVEMENT OF UNSYMMETRICAL THREE-PHASE LOADS

The third objective of load compensation is the balancing of unbalanced three-phase loads. We first model the load as a delta-connected admittance network for a general unbalanced three-phase load as shown in Fig. 9.4 in which the admittances Y_L^{ab}, Y_L^{bc}, and Y_L^{cd} are complex and unequal.

In this case, supply voltages are assumed to be balanced. Any ungrounded Y-connected load can be represented by a delta-connected load by means of the Y-Δ transformation.

A compensator can be any passive three-phase admittance network, which when combined in parallel with the load will present a real and balanced load with respect to the supply.

9.6.1 p.f. Correction

Each load admittance can be made purely real by connecting, in parallel, a compensating susceptance equal to the negative of the load susceptance in that branch of the delta-connected load as shown in Fig. 9.5(a).

If load admittance, $Y_L^{ab} = G_L^{ab} + jB_L^{ab}$, then the compensating susceptance $B_C^{ab} = -B_L^{ab}$ is connected across Y_L^{ab}:

$$B_C^{bc} = \frac{G_L^{ab}}{\sqrt{3}}$$

An inductive susceptance between phases 'c' and 'a' as shown in Fig. 9.5(a) is

$$B_C^{ca} = \frac{G_L^{ab}}{\sqrt{3}}$$

Now, the line currents will be balanced and are in phase with their respective phase voltages. The compensated single-phase load with a positive sequence equivalent circuit is shown in Fig. 9.5(a).

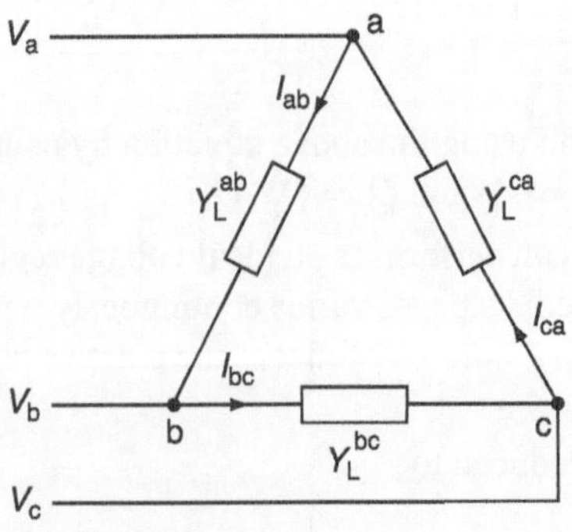

FIG. 9.4   *Unbalanced three-phase load*

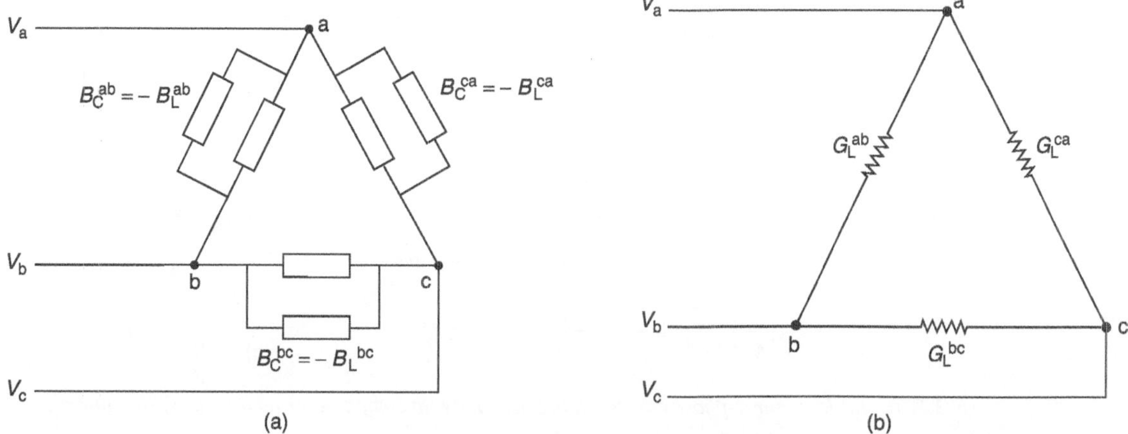

FIG. 9.5   *(a) Connection of p.f. correcting susceptance; (b) resultant unbalanced real load with unity p.f.*

Similarly, the compensating susceptance, $B_c^{bc} = -B_L^{bc}$ and $B_c^{cd} = -B_L^{ca}$ are connected across Y_L^{bc} and Y_L^{ca}, respectively, as shown in Fig. 9.5(a).

Figure 9.5(b) shows the resultant unbalanced real load with unity p.f.

9.6.2 Load balancing

Now, we make this real unbalanced load to a balanced one. To do this, let us consider a single-phase load (G_L^{ab}) (as shown in Fig. 9.6(a)) of the Δ-connected load as shown in Fig. 9.5(b). The three-phase positive sequence line currents can be balanced by connecting capacitive susceptance between phases 'b' and 'c' and together with the inductive susceptance between phases 'c' and 'a' as shown in Fig. 9.6(b).

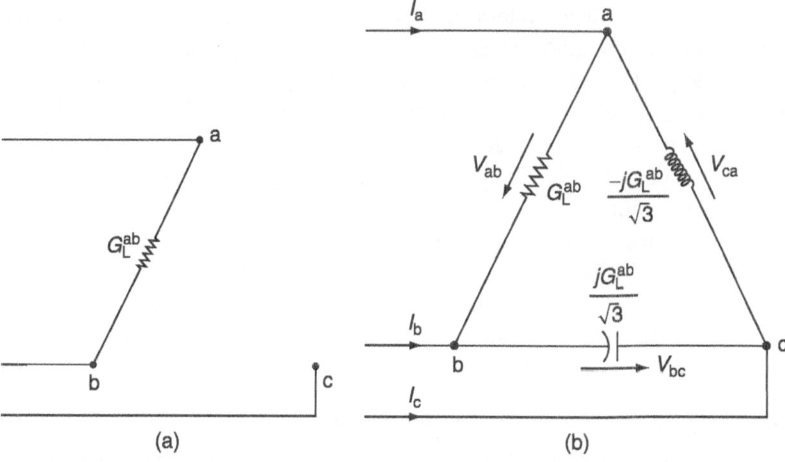

FIG. 9.6   *(a) Single-phase unity p.f. load before positive sequence balancing;*
(b) positive sequence balancing of a single-phase u.p.f. load

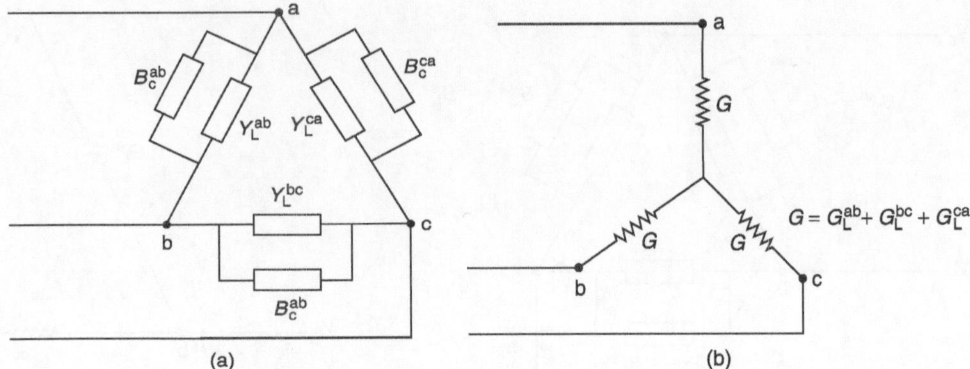

FIG. 9.7   *(a) Ideal three-phase compensating network with compensator admittances; (b) equivalent circuit of real and balanced compensated load admittance*

Similarly, the real admittances in the remaining phases 'bc' and 'ca' can be balanced.

The resultant compensator admittance (susceptance) represented by an equivalent circuit is shown in Fig. 9.7(a).

$$B_c^{ab} = -B_L^{ab} \text{ (p.f. correction)} + (G_L^{ca} - G_L^{bc})/\sqrt{3} \qquad \text{(load balancing)}$$

$$B_c^{bc} = -B_L^{bc} + (G_L^{ab} - G_L^{ca})/\sqrt{3}$$

$$B_c^{ca} = -B_L^{ca} + (G_L^{bc} - G_L^{ab})/\sqrt{3}$$

The resulting compensated load admittance is purely real and balanced, as shown in the equivalent circuit of Fig. 9.7(b).

9.7 UNCOMPENSATED TRANSMISSION LINES

An electric transmission line has four parameters, which affect its ability to fulfill its function as part of a power system and these are a series combination of resistance, inductance, shunt combination of capacitance, and conductance. These parameters are symbolized as R, L, C, and G, respectively. These parameters are distributed along the whole length at any line. Each small length at any section of the line will have its own values and concentration of all such parameters for the complete length of line into a single one is not possible. These are usually expressed as resistance, inductance, and capacitance per kilometer.

Shunt conductance that is mostly due to the breakage over the insulators is almost always neglected in a power transmission line. The leakage loss in a cable is uniformly distributed over the length of the cable, whereas it is different in the case of overhead lines. It is limited only to the insulators and is very small under normal operating conditions. So, it is neglected for an overhead transmission line.

9.7.1 Fundamental transmission line equation

Consider a very small element of length Δx at a distance of x from the receiving end of the line. Let z be the series impedance per unit length, y the shunt admittance per unit length, and l the length of the line.

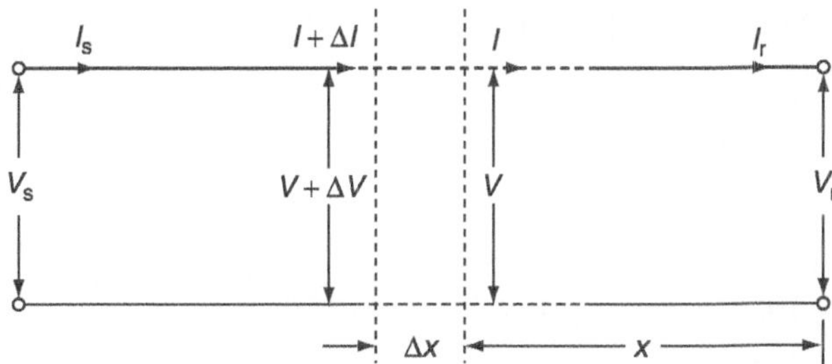

FIG. 9.8 *Representation of a transmission line on a single-phase basis*

Then,

$Z = zl =$ total series impedance of the line

$Y = yl =$ total shunt admittance of the line

The voltage and current at a distance x from the receiving end are V and I, and at distance $x + \Delta x$ are $V + \Delta V$ and $I + \Delta I$, respectively (Fig. 9.8).

So, the change of voltage, $\Delta V = Iz\Delta x$, where $z\Delta x$ is the impedance of the element considered:

$$\frac{\Delta V}{\Delta x} = Iz$$

$$\underset{\Delta x \to 0}{\text{Lt}} \frac{\Delta V}{\Delta x} = \frac{dV}{dx} = Iz \tag{9.6}$$

Similarly, the change of current, $\Delta I = Vy\Delta x$, where $y\Delta x$ is the admittance of the element considered:

$$\frac{\Delta I}{\Delta x} = Vy$$

$$\underset{\Delta x \to 0}{\text{Lt}} \frac{\Delta I}{\Delta x} = \frac{dI}{dx} = Vy \tag{9.7}$$

Differentiating Equation (9.6) with respect to x, we get

$$\frac{d^2V}{dx^2} = z\frac{dI}{dx} \tag{9.8}$$

Substituting the value of $\dfrac{dI}{dx}$ from Equation (9.7) in Equation (9.8), we get

$$\frac{d^2V}{dx^2} = zyV \tag{9.9}$$

Equation (9.9) is a second-order differentiatial equation and its solution is

$$V_{(x)} = Ae^{\sqrt{yz}\,x} + Be^{-\sqrt{yz}\,x} \tag{9.10}$$

Differentiating Equation (9.10) with respect to x, we get

$$\frac{dV_{(x)}}{dx} = A\sqrt{yz}\,e^{\sqrt{yz}\,x} - B\sqrt{yz}\,e^{-\sqrt{yz}\,x} \tag{9.11}$$

From Equations (9.6) and (9.11), we have

$$Iz = A\sqrt{yz}\, e^{\sqrt{yz}\,x} - B\sqrt{yz}\, e^{-\sqrt{yz}\,x}$$

$$\therefore I(x) = A\sqrt{\frac{y}{z}}\, e^{\sqrt{zy}\,x} - B\sqrt{\frac{y}{z}}\, e^{-\sqrt{zy}\,x} \tag{9.12}$$

From Equation (9.10), we have

$$V(x) = Ae^{\gamma x} + Be^{-\gamma x} \tag{9.13}$$

From Equation (9.12), we have

$$I_{(x)} = \frac{A}{Z_C}e^{-\gamma x} + \frac{B}{Z_C}e^{-\gamma x} \tag{9.14}$$

where $Z_c = \sqrt{\dfrac{z}{y}}$ is known as characteristic impedance or surge impedance and $\gamma = \sqrt{yz}$

is known as the propagation constant.

The constants A and B can be evaluated by using the conditions at the receiving end of the line.

The conditions are

at $x = 0$, $V = V_r$ and $I = I_r$

Substituting the above conditions in Equations (9.11) and (9.12), we get

$$\therefore V_r = A + B \tag{9.15}$$

and $I_r = \dfrac{1}{Z_C}(A - B)$ (9.16)

Solving Equations (9.15) and (9.16), we get

$$A = \frac{V_r + I_r Z_C}{2} \quad \text{and} \quad B = \frac{V_r - I_r Z_C}{2}$$

Now, substituting the values of A and B in Equations (9.13) and (9.14), then we get

$$V_{(x)} = \frac{V_r + I_r Z_C}{2}e^{\gamma x} + \frac{V_r - I_r Z_C}{2}e^{-\gamma x}$$

$$= V_r\left[\frac{e^{\gamma x} + e^{-\gamma x}}{2}\right] + I_r Z_C\left[\frac{e^{\gamma x} - e^{-\gamma x}}{2}\right]$$

$$\therefore V_{(x)} = V_r \cos h\,\gamma x + I_r Z_C \sin h\,\gamma x \tag{9.17}$$

$$I_{(x)} = \frac{1}{Z_C}\left[\frac{V_r + I_r Z_C}{2}e^{\gamma x} - \frac{V_r - I_r Z_C}{2}e^{-\gamma x}\right]$$

$$= \frac{1}{Z_C}V_r\left(\frac{e^{\gamma x} - e^{-\gamma x}}{2}\right) + I_r\left(\frac{e^{\gamma x} + e^{-\gamma x}}{2}\right)$$

$$\therefore I_{(x)} = \frac{V_r}{Z_C} \sin h \, \gamma x + I_r \cos h \, \gamma x \tag{9.18}$$

where $V_{(x)}$ and $I_{(x)}$ are the voltages and currents at any distance x from the receiving end.

For a lossless line $\gamma = j\beta$ and the hyperbolic functions, i.e., $\cos h \, \gamma x = \cos h \, j\beta x = \cos \beta x$ and $\sin h \, \gamma x = \sin h \, j\beta x = j \sin \beta x$.

Therefore, Equations (9.17) and (9.18) can be modified as

$$V_{(x)} = V_r \cos \beta x + jI_r Z_C \sin \beta x \tag{9.19}$$

and

$$I_{(x)} = j\frac{V_r}{Z_C} \sin \beta x + I_r \cos \beta x \tag{9.20}$$

where β is the electrical length of the line (radians or wavelength)

$$\beta = \omega\sqrt{LC} = \frac{2\pi f}{v} = \frac{2\pi}{\lambda}$$

v = velocity of light = 3×10^8 m/s
λ = wavelength of light

9.7.2 Characteristic impedance

The quantity $\sqrt{\dfrac{z}{y}}$ is a complex number as y and z are in complex.

It is denoted by Z_C or Z_0. It has the dimension of impedance, since

$$\sqrt{\frac{z}{y}} = \sqrt{\frac{\text{ohms/unit length}}{\text{ohms/unit length}}} = \text{ohms}$$

$$Z_c = \sqrt{\frac{z}{y}} = \sqrt{\frac{r + j\omega L}{g + j\omega C}}$$

This quantity depends upon the characteristic of the line per unit length. It is, therefore, called characteristic impedance of the line. It also depends upon the length of the line, radius, and spacing between the conductors. For a lossless line, $r = g = 0$, the characteristic impedance becomes

$$Z_C = \sqrt{\frac{L}{C}}$$

The characteristic impedance is also called the surge or natural impedance of the line.

The approximate value of the surge impedance for overhead lines is 400 Ω and that for underground cables is 40 Ω, and the transformers have several thousand ohms as their surge impedance. Surge impedance is the impedance offered to the propagation of a voltage or current wave during its travel along the line.

9.7.3 Surge impedance or natural loading

The surge impedance loading (SIL) of a transmission line is the MW loading of a transmission line at which a natural reactive power balance occurs (zero resistance).

Transmission lines produce reactive power (*MVAr*) due to their natural capacitance. The amount of *MVAr* produced is dependent on the transmission line's capacitive reactance (X_C) and the voltage (*kV*) at which the line is energized.

Now the *MVAr* produced is

$$MVAr = \frac{(kV)^2}{X_c} \tag{9.21}$$

Transmission lines also utilize reactive power to support their magnetic fields. The magnetic field strength is dependent on the magnitude of the current flow in the line and the natural inductive reactance (X_L) of the line. The amount of *MVAr* used by a transmission line is a function of the current flow and inductive reactance.

The *MVAr* used by a transmission line $= I^2 X_L$ $\tag{9.22}$

Transmission line SIL is simply the MW loading (at a unity p.f.) at which the line *MVAr* usage is equal to the line *MVAr* production. From the above statement, the SIL occurs when

$$I^2 X_L = \frac{(kV)^2}{X_C}$$

$$X_L X_C = \frac{(kV)^2}{I^2} \tag{9.23}$$

And the Equation (9.23) can be rewritten as

$$\sqrt{\frac{V^2}{I^2}} = \frac{2\pi fL}{2\pi fC}$$

or $\quad \dfrac{V}{I} = \sqrt{\dfrac{L}{C}} = \text{surge impedance}$ $\tag{9.24}$

The term $\sqrt{\dfrac{L}{C}}$ is the 'surge impedance'.

The theoretical significance of the surge impedance is that if a purely resistive load that is equal to the surge impedance were connected to the end of a transmission line with no resistance, a voltage surge introduced to the sending end of the line would be absorbed completely at the receiving end. The voltage at the receiving end would have the same magnitude as the sending-end voltage and would have a voltage phase angle that is lagging with respect to the sending end by an amount equal to the time required to travel across the line from the sending to the receiving end.

The concept of surge impedance is more readily applied to telecommunication systems rather than to power systems. However, we can extend the concept to the power transferred across a transmission line. The surge impedance loading (power transmitted at this condition) or SIL (in MW) is equal to the voltage squared (in kV) divided by the surge impedance (in ohms):

$$\therefore \text{SIL (MW)}, \quad P_0 = \frac{V_{L-L}^2}{\text{surge impedance}}$$

Note: In this formula, the SIL is dependent only on the voltage (kV) of the line is energized and the line surge impedance. The line length is not a factor in the SIL or surge impedance calculations. Therefore, the SIL is not a measure of a transmission line power transfer capability as it neither takes into account the line length nor considers the strength of the local power system.

The value of the SIL to a system operator is realized when a line is loaded above its SIL, it acts like a shunt reactor absorbing $MVAr$ from the system and when a line is loaded below its SIL, it acts like a shunt capacitor supplying $MVAr$ to the system.

9.8 UNCOMPENSATED LINE WITH OPEN CIRCUIT

In this section, we shall discuss the cases: (a) voltage and current profiles, (b) symmetrical line at no-load, and (c) underexcited operation of generators.

9.8.1 Voltage and current profiles

A lossless line is energized at the sending end and is open-circuited at the receiving end.

From Equations (9.19) and (9.20) with $I_r = 0$

$$\therefore V(x) = V_r \cos \beta x \tag{9.25}$$

$$I(x) = j \left[\frac{V_r}{Z_c} \right] \sin \beta x \tag{9.26}$$

Voltage and current at the sending end are given by equations with $x = l$ as

$$V_{(x)} = V_s, \quad I_{(x)} = I_{(s)}$$

Equations (9.25) and (9.26) are modified as

$$V_s = V_r \cos \theta \tag{9.27}$$

where $\theta = \beta l$

$$\text{and } I_s = j \left[\frac{V_r}{Z_c} \right] \sin \theta = j \left[\frac{V_s}{Z_c} \right] \tan \theta \quad \left(\because V_r = \frac{V_s}{\cos \theta} \right) \tag{9.28}$$

From Equations (9.25) and (9.26), the voltage profile equation is

$$V(x) = V_s \frac{\cos \beta x}{\cos \theta} \quad \left(\because V_r = \frac{V_s}{\cos \theta} \right) \tag{9.29}$$

And the current profile equation, $\quad I(x) = j \dfrac{V_s}{Z_c} \dfrac{\sin \beta x}{\cos \theta} \tag{9.30}$

9.8.2 The symmetrical line at no-load

This is similar to an open-circuited line energized at one end. This is a line identical at both ends, but with no power transfer. Suppose the terminal voltages are maintained as same values,

i.e., $\quad V_s = V_r$

From Equations (9.19) and (9.20) with $x = l$, we have

$$V_s = V_r \cos \theta + j Z_c I_r \sin \theta \tag{9.31}$$

$$I_s = j \left[\frac{V_r}{Z_c} \right] \sin \theta + I_r \cos \theta \tag{9.32}$$

The electrical conditions are the same $(V_\mathrm{s} = V_\mathrm{r})$; there would not be any power transfer. Therefore, by symmetry $I_\mathrm{s} = I_\mathrm{r}$.

Substituting the above condition in Equations (9.32), we get

$$\therefore -I_\mathrm{r} = j\left|\frac{V_\mathrm{r}}{Z_\mathrm{c}}\right|\frac{\sin\theta}{1+\cos\theta} \tag{9.33}$$

$$= j\frac{V_\mathrm{r}}{Z_\mathrm{c}}\tan\frac{\theta}{2} \tag{9.34}$$

Substituting Equation (9.33) in Equation (9.31), we get

$$V_\mathrm{s} = V_\mathrm{r}\cos\theta + jZ_\mathrm{c}\left\{-j\left|\frac{V_\mathrm{r}}{Z_\mathrm{c}}\right|\frac{\sin\theta}{1+\cos\theta}\right\}\sin\theta$$

$$= \frac{V_\mathrm{r}(\cos\theta + \cos^2\theta + \sin^2\theta)}{1+\cos\theta}$$

$$\therefore V_\mathrm{s} = V_\mathrm{r} \tag{9.35}$$

From Equation. (9.34), we have

$$I_\mathrm{s} = j\frac{V_\mathrm{s}}{Z_\mathrm{c}}\tan\frac{\theta}{2} \quad (\therefore V_\mathrm{s} = V_\mathrm{r};\, I_\mathrm{s} = -I_\mathrm{r}) \tag{9.36}$$

A comparison of Equations (9.35) and (9.36) with Equation (9.28) shows that the line is equivalent to two equal halves connected back-to-back. Half the line-charging current is supplied from each end. By symmetry, the midpoint current is zero, whereas the midpoint voltage is equal to the open-circuit voltage of the line having half the total length.

From Equation (9.31) the midpoint voltage is

$$V_\mathrm{m} = \frac{V_\mathrm{s}}{\cos\left(\dfrac{\theta}{2}\right)} \quad (\because V_\mathrm{r} = V_\mathrm{m};\quad I_\mathrm{r} = 0)$$

9.8.3 Underexcited operation of generators due to line-charging

No load at the receiving ends, i.e., $I_\mathrm{r} = 0$. The charging reactive power at the sending end is

$$Q_\mathrm{s} = \mathrm{Im}(V_\mathrm{s}I_\mathrm{s}^*)$$

From Equation (9.28), we have

Line-charging current at the sending end, $\quad I_\mathrm{s} = j\left[\dfrac{V_\mathrm{s}}{Z_\mathrm{c}}\right]\tan\theta$

$\therefore$ Line-charging power at the sending end, $\quad Q_\mathrm{s} = -P_0\tan\theta$

For a 300-km line, Q_s is nearly 43% of the natural load expressed in MVA. At 400 kV, the generators would have to absorb 172 *MVAr*.

The reactive power absorption capability of a synchronous machine is limited for two reasons:

- The heating of the ends of the stator core increases during underexcited operation.
- The reduced field current results in reduced internal e.m.f of the machine and this weakens the stability.

Using a compensator, this problem can be reduced by two ways:

- If the line is made up of two (or) more parallel circuits, one (or) more of the circuits can be switched off under light load (or) open-circuit conditions.

- If the generator absorption is limited by stability and not by core-end heating, the absorption limit can be increased by using a rapid response excitation system, which restores the stability margins when the steady-state field current is low. The underexcited operation of generators can set a more stringent limit to the maximum length of an uncompensated line than the open-circuit voltage rise.

9.9 THE UNCOMPENSATED LINE UNDER LOAD

In this section, the effects of line length, load power, and p.f. on voltage as well as reactive power are discussed.

9.9.1 Radial line with fixed sending-end voltage

A load $(P_r + j\,Q_r)$ at the receiving end of a radial line draws the current.

$$\text{i.e.,} \quad I_r = \frac{P_r - jQ_r}{V_r}$$

From Equation (9.19), with $x = l$, for a lossless line, the sending-and receiving-end voltages are related as

$$V_s = V_r \cos\theta + jZ_c \sin\theta \left(\frac{P_r - jQ_r}{V_r} \right) \tag{9.37}$$

If V_s is fixed, this quadratic equation can be solved for V_r. The solution gives how V_r varies with the load and its p.f. as well as with the line length.

Several fundamental important properties of AC transmission are evident from Fig. 9.9

- For each load p.f., there is a maximum transmissible power.

- The load p.f. has a strong influence on the receiving-end voltage.

- Uncompensated lines between about 150-km and 300-km long can be operated at normal voltage provided that the load p.f. is high. Longer lines, with large voltage variations, are impractical at all p.f.'s unless some means of voltage compensation is provided.

The midpoint voltage variation on a symmetrical 300-km line is the same as the receiving-end voltage variations on a 150-km line with a unity p.f. load.

9.9.2 Reactive power requirements

From the line voltage and the level of power transmission, the reactive power requirements can be determined. It is very important to know the reactive power requirements because they determine the reactive power ratings of the synchronous machines as well as those of any compensating equipment. If any inductive load is connected at the sending end of the line, it will support the synchronous generators to absorb the line-charging reactive power. With the absence of the compensating equipment, the

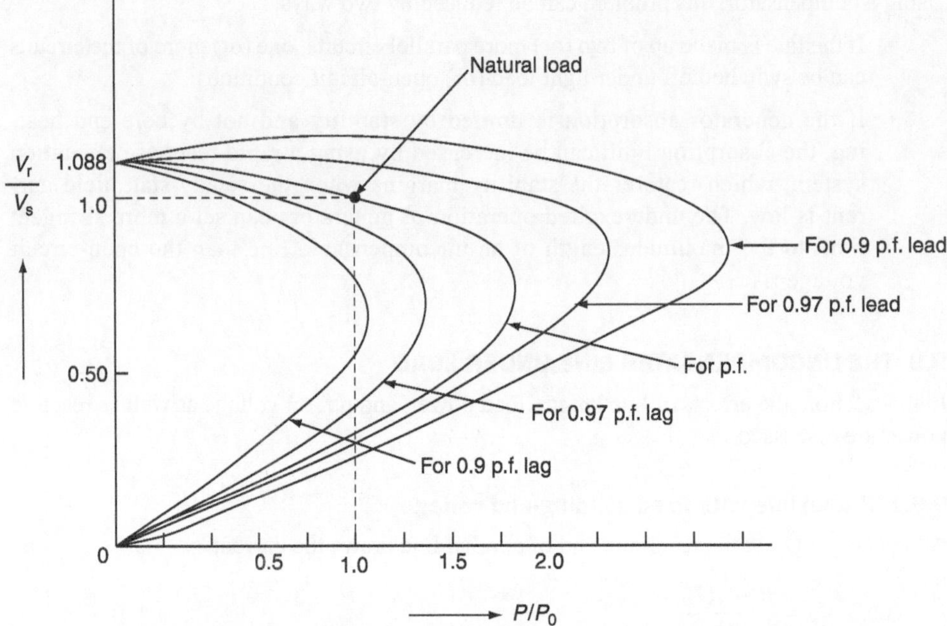

FIG. 9.9   *Magnitude of receiving-end voltage as a function of load and load p.f.*

synchronous machines must absorb or generate the difference between the line and the local load reactive powers.

The equations of voltage and current for the sending-end half of the symmetrical line is

$$V_s = V_m \cos\frac{\theta}{2} + jZ_0 I_m \sin\frac{\theta}{2} \quad (\because V_r = V_m;\ I_r = I_m) \tag{9.38}$$

$$I_s = j\frac{V_m}{Z_0}\sin\frac{\theta}{2} + I_m \cos\frac{\theta}{2} \tag{9.39}$$

The power at midpoint is

$$P_m + j\,Q_m = V_m I_m^* = P = \text{transmitted power}$$

Since $Q_m = 0$, because no reactive power flows at the midpoint.
The power at the sending end is

$$P_s + j\,Q_s = V_s I_s^*$$

Substituting V_s and I_s from Equations (9.38) and (9.39) in the above equation, we get

$$P_s + j\,Q_s = P + j\frac{\sin\theta}{2}\left[Z_0 I_m^2 + \frac{V_m^2}{Z_0}\right]$$

If the line is assumed to be lossless, then $P_s = P_r = P$:

$$\therefore Q_s = P_0\frac{\sin\theta}{2}\left[\left(\frac{P}{P_0}\right)^2\left(\frac{V_0}{V_m}\right)^2 - \left(\frac{V_m}{V_0}\right)^2\right]$$

The above expression gives the relation between the midpoint voltage and the reactive power requirements of the symmetrical line.

If the terminal voltages are continuously adjusted so that the midpoint voltage, $V_{\mathrm{m}} = V_0 = 1.0$ p.u. at all levels of power transmission

$$\therefore Q_{\mathrm{s}} = P_0 \frac{\sin \theta}{2} \left(\left(\frac{P}{P_0} \right)^2 - 1 \right)$$

9.9.3 The uncompensated line under load with consideration of maximum power and stability

Consider Equation (9.37) as

$$V_{\mathrm{s}} = V_{\mathrm{r}} \cos \theta + j Z_{\mathrm{c}} \left(\frac{P_{\mathrm{r}} - j Q_{\mathrm{r}}}{V_{\mathrm{r}}} \right) \sin \theta \qquad (9.40)$$

If V_{r} is taken as reference phasor, then:

$$V_{\mathrm{s}} = V_{\mathrm{s}} \, e^{j\delta} = V_{\mathrm{s}} (\cos \delta + j \sin \delta) \qquad (9.41)$$

where δ is the phase angle between V_{s} and V_{r} and is called the load angle (or) the transmission angle.

Equating real and imaginary parts of Equations (9.40) and (9.41), we get

$$V_{\mathrm{s}} \cos \delta = V_{\mathrm{r}} \cos \theta + Z_{\mathrm{c}} \frac{Q_{\mathrm{r}}}{V_{\mathrm{r}}} \sin \theta$$

$$V_{\mathrm{s}} \sin \delta = Z_{\mathrm{c}} \frac{P_{\mathrm{r}}}{V_{\mathrm{r}}} \sin \delta$$

$$\therefore P_{\mathrm{r}} = \frac{V_{\mathrm{s}} V_{\mathrm{r}}}{Z_{\mathrm{c}} \sin \theta} \sin \delta \quad \text{(since neglecting the losses)}$$

For an electrically short line, $\sin \theta = \theta = \beta l$:

$$\beta l = \omega l \sqrt{LC}$$

Then, $Z_{\mathrm{c}} \theta = \sqrt{\dfrac{L}{C}} \left(\omega l \sqrt{LC} \right) = \omega L = X_L$, the series reactance of the line:

$$\therefore P_{\mathrm{r}} \approx \frac{V_{\mathrm{s}} V_{\mathrm{r}}}{X_{\mathrm{L}}} \sin \delta$$

$$= P_{\max} \sin \delta$$

where $P_{\max} = \dfrac{V_{\mathrm{s}} V_{\mathrm{r}}}{X_{\mathrm{L}}}$

9.10 COMPENSATED TRANSMISSION LINES

The change in the electrical characteristics of a transmission line in order to increase its power transmission capability is known as line compensation. While satisfying the requirements for a transmission system (i.e., synchronism, voltages must be kept near their rated values, etc.), a compensation system ideally performs the following functions:

- It provides the flat voltage profile at all levels of power transmission.

- It improves the stability by increasing the maximum transmission capacity.

- It meets the reactive power requirements of the transmission system economically.

The following types of compensations are generally used for transmission lines:

    (i)   Virtual-Z_0.

    (ii)   Virtual-θ.

    (iii)   Compensation by sectioning.

The effectiveness of a compensated system is gauged by the product of the line length and maximum transmission power capacity. Compensated lines enable the transmission of the natural load over larger distances, and shorter compensated lines can carry loads more than the natural load.

The flat voltage profile can be achieved if the effective surge impedance of the line is modified as to a virtual value Z_0', for which the corresponding virtual natural load $\left(\dfrac{V^2(kV)}{Z_0'}\right)$ is equal to the actual load. The surge impedance of the uncompensated line is $Z_0 = \sqrt{\dfrac{L}{C}}$, which can be written as $\sqrt{X_L X_c}$, if the series and/or the shunt reactance X_L and/or X_c are modified, respectively. Then, the line can be made to have virtual surge impedance Z_0' and a virtual natural load P' for which

$$P' = \frac{V^2(kV)}{Z_0'}$$

Compensation of line, by which the uncompensated surge impedance Z_0 is modified to virtual surge impedance Z_0', is called *virtual surge impedance compensation* or *virtual Z_0 compensation.*

Once a line is computed for Z_0, the only way to improve stability is to reduce the effective value of θ. Two alternative compensation strategies have been developed to achieve this.

- Apply series capacitors to reduce X_L and thereby reduce θ, since $\theta = \beta l = \omega\sqrt{LC}\,l = \sqrt{\dfrac{X_L}{X_c}}$ at fundamental frequency. This method is called line-length compensation (or) θ compensation.

- Divide the line into shorter sections that are more (or) less independent of one another. This method is called compensation by sectioning. It is achieved by connecting constant voltage compensations at intervals along the line.

9.11 SUB-SYNCHRONOUS RESONANCE

In this section, various effects due to sub-synchronous resonance are discussed in detail.

9.11.1 Effects of series and shunt compensation of lines

The objective of series compensation is to cancel part of the series inductive reactance of the line using series capacitors, which results in the following factors.

(i) Increase in maximum transferable power capacity.

(ii) Decrease in transmission angle for considerable amount of power transfer.

(iii) Increase in virtual surge impedance loading.

From a practical point of view, it is desirable not to exceed series compensation beyond 80%. If the line is compensated at 100%, the line behaves as a purely resistive element and would result in series resonance even at fundamental frequency since the capacitive reactance equals the inductive reactance, and it would be difficult to control voltages and currents during disturbances. Even a small disturbance in the rotor angles of the terminal synchronous machine would result in flow of large currents.

The location of series capacitors is decided partly by economical factors and partly by the selectivity of fault currents as they would depend upon the location of the series capacitor. The voltage rating of the capacitor will depend upon the maximum fault current that likely flows through the capacitor.

The net inductive reactance of the line becomes

$$X_{l\,net} = X_l - X_{sc}$$

The connection of the transmission line and the series capacitor behaves like a series resonance circuit with inductive reactance of line in series with the capacitance of the series capacitor.

The effects of series and shunt compensation of overhead transmission lines are as follows:

- For a fixed degree of series compensation, the capacitive shunt compensation decreases the virtual surge impedance loading of the line. However, the inductive shunt compensation increases the virtual surge impedance and decreases the virtual surge impedance loading of the line.

- If the inductive shunt compensation is 100%, then the virtual surge impedance becomes infinite and the loading is zero, which implies that a flat voltage profile exists at zero loads and the Ferranti effect can be eliminated by the use of shunt reactors.

- Under a heavy-load condition, the flat voltage profile can be obtained by using shunt capacitors.

- A flat voltage profile can be obtained by series compensation for heavy loading condition.

- Voltage control using series capacitors is not recommended due to the lumped nature of series capacitors, but normally they are preferred for improving the stability of the system.

- Series compensation has no effect on the load-reactive power requirements of the generator and, therefore, the series-compensated line generates as much line-charging reactive power at no load as completely uncompensated line of the same length.

- If the length of the line is large and needs series compensation from the stability point of view, the generator at the two ends will have to absorb an excessive reactive power and, therefore, it is important that the shunt compensation (inductive) must be associated with series compensation.

9.11.2 Concept of SSR in lines

Consider a transmission line compensated by a series capacitors connection.

Let L_L, L_g, and L_t be the inductance of the line, generator, and transformer, respectively. Let C_{sc} be the capacitance of the series capacitor, X_L the total inductive reactance $(X_L + X_g + X_t)$, and X_{sc} the reactance of the series capacitor.

The natural frequency of oscillation of the above-said series resonance circuit is given by the relation

$$f_0 = \frac{1}{2\pi\sqrt{LC}} = \frac{1}{2\sqrt{(L_L + L_g + L_t)C_{sc}}}$$

The inductive reactance of the system is $X_l = 2\pi f L$

Capacitive reactance of the capacitor is $X_c = \dfrac{1}{2\pi f C_{sc}}$

Therefore, the natural frequency of oscillation in terms of X_l and X_c is expressed as

$$f_0 = \frac{1}{2\pi\sqrt{\dfrac{X_l}{2\pi f} \times \dfrac{1}{2\pi X_c}}}$$

$$= f\sqrt{\frac{X_l}{X_c}}$$

where f is the rated frequency.

The term $\dfrac{X_L}{X_c}$ represents the degree of series compensation and it varies between 25% and 65%; therefore, the natural frequency of oscillation becomes less when compared to the natural frequency $(f_0 < f)$, i.e., the series resonance will occur at sub-synchronous frequency.

There are three types of sub-synchronous oscillations that have been identified due to SSR conditions.

9.11.2.1 *SSR due to induction generator effect*

The transient currents at the sub-harmonic frequency resulted in a series-compensated network due to a switching operation or a fault. These sub-harmonic frequency currents assume dangerously high values and even become unstable. The unstable operation is exhibited in the form of a negative resistance in the equivalent circuit of synchronous and induction motors.

Considering a round-rotor synchronous machine, the sub-harmonic frequency operation can be studied with the help of its equivalent circuit as shown in Fig. 9.10.

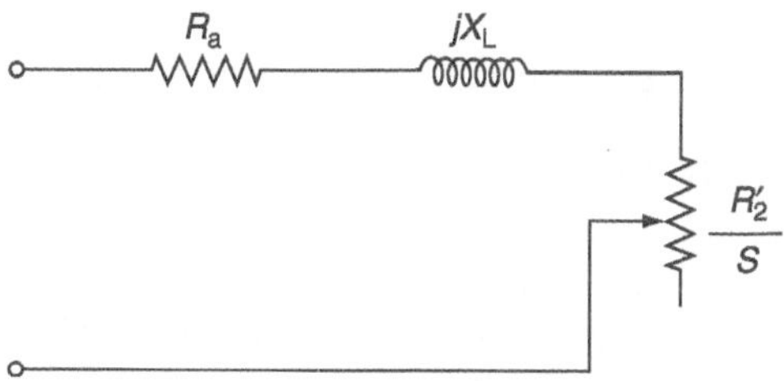

FIG. 9.10   *Equivalent circuit of a synchronous machine for sub-harmonic operation*

The sub-harmonic currents are excited in the stator winding of the synchronous machine due to some disturbances and these sub-harmonic currents would generally be unbalanced.

The positive sequence component of these unbalanced sub-harmonic currents will produce a magnetic field, which rotates in the same direction of rotation of the rotor but with a speed $N < N_s$. The machine behaves as an induction generator as far as sub-harmonic currents are considered. Due to this speed deviation between the rotor and the magnetic field, the slip $S = \dfrac{f_0 - f}{f_0}$ will be present. Since $f_0 < f$, the slip S becomes negative.

Therefore, the equivalent resistance of the damper winding and the solid rotor resistance when referred to the stator side, i.e., $\dfrac{R_2'}{S}$ becomes negative and therefore provides negative damping.

If the series compensation is very high, the slip S would turn out to be very small and hence the equivalent resistance becomes very high and may become large enough to have total resistance of the system, which is negative. Therefore, it provides negative damping of the sub-harmonic currents, and voltage may build up to dangerously high values. Several measures are to be taken to prevent SSR in the system.

9.11.2.2 SSR due to torsional interaction between electrical and mechanical systems

The sub-harmonic currents produce field rotations in the direction with respect to the rotor and main field and which in turn produces an alternating torque on the rotor at the frequency $(f - f_0)$. If this frequency difference coincides with one of the natural tensional resonances of the machine shaft system, tensional oscillations may be excited and this operation is entirely known as SSR. It means that whenever the natural frequency of the mechanical oscillation of the rotor equals $(f - f_0)$, mechanical resonance would take place. Hence, SSR is treated as a combined electrical–mechanical resonance phenomenon.

The currents of high frequency may produce torque in some of the shafts, which may have the same natural frequency as the torque frequency (called swing frequency, which is the lowest frequency of natural oscillation of an equivalent system of a turbine cylinder and a generator) such that these shafts may break down due to the twisting action. Hence, resonant frequencies may be extended upto several hundreds of Hz. The large multiple-stage steam turbines that have more than one tensional modes in the frequency range of 0.5 Hz are more susceptible to SSR.

In the SSR phenomenon, if the resonance frequency coincides with the swing frequency, the whole turbine-generator assembly may come out from its foundation, and/or if the frequency of the torque developed coincides with the natural frequency of oscillation of some shafts and if oscillations build up sufficiently, it results in the breaking of the shaft.

9.11.2.3 *SSR due to large disturbances*

Due to the large disturbances like any switching operation or any fault condition, the condition of SSR occurs in the system even if oscillations are damped out. This SSR condition results in a 'low cycle fatigue' condition in a mechanical system or slow deterioration of the mechanical system due to the reduction in life of the shafts.

The corrective measures for SSR are:

(i)   By passing a series of capacitors.

(ii) Use of very sensitive relays to detect even small levels of sub-harmonic currents.

(iii) Modulation of generator field current to provide an increased positive damping sub-harmonic frequency.

9.12 SHUNT COMPENSATOR

A shunt-connected static VAr compensator, composed of thyristor-switched capacitors (TSCs) and thyristor-controlled reactors (TCRs), is shown Fig. 9.11. With proper co-ordination of the capacitor switching and reactor control, the VAr output can be varied continuously between the capacitive and inductive rating of the equipment. The compensator

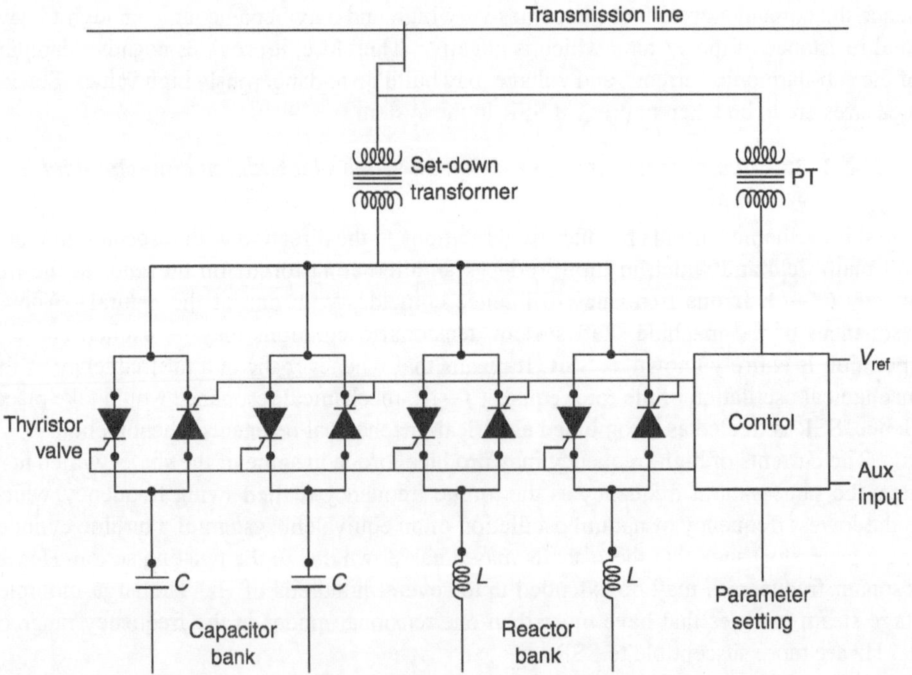

FIG. 9.11   *Static VAr compensator employing TSCs and TCR*

is normally operated to regulate the voltage of the transmission system at a selected terminal, often with an appropriate modulation option to provide damping if power oscillation is detected.

9.12.1 Thyristor-controlled reactor

A shunt-connected thyristor-controlled inductor has an effective reactance, which is varied in a continuous manner by partial-conduction control of the thyristor valve.

With the increase in the size and complexity of a power system, fast reactive power compensation has become necessary in order to maintain the stability of the system. The thyristor-controlled shunt reactors have made it possible to reduce the response time to a few milliseconds. Thus, the reactive power compensator utilizing the thyristor-controlled shunt reactors become popular. An elementary single-phase TCR is shown Fig.9.12.

It consists of a fixed reactor of inductance L and a bidirectional thyristor valve. The thyristor valve can be brought into conduction by the application of a gate pulse to the thyristor, and the valve will be automatically blocked immediately after the AC current crosses zero. The current in the reactor can be controlled from maximum to zero by the method of firing angle control. Partial conduction is obtained with a higher value of firing angle delay. The effect of increasing the gating angle is to reduce the fundamental component of current. This is equivalent to an increase in the inductance of the reactor, reducing its current. As far as the fundamental component of current is concerned, the TCR is a controllable susceptance, and can, therefore, be used as a static compensator.

The current in this circuit is essentially reactive, lagging the voltage by 90° and this is continuously controlled by the phase control of the thyristors. The conduction angle control results in a non-sinusoidal current wave form in the reactor. In other words, the TCR generates harmonics. For identical positive and negative current half-cycle time, only odd harmonics are generated as shown in Fig. 9.13. By using filters, we can reduce the magnitude of harmonics.

TCR's characteristics are:

- Continuous control.
- No transients.
- Generation of harmonics.

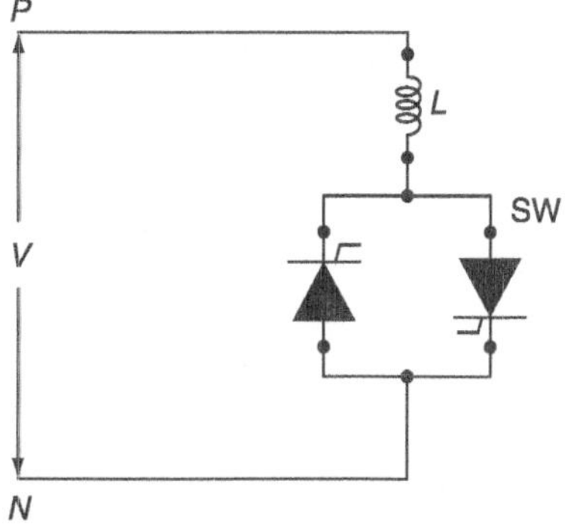

FIG. 9.12   *TCR*

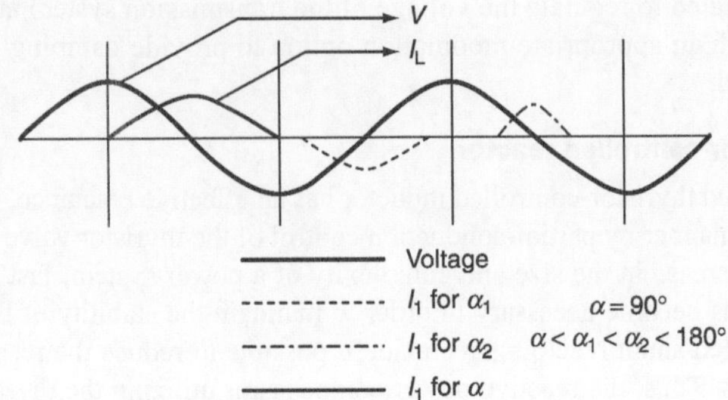

FIG. 9.13 *TCR waveform*

9.12.2 Thyristor-switched capacitor

A shunt-connected TSC shows that an effective reactance is varied in a step-wise manner by full- or zero-conduction operation of the thyristor valve.

The TSC is also a sub-set of SVC in which thyristor-based AC switches are used to switch in and switch out shunt capacitors units in order to achieve the required step change in the reactive power supplied to the system. Unlike shunt reactors, shunt capacitors cannot be switched continuously with a variable firing angle control.

Depending on the total VAr requirement, a number of capacitors are used, which can be switched into or out of the system individually. The control is done continuously by sensing the load VArs. A single-phase TSC is shown in Fig. 9.14.

It consists of a capacitor, a bidirectional thyristor valve, and relatively small surge current in the thyristor valve under abnormal operating conditions (e.g., control malfunction causing capacitor switching at a 'wrong time'); it may also be used to avoid resonance with system impendence at particular frequencies. The problem of achieving transient-free switching of the capacitors is overcome by keeping the capacitors charged to the positive or negative peak value of the fundamental frequency network voltage at all times when they are in the stand-by state. The switching-on-transient is then selected at the time when the same polarity exists in the capacitor voltage. This

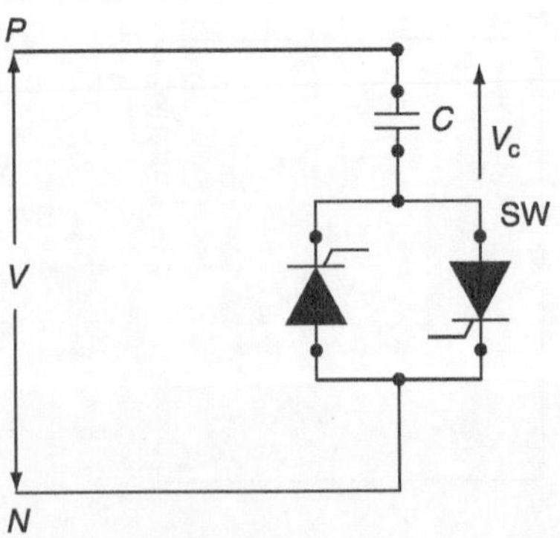

FIG. 9.14 *TSC*

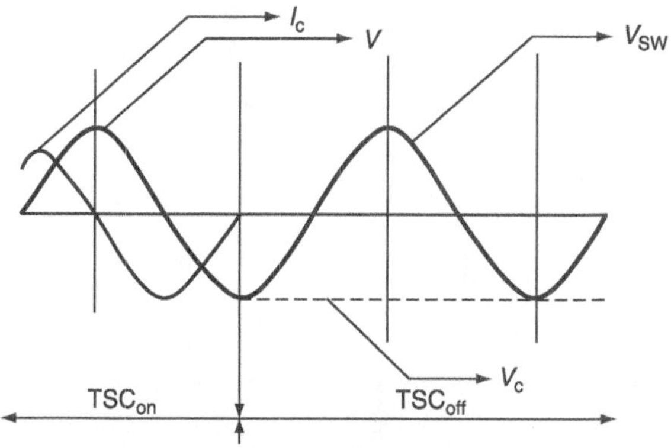

FIG. 9.15   *TSC waveforms*

ensures that switching on takes place at the natural zero passage of the capacitor current. The switching thus takes place with practically no transients. This is called zero-current switching.

Switching off a capacitor is accomplished by suppression-offering pulses to the anti-parallel thyristors so that the thyristors will switch off as soon as the current becomes zero. In principle, the capacitor will then remain charged to the positive or negative peak voltage and be prepared for a new transient-free switching-on as shown in Fig. 9.15.

TSC's characteristics are:

- Steeped control.
- No transients.
- No harmonics.
- Low losses.
- Redundancy and flexibility.

9.13 SERIES COMPENSATOR

In the TSC scheme, increasing the number of capacitor banks in series, controls the degree of series compensation. To accomplish this, each capacitor bank is controlled by a thyristor bypass switch or valve. The operation of the thyristor switches is co-ordinated with voltage and current zero-crossing; the thyristor switch can be turned on to bypass the capacitor bank when the applied AC voltage crosses zero, and its turn-off has to be initiated prior to a current zero at which it can recover its voltage-blocking capability to activate the capacitor bank. Initially, capacitor is charged to some voltage, while switching the SCR's, they may get damaged because of this initial voltage. In order to protect the SCR's from this kind of damage, resistor is connected in series with capacitor as shown in Fig. 9.16.

In a fixed capacitor, the TCR scheme as shown in Figs. 9.17 and 9.18, the degree of series compensation in the capacitive operating region is increased (or decreased) by increasing (or decreasing) the current in the TCR. Minimum series compensation is reached when the TCR is switched off. The TCR may be designed for a substantially higher maximum admittance at

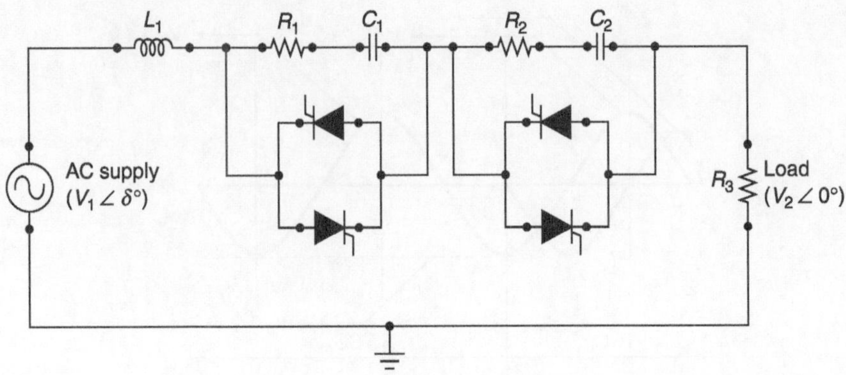

FIG. 9.16   *Series compensator*

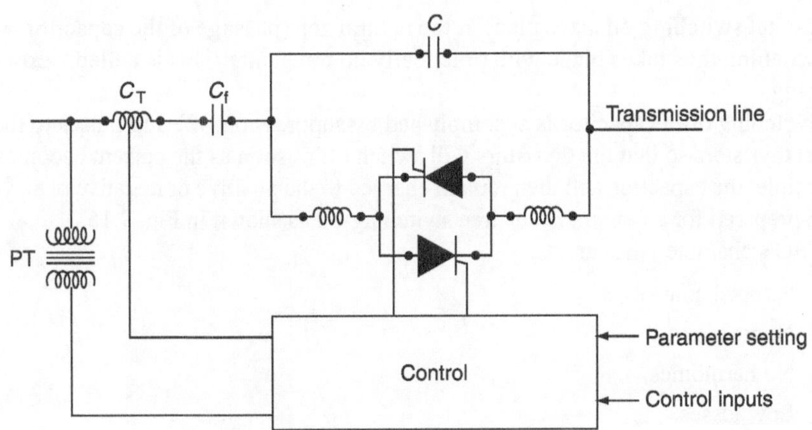

FIG. 9.17   *Thyristor-controlled capacitor*

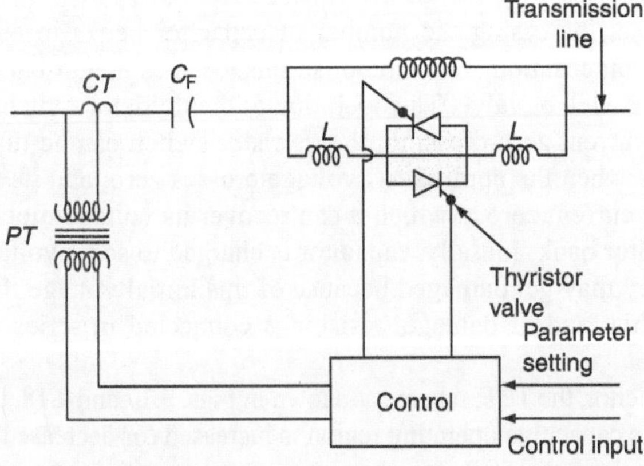

FIG. 9.18   *TCR*

full thyristor conduction than that of the fixed shunt-connected capacitor. In this case, the TCR, time with an appropriate surge-current rating can be used essentially as a bypass switch to limit the voltage across the capacitor during faults and the system contingencies of similar effects.

Controllable series compensation can be highly effective in damping power oscillation and preventing loop flows of power.

The expression for power transferred is given by

$$P = \frac{V_s V_r}{X} \sin \delta$$

where V_s is the sending-end voltage, V_r the receiving-end voltage, δ the angle between V_s and V_r, and $X = X_L - X_C$.

In interconnected power systems, the actual transfer of power from one region to another might take unintended routes depending on impedances of transmission lines connecting the areas. Controlled series compensation is a useful means for optimizing power flow between regions for varying loading and network configurations. It becomes possible to control power flows in order to achieve a number of goals that are listed below:

- Minimizing system losses.
- Reduction of loop flows.
- Elimination of line overloads.
- Optimizing load sharing between parallel circuits.
- Directing power flows along contractual paths.

9.14 UNIFIED POWER-FLOW CONTROLLER

In the UPFC, an AC voltage generated by a thyristor-based inverter is injected in series with the phase voltage. In Fig. 9.19, Converter-2 performs the main function of the UPFC by injecting, via., a series transformer, an AC voltage with controllable magnitude and a phase angle in series with the transmission line. The basic function of Converter-1 is to supply or

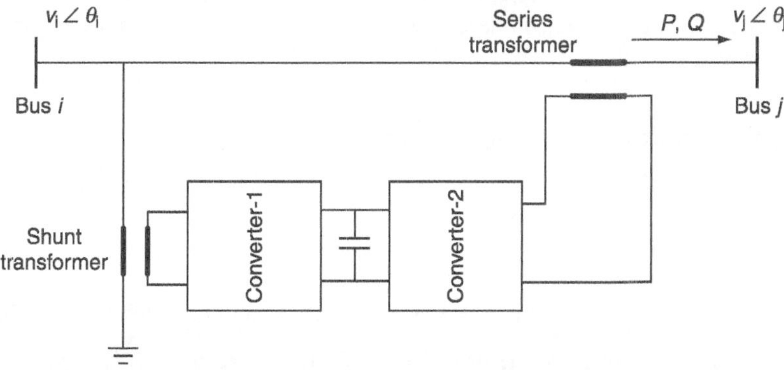

FIG. 9.19   *The UPFC*

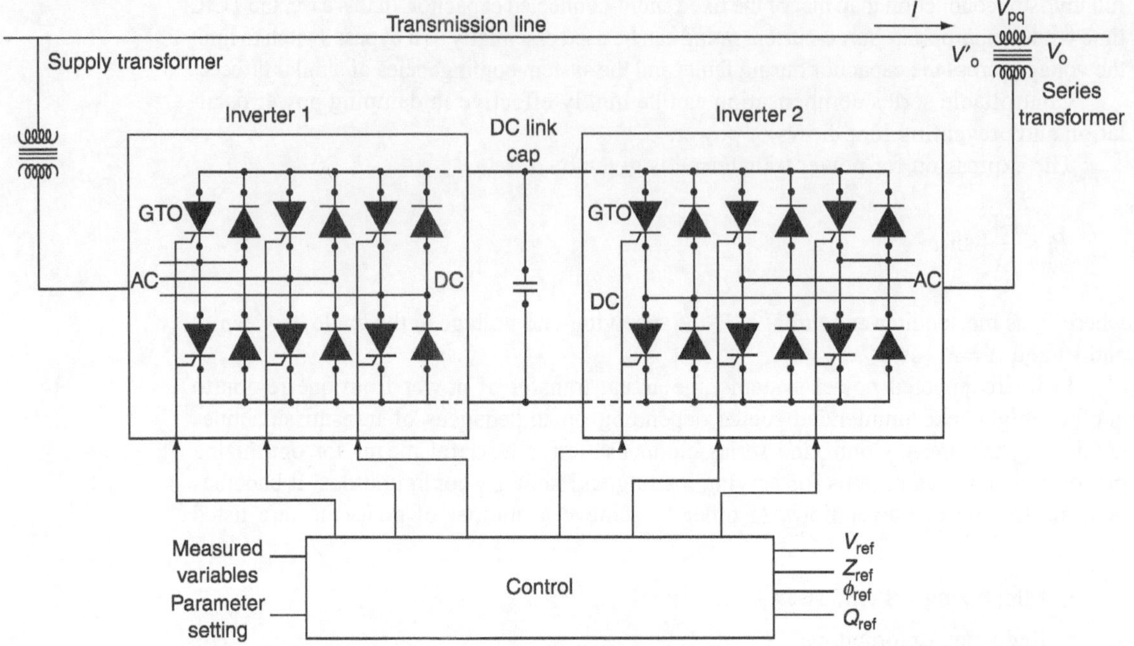

FIG. 9.20   *Implementation of the UPFC using two voltage source inverters with a direct voltage link simultaneously*

absorb the real power demanded by Converter-2 at the common DC link. It can also generate or absorb controllable reactive power and provide an independent shunt-reactive compensation for the line. Converter-2 either supplies or absorbs the required reactive power locally and exchanges the active power as a result of the series injection voltage.

Generally, the impedance control would cost less and be more effective than the phase angle control, except where the phase angle is very small or very large or varies widely.

In general, it has three control variables and can be operated in different modes. The shunt-connected converter regulates the voltage bus 'i' in Fig. 9.20 and the series-connected converter regulates the active and reactive power or active power and the voltage at the series-connected node. In principle, a UPFC is able to perform the functions of the other FACTS devices, which have been described, namely voltage support, power-flow control, and an improved stability.

9.15 BASIC RELATIONSHIP FOR POWER-FLOW CONTROL

The basic concept of controlling power transmission in real time assumes the available means for rapidly changing those parameters of the power system, which determine the power flow. To consider the possibilities for a power-flow control, power relationships for the simple two-machine model are shown in Figs. 9.21 and 9.22.

Figure 9.22 shows the sending- and receiving-end generators with voltage phasors V_s and V_r, inductive transmission line impedance (X_L) in two sections $\dfrac{X_L}{2}$, and generalized power-flow controller operated (for convenience) at the middle of the line. The power-flow

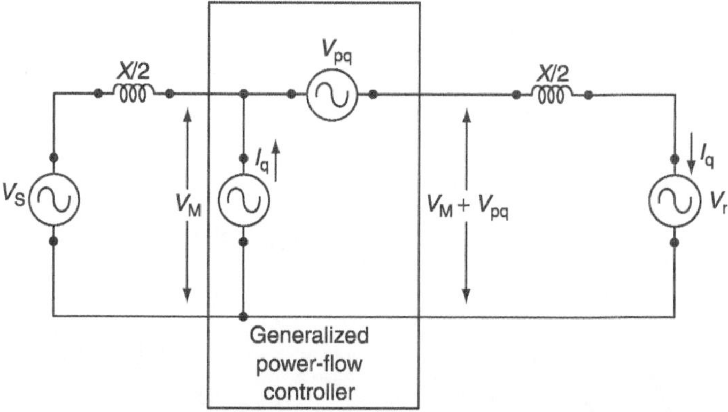

FIG. 9.21   *Simple two-machine power system with a generalized power-flow controller*

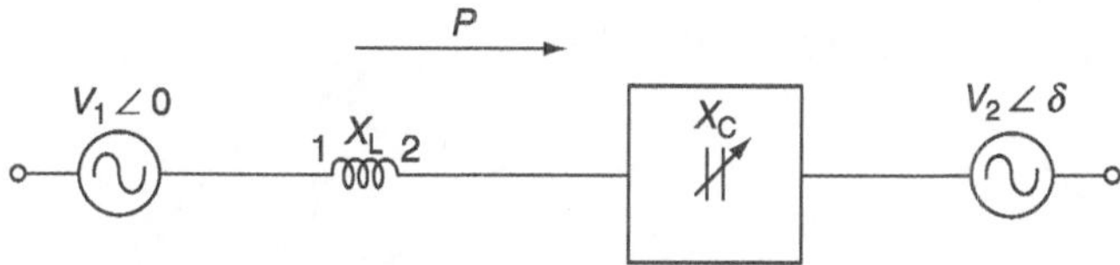

FIG. 9.22   *Power-flow relation*

controller consists of two controllable elements, i.e., a voltage source (V_{xy}) and a current source (I_x) are connected in series and shunt, respectively, with the line at the midpoint. Both the magnitude and the angle of the voltage V_{xy} are freely variables, whereas only the magnitude of current I_x is variable; its phase angle is fixed at 90° with respect to the reference phasor of midpoint voltage V_m. The basic power-flow relation is shown in Fig. 9.22 by using FACTS controller in a normal transmission system.

The four classical cases of power transmission are as follows:

(i)   Without line compensation.

(ii)   With series capacitive compensation.

(iii)   With shunt compensation.

(iv)   With phase angle control,

They can be obtained by appropriately specifying V_{xy} and I_x in the generalized power-flow controller.

9.15.1 Without line compensation

Consider that the power-flow controller is off, i.e., both V_{xy} and I_x are zero. Then, the power transmitted between the sending- and receiving-end generators can be expressed by the well-known formula:

$$P_{(1)} = \frac{V^2}{X}\sin\delta \quad (\text{assume } V_r = V_s = V)$$

where δ is the angle between the sending- and receiving-end voltage phasors. Power $P_{(1)}$ is plotted against angle δ in Fig. 9.23.

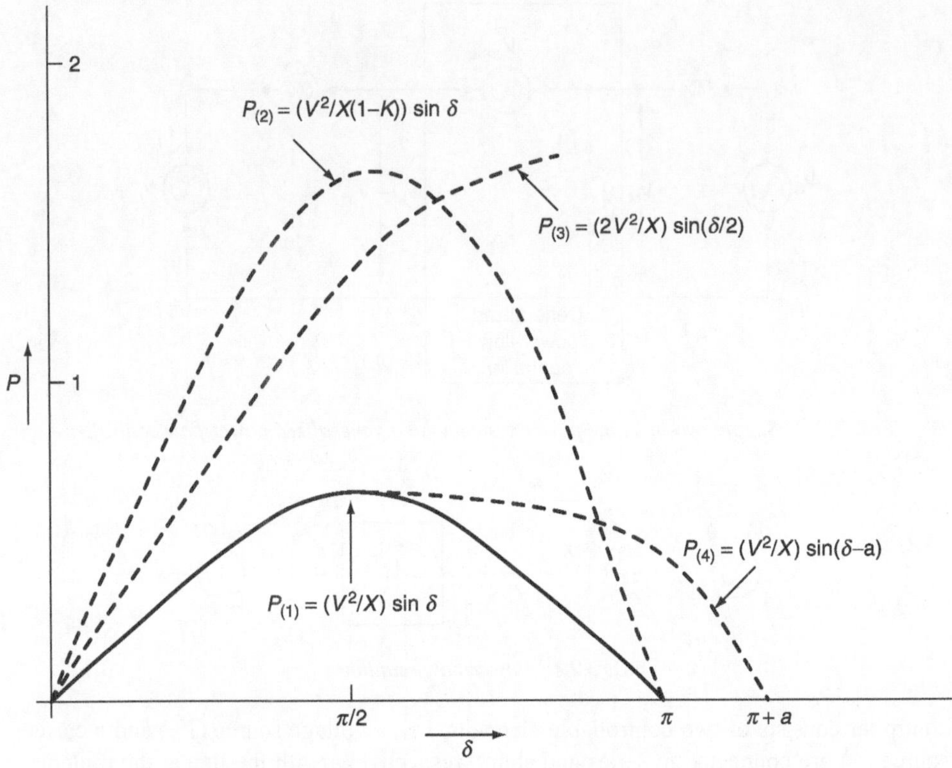

FIG. 9.23   *The basic power transmission on characteristics for four different cases*

9.15.2 With series capacitive compensation

Assume parallel current source, $I_x = 0$, and series voltage source, $V_{xy} = -jnX_L I$, i.e., the voltage inserted in series with the line lags the line current by 90° with an amplitude that is proportional to the magnitude of the line current and that of the line impedance. In other words, the voltage source acts at the fundamental frequency precisely as a series-compensating capacitor. The degree of series compensating is defined by coefficient n (i.e., $0 \leq n \leq 1$). With this, P against δ relationship becomes

$$P_{(2)} = \frac{V^2}{X_L(1-n)}\sin\delta$$

9.15.3 With shunt compensation

Consider that series voltage source $V_{xy} = 0$ and parallel current source $I_x = -j\dfrac{4V}{X_L}\left(1-\cos\dfrac{\delta}{2}\right)$,

i.e., the current source I_x draws just enough capacitive current to make the magnitude of the midpoint voltage V_m equal to V. In other words, the reactive current source acts like an ideal shunt compensator, which segments the transmission line into two independent parts, each with an impedance of $\dfrac{X_L}{2}$ by generating the reactive power necessary to keep the midpoint voltage constant, independently of angle δ. For this case of ideal midpoint compensation, the P against δ relation can be written as

$$P_{(3)} = \frac{2V^2}{X_L} \sin\frac{\delta}{2}$$

9.15.4 With phase angle control

Assume that $I_x = 0$ and $V_{xy} = \pm jV_m \tan\alpha$, i.e., a voltage ($V_{xy}$) with the amplitude $\pm jV_m \tan\alpha$, is added in quadrature to the midpoint voltage (V_m) to produce the desired α phase shift. The basic idea behind the phase shifter is to keep the transmitted power at a desired level independently of angle δ in a pre-determined operating range. Thus, for example, the power can be kept at its peak value after angle δ is $\pi/2$ by controlling the amplitude of the quadrature voltage V_{xy} so that the effective phase angle $(\delta - \alpha)$ between the sending- and receiving-end voltages stays at $\pi/2$. In this way, the actual transmitted power may be increased significantly even though the phase shifter does not increase the steady-state power transmission limit. Considering $(\delta - \alpha)$ as the effective phase angle between the sending-end and the receiving-end voltage, the transmitted power can be expressed as

$$P_{(4)} = \frac{V^2}{X_L} \sin(\delta - \alpha)$$

From Fig. 9.23, it can be seen that the power in without compensating is less shown in the $P_{(1)}$ curve. Power is increased by using the series capacitor compensation shown in the $P_{(2)}$ curve. The power angle curve with the shunt compensator is shown in the $P_{(3)}$ curve, in this case, power is increased and it seems that voltage is increased. The concept of the phase angle control is shown in the $P_{(4)}$ curve by shifting the curve higher, and power can be obtained.

9.16 COMPARISON OF DIFFERENT TYPES OF COMPENSATING EQUIPMENT FOR TRANSMISSION SYSTEMS

The comparison among different types of compensating equipment for transmission systems is tabulated below (Table 9.1).

TABLE 9.1 *Comparison of different types of compensating equipment for transmission systems*

Compensating equipment	Advantages	Disadvantages
Switched shunt reactor	Simple in principle and construction	Fixed in value
Switched shunt capacitor	Simple in principle and construction	Fixed in value-switching transients. Required overvoltage protection and sub-harmonic filters. Limited overload capacity
Series capacitor	Simple in principle. Performance relatively sensitive to location. Has useful overload capability	High-maintenance requirements. Slow control response

(Continued)

TABLE 9.1 *(Continued)*

Compensating equipment	Advantages	Disadvantages
Synchronous condenser	Fully controllable. Low harmonics	Performance sensitive to location. Requires strong foundations
Polyphase-saturated reactor	Very rugged construction. Large overload capability. No effect on fault level. Low harmonics	Essentially fixed in value. Performance sensitive to location and noisy
TCR	Fast response. Fully controllable. No effect on fault level. Can be rapidly repaired after failures	Generator harmonics performance sensitive to location
TSC	Can be rapidly repaired after failures. No harmonics	No inherent absorbing capability to limit overvoltages. Complex bus work and controls low-frequency resonance with system. Performance sensitive to location

9.17 VOLTAGE STABILITY—WHAT IS IT?

Voltage instability does not mean the problem of low voltage in steady-state condition. As a matter of fact, it is possible that the voltage collapse may be precipitated even if the initial operating voltages may be at acceptable levels.

Voltage collapse may be either fast or slow. Fast voltage collapse is due to induction motor loads or HVDC converter stations and slow voltage collapse is due to on-load tap changer and generator excitation limiters.

Voltage stability is also sometimes termed load stability. The terms voltage instability and voltage collapse are often used interchangeably.

It is to be understood that the voltage stability is a sub-set of overall power system stability and is a dynamic problem. The voltage instability generally results in monotonically (or a periodically) decreasing voltages. Sometimes, the voltage instability may manifest as undamped (or negatively damped) voltage oscillations prior to voltage collapse.

9.17.1 Voltage stability

Definition: A power system at a given operating state and subjected to a given disturbance is voltage stable if voltages near the loads approach post-disturbance equilibrium values. The disturbed state is within the regions of attractions of stable post-disturbance equilibrium.

The concept of voltage stability is related to the transient stability of a power system.

9.17.2 Voltage collapse

Following voltage instability, a power system undergoes voltage collapse if the post-disturbance equilibrium voltages near the load are below acceptable limits. The voltage collapse may be either total or partial.

The absence of voltage stability leads to voltage instability and results in progressive decrease of voltages. When destabilizing controls (such as OLTC) reach limits or due to other control actions (under voltage load shedding), the voltages are stabilized (at acceptable or unacceptable levels). Thus, abnormal voltage levels in the steady state may be the result of voltage instability, which is a dynamic phenomenon.

9.18 VOLTAGE-STABILITY ANALYSIS

The voltage-stability analysis is carried out by load flow methods, which are basically post-disturbance power-flow methods. Besides these methods, P–V curves and Q–V curves are the other power-flow-based methods generally used for voltage-stability analysis. By these two methods, the steady-state loadability limits are determined, which are related to voltage stability.

9.18.1 *P–V curves*

The conceptual analysis of voltage stability is useful carried out by using P–V curves. These are useful for the study of analysis of radial systems.

This method is also applicable for a large interconnected network to which the total load connected is P and the voltage of the critical bus is V. The total load P may be the power transferred over a transmission line. The voltage at several buses can be plotted.

9.18.1.1 *Interpretation of P–V curves*

Consider a radial line with an asynchronous load is as shown in Fig. 9.24(a).

The load is $P_L + jQ_L$ at the receiving end keeping the sending-end voltage V_s constant.

Even the radial line is connected with the asynchronous load, the maximum power can be transmitted over the line.

Let us consider that the load is u.p.f load and the sending-end voltage source and line form a voltage source with an open-circuit voltage V_{oc} and impedance $(R + jX)$ and at the receiving end a variable resistive load R_L is connected such that the p.f. is unity (Fig. 9.24(b)).

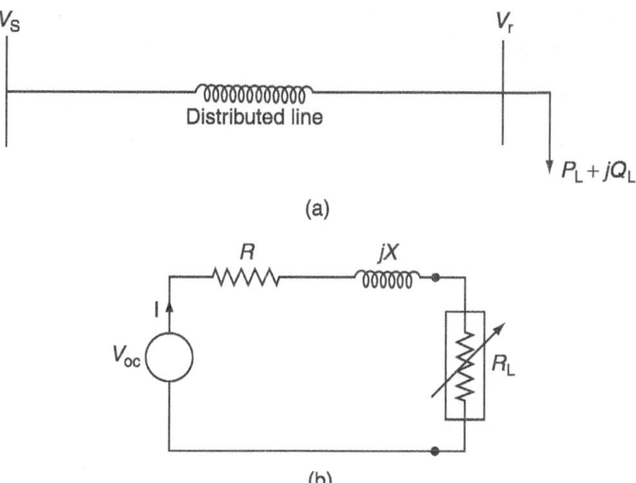

FIG. 9.24 *(a) A radial line terminated through an asynchronous load $P_L + jQ_L$; (b) equivalent circuit of Fig. 9.24 with u.p.f. load*

The short-circuit current, $I_{sc} = \dfrac{V_{oc}}{Z} = \dfrac{V_{oc}}{R + jX}$

Short-circuit p.f., $\cos \phi_{sc} = \dfrac{R}{Z}$

The total load current, $I_L = \dfrac{V_{oc}}{(R + jX) + R_L} = \dfrac{V_{oc}}{(R + R_L) + jX}$

The power delivered to the load, $P = I_L^2 R_L = \dfrac{V_{oc}^2 R_L}{(R + R_L)^2 + X^2}$ **(9.42)**

Condition for maximum power delivered is $\dfrac{dP}{dR_L} = 0$

$$\frac{d}{dR_L}\left[\frac{V_{oc}^2}{(R + R_L)^2 + X^2} R_L \right] = 0$$

$$\left[(R + R_L)^2 + X^2\right] - R_L \times 2(R + R_L) = 0$$

$$R_L^2 + R^2 + 2RR_L + X^2 - 2RR_L - 2R_L^2 = 0$$

$$\text{or} \quad R_L^2 + R_2 + X^2 - 2R_L^2 = 0$$

$\therefore R_L = Z$, is the condition for maximum power delivered

Substituting this condition in Equation (9.42), we get the maximum power as

$$P_m = \frac{V_{oc}^2 Z}{(R + Z)^2 + X^2}$$

$$= \frac{V_{oc}^2 I_{sc} Z^2}{R^2 + Z^2 + 2RZ + X^2}$$

$$= \frac{V_{oc}^2 I_{sc} Z^2}{2Z^2 + 2RZ}$$

$$= \frac{V_{oc}^2 I_{sc} Z^2}{2Z(Z + R)}$$

$$= \frac{V_{oc}^2 I_{sc} Z^2}{2Z^2 \left(1 + \dfrac{R}{Z}\right)}$$

$$\therefore P_m = \frac{V_{oc}^2 I_{sc}}{2(1 + \cos \phi_{sc})}$$

Now V_{oc} is the open-circuit voltage, i.e., V_r when $I_r = 0$.

Let x be the distance from the sending end and l be the length of the line

For a lossless line, $r = 0$ and $g = 0$, then the voltage at distance x from the sending end becomes

$$V_{(x)} = V_r \cos \beta(l - x) + jZ_c I_r \sin \beta(l - x)$$

where β is the phase constant

Suppose the line is open circuited at the receiving end, i.e., $I_r = 0$,

$$\therefore\ V_{(x)} = V_r \cos \beta(l - x)$$

At $x = 0$, $V_{(x)} = V_s$, and $I_{(x)} = I_s$

$$V_s = V_r \cos \beta l$$

$$= V_0 \cos \beta l$$

or $V_0 = \dfrac{V_s}{\cos \beta l}$

Similarly, the short-circuit current I_{sc} is the value of I_r when $V_r = 0$

$$V_s = jZ_c I_{sc} \sin \beta l$$

or $I_{sc} = \dfrac{V_s}{\sin \beta l}$

Assuming the line to be lossless, $\cos \phi_{sc} = 0$

$$P_m = \frac{V_0 I_{sc}}{2} = \frac{V_s}{2\cos \beta l} \times \frac{V_s}{Z_c \sin \beta l} = \frac{V_s^2}{2Z_c \cos \beta l \, \sin \beta l} \tag{9.43}$$

Equation (9.43) represents loci of maximum power for different line lengths at unity p.f.

The receiving-end current, $I_r = \dfrac{(P_r - jQ_r)}{V_r}$

The sending-end voltage of the line, if assuming the line to be lossless, now becomes

$$V_s = V_r \cos \beta l + jZ_c \left(\frac{P_r - jQ_r}{V_r} \right) \sin \beta l \tag{9.44}$$

For a fixed sending-end voltage V_s and the fixed line length, Equation (9.44) is quadratic in V_r and thus will have two roots. Figure 9.25 shows a graphical relation between $\dfrac{V_r}{V_s}$ as a function of normalized loading $\dfrac{P_L}{P_c}$.

From Fig. 9.25, it is observed that the maximum power can be transmitted for each load p.f. and for any loading, there are two different values of V_r.

The normal operation of the power system is along the upper part of the curve where the receiving-end voltage is nearly 1.0 p.u. The load is increased by decreasing the effective resistance of the load upto the maximum power; the product of load voltage and current increases as the system is stable. As the point of maximum power is reached, a further reduction in effective load resistance reduces the voltage more than the increase in current and therefore, there is an effective reduction in power transmission. The voltage finally collapses to zero and the system at the receiving end is effectively short-circuited and therefore, the power transmitted is zero (point at orgin in Fig. 9.25).

It is observed from Fig. 9.25 that the power transmitted is zero both at Point K and Point 0. Point K corresponding to the open circuit and Point 0 corresponding to the short circuit and in either case the power transmitted is zero.

Figure 9.25 shows the relation between $\dfrac{V_r}{V_s}$ as a function of normalized loading $\dfrac{P_L}{P_c}$, where P_c is the surge impedance load. These relation curves between $\dfrac{V_r}{V_s}$ and $\dfrac{P_L}{P_c}$ are

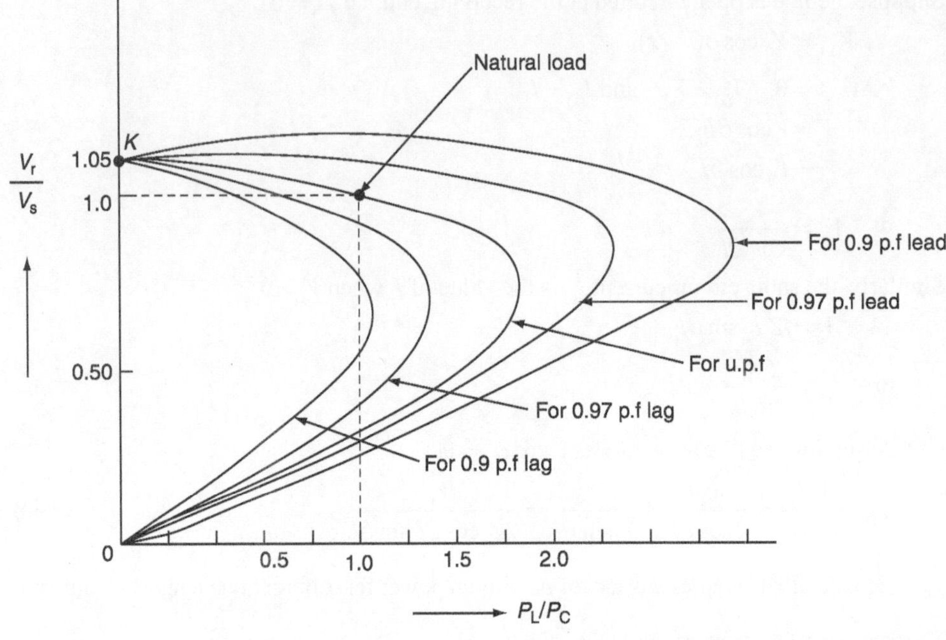

FIG. 9.25 $\dfrac{V_r}{V_s}$ *as a function of normalized loading* $\dfrac{P_L}{P_c}$

known as *normalized P–V curves*. These *P–V* curves are different for different p.f.'s. At more leading p.f.'s, the maximum power is higher and for that the shunt compensation is provided. The nose voltage of the *P–V* curve has the critical voltage at the receiving end for maximum power transfer. With leading p.f., the critical voltage is higher, which is a very important aspect of voltage stability.

The main disadvantage of the load-flow solution for *P–V* curves is that it is likely to diverge near the maximum power-transfer point or the nose point of the *P–V* curve. A load-flow solution is at various *P–V* buses or generators buses for particular generations. However, when the load changes, the scheduling of generation at various generator buses also changes. This is yet another disadvantage of the load-flow solution method.

9.18.2 Concept of voltage collapse proximate indicator

Consider that a purely inductive load is connected to a source through a lossless line as shown in Fig. 9.26.

Here, $P = 0$ and $\delta = 0$ since the load is purely inductive; therefore, the reactive power at the receiving end is expressed as $Q_r = \dfrac{V_s V_r}{X} - \dfrac{V_r^2}{X}$.

The condition for maximum reactive power transfer is $\dfrac{\mathrm{d}Q_r}{\mathrm{d}V_r} = 0$.

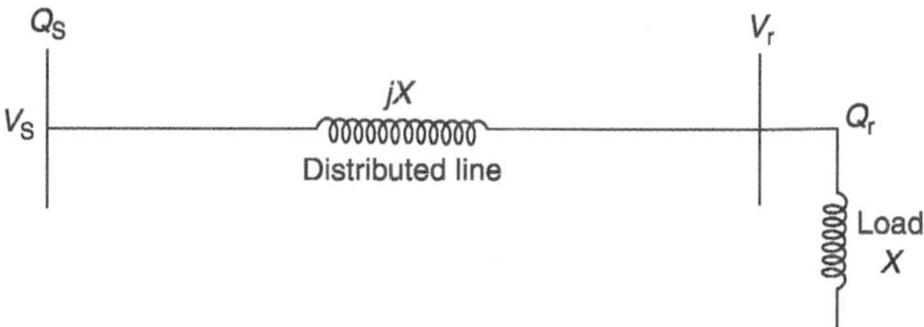

FIG. 9.26   *Radial line connected with purely inductive load to source*

$$\frac{d}{dV_r}\left(\frac{V_s V_r}{X} - \frac{V_r^2}{X}\right) = 0$$

$$\frac{1}{X}(V_s - 2V_r) = 0$$

$$\therefore V_r = \frac{V_s}{2} = V_{critical}$$

where $V_{critical}$ is the receiving-end voltage for maximum reactive power transfer.

Therefore, the maximum reactive power is expressed as

$$Q_{r\,max.} = \frac{2V_r V_r}{X} - \frac{V_r^2}{X} = \frac{V_r^2}{X}$$

As half of the drop will be across the line and another half across the load, $X_{load} = X$; hence, $Q_{r\,max} = \dfrac{V_r^2}{X_{load}}$; i.e., the maximum reactive power is transferred when the load reactance is equal to the line reactance or the source reactance.

The short-circuit reactive power of the line is $Q_{sc} = \dfrac{V_s^2}{X}$; hence, the normalized maximum reactive power is

$$q_{max} = \frac{Q_{r\,max}}{Q_{sc}} = \frac{V_r^2}{X} \times \frac{X}{V_s^2} = \frac{V_r^2}{4V_r^2} = 0.25$$

Also $V_{critical} \dfrac{V_{r\ critical}}{V_s} = \dfrac{V_s}{2V_s} = 0.5$

Now $Q_s = Q_r + I^2 X$

Since $P = 0,\quad I^2 = \dfrac{Q_s^2}{V_s^2}$

Hence, $Q_s = Q_r + X\dfrac{Q_s^2}{V_s^2}$ **(9.45)**

Multiplying both sides of Equation (9.45) by $\dfrac{V_s^2}{X}$, we get

$$Q_s \frac{V_s^2}{X} = Q_r \frac{V_s^2}{X} + X \frac{Q_s^2}{V_s^2} \times \frac{V_s^2}{X}$$

$$Q_s^2 - \frac{V_s^2}{X} Q_s + \frac{V_s^2}{X} Q_r = 0$$

$$Q_s = \frac{1}{2}\left(\frac{V_s^2}{X} \pm V_s \sqrt{\frac{V_s^2}{X} - \frac{4Q_r}{X}} \right)$$

Now $\quad \dfrac{dQ_s}{dQ_r} = \dfrac{V_s}{2} \dfrac{1}{2} \dfrac{1}{\sqrt{\dfrac{V_s^2}{X} - \dfrac{4Q_r}{X}}} \dfrac{4}{X}$

$$= \frac{1}{\sqrt{\dfrac{V_s^2}{X}\dfrac{X}{V_s^2} - 4\dfrac{Q_r}{X} \times \dfrac{X_2}{V_s^2}}}$$

$$= \frac{1}{\sqrt{1 - 4Q_r \dfrac{X}{4V_r^2}}} \quad (\because V_s = 2V_r \ \text{at maximum power})$$

$$\therefore \frac{dQ_s}{dQ_r} = \frac{1}{\sqrt{1 - \dfrac{Q_r}{Q_{max}}}}$$

The differentiation of the sending-end reactive power Q_s with respect to the receiving-end reactive power Q_r, i.e., $\dfrac{dQ_s}{dQ_r}$, is known as the voltage collapse proximate indicator (VCPI) of a radial line.

The receiving-end voltage varies from V_s at no load to $\dfrac{V_s}{2}$ at maximum load Q_{max}.

However, VCPI is unity at no load since at no load $Q_r = 0$ and $\left.\dfrac{dQ_s}{dQ_r}\right|_{Q_r=0} = 1$ and it is infinity

at maximum load since at this load $Q_r = Q_{r_{max}}$ and hence $\dfrac{dQ_s}{dQ_r} = \infty$.

It is clear that near the maximum load, an extremely large amount of reactive power is required at the sending end to supply an incremental increase in load. Thus, VCPI is a very sensitive indicator of impeding voltage collapse. There, lated quantities reactive reserve activation and reactive losses are also sensitive indicators.

9.18.3 Voltage-stability analysis: Q–V curves

Q–V curves can be obtained from the normalized P–V curves.

$$\text{Let} \quad p = \frac{PX}{V_s^2}, \quad q = \frac{Q_r X}{V_s^2} \quad \text{and} \quad v = \frac{V_r}{V_s}.$$

For constant values of p, we note the two pairs of values of v and q for each p.f. and report these values. The result of these plots is shown in Fig. 9.27.

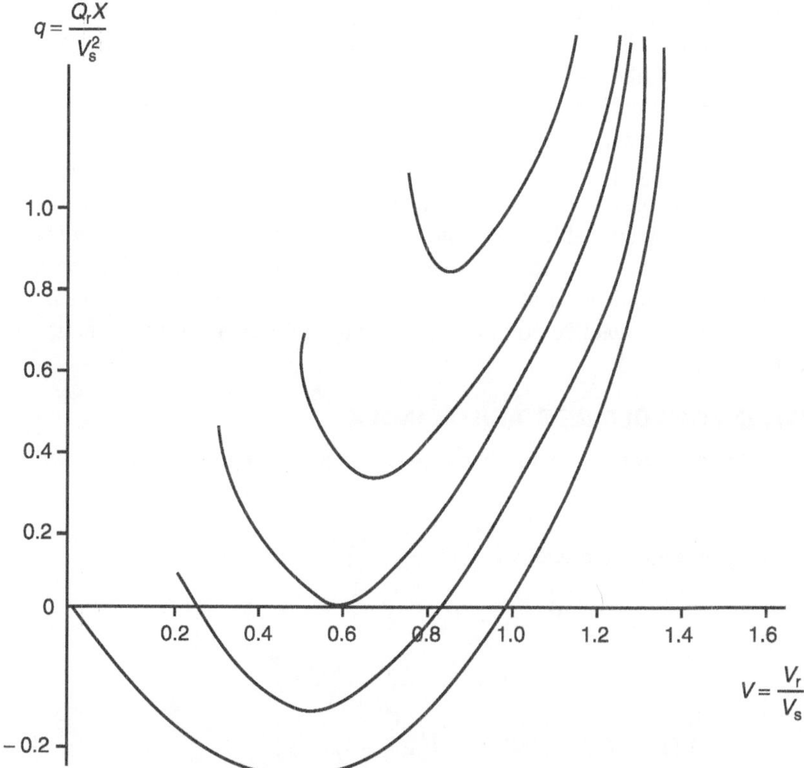

FIG. 9.27   *Normalized Q–V curves for fixed source and reactive network
loads are at constant power*

The critical voltage is high for loadings (i.e., v is above 1 p.u. for $p = 1$ p.u.). The right-hand side of the curves indicates the normal operating conditions where the application of shunt capacitors raises the voltage. The steep-sloped linear portions of the right side of the curve are equivalent to the figure below (rotate Fig. 9.28 clock-wise by 90°).

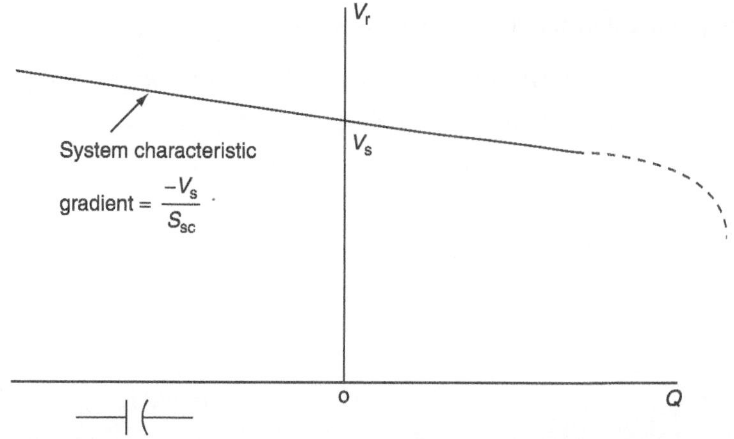

FIG. 9.28   *System approximate voltage-reactive power characteristic*

The Q–V curves for large systems are obtained by a series of power-flow simulations. Q–V curves result from the plot of voltage at a test or critical bus versus reactive power on the same bus. Generally, the P–V bus or generator bus has reactive power constraints for a load-flow solution. A fictitious synchronous condenser is represented at the test bus and allows the bus to have any reactive power for a fixed p and v. The value of v at the bus changes for obtaining another point on the Q–V curve, and obtains reactive power flow for different scheduled voltages at the bus. Scheduled voltage at the bus is an independent variable and forms an abscissa variable. The capacitive reactive power required to maintain the scheduled voltage at the bus is a dependent variable and is plotted in the positive vertical direction. Without the application of shunt reactive compensation at the test bus, the operating point is at the zero reactive point corresponding to the removal of the fictitious synchronous condenser.

9.19 DERIVATION FOR VOLTAGE-STABILITY INDEX

Consider a typical branch consisting of sending- and receiving-end buses as shown in Fig. 9.29.

$$\text{Current flowing through the branch, } I = \frac{V_s \angle \delta_1 - V_R \angle \delta_2}{R + jX}$$

$$P - jQ = V_R^* I$$

$$\therefore P - iQ = \frac{V_s V_R \angle(\delta_1 - \delta_2) - V_R^2}{R + iX}$$

$$\Rightarrow (P - jQ)(R + jX) = V_s V_P \lfloor(\delta_1 - \delta_2) - V_R^2$$

$$\Rightarrow (RP + XQ) + j(XP - RQ) = V_s V_P \lfloor(\delta_1 - \delta_2) - V_R^2$$

$$V_R^2 + (RP + XQ) + j(XP - RQ) = V_s V_R \lfloor \delta_1 - \delta_2$$

The real term of the above equation is

$$V_S V_R \cos(\delta_1 - \delta_2) = V_R^2 + (RP + XQ)$$

and the imaginary part is

$$V_S V_R \sin(\delta_1 - \delta_2) = XP - RQ$$

Squaring and adding the above two terms, we get

$$V_R^4 + R^2 P^2 + X^2 Q^2 + 2R \times PQ + 2V_R^2(RP + XQ) + X^2 P^2 + R^2 Q^2 - 2R \times PQ = V_S^2 V_R^2$$

$$\Rightarrow V_R^4 + P^2(R^2 + X^2) + Q^2(R^2 + X^2) + 2V_R^2(RP + XQ) - V_S^2 V_R^2 = 0$$

$$\Rightarrow V_R^4 + V_R^2(2RP + 2XQ - V_S^2) + (P^2 + Q^2)(R^2 \times X^2) = 0$$

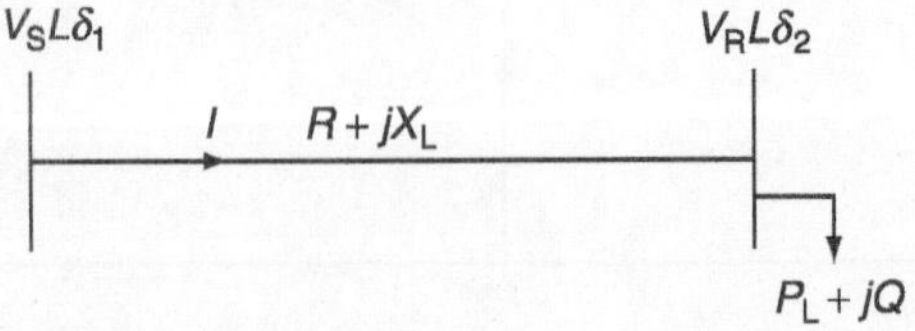

FIG. 9.29 *Single-line model of typical branch*

The above equation is a quadratic equation of V_R^2. The system is to be stable if $V_R^2 \geq 0$. It is possible when

$$b^2 - 4ac \geq 0$$

i.e., $\quad [2(RP + XQ) - V_S^2)]^2 - 4(P^2 + Q^2)(R^2 + X^2) \geq 0$

or $\quad 4R^2 p^2 + 4X^2 Q^2 + 4R \times PQ + V_S^4 - 4V_S^2(RP + XQ)$

$\qquad - 4R^2 p^2 - 4X^2 p^2 - 4R^2 Q^2 - 4X^2 Q^2 \geq 0$

Simplifying the above equation, we get

$$V_S^4 - 4V_S^2(RP + XQ) - 4(PX - RQ)^2 \geq 0$$

or $\quad 4(PX - RQ)^2 + 4V_S^2(RP + XQ) \leq V_S^4$

Dividing both sides of the above equation by V_S^4, we get

$$4\left(\frac{PX - RQ}{V_S^2}\right)^2 + 4\left(\frac{RP + XQ}{V_S^2}\right) \leq 1$$

$$\therefore \ L_{(p.u.)} = 4\left\{\left(\frac{PX - RQ}{V_S^2}\right)^2 + \left(\frac{RP + XQ}{V_S^2}\right)\right\}$$

where L = stability index
For stable systems, $L \leq 1$.

Example 9.1: A 440 V, 3-ϕ distribution feeder has a load of 100 kW at lagging p.f. with the load current of 200 A. If the p.f. is to be improved, determine the following:

    (i)   uncorrected p.f. and reactive load and

    (ii) new corrected p.f. after installing a shunt capacitor of 75 kVAr.

Solution:

    (i)   Uncorrected p.f. $= \cos\phi = \dfrac{P}{\sqrt{3}V_L I_L} = \dfrac{100 \times 10^3}{\sqrt{3} \times 440 \times 200} = 0.656$ lagging

$$Q_L = P \tan\phi = 115.055 \text{ kVAr}$$
$$Q_c = 75 \text{ kVAr}$$

    (ii) Corrected p.f. $= \dfrac{P}{\sqrt{(P^2 + (Q - Q_c)^2)}}$

$$= \dfrac{100}{\sqrt{(100)^2 + (115.055 - 75)^2}} = 0.928 \text{ lagging}$$

Example 9.2: A synchronous motor having a power consumption of 50 kW is connected in parallel with a load of 200 kW having a lagging p.f. of 0.8. If the combined load has a p.f. of 0.9, what is the value of leading reactive kVA supplied by the motor and at what p.f. is it working?

Solution:

    Let:

    p.f. angle of motor $= \phi_1$

p.f. angle of load $= \phi_2 = \cos^{-1}(0.8) = 36.87°$

Combined p.f. angle (both motor and load), $\phi = \cos^{-1}(0.9) = 25.84°$

$\tan \phi_2 = \tan 36° \, 87' = 0.75$; $\tan \phi = \tan 25°84' = 0.4842$

Combined power, $P = 200 + 50 = 250 \text{ kW}$

Total $kVAr$ of a combined system $= P \tan \phi_1 = 250 \times 0.4842 = 121.05$

Load $kVAr = 200 \times \tan \phi_2 = 200 \times 0.75 = 150$

$\therefore$ Leading $kVAr$ supplied by synchronous motor $= 150 - 121.05 = 28.95$

p.f. angle at which the motor is working, $\phi_1 = \tan^{-1} 28.95/50 = 30.07°$

p.f. at which the motor is working $= \cos \phi_1 = 0.865$ (lead)

Example 9.3: A 3-ϕ, 5-kW induction motor has a p.f. of 0.85 lagging. A bank of capacitor is connected in delta across the supply terminal and p.f. raised to 0.95 lagging. Determine the kVAr rating of the capacitor in each phase.

Solution:

The active power of the induction motor, $P = 5 \text{ kW}$

When the p.f. is changed from 0.85 lag to 0.95 lag by connecting a condenser bank, the leading $kVAr$ taken by the condenser bank $= P (\tan \phi_2 - \tan \phi_1)$

$$= 5(0.6197 - 0.3287) = 1.455$$

$\therefore$ Rating of capacitor connected in each phase $= 1.455/3 = 0.485 \text{ kVAr}$

Example 9.4: A 400 V, 50 Hz, 3-ϕ supply delivers 200 kW at 0.7 p.f. lagging. It is desired to bring the line p.f. to 0.9 by installing shunt capacitors (Fig. 9.30). Calculate the capacitance if they are: (a) star connected and (b) delta connected.

Solution:

(a) For star connection:

Phase voltage, $V_{\text{ph}} = 400/\sqrt{3} = 230.94 \text{ V/ph}$

Load current, $I = \dfrac{200 \times 10^3}{\sqrt{3} \times 400 \times 0.7} = 412.39 \text{ A}$

The active component of current, $I_a = I \cos \phi_1 = 412.39 \times 0.7 = 288.68 \text{ A}$

Reactive component of current, $I_r = I \sin \phi_1 = \dfrac{I_a}{\cos \phi_1} \times \sin \phi_1 = I_a \tan \phi_1$

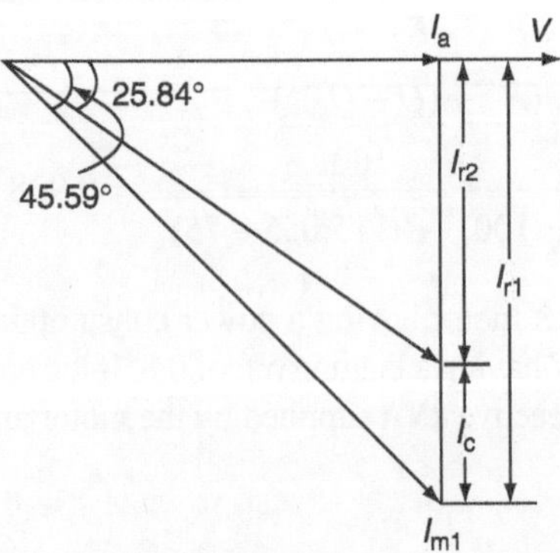

FIG. 9.30 *Phasor diagram*

For a fixed load, let us take:

The reactive component of load current without capacitor $= I_\mathrm{a} \tan \phi_1$

The reactive component of load current with capacitor $= I_\mathrm{a} \tan \phi_2$

Current taken by the capacitor installed for improving p.f., $I_\mathrm{C} = I_\mathrm{a} (\tan \phi_1 - \tan \phi_2)$

$$= 288.68 \left(\tan \left(\cos^{-1} 0.7 \right) - \tan \left(\cos^{-1} 0.9 \right) \right)$$

$$= 154.7 \, \mathrm{A}$$

The value of capacitor to be connected, $C = \dfrac{I_\mathrm{C}}{2\pi f V}$

$$= \frac{154.7}{2 \times \pi \times 50 \times 230.94} = 2{,}132.26 \, \mu\mathrm{F}$$

(b) For delta connection:

Phase voltage, $V_\mathrm{ph} = 400 \, \mathrm{V}$

Load current, $I = \dfrac{200 \times 10^3}{\sqrt{3} \times 400 \times 0.7} = 412.39 \, \mathrm{A}$

Phase current, $I_\mathrm{ph} = \dfrac{412.39}{\sqrt{3}} = 238.09 \, \mathrm{A}$

The active component of phase current, $I_\mathrm{a} = I \cos \phi_1 = 238.09 \times 0.7 = 166.663 \, \mathrm{A}$

Current taken by the capacitor installed for improving p.f., $I_\mathrm{C} = I_\mathrm{a} (\tan \phi_1 - \tan \phi_2)$

$$= 166.663 \left(\tan \left(\cos^{-1} 0.7 \right) - \tan \left(\cos^{-1} 0.9 \right) \right)$$

$$= 89.312 \, \mathrm{A}$$

The value of capacitor to be connected, $C = \dfrac{I_\mathrm{C}}{2\pi f V}$

$$= \frac{89.312}{2 \times \pi \times 50 \times 400} = 710.7 \, \mu\mathrm{F}$$

Example 9.5: A 3-ϕ 500 HP, 50 Hz, 11 kV star-connected induction motor has a full load efficiency of 85% at lagging p.f. of 0.75 and is connected to a feeder. If the p.f. of load is desired to be corrected to 0.9 lagging, determine the following:

(a) size of the capacitor bank in kVAr and

(b) capacitance of each unit if the capacitors are connected in Δ as well as in Y.

Solution:

Induction motor output $= 500 \, \mathrm{HP}$

Efficiency $\eta = 85\%$,

$\eta = \text{output/input}$

Input of the induction motor, $P = \text{output}/\eta = 500/0.85 = 588.235 \, \mathrm{HP}$

$$= 588.235 \times 746 = 438.82 \, \mathrm{kW}$$

Initial p.f. $(\cos \phi_1) = 0.75 \Rightarrow \tan \phi_1 = 0.88$

Corrected p.f. $(\cos \phi_2) = 0.9 \Rightarrow \tan \phi_2 = 0.48$

Leading $kVAr$ taken by the capacitor bank, $Q_c = P\,(\tan\phi_1 - \tan\phi_2)$
$$= 438.82\,(0.88 - 0.48) = 175.53 \text{ kVAr}$$

Line current drawn, $I_L = \dfrac{Q_c}{\sqrt{3}V_{L-L}} = \dfrac{175.53}{\sqrt{3}\times 11} = 9.213 \text{ A}$

Case 1: Delta connection:

Charging current per phase, $I_c = \dfrac{I_L}{\sqrt{3}} = 5.319 \text{ A}$

Reactance of capacitor bank per phase, $X_c = \dfrac{V_{L-L}}{I_c} = \dfrac{11\times 10^3}{5.319} = 2.068 \text{ k}\Omega$

$$X_c = \frac{1}{2\pi fc} \Rightarrow c = \frac{1}{2\pi fX_c}$$

Capacitance of capacitor bank, $C = \dfrac{1}{2\pi \times 50\times 2.068 \times 10^3} = 1.539 \text{ }\mu\text{F}$

Case 2: Star connection:

$I_L = I_c = 9.213 \text{ A}$

Reactance of capacitor bank per phase, $X_c = \dfrac{V_{L-N}}{I_c} = \dfrac{11\times 10^3}{\sqrt{3}\times 9.213} = 0.689 \text{ k}\Omega$

Capacitance of capacitor bank, $C = \dfrac{1}{2\pi fX_C} = 4.619 \text{ }\mu\text{F}$

Example 9.6: A star-connected 400 HP (metric), 2,000 V, and 50 Hz motor works at a p.f. of 0.7 lag. A bank of mesh-connected condensers is used to raise the p.f. to 0.9 lag. Calculate the capacitance of each unit and total number of units required, if each unit is rated 400 V, 50 Hz. The motor efficiency is 90% (Fig. 9.31).

Solution:

Motor output $= 400$ HP
Supply voltage $= 2,000$ V
i.e., $V = 2,000$ V
p.f. without condenser $= 0.7$ lag

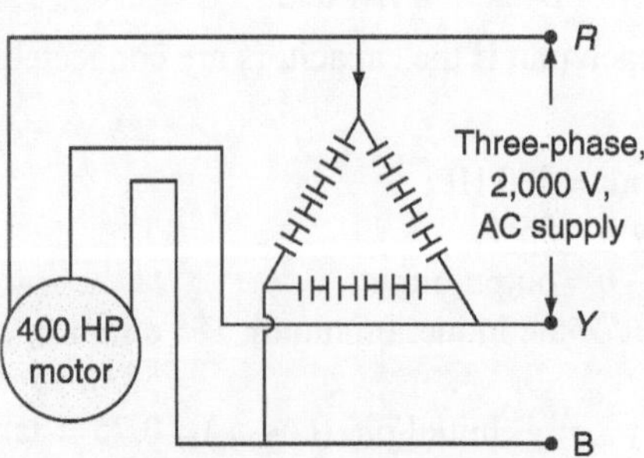

FIG. 9.31   *Circuit diagram*

Efficiency of motor, $\eta = 0.9$;

$$\therefore \text{ Motor current} = \frac{400 \times 746}{\sqrt{3} \times 2,000 \times 0.7 \times 0.9} = 136.73 \ (\because 1 \text{ HP} = 746 \text{ W})$$

For fixed loads, the active component of current is the same for improved p.f., whereas the reactive component will be changed.

$\therefore$ The active component of current at 0.7 p.f. lag., $I_{a_1} = I \cos \phi_1$

$$= 136.73 \times 0.7 = 95.71 \text{ A}$$

Line current taken by the capacitor installed for improving p.f., $I_C = I_a (\tan \phi_1 - \tan \phi_2)$

$$= 95.71 \left(\tan \left(\cos^{-1} 0.7 \right) - \tan \left(\cos^{-1} 0.9 \right) \right)$$

$$= 51.28 \text{ A}$$

The bank of condensers used to improve the p.f. is connected in delta. The voltage across each phase is $2,000$ V, but each unit of condenser bank is of 400 V. So, each phase of the bank will have five condensers connected in series as shown in Fig. 9.31.

$$\text{The current in each phase of the bank} = \frac{51.28}{\sqrt{3}} = 29.61 \text{ A}$$

Let X_c be the reactance of each condenser.

$$\text{Then the charge current, } I_c = \frac{400}{X_c} = 29.61$$

$$\text{or} \quad X_c = \frac{400}{29.61} = 13.51 \ \Omega$$

$$C = \frac{1}{2\pi \times 50 \times 13.51} = 235.61 \ \mu\text{F}$$

$$\text{Capacitance of each phase of the bank} = \frac{235.61}{5} = 47.12 \ \mu\text{F}$$

Example 9.7: A 3-ϕ, 50 Hz, 2,500 V motor develops 600 HP, the p.f. being 0.8 lagging and the efficiency 0.9. A capacitor bank is connected in delta across the supply terminals and the p.f. is raised to unity. Each of the capacitance units is built of five similar 500 V capacitors (Fig. 9.32). Determine the capacitance of each capacitor.

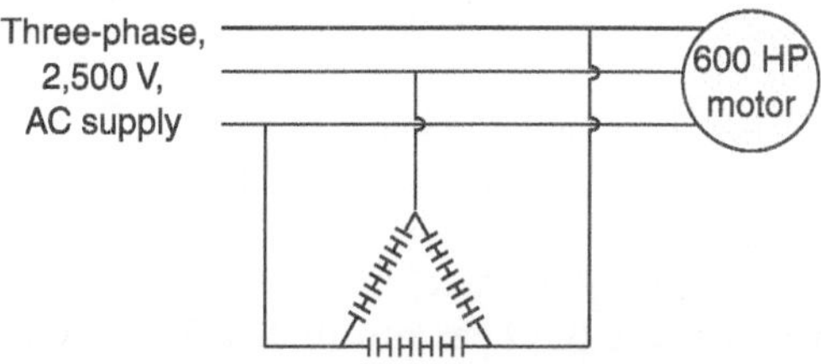

FIG. 9.32   *Circuit diagram*

Solution:

$$\text{Motor input } (P_{\text{i}}) = \frac{\text{output}}{\eta}$$

$$= \frac{600 \times 746}{0.9}$$

$$= 497.33 \text{ kW}$$

$$\text{Leading } kVAr \text{ supplied by the capacitor bank} = P(\tan \phi_1 - \tan \phi_2)$$

$$= 497.33 \ (0.75 - 0)$$

$$= 373 \text{ kVAr}$$

$$\text{Leading } kVAr \text{ supplied by each of three sets} = \frac{373}{3}$$

$$= 124.33 \text{ kVAr}$$

$$\text{Current per phase of capacitor bank, } I_{\text{C}} = \frac{V_{\text{Ph}}}{X_{\text{C}}}$$

$$= 2\pi f C V_{\text{ph}}$$

$$= 2\pi \times 50 \times C \times 2,500 \text{ A}$$

$$kVAr \text{ required/phase} = \frac{V_{\text{Ph}} I_{\text{C}}}{1,000}$$

$$= \frac{2,500 \times 2\pi \times 50 \times C \times 2,500}{1,000}$$

$$= 19,63,495.41 \, C$$

But leading $kVAr$ supplied by each phase $= 124.33$ kVAr

$$\therefore \ 19,63,495.41 C = 124.33$$

$$C = \frac{124.33}{19,63,495.41} = 63.32 \ \mu\text{F}$$

Since it is the combined capacitance of five equal capacitors joined in series,
The capacitance of each unit $= 5 \times 63.32 \ \mu\text{F}$

$$= 3,116.6 \ \mu\text{F}$$

Example 9.8: A 3-ϕ, 50 Hz, 30-km transmission line supplies a load of 5 MW at p.f. 0.7 lagging to the receiving end where the voltage is maintained constant at 11 kV. The line resistance and inductance are 0.02 Ω and 0.84 mH per phase per km, respectively. A capacitor is connected across the load to raise the p.f. to 0.9 lagging (Fig. 9.33). Calculate: (a) the value of the capacitance per phase and (b) the voltage regulation.

Solution:

$$\text{Length of the line} = 30 \text{ km}$$

$$\text{Frequency} = 50 \text{ Hz}$$

$$\text{Load} = 5 \text{ MW at 0.7 lagging p.f.}$$

$$\text{Receiving-end voltage, } V_{\text{r}} = 11 \text{ kV}$$

$$\text{Line resistance per phase} = 0.02 \ \Omega \text{ per km} = 0.02 \times 30 = 0.6 \ \Omega$$

$$\text{Reactance of 30-km length per phase, } X = 2 \times \pi \times f \times L \times 30$$

$$= 2 \times \pi \times 50 \times 0.84 \times 10^{-3} \times 30 = 7.92 \ \Omega$$

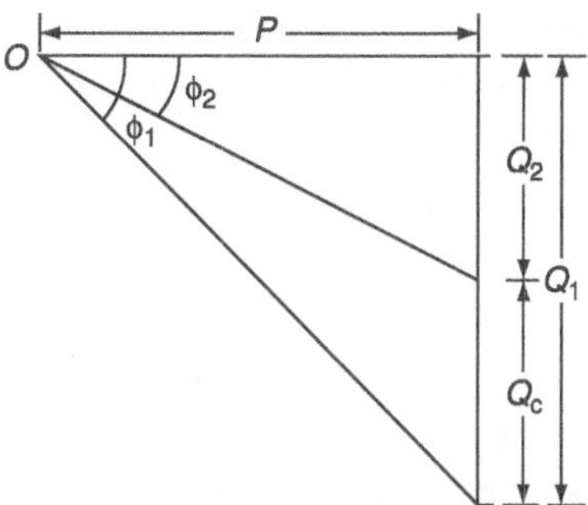

FIG. 9.33   *Phasor diagram*

For fixed loads, the active component of power is the same for improved p.f., whereas the reactive component of power will be changed.

$\therefore$ The active component of current at 0.7 p.f. lag, $P = 5$ MW

Reactive power ($MVAr$) supplied by the capacitor bank $= P(\tan \phi_1 - \tan \phi_2)$

$$= 5\,(1.02 - 0.484)$$
$$= 2.679 \text{ MVAr}$$

Reactive power ($MVAr$) supplied by the capacitor bank $= \dfrac{2.679}{3}$ MVAr

$$= 0.893 \text{ MVAr} = 893 \text{ kVAr}$$

Let C be the capacitance to be connected per phase across the load.

$$kVAr \text{ required/phase} = \frac{V_{Ph} I_C}{1,000} = \frac{V_{Ph}\left(V_{ph}\,\omega C\right)}{1,000}$$

$$= \frac{11,000 \times 2\pi \times 50 \times C \times 11,000}{1,000}$$

$$= 38,013,271.11 C$$

But leading $kVAr$ supplied by each phase $= 893$ kVAr

$$\therefore 3,80,13,271.11 C = 893$$

$$C = \frac{893}{3,80,13,271.11} = 23.49 \text{ μF}$$

Sending-end voltage with improved p.f. $= V_r + I(R \cos \phi_2 + jX \sin \phi_2)$

Receiving-end voltage, $V_r = 11 \text{ kV}(L-L) = \dfrac{11}{\sqrt{3}} = 6,350.85$ V

Current with improved p.f., $I = \dfrac{5 \times 10^6}{\sqrt{3} \times 11 \times 10^3 \times 0.9} = 291.59$ A

$\therefore$ Sending-end voltage, $V_s = 6,350.85 + 291.59(0.6 \times 0.9 + j\,7.92 \times 0.436)$

$$= 6,508.31 + j1006.9 \text{ V/ph}$$
$$= 6,585.74 \angle 8.79 \text{ V/ph}$$
$$= 11,406.84 \text{ V(L-L)}$$

$$\% \text{ regulation} = \frac{11,406.84 - 11,000}{11,000} \times 100 = 3.7$$

Example 9.9: A synchronous motor improves the p.f. of a load of 200 kW from 0.7 lagging to 0.9 lagging and at the same time carries an additional load of 100 kW (Fig. 9.34). Find: (i) The leading kVAr supplied by the motor, (ii) kVA rating of motor, and (iii) p.f. at which the motor operates.

Solution:

$$\text{Load, } P_1 = 200 \text{ kW}$$

$$\text{Motor load, } P_2 = 100 \text{ kW}$$

$$\text{p.f. of the load } 200 \text{ kW} = 0.7 \text{ lag}$$

$$\text{p.f. of the combined load } (200 + 100) \text{ kW} = 0.9 \text{ lag}$$

$$\text{Combined load} = P_1 + P_2 = 200 + 100 = 300 \text{ kW}$$

$\triangle$OAB is a power triangle without additional load, $\triangle$ODC the power triangle for combined load, and $\triangle$BEC for the motor load.

From Fig. 9.34, we have

(i)   Leading *kVAr* taken by the motor = CE

$$= DE - DC = AB - DC$$

$$= 200 \tan(\cos^{-1}0.7) - 300 \tan(\cos^{-1}0.9)$$

$$= 200 \times 1.02 - 300 \times 0.4843$$

$$= 58.71 \text{ KVAr}$$

(ii) *kVA* rating of motor = BE

$$= \sqrt{(BC)^2 + (EC)^2}$$

$$= \sqrt{(100)^2 + (58.71)^2}$$

$$= 115.96 \text{ kVA}$$

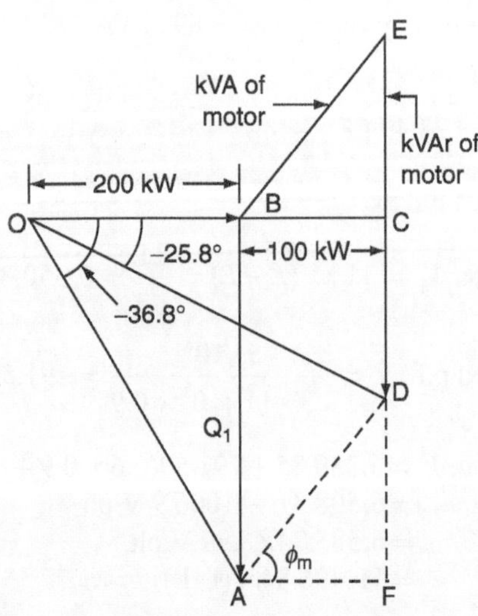

FIG. 9.34   *Phasor diagram*

(iii)   p.f. of motor at which it operates, $\cos \phi_m = \dfrac{BC}{BE}$

$$= \dfrac{100}{115.96}$$

$$= 0.862 \text{ leading}$$

Example 9.10: A 37.3 kW induction motor has p.f. 0.9 and efficiency 0.9 at full-load, p.f. 0.6, and efficiency 0.7 at half-load. At no-load, the current is 25% of the full-load current and p.f. 0.1. Capacitors are supplied to make the line p.f. 0.8 at half-load. With these capacitors in circuit, find the line p.f. at (i) full-load and (ii) no-load.

Solution:

Full-load current, $I_1 = 37.3 \times 10^3/(\sqrt{3} \; V_L \times 0.9 \times 0.9) = 26{,}586/V_L$

At full load:

Motor input, $P_1 = 37.3/0.9 = 41.44$ kW

Lagging $kVAr$ drawn by the motor, $kVAr_1 = P_1 \tan \phi_1 = 41.44 \tan(\cos^{-1} 0.9) = 20.07$

At half load:

Motor input, $P_2 = (0.5 \times 37.3)/0.7 = 26.64$ kW

Lagging $kVAr$ drawn by the motor, $kVAr_2 = P_2 \tan \phi_2 = 26.64 \tan(\cos^{-1} 0.6) = 35.52$

At no load:

No-load current, $I_0 = 0.25 \text{ (full-load current)} = 0.25 \times 26{,}586/V_L = 6{,}646.5/V_L$

Motor input at no-load, $P_o = \sqrt{3} \; V_L I_0 \cos \phi_0$

$$= \sqrt{3} \times 6{,}646.5 \times V_L \times 0.1/V_L = 1.151 \text{ kW}$$

Lagging $kVAr$ drawn by the motor, $kVAr_0 = 1.151 \tan(\cos^{-1} 0.1)$

$$= 11.452$$

Lagging $kVAr$ drawn from the mains at half-load with capacitors,

$$kVAr_{2c} = 26.64 \tan(\cos^{-1} 0.8) = 19.98$$

$kVAr$ supplied by capacitors, $kVAr_c = kVAr_2 - kVAr_{2c} = 35.52 - 19.98 = 15.54$

$kVAr$ drawn from the main at full load with capacitors $kVAr_{1c} = kVAr_1 - kVAr_c$

$$= 20.07 - 15.54 = 4.53$$

(i)   Line p.f. at full load $= \cos\left(\tan^{-1} \dfrac{kVAr_{1c}}{P_1}\right) = \cos(\tan^{-1} 4.53/41.44)$

$$= 0.994 \text{ lagging}$$

(ii) $kVAr$ drawn from mains at no-load with capacitors $= 11.452 - 15.54 = -4.088$

Line p.f. at no-load $= \cos(\tan^{-1} -4.088/1.151) = \cos(-74.27°) = 0.271 \text{ leading}$.

Example 9.11: A single-phase system supplies the following loads:

(i)   Lighting load of 50 kW at unity p.f.

(ii)   Induction motor load of 125 kW at p.f. 0.707 lagging.

(iii)   Synchronous motor load of 75 kW at p.f. 0.9 leading.

(iv)   Other miscellaneous loads of 25 kW at p.f. 0.8 lagging.

Determine the total kW and kVA delivered by the system and the p.f. at which it works.

Solution:

$$\text{Total kW of the load} = 50 + 125 + 75 + 25 = 275 \text{ kW}$$
$$kVAr \text{ of lighting load} = 50 \times 0 = 0$$
$$kVAr \text{ of induction motor} = -125 \tan(\cos^{-1} 0.707) = -125.04$$
$$kVAr \text{ of synchronous motor} = 75 \tan(\cos^{-1} 0.9) = 36.32$$
$$kVAr \text{ of miscellaneous loads} = 25 \tan(\cos^{-1} 0.8) = -18.75$$
$$\therefore \text{Total } kVAr \text{ of the load} = 0 - 125.04 + 36.32 - 18.75 = -107.47$$

$$\text{Total } kVA \text{ load} = \sqrt{(kW)^2 + (kVAr)^2} = \sqrt{(275)^2 + (107.47)^2} = 295.25$$

$$\text{p.f.} = \frac{\text{Total } kW}{\text{Total } kVA} = \frac{275}{295.25} = 0.93 \text{ lagging}$$

KEY NOTES

- For quality power, i.e., voltage and frequency at every supply point would remain constant, free from harmonics, and the power factor would remain unity, compensation is required.

- The objectives of load compensation are:

  (i) Power-factor correction.

  (ii) Voltage regulation improvement.

  (iii) Balancing of load.

- Characteristics of the ideal compensator are to:

  (i) Provide a controllable and variable amount of reactive power without any delay according to the requirements of the load.

  (ii) Maintain constant-voltage characteristics at its terminals.

  (iii) Operate independently in the three phases.

- **Voltage regulation** is defined as the proportional change in supply-voltage magnitude associated with a defined change in load current, i.e., from no-load to full load.

- The SIL of a transmission line is the MW loading of a transmission line at which a natural reactive power balance occurs (zero resistance).

- **Voltage stability:** A power system at a given operating state and subjected to a given disturbance is voltage stable if voltages near loads approach post-disturbance equilibrium values. The disturbed state is within the regions of attractions of stable post-disturbance equilibrium.

- **Voltage collapse:** Following voltage instability, a power system undergoes voltage collapse if the post-disturbance equilibrium voltages near the load are below the acceptable limits. The voltage collapse may be either total or partial.

SHORT QUESTIONS AND ANSWERS

(1)   Define the need of compensation.

For maintaining the quality power, i.e., voltage and frequency at every supply point would remain constant, free from harmonics and the p.f. would remain unity and compensation is needed.

(2)   What are the objectives of load compensation?

The objectives of load compensation are:

  (i) p.f. correction.

  (ii) Voltage regulation improvement.

  (iii) Balancing of load.

(3)   What are the characteristics of an ideal compensator?

The characteristics of the ideal compensator are:

(i) To provide a controllable and variable amount of reactive power without any delay according to the requirements of the load.

(ii) To maintain a constant-voltage characteristic at its terminals and

(iii) Should operate independently in the three phases.

(4) Define the voltage regulation.

It is defined as the proportional change in supply voltage magnitude associated with a defined change in load current, i.e., from no load to full load.

(5) Define the surge impedance loading (SIL) of a transmission line.

It is the MW loading of a transmission line at which a natural reactive power balance occurs (zero resistance).

(6) What is meant by voltage stability?

A power system at a given operating state and subjected to a given disturbance is voltage stable if voltages near loads approach post-disturbance equilibrium values. The disturbed state is within the regions of attractions of stable post-disturbance equilibrium.

(7) What is meant by voltage collapse?

Following voltage instability, a power system undergoes voltage collapse if the post-disturbance equilibrium voltages near the load are below acceptable limits. The voltage collapse may be either total or partial.

MULTIPLE-CHOICE QUESTIONS

(1) The major reason for low lagging p.f. of supply system is due to the use of ___________ motors.

(a) Induction.

(b) Synchronous.

(c) DC.

(d) None of these.

(2) The maximum value of p.f. can be ___________.

(a) 1.

(b) 0.9.

(c) 0.8.

(d) 0.7.

(3) By improving the p.f. of the system, the kilowatts delivered by generating stations are ___________.

(a) Decreased.

(b) Increased.

(c) Not changed.

(d) None of these.

(4) Power factor can be improved by installing such a device in parallel with load, which takes:

(a) Lagging reactive power.

(b) Leading reactive power.

(c) Apparent power.

(d) None of these.

(5) The main reason for low p.f. of supply system is due to the use of ___________.

(a) Resistive load.

(b) Inductive load.

(c) Synchronous motor.

(d) All of these.

(6) The only motor that can also be worked at leading p.f. and can supply mechanical power ___________.

(a) Synchronous induction generator.

(b) Synchronous motor.

(c) Alternator.

(d) None of these.

(7) An over-excited synchronous motor on no-load is known as ___________.

(a) Synchronous induction generator.

(b) Synchronous condenser.

(c) Alternator.

(d) None of these.

(8) For synchronous condensers, the p.f. improvement apparatus should be located at ___________.

(a) Sending end.

(b) Receiving end.

(c) Both (a) and (b).

(d) None of these.

(9) A disadvantage of synchronous condenser is:

(a) Continuous losses in motor.

(b) High maintenance cost.

(c) Noisy.

(d) All of the above.

(10) The smaller the lagging reactive power drawn by a circuit, its p.f. will be ___________.

(a) Better.

(b) Poorer.

(c) Unity.

(d) None of these.

(11) kVAR is equal to ___________.

(a) kW tan ϕ.

(b) kW sin ϕ.

(c) kVA cos ϕ.

(d) None of these.

(12) For a particular power, the current drawn by the circuit is minimum when the value of p.f. is ___________.

(a) 0.8 lagging.

(b) 0.8 leading.

(c) Unity.

(d) None of these.

(13) Synchronous capacitors are normally ___________ cooled.

(a) Air.

(b) Oil.

(c) Water.

(d) None of these.

(14) To improve the p.f. of 3-ϕ circuits, the size of each capacitor when connected in delta with respect to when connected in star is ___________.

(a) 1/6th.

(b) 1/4th.

(c) 3 times.

(d) 1/3rd.

(15) The p.f. improvement equipment is always placed ___________.

(a) At the generating station.

(b) Near the transformer.

(c) Near the apparatus responsible for low p.f.

(d) Near the bus bar.

(16) A synchronous machine has higher capacity for:

(a) Leading p.f.

(b) Lagging p.f.

(c) It does not depend upon the p.f. of the machine.

(d) None of these.

(17) If a synchronous machine is underexcited, it takes lagging VARs from the system when it is operated as a ___________.

(a) Synchronous motor.

(b) Synchronous generator.

(c) Synchronous motor as well as generator.

(d) None of these.

(18) A synchronous phase modifier as compared to synchronous motor used for mechanical loads has ___________.

(a) Larger shaft and higher speed.

(b) Smaller shaft and higher speed.

(c) Larger shaft and smaller speed.

(d) Smaller shaft and smaller speed.

(19) The phase advancer is mounted on the main motor shaft and is connected in the ___________ motor.

(a) Rotor.

(b) Stator.

(c) Core.

(d) None of these.

(20) Industrial heating furnaces such as arc and induction furnaces operate on ___________.

(a) Very low lagging p.f.

(b) Very low leading p.f.

(c) Very high leading p.f.

(d) None of these.

(21) If a synchronous machine is overexcited, it takes lagging VARs from the system when it is operated as:

    (a) Synchronous motor.

    (b) Synchronous generator.

    (c) Synchronous motor as well as generator.

    (d) None of these.

(22) A machine designed to operate at full load is physically heavier and is costlier if the operating p.f. is:

    (a) Lagging.

    (b) Leading.

    (c) The size and cost do not depend on p.f.

    (d) None of these.

(23) Unit of reactive power is:

    (a) MW.

    (b) MVAr.

    (c) MVA.

    (d) KVA.

(24) Reactive power is _____________ power.

    (a) Wattfull.

    (b) Wattless.

    (c) Loss.

    (d) None of these.

(25) Transmission line parameters are:

    (a) R.

    (b) L.

    (c) C.

    (d) All of these.

(26) On fundamental $T\gamma$. line expression $V(x) = Ae^{\gamma x} + Be^{-\gamma x}$, γ represents:

    (a) Distance (or) length.

    (b) Velocity of light.

    (c) Propagation constant.

    (d) None of these.

(27) Characteristic impedance is _____________.

    (a) $\sqrt{L-C}$.

    (b) $\omega\sqrt{LC}$.

    (c) $\omega/\sqrt{LC}$.

    (d) $\sqrt{\dfrac{L}{C}}$.

(28) β is _____________.

    (a) $\omega\sqrt{L/C}$.

    (b) $\omega\sqrt{LC}$.

    (c) $\omega/\sqrt{LC}$.

    (d) All of these.

(29) Advantage of operating at natural load is:

    (a) Insulation is uniformly stressed.

    (b) Reactive power balance is achieved.

    (c) Both (a) and (b).

    (d) None of these.

(30) An uncompensated line on open-circuit leads to _____________.

    (a) Ferranti effect.

    (b) Line-charging current flowing into generators is more.

    (c) Both (a) and (b).

    (d) None of these.

(31) A symmetrical line at no-load means _____________.

    (a) No power transmission.

    (b) $V_s = V_r$.

    (c) Both (a) and (b).

    (d) None of these.

(32) During the underexcited operation of a synchronous generator:

    (a) Heating of the ends of the stator core increases.

    (b) Reduces field current, results in the internal emf, which causes weak stability.

    (c) Both (a) and (b).

    (d) None of these.

(33) For a symmetrical line with $V_s = V_0$, the maximum voltage occurs at:

    (a) Sending end.

    (b) Receiving end.

    (c) Midpoint.

    (d) None of these.

(34) Unit of p.f. is:

    (a) s.

    (b) m.

    (c) No units.

    (d) None of these.

(35) Unit of time constant is:

    (a) m.

    (b) kg.

    (c) s.

    (d) miles.

(36) Power transmission through a line is improved by:

    (a) Increasing the line voltage.

    (b) Decreasing the line reactance.

    (c) Both (a) and (b).

    (d) None of these.

(37) A linear device must satisfy:

    (a) Homogeneity.

    (b) Additivity.

    (c) Both (a) and (b).

    (d) None of these.

(38) Fundamental requirements of AC-power transmission is:

    (a) Synchronous machines must remain stably in synchronizer.

    (b) Voltages must be kept near to their rated values.

    (c) Both (a) and (b).

    (d) None of these.

(39) Load compensation is:

    (a) The control of reactive power to improve quality of supply.

    (b) The control of real power to improve quality of supply.

    (c) The control of voltage and its angle to improve the quality of supply.

    (d) Both (a) and (b).

(40) Power factor under natural load is:

    (a) Lagging.

    (b) Leading.

    (c) Unity.

    (d) None of these.

(41) Steady-state stability of unit occurs when $\delta =$ __________.

    (a) 30°.

    (b) 20°.

    (c) 90°.

    (d) 0°.

(42) 'θ' in fundamental transmission line equation is:

    (a) β.

    (b) ax.

    (c) βl.

    (d) β/a.

(43) Rating of a compensator is:

    (a) MVAr.

    (b) Time of response.

    (c) Both (a) and (b).

    (d) None of these.

(44) Load compensation includes:

    (a) p.f. correction.

    (b) Voltage regulation.

    (c) Load balancing.

    (d) All of these.

(45) For a symmetrical line, the voltage is more at:

    (a) Sending end.

    (b) Receiving end.

    (c) Midpoint.

    (d) All of these.

(46) Load compensation can be achieved by:

    (a) Installing the compensating equipment near the source.

    (b) Installing the compensating equipment near the load.

    (c) Either (a) or (b).

    (d) Both (a) and (b).

(47) pf correction of load is achieved by:

    (a) Generating reactive power as close as possible to the source.

    (b) Generating reactive power as close as possible to the load.

(c) Generating real power as close as possible to the load.

(d) Generating real power as close as possible to the source.

(48) The main function of an ideal compensator is:

(a) Instantaneous pf correction to unity.

(b) Elimination (or) reduction of voltage regulation.

(c) Phase balance of the load currents and voltages.

(d) All.

(49) The important characteristic of an ideal compensator is:

(a) To provide a controllable and variable amount of reactive power without any delay.

(b) To maintain a constant voltage characteristic at its terminals.

(c) Should operate independently in the three phases.

(d) All the above.

(50) Characteristic impedance of the line depends upon:

(a) The characteristic of the line per unit length.

(b) Length of the line.

(c) Radius and spacing between conductors.

(d) All.

(51) The surge impedance loading (SIL) is expressed as:

(a) $\text{SIL} = \dfrac{\left(V_{L\text{-}L}\right)^2}{\text{surge impedance}}$.

(b) $\text{SIL} = \dfrac{V_{L\text{-}L}}{\text{surge impedance}}$.

(c) $\text{SIL} = (V_{L\text{-}L})^2 \times \text{surge impedance}$.

(d) None.

(52) When a line is loaded above its SIL, it acts like:

(a) Shunt reactor absorbing MUAR from the system.

(b) Shunt capacitor supplying MUAR to the system.

(c) Series capacitor supplying MUAR to the system.

(d) Series reactor absorbing MUAR from the system.

(53) When a line is loaded below its SIL, it acts like:

(a) Shunt reactor absorbing MUAR from the system.

(b) Shunt capacitor supplying MUAR to the system.

(c) Shunt capacitor supplying MUAR to the system.

(d) Shunt reactor absorbing MUAR from the system.

(54) If any inductive load is connected at the sending end of the line, it will support the synchronous generators:

(a) To absorb the line-charging reactive power.

(b) To absorb the load-charging reactive power.

(c) To supply the line-charging reactive power.

(d) To supply the load-charging reactive power.

(55) The change in electrical properties of a transmission line in order to increase its power transmission capability is known as:

(a) Load compensation.

(b) Line compensation.

(c) Load synchronism.

(d) Line synchronism.

(56) Apply series capacitors to reduce X_L and thereby reduce θ at the fundamental frequency. This method is called:

(a) Line-length compensation (or) θ-compensation.

(b) Compensation by sectioning.

(c) Load balancing.

(d) All the above.

(57) Series compensation results in:

(a) Increase in maximum transferable power capacity.

(b) Decrease in transmission angle for considerable amount of power transfer.

(c) Increase in virtual surge impedance loading.

(d) All the above.

(58) For a heavy loading condition, a flat voltage profile can be obtained by:

(a) Series compensation.

(b) Shunt compensation.

(c) (a) or (c).

(d) None.

(59) Inductive shunt compensation ___________ the virtual surge impedance and ___________ the virtual SIL of the line:

(a) Decreases, decreases.

(b) Decreases, increases.

   (c) Increases, decreases.

   (d) Decreases, increases.

(60) If the inductive shunt compensation is 100% then:

   (a) Flat voltage profile exists at zero loads.

   (b) Ferranti effect can be eliminated.

   (c) Both (a) and (b).

   (d) None.

(61) Sub-synchronous resonance (SSR) is treated as __________ type of phenomenon.

   (a) Electrical.

   (b) Mechanical.

   (c) Combined electrical–mechanical.

   (d) Damped frequency resonance.

(62) UPFC is able to perform:

   (a) Voltage support.

   (b) Power flow control.

   (c) Improved stability.

   (d) All.

(63) The voltage stability analysis is carried out by which power flow-based method?

   (a) P–V curves.

   (b) Q–V curves.

   (c) Both (a) and (b).

   (d) None.

(64) Voltage collapse proximate indicator (VCPI) for a radial line is defined as:

   (a) $\dfrac{dQ_s}{dQ_r}$.

   (b) $\dfrac{dQ_r}{dQ_s}$.

   (c) $\dfrac{dQ_s}{dV_s}$.

   (d) $\dfrac{dQ_r}{dV_r}$.

REVIEW QUESTIONS

(1) Explain the objectives of load compensation.

(2) Explain the voltage regulation with and without compensators.

(3) What are the specifications of load compensation?

(4) Explain the effects on uncompensated line under no-load and load conditions.

(5) Explain the effects on compensated line.

(6) Explain the concept of sub-synchronous resonance.

(7) Compare the different types of compensating equipment for transmission systems.

(8) Explain the concepts of voltage stability and voltage collapse.

(9) Derive the voltage stability index of a typical branch of a power system.

PROBLEMS

(1) A 3-ϕ, 5 kW induction motor has a p.f. of 0.8 lag. A bank of capacitors is connected in delta across the supply terminals and p.f. is raised to 0.95 lag. Determine the kVAr rating of the capacitors connected in each phase.

(2) A 3-ϕ, 50 Hz, 400 V motor develops 100 HP, the p.f. being 0.7 lag and efficiency 93%. A bank of capacitors is connected in delta across the supply terminals and p.f. is raised to 0.95 lag. Each of the capacitance units is built of four similar 100 V capacitors. Determine the capacitance of each capacitor.

(3) A star-connected 400 HP, 2,000 V, 50 Hz motor works at a p.f. of 0.75 lagging. A bank of mesh-connected condensers is used to raise the p.f. to 0.98 lagging. Determine the capacitance of each unit and total number of units required; if each is rated 500 V, 50 Hz. The motor efficiency is 85%.

(4) A 3-ϕ, 50 Hz, 3,000 V motor develops 600 HP, the p.f. being 0.75 lagging and the efficiency 0.95. A bank of capacitors is connected in delta across the supply terminals and the p.f. raised to 0.98 lagging. Each of the capacitance units is built of five similar 600 V capacitors. Determine the capacitance of each capacitor.

10 Voltage Control

O B J E C T I V E S

After reading this chapter, you should be able to:

- obtain an overview of voltage control
- discuss the parameters or equipments causing reactive power
- understand the methods of voltage control and
- calculate the rating of synchronous phase modifier

10.1 INTRODUCTION

A power system must be designed in such a way so as to maintain the voltage variations at the **consumer terminals** within specified limits. In practice, all the equipments on the power system are designed to operate satisfactorily at the **rated voltages** or within speci-fied limits, at most $\pm 6\%$ at the **consumer terminals**. The main reason for the variation in voltage at the consumer terminals is the variation in load on the supply power system. In case load on the supply system increases, the voltage at the consumer terminals decreases due to an increase in voltage drop in power system components and vice versa when load is decreased. Most of the electronic equipments are sensitive to voltage variations; hence, the voltage must be maintained constant. It can be maintained within the limits by providing voltage-control equipment.

10.2 NECESSITY OF VOLTAGE CONTROL

The voltage at the consumer terminals changes with the variation in load on the supply system, which is undesirable due to the following reasons:

(i) In case of **lighting load**, for example, incandescent lamp is acutely sensitive to voltage changes. Fluctuations in voltage beyond a certain level may even decrease the life of the lamp.

(ii) In case of **power load** consisting of **induction motors,** the voltage variations may cause a variation in the torque of an induction motor, as the torque is pro-portional to the square of the terminal voltage. If the supply voltage is low, then the starting torque of the motor will be too low.

(iii) If the voltage variation is more than a specified value, then the performance of the equipments suffers and the life of the equipment is reduced.

(iv) The picture on a **television** set starts rolling if the voltage is below a certain level because the **fluorescent tube** refuses to glow at low voltages. Hence, volt-age variations must be regulated and kept to a minimum level.

Before discussing the various methods of voltage control, it is very important to know about the various sources and sinks of reactive power in a power system.

Test Yourself

Why is voltage tolerance more than frequency tolerance?

10.3 GENERATION AND ABSORPTION OF REACTIVE POWER

(i) **Synchronous machine:** These can be used either to generate or absorb reactive power. The ability to supply reactive power is determined by the short-circuit ratio $\left(\dfrac{1}{\text{synchronous reactance}}\right)$. An **overexcited synchronous machine** generates $kVAr$ and acts as a shunt capacitor, while a **underexcited synchronous machine** absorbs it and acts as a shunt reactor. The machine is the main source of supply to the system of both positive and negative $VArs$.

(ii) **Overhead lines:** When fully loaded, lines absorb reactive power with a current I amperes for a line of reactance per phase X ohms, the $VArs$ absorbed are I^2X per phase. On light loads, the shunt capacitances of longer lines may become predominant and the lines then become VAr generators.

(iii) **Transformers:** Transformers absorb reactive power. The mathematical expression for the reactive power absorbed by a transformer is $Q_T = 3\,|I|^2 X_T\ VAr$.
where X_T is the transformer reactance per phase in ohms and $|I|$ is the current flowing through in amperes.

(iv) **Cables:** Cables act as VAr generators because they have a very small inductance and relatively very large capacitance due to the nearness of the conductors.

(v) **Loads:** A load at 0.8 p.f. implies a reactive power demand of 0.75 kVAr per kW of power, which is more significant than the simple quoting of the p.f. In planning power systems, it is required to consider reactive power requirements to ascertain whether the generator is able to operate any range of p.f.

10.4 LOCATION OF VOLTAGE-CONTROL EQUIPMENT

The consumer apparatus should operate satisfactorily. This is achieved by installing voltage-control equipment at suitable places.

The voltage-control equipment is placed in two or more than two places in a power system because of the following reasons:

(i) The power system is a combination of wide-ranging networks and there is a voltage drop in different sections of the distribution and transmission systems.

(ii) The various circuits of a power system have different load characteristics.

The voltage-control equipment is placed at:

(i) Generating stations.

(ii) Transformer stations.

(iii) The feeders.

When power is supplied to a load through a transmission line keeping the sending-end voltage constant, the receiving-end voltage varies with magnitude of load and p.f. of the load. The higher the load with smaller p.f., the greater is the voltage variation.

10.5 METHODS OF VOLTAGE CONTROL

The different voltage-control methods are:

   (i)   Excitation control.

  (ii)   Shunt capacitors.

 (iii)   Series capacitors.

 (iv)   Tap-changing transformers.

  (v)   Boosters.

 (vi)   Synchronous condensers.

10.5.1 Excitation control

This method is used only at the **generating station**. Due to the voltage drop in the synchronous reactance of armature, the alternator terminal voltage changes and hence the load on the supply system also undergoes a change. This can be maintained constant by changing the field current of the alternator. This process is called excitation control. By using an automatic or a hand-operated regulator, the excitation of the alternator can be controlled.

In modern systems, automatic regulator is preferred. The two main types of automatic voltage regulators are:

  (a)   Tirril regulator.

  (b)   Brown—Boveri regulator

(a) Tirril automatic regulator: Tirril regulator is a fast-acting **electromagnetical regulator** and it gives $\pm 0.5\%$ regulating deviation between no-load and full load of an alternator.

Construction: Tirril voltage regulator is a **vibrating-type voltage regulator** in which a resistance R is connected in the exciter circuit to get the required value of voltage by adjusting the proper value of resistance. Figure 10.1 shows the main parts of the Tirril voltage regulator.

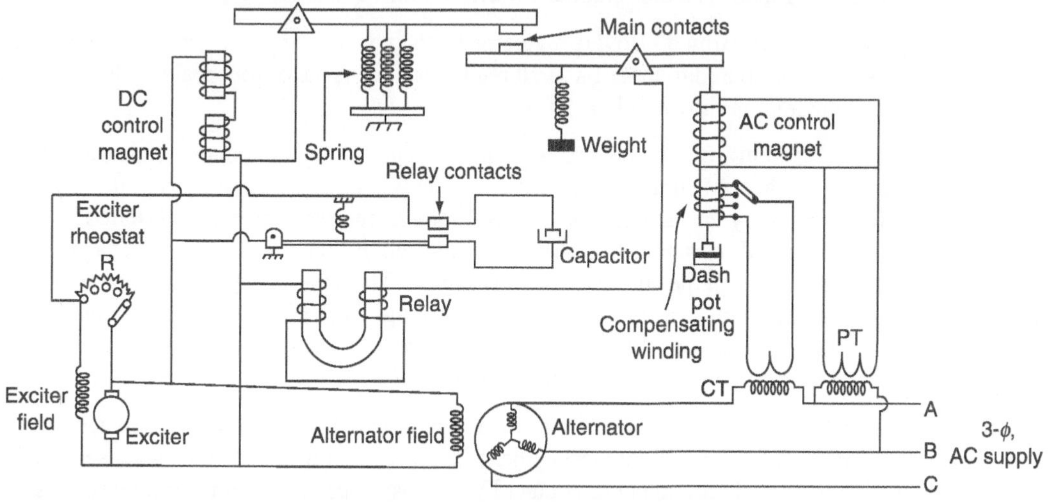

FIG. 10.1   *Tirril automatic voltage regulator*

Differential relay: It is a 'U'-shaped (horseshoe) relay magnet. It has two identical windings on both limbs as shown in Fig. 10.1, which are connected across the armature of the exciter only when the main contacts are closed. A **capacitor** is connected in parallel to the relay for reducing the spark when the relay contacts are opened.

Excitation system: It consists of a solenoid energized by the voltage equal to the exciter terminal voltage. The counter-balance force of an excitation solenoid is provided by three springs, which are acting in sequence and are shown in Fig. 10.1.

Main control unit: It is a **solenoid** excited from an AC supply. The lower part of this solenoid is connected with a dashpot, which provides damping to the measuring unit.

Main contacts: These are attached to the levers that are operated by measuring and excitation solenoids as shown in Fig. 10.1. The lever on the left side is controlled by the exciter control magnet and the lever on the right side is controlled by the main control magnet.

Principle of operation: Under normal operating conditions, i.e., the system is operating at pre-set load and voltage conditions, the main contacts are open. The **field rheostat** is in the circuit. If the load on the alternator increases, the terminal voltage decreases. When the pre-set excitation settings of the device is low, the m.m.f developed by the measuring system or the solenoid is low, causing a disturbance in the equilibrium and, therefore, main contacts are closed. These results in de-energization of differential relay and relay contacts are closed. So, the resistance 'R' in the field is short-circuited. When this is out of circuit, total field current flows through the exciter, and the exciter terminal voltage increases. Thus, the voltage across the alternator terminals increases due to the increase in alternator field current.

Due to this increased voltage, the pull of the solenoid exceeds the spring force and so the main contacts are opened again and the resistance is inserted in the exciter field. A similar process is repeated if the terminal voltage is reduced.

(b) Brown–Boveri regulator: This differs from the Tirril regulator. In this, the resistance of regulator is either gradually varied or varied in small steps.

Construction: Brown–Boveri regulator is not a vibrating type; hence, wear and tear is less when compared to that of a Tirril regulator. It consists of four main parts and its schematic diagram is shown in Fig. 10.2(a).

Control system: It contains two windings 'P' and 'Q' wound on an annular core of laminated steel sheet as shown in Fig. 10.2(a). The windings are excited from the three-phase alternator supply through the resistances R_c and R_f and resistance R_{se} is inserted in winding 'P'. The ratio of resistance to reactance is adjusted in such a way so as to get a phase angle difference between the currents in two windings. This results in the formation of a rotating magnetic field and hence develops an electromagnetic torque on the aluminium drum D. This torque depends on the terminal voltage of the alternator and on the resistances R_c and R_f. The torque decreases with increased values of R_f.

Operating system: It consists of two resistance sectors made up of contact blocks on the inner surface of roll-contact segments as shown in Fig. 10.2(b). Contact segments and resistance sectors are made to contact by using springs. The two resistance sectors R and R are connected in series, and this combination is connected in series with exciter field circuits. If the alternator voltage changes from its pre-determined value, the contact segments

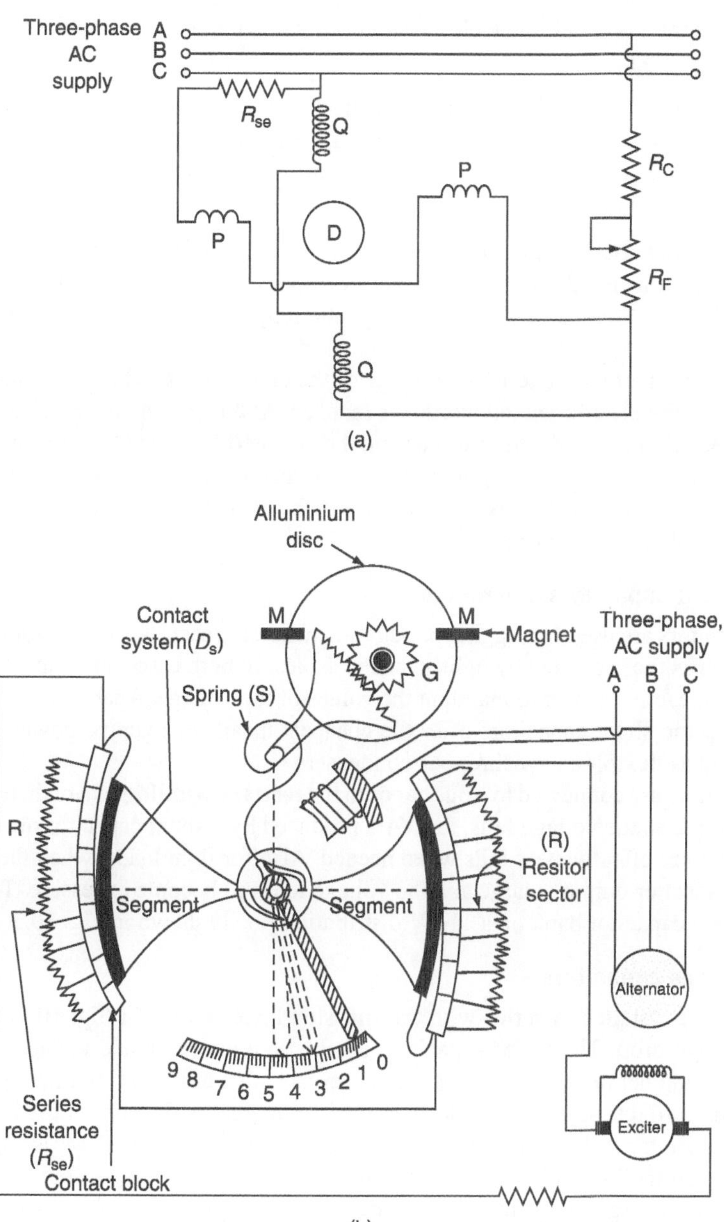

FIG. 10.2 *(a) Schematic diagram of Brown–Boveri regulator; (b) detailed diagram of Brown–Boveri regulator*

roll on the inside of resistance sectors, rotates clock-wise or anti-clock-wise under the action of the two windings P and Q.

Mechanical control torque: Mechanical torque is produced by **springs** (main and auxiliary) and is independent of the position of the control system. In a steady deflection state, the mechanical torque is equal to the electrical torque, which is produced by the current in the split-phase winding.

Damping torque: It consists of an aluminium disc, which is rotated in between two **magnets** M and M, and a spring S is attached to it. When there is a change in the alternator voltage, eddy currents are produced in the disc and torque is developed; thereby controlling the response of the moving system.

Principle of operation: Suppose the voltage of the alternator terminals is set to the normal value by adjusting R_c and R_f and is in Position-3 on the scale. In this position, the mechanical torque is equal to the electromagnetic torque and the moving system is under equilibrium.

Let us assume that the terminal voltage of the alternator is reduced due to the rise in load, and then the electromagnetic torque is reduced. At this instant, the mechanical torque is greater than the electromagnetic torque and the disc starts to rotate (assume in anti-clockwise direction). Due to this, the pointer moves to Position-1. The resistance in the exciter field will be reduced, which causes an increase in the exciter field current. So, the terminal voltage of an alternator increases.

10.5.2 Shunt capacitors and reactors

Shunt capacitors are used for lagging p.f. circuits; whereas reactors are used for leading p.f. circuits such as those created by lightly loaded cables. In both cases, the effect is to supply the required reactive power to maintain the values of the voltage. Apart from synchronous machines, static shunt capacitors offer the cheapest means of reactive power supply but these are not as flexible as synchronous condensers.

Capacitors are connected to a bus bar or to the **tertiary winding** of a main transformer. In this method, as the voltage falls, the VArs produced by a shunt capacitor or reactor also falls. Thus, their effectiveness falls when needed. Also for light loads, when the voltage is high, the capacitor output is large and the voltage tends to become excessive. The view of a three-phase capacitor bank on a 11-kV distribution line is shown in Fig. 10.3.

11.5.3 Series capacitors

Capacitors are installed in series with transmission lines (shown in Fig. 10.4) in order to reduce voltage drop. The series capacitors compensate the reactance voltage drop in the line by reducing net reactance. A capacitor in series with a transmission line serving a lagging p.f. load will cause a rise in voltage as the load increases. The p.f. of the load through the series capacitor and line must be lagging if the voltage drop is to decrease appreciably. The voltage on the load side of the series capacitor is raised above the source side, acting to improve the voltage regulation of the feeder. Since the voltage rise or drop occurs instantaneously with variations in the load, the series capacitor response as a voltage regulator is faster and smoother than the regulators.

The main drawback of this capacitor is the high voltage produced across the capacitor terminals under short-circuit conditions. The drop across the capacitor is $I_f X_c$, where I_f is the fault current which is many times the full-load current under certain circuit conditions. It is essential, therefore, that the capacitor is taken out of service as quickly as possible. A spark gap with a high-speed contactor can be used to protect the capacitor under these conditions.

FIG. 10.3 *View of a three-phase capacitor bank on a 11-kV distribution line*

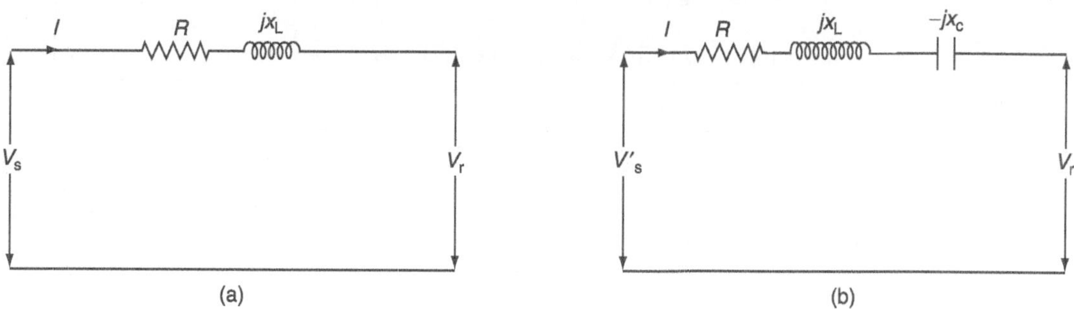

FIG. 10.4 *Circuit diagram without and with series compensation*

Figures 10.4 and 10.5 show the line and its voltage phasor diagrams without and with series compensation. The voltage drop of the line without a series capacitor is approximately given by

$$V_d = I_r R \cos \phi + I_r X_L \sin \phi$$

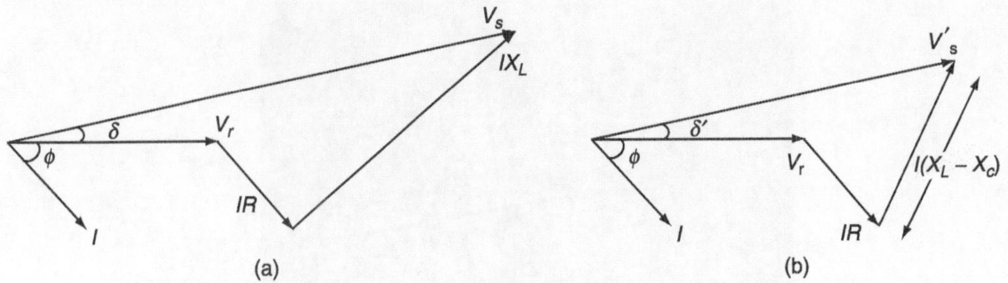

FIG. 10.5   *Phasor diagrams of Fig. 10.4*

and the voltage drop with a series capacitor,

$$V_{\mathrm{d}} = I_{\mathrm{r}} R \cos \phi + I_{\mathrm{r}} (X_{\mathrm{L}} - X_{\mathrm{C}}) \sin \phi$$

where X_{c} is the capacitive reactance of the series capacitor. A comparison between shunt and series capacitors is tabulated below (Table 10.1).

10.5.4 Tap-changing transformers

A tap-changing transformer is a static device having a number of tap settings on its secondary side for obtaining different secondary voltages. The basic function of this device is to change the **transformation ratio**, whereby the voltage in the secondary circuit is varied making possible voltage control at all voltage levels at any load. The supply may not be interrupted when tap changing is done with and without load.

Types of tap-changing transformers are:

(a)   Off-load tap-changing transformer.

(b)   On-load tap-changing transformer.

TABLE 10.1   *Comparison of shunt and series capacitors*

Shunt capacitor	Series capacitor
1. Supplies fixed amount of reactive power to the system at the point where they are installed. Its effect is felt in the circuit from the location towards supply source only	1. Quantum of compensation is independent of load current and instantaneous changes occur. Its effect is from its location towards the load end
2. It reduces the reactive power flowing in the line and causes: (a) Improvement of p.f. of a system (b) Voltage profile improvement (c) Decreases kVA loading on source, i.e., generators, transformers, and line upto location and thus provides an additional capacity	2. It is effective: (a) On tie lines, the power transfer is greater (b) Specifically, suitable for situations when flickers due to respective load functions occur

3. The location has to be as near to the load point as possible. In practice, due to the high compensation required, it is found to be economical to provide group compensation on lines and sub-stations

3. As a thumb rule, the best location is 1/3rd of electrical impedance from the source bus

4. As fixed kVAr is supplied, this may sometimes result in overcompensation in the light-load period. Switched kVAr banks are comparatively costlier than fixed kVAr and become necessary

4. As full-load current is to pass through, the capacity (current rating) should be more than the load current

5. As the p.f. approaches unity, larger compensation is required for the improvement of p.f.

5. As series capacitors carry fault current, special protection is required to protect from fault current

6. Where lines are heavily loaded, compensation required will be more

6. Causes sudden rises in voltage at the location

7. Cost of compensation is lower than that of the cost required for series capacitor

7. Cost of a series capacitor is higher than that of a shunt capacitor

10.5.4.1 *Off-load tap-changing transformers*

The simple tap-changing arrangement of a transformer is shown in Fig. 10.6. The voltage can be varied by varying a number of tappings on the secondary side of the transformer as shown in Fig. 10.6.

Figure 10.6 refers to the off-load tap-changing transformer, which requires the disconnection of the transformer from the load when the tap setting is to be changed.

The output of the secondary side of the transformer changes with the change in the tap position of the secondary winding. The secondary voltage is minimum when the movable arm makes contact with stud 1, whereas it is maximum when it is in position N. When the load on the transformer increases, the voltage across secondary terminals decreases. This can be increased to the desired value by adding the number of turns on the secondary of the transformer by changing taps.

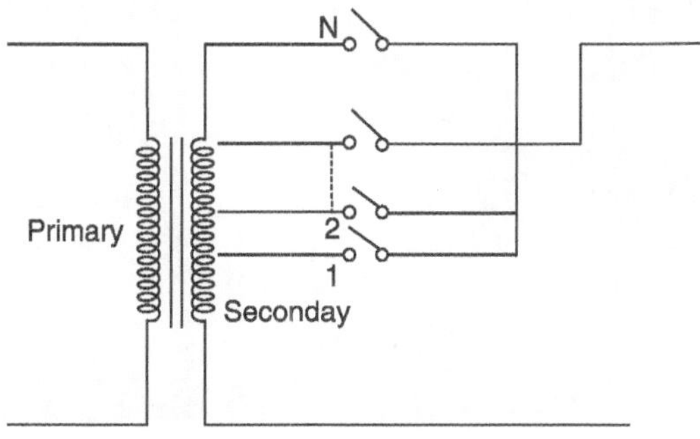

FIG. 10.6   *Off-load tap-changing transformer arrangement*

Thus, in the case of tap-changing transformers, the main drawback is that the taps are changed only after the removal of the load. This can be overcome by using an on-load tap-changing transformer with reactors.

10.5.4.2 On-load tap-changing transformer

To supply uninterrupted power to the load (consumer), tap changing has to be performed when the system is on load. The secondary winding in a tap-changing transformer consists of two identical parallel windings with similar tappings. For example, 1, 2, ..., N and $1'$, $2'$, ..., N' are the tappings on both the parallel windings of such a transformer. These two parallel windings are controlled by switches S_a and S_b as shown in Fig. 10.7(a). In the

FIG. 10.7   *(a) On-load tap-changing transformer arrangement; (b) on-line tap-changing transfomer*

normal operating conditions, switches S_a, S_b, and tappings 1 and 1′ are closed, i.e., both the secondary windings of the transformer are connected in parallel, and each winding carries half of the total load current by an equal sharing. The secondary side of the transformer is at a rated voltage under no load, when the switches S_a and S_b are closed and movable arms make contact with stud 1 and 1′, whereas it is maximum (above the rated value) under no load, when the movable arms are in position N and N′. The voltage at the secondary terminal decreases with an increase in the load. To compensate for the decreased voltages, it is required to change switches from positions 1 and 1′ to positions 2 and 2′ (number of turns on the secondary is increased). For this, open any one of the switches S_a and S_b, assuming that S_a is opened. At this instant, the secondary winding controlled by switch S_b carries full-load current through one winding. Then, the tapping is changed to position 2 on the winding of the disconnected transformer and close the switch S_a. After this, switch S_b is opened for disconnecting its winding, and change the tapping position from 1′ to 2′ and then switch S_b is closed. Similarly, tapping positions can be changed without interrupting the power supply to the consumers. The online tap-changing transformer is shown in Fig. 10.7(b).

This method has the following disadvantages:

(i)   It requires two windings with rated current-carrying capacity instead of one winding.

(ii) It requires two operations for the change of a single step.

(iii) Complications are introduced in the design in order to obtain a high reactance between the parallel windings.

10.5.5 Booster transformers

The booster transformer performs the function of boosting the voltage. It can be installed at a sub-station or at any intermediate point of line.

In the circuit shown in Fig. 10.8(a), P and Q are the two relays. The secondary of the booster transformer is connected in series with the line whose voltage is to be controlled, and the primary of the booster transformer is supplied from a regulating transformer with on-load tap-changing gear. The booster can be brought into the circuit by the closure of relay Q and the opening of relay P, and vice versa as shown in Fig. 10.8(a). The secondary of the booster transformer injects a voltage in phase with the line voltages. By changing the tapping on the regulating transformer, the magnitude of V_Q can be changed and thus the feeder voltage V_F can be regulated. The view of booster and distribution transformer connection (left to right) is shown in Fig. 10.8(b).

Advantages

(i)   It can be installed at any intermediate point in the system.

(ii) The rating of the booster transformer is about 10% that of the main transformer (product of current and injected voltage).

Disadvantages

When used in conjunction with main transformer:

(i)   More expensive than a transformer with on-load tap changes.

(ii) Less efficient due to losses in booster.

(iii) Requires more space.

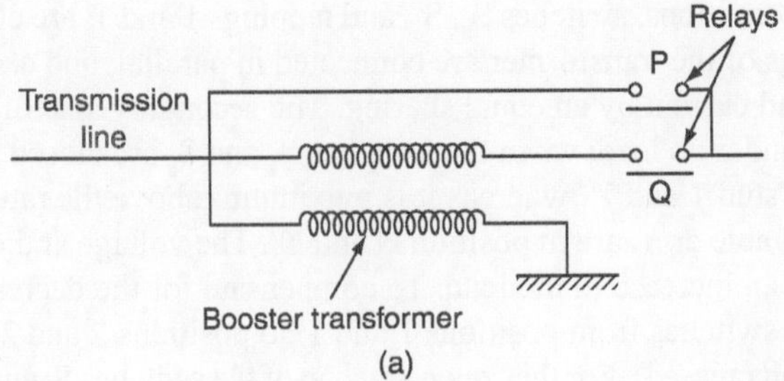

FIG. 10.8   *(a) Booster transformer; (b) view of booster and distribution transfomer connection (left to right)*

10.5.6 Synchronous condensers

A **synchronous condenser (synchronous phase modifier)** is a synchronous motor running without mechanical load. It is connected in parallel with the load at the receiving end of the line. Depending upon its excitation, it either generates or absorbs the reactive power. It takes leading current when its field is **overexcited**, i.e., above normal speed and takes lagging current when it is **underexcited**. Thus, the current drawn by a synchronous phase modifier can be varied from lagging to leading by varying its excitation. It is a very convenient device to keep

TABLE 10.2 *Comparison of synchronous condenser and static capacitors*

Synchronous condenser	Static capacitors
1. Harmonics in the voltage does not exist	1. Large harmonics are produced in the system
2. Power factor variation is stepless (uniform)	2. Power factor varies in steps
3. It allows overloading for a short period	3. It does not allow any overloading
4. Power loss is more	4. Power loss is less
5. It is more economical in the case of large kVAr	5. It is more economical for small kVAr requirement
6. Failure rate is less and, therefore, this is more reliable	6. Failure rate is more and, therefore, it is less reliable

the receiving-end voltage constant under any condition of load. It also improves the p.f. and the output can vary smoothly.

A synchronous phase modifier has a smaller shaft and bearing and higher speed as compared to a synchronous motor used for mechanical loads. A synchronous phase modifier has higher overall efficiency as compared with a synchronous motor.

Advantages

(i) Flexibility for use in all load conditions because when the machine is under excited, it consumes reactive power.

(ii) There is a smooth variation of reactive VArs by synchronous capacitors.

(iii) It can be overloaded for short periods.

Disadvantages

(i) Possibility of falling out of control in case of sudden changes in voltage.

(ii) These machines add to short-circuit capacity of the system during fault condition.

A comparison between synchronous condenser and static capacitors is presented in Table 10.2.

10.6 RATING OF SYNCHRONOUS PHASE MODIFIER

An expression of sending-end voltage in terms of transmission line constants is

$$\overline{V}_s = \overline{A}\,\overline{V}_r + \overline{B}\,\overline{I}_r \qquad (10.1)$$

where

$\overline{V}_s = V_s \angle \delta =$ sending-end voltage

$\overline{V}_r =$ receiving-end voltage (reference phasor)

$\overline{I}_r = I_r \angle -\phi_r =$ receiving-end current

$\overline{A} = A \angle \alpha$

$\overline{B} = B \angle \beta$ are the line constants

Equation (10.1) can be written in a phasor form as

$$V_s \angle \delta = AV_r \angle \alpha + BI_r \angle (\beta - \phi_r)$$

$$= AV_r \cos \alpha + jAV_r \sin \alpha + BI_r \cos(\beta - \phi_r) + jBI_r \sin(\beta - \phi_r) \tag{10.2}$$

The real part of Equation (10.2) is

$$V_s \cos \delta = AV_r \cos \alpha + BI_r \cos(\beta - \phi_r) \tag{10.3}$$

and the imaginary part is

$$V_s \sin \delta = AV_r \sin \alpha + BI_r \sin(\beta - \phi_r) \tag{10.4}$$

Squaring and adding Equations (10.3) and (10.4), we get

$$V_s^2 = A^2 V_r^2 + B^2 I_r^2 + 2ABV_r I_r \cos \alpha \cos(\beta - \phi_r) + 2ABV_r I_r \sin \alpha \sin(\beta - \phi_r)$$

$$= A^2 V_r^2 + B^2 I_r^2 + 2ABV_r I_r \cos(\alpha - \beta + \phi_r)$$

$$= A^2 V_r^2 + B^2 I_r^2 + 2ABV_r I_r \left[\cos(\alpha - \beta) \cos \phi_r - \sin(\alpha - \beta) \sin \phi_r \right] \tag{10.5}$$

Real power at receiving end, $P_r = V_r I_r \cos \phi_r$
Reactive power at receiving end, $Q_r = V_r I_r \sin \phi_r$
Receiving-end current can be written as

$$I_r = I_r \cos \phi_r - jI_r \sin \phi_r \quad (\because \text{lagging p.f.})$$

$$= I_p - jI_q$$

$$\therefore \ I_r^2 = I_p^2 + I_q^2$$

where $\ I_p = \dfrac{P_r}{V_r}, \quad I_q = \dfrac{Q_r}{V_r}$

Substituting the above quantities in Equation (10.5), we have

$$V_s^2 = A^2 V_r^2 + B^2 I_r^2 + 2ABP_r \cos(\alpha - \beta) - 2ABQ_r \sin(\alpha - \beta) \tag{10.6}$$

In Equation (10.6), I_r^2 is replaced by I_p^2 and I_q^2 expressions,

$$\therefore V_s^2 = A^2 V_r^2 + B^2 \left[\frac{P_r^2}{V_r^2} + \frac{Q_r^2}{V_r^2} \right] + 2ABP_r \cos(\alpha - \beta) - 2ABQ_r \sin(\alpha - \beta) \tag{10.7}$$

Equation (10.7) is useful for calculating the sending-end voltage by knowing the values of A, B, α, β, P_r, Q_r, and V_r (or) sometimes the sending-end and receiving-end voltages are fixed and A, B, α, β, P_r, and Q_r are given. It is required to find out the rating of the phase modifier. In this case, the required quantity is Q_r, where Q_r is the net reactive power at the receiving end and not the reactive power for the load. So, if the net reactive power required to maintain certain voltages at the two ends is known, the rating of the phase modifier can be found.

Example 10.1: A 3-ϕ overhead line has resistance and reactance per phase of 25 and 90 Ω, respectively. The supply voltage is 145 kV while the load-end voltage is maintained at 132 kV for all loads by an automatically controlled synchronous phase modifier. If the kVAr rating of the modifier has the same value for zero loads as for a load of 50 MW, find the rating of the synchronous phase modifier.

Solution:

$$\therefore V_s^2 = A^2 V_r^2 + B^2 \left[\frac{P_r^2}{V_r^2} + \frac{Q_r^2}{V_r^2} \right] + 2ABP_r \cos(\alpha - \beta) - 2ABQ_r \sin(\alpha - \beta) \qquad \textbf{(10.8)}$$

We have $V_s = AV_r + BI_r$ $\qquad\qquad\qquad$ **(10.9)**

From given data:

Sending-end voltage, $V_s = \dfrac{145}{\sqrt{3}}$

Receiving-end voltage, $V_r = \dfrac{132}{\sqrt{3}}$

Line impedance, $Z = 25 + j90$

Assuming short-line model, $V_s = V_r + I_r Z$

$$\therefore \frac{145}{\sqrt{3}} = \frac{132}{\sqrt{3}} + I_r \left(25 + j90 \right)$$

$$83.71 = 76.2 + I_r \times 93.4 \angle 74.47° \qquad\qquad \textbf{(10.10)}$$

Comparing Equations (10.9) and (10.10), we have

$\qquad A = 1; \qquad \alpha = 0,$
$\qquad B = 93.4 \quad \beta = 74.47°$
$\qquad P_r = V_r I_r \cos \phi_r = 50$ MW (given)

Substituting these values in Equation (10.8), we get

$$(83.71)^2 = 1^2 (76.2)^2 + (93.4)^2 \left[\frac{50^2}{(76.2)^2} + \frac{Q_r^2}{(76.2)^2} \right] + 2 \times 1 \times 93.4 \times \cos\left(0 - 74.47° \right) \times 50$$

$$- 2 \times 1 \times 93.4 \times Q_r \sin\left(0 - 74.47° \right)$$

$$1,200.92 \times 10^6 = 3,755.98 \times 10^6 + 1.50 Q_r^2 \times 10^6 + 2,500.7 \times 10^6 + 179.98 Q_r$$

$$1,200.92 = 3,755.98 + 1.50 Q_r^2 + 2,500.7 + 179.98 Q_r$$

$$1.50 Q_r^2 + 179.98 Q_r + 5,055.76 = 0$$

Solving the above equation, we get

$\qquad Q_r = -44.87$ MVAr

$$\frac{Q_r}{P_r} = \frac{V_r I_r \sin \phi_r}{V_r I_r \cos \phi_r}$$

$$= \frac{-44.87}{50}$$

$$\tan \phi_r = -0.8974$$

Power factor angle at the receiving-end voltage, $\phi_r = -41.9°$

$\therefore$ The power factor is $\cos\phi_r = 0.7442$ leading

$\therefore$ The rating of the synchronous modifier $= 44.87$ MVArs

Example 10.2: A 3-ϕ feeder having a resistance of 3 Ω and a reactance of 10 Ω supplies a load of 2 MW at 0.85 p.f. lag. The receiving-end voltage is maintained at 11 kV by means of a static condenser drawing 2.1 MVAr from the line (Fig. 10.9). Calculate the sending-end voltage and p.f. What is the regulation and efficiency of the feeder?

Solution:

$$\text{Load current,} \quad I_L = \frac{2,000}{\sqrt{3} \times 11 \times 0.85}$$

$$= 123.5 \text{ A}$$

$$\text{Shunt branch current,} \quad I_C = \frac{2,100}{\sqrt{3} \times 11}$$

$$= 110.22 \text{ A}$$

$$\text{Receiving-end current,} \quad I_r = I_L + I_C$$

$$= 123.5\angle -31.79° + 110.22\angle 90°$$

$$= 105 - j65 + j110.22$$

$$= 105 + j45.22$$

$$= 114.3\angle 23.3° \text{ A}$$

The vector diagram is shown in Fig. 10.10. Here, the current is a leading current. From the circuit diagram shown in Fig. 10.9,

The sending-end voltage, $V_s = V_r + I_r Z$

$$= \frac{11,000}{\sqrt{3}} \angle 0 + 114.3\angle 23.3 \times (3 + j10)$$

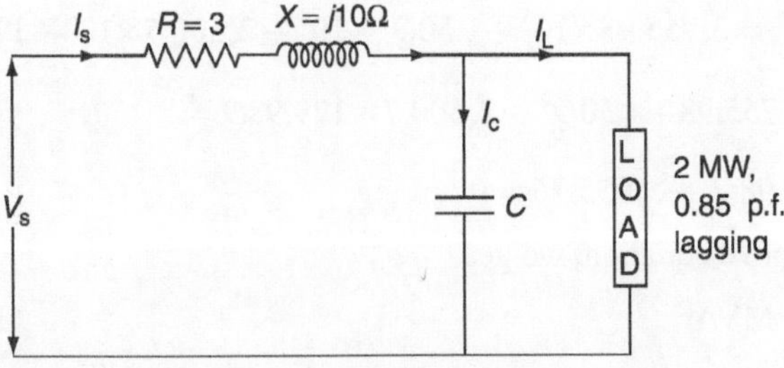

FIG. 10.9   *Circuit diagram*

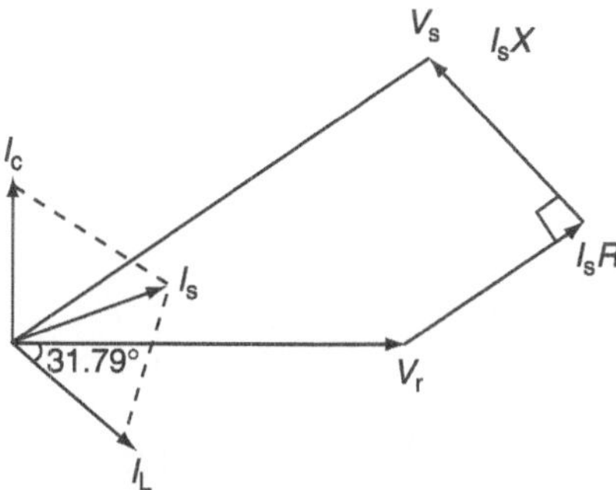

FIG. 10.10   *Phasor diagram*

$$= 6,350.85\angle 0 + 1,193.33\angle 96.6$$
$$= 6,350.85 + j0 - 137.16 + j1,185.42$$
$$= 6,213.69 + j1,185.42$$
$$= 6,325.74\angle 10.8 \text{ V}$$

$\therefore$ The sending-end voltage, $V_s = 6,325.75 \times \sqrt{3}\ (L-L)$

$$= 10.96 \text{ kV}$$

From the phasor diagram shown in Fig. 10.10,

sending-end p.f. $= \cos(23.3 - 10.8) = \cos 12.5 = 0.976$

$$\% \text{ regulation} = \frac{10.96 - 11}{11} \times 100$$

$$= -0.364\%$$

$$\text{Efficiency} = \frac{\text{output}}{\text{output} + \text{losses}}$$

$$= \frac{2,000 \times 10^3}{2,000 \times 10^3 + 3 \times (114.3)^2 \times 3} \times 100$$

$$= 94.44\%$$

Example 10.3: A single-circuit 3-ϕ 220-kV line runs at no-load. Voltage at the receiving end of the line is 205 kV. Find the sending-end voltage if the line has a resistance of 21.7 Ω, a reactance of 85.2 Ω, and the total susceptance of 5.32×10^{-4} $\mho$ (Fig. 10.11). The transmission line is to be represented by Π-model.

Solution:

The sending-end voltage V_1 differs from the receiving-end voltage V_2 by the value of voltage drop due to charging current in the line impedance, as shown in Fig. 10.11. With the quadrate-axis component of the voltage drop being neglected as shown in Fig. 10.12, we find $|V_1|$

$$|V_1| = |V_2| - \frac{Q_c}{|V_2|} X$$

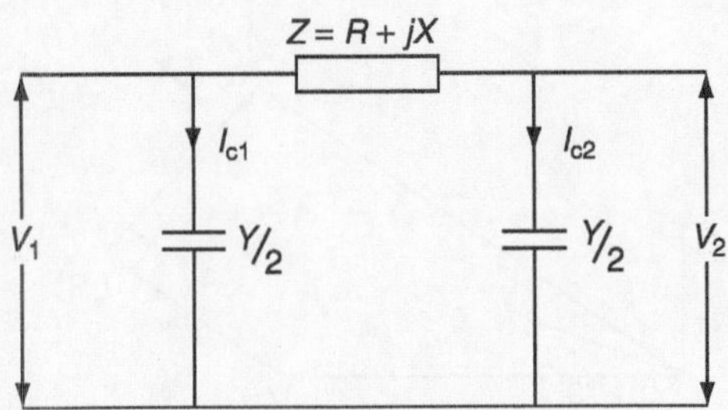

FIG. 10.11   *Circuit diagram*

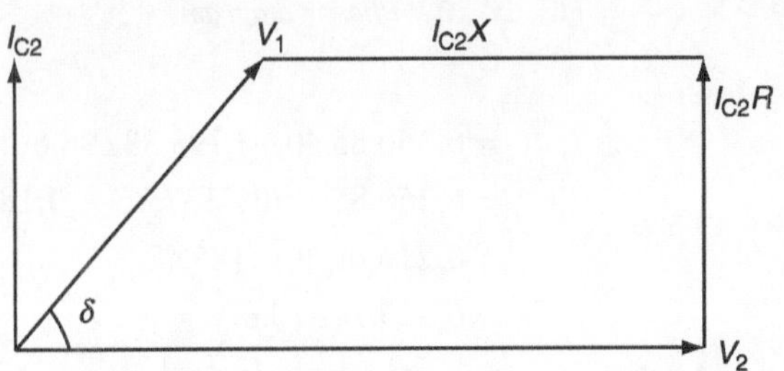

FIG. 10.12   *Phasor diagram*

We can also find $|V_1|$ from expression $|V_1| = |V_2| + \dfrac{P_2 R + Q_2 X}{|V_2|}$, where $Q_2 = -Q_c$ because

the current is leading and $P_2 = 0$.

We are given line voltage at the receiving end, therefore, on per phase, we have

$$|V_2| = \frac{205}{\sqrt{3}} = 118.35 \text{ kV}$$

$$Q_c = |V_2|^2 \frac{B}{2} = |V_2|^2 \frac{\omega c}{2}$$

$$= (118.35)^2 \times \frac{5.32}{2} \times 10^{-4} = 3.725 \text{ MVAr}$$

$X = 85.2$ (given)

Hence,$\quad |V_1| = |V_2| - \dfrac{Q_c}{|V_2|} X = 118.35 - \dfrac{3.725}{118.35} \times 85.2$

$$= 118.35 - 2.69 = 115.67 \text{ kV}$$

Sending-end voltage, line to line

$$= \sqrt{3}\,|V_1| = 200.34 \text{ kV}$$

KEY NOTES

- Sources and sinks of a reactive power are synchronous machine, overhead lines, transformers, cables, and loads.

- The voltage-control equipment is located at generating stations, transformer stations, and feeders.

- The various methods for voltage control are: excitation control, shunt capacitors, series capacitors by using tap-changing transformers, boosters, and synchronous condensers.

- **Excitation control:** This method is used only at the generating station. Due to the voltage drop in the synchronous reactance of armature, whenever the load on the supply system changes, the terminal voltage of the alternator also changes. This can be kept constant by changing the field current of the alternator according to the changes in load. This is known as excitation control.

- **Shunt capacitors and reactors:** Shunt capacitors are used for lagging p.f. circuits; whereas reactors

are used for leading p.f. circuits such as created by lightly loaded cables.

- **Series capacitor:** It is installed in series with transmission lines to reduce the frequency of voltage drops.

- **Tap-changing transformers:** The basic operation of a tap-changing transformer is by changing the transformation ratio, the voltage in the secondary circuit is varied.

- **Booster transformers:** The booster transformer performs the function of boosting the voltage. It can be installed at a sub-station or any intermediate point of line.

- **Synchronous condensers:** It is connected in parallel with the load at the receiving end of the line. It can either generate or absorb reactive power by varying the excitation of its field winding.

SHORT QUESTIONS AND ANSWERS

(1) What are the different methods of voltage control?

The following methods are generally employed for controlling the receiving-end voltage.

  (a) By excitation control.

  (b) By using tap-changing transformer.

  (c) Auto-transformer tap changing.

  (d) Booster transformer.

  (e) Induction regulators.

  (f) By synchronous condensers.

(2) What is meant by excitation voltage control?

Due to the voltage drops in the synchronous reactance of the armature, whenever the load on the supply system changes, the terminal voltage of the alternator changes correspondingly. This can be kept constant by changing the field current of the alternator according to the changes in load.

(3) What are the disadvantages of tap-changing transformers?

  (a) During switching, the impedance of transformer is increased and there will be a voltage surge.

  (b) There are twice as many tappings as the voltage steps.

(4) What is the synchronous condenser?

A synchronous motor takes a leading current when overexcited and therefore behaves as a capacitor. An overexcited synchronous motor running on no-load is known as a synchronous condenser.

(5) What is a booster transformer?

The transformer, which is used to control the voltage of the transmission line at a point far away from the main transformer, is known as booster transformer.

(6) How does a shunt-capacitor bank control the system voltage under light loads and heavy loads?

The shunt-capacitor bank provided with fixed and variable elements may be either removed from or added to the bank to decrease or increase the capacitance under no-load and heavy-load conditions, respectively.

(7) Under what condition does a synchronous motor take a leading current?

The synchronous motor takes leading current when its field is overexcited under high-load conditions.

(8) When is the shunt-inductor compensation required?

The shunt-inductor compensation is required whenever the loading is less than the surge impedance loading.

MULTIPLE-CHOICE QUESTIONS

(1) The voltage of the power supply at the consumer's service must be held substantially __________

  (a) Constant.

  (b) Smooth variation.

  (c) Random variation.

  (d) None of these.

(2) Low voltage reduces the __________ from incandescent lamps.

  (a) Power output.

  (b) Power input.

  (c) Light output.

  (d) Current.

(3) Motors operated at below normal voltage draw abnormally __________ currents.

  (a) Low.

  (b) High.

  (c) Medium.

  (d) None of these.

(4) Permissible voltage variation is __________.

  (a) $\pm 10\%$.

  (b) $\pm 20\%$.

  (c) $\pm 50\%$.

  (d) $\pm 5\%$.

(5) By drawing high currents at low voltages, the motors get __________.

  (a) Overheated.

  (b) Cool.

  (c) Constant heat.

  (d) None of these.

(6) Domestic circuits' supply voltage is __________.

  (a) 230 V.

  (b) 110 V.

  (c) 240 V.

  (d) 220 V.

(7) The voltage may normally vary between the limits of __________.

  (a) 210 and 230 V.

  (b) 230 and 240 V.

  (c) 230 and 520 V.

  (d) 210 and 235 V.

(8) Above normal voltages reduces the __________ of the lamps.

  (a) Life.

  (b) Strength.

  (c) Lighting.

  (d) Color.

(9) The voltage at the bus can be controlled by the injection of __________ power of the correct sign.

  (a) Real.

  (b) Reactive.

  (c) Complex.

  (d) Both real and reactive.

(10) General methods of voltage control are __________.

  (a) Use of tap-changing transformer.

  (b) Synchronous condensers.

  (c) Static capacitors.

  (d) All of these.

(11) Use of thyristor-controlled static compensators is _______________.

    (a) Voltage control.

    (b) Power control.

    (c) Current control.

    (d) None of these.

(12) An overexcited synchronous machine operated as generator or motor generates _______________

    (a) kVA.

    (b) kVAr.

    (c) kW.

    (d) kI.

(13) Synchronous motor running at no-load and overexcited load is known as _______________

    (a) Synchronous condenser.

    (b) Shunt capacitor.

    (c) Series capacitor.

    (d) None of these.

(14) The excitation-control method is only suitable for _______________ lines.

    (a) Short.

    (b) Medium.

    (c) Long.

    (d) All of these.

(15) It is _______________ to maintain the same voltage at both ends of transmission lines by the synchronous-condenser method.

    (a) Economical.

    (b) Not economical.

    (c) Difficult.

    (d) Easy.

(16) Shunt capacitors and reactors are used across lightly loaded lines to absorb some of the leading _______________ again to control the voltage.

    (a) VArs.

    (b) VA.

    (c) VBRS.

    (d) None of these.

(17) Disadvantages of shunt capacitors are _______________.

    (a) Fall of voltage.

    (b) Reduction in VArs.

    (c) Reduction in effectiveness.

    (d) All of these.

(18) _______________ reduces the inductive reactance between the load and the supply point.

    (a) Shunt capacitor.

    (b) Shunt reactor.

    (c) Series capacitor.

    (d) Transformer.

(19) The disadvantage of a series capacitor is that it produces _______________ voltage across the capacitor under short-circuit condition.

    (a) Low.

    (b) High.

    (c) Very low.

    (d) Either (a) or (b).

(20) A spark gap with a high-speed contactor is the _______________ used for shunt capacitor.

    (a) Protective device.

    (b) Control.

    (c) Fuse.

    (d) Circuit breaker.

(21) The different types of tap-changing transformers are _______________.

    (a) Off-load.

    (b) On-load.

    (c) Both (a) and (b).

    (d) Either (a) or (b).

(22) The purpose of using booster transformers is _______________ the voltage.

    (a) Transforming.

    (b) Bucking .

    (c) Boosting.

    (d) Bucking and boosting.

(23) More expensive, less efficient, and take more floor area are the disadvantages of the _______________ transformer.

    (a) Off-load tap.

(b) On-load.

(c) Booster.

(d) Induction.

(24) If a synchronous machine gets overexcited, takes lagging VArs from the system when it is operated as a ___________.

(a) Synchronous motor.

(b) Synchronous generator.

(c) Either (a) or (b).

(d) Synchronous phase modifier.

(25) For a synchronous phase modifier, the load angle is ___________.

(a) 0°.

(b) 25°.

(c) 30°.

(d) 50°.

REVIEW QUESTIONS

(1) Why is voltage control required in power systems? Mention the different methods of voltage control employed in power system. Explain one method of voltage control in detail giving a neat connection diagram.

(2) Why is excitation control necessary in an alternator?

(3) Describe 'off-load' and 'on-load' tap-changing transformers.

(4) Explain the function of a synchronous phase modifier placed at the receiving end of the transmission line.

(5) Show with the aid of a vector diagram, how the voltage at the receiving end of a transmission line can be maintained constant by the use of a synchronous phase modifier.

PROBLEMS

(1) A 3-ϕ 33-kV overhead transmission line has a resistance of 5 Ω/phase and a reactance of 18 Ω/phase with the help of a synchronous modifier, the receiving-end voltage is kept constant at 33 kV. Calculate the kVA of the phase modifier if the load at the receiving end is 60 MW at 0.85 p.f. lagging. What will be the maximum load that can be transmitted?

(2) If the voltage at the sending end is to be maintained at 66 kV, determine the MVAr of the phase modifier to be installed for a 3-ϕ overhead transmission line having an impedance of (7 + j 19) Ω/phase, delivering a load of 80 MW at 0.85 p.f. lagging and with voltage 66 kV.

(3) A 3-ϕ induction motor delivers 450 HP at an efficiency of 95% when the operating p.f. is 0.85 lag. A loaded synchronous motor with a power consumption of 110 kW is connected in parallel with the induction motor. Calculate the necessary kVA and the operating p.f. of the synchronous motor if the overall p.f. is to be unity.

11 Modeling of Prime Movers and Generators

OBJECTIVES

After reading this chapter, you should be able to:

- develop the modeling of hydraulic and steam turbines
- discuss reheat and non-reheat type of steam turbine configurations
- develop the simplified models of a synchronous machine
- discuss the application of Park's transformation to synchronous machine modeling
- study the swing equation model of a synchronous machine

11.1 INTRODUCTION

For the study of power system dynamics, the simple equivalent methods of the synchronous generators are not adequate. The accurate description of power system dynamics requires the detailed models of synchronous machines.

In this work, the above-required detailed models of synchronous machines are developed from the basic equations. The time-invariant synchronous machine equations are developed through the application of PARK'S transformation and with the use of phase variables.

The detailed synchronous machine model is derived accompanied by its representation using per unit quantities and the consideration of d-axis and q-axis equivalent circuits.

First, the most simplified model of the synchronous machine is discussed and later the detailed model is developed.

The effect of saliency is discussed. The steady-state model and dynamic model representations of a synchronous machine are discussed. Finally, the mechanical behavior of a synchronous machine is studied in the form of derivation of swing equation.

The speed control of a prime mover is essential for the frequency regulation of a power system network. This is achieved by providing a speed-governor mechanism. The parallel operation for generators requires droop characteristics incorporated in the speed governing system to secure stability and economic division of load. Hence, to maintain constant frequency, it is necessary for the prime mover control to adjust the power generation according to economic dispatch of load among various units.

The prime mover controls are classified into three different categories as:

(i) Primary control (speed-governor control).

(ii) Secondary control (load frequency control (LFC)).

(iii) Tertiary control involving economic dispatch.

In this unit, models of a hydraulic turbine with a penstock system and a steam turbine are developed. In modeling the steam turbine, six common system configurations in the form of non-reheat type and reheat type are considered.

11.2 HYDRAULIC TURBINE SYSTEM

The representation of a hydro-turbine is highly dependent on the type of prime mover because each type has different speed control mechanisms.

According to the type of head conditions, there are three types of hydro-turbines.

(i) **Low head:** Up to 100" height, specific speed (90−180 rpm), speed (100−400 rpm).

These are propeller type of reaction turbines.

(ii) **Medium head:** 50"−1,000" height, specific speed (90−200 rpm), speed (100−400 rpm).

These are Francis type of reaction turbines.

(iii) **High speed:** From 800" and above height, specific speed (3−7 rpm), speed (120−720 rpm). These are of impulse type of turbines (Pelton wheel).

11.2.1 Modeling of hydraulic turbine

The transient characteristics of hydro-turbines are obtained by the dynamics of water flow in the penstock. A hydraulic turbine with a penstock system is as shown in Fig. 11.1.

Let l be the length of the penstock in m, Q the discharge of water to the turbine in m³/s, v the velocity of water discharge in m/s, and H the operating head in m.

For a particular change in load, let ΔH be the p.u. change in head, ΔN the p.u. change in speed, ΔX the p.u. change in turbine gate opening, ΔQ the p.u. change in water discharge, and ΔT the p.u. change in turbine torque.

It has been proved that

$$\frac{\Delta H}{\Delta Q} = -Z\frac{1-e^{-2\tau_e s}}{1+e^{-2\tau_e s}} \tag{11.1}$$

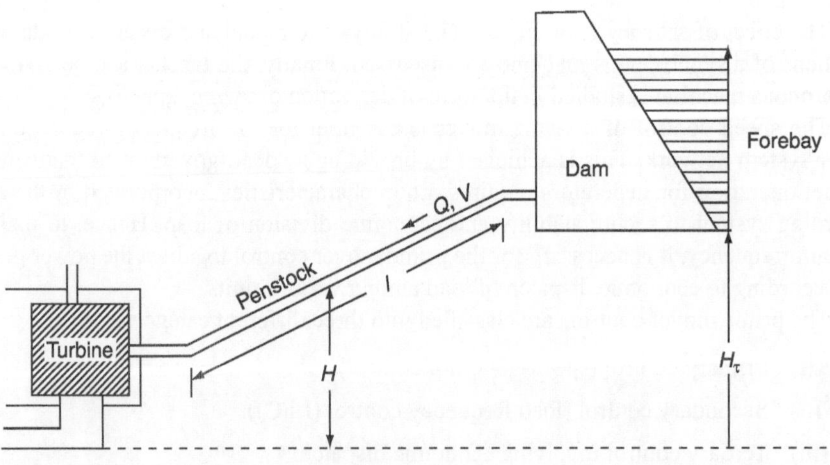

FIG. 11.1   *Hydraulic turbine with penstock system*

where Z is the normalized penstock impedance, τ_e the elastic limit of penstock, and s the complex frequency $= \sigma + j\omega$.

According to first-order PADE's approximation, we have

$$e^{-x} = \frac{1 - \dfrac{x}{2}}{1 + \dfrac{x}{2}}$$

$$\frac{\Delta H}{\Delta Q} = -\tau_e Z s = -s\tau_\omega \tag{11.2}$$

where the time constant τ_w is called the water starting time or water time constant.

The changes in flow and torque of the turbine about a steady-state condition can be represented by the following linearized equations:

$$\Delta Q = a_{11}\Delta H + a_{12}\Delta N + a_{13}\Delta x \tag{11.3}$$

$$\Delta T = a_{21}\Delta H + a_{22}\Delta N + a_{23}\Delta x \tag{11.4}$$

where a_{11}, a_{12}, a_{13}, a_{21}, a_{22}, and a_{23} are constants and are expressed as

$$a_{11} = \frac{\partial \Delta Q}{\partial \Delta H}, \quad a_{21} = \frac{\partial \Delta T}{\partial \Delta H}$$

$$a_{12} = \frac{\partial \Delta Q}{\partial \Delta N}, \quad a_{22} = \frac{\partial \Delta T}{\partial \Delta N}$$

$$a_{13} = \frac{\partial \Delta Q}{\partial \Delta X}, \quad a_{23} = \frac{\partial \Delta T}{\partial \Delta X}$$

The change in turbine torque due to the speed changes can be neglected in comparison to other changes since the change in speed is relatively small.

$$\therefore \Delta Q = a_{11}\Delta H + a_{13}\Delta X$$

$$\Delta T = a_{21}\Delta H + a_{23}\Delta X$$

$\therefore$ By using Equations (11.2) and (11.3), we have

$$\frac{\Delta T}{\Delta X} = \frac{a_{21}\Delta H + a_{23}\Delta X}{\Delta X}$$

$$= 3a_{23} \left| \frac{1 - \dfrac{a_{13}a_{21}}{a_{23}} - a_{11}\tau_\omega s}{1 + a_{11}\tau_\omega s} \right|$$

For an ideal turbine with valve opening X_0,

$$a_{11} = 0.5X_0$$
$$a_{12} = 0$$
$$a_{13} = 1.0$$
$$a_{21} = 1.5X_0$$
$$a_{22} = -1.0$$
$$a_{23} = 1.0$$

$$\therefore \frac{\Delta T}{\Delta X} = \frac{1-\left(1.5X_0 + 0.5X_0\right)\tau_\omega s}{1+0.5X_0\tau_\omega s} = \frac{1-X_0\tau_\omega s}{1+0.5X_0\tau_\omega s}$$

At full-load, $X_0 = 1.0$ p.u.

$$\therefore \frac{\Delta T}{\Delta X} = \frac{1-\tau_\omega s}{1+0.5\tau_\omega s} \tag{11.5}$$

Equation (11.5) is the transfer function of classical hydraulic turbine of penstock model.

i.e., $$G_{\text{HT}}(s) = \frac{1-\tau_\omega s}{1+0.5\tau_\omega s} = \frac{\Delta T}{\Delta X}$$

The approximate linear models for hydro-turbines are as shown in Fig. 11.2.

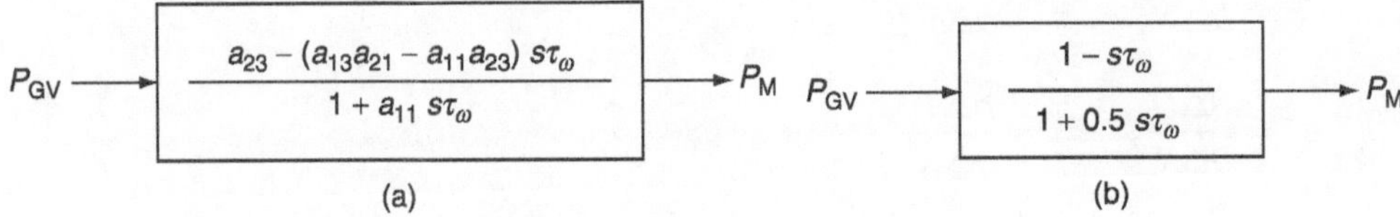

FIG. 11.2   *(a) Linearized model of hydraulic turbine; (b) linearized model of an ideal hydraulic turbine*

The input P_{GV} for the hydraulic turbine is given from the speed governor. It is the gate opening expressed in p.u.

The value of τ_ω lies in the range of 0.5 s–5.0 s. The typical value of τ_ω is around 1.0 s.

11.2.1.1 *Calculation of water time constant* (τ_ω)

Water time constant is associated with the acceleration time for water in the penstock between the turbine inlet and the forebay or between the turbine inlet and the surge tank if it exists.

The water time constant τ_ω is given by

$$\tau_\omega = \frac{lv}{H_T g}$$

where l is the length of penstock in m, v the velocity of water flow in m/s, H_T the total head in m, and g the acceleration due to gravity in m/s².

In terms of power generated by the plant 'P', the water time constant τ_ω is expressed as

$$\tau_\omega = \frac{11.8Pl}{H_T^2\,Aeg}$$

or

$$\tau_\omega = \frac{0.366Pl}{H_T^2\,Ae}$$

where P is the generated electrical power in kW and is given as

$$P = \frac{vH_T Ae}{11.8}$$

where A is the average penstock area in m² and e is the product of efficiencies of turbine and generator:

i.e., $\quad e = \eta_{\text{turbine}} \times \eta_{\text{generator}}$

11.3 STEAM TURBINE MODELING

The two common steam turbine system configurations are:

(1) Non-reheat type.

(2) Reheat type.

11.3.1 Non-reheat type

A simple non-reheat type turbine is modeled by a single time constant.

The functional block diagram representation of a non-reheat type of steam turbine is as shown in Fig. 11.3.

The approximate linear model of the non-reheat steam turbine is shown in Fig. 11.4.

Here, P_{GV} represents the power at the gate of the valve outlet, τ_{CH} the steam-chest time constant, and P_{m} the mechanical power at the turbine shaft.

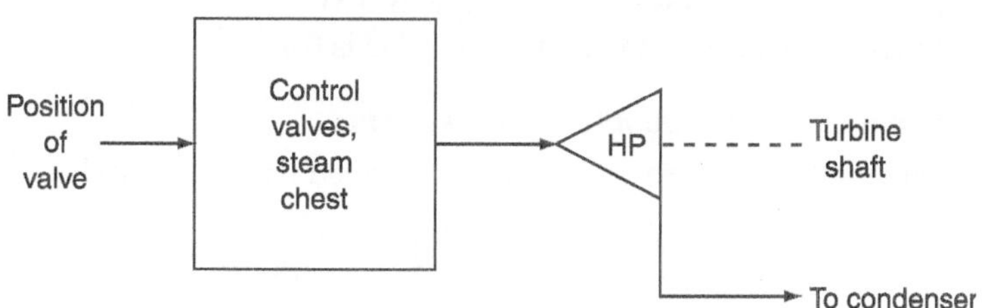

FIG. 11.3   *Block diagram representation of a non-reheat type of steam turbine*

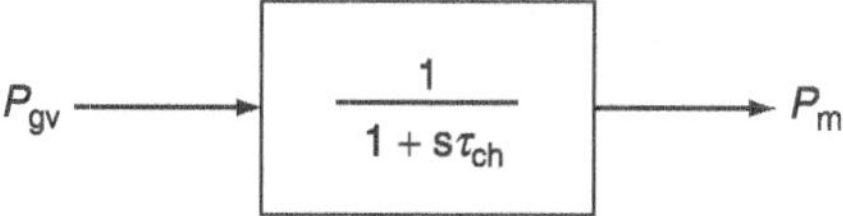

FIG. 11.4   *Approximate linear model of a non-reheat steam turbine*

11.3.2 Reheat type

There are mainly two configurations and they are:

    (i)   Tandem compound system configuration.

    (ii) Cross-compound system configuration.

These two configurations are further classified into the following types:

    (a) Tandem compound, single reheat type.

    (b) Tandem compound, double reheat type.

    (c) Cross-compound, single reheat type with two low-pressure (LP) turbines.

    (d) Cross-compound, single reheat type with single LP turbine.

    (e) Cross-compound, double reheat type.

A tandem compound system has only one shaft on which all the turbines are mounted. The turbines are of high pressure (HP), low pressure LP, and intermediate pressure (IP) turbines. Sometimes, there may be also a very high pressure (VHP) turbine mounted on the shaft.

The functional block diagram representations of tandem compound reheat system configurations and their linear model representations are shown in Figs. 11.5–11.8.

11.3.2.1 *Tandem compound single reheat system*

The tandem compound single reheat system is shown in Figs. 11.5 and 11.6.

All compound steam turbines use governor-controlled valves, at the inlet to the HP turbine, to control the steam flow. The steam chest, reheater, and cross-over piping introduce delays. These time delays are represented by:

τ_{CH} = Steam-chest time constant (from 0.1 to 0.4 s)

τ_{RH} = Reheat time constant (from 4 to 11 s)

τ_{CO} = Cross-over time constant (from 0.3 to 0.5 s)

The fractions of total turbine power are represented by:

F_{HP} = Fraction of HP turbine power (typical value is 0.3)

F_{IP} = Fraction IP turbine power (typical value is 0.3)

F_{LP} = Fraction of LP turbine power (typical value is 0.4)

11.3.2.2 *Tandem compound double reheat system*

The tandem compound double reheat system is shown in Figs. 11.7 and 11.8.

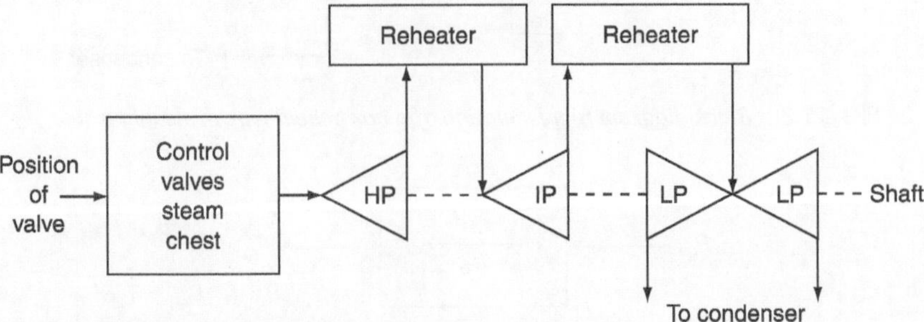

FIG. 11.5   *Functional block diagram representation—tandem compound single reheat system*

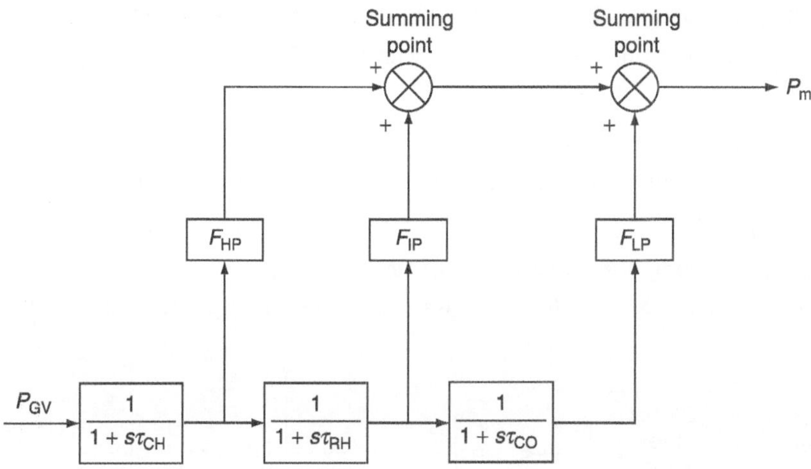

FIG. 11.6   *Approximate linear model—tandem compound single reheat system*

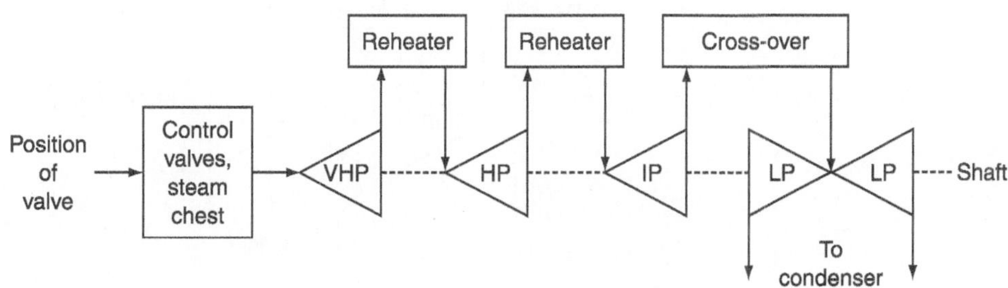

FIG. 11.7   *Functional block diagram representation—tandem compound double reheat system*

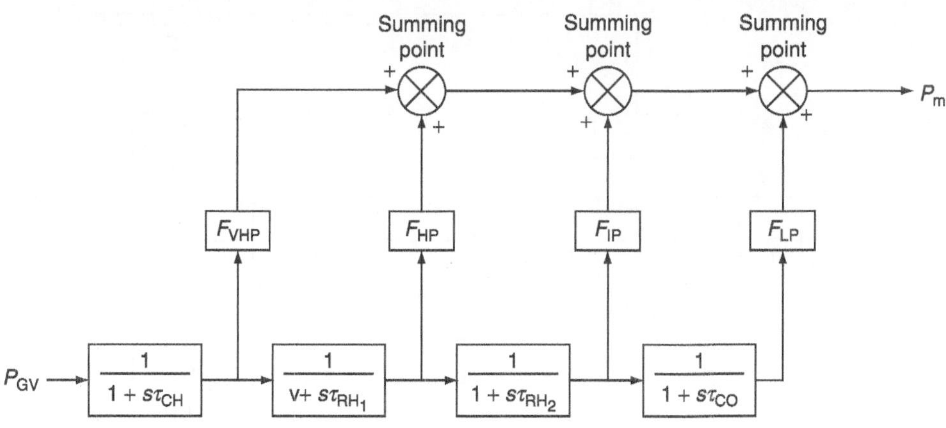

FIG. 11.8   *Approximate linear model—tandem compound double reheat system*

The time delays are represented by:

τ_{RH_1} = first reheat time constant

τ_{RH_2} = second reheat time constant

11.3.2.3 *Cross-compound single reheat system (with two LP turbines)*

The cross-compound single reheat system with two LP turbines is shown in Figs. 11.9 and 11.10.

11.3.2.4 *Cross-compound single reheat system (with single LP turbine)*

The cross-compound single reheat system with a single LP turbine is shown in Figs. 11.11 and 11.12.

11.3.2.5 *Cross-compound double reheat type*

The cross-compound double reheat-type system is shown in Figs. 11.13 and 11.14.

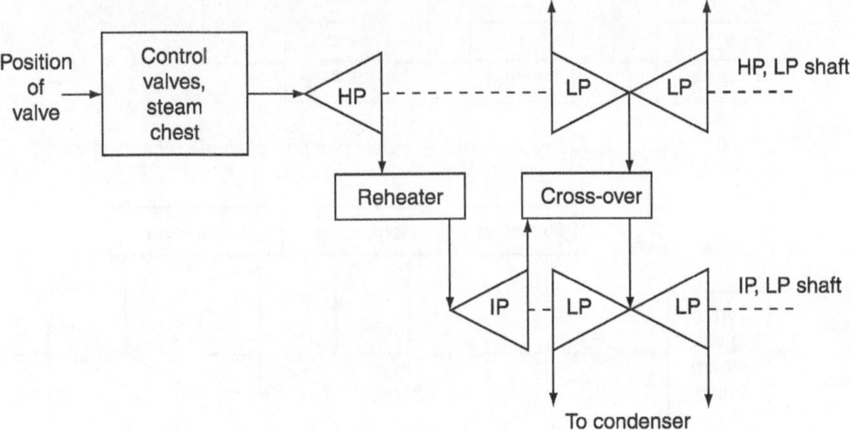

FIG. 11.9   *Functional block diagram representation*

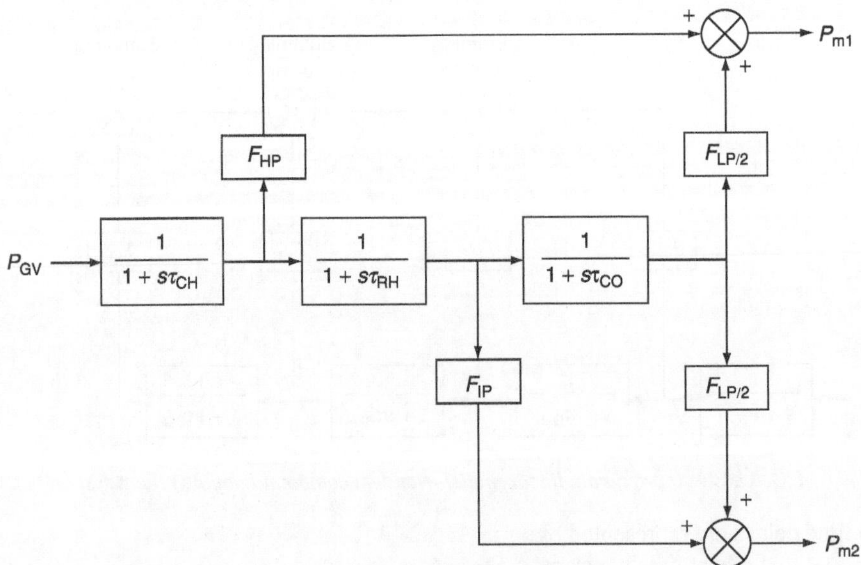

FIG. 11.10   *Approximate linear model*

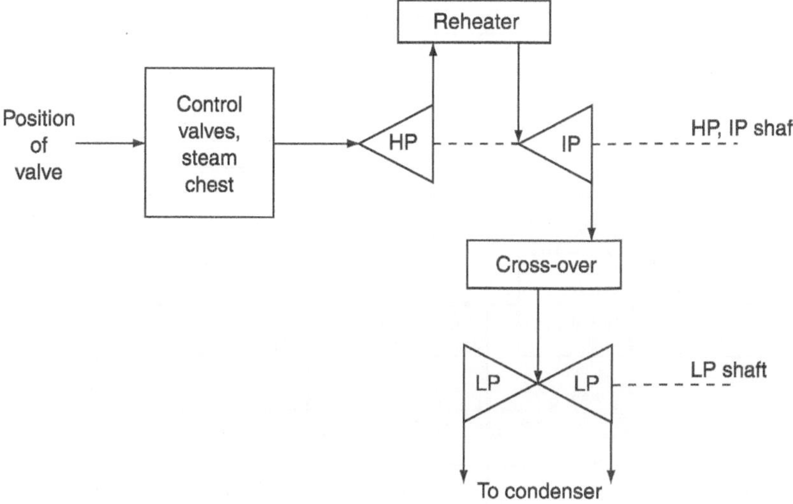

FIG. 11.11   *Functional block diagram representation*

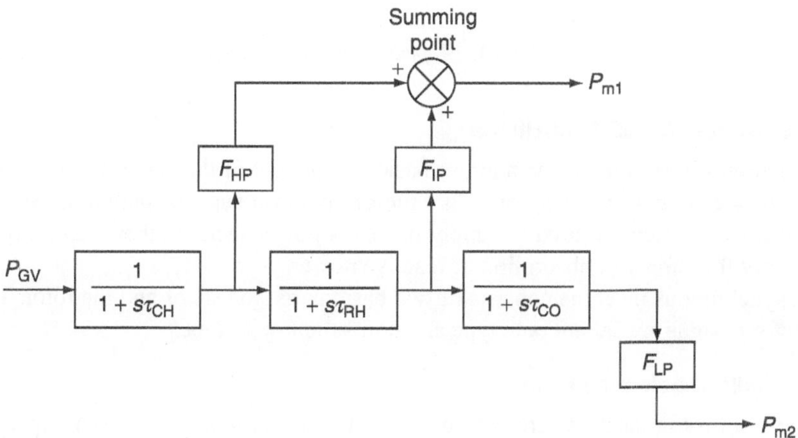

FIG. 11.12   *Approximate linear model*

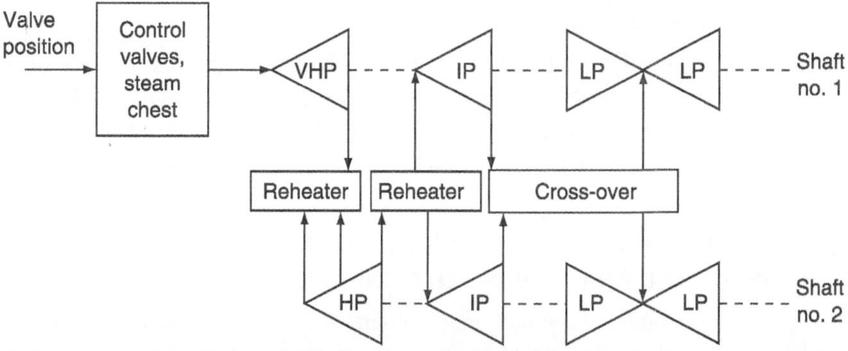

FIG. 11.13   *Functional block diagram representation*

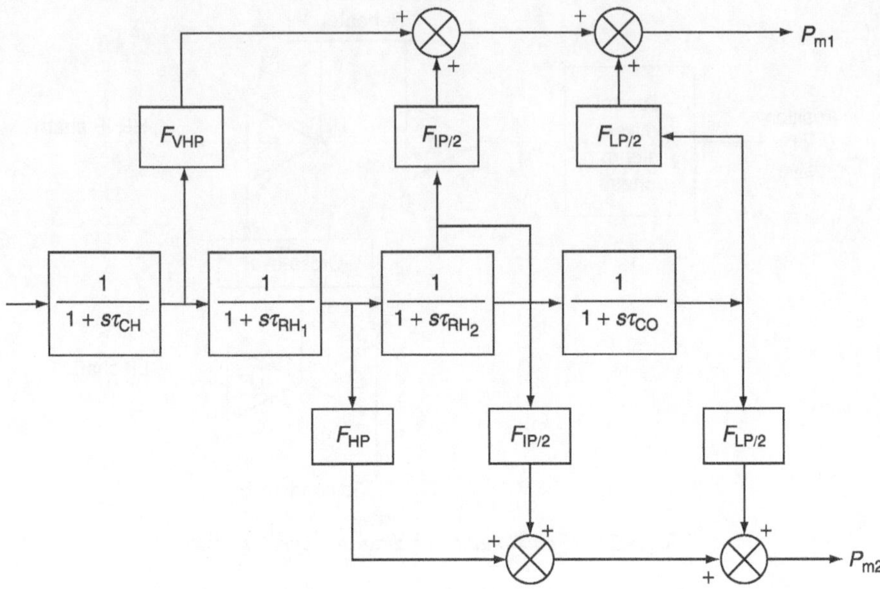

FIG. 11.14   *Approximate linear model*

11.4 SYNCHRONOUS MACHINES

The synchronous machine is the main or basic component of the electrical power system. It may operate either as a generator or as a motor. In power system operation, the synchronous machine is often required to supply power at power factors other than unity, which necessitates the supply or absorption of reactive power.

A synchronous machine consists of two basic parts: the stator and the rotor. The two basic rotor designs are salient pole type and non-salient-pole type.

11.4.1 Salient-pole-type rotor

In this type of rotor, the poles project from the rotor and exhibit a narrow air gap under the pole structure and a wider air gap between the poles. This type of rotor structure is shown in Fig. 11.15.

11.4.2 Non-salient-pole-type rotor

It consists of a cylindrical rotor as shown in Fig. 11.16, often made from a single steel forging, in which the field winding is embedded in longitudinal slots machined in its structure.

The mathematical modeling of a synchronous machine is complicated because of its multitude of windings, all characterized by time-varying self-inductances and mutual inductances.

11.5 SIMPLIFIED MODEL OF SYNCHRONOUS MACHINE (NEGLECTING SALIENCY AND CHANGES IN FLUX LINKAGES)

The most simplified model of a synchronous generator for the purpose of transient stability studies is a constant voltage source behind proper reactance. The voltage source may be subtransient, transient, or steady state and the reactance may be the corresponding reactances.

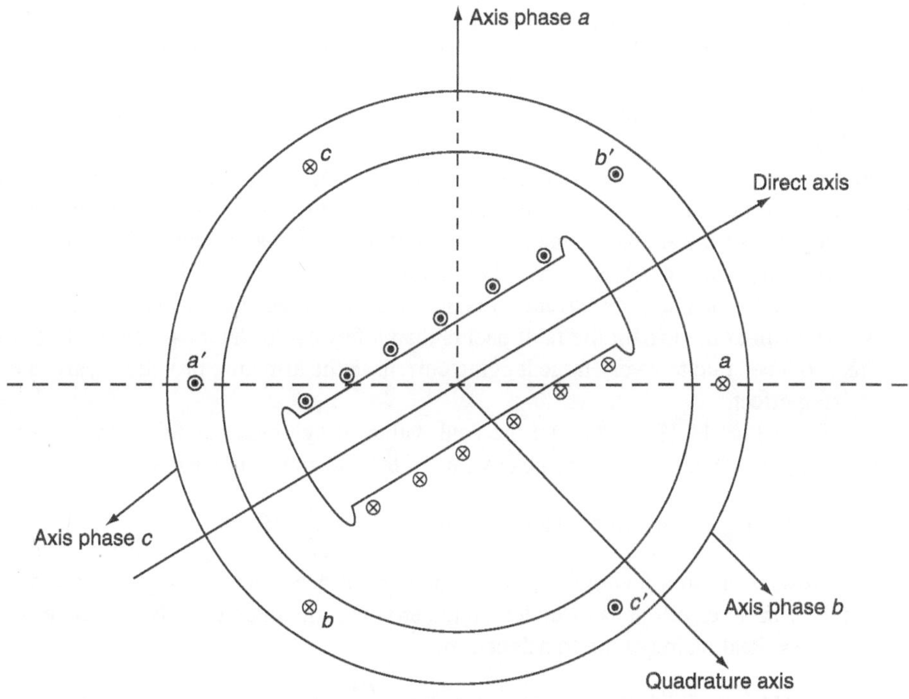

FIG. 11.15 *Salient-pole-type rotor structure*

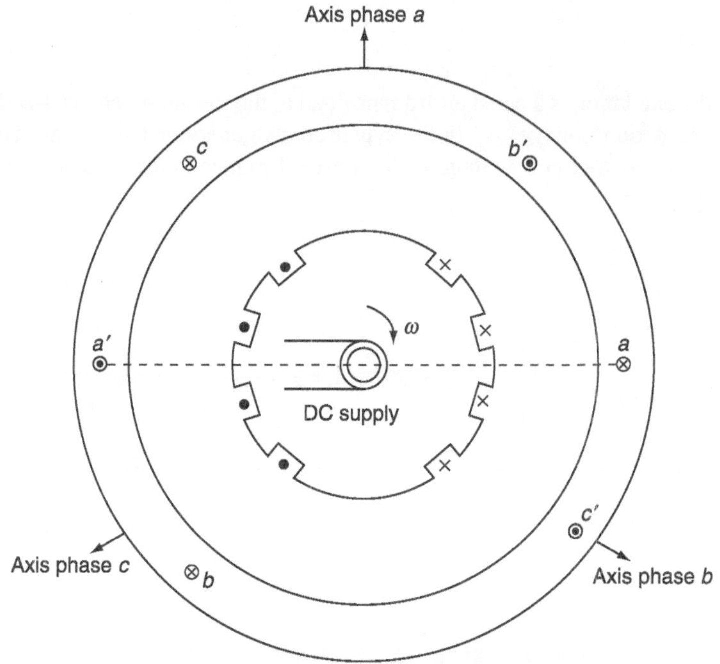

FIG. 11.16 *Non-salient-pole-type rotor structure*

In this model, saliency and changes in the flux linkages are neglected. However to understand this model, let us consider a synchronous generator operating at no-load before a 3-ϕ short circuit is applied at its terminals. The current flowing in the synchronous generator just after the 3-ϕ short circuit occurs at its terminals is similar to the current flows in an R–L circuit when an AC voltage is suddenly applied. Hence, the current will have both the AC (i.e., steady state) component as well as the DC (i.e., transient) component, which decays exponentially with the time constant L/R. If the DC component is neglected, the oscillograph of the AC component of the current that flows in the synchronous generator just after the fault occurs will have the shape as shown in Fig. 11.17.

Just after the fault, the current is maximum as the air gap flux, which generates voltage, is maximum at the instant the fault occurs than a few cycles later as the armature reaction flux produced due to a very large lagging current in the armature provides nearly a demagnetizing effect.

From Fig. 11.17, let OA be the peak value of symmetrical AC current (neglecting DC component), also known as peak value of the sub-transient current:

$$\therefore \text{ RMS value of sub-transient current} = I'' = \frac{\text{OA}}{\sqrt{2}} \tag{11.6}$$

Now, if the first few cycles, where the current decrement is very fast, are neglected and the current envelope is extended up to zero time, the intercept OB is obtained:

OB = Peak value of the transient current

$$\therefore \text{ RMS value of transient current} = I' = \frac{\text{OB}}{\sqrt{2}} \tag{11.7}$$

However, the steady-state value of the short-circuit current (i.e., sustained value of

$$\text{short-circuit current)} = \frac{\text{OC}}{\sqrt{2}} = I \tag{11.8}$$

Since the excitation is a constant from no-load to the instant when the 3-ϕ short circuit occurs, the excitation voltage 'E_g' in the synchronous generator will remain constant and is known as an 'open-circuit voltage or the no-load induced emf', and is represented as shown in Fig. 11.18.

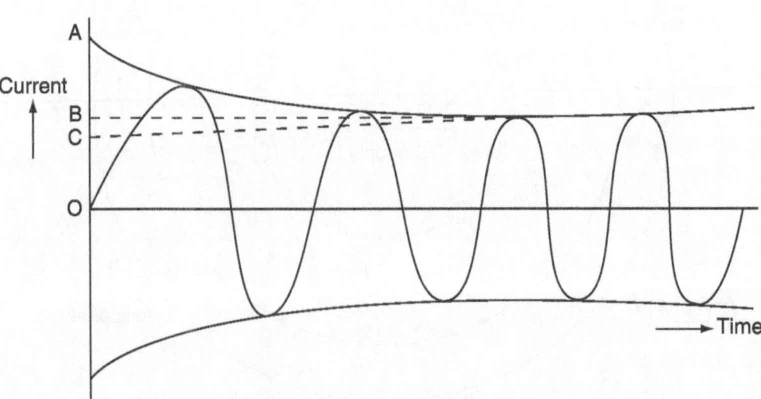

FIG. 11.17   *Oscillograph of the current in the synchronous generator*

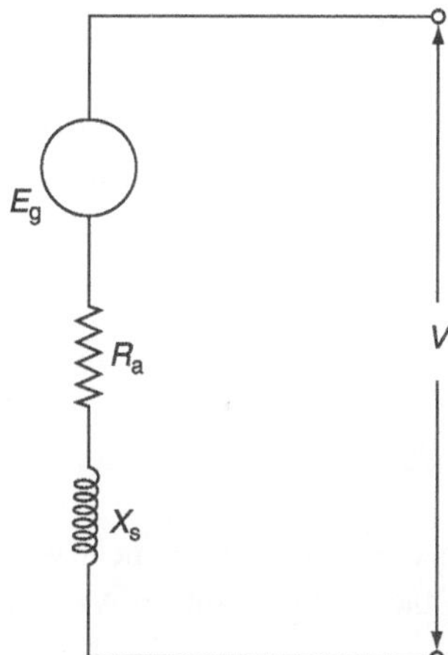

FIG. 11.18   *Equivalent circuit of the synchronous generator*

The phasor diagram of a non-salient-pole synchronous generator for steady-state analysis is as shown in Fig. 11.19.

Now, the machine equation becomes

$$E_g = V + I_a R_a + j I_a X_s \tag{11.9}$$

where E_g is the excitation voltage (or) open-circuit voltage,

V is the full-load terminal voltage,

I_a is the armature current,

R_a is the armature resistance/phase,

X_s is the synchronous reactance/phase,

ϕ is the phase angle between V and I_a,

δ is the torque angle or power angle, and

θ the impedance angle $\left(\tan \theta = \dfrac{X_s}{R_a} \right)$.

It is seen that the current in the synchronous generator changing from sub-transient state (I'') to transient state (I') and to steady state (I) and hence the synchronous reactance of the generator must change, as E_g is constant, from sub-transient reactance (X'') to transient reactance (X') to steady-state reactance (X).

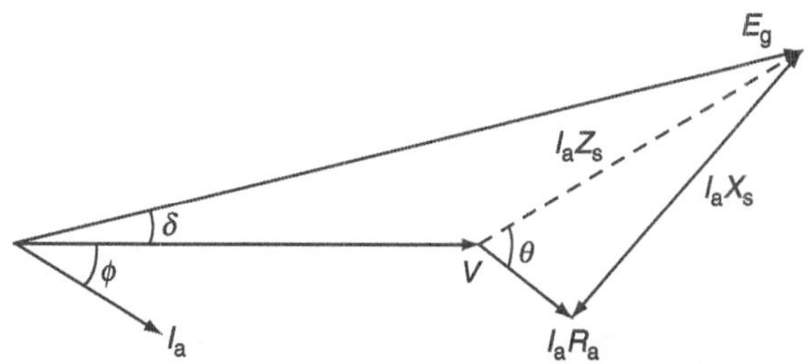

FIG. 11.19   *Phasor diagram of a non-salient-pole synchronous generator for steady-state analysis*

i.e., $\quad X'' = \dfrac{E_g}{I''}$

$$X' = \dfrac{E_g}{I'}$$

$$X = \dfrac{E_g}{I} \tag{11.10}$$

The armature reaction flux is produced by a large lagging current as this current is limited only by armature impedance, where winding resistance is negligible compared to synchronous reactance, X_s.

This armature reaction flux at this instant is nearly demagnetizing in nature because it acts along the direct axis of the machine; the above reactances are referred to as direct axis reactances, i.e.,

$$X_d'' = \dfrac{E_g}{I''} = \text{direct axis sub-transient reactance}$$

$$X_d' = \dfrac{E_g}{I'} = \text{direct axis transient reactance}$$

$$X_d = \dfrac{E_g}{I} = \text{direct axis component of steady state or} \tag{11.11}$$
$$\text{synchronous reactance}$$

Here I'', I', and I are the sub-transient, transient, and steady-state value of short-circuit currents, respectively, and E_g is excitation (or) open-circuit voltage (or no-load induced emf) in the armature. Hence, the simplest model of the synchronous generator is a constant voltage 'E_g' in series with the proper impedance (or) reactance, i.e., X''_d or X'_d or X_d as shown in Fig. 11.20.

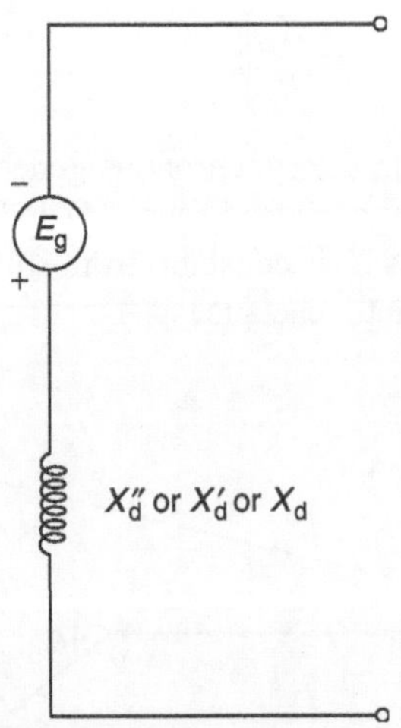

FIG. 11.20   *Simplified model of synchronous generator by neglecting the saliency and flux linkage changes*

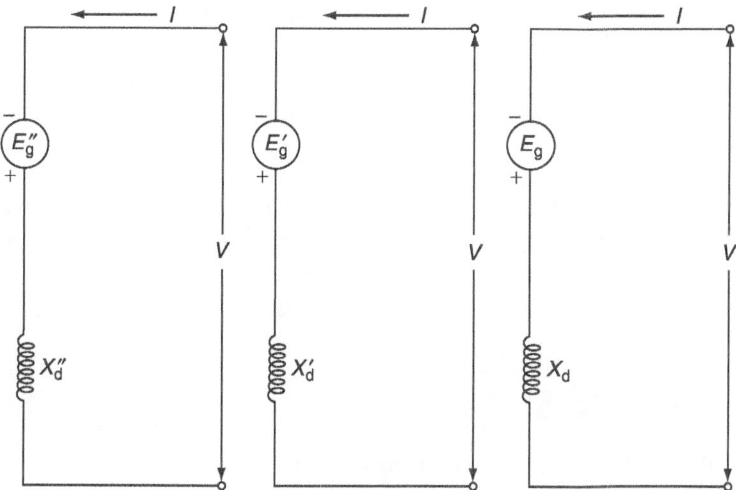

FIG. 11.21   *Representation of synchronous generator*

When the synchronous machine connected to the power system is operating at load before the fault occurs, the synchronous generator is represented by an appropriate voltage source behind the respective reactances as shown in Fig. 11.21.

This modeling (or) representation can easily be obtained for any fault in the power system with the help of Thevenin's equivalent circuit, from which it is clear that the flux linkages and hence the internal voltage of the machine remain constant, but only its phase angle changes.

Now, the machine equations of the model of Fig. 11.21 will be expressed as

$$E''_g = V + jIX''_d$$

$$E'_g = V + jIX'_d$$

$$E_g = V + jIX_d \qquad\qquad (11.12)$$

where E''_g, E'_g, and E_g are sub-transient, transient or steady-state excitation voltages of the synchronous generator, respectively. V is the terminal voltage of the synchronous generator, and I is the current in the synchronous generator.

For the synchronous motor, Equations (11.12) may obtain the following formation.

$$E''_g = V - jIX''_d$$

$$E'_g = V - jIX'_d$$

$$E_g = V - jIX_d \qquad\qquad (11.13)$$

11.6 EFFECT OF SALIENCY

The salient-pole rotor is shown in Fig. 11.15. It has a direct axis of rotor-field winding, i.e., d-axis and quadrature axis, i.e., the q-axis of the rotor-field winding. The d-axis and the q-axis revolve with the rotor, while the magnetic axes of the three stator phases remain fixed.

At the instant of time, θ is the angle from the axis of Phase-a to the d-axis. The corresponding angle from the Phase-b axis to the d-axis is $\theta + 240°$ or $\theta - 120°$. The angle from the Phase-c axis is $\theta + 120°$. As the rotor turns, θ varies with time, with a constant rotor angular velocity, ω, i.e., $\theta = \omega t$.

In order to include the effect of saliency, the simplest model of synchronous machine can be represented by a fictitious voltage 'E_q' located at the q-axis. The d-axis is taken along the main pole axis while the q-axis lags the d-axis by 90°. Then the voltage E_q is expressed in terms of full-load terminal voltage V and full-load armature current I_a as

$$E_q = V + I_a R_a + j I_a X_q \tag{11.14}$$

where X_q is the quadrature axis synchronous reactance.

The equivalent circuit and phasor diagram for the case of effect of saliency are given in Figs. 11.22(a) and (b).

The excitation voltage (or) open-circuit voltage will be calculated as

$$E_g = V + I_a R_a + j I_{ad} X_d + j I_{aq} X_q \tag{11.15}$$

where $X_d = X_l + X_{ad}$ and $X_q = X_l + X_{aq}$. X_d is the direct axis armature synchronous reactance,

X_q the quadrature axis armature synchronous reactance,

X_{ad} the d-axis component of armature magnetizing reactance, and

X_{aq} the q-axis component of the armature magnetizing reactance.

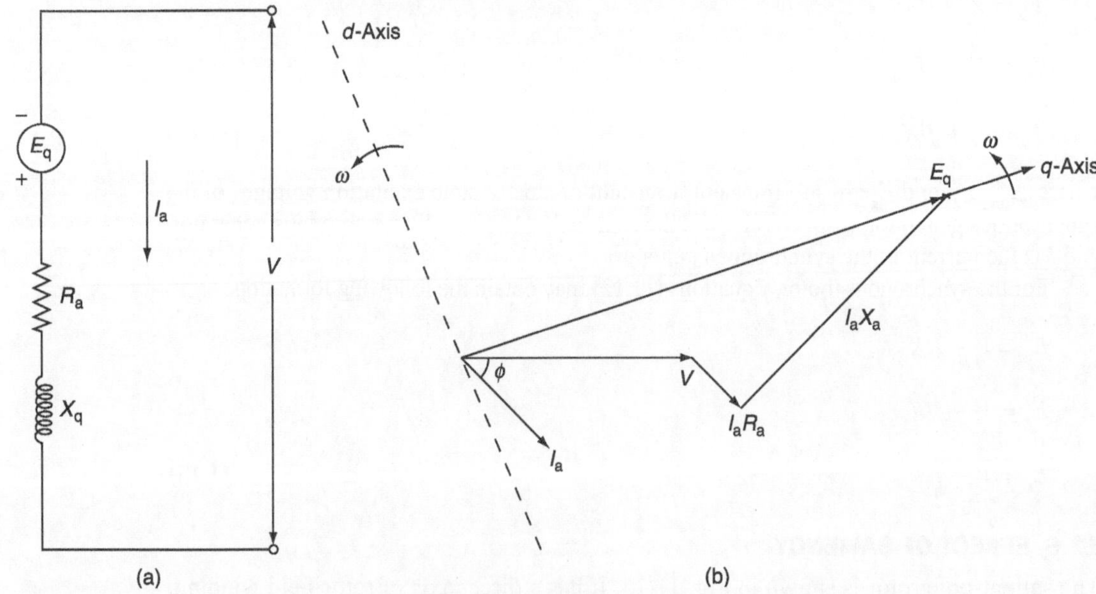

FIG. 11.22 *Effect of saliency; (a) equivalent circuit; (b) phasor diagram*

X_{ad} corresponds to the *d*-axis component of the armature reaction flux, ϕ_{ad}, and X_{aq} corresponds to the *q*-axis component of the armature reaction flux ϕ_{aq}.

Now, the phasor diagram including the effect of saliency is drawn as shown in Fig. 11.23.

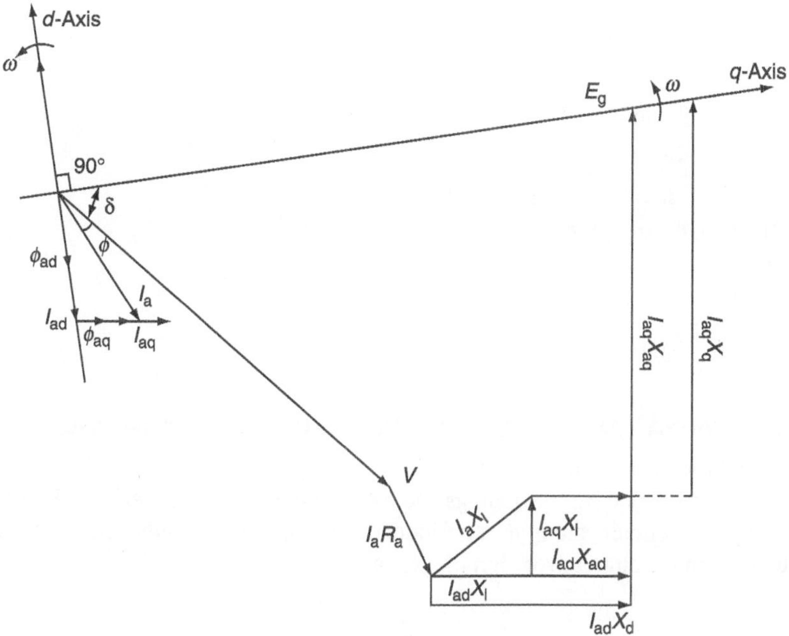

FIG. 11.23   *Phasor diagram of synchronous machine including the effect of saliency*

11.7 GENERAL EQUATION OF SYNCHRONOUS MACHINE

The synchronous machine has at least four windings, three on stator carrying AC and one on rotor with DC excitation.

When a coil has an instantaneous voltage V applied across its terminals and a current 'i' that flows from a positive terminal into the coil, the governing equation becomes

$$V = ir + \dot{\lambda}$$

$$= ir + \frac{d\lambda}{dt} \tag{11.16}$$

Hence, the instantaneous terminal voltage V of any winding will be in the form,

$$V = \pm \Sigma\, ir \pm \Sigma\, \dot{\lambda}$$

where λ is the flux linkage (it may be represented by the symbol Ψ), r the resistance of the winding, and i the current with positive direction of stator current flowing out of the generator terminal.

For the three stator windings a, b, and c, the voltage equations are

$$V_a = i_a r_a + \frac{d\lambda_a}{dt}$$

$$V_b = i_b r_b + \frac{d\lambda_b}{dt}$$

$$V_c = i_c r_c + \frac{d\lambda_c}{dt}$$

(11.17)

and for rotor-field winding,

$$V_f = i_f r_f + \frac{d\lambda_f}{dt}$$

In practice, $r_a = r_b = r_c = r$

$\therefore \ \lambda = Li$,

Equation (11.16) can be written as

$$V = ir + \frac{d}{dt}(Li)$$

(11.18)

11.8 DETERMINATION OF SYNCHRONOUS MACHINE INDUCTANCES

The 3-ϕ synchronous machine without damper windings may be considered as a set of coupled circuits formed by the 3-ϕ windings and rotor-field windings as shown in Fig. 11.24.

In the above circuit, there are self-inductances and mutual inductances, which vary periodically with the angular rotation of a rotor.

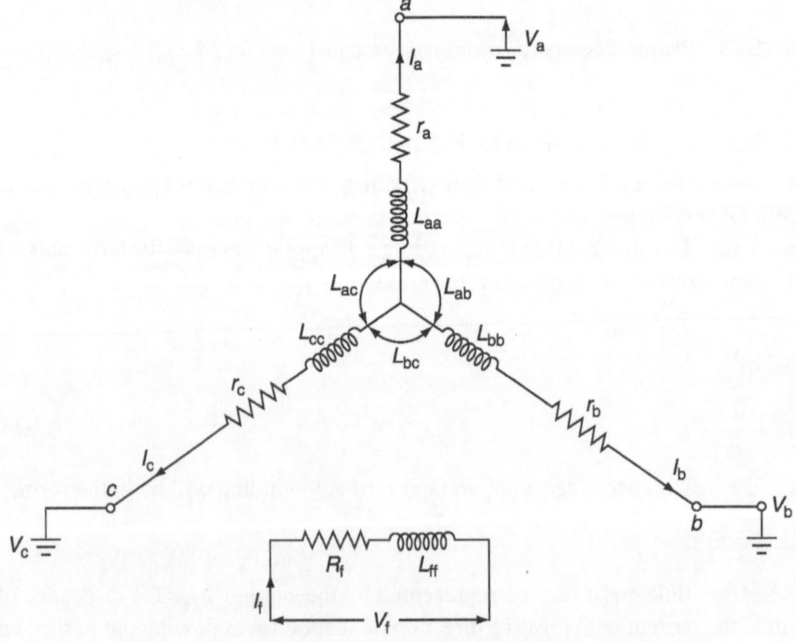

FIG. 11.24   *Circuit diagram of 3-ϕ, synchronous machine*

11.8.1 Assumptions

The following assumptions are usually made to determine the nature of the machine inductances to develop the detailed model of the synchronous machines:

(i) The self-inductance and mutual inductance of the machine are independent of the magnitude of winding currents because the magnetic saturation is neglected. Thus, the machine is assumed to be magnetically linear.

(ii) The shape of the air gap and the distribution of windings are such that all the above inductances may be represented as constants plus sinusoidal functions of electrical rotor positions.

(iii) Slotting effects are ignored. Distributed windings comprise finely spread conductors of negligible diameter.

(iv) Magnetic materials are free from hysterisis and eddy current losses.

(v) The machine may be considered without damper windings. If the damper winding is present, then its influence may be neglected.

(vi) Higher order time and space harmonics are neglected.

11.9 ROTOR INDUCTANCES

In this section, we shall discuss rotor self-inductance and stator to rotor mutual inductances in detail.

11.9.1 Rotor self-inductance

The stator, i.e., armature has a cylindrical structure. Hence, the self-inductance of the rotor field winding 'f' will not depend upon the position of the rotor and will be a constant one.

i.e., L_{ff} = self-inductance of rotor-field winding = constant

11.9.2 Stator to rotor mutual inductances

The stator to rotor mutual inductances will vary periodically with β. The mutual inductance between the field winding and any armature phase is the greatest when the d-axis coincides with the axis of that phase. Figure 11.25 shows the effect of field winding on mutual inductances.

Consider the example, the mutual inductance between the field winding and Phase-a (M_f) will be maximum at $\beta = 0$ and at $\beta = 90°$ and negative maximum ($-M_f$) at $\beta = 180°$ and zero again at $\beta = 270°$.

Accordingly, with space m.m.f. and flux distribution assumed to be sinusoidal, the mutual inductance between the field winding and Phase-a (L_{af}) can be expressed as

$$L_{af} = L_{fa} = M_f \cos \beta$$

Based on the above, the similar expressions for Phases-b and c can be obtained directly by replacing β with ($\beta - 120°$) and ($\beta + 120°$), respectively,

i.e., $L_{af} = L_{fa} = M_f \cos \beta$

$$L_{bf} = L_{fb} = M_f \cos(\beta - 120°)$$

$$L_{cf} = L_{fc} = M_f \cos(\beta + 120°) \qquad (11.19)$$

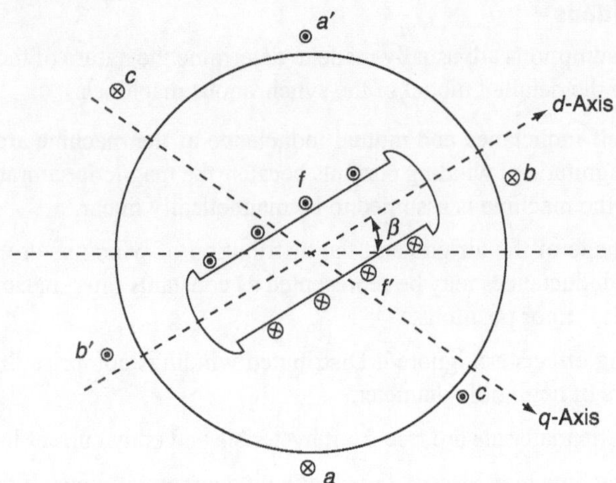

FIG. 11.25   *Effect of field winding on mutual inductances*

11.10 STATOR SELF-INDUCTANCES

The self-inductance of any stator phase is always positive but varies with the position of the rotor. It is the greatest when the d-axis of the field coincides with the axis of the armature phase and being least when the q-axis coincides with it. There will be a second-harmonics variation because of different air-gap geometry along the d and q-axes. For example, the self-inductance of Phase-a (L_{aa}) will be a maximum for $\beta = 0$ and a minimum for $\beta = 90°$ and maximum again for $\beta = 180°$ and so on.

When Phase 'a' is excited, with space harmonics ignored, the m.m.f. wave of Phase 'a' will be a cosine wave (space distribution) centered on the Phase-a-axis as shown in Fig. 11.26.

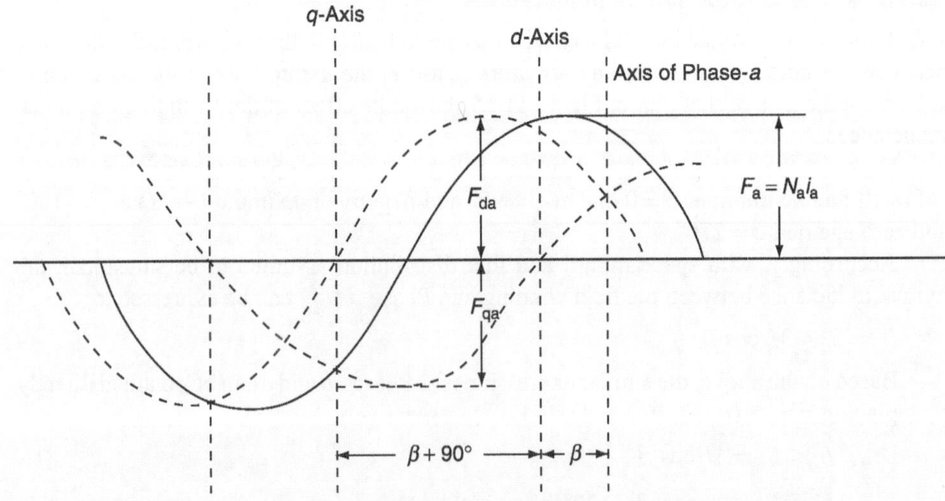

FIG. 11.26   *The m.m.f. wave of Phase-a with its d-axis and q-axis components*

The peak amplitude of this m.m.f. wave of Phase 'a' is

$$F_a = N_a\, i_a \tag{11.20}$$

where N_a is the effective turns/phase and i_a the instantaneous current in Phase 'a'.

Let us resolve this m.m.f. wave into two-component sinusoidal space distributions, one centered on the d-axis (F_{da}) and the other on the q-axis (F_{qa}).

The peak amplitudes of these two resolved components are

$$F_{da} = F_a \cos \beta \tag{11.21}$$

$$F_{qa} = F_a \cos (\beta + 90°) = -F_a \sin \beta \tag{11.22}$$

The advantage of resolving m.m.f. is that two components m.m.f. waves act on specific air-gap geometry in their respective axes.

The fundamental air-gap fluxes per pole along the two axes are, accordingly,

$$\phi_{da} = \frac{F_{da}}{R_d} = F_{da} P_d = F_a P_d \cos \beta \tag{11.23}$$

$$\phi_{qa} = \frac{F_{qa}}{R_q} = F_{qa} P_q = -F_a P_q \sin \beta \tag{11.24}$$

where P_d is the permeance along the d-axis and P_q the permeance along the q-axis.

These are known as machine constants and their values can be found from a flux plot for specific machine geometry.

Let ϕ_{ga} be the air-gap flux linking with Phase-a and be expressed as

$$\phi_{ga} = \phi_{da} \cos \beta - \phi_{qa} \sin \beta \tag{11.25}$$

$$= F_a (P_d \cos^2 \beta + P_q \sin^2 \beta)$$

$$\therefore \phi_{ga} = N_a i_a \left[\frac{P_d + P_q}{2} + \frac{P_d - P_q}{2} \cos 2\beta \right] \quad [\because F_a = N_a\, i_a] \tag{11.26}$$

Since the air-gap flux linkage, $\lambda_{ga} = \phi_{ga} N_a$

$$\lambda_{ga} = N_a^2 i_a \left[\frac{P_d + P_q}{2} + \frac{P_d - P_q}{2} \cos 2\beta \right] \tag{11.27}$$

If ϕ_{la} represents the leakage flux of Phase 'a', which does not cross the air gap, then the flux linkages of Phase 'a' due to leakage flux only are

$$\lambda_{la} = \phi_{la} N_a = F_a P_l N_a = N_a^2 i_a P_l$$

where P_l is the constant leakage permeance of armature Phase 'a'.

Therefore, the total flux linkage of Phase a can be expressed as

$$\lambda_a = \lambda_{ga} + \lambda_{la}$$

$$= \left\{ \left[\left(N_a^2 i_a\right) \frac{P_d + P_q}{2} + \frac{P_d - P_q}{2} \cos 2\beta \right] + N_a^2 i_a P_l \right\}$$

$$= N_a^2 i_a \left[P_l + \frac{P_d + P_q}{2} + \frac{P_d - P_q}{2} \cos 2\beta \right]$$

Since by definition, the inductance is the proportionality factor relating flux linkages to current, the self-inductance of Phase '*a*' due to the air-gap flux when only Phase '*a*' is excited, will be

$$L_{aa} = \frac{\lambda_a}{i_a} = N_a^2 \left[P_l + \frac{P_d + P_q}{2} + \frac{P_d - P_q}{2} \cos 2\beta \right] \tag{11.28}$$

$$\Rightarrow L_{aa} = L_s + L_m \cos 2\beta \tag{11.29}$$

where $L_s = N_a^2 \left[P_l + \dfrac{P_d + P_q}{2} \right]$ is a constant term,

$L_m = N_a^2 \left[\dfrac{P_d - P_q}{2} \right]$ is the amplitude of second-harmonics variation

For Phase '*b*', the variation of self-inductance is similar, except that the maximum value occurs when the *d*-axis coincides with the Phase *b*-axis. The self-inductances of Phase '*b*' and Phase '*c*' can be obtained by replacing β by $(\beta - 120°)$ and $(\beta + 120°)$, respectively:

$$\begin{aligned} L_{bb} &= L_s + L_m \cos 2(\beta - 120°) \\ &= L_s + L_m \cos(2\beta - 240°) \\ &= L_s + L_m \cos(2\beta + 120°) \end{aligned} \tag{11.30}$$

$$\begin{aligned} L_{cc} &= L_s + L_m \cos 2(\beta + 120°) \\ &= L_s + L_m \cos(2\beta + 240°) \\ &= L_s + L_m \cos(2\beta - 120°) \end{aligned} \tag{11.31}$$

i.e., the stator self-inductances are obtained as

$$\begin{aligned} L_{aa} &= L_s + L_m \cos 2\beta \\ L_{bb} &= L_s + L_m \cos(2\beta + 120°) \\ L_{cc} &= L_s + L_m \cos(2\beta - 120°) \end{aligned}$$

11.11 STATOR MUTUAL INDUCTANCES

The mutual inductances between stator phases will also exhibit a second-harmonics variation with β because of the rotor shape. The mutual inductance between two phases can be found by evaluating the air-gap flux linking one phase when another phase is excited. For example, the mutual inductance between Phases a and b, $L_{ab} = L_{ba}$, can be obtained by evaluating ϕ_{gba} linking Phase '*b*' when only Phase '*a*' is *excited*. ϕ_{gba} can be computed from Equation (11.24) by replacing β with $(\beta - 120°)$ as

$$\begin{aligned} \phi_{gba} &= \phi_{da} \cos(\beta - 120°) - \phi_{qa} \sin(\beta - 120°) \\ &= F_a \left[P_d \cos \beta \cos(\beta - 120°) + P_q \sin \beta \sin(\beta - 120°) \right] \end{aligned}$$

$$= N_a i_a \left[-\frac{P_d + P_q}{4} + \frac{P_d - P_q}{2} \cos(2\beta - 120)° \right] \tag{11.32}$$

The mutual inductance between Phase-a and Phase-b due to the air-gap flux is then

$$L_{ab} = \frac{N_a \phi_{gba}}{i_a} = \frac{\lambda_{ba}}{i_a} = N_a^2 \left[-\frac{P_d - P_q}{4} + \frac{P_d - P_q}{2} \cos(2\beta - 120°) \right] \tag{11.33}$$

$$L_{ab} = -M_s + L_m \cos(2\beta - 120°) \tag{11.34}$$

where $\quad M_s = N_a^2 \left[\dfrac{P_d + P_q}{4} \right]$

Similarly, the mutual inductances of stator L_{bc} and L_{ca} can be obtained as

$$L_{bc} = -M_s + L_m \cos 2\beta \tag{11.35}$$

$$L_{ca} = -M_s + L_m \cos(2\beta + 120°) \tag{11.36}$$

i.e., the stator mutual inductances are expressed as

$$L_{ab} = -M_s + L_m \cos(2\beta - 120°)$$
$$L_{bc} = -M_s + L_m \cos(2\beta)$$
$$L_{ca} = -M_s + L_m \cos(2\beta + 120°)$$

Here, L_s, L_m, and M_s are regarded as known machine constants and are determined either by tests or calculated by designed. All the inductances except L_{ff} are functions of β and thus they are time-varying.

11.12 DEVELOPMENT OF GENERAL MACHINE EQUATIONS — MATRIX FORM

From the knowledge of the above time-varying inductances, the general machine equations are developed. If a motoring mode is considered, then for any winding, the general equation relating the applied voltage of the input current is

$$v = ir + \frac{d\lambda}{dt}$$

In terms of self-inductance and mutual inductance, the flux linkages are expressed as

$$\lambda_a = L_{aa} i_a + L_{ab} i_b + L_{ac} i_c + L_{af} i_f$$
$$\lambda_b = L_{ba} i_a + L_{bb} i_b + L_{bc} i_c + L_{bf} i_f$$
$$\lambda_c = L_{ca} i_a + L_{cb} i_b + L_{cc} i_c + L_{cf} i_f$$
$$\lambda_f = L_{fa} i_a + L_{fb} i_b + L_{fc} i_c + L_{ff} i_f \tag{11.37}$$

Equations (11.17) can be modified as

$$v_a = i_a r_a + \frac{d}{dt}\left(L_{aa} i_a + L_{ab} i_b + L_{ac} i_c + L_{af} i_f \right)$$

$$v_{b} = i_{b}r_{b} + \frac{\mathrm{d}}{\mathrm{d}t}\left(L_{ba}i_{a} + L_{bb}i_{b} + L_{bc}i_{c} + L_{bf}i_{f}\right)$$

$$v_{c} = i_{c}r_{c} + \frac{\mathrm{d}}{\mathrm{d}t}\left(L_{ca}i_{a} + L_{cb}i_{b} + L_{cc}i_{c} + L_{cf}i_{f}\right)$$

$$v_{f} = i_{f}r_{f} + \frac{\mathrm{d}}{\mathrm{d}t}\left(L_{fa}i_{a} + L_{fb}i_{b} + L_{fc}i_{c} + L_{ff}i_{f}\right) \tag{11.38}$$

Write the above equations in a compact matrix form:

$$[V] = [i][r] + \frac{\mathrm{d}}{\mathrm{d}t}[\lambda] \tag{11.39}$$

$$\text{where} \quad [V] \triangleq \begin{bmatrix} v_{a} \\ v_{b} \\ v_{c} \\ \cdot \\ \cdot \\ \cdot \\ v_{f} \end{bmatrix}$$

$$[i] = \begin{bmatrix} i_{a} \\ i_{b} \\ i_{c} \\ \cdot \\ \cdot \\ \cdot \\ i_{f} \end{bmatrix}$$

$$[r] = \begin{bmatrix} r_{a} & 0 & 0 & 0 \\ 0 & r_{b} & 0 & 0 \\ 0 & 0 & r_{c} & 0 \\ \cdot & \cdot & \cdot & \cdot \\ 0 & 0 & 0 & r_{f} \end{bmatrix}$$

$$\text{and} \quad [\lambda] = \begin{bmatrix} \lambda_{a} \\ \lambda_{b} \\ \lambda_{c} \\ \cdot \\ \cdot \\ \cdot \\ \lambda_{f} \end{bmatrix}$$

The vectors of the flux linkages are proportional to the currents with matrix of self-inductance and mutual inductance as the proportionality factor.

Hence, $[\lambda] = [L]\,[i]$ **(11.40)**

$$
\text{where}\quad [L] \triangleq
\begin{bmatrix}
L_{aa} & L_{ab} & L_{ac} & L_{af} \\
L_{ba} & L_{bb} & L_{bc} & L_{bf} \\
L_{ca} & L_{cb} & L_{cc} & L_{cf} \\
\cdot & \cdot & \cdot & \cdot \\
L_{fa} & L_{fb} & L_{fc} & L_{ff}
\end{bmatrix}
$$

i.e.,

$$
\begin{bmatrix}
\lambda_a \\
\lambda_b \\
\lambda_c \\
\cdot \\
\lambda_f
\end{bmatrix}
=
\begin{bmatrix}
L_{aa} & L_{ab} & L_{ac} & L_{af} \\
L_{ba} & L_{bb} & L_{bc} & L_{bf} \\
L_{ca} & L_{cb} & L_{cc} & L_{cf} \\
\cdot & \cdot & \cdot & \cdot \\
L_{fa} & L_{fb} & L_{fc} & L_{ff}
\end{bmatrix}
\begin{bmatrix}
i_a \\
i_b \\
i_c \\
\cdot \\
i_f
\end{bmatrix}
\qquad\qquad
\textbf{(11.41)}
$$

The inductance L is symmetric, i.e., $L_{ab} = L_{ba}$, etc.

If the expressions of inductances of Equations (11.19) – (11.36) are substituted in Equations (11.38), the solution of resulting equations can be simplified by a transformation, known as Blondel's transformation, which is also generally called *Park's transformation.*

11.13 BLONDEL'S TRANSFORMATION (OR) PARK'S TRANSFORMATION TO '*dqo*' COMPONENTS

Equations (11.37)–(11.41) are a set of differential equations describing the behavior of machines. However, the solution of these equations is complicated since the inductances are the functions of rotor angle 'β', which in turn, is a function of time.

The complication in getting the solution can be avoided by transferring the physical quantities in the armature windings through a linear, time-dependent, and power-invariant transformation called *Park's transformation.*

This transformation is based on the fact that the rotating field produced by 3-ϕ stator currents in the synchronous machine can be equally produced by 2-ϕ currents in a 2-ϕ winding. Let us consider the 3-windings for the three phases a, b, and c of a synchronous machine as shown in Fig. 11.27.

When all the three phases are excited, the total stator m.m.f.'s that act along the d-axis and q-axis are

$$
\begin{aligned}
F_{ds} &= F_a \cos\beta + F_b \cos(\beta - 120°) + F_c \cos(\beta + 120°) \\
&= N_a\,[i_a \cos\beta + i_b \cos(\beta - 120°) + i_c \cos(\beta + 120°)]
\end{aligned}
\qquad \textbf{(11.42)}
$$

and

$$
\begin{aligned}
F_{qs} &= -F_a \sin\beta - F_b \sin(\beta - 120°) - F_c \sin(\beta + 120°) \\
&= N_a\,[-i_a \sin\beta - i_b \sin(\beta - 120°) - i_c \sin(\beta + 120°)]
\end{aligned}
\qquad \textbf{(11.43)}
$$

Let us consider two fictitious windings, one placed on the d-axis and the other on the q-axis, as shown in Fig. 11.28.

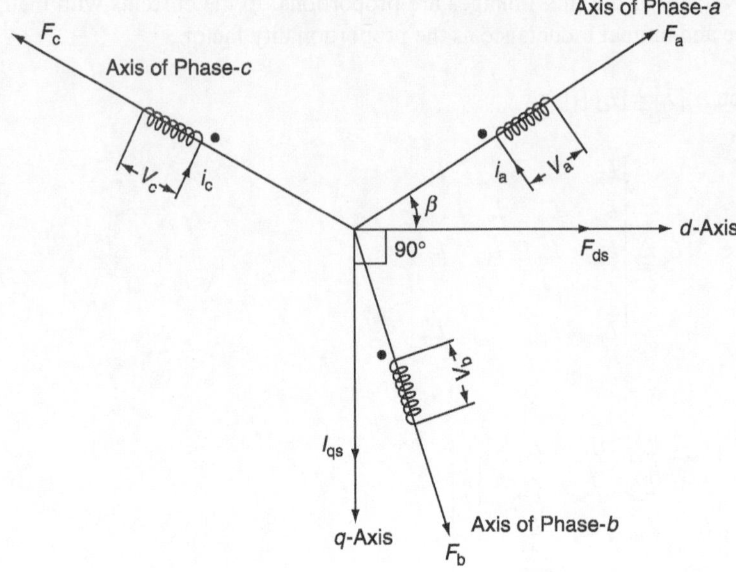

FIG. 11.27   *The m.m.f.'s of 3-φ windings and their resultant m.m.f.'s along the d-axis and the q-axis*

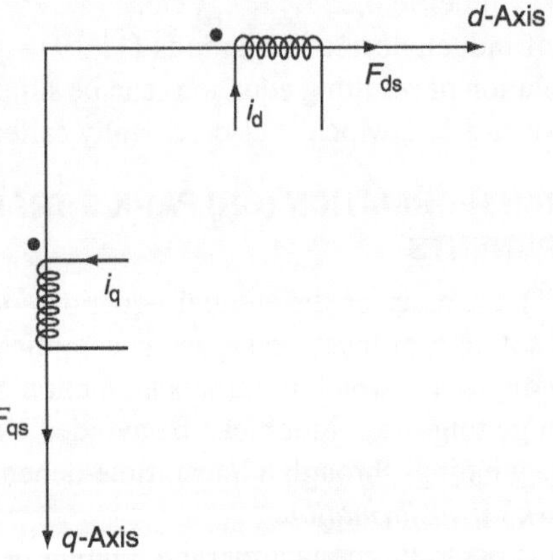

FIG. 11.28   *Fictitious coils on d and q-axes*

The winding on the d-axis rotates at the same speed as the rotor-field winding and remains in such a position that its axis always coincides with the direct axis of the rotor field. Hence, the instantaneous current i_d gives the same m.m.f. F_{ds} on this axis as do the actual three instantaneous armature phase currents flowing in the actual armature windings. Similarly, the current i_q flowing in the winding on the q-axis gives the same m.m.f. F_{qs} on this axis as do the actual three instantaneous armature phase currents flowing in the actual armature windings.

In order to transform quantities in the f-a-b-c axes into the f-d-q-0 axes, the constraints of the system forming the basis of the transformation may be obtained by viewing the currents, m.m.f.'s, voltages, and flux linkages in the two axes. The currents in view of Equations (11.42) and (11.43) are

$$i_f = i_f$$

$$i_\mathrm{d} = K_\mathrm{d}\left[i_\mathrm{a}\cos\beta + i_\mathrm{b}\cos(\beta-120^\circ) + i_\mathrm{c}\cos(\beta+120^\circ)\right]$$

$$i_\mathrm{q} = K_\mathrm{q}\left[i_\mathrm{a}\cos(\beta+90^\circ) + i_\mathrm{b}\cos(\beta-120^\circ+90^\circ) + i_\mathrm{c}\cos(\beta+120^\circ+90^\circ)\right]$$

$$= -K_\mathrm{q}\left[i_\mathrm{a}\sin\beta + i_\mathrm{b}\sin(\beta-120^\circ) + i_\mathrm{c}\sin(\beta+120^\circ)\right] \tag{11.44}$$

A new current, known as zero-sequence current, which does not produce any rotating field is introduced and is expressed as

$$i_\mathrm{o} = \frac{i_\mathrm{a}+i_\mathrm{b}+i_\mathrm{c}}{3} = K_0\left(i_\mathrm{a}+i_\mathrm{b}+i_\mathrm{c}\right) \tag{11.45}$$

where K_d, K_q, and K_0 are constants.

Now in this system, there are three original phase currents i_a, i_b, and i_c and three field currents i_d, i_q, and i_o. For the ungrounded star connection or for the balanced 3-ϕ condition, the sum of phase currents is zero and hence i_o must also be zero.

Equations (11.44) and (11.45) are expressed in a matrix form as

$$\begin{bmatrix} i_\mathrm{d} \\ i_\mathrm{q} \\ i_\mathrm{o} \\ i_\mathrm{f} \end{bmatrix} = \begin{bmatrix} K_\mathrm{d}\cos\beta & K_\mathrm{d}\cos(\beta-120^\circ) & K_\mathrm{d}\cos(\beta+120^\circ) & 0 \\ -K_\mathrm{q}\sin\beta & -K_\mathrm{q}\sin(\beta-120^\circ) & -K_\mathrm{q}\sin(\beta+120^\circ) & 0 \\ K_0 & K_0 & K_0 & 0 \\ 0 & 0 & 0 & 1 \end{bmatrix} \begin{bmatrix} i_\mathrm{a} \\ i_\mathrm{b} \\ i_\mathrm{c} \\ i_\mathrm{f} \end{bmatrix} \tag{11.46}$$

In a more compact matrix form, Equation (11.46) can be expressed as

$$[i]_\mathrm{dqof} \triangleq [P]_\mathrm{dqof}\,[i]_\mathrm{abcf} \tag{11.47}$$

This is known as **PARK's transformation (or Blondel's transformation).**

$$\text{where}\quad [i]_\mathrm{dqof} = \begin{bmatrix} i_\mathrm{d} \\ i_\mathrm{q} \\ i_\mathrm{o} \\ i_\mathrm{f} \end{bmatrix}; \quad [i]_\mathrm{abcf}\begin{bmatrix} i_\mathrm{a} \\ i_\mathrm{b} \\ i_\mathrm{c} \\ i_\mathrm{f} \end{bmatrix}$$

$$\text{and}\quad [P]_\mathrm{dqof} = \begin{bmatrix} K_\mathrm{d}\cos\beta & K_\mathrm{d}\cos(\beta-120^\circ) & K_\mathrm{d}\cos(\beta+120^\circ) & 0 \\ -K_\mathrm{q}\sin\beta & -K_\mathrm{q}\sin(\beta-120^\circ) & -K_\mathrm{q}\sin(\beta+120^\circ) & 0 \\ K_0 & K_0 & K_0 & 0 \\ 0 & 0 & 0 & 1 \end{bmatrix} \begin{matrix} d \\ q \\ o \\ f \end{matrix} \tag{11.48}$$

(columns labelled a, b, c, f)

and is known as **PARK's transformation matrix (or Blondel's transformation matrix).**

The effect of Park's transformation is simply to transform all stator quantities from phases a, b, and c into new variables, the frame of reference of which moves with the rotor, i.e., the Park's transformation matrix $[P]_\mathrm{dqof}$ transforms the field of phasors to the field of d-q-o-f components and it is a linear, time-dependent matrix.

11.14 INVERSE PARK'S TRANSFORMATION

The inverse transformation, which transforms the d-q-o-f quantities into the phase quantities, is expressed as

$$[i]_\mathrm{abcf} = [P]^{-1}[i]_\mathrm{dqof} \tag{11.49}$$

$$\text{where} \quad [P]^{-1} = \begin{array}{c} a \\ b \\ c \\ f \end{array} \begin{bmatrix} \dfrac{2}{3}\dfrac{1}{K_d}\cos\beta & \dfrac{-2}{3}\dfrac{1}{K_q}\sin\beta & \dfrac{1}{3K_0} & 0 \\[3mm] \dfrac{2}{3}\dfrac{1}{K_d}\cos(\beta-120°) & \dfrac{-2}{3}\dfrac{1}{K_q}\sin(\beta-120°) & \dfrac{1}{3K_0} & 0 \\[3mm] \dfrac{2}{3}\dfrac{1}{K_d}\cos(\beta+120°) & \dfrac{-2}{3}\dfrac{1}{K_q}\sin(\beta+120°) & \dfrac{1}{3K_0} & 0 \\[3mm] 0 & 0 & 0 & 1 \end{bmatrix} \qquad (11.50)$$

11.15 POWER-INVARIANT TRANSFORMATION IN 'f-d-q-o' AXES

The total instantaneous power delivered to the motor is

$$P_d = V_a i_a + V_b i_b + V_c i_c \qquad (11.51)$$

When *f-d-q-o* axes' quantities are substituted for the phase quantities by using Equation (11.50) and by further manipulation, these results give power in terms of new quantities as

$$P_d = \frac{2}{3K_d^2}V_d i_d + \frac{2}{3K_q^2}V_q i_q + \frac{1}{3K_0^2}V_o i_o \qquad (11.52)$$

Equation (11.52) gives the true power associated with the armature of the new system. This power invariance is preserved if

$$\frac{2}{3}\frac{1}{K_d^2}=1, \quad \frac{2}{3}\frac{1}{K_q^2}=1, \quad \text{and} \quad \frac{1}{3K_o^2}=1$$

$$\Rightarrow K_d = \pm\sqrt{\frac{2}{3}}\,; \quad K_q = \pm\sqrt{\frac{2}{3}} \quad \text{and} \quad K_o = \pm\frac{1}{\sqrt{3}}$$

Since the angle between the *d*-axis and the Phase-*a* axis is β, and the *q*-axis has been chosen ahead of the *d*-axis, the values of K_d and K_q must be positive; hence, the selected values of the three quantities are

$$K_d = \sqrt{\frac{2}{3}};\ K_q = \sqrt{\frac{2}{3}}, \quad \text{and} \quad K_o = \frac{1}{\sqrt{3}}$$

With these values, Park's transformation matrix and its inverse become

$$[P]_{dqof} = \begin{array}{c} d \\ q \\ o \\ f \end{array} \begin{bmatrix} \sqrt{\dfrac{2}{3}}\cos\beta & \sqrt{\dfrac{2}{3}}\cos(\beta-120°) & \sqrt{\dfrac{2}{3}}\cos(\beta+120°) & 0 \\[3mm] -\sqrt{\dfrac{2}{3}}\sin\beta & -\sqrt{\dfrac{2}{3}}\sin(\beta-120°) & -\sqrt{\dfrac{2}{3}}\sin(\beta+120°) & 0 \\[3mm] \dfrac{1}{\sqrt{3}} & \dfrac{1}{\sqrt{3}} & \dfrac{1}{\sqrt{3}} & 0 \\[3mm] 0 & 0 & 0 & 1 \end{bmatrix} \qquad (11.53)$$

$$
[P]^{-1}_{\text{dqof}} = \begin{array}{c} \\ a \\ b \\ c \\ f \end{array}
\begin{array}{cccc} d & q & o & f \\ \end{array}
\left|
\begin{array}{cccc}
\sqrt{\dfrac{2}{3}}\cos\beta & -\sqrt{\dfrac{2}{3}}\sin\beta & \dfrac{1}{\sqrt{3}} & 0 \\[3mm]
\sqrt{\dfrac{2}{3}}\cos(\beta-120°) & -\sqrt{\dfrac{2}{3}}\sin(\beta-120°) & \dfrac{1}{\sqrt{3}} & 0 \\[3mm]
\sqrt{\dfrac{2}{3}}\cos(\beta+120°) & -\sqrt{\dfrac{2}{3}}\sin(\beta+120°) & \dfrac{1}{\sqrt{3}} & 0 \\[3mm]
0 & 0 & 0 & 1
\end{array}
\right|
\tag{11.54}
$$

Now, the transformation is called unitary transformation because the inverse of the transformation matrix is the transpose of the matrix, i.e., $[P]^{-1} = [P^*_{\text{T}}]$

i.e., Inverse of the matrix = transpose of conjugate of the matrix.

∴ The current equation of a synchronous machine Equation (11.46) can be expressed as

$$
\begin{bmatrix} i_d \\ i_q \\ i_o \\ \cdot \\ \cdot \\ \cdot \\ i_f \end{bmatrix}
\triangleq
\begin{array}{c} \\ d \\ q \\ o \\ \\ \\ \\ f \end{array}
\begin{array}{cccc} a & b & c & f \\ \end{array}
\left|
\begin{array}{cccc}
\sqrt{\dfrac{2}{3}}\cos\beta & \sqrt{\dfrac{2}{3}}\cos(\beta-120°) & \sqrt{\dfrac{2}{3}}\cos(\beta+120°) & 0 \\[3mm]
-\sqrt{\dfrac{2}{3}}\sin\beta & -\sqrt{\dfrac{2}{3}}\sin(\beta-120°) & -\sqrt{\dfrac{2}{3}}\sin(\beta+120°) & 0 \\[3mm]
\dfrac{1}{\sqrt{3}} & \dfrac{1}{\sqrt{3}} & \dfrac{1}{\sqrt{3}} & 0 \\[3mm]
\cdot & \cdot & \cdot & \cdot \\
\cdot & \cdot & \cdot & \cdot \\
\cdot & \cdot & \cdot & \cdot \\
0 & 0 & 0 & 1
\end{array}
\right|
\begin{bmatrix} i_a \\ i_b \\ i_c \\ \cdot \\ \cdot \\ \cdot \\ i_f \end{bmatrix}
\tag{11.55}
$$

and in terms of inverse Park's transformation, the current equations of synchronous machine can be expressed as

$$
\begin{bmatrix} i_a \\ i_b \\ i_c \\ \cdot \\ \cdot \\ \cdot \\ i_f \end{bmatrix}
\triangleq
\begin{array}{c} \\ a \\ \\ b \\ \\ c \\ \\ \\ f \end{array}
\begin{array}{cccc} d & q & o & f \\ \end{array}
\left|
\begin{array}{cccc}
\sqrt{\dfrac{2}{3}}\cos\beta & -\sqrt{\dfrac{2}{3}}\sin\beta & \dfrac{1}{\sqrt{3}} & 0 \\[3mm]
\sqrt{\dfrac{2}{3}}\cos(\beta-120°) & -\sqrt{\dfrac{2}{3}}\sin(\beta-120°) & \dfrac{1}{\sqrt{3}} & 0 \\[3mm]
\sqrt{\dfrac{2}{3}}\cos(\beta+120°) & -\sqrt{\dfrac{2}{3}}\sin(\beta+120°) & \dfrac{1}{\sqrt{3}} & 0 \\[3mm]
\cdot & \cdot & \cdot & \cdot \\
\cdot & \cdot & \cdot & \cdot \\
\cdot & \cdot & \cdot & \cdot \\
0 & 0 & 0 & 1
\end{array}
\right|
\begin{bmatrix} i_d \\ i_q \\ i_o \\ \cdot \\ \cdot \\ \cdot \\ i_f \end{bmatrix}
\tag{11.56}
$$

11.16 FLUX LINKAGE EQUATIONS

The flux linkage equations in matrix form are $[\lambda] = [L][i]$. In terms of self-inductance and mutual inductance, the flux linkages from Equation (11.37) are expressed as

$$\lambda_a = L_{aa} i_a + L_{ab} i_b + L_{ac} i_c + L_{af} i_f$$

$$\lambda_b = L_{ba} i_a + L_{bb} i_b + L_{bc} i_c + L_{bf} i_f$$

$$\lambda_c = L_{ca} i_a + L_{cb} i_b + L_{cc} i_c + L_{cf} i_f$$

$$\lambda_f = L_{fa} i_a + L_{fb} i_b + L_{fc} i_c + L_{ff} i_f$$

In matrix form, it is expressed as

$$
\begin{bmatrix} \lambda_a \\ \lambda_b \\ \lambda_c \\ \cdot \\ \cdot \\ \cdot \\ \lambda_f \end{bmatrix}
=
\begin{bmatrix}
L_{aa} & L_{ab} & L_{ac} & L_{af} \\
L_{ba} & L_{bb} & L_{bc} & L_{bf} \\
L_{ca} & L_{cb} & L_{cc} & L_{cf} \\
\cdot & \cdot & \cdot & \cdot \\
\cdot & \cdot & \cdot & \cdot \\
\cdot & \cdot & \cdot & \cdot \\
L_{fa} & L_{fb} & L_{fc} & L_{ff}
\end{bmatrix}
\begin{bmatrix} i_a \\ i_b \\ i_c \\ \cdot \\ \cdot \\ \cdot \\ i_f \end{bmatrix}
$$

In vector form, it can be expressed as

$$[\lambda]_{abcf} = [L]_{abcf}[i]_{abcf} \tag{11.57}$$

The flux linkages of f-d-q-o coils in terms of four currents are obtained upon transforming both sides by using the transformation matrix $[P]$ and its inverse $[P]^{-1}$ as follows:

$$[\lambda]_{dqof} = [P][\lambda]_{abcf} \tag{11.58}$$

$$= [P]\{[L]_{abcf}[i]_{abcf}\}$$

$$= [P][L]_{abcf}[P]^{-1}[i]_{dqof} \tag{11.59}$$

$$\left[\because \text{ from Equation (11.56); } [i]_{abcf} = [P]^{-1}[i]_{dqof} \right]$$

$$
\begin{bmatrix} \lambda_d \\ \lambda_q \\ \lambda_o \\ \cdot \\ \cdot \\ \cdot \\ \lambda_f \end{bmatrix}
=
\begin{array}{c}
 \\ d \\ q \\ o \\ \\ \\ \\ f
\end{array}
\begin{array}{cccc}
d & q & o & f \\
\end{array}
\begin{bmatrix}
L_d & 0 & 0 & \frac{3}{2} k_d L_{af} \\
0 & L_q & 0 & 0 \\
0 & 0 & L_0 & 0 \\
\cdot & \cdot & \cdot & \cdot \\
\cdot & \cdot & \cdot & \cdot \\
\cdot & \cdot & \cdot & \cdot \\
\frac{1}{K_a} L_{af} & 0 & 0 & L_{ff}
\end{bmatrix}
\begin{bmatrix} i_d \\ i_q \\ i_o \\ \cdot \\ \cdot \\ \cdot \\ i_f \end{bmatrix}
\tag{11.60}
$$

By substituting $K_d = \sqrt{\dfrac{2}{3}}$, we get

$$
\begin{bmatrix} \lambda_{\mathrm{d}} \\ \lambda_{\mathrm{q}} \\ \lambda_{\mathrm{o}} \\ \cdot \\ \cdot \\ \cdot \\ \lambda_{\mathrm{f}} \end{bmatrix} = \begin{bmatrix} L_{\mathrm{d}} & 0 & 0 & \sqrt{\tfrac{3}{2}}L_{\mathrm{af}} \\ 0 & L_{\mathrm{q}} & 0 & 0 \\ 0 & 0 & L_{0} & 0 \\ \cdot & \cdot & \cdot & \cdot \\ \cdot & \cdot & \cdot & \cdot \\ \cdot & \cdot & \cdot & \cdot \\ \sqrt{\tfrac{3}{2}}L_{\mathrm{af}} & 0 & 0 & L_{\mathrm{ff}} \end{bmatrix} \begin{bmatrix} i_{\mathrm{d}} \\ i_{\mathrm{q}} \\ i_{\mathrm{o}} \\ \cdot \\ \cdot \\ \cdot \\ i_{\mathrm{f}} \end{bmatrix}
\tag{11.61}
$$

where $L_{\mathrm{d}} =$ direct axis synchronous inductance $= L_{\mathrm{S}} + M_{\mathrm{S}} + \dfrac{3}{2}L_{\mathrm{m}}$

$L_{\mathrm{q}} =$ quadrature axis synchronous inductance $= L_{\mathrm{S}} + M_{\mathrm{S}} - \dfrac{3}{2}L_{\mathrm{m}}$

$L_{\mathrm{o}} =$ zero-sequence inductance $= L_{\mathrm{S}} - 2M_{\mathrm{S}}$

Note: The mutual inductance L_{af} can also be represented by M_{af} or M.
From Equation (11.61), we have

$$
\left.\begin{aligned}
\lambda_{\mathrm{d}} &= L_{\mathrm{d}}i_{\mathrm{d}} + \sqrt{\tfrac{3}{2}}M_{\mathrm{f}}\, i_{\mathrm{f}} \\
\lambda_{\mathrm{q}} &= L_{\mathrm{q}}i_{\mathrm{q}} \\
\lambda_{\mathrm{o}} &= L_{\mathrm{o}}i_{\mathrm{o}} \\
\lambda_{\mathrm{f}} &= \sqrt{\tfrac{3}{2}}M_{\mathrm{f}}i_{\mathrm{d}} + L_{\mathrm{ff}}i_{\mathrm{f}}
\end{aligned}\right\}
\tag{11.62}
$$

From Equation (11.62), it is noticed that the inductances are not functions of rotor position β. Hence, it is a greater advantage of transforming phase quantities to a suitable set of quantities of f-d-q-o-axes.

11.17 VOLTAGE EQUATIONS

The original voltage equation of a synchronous machine is

$$
[v(t)] = -[R][i(t)] - \frac{d}{dt}[\lambda]
\tag{11.63}
$$

i.e.,
$$
\begin{bmatrix} v_{\mathrm{a}} \\ v_{\mathrm{b}} \\ v_{\mathrm{c}} \\ v_{\mathrm{f}} \end{bmatrix} = - \begin{bmatrix} R_{\mathrm{a}} & 0 & 0 & 0 \\ 0 & R_{\mathrm{b}} & 0 & 0 \\ 0 & 0 & R_{\mathrm{c}} & 0 \\ 0 & 0 & 0 & R_{\mathrm{f}} \end{bmatrix} \begin{bmatrix} i_{\mathrm{a}} \\ i_{\mathrm{b}} \\ i_{\mathrm{c}} \\ i_{\mathrm{f}} \end{bmatrix} - \frac{d}{dt} \begin{bmatrix} \lambda_{\mathrm{a}} \\ \lambda_{\mathrm{b}} \\ \lambda_{\mathrm{c}} \\ \lambda_{\mathrm{f}} \end{bmatrix}
$$

For transformation, let us pre-multiply both sides of Equation (11.63) by $[P]$,

$$
[P][v(t)]_{\mathrm{abcf}} = -[P][R][i(t)]_{\mathrm{abcf}} - [P]\frac{d}{dt}[\lambda]_{\mathrm{abcf}}
$$

$$[v(t)]_{\text{dqof}} = [-R][i(t)]_{\text{dqof}} - [P]\frac{d}{dt}[P]^{-1}[\lambda]_{\text{dqof}} \quad \left(\because [P]^{-1}[\lambda]_{\text{dqof}} = [\lambda]_{\text{abcf}} \right) \tag{11.64}$$

Hence, $\quad [v]_{\text{dqof}} = -[R][i]_{\text{dqof}} - [P]\frac{d}{dt}[P][\lambda]_{\text{dqof}}$

$$[v]_{\text{dqof}} = [-R][i]_{\text{dqof}} - [P]\left[[P]^{-1}\frac{d}{dt}[\lambda]_{\text{dqof}} - [P]\left[\frac{d}{dt}[P]^{-1}[\lambda]_{\text{dqof}} \right] \right]$$

$$-[R][i]_{\text{dqof}} - \frac{d}{dt}[\lambda]_{\text{dqof}} - P\left[\frac{d}{dt}[P]^{-1}[\lambda]_{\text{dqof}} \right] \tag{11.65}$$

It can be shown that $-[P]\frac{d}{dt}[P]^{-1}$ is a matrix with zero entries except for $-\omega$ in the first row, second column and $+\omega$ in the second row, first column,

i.e.,

$$-[P]\frac{d}{dt}[P]^{-1} = \begin{vmatrix} \sqrt{\frac{2}{3}}\cos\beta & \sqrt{\frac{2}{3}}\cos(\beta-120°) & \sqrt{\frac{2}{3}}\cos(\beta+120°) & 0 \\ -\sqrt{\frac{2}{3}}\sin\beta & -\sqrt{\frac{2}{3}}\sin(\beta-120°) & \sqrt{\frac{2}{3}}\sin(\beta+120°) & 0 \\ \frac{1}{\sqrt{3}} & \frac{1}{\sqrt{3}} & \frac{1}{\sqrt{3}} & 0 \\ 0 & 0 & 0 & 1 \end{vmatrix}$$

$$\times \frac{d}{dt} \begin{vmatrix} \sqrt{\frac{2}{3}}\cos\beta & -\sqrt{\frac{2}{3}}\sin\beta & \frac{1}{\sqrt{3}} & 0 \\ \sqrt{\frac{2}{3}}\cos(\beta-120°) & -\sqrt{\frac{2}{3}}\sin(\beta-120°) & \frac{1}{\sqrt{3}} & 0 \\ \sqrt{\frac{2}{3}}\cos(\beta+120°) & -\sqrt{\frac{2}{3}}\sin(\beta+120°) & \frac{1}{\sqrt{3}} & 0 \\ 0 & 0 & 0 & 1 \end{vmatrix}$$

$$= \begin{vmatrix} 0 & -\omega & 0 & 0 \\ \omega & 0 & 0 & 0 \\ 0 & 0 & 0 & 0 \\ 0 & 0 & 0 & 0 \end{vmatrix} \tag{11.66}$$

After the transformation, synchronous machine equations in matrix form become

$$\begin{bmatrix} v_d \\ v_q \\ v_o \\ v_f \end{bmatrix} = -\begin{bmatrix} R & 0 & 0 & 0 \\ 0 & R & 0 & 0 \\ 0 & 0 & R & 0 \\ 0 & 0 & 0 & R_f \end{bmatrix}\begin{bmatrix} i_d \\ i_q \\ i_o \\ i_f \end{bmatrix} - \begin{bmatrix} \frac{d\lambda_d}{dt} \\ \frac{d\lambda_q}{dt} \\ \frac{d\lambda_o}{dt} \\ \frac{d\lambda_f}{dt} \end{bmatrix} + \begin{bmatrix} -\omega\lambda_q \\ \omega\lambda_d \\ 0 \\ 0 \end{bmatrix} \tag{11.67}$$

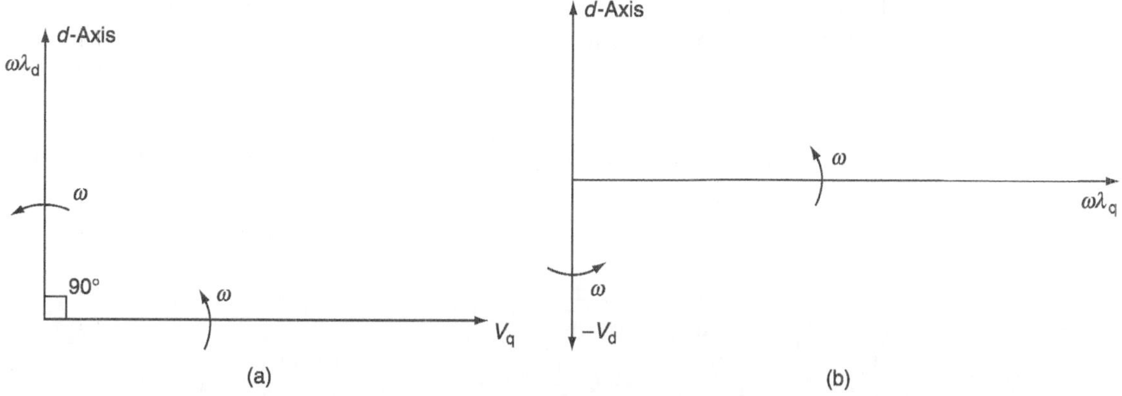

FIG. 11.29   *(a) Phasor diagrams V_q versus $\omega\lambda_d$; (b) phasor diagrams $\omega\lambda_q$ versus $-V_d$*

$$\because R_{\mathrm{a}} = R_{\mathrm{b}} = R_{\mathrm{c}} = R$$

where $\omega =$ angular velocity of rotation $= \dfrac{\mathrm{d}\beta}{\mathrm{d}t}$

From Equation (11.67), the voltage equations obtained are

$$\left.\begin{aligned}
v_{\mathrm{d}} &= -Ri_{\mathrm{d}} - \frac{\mathrm{d}\lambda_{\mathrm{d}}}{\mathrm{d}t} - \omega\lambda_{\mathrm{q}} \\[4pt]
v_{\mathrm{q}} &= -Ri_{\mathrm{q}} - \frac{\mathrm{d}\lambda_{\mathrm{q}}}{\mathrm{d}t} - \omega\lambda_{\mathrm{d}} \\[4pt]
v_{0} &= Ri_{\mathrm{o}} - \frac{\mathrm{d}\lambda_{\mathrm{o}}}{\mathrm{d}t} \\[4pt]
v_{\mathrm{f}} &= -R_{\mathrm{f}}i_{\mathrm{f}} - \frac{\mathrm{d}\lambda_{\mathrm{f}}}{\mathrm{d}t}
\end{aligned}\right\} \qquad (11.68)$$

For sinusoidal steady-state conditions, the flux phasor leads the voltage phasor by 90°. This means that v_{q} will be induced by the flux in the direct axis $(\omega\lambda_{\mathrm{d}})$; i.e., v_{q} will be induced by $(\omega\lambda_{\mathrm{d}})$ and similarly $-v_{\mathrm{d}}$ will be induced by $(\omega\lambda_{\mathrm{q}})$ as shown in Figs. 11.29(a) and (b).

11.18 PHYSICAL INTERPRETATION OF EQUATIONS (11.62) AND (11.68)

The terms $\omega\lambda_{\mathrm{d}}$ and $\omega\lambda_{\mathrm{q}}$ are speed voltages (flux changes in space) and the terms $\dfrac{\mathrm{d}\lambda_d}{\mathrm{d}t}$ and $\dfrac{\mathrm{d}\lambda_q}{\mathrm{d}t}$ are transformer voltages (flux changes in time). Usually, these transformer voltages are small compared with speed voltages and may be neglected. The neglected transformer voltages correspond to negligence of the harmonics and DC components in transient solution for stator voltages and currents. Negligence of harmonics and DC components in the phase current is very common in machine analysis. Neither harmonics nor DC components have a significant effect on the average torque of the machine since harmonics are usually small and DC components die away very rapidly.

The solutions of network equations become extremely difficult and complex when the harmonics and DC components are present in electrical quantities if the transformer voltages are included. Hence, it is preferable to approximate the associated damping torques by additional terms in the swing equation.

11.19 GENERALIZED IMPEDANCE MATRIX (VOLTAGE–CURRENT RELATIONS)

By combining Equations (11.62) and (11.68), we get

$$\begin{bmatrix} v_d \\ v_q \\ v_o \\ v_f \end{bmatrix} = \begin{bmatrix} r+L_d\dfrac{\mathrm{d}}{\mathrm{d}t} & -L_q\dfrac{\mathrm{d}\beta}{\mathrm{d}t} & 0 & \sqrt{\dfrac{3}{2}}M_{af}\dfrac{\mathrm{d}}{\mathrm{d}t} \\[2mm] L_d\dfrac{\mathrm{d}\beta}{\mathrm{d}t} & r+L_q\dfrac{\mathrm{d}}{\mathrm{d}t} & 0 & M_{af}\dfrac{\mathrm{d}\beta}{\mathrm{d}t} \\[2mm] 0 & 0 & r+L_o\dfrac{\mathrm{d}}{\mathrm{d}t} & 0 \\[2mm] -\sqrt{\dfrac{3}{2}}M_{af}\dfrac{\mathrm{d}}{\mathrm{d}t} & 0 & 0 & r_f+L_{ff}\dfrac{\mathrm{d}}{\mathrm{d}t} \end{bmatrix} \begin{bmatrix} i_d \\ i_q \\ i_o \\ i_f \end{bmatrix} \qquad (11.69)$$

In compact form, the above matrix can be expressed as

$$[V]_{dqof} = [Z]_{dqof}[i]_{dqof} \qquad (11.70)$$

It is observed that the impedance matrix is symmetrical in f and d-axes. It consists of two terms, one relating to transformer voltages and the second $\left(\dfrac{\mathrm{d}\beta}{\mathrm{d}t}=\omega\right)$ relating to speed voltages, as given below:

$$[Z]_{dqof} = \begin{bmatrix} r+L_d\dfrac{\mathrm{d}}{\mathrm{d}t} & 0 & 0 & \sqrt{\dfrac{3}{2}}M_{af}\dfrac{\mathrm{d}}{\mathrm{d}t} \\[2mm] 0 & r+L_q\dfrac{\mathrm{d}}{\mathrm{d}t} & 0 & 0 \\[2mm] 0 & 0 & r+L_o\dfrac{\mathrm{d}}{\mathrm{d}t} & 0 \\[2mm] -\sqrt{\dfrac{3}{2}}M_{af}\dfrac{\mathrm{d}}{\mathrm{d}t} & 0 & 0 & r_f+L_{ff}\dfrac{\mathrm{d}}{\mathrm{d}t} \end{bmatrix}$$

$$+\begin{bmatrix} 0 & -L_q\dfrac{\mathrm{d}\beta}{\mathrm{d}t} & 0 & 0 \\[2mm] L_q & 0 & 0 & M_{af}\dfrac{\mathrm{d}\beta}{\mathrm{d}t} \\[2mm] 0 & 0 & 0 & 0 \\[2mm] 0 & 0 & 0 & 0 \end{bmatrix} \qquad (11.71)$$

where $\dfrac{\mathrm{d}\beta}{\mathrm{d}t}=\omega=$ angular speed of rotation

11.20 TORQUE EQUATION

$$\text{Torque, } T = \frac{\text{power associated with speed (rotational) voltages}}{\text{mechanical angular velocity of rotor}}$$

The speed voltages are:

(i) In the direct axis, $\lambda_d \omega$ and

(ii) In the quadrature axis, $-\lambda_q \omega$

$$\text{Mechanical angular velocity} = \frac{2}{\text{poles}} \times \text{electrical angular velocity}$$

$$= \frac{2}{p}\frac{d\theta}{dt} = \frac{2}{p}\omega$$

The total 3-ϕ power output of a synchronous machine is given by

$$P_{out} = v_a i_a + v_b i_b + v_c i_c \tag{11.72}$$

$$= \left[v_{abc}\right]^T \left[i\right]_{abc} \quad \text{p.u.}$$

$$= \left[\left[P\right]^{-1}\left[V\right]_{dqof} \right]^T \left[\left[P\right]^{-1}\left[i\right]_{dqof} \right]$$

$$= \left[V\right]^T_{dqof} \left[\left[P\right]^{-1}\right]^T \left[P\right]^{-1}\left[i\right]_{dqof}$$

$$= \left[V\right]^T_{dqof} \left[P\right]\left[P\right]^{-1}\left[i\right]_{dqof}$$

$$= \left[V\right]^T_{dqof} \left[i\right]_{dqof}$$

$$P_{out} = v_d i_d + v_q i_q + v_o i_o \tag{11.73}$$

Assume balanced but not necessarily steady-state conditions, thus $v_0 = 0$ and $i_0 = 0$.

$$\therefore P_{out} = v_d i_d + v_q i_q \tag{11.74}$$

From Equation (11.68),

$$V_d = -Ri_d - \frac{d\lambda_d}{dt} - \omega\lambda_q \qquad \text{and}$$

$$V_q = -Ri_q - \frac{d\lambda_q}{dt} + \omega\lambda_d$$

Substituting v_d and v_q expressions in Equation (11.74), we get

$$p = \left(-Ri_d - \frac{d\lambda_d}{dt} - \omega\lambda_q\right)i_d + \left(-Ri_q - \frac{d\lambda_q}{dt} + \omega\lambda_d\right)i_q$$

$$= -Ri_d^2 - \frac{d\lambda_d}{dt}i_d - \omega\lambda_q i_d - Ri_q^2 - \frac{d\lambda_q}{dt}i_q + \omega\lambda_d i_q$$

$$= -R(i_d^2 + i_q^2) + \omega(\lambda_d i_q - \lambda_q i_d) - \left(\frac{d\lambda_d}{dt}i_d + \frac{d\lambda_q}{dt}i_q\right)$$

$$p = -\left(\frac{d\lambda_q}{dt} i_q + \frac{d\lambda_d}{dt} i_d\right) + \omega(\lambda_d i_q - \lambda_q i_d) - r(i_d{}^2 + i_q{}^2) \tag{11.75}$$

The above expression consists of three terms and they are:

- The first term represents the rate of change of stator magnetic field energy.
- The second term represents the power transferred across the air gap.
- The third term represents the stator ohmic losses.

The machine torque is obtained from the second term,

$$T_e = \frac{\partial p}{\partial \omega}$$

$$= \frac{\partial}{\partial \omega}\left[\omega\left(\lambda_d i_q - \lambda_q i_d\right)\right]$$

$$= (\lambda_d i_q - \lambda_q i_d) \ \text{p.u.} \tag{11.76}$$

Substituting for λ_d and λ_q from Equation (11.62) in Equation (11.76), we get

$$\lambda_d = L_d i_d + \frac{\sqrt{3}}{2} M_f \ i_f$$

$$\lambda_q = L_q i_q$$

$$T_e = \left(L_d \ i_d + \frac{\sqrt{3}}{2} M_f i_f\right) i_q - \left(L_q i_q\right) i_d$$

$$= L_d i_q i_d - \left(L_q i_q\right) i_d + \frac{\sqrt{3}}{2} M_f i_q i_f$$

$$= \frac{\sqrt{3}}{2} M_f i_q i_f + L_d i_d i_q - L_q i_d i_q$$

$$= \frac{\sqrt{3}}{2} M_f i_q i_f + \left(L_d - L_q\right) i_d i_q \tag{11.77}$$

For a cylindrical rotor synchronous machine, the direct axis and quadrature axis inductances are equal, i.e., $L_d = L_q$:

$$\therefore \text{Torque, } T_e = \frac{\sqrt{3}}{2} M_f i_f i_q \tag{11.78}$$

For a salient-pole machine, there is a saliency torque:

$$T_{\text{saliency}} = \left(L_d - L_q\right) i_d i_q \tag{11.79}$$

This torque exists only because of non-uniformity in the permeance of the air gap along the d- and q-axes. This is the reluctance torque of a salient-pole machine and exists even when the field excitation is zero.

11.21 SYNCHRONOUS MACHINE—STEADY-STATE ANALYSIS

Consider a 3-ϕ synchronous machine that has three armature (stator) windings a, b, and c, one field winding 'f' on the rotor with its flux in the direction of the d-axis, and one fictitious winding 'g' on the rotor with its flux in the quadrature axis as shown in Fig. 11.30.

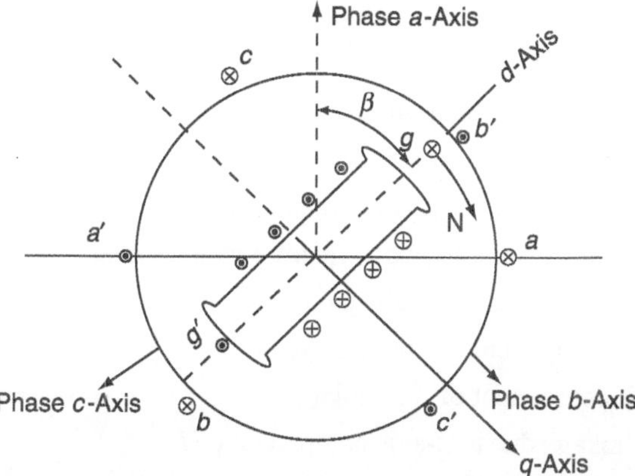

FIG. 11.30 *Three-phase synchronous machine with stator and rotor windings*

The fictitious winding 'g' approximates the effect of eddy currents circulating in the iron (rotor iron in round-rotor machine and negligible in salient-pole machine) and to some extent the effect of damper windings. This fictitious winding is short circuited since it is not connected to any voltage source.

Since the electromagnetic transients in the network are much faster than the mechanical transients, the steady-state phasor solutions on the network side are performed.

11.21.1 Salient-pole synchronous machine

The phasor diagram of an overexcited salient-pole synchronous generator for lagging p.f. is shown in Fig. 11.31.

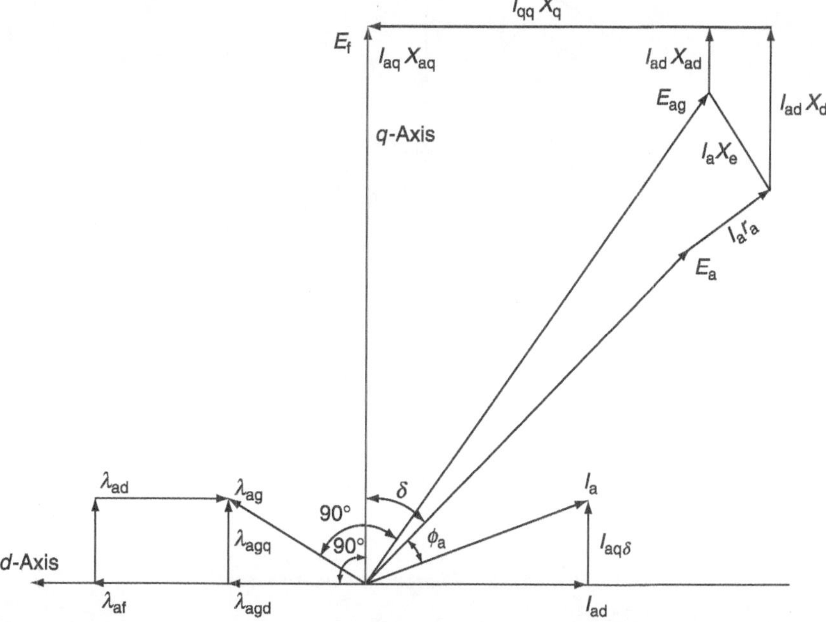

FIG. 11.31 *Phasor diagram of a salient-pole synchronous generator*

δ is the power angle or torque angle

E_a the terminal voltage $= v_a$

E_{ag} the voltage due to air-gap flux

E_{af} the voltage due to flux produced by main rotor-field current

$I_{ad}X_{ad}$ the voltage drop across d-axis armature magnetizing reactance

$I_{aq}X_{aq}$ the voltage drop across q-axis armature magnetizing reactance

λ_{ag} the flux linkage due to net air-gap flux

λ_{agd} the d-axis component of flux linkage

λ_{agq} the q-axis component of flux linkage

λ_{ad} the flux linkage due to d-axis component of I_a

λ_{aq} the flux linkage due to q-axis component of I_a

x_d the d-axis component of synchronous reactance $= x_1 + x_{ad}$

x_q the q-axis component of synchronous reactance $= x_1 + x_{aq}$

Figure 11.32 represents the phasors and their speeds:

δ is the angle between synchronously rotating reference phasor axis and q-axis

ω_s the synchronous speed

ω the speed of rotor

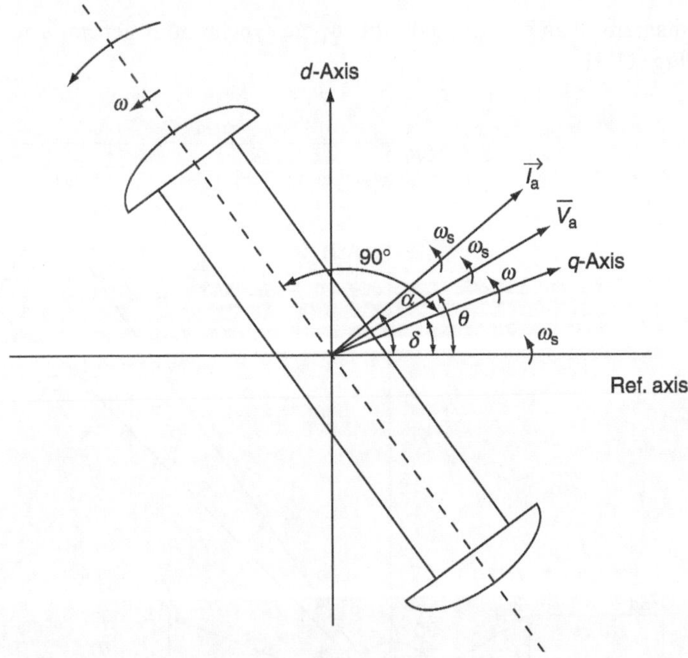

FIG. 11.32   *Phasors and their speeds*

From the phasor diagram shown in Fig. 11.32, we have

$$\left.\begin{array}{l} i_a = I\cos(\omega_s t + \alpha) \\ i_b = I\cos(\omega_s t + \alpha - 120°) \\ i_c = I\cos(\omega_s t + \alpha + 120°) \end{array}\right\}$$
(11.80)

By applying Park's transformation, we get

$$\left.\begin{array}{l} i_q = I\cos(\alpha - \delta) = I_q \\ i_d = I\sin(\alpha - \delta) = I_d \\ i_o = 0 = I_0 \end{array}\right\}$$
(11.81)

Equations (11.81) can be expressed as a phasor equation:

$$I_q + jI_d = Ie^{j(\alpha - \delta)}$$
$$= \bar{I}e^{-j\delta} \text{ as } \bar{I} = I\angle\alpha$$
(11.82)

For the steady-state analysis, the q-axis will be considered as the real axis and the d-axis as the imaginary axis since the voltage induced in a normal steady-state operation lies on the q-axis.

Similarly, by Park's transformation, we get

$$v_q = v\cos(\theta - \delta)$$
$$v_d = v\sin(\theta - \delta)$$
$$V_o = 0$$

In complex notation, $\quad v_q + jv_d = ve^{-j(\theta - \delta)} = \bar{v}e^{-j\delta}$

$$\text{as } \bar{v} = v\angle\theta$$
(11.83)

To get the steady-state analysis, we make use of the following assumptions:

(i) Transformer voltages, $\dfrac{d\lambda_d}{dt}$ and $\dfrac{d\lambda_q}{dt}$, being small and are therefore neglected.

(ii) Balanced network currents and voltages are assumed.

The reasons for the above assumptions are that changes in λ_d and λ_q are very slow in time with the oscillations of angle δ and hence $\dfrac{d\lambda_d}{dt}$ and $\dfrac{d\lambda_q}{dt}$ are very small compared with $\omega\lambda_d$ and $\omega\lambda_q$.

Due to the above assumptions, Equations (11.62) and (11.68) can be rewritten by dropping the transformer voltage terms, zero-sequence currents, and voltages:

$$\left.\begin{array}{l} v_d = -R_a i_d - \omega\lambda_q \\ v_q = -R_a i_q - \omega\lambda_d \\ \lambda_d = L_d i_d + \sqrt{\dfrac{3}{2}}M_f i_f \\ \lambda_q = L_q i_q - \sqrt{\dfrac{3}{2}}M_g i_g \end{array}\right\}$$
(11.84)

Substituting for λ_d and λ_q in the above equations of voltages, we get

$$\left.\begin{aligned}
v_q &= -R_a i_q + \omega L_d i_d + \sqrt{\tfrac{3}{2}}\,\omega M_f i_f \\
v_d &= -R_a i_d - \omega L_d i_d + \sqrt{\tfrac{3}{2}}\,\omega M_g i_g \\
\lambda_f &= \sqrt{\tfrac{3}{2}}\,M_f i_d + L_{ff} i_f \\
v_f &= -R_f i_f - \frac{d\lambda_f}{dt}
\end{aligned}\right\} \tag{11.85}$$

$$\left.\begin{aligned}
\Rightarrow \frac{d\lambda_f}{dt} &= -v_f - R_f i_f \\
0 &= -R_g i_g - \frac{d\lambda_g}{dt}
\end{aligned}\right\} \tag{11.86}$$

Hence, $\quad \dfrac{d\lambda_g}{dt} = -R_g i_g$

$$I_q + j I_d = I\,e^{j(\alpha-\delta)}$$

$$v_q + j v_d = v\,e^{j(\theta-\delta)}$$

11.21.2 Non-salient-pole synchronous (cylindrical rotor) machine

For this case, $X_d = X_q$

$$\therefore E_{af} = E_a + i_a R_a + j i_a X_l + j i_a X_{ad}$$

$$= E_a + i_a R_a + j i_d X_d \tag{11.87}$$

The equivalent circuit of a non-salient-pole synchronous machine is represented by a source E_{af} (induced emf) in series with the internal impedance $R_a + jX_d$ as shown in Fig. 11.33.

The phasor diagram is shown in Fig. 11.34. Now, the synchronous reactance is defined as $X_s = X_{aq} + X_l$, and if resistance 'r' is neglected, the cylindrical rotor synchronous generator is represented by the equivalent circuit as shown in Fig. 11.35.

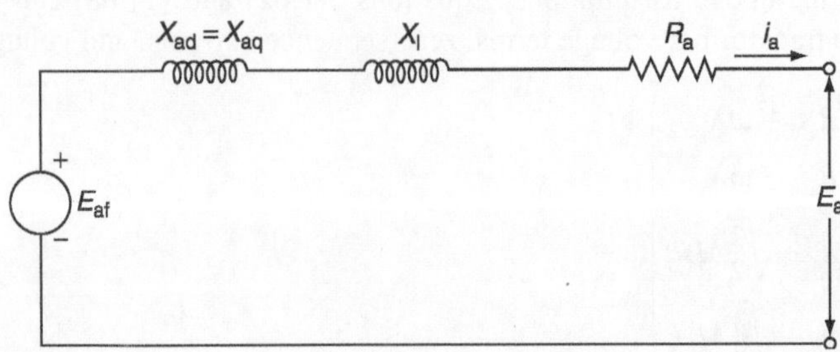

FIG. 11.33   *Equivalent circuit of non-salient-pole synchronous generator*

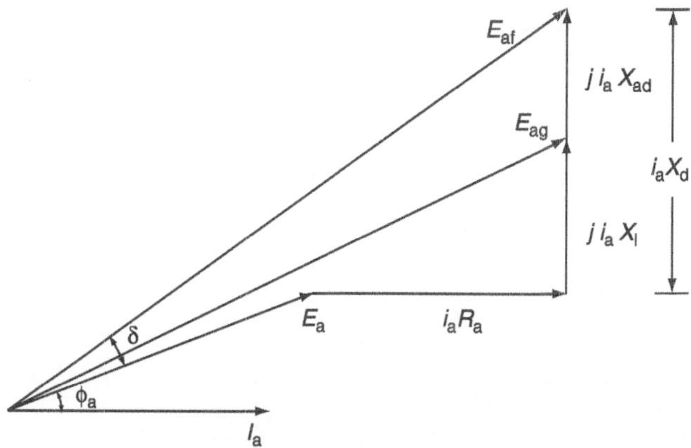

FIG. 11.34 *Phasor diagram of non-salient-pole synchronous generator*

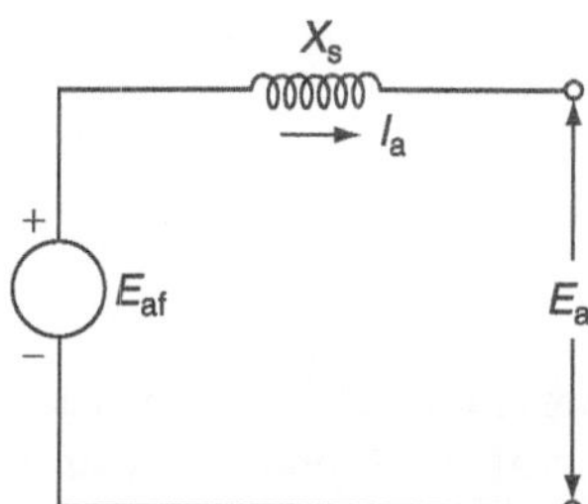

FIG. 11.35 *Equivalent circuit*

Now, Equation (11.87) becomes

$$E_{af} = E_a + ji_a X_s \qquad\qquad (11.88)$$

The above relationship represents the model of the cylindrical rotor (non-salient pole) generator under steady-state conditions and of which a very useful equivalent circuit is shown in Fig. 11.35.

11.22 DYNAMIC MODEL OF SYNCHRONOUS MACHINE

In this section, we shall discuss the dynamic model of synchronous machines—salient pole synchronous generator, dynamic equations of synchronous machine, and equivalent circuit of synchronous generator—in detail.

11.22.1 Salient-pole synchronous generator—sub-transient effect

During normal steady-state conditions, there is no transformer action between stator and rotor windings of synchronous machines, as the resultant field produced by stator windings and rotor windings revolves with the same (synchronous) speed and in the same direction. However, during disturbances, the rotor speed is no longer the same as that of the revolving field produced by stator windings, which always rotates with synchronous speed. Hence, the synchronous generator becomes a transformer.

Synchronous machine dynamic equations

We have

$$v_q = -R_a I_q + \omega L_d I_d + \sqrt{\frac{3}{2}}\,\omega M_f I_f$$

$$v_d = -R_a I_d - \omega L_q I_q + \sqrt{\frac{3}{2}}\,\omega M_g I_g$$

$$\lambda_f = \sqrt{\frac{3}{2}}\,M_f I_d + L_{ff} I_f$$

$$\lambda_g = -\sqrt{\frac{3}{2}}\,M_g I_g + L_{gg} I_g$$

$$\frac{d\lambda_f}{dt} = -v_f - R_f I_f$$

$$\frac{d\lambda_g}{dt} = -R_g I_g$$

$$i_q = I\cos(\alpha - \delta)$$

$$i_d = I\sin(\alpha - \delta)$$

$$v_q = v\cos(\theta - \delta)$$

$$v_d = v\sin(\theta - \delta)$$

(11.89)

Let $X_d = \omega L_d = d$-axis component of synchronous reactance,
$X_q = \omega L_q = q$-axis component of synchronous reactance:

$$\therefore v_q = -R_a I_q + X_d I_d + \sqrt{\frac{3}{2}}\,\omega M_f I_f$$

(11.90)

and

$$\lambda_f = \sqrt{\frac{3}{2}}\,M_f I_d + L_{ff} I_f$$

$$\Rightarrow I_f = \frac{\lambda_f - \sqrt{\frac{3}{2}}\,M_f I_d}{L_{ff}}$$

Substituting the I_f value in Equation (11.90), we get

$$v_q = -R_a I_q + X_d I_d + \sqrt{\frac{3}{2}}\,\omega M_f \left(\frac{\lambda_f - \sqrt{\frac{3}{2}} M_f I_d}{L_{ff}} \right)$$

$$= -R_a I_q + X_d I_d + \sqrt{\frac{3}{2}}\,\frac{\omega M_f \lambda_f}{L_{ff}} - \frac{\frac{3}{2}\omega M_f^2 I_d}{L_{ff}}$$

(11.91)

Let $\sqrt{\frac{3}{2}}\,\dfrac{\omega M_f \lambda_f}{L_{ff}} = E_q' =$ sub-transient voltage along the q-axis

Since at this time, both windings 'f' and 'g' are present along the d-axis then, Equation (11.91) becomes

$$v_q = -R_a I_q + X_d I_d + E_q' - \frac{\frac{3}{2}\omega M_f^2 I_d}{L_{ff}}$$

$$\Rightarrow v_q = -R_a I_q + \left(X_d - \frac{3}{2}\frac{\omega M_f^2}{L_{ff}}\right)I_d + E_q' \tag{11.92}$$

Let $\left(X_d - \frac{3}{2}\frac{\omega M_f^2}{L_{ff}}\right) = X_d' = $ transient d-axis reactance

Since at this time, both the d-axis component of the armature windings and the f-winding are present, Equation (11.92) becomes

$$v_q = -R_a I_q + X_d' I_d + E_q' \tag{11.93}$$

Similarly, the equation of v_d is given by

$$v_d = -R_a I_d - \omega L_q I_q + \sqrt{\frac{3}{2}}\,\omega M_g I_g$$

Substituting, $\omega L_q = X_q$, we get

$$v_d = -R_a I_d - X_q I_q + \sqrt{\frac{3}{2}}\,\omega M_g I_g \tag{11.94}$$

We know that $\quad \lambda_g = -\sqrt{\frac{3}{2}}M_g I_q + L_{gg}I_g$

$$\Rightarrow I_g = \frac{\lambda_g + \sqrt{\frac{3}{2}}M_g I_q}{L_{gg}}$$

Substituting I_g expression in Equation (11.94), we get

$$v_d = -R_a I_d - X_q I_q + \sqrt{\frac{3}{2}}\,\omega M_g \left[\frac{\lambda_g + \sqrt{\frac{3}{2}}M_g I_q}{L_{gg}}\right]$$

$$= -R_a I_d - X_q I_q + \sqrt{\frac{3}{2}}\,\frac{\omega M_g \lambda_g}{L_{gg}} + \frac{3}{2}\frac{\omega M_g^2}{L_{gg}}I_q$$

$$= -R_a I_d - \left[X_q - \frac{3}{2}\frac{\omega M_g^2}{L_{gg}}\right]I_q + \sqrt{\frac{3}{2}}\,\frac{\omega M_g \lambda_g}{L_{gg}} \tag{11.95}$$

Let $\sqrt{\frac{3}{2}}\,\frac{\omega M_g \lambda_g}{L_{gg}} = E_d' = $ Transient voltage along the d-axis (since both the q-axis component and g-windings are present to give transient state) and also let

$$X_q - \frac{3}{2}\frac{\omega M_g^2}{L_{gg}} = X_q' = q\text{-axis component of transient reactance}$$

Hence, Equation (11.95) becomes

$$v_d = -R_a I_d - X_q' I_q + E_d' \tag{11.96}$$

Let $\quad \tau_{do}' = \frac{L_{ff}}{R_f} = $ direct axis open-circuit transient time constant

$$\tau_{qo}' = \frac{L_{gg}}{R_g} = \text{quadrature axis open-circuit transient time constant}$$

and also,

$$E_\mathrm{d} = \sqrt{\frac{3}{2}}\,\omega\,M_\mathrm{g}\,I_\mathrm{g} = \omega\lambda_\mathrm{g} \left.\vphantom{\begin{matrix}a\\a\\a\\a\end{matrix}}\right\}$$

$$E_\mathrm{q} = \sqrt{\frac{3}{2}}\,\omega\,M_\mathrm{f}\,I_\mathrm{f} = \omega\lambda_\mathrm{f}$$

$$\text{i.e.,}\quad E = E_\mathrm{q} + jE_\mathrm{d}$$

(11.97)

Substituting Equation (11.97) in Equations (11.90) and (11.94), we get

$$v_\mathrm{q} = -R_\mathrm{a}I_\mathrm{q} + \omega L_\mathrm{d}I_\mathrm{d} + \sqrt{\frac{3}{2}}\,\omega\,M_\mathrm{f}I_\mathrm{f}$$

$$\Rightarrow v_\mathrm{q} = -R_\mathrm{a}I_\mathrm{q} + X_\mathrm{d}I_\mathrm{d} + E_\mathrm{q}$$

(11.98)

$$v_\mathrm{d} = -R_\mathrm{a}I_\mathrm{d} - X_\mathrm{q}I_\mathrm{q} + E_\mathrm{d}$$

(11.99)

From Equations (11.93) and (11.98)

$$v_\mathrm{q} = -R_\mathrm{a}I_\mathrm{q} + X_\mathrm{d}'I_\mathrm{d} + E_\mathrm{q}' = -R_\mathrm{a}I_\mathrm{q} + X_\mathrm{d}I_\mathrm{d} + E_\mathrm{q}$$

$$\text{i.e.,}\quad E_\mathrm{q}' = E_\mathrm{q} + \left(X_\mathrm{d} - X_\mathrm{d}'\right)I_\mathrm{d}$$

(11.100)

Similarly from Equations (11.96) and (11.99), we have

$$v_\mathrm{d} = -R_\mathrm{a}I_\mathrm{d} - X_\mathrm{q}'I_\mathrm{q} + E_\mathrm{d}' = -R_\mathrm{a}I_\mathrm{d} - X_\mathrm{q}I_\mathrm{q} + E_\mathrm{d}$$

$$\text{i.e.,}\quad E_\mathrm{d}' = E_\mathrm{d} - \left(X_\mathrm{q} - X_\mathrm{q}'\right)I_\mathrm{q}$$

(11.101)

Figure 11.36 shows the phasor diagram of a synchronous machine under the transient state.

From the phasor diagram also we get

$$E_\mathrm{q}' = E_\mathrm{q} + I_\mathrm{d}\left(X_\mathrm{d} - X_\mathrm{d}'\right) = \text{voltage behind the } d\text{-axis component of transient reactance}$$

$$E_\mathrm{d}' = E_\mathrm{d} - I_\mathrm{q}\left(X_\mathrm{q} - X_\mathrm{q}'\right) = \text{voltage behind the } q\text{-axis component of transient reactance}$$

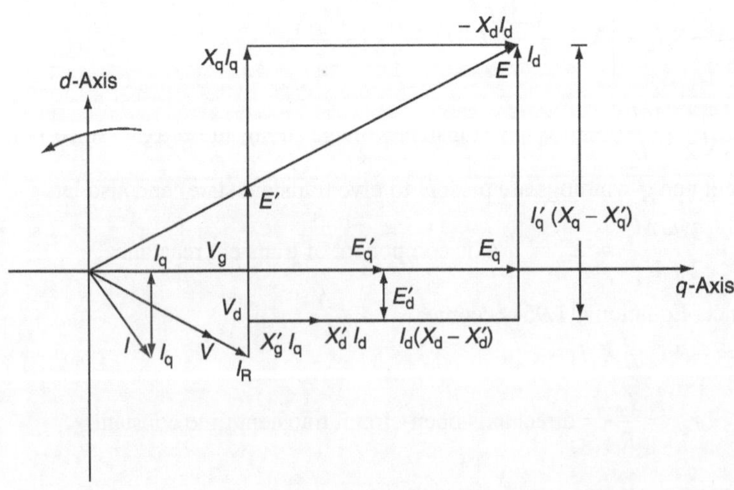

FIG. 11.36   *Phasor diagram*

$$E'_q = E_q + I_d\left(X_d - X'_d\right)$$

$$= \sqrt{\frac{3}{2}}\,\omega M_f I_f + \frac{3}{2}\frac{\omega M_f^2}{L_{ff}}I_d$$

$$= \sqrt{\frac{3}{2}}\frac{\omega M_f}{L_{ff}}\left(\sqrt{\frac{3}{2}}\,M_f I_d + L_{ff} I_f\right)$$

$$= \sqrt{\frac{3}{2}}\frac{\omega M_f}{L_{ff}}\lambda_f \tag{11.102}$$

$$E'_d = E_d - (X_q - X'_q)I_q$$

$$= \sqrt{\frac{3}{2}}\,\omega M_g I_g - \frac{3}{2}\frac{\omega M_g^2}{L_{gg}}I_q$$

$$= \sqrt{\frac{3}{2}}\frac{\omega M_g}{L_{gg}}\left(-\sqrt{\frac{3}{2}}\,M_g I_q + L_{gg} I_g\right)$$

$$= \sqrt{\frac{3}{2}}\frac{\omega M_g}{L_{gg}}\lambda_g \tag{11.103}$$

From Equations (11.102) and (11.103), we have

$$\left.\begin{aligned}
E'_q &= \sqrt{\frac{3}{2}}\frac{\omega M_f}{L_{ff}}\lambda_f \\[2mm]
E'_d &= \sqrt{\frac{3}{2}}\frac{\omega M_g}{L_{gg}}\lambda_g
\end{aligned}\right\} \tag{11.104}$$

and $\qquad v_f = \sqrt{\dfrac{3}{2}}\dfrac{\omega M_f}{R_f}\lambda_f$

Therefore, from the above analysis, we get

$$E'_q = E_q + \left(X_d - X'_d\right)I_d = \sqrt{\frac{3}{2}}\frac{\omega M_f}{L_{ff}}\lambda_f \tag{11.105}$$

$$E'_d = E_d + \left(X_q - X'_q\right)I_q = \sqrt{\frac{3}{2}}\frac{\omega M_g}{L_{gg}}\lambda_g \tag{11.106}$$

Taking the derivative for Equation (11.105), we get

$$\frac{dE'_q}{dt} = \sqrt{\frac{3}{2}}\frac{\omega M_f}{L_{ff}}\frac{d\lambda_f}{dt}$$

$$= \sqrt{\frac{3}{2}}\frac{\omega M_f}{L_{ff}}(-v_f - R_f I_f) \quad \left[\because \frac{d\lambda_f}{dt} = -v_f - R_f I_f\right]$$

$$= -\frac{1}{\tau'_{do}} \sqrt{\frac{3}{2}} \left(\frac{\omega M_f v_f}{R_f} + \omega M_f I_f \right)$$

$$\left[\because \tau'_{do} = \frac{L_{ff}}{R_f} = d\text{-axis open-circuit transient reactance time constant} \right]$$

$$= -\frac{1}{\tau'_{do}} \left(V_f + E_q \right) \tag{11.107}$$

and also taking the derivative of Equation (11.106), we have

$$\frac{dE'_d}{dt} = \sqrt{\frac{3}{2}} \frac{\omega M_g}{L_{gg}} \left(-R_g I_g \right) \qquad \left[\because \frac{d\lambda_g}{dt} = -R_g I_g \right]$$

$$= -\frac{1}{\tau'_{qo}} \sqrt{\frac{3}{2}} \left(\omega M_g I_g \right)$$

$$\because \tau'_{qo} = \frac{L_{gg}}{R_g} = q\text{-axis component of open-circuit transient reactance time constant}$$

$$\frac{dE'_d}{dt} = -\frac{1}{\tau'_{qo}} E_d \left[\because \sqrt{\frac{3}{2}} \omega M_g I_g = E_d \right] \tag{11.108}$$

i.e., $$\frac{dE'_q}{dt} = -\frac{1}{\tau'_{do}} \left(v_f + E_q \right)$$

$$\frac{dE'_d}{dt} = -\frac{1}{\tau'_{qo}} \left(E_d \right)$$

Equations (11.93) and (11.96) can be written in the matrix form as

$$\begin{bmatrix} v_q \\ v_d \end{bmatrix} = \begin{bmatrix} -R_a & X'_d \\ -X'_q & -R_a \end{bmatrix} \begin{bmatrix} I_q \\ I_d \end{bmatrix} + \begin{bmatrix} E'_q \\ E'_d \end{bmatrix} \tag{11.109}$$

11.22.2 Dynamic model of synchronous machine including damper winding

The 'f' and 'g' coils in the rotor winding produce transient effect in terms of X'_d and X'_q in the synchronous machine. The field coil 'f' exists physically whereas the 'g' coil is hypothetical for representing the rotor eddy currents in the q-axis. However, it is quite difficult to calculate g-coil inductance.

The more accurate representation of synchronous machine is obtained by adding two more fictitious windings on the rotor, one along the d-axis known as 'K_d' winding and the other along the q-axis, known as 'K_q' winding. These damper windings can be approximated by two hypothetical coils, both short-circuited as there is no voltage source connected to them.

The dynamic machine equations will now be modified to include the damper windings by substituting the scalar (λ_f and λ_g) vectors:

$$\begin{bmatrix} \lambda_f \\ \lambda_{kd} \end{bmatrix} \text{ and } \begin{bmatrix} \lambda_g \\ \lambda_{kg} \end{bmatrix}$$

Now in this model, the Park's transformation will be applied with the following assumptions:

(a) Mutual inductances from stator coils a, b, and c (or its component 'd' and 'q' axes) to the 'K_d' coil is the same as 'f' coil, and to the K_g coil the same as 'g' coil and

(b) Mutual inductance between 'K_d' coil and 'f' coil (and 'K_g' coil and 'g' coil) is the same as 'd' component of stator coils and f-coils (or 'q' component of stator to 'g' coil).

The resultant equations after Park's transformation are

$$\begin{bmatrix} v_d \\ v_q \\ v_{oh} \\ v_f \\ 0 \\ 0 \\ 0 \end{bmatrix} = -\begin{bmatrix} R_a & & & & & & \\ & R_a & & & & & \\ & & R_a & & & & \\ & & & R_f & & & \\ & & & & R_{kd} & & \\ & & & & & R_g & \\ & & & & & & R_{kg} \end{bmatrix}\begin{bmatrix} i_d \\ i_q \\ i_o \\ i_f \\ i_{kd} \\ i_g \\ i_{kg} \end{bmatrix} - \begin{bmatrix} \frac{d\lambda_d}{dt} \\ \frac{d\lambda_q}{dt} \\ \frac{d\lambda_o}{dt} \\ \frac{d\lambda_f}{dt} \\ \frac{d\lambda_{kd}}{dt} \\ \frac{d\lambda_g}{dt} \\ \frac{d\lambda_{kg}}{dt} \end{bmatrix} + \begin{bmatrix} -\omega\lambda_q \\ \omega\lambda_d \\ 0 \\ 0 \\ 0 \\ 0 \\ 0 \end{bmatrix} \qquad \textbf{(11.110)}$$

From the above matrix representation, the modified voltage equations with the inclusion of damper windings are

$$\left.\begin{aligned} v_d &= -R_a I_d - \frac{d\lambda_d}{dt} - \omega\lambda_q \\[6pt] v_q &= -R_a I_q - \frac{d\lambda_q}{dt} + \omega\lambda_d \\[6pt] v_o &= -R_a I_0 - \frac{d\lambda_0}{dt} \\[6pt] v_f &= -R_f I_f - \frac{d\lambda_f}{dt} \\[6pt] 0 &= -R_{kd} I_{kd} - \frac{d\lambda_{kd}}{dt} \\[6pt] 0 &= -R_{kg} I_{kg} - \frac{d\lambda_{kg}}{dt} \end{aligned}\right\} \qquad \textbf{(11.111)}$$

The modified flux linkages with the inclusion of damper windings are obtained in the form of matrix as

$$\begin{bmatrix} \lambda_d \\ \lambda_q \\ \lambda_o \\ \lambda_f \\ \lambda_{kd} \\ \lambda_g \\ \lambda_{kg} \end{bmatrix} = \begin{bmatrix} L_d & 0 & 0 & \left[\sqrt{\tfrac{3}{2}}M_f \;\; \sqrt{\tfrac{3}{2}}M_f\right] & 0 & 0 \\ 0 & L_q & 0 & 0 & 0 & \left[-\sqrt{\tfrac{3}{2}}M_g \;\; -\sqrt{\tfrac{3}{2}}M_g\right] \\ 0 & 0 & L_0 & 0 & 0 & 0 \\ \left[\sqrt{\tfrac{3}{2}}M_f \atop \sqrt{\tfrac{3}{2}}M_f\right] & 0 & 0 & \left[\begin{matrix}L_{ff} & \sqrt{\tfrac{3}{2}}M_f \\ \sqrt{\tfrac{3}{2}}M_f & L_{kd\,Kd}\end{matrix}\right] & 0 & 0 \\ 0 & \left[-\sqrt{\tfrac{3}{2}}M_g \atop -\sqrt{\tfrac{3}{2}}M_g\right] & 0 & 0 & 0 & \left[\begin{matrix}L_{gg} & -\sqrt{\tfrac{3}{2}}M_g \\ -\sqrt{\tfrac{3}{2}}M_g & L_{kg\,kg}\end{matrix}\right] \end{bmatrix}\begin{bmatrix} i_d \\ i_q \\ i_o \\ \left[i_f \atop i_{kd}\right] \\ \left[i_g \atop i_{kg}\right] \end{bmatrix} \qquad \textbf{(11.112)}$$

i.e., the flux linkage equations are

$$
\left.
\begin{aligned}
\lambda_{\mathrm{d}} &= L_{\mathrm{d}}I_{\mathrm{d}} + \sqrt{\tfrac{3}{2}}\,M_{\mathrm{f}}I_{\mathrm{f}} + \sqrt{\tfrac{3}{2}}\,M_{\mathrm{f}}I_{\mathrm{kd}} \\[4pt]
\lambda_{\mathrm{q}} &= L_{\mathrm{q}}I_{\mathrm{q}} - \sqrt{\tfrac{3}{2}}\,M_{\mathrm{g}}I_{\mathrm{g}} - \sqrt{\tfrac{3}{2}}\,M_{\mathrm{g}}I_{\mathrm{kg}} \\[4pt]
\lambda_{0} &= L_{0}I_{0} \\[4pt]
\lambda_{\mathrm{f}} &= \sqrt{\tfrac{3}{2}}\,M_{\mathrm{f}}I_{\mathrm{d}} + L_{\mathrm{ff}}I_{\mathrm{f}} + \sqrt{\tfrac{3}{2}}\,M_{\mathrm{f}}I_{\mathrm{kd}} \\[4pt]
\lambda_{\mathrm{kd}} &= \sqrt{\tfrac{3}{2}}\,M_{\mathrm{f}}I_{\mathrm{d}} + \sqrt{\tfrac{3}{2}}\,M_{\mathrm{f}}I_{\mathrm{f}} + L_{\mathrm{kd\,kd}}I_{\mathrm{kd}} \\[4pt]
\lambda_{\mathrm{g}} &= -\sqrt{\tfrac{3}{2}}\,M_{\mathrm{g}}I_{\mathrm{q}} + L_{\mathrm{gg}}I_{\mathrm{g}} - \sqrt{\tfrac{3}{2}}\,M_{\mathrm{g}}I_{\mathrm{kg}} \\[4pt]
\lambda_{\mathrm{kg}} &= -\sqrt{\tfrac{3}{2}}\,M_{\mathrm{g}}I_{\mathrm{q}} - \sqrt{\tfrac{3}{2}}\,M_{\mathrm{g}}I_{\mathrm{g}} + L_{\mathrm{kg\,kg}}I_{\mathrm{kg}}
\end{aligned}
\right\}
\qquad (11.113)
$$

11.22.3 Equivalent circuit of synchronous generator—including damper winding effect

11.2.3.1 Along the d-axis

The equivalent circuit of the synchronous generator along the d-axis excluding resistances is as shown in Fig. 11.37, where X_{l} is the armature leakage reactance, X_{ad} the armature magnetizing reactance, X_{f} the field reactance, and X_{g} the fictitious winding reactance.

Initially, all the reactances are in the circuit (i.e., just at the instant when the fault has occurred) and therefore, initial or sub-transient reactance, are the lowest. After some time, the g-winding (damper winding) is out of circuit as it has a very low time constant and hence we have only field winding and armature reactances in parallel. This reactance is known as transient reactance and is larger than the previous one. However, after some time, when the disturbance altogether disappears, field winding is also out of circuit and hence we have only armature reactance, $(X_{\mathrm{d}} = X_{\mathrm{l}} + X_{\mathrm{ad}})$ called the steady-state reactance of the circuit.

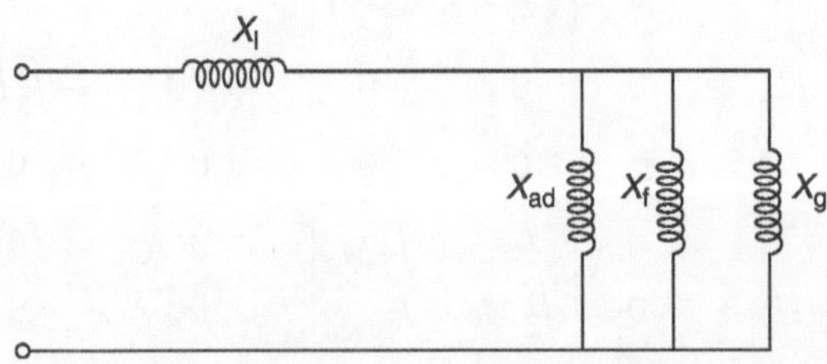

FIG. 11.37   *Equivalent circuit of synchronous generator*

For a more accurate representation, two more fictitious windings 'K_d' winding and 'K_q' winding are added. Hence, the equivalent circuit of the synchronous generator along the d-axis can be represented as shown in Fig. 11.38.

The parallel combination of X_ad, X_f, and X_kd is known as the d-axis component of sub-transient reactance X''_ad and is represented as $X''_\mathrm{ad} = X_\mathrm{ad}/X_\mathrm{f}/X_\mathrm{kd}$.

Hence, the d-axis component of the sub-transient synchronous reactance is given by

$$X''_\mathrm{d} = X_\mathrm{l} + X''_\mathrm{ad}$$

This reactance is very small. After some time, as the hunting becomes less, the winding K_d is also out of circuit since it has a low time constant.

Thus, the resultant equivalent circuit becomes as shown in Fig. 11.39.

The parallel combination of reactances X_ad and X_f is known as X'_ad, d-axis component of transient armature reactance.

The d-axis component of transient synchronous reactance is $X'_\mathrm{d} = X_\mathrm{l} + X'_\mathrm{ad}$

Generally, $X'_\mathrm{d} > X''_\mathrm{d}$

Finally, when the disturbance is altogether over, there will not be hunting of the rotor and hence there will not be any transformer action between the stator and the rotor. Hence, the equivalent circuit of synchronous generator becomes as shown in Fig. 11.40.

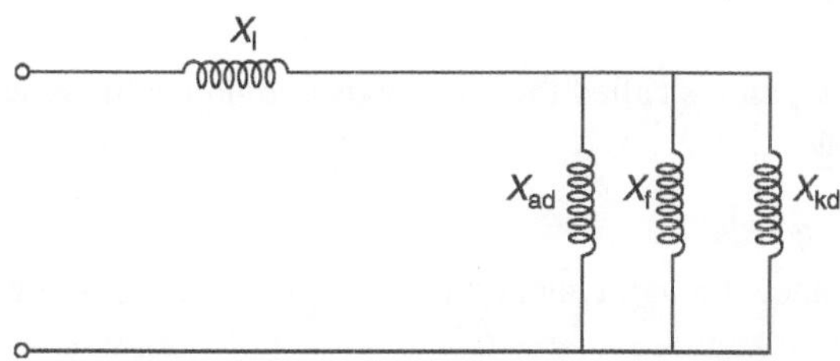

FIG .11.38   *Accurate representation of equivalent circuit of synchronous generator*

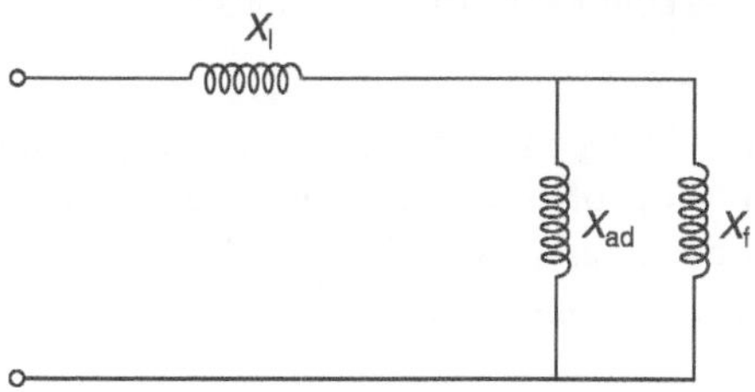

FIG. 11.39   *Resultant equivalent circuit of synchronous generator*

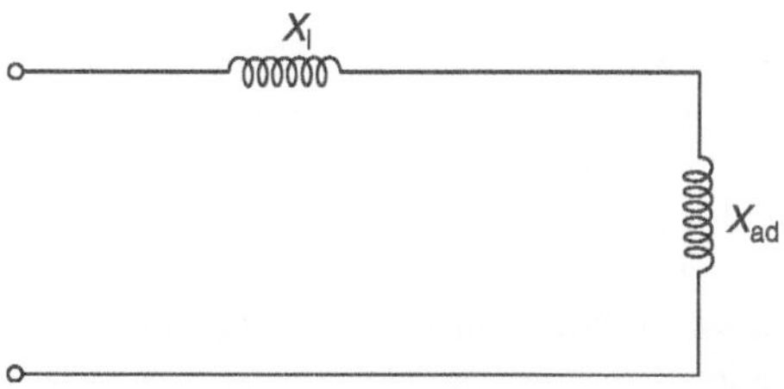

FIG. 11.40   *Equivalent circuit*

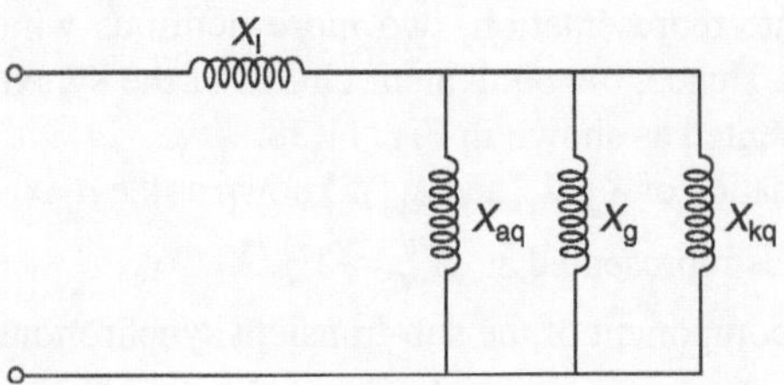

FIG. 11.41   *Equivalent circuit of synchronous generator along the q-axis*

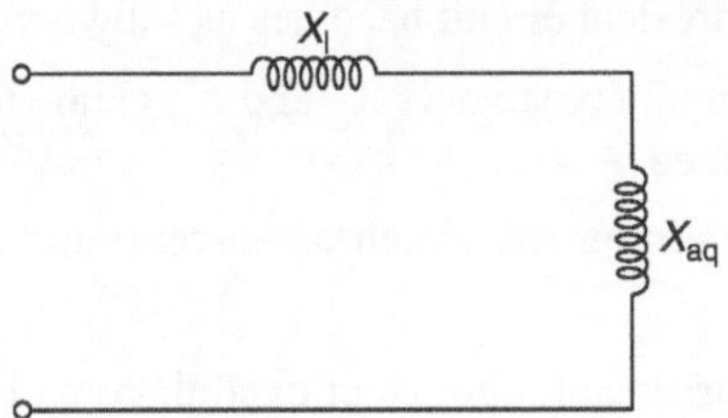

FIG. 11.42   *Resultant equivalent circuit of the synchronous generator*

Here, $X_d = X_1 + X_{ad}$ and is called the direct axis component of synchronous reactance. It is obvious that $X_d'' < X_d' < X_d$.

11.2.3.2 *Along the q-axis*

Just after the disturbance, the equivalent circuit of the machine will become as shown in Fig. 11.41.

Here, $X_q'' = X_1 + (X_{aq} \text{ // } X_g \text{ // } X_{kq})$

where X_q'' is the q-axis component of sub-transient synchronous reactance. The value of X_q'' is very small.

After some time, hunting becomes less and less, both g-winding and K_q, which have a low time constant, will be out of circuit and hence the resultant equivalent circuit of the synchronous generator becomes as shown in Fig. 11.42.

Here, $X_q' = X_q = X_1 + X_{aq}$, where X_q' is the q-axis component of transient synchronous reactance.

11.23 MODELING OF SYNCHRONOUS MACHINE—SWING EQUATION

The mechanical behavior of a synchronous machine can be established by interconnecting the electrical and mechanical sides of a synchronous machine in terms of electrical and mechanical torque. This is provided by the dynamic equation for the acceleration or deceleration of the rotor of a combined turbine and synchronous generator system, which is usually called the *swing equation*.

While developing a swing equation or a mechanical equation, the following basic assumptions are to be made:

(i) Synchronous machine rotor speed must be synchronous speed.

(ii) The rotational power losses due to friction and windage are neglected.

(iii) Mechanical shaft power is smooth, i.e., the shaft power is constant.

Let us consider a single rotating machine with steady-state angular speed ω_s and phase angle δ. Due to various electrical or mechanical disturbances, the machine will be subjected to differences in mechanical and electrical torque, causing it to accelerate or decelerate. Hence, during disturbance, the rotor will accelerate or decelerate with respect to the synchronously rotating air-gap m.m.f. and a relative motion begins.

Let $\quad\theta$ be the angular position of the rotor at any instant 't'

$\qquad\omega$ the angular velocity (rad/s)

$\qquad\alpha$ the acceleration

$\qquad\delta$ the phase angle of a rotation machine

$\qquad T_{net}$ the net accelerating torque in a machine

$\qquad T_{elec}$ the electrical torque exerted on the machine by the generator

$\qquad P_{net}$ the net accelerating power

$\qquad P_{mech}$ the mechanical power input

$\qquad P_{elec}$ the electrical power output

$\qquad J$ the moment of inertia for the machine

$\qquad M = J\omega$; angular momentum of the machine in kg-m^2

$$J\alpha = T_{net}$$

$$P_{net} = \omega T_{net} = \omega(J\alpha) = M\alpha$$

Consider a synchronous generator developing an electromagnetic torque T_e and running at the synchronous speed ω_s. If T_m is the driving mechanical torque, then under steady-state conditions, with negligible losses,

$$T_m = T_e$$

A departure from the steady state due to a disturbance results in an accelerating ($T_m > T_e$) or decelerating ($T_e > T_m$) torque T_a on the rotor:

$$T_a = T_m - T_e$$

Neglecting the frictional and damping torque, from the law of rotation, we have

$$J\frac{d^2\theta}{dt^2} = T_a = T_m - T_e$$

Now, the value of θ is continuously changing with time 't'. It is convenient to measure θ with respect to a reference axis, which is rotating at synchronous speed.

If δ is the angular displacement of a rotor in electrical degree from the synchronous rotating reference axis and ω_s, the synchronous speed in electrical degrees, then θ can be expressed as the sum of: (i) time-varying angle $\omega_s t$ on the rotating axis and (ii) the torque angle δ of the rotor with respect to the rotating reference axis as shown in Fig. 11.43:

$$\theta = \omega_s t + \delta$$

$$\frac{d\theta}{dt} = \omega_s + \frac{d\delta}{dt}$$

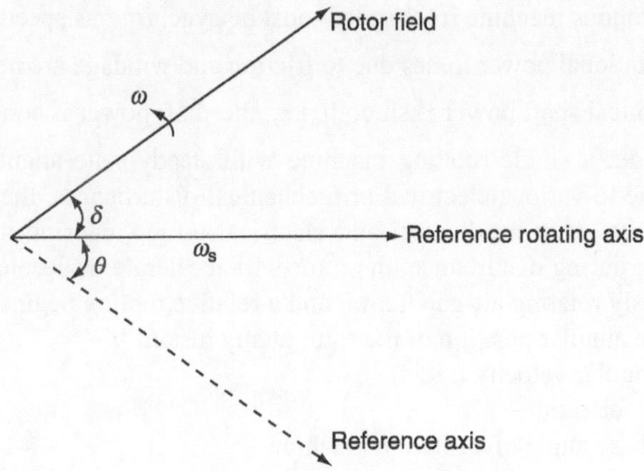

FIG. 11.43   *Phasor representation of rotor-field position*

and the rotor angular acceleration is obtained by differentiating the above equation again:

$$\frac{\mathrm{d}^2\theta}{\mathrm{d}t^2} = \frac{\mathrm{d}^2\delta}{\mathrm{d}t^2} \text{ is electrical rad/sec}$$

$$\therefore (T_\mathrm{m} - T_\mathrm{e}) = J\frac{\mathrm{d}^2\theta}{\mathrm{d}t^2}$$

The torque acting on the rotor of a synchronous generator includes the mechanical input torque from the prime mover, torque due to rotational losses (i.e., friction, windage, and core loss), electrical output torque, and damping torques due to prime mover, generator, and power system.

The electrical and mechanical torques acting on the rotor are of opposite sign and are of a result of electrical input and mechanical load. By neglecting damping and rotational losses, so that accelerating torque is

$$T_\mathrm{a} = T_\mathrm{m} - T_\mathrm{e}$$

and multiplying with ω, we get

$$J\omega\frac{\mathrm{d}^2\theta}{\mathrm{d}t^2} = \omega(T_\mathrm{m} - T_\mathrm{e}) = P_\mathrm{m} - P_\mathrm{e}$$

i.e., $$P_\mathrm{a} = \omega T_\mathrm{a} = J\omega\frac{\mathrm{d}^2\delta}{\mathrm{d}t^2} = M\frac{\mathrm{d}^2\delta}{\mathrm{d}t^2}$$

The swing equation in terms of moment of inertia or angular momentum is

$$M\frac{\mathrm{d}^2\delta}{\mathrm{d}t^2} = (P_\mathrm{m} - P_\mathrm{e}) \qquad (\because J\omega = M)$$

$\therefore$ Swing equation is also expressed as

$$M\frac{\mathrm{d}^2\delta}{\mathrm{d}t^2} = P_\mathrm{m} - \frac{EV}{X}\sin\delta$$

where $$P_\mathrm{e} = \frac{EV}{X}\sin\delta$$

KEY NOTES

- The prime mover controls are classified as:

  (i) Primary control (speed governor control).

  (ii) Secondary control (load frequency control (LFC)).

  (iii) Tertiary control involving economic dispatch.

- The transient characteristics of hydro-turbines are obtained by the dynamics of water flow in the penstock.

- The water starting time or water time constant value lies in the range of 0.5 – 5.0 s.

- Steam turbine system configurations are:

  (i) Non-reheat type.

  (ii) Reheat type.

- Reheat type steam turbines are classified as:

  (i) Tandem compound, single reheat type.

  (ii) Tandem compound, double reheat type.

  (iii) Cross-compound, single reheat type with two LP turbines.

  (iv) Cross-compound, single reheat type with single LP turbine.

  (v) Cross-compound, double reheat type.

- Most simplied model of a synchronous generator for the purpose of transient stability studies is a constant voltage source behind proper reactance.

- In order to include the effect of saliency, the simplest model of a synchronous machine can be represented by a fictitious voltage 'E_q' located at the q-axis. The d-axis is taken along the main pole axis while the q-axis lags the d-axis by 90°.

- The stator to rotor mutual inductances will vary periodically with the angle between the q-axis and the d-axis of a synchronous machine.

- The self-inductance of any stator phase is always positive but varies with the position of the rotor. It is the greatest when the d-axis of the field coincides with the axis of the armature phase and is the least when the q-axis coincides with it.

- The effect of Park's transformation is simply to transform all stator quantities from phases a, b, and c into new variables, the frame of reference of which moves with the rotor, i.e., Park's transformation matrix $[P]_{dqof}$ transforms the field of phasors to the field of d-q-o-f components and it is a linear, time-dependent matrix.

SHORT QUESTIONS AND ANSWERS

(1) What is the significance of water time constant, τ_ω?

$$\frac{\Delta H}{\Delta Q} = -s\tau_e z$$

where ΔH is the p.u. change in water head, ΔQ the p.u. change in the water discharge, τ_ω the water time constant, τ_e the elastic limit of penstock, z the normalized penstock impedance, and τ_e known as the water time constant or water starting time.

    The value of τ_ω lies in the range of 0.5 – 5.0 s.

    The typical value of τ_ω is around 1.0 s.

(2) How are the transient characteristics of hydro-turbines obtained?

    By the dynamics of water flow in the penstock.

(3) The water time constant τ_ω is associated with what time?

    τ_ω is associated with acceleration time for water in the penstock between the turbine inlet and the forebay or between the turbine inlet and the surge tank if it exists.

(4) Write the expression for water time constant τ_ω in terms of velocity of flow of water.

$$\tau_\omega = \frac{L\upsilon}{H_T g}$$

where L is the length of penstock in m, v the velocity of water flow in m/s, H_T the total head in m, and g the acceleration due to gravity in m/s².

(5) Write the expression for water time constant τ_ω in terms of power generation of the plant.

$$\tau_\omega = \frac{11.8\,PL}{H_T^2\, Aeg} \quad \text{or} \quad \tau_\omega = \frac{0.366\,PL}{H_T^2\, Ae}$$

where P is the power generation in $\text{kW} = \dfrac{VH_\text{T}Ae}{11.8}$

where $e = \eta_\text{turbine} \times \eta_\text{generation}$

(6) What are the common steam turbine system configurations?

- Non-reheat system.
- Reheat type.

(7) What are the compound system configurations of a steam turbine?

Tandem compound and cross-compound system configuration.

(8) What are the types of tandem compound system configuration?

Single reheat type and double reheat type.

(9) What are the types of cross-compound system configuration?

- Single reheat type with two LP turbines.
- Single reheat type with single LP turbines.
- Double reheat type.

(10) What do you mean by tandem compound reheat-type steam turbine?

Tandem compound system configuration has only one shaft on which all the turbine (are of HP, LP, and IP) types are mounted.

(11) What are the components that introduce the time delays and how can these delays be represented?

Steam chest, reheat, and cross-over piping are the components that introduce the time delays in the operation of steam turbines.

The time delays can be represented by:

$\tau_\text{CH} = $ steam-chest time constant (0.1–0.4 s).

$\tau_\text{RH} = $ reheat time constant (4–11 s).

$\tau_\text{CO} = $ cross-piping time constant (0.3–1.5 s).

(12) What is the most simplified model of a synchronous generator for the purpose of a transient stability study?

A constant voltage source behind a proper reactance, the voltage source may be sub-transient or steady state and the reactance may be corresponding reactance.

(13) Why is the reactance of the synchronous generator equivalent circuit referred to as direct axis reactance when the $3\text{-}\phi$ short-circuit fault occurs?

The armature reaction flux at that instant is nearly demagnetizing in nature and because it acts along the direct axis of the machine, the equivalent circuit reactance is referred to as direct axis reactance.

(14) The modeling of a synchronous machine is easily obtained for any fault in a power system by which circuit?

Thevenin's equivalent circuit.

(15) Write the expression for excitation voltage or open-circuit voltage of a synchronous machine for the effect of saliency.

$E_\text{q} = V + I_\text{a} R_\text{a} + jI_\text{ad}X_\text{d} + jI_\text{aq}X_\text{q}.$

(16) Write the voltage equations for the three stator windings and rotor windings in terms of flux linkages.

$V_\text{a} = i_\text{a}r_\text{a} + d\lambda_\text{a}/dt.$

$V_\text{b} = i_\text{b}r_\text{b} + d\lambda_\text{b}/dt.$

$V_\text{c} = i_\text{c}r_\text{c} + d\lambda_\text{c}/dt.$

$V_\text{f} = i_\text{f}r_\text{f} + d\lambda_\text{f}/dt.$

(17) What are the assumptions usually made to determine the nature of the machine inductance to help the detailed model of synchronous machines?

- The self-inductance and mutual inductance of the machine are independent of the magnitude of winding currents because of neglecting magnetic saturation.

- The shape of the air gap and the distribution of windings are such that all the machine inductances may be represented as constant plus sinusoidal functions of electrical rotor positions.

- Slottings are ignored.

- Magnetic materials are free from hysteresis and eddy current losses.

- The machine may be considered without damper windings. If a damper winding is presented, its influence may be neglected.

- Higher time and space harmonics are neglected.

(18) Write the expressions for stator to rotor mutual inductances of a synchronous machine.

$L_\text{af} = L_\text{fa} = M_\text{f} \cos \beta$

$L_{bf} = L_{fb} = M_f \cos(\beta - 120°)$

$L_{cf} = L_{fc} = M_f \cos(\beta + 120°).$

(19) Write the expressions for stator self-inductances of a synchronous machine.

$L_{aa} = L_s + L_m \cos 2\beta$

$L_{bb} = L_s + L_m \cos(2\beta + 120°)$

$L_{cc} = L_s + L_m \cos(2\beta - 120°).$

(20) Write the expressions for rotor self-inductances of a synchronous machine.

$L_{ab} = -M_s + L_m \cos(2\beta - 120°)$

$L_{bc} = -M_s + L_m \cos 2\beta$

$L_{ca} = -M_s + L_m \cos(2\beta + 120°).$

(21) What is Park's transformation and what is its requirement?

The behavior of a synchronous machine can be described by a set of differential equations. The solution of these differential equations is complicated since the inductances are the functions of rotor angle β, which in turn is a function of time. The complication in getting the solution can be avoided by transferring the physical quantity in the armature windings through a linear, time-dependent, and power-invariant transformer called Park's transformation.

(22) In Park's transformation, which quantities are transformed?

Quantities such as currents, m.m.f.s, voltages, and flux linkages in *a b c f* axes to *d q o f* axes are transformed.

(23) Express Park's transformation matrix.

$$[p]_{dqof} = \begin{array}{c} d \\ q \\ o \\ f \end{array} \begin{bmatrix} k_d \cos\beta & k_d \cos(\beta - 120°) & k_d \cos(\beta + 120°) & 0 \\ -k_q \sin\beta & -k_q \sin(\beta - 120°) & -k_q \sin(\beta + 120°) & 0 \\ k_0 & k_0 & k_0 & 0 \\ 0 & 0 & 0 & 1 \end{bmatrix}.$$

(24) What is the function of Park's transformation matrix?

Park's transformation matrix $[p]$ transforms the field of *a b c f* components to the field of *d q o f* components.

(25) Park's transformation matrix is called unitary matrix. Why?

Because the inverse of Park's transformation matrix is the transpose of conjugate of the matrix.

$[p]^{-1} = [p^*]^T$

(26) Write the expression for the flux linkages of *f d q o* coils in terms of coil currents by using transformation matrix and its inverse.

$[\lambda]_{dqof} = [p][L]_{abcf}[p]^{-1}[i]_{dqof}.$

(27) Write the expression for 3-ϕ power output of a synchronous machine in terms of rate of flux linkages *d*-axis and *q*-axis currents.

$$P = -\left(\frac{d\lambda_q}{dt}i_q + \frac{d\lambda_d}{dt}i_d\right) + \omega\left(\lambda_d i_d - \lambda_q i_q\right) - r\left(i_d^2 + i_q^2\right)$$

(28) In the expression for 3-ϕ power output of a synchronous machine in terms of rate of flux linkages *d*-axis and *q*-axis currents, what parameters do the three terms indicate?

First term: $-\left(\dfrac{d\lambda_q}{dt}i_q + \dfrac{d\lambda_d}{dt}i_d\right)$ represents the rate

of change of stator magnetic field energies.

Second term $\omega(\lambda_d i_d - \lambda_q i_q)$ represents the power transferred across the air gap.

Third term $r\left(i_d^2 + i_q^2\right)$ represents the stator ohmic losses.

(29) What is the significance of saliency torque $T_e = (L_d - L_q)i_d i_q$?

The saliency torque exists only because of non-uniformity in the performance of the air gap along the *d*-axis and the *q*-axis. This is the reluctance torque of a salient-pole machine and exists even when the field excitation is zero.

(30) What is the significance of fictitious winding 'g' on rotor in detailed modeling of a synchronous machine?

The fictitious winding 'g' on the rotor of the synchronous machine approximates the effects of eddy currents circulating in the iron currents and to some extent the effect of damper windings. This fictitious winding is short-circuited since it it not connected to any voltage source.

(31) What is the significance of 'f' and 'g' coils in rotor winding in the modeling of a synchronous machine?

The 'f' and 'g' coils in the rotor winding produce transient effect in terms of X'_d and X'_q in a synchronous machine; the field coil 'f' exists physically, whereas 'g' coil is hypothetical for representing the rotor eddy currents in the q-axis.

(32) What is the significance of k_d winding and k_g winding in rotor winding in the modeling of a synchronous machine?

The more accurate representation of a synchronous machine is obtained by adding two more fictitious windings, f-coil and g-coil windings. One fictitious winding is known as the k_d winding along the d-axis and the other along the q-axis is known as the k_q axis. These two damper windings can be approximated by two hypothetical coils both short circuited as there is no voltage source connected to them.

MULTIPLE-CHOICE QUESTIONS

(1) The transient characteristics of hydro-turbines are obtained by:

(a) The dynamics of water in the reservoir.

(b) The dynamics of water in the penstock.

(c) The water head.

(d) None of these.

(2) The water time constant of hydro-turbine T_w is associated with:

(a) The acceleration time for water in the penstock between the turbine inlet and the forebay.

(b) The acceleration time for water in the penstock between the turbine inlet and the surge tank if it exists.

(c) Either (a) or (b).

(d) None of these.

(3) Expression for water time constant T_w in terms of power generation of the plant P is:

(a) $T_w = \dfrac{11.8PL}{H_T^2 Aeg}$.

(b) $T_w = \dfrac{0.366PL}{H_T^2 Ae}$.

(c) Either (a) or (b).

(d) None of these.

(4) In the expression of water time constant

$$T_w = \dfrac{11.8PL}{H_T^2 Aeg}\text{, the term e is given as:}$$

(a) $e = \eta_{turbine} \times \eta_{generator}$.

(b) $e = \eta_{turbine} / \eta_{generator}$.

(c) $e = \eta_{turbine} + \eta_{generator}$.

(d) $e = \eta_{turbine} - \eta_{generator}$.

(5) The steam turbines are mainly classified into:

(a) HP turbines and LP turbines.

(b) Single and double-type turbines.

(c) Non-reheat and reheat-type turbines.

(d) None of these.

(6) The system where only one shaft on which all the turbines are mounted are:

(a) Tandem compound system.

(b) Cross-compound system.

(c) Either (a) or (b).

(d) Both (a) and (b).

(7) In the tandem compound system, the turbines mounted on a shaft are:

(a) Only HP type.

(b) IP type.

(c) Only LP type.

(d) All of these.

(8) In _____________ type of steam turbines, the governor-controlled values are used at the inlet to control steam flow.

(a) Tandem compound system.

(b) Cross-compound system.

(c) Either (a) or (b).

(d) All compound.

(9) In controlling the steam flow, the time delays are introduced due to:

    (a) Steam chest.

    (b) Reheater.

    (c) Cross-over piping.

    (d) All of these.

(10) Match the following:

A		B
(a) Steam-chest time constant (τ_{ch}).	(i)	0.3–0.5 s.
(b) Reheat time constant (τ_{Rh}).	(ii)	4–11 s.
(c) Cross-over time constant (τ_{co}).	(iii)	0.1–0.4 s.
(d) Water time constant (τ_{w}).	(iv)	0.5–5.0 s.

(11) A non-reheat type steam turbine is modeled by:

    (a) A single time constant.

    (b) Two time constants.

    (c) Without time constant.

    (d) Either (a) or (b).

(12) The most simplified model of a synchronous generator is:

    (a) A constant voltage source behind proper reactance.

    (b) A constant current source behind proper reactance.

    (c) A variable voltage source behind reactance.

    (d) A variable current source behind reactance.

(13) The armature reaction flux at the instant of fault occurs due to a very large lagging current, which is nearly ____________ in nature.

    (a) Magnetizing.

    (b) Demagnetizing.

    (c) Either (a) or (b).

    (d) None of these.

(14) The armature reaction flux at the instant of fault occurs acts along which axis of the machine?

    (a) Direct axis.

    (b) Quadrature axis.

    (c) Both (a) and (b).

    (d) None of these.

(15) The most specified model representation of a synchronous generator can easily be obtained for any fault in the power system with the help of ____________ circuit.

    (a) Norton's equivalent.

    (b) Thevenin's equivalent.

    (c) Maximum power transfer theorem equivalent.

    (d) None of these.

(16) In order to include the effect of saliency, the simplest model of a synchronous machine can be represented as:

    (a) A fictitious voltage E_q located at the q-axis.

    (b) A fictitious voltage E_d located at the d-axis.

    (c) A fictitious voltage $E_d E_q$ along both the axes.

    (d) None of these.

(17) The expression for E_q in terms of a full-load terminal voltage V and full-load armature current I_a is:

    (a) $E_q = V + j I_a (R_a + X_q)$.

    (b) $E_q = V + j I_a R_a + I_a X_q$.

    (c) $E_q = V - I_a R_a + j I_a X_q$.

    (d) $E_q = V + I_a R_a + j I_a X_q$.

(18) For the synchronous generator, without the effect of saliency, the machine equation can be represented as:

    (a) $E_g = V + j I X_q$.

    (b) $E_g = V + j I X_d$.

    (c) $E_g = V - j I X_q$.

    (d) $E_g = V - j I X_d$.

(19) The expression for excitation or open-circuit voltage of the synchronous generator with the effect of saliency is:

    (a) $E_g = V + j I X_d$.

    (b) $E_g = V + I_a R_a + j I_a X_q$.

    (c) $E_q = V + I_a R_a + j I_{ad} X_d + j I_{aq} X_q$.

    (d) None of these.

(20) The synchronous machine has:

    (a) Three windings on stator carrying AC.

    (b) One winding on rotor carrying DC excitation.

(c) Either (a) or (b).

(d) Both (a) and (b).

(21) The instantaneous terminal voltage of synchronous machine of any winding expressed in terms of flux linkages is:

(a) $V = ir + \lambda$.

(b) $V = ir + j\lambda$.

(c) $V = ir + d\lambda/dt$.

(d) None of these.

(22) To develop the detailed model of synchronous machine, which of the following assumptions are usually made to determine the nature of the machine inductance?

(i) The self-inductance and mutual inductance of machine are independent of magnitudes of winding current.

(ii) The self-inductance and mutual inductance may be represented as constants plus sinusoidal functions of electrical rotor positions.

(iii) Slotting effects are ignored.

(iv) Magnetizing materials are free from hysteresis and eddy current losses

    (a) (i) and (ii)

    (b) All except (iv)

    (c) All except (i)

    (d) All of these

(23) The self-inductance of any stator phase of a synchronous machine is always ___________, but varies the position of ___________ .

(a) Positive, rotor.

(b) Negative, rotor.

(c) Positive, stator.

(d) Negative, stator.

(24) The self-inductance of any stator phase of a synchronous machine is greater when:

(a) The q-axis of field coincides with the axis of armature phase.

(b) The d-axis of field coincides with the axis of armature phase.

(c) Either (a) or (b).

(d) None of these.

(25) The self-inductance of any stator phase of a synchronous machine is least when:

(a) The q-axis of field coincides with the axis of armature phase.

(b) The d-axis of field coincides with the axis of armature phase.

(c) Either (a) or (b).

(d) None of these.

(26) The expressions for self-inductances of stator phases of a synchronous machine are:

(a) $L_{aa} = L_m + L_s \cos 2\beta$

$L_{bb} = L_m + L_s \cos(2\beta + 120°)$

$L_{cc} = L_m + L_s \cos(2\beta - 120°)$.

(b) $L_{aa} = L_m - L_s \cos 2\beta$

$L_{bb} = L_m - L_s \cos(2\beta + 120°)$

$L_{cc} = L_m - L_s \cos(2\beta - 120°)$.

(c) $L_{aa} = L_s + L_m \cos 2\beta$

$L_{bb} = L_s + L_m \cos(2\beta + 120°)$

$L_{cc} = L_s + L_m \cos(2\beta - 120°)$.

(d) $L_{aa} = L_s - L_m \cos 2\beta$

$L_{bb} = L_s - L_m \cos(2\beta + 120°)$

$L_{cc} = L_s - L_m \cos(2\beta - 120°)$.

(27) The stator mutual inductances of a synchronous machine are:

(a) $L_{ab} = M_s + L_m \cos(2\beta - 120°)$

$L_{bc} = M_s + L_m \cos 2\beta$

$L_{ca} = M_s + L_m \cos(2\beta + 120°)$.

(b) $L_{ab} = M_s + L_m \cos 2\beta$

$L_{bc} = M_s + L_m \cos(2\beta - 120°)$

$L_{ca} = M_s + L_m \cos(2\beta + 120°)$.

(c) $L_{ab} = -M_s + L_m \cos 2\beta$

$L_{bc} = -M_s + L_m \cos(2\beta - 120°)$

$L_{ca} = -M_s + L_m \cos(2\beta + 120°)$.

(d) $L_{ab} = -M_s + L_m \cos(2\beta - 120°)$

$L_{bc} = -M_s + L_m \cos 2\beta$

$L_{ca} = -M_s + L_m \cos(2\beta + 120°)$.

(28) Park's transformation matrix is:

   (a) Linear.

   (b) Time-dependent.

   (c) Power-invariant.

   (d) All of these.

(29) The fact on which Park's transformation based is:

   (a) The rotating field produced by 3-ϕ stator currents in the synchronous machine can be equally produced by 2-ϕ currents in 2-ϕ winding.

   (b) The rotating field produced by 3-ϕ stator currents in the synchronous machine can be equally produced by 1-ϕ currents in 1-ϕ winding.

   (c) Either (a) or (b).

   (d) None of these.

(30) The matrix form of representation of Park's transformation is:

   (a) $[i]_{dqof} \cong [P]_{dqof} [i]_{abcf}.$

   (b) $[i]_{abcf} \cong [P]_{dqof} [i]_{abcf}.$

   (c) $[i]_{dqof} \cong [P]_{abcf} [i]_{abcf}.$

   (d) $[i]_{dqof} \cong [P]_{dqof} [i]_{abcf}.$

(31) Park's transformation matrix is:

   (a) $$[P]_{dqof} = \begin{matrix} d \\ q \\ o \\ f \end{matrix} \begin{bmatrix} k_q \cos\beta & k_q \cos(\beta-120°) & k_q \cos(\beta+120°) & 0 \\ k_d \sin\beta & -k_d \sin(\beta-120°) & -k_d \sin(\beta+120°) & 0 \\ 0 & 0 & 0 & 0 \\ k_o & k_o & k_o & 1 \end{bmatrix}.$$

   (b) $$[P]_{dqof} = \begin{matrix} d \\ q \\ o \\ f \end{matrix} \begin{bmatrix} k_d \cos\beta & k_d \cos(\beta-120°) & k_d \cos(\beta+120°) & 0 \\ -k_q \sin\beta & -k_q \sin(\beta-120°) & -k_q \sin(\beta+120°) & 0 \\ k_o & k_o & k_o & 0 \\ 0 & 0 & 0 & 1 \end{bmatrix}.$$

   (c) $$[P]_{dqof} = \begin{matrix} d \\ q \\ o \\ f \end{matrix} \begin{bmatrix} k_q \cos\beta & k_q \cos(\beta-120°) & k_q \cos(\beta+120°) & 0 \\ -k_d \sin\beta & -k_d \sin(\beta-120°) & -k_d \sin(\beta+120°) & 0 \\ k_o & k_o & k_o & 0 \\ 0 & 0 & 0 & 1 \end{bmatrix}.$$

   (d) None of these.

(32) Park's transformation matrix $[p]_{dqof}$ transforms:

   (a) Field of stator phasors to the field of *d-q-o-f* components.

   (b) Field of rotor phasors to the field of *d-q-o-f* components.

   (c) Field of stator phasors to the field of *a-b-c-f* components.

   (d) Field of rotor phasors to the field of *a-b-c-f* components.

(33) Park's transformation matrix $[p]_{dqof}$ is:

   (a) A linear matrix.

   (b) A time-dependent matrix.

   (c) Non-linear and time-invariant matrix.

   (d) Both (a) and (b).

(34) Park's transformation matrix $[p]_{dqof}$ is:

(a) $[p]_{dqof} = \begin{array}{c} d \\ q \\ o \\ f \end{array} \begin{bmatrix} \sqrt{\dfrac{2}{3}}\cos\beta & \sqrt{\dfrac{2}{3}}\cos(\beta-120°) & \sqrt{\dfrac{2}{3}}\cos(\beta+120°) & 0 \\ \sqrt{\dfrac{2}{3}}\sin\beta & \sqrt{\dfrac{2}{3}}\sin(\beta-120°) & \sqrt{\dfrac{2}{3}}\sin(\beta+120°) & 0 \\ 0 & 0 & 0 & 0 \\ \sqrt{\dfrac{1}{3}} & \sqrt{\dfrac{1}{3}} & \sqrt{\dfrac{1}{3}} & 1 \end{bmatrix}$
$\qquad\qquad\quad\; a \qquad\qquad b \qquad\qquad c \qquad\quad f$

(b) $[p]_{dqof} = \begin{array}{c} d \\ q \\ o \\ f \end{array} \begin{bmatrix} \sqrt{\dfrac{2}{3}}\cos\beta & \sqrt{\dfrac{2}{3}}\cos(\beta-120°) & \sqrt{\dfrac{2}{3}}\cos(\beta+120°) & 0 \\ -\sqrt{\dfrac{2}{3}}\sin\beta & -\sqrt{\dfrac{2}{3}}\sin(\beta-120°) & -\sqrt{\dfrac{2}{3}}\sin(\beta+120°) & 0 \\ \dfrac{1}{\sqrt{3}} & \dfrac{1}{\sqrt{3}} & \dfrac{1}{\sqrt{3}} & 0 \\ 0 & 0 & 0 & 1 \end{bmatrix}$
$\qquad\qquad\quad\; a \qquad\qquad b \qquad\qquad c \qquad\quad f$

(c) $[p]_{dqof} = \begin{array}{c} d \\ q \\ o \\ f \end{array} \begin{bmatrix} -\sqrt{\dfrac{2}{3}}\cos\beta & -\sqrt{\dfrac{2}{3}}\cos(\beta-120°) & -\sqrt{\dfrac{2}{3}}\cos(\beta+120°) & 0 \\ \sqrt{\dfrac{2}{3}}\sin\beta & \sqrt{\dfrac{2}{3}}\sin(\beta-120°) & \sqrt{\dfrac{2}{3}}\sin(\beta+120°) & 0 \\ 0 & 0 & 0 & 0 \\ \sqrt{\dfrac{1}{3}} & \sqrt{\dfrac{1}{3}} & \sqrt{\dfrac{1}{3}} & 1 \end{bmatrix}$
$\qquad\qquad\quad\; a \qquad\qquad b \qquad\qquad c \qquad\quad f$

(d) None of these.

(35) Which of the following is correct regarding to Park's transformation?

(a) $[P] = [P]^{-1}$.

(b) $[P]^{-1} = [P]^{T}$.

(c) $[P]^{-1} = [P^{*}]$.

(d) $[P]^{-1} = [P^{*}]^{T}$.

(36) Park's transformation matrix is called unitary transformation since:

(a) Inverse of Park's transformation matrix is equivalent to transpose of the matrix.

(b) Park's transformation matrix is equal to transpose of the matrix.

(c) Inverse of Park's transformation matrix is equal to transpose of conjugate of the matrix.

(d) None of these.

(37) The flux linkages of $f\,d\,q\,o$ coils are expressed in terms of Park's transformation matrix and its inverse as:

(a) $[\lambda]_{dqof} = [P][L]_{abcf}[P]^{-1}[i]_{dqof}$.

(b) $[\lambda]_{dqof} = [P][L]_{dqof}[P]^{-1}[i]_{abcf}$.

(c) $[\lambda]_{dqof} = [P]^{-1}[L]_{dqof}[P][i]_{dqof}$.

(d) None of these.

(38) The expression for 3-ϕ power output of a synchronous machine is:

$$P = -\left(\frac{d\lambda_q}{dt}i_q + \frac{d\lambda_d}{dt}i_d\right) + \omega\left(\lambda_d i_q - \lambda_q i_d\right) - r\left(i_d^2 + i_q^2\right)$$

Which of the following is correct?

(a) First term represents the rate of change of stator magnetic field changes.

(b) Second term represents the power transferred across the air gap.

(c) Third term represents the stator ohmic losses

(d) All of these.

(39) The torque expression of a salient-pole synchronous machine is:

(a) $T_e = \dfrac{\sqrt{3}}{2}M_f i_f i_q + (L_d - L_q)i_d i_q.$

(b) $T_e = \dfrac{\sqrt{3}}{2}M_f i_f i_q.$

(c) $T_e = \dfrac{\sqrt{3}}{2}M_f i_f i_q + (L_d - L_q)i_d i_q.$

(d) $T_e = \dfrac{\sqrt{3}}{2}M_f i_f i_q + (i_d - i_q)L_d L_q.$

(40) The torque expression of a cylindrical rotor synchronous machine is:

(a) $T_e = \dfrac{\sqrt{3}}{2}M_f i_f i_q + (L_d - L_q)i_d i_q.$

(b) $T_e = \dfrac{\sqrt{3}}{2}M_f i_f i_q.$

(c) $T_e = \dfrac{\sqrt{3}}{2}M_f i_f i_q + (L_d - L_q).$

(d) $T_e = \dfrac{\sqrt{3}}{2}M_f i_f i_q + (i_d - i_q)L_d L_q.$

(41) In the dynamic model of a synchronous machine including damper windings, the more accurate representation is obtained by adding two fictitious windings on rotor, k_d winding and k_q winding along the d-axis and the q-axis. These damper windings are approximated as:

(a) Two hypothetical coils both open-circuited.

(b) Two hypothetical coils with voltage sources connected to them.

(c) Two hypothetical coils both short-circuited as there is no voltage source connected to them.

(d) None of these.

REVIEW QUESTIONS

(1) Develop the linearized modeling of a hydraulic turbine.

(2) Discuss the different configurations of reheat type of steam turbines with a representation of their functional block diagrams and approximate their linear models.

(3) Explain the simplified model of a synchronous machine.

(4) Describe the effect of saliency in synchronous machine modeling.

(5) Derive the self-inductance and mutual inductance stator and rotor of synchronous machines.

(6) Explain Park's transformation and inverse Park's transformation.

(7) Develop the steady-state analysis of salient and non-salient-pole synchronous machines.

(8) Develop the dynamic analysis of salient and non-salient-pole synchronous machines, with and without damper windings.

12 Modeling of Speed Governing and Excitation Systems

OBJECTIVES

After reading this chapter, you should be able to:

- develop the modeling of speed-governor systems for steam and hydraulic turbines
- develop the modeling of speed-governor systems with limiters
- study the effect of excitation variation on synchronous machines
- discuss the methods of providing excitation of synchronous machines
- study the structure of a general excitation system
- develop the transfer functions of various components of an excitation system

12.1 INTRODUCTION

Two important control loops are needed for the economic and reliable operation of a power system. They are:

(a) Load frequency control (LFC) loop (p.f. control loop) for the regulation of system frequency.

(b) Automatic voltage control loop (Q–V control loop) for the regulation of system voltage magnitude.

These control loops indirectly influence the real and reactive power balances in the power system network.

The LFC is achieved by the speed-governor mechanism. The basic principle of the speed-governor mechanism is that according to the load variation, the speed of the rotor shaft of the synchronous machine is varied and hence the frequency of the system is varied. This change in frequency is sensed and compared with a reference and produces a feedback signal. This feedback signal makes the variation of generated power of synchronous generator by adjusting the opening of the steam inlet valve to steam turbine or water gates in the case of a hydro-turbine. Hence, the real power balance between real power generation and real power demand is achieved. This is the basic principle of the speed-governor mechanism.

The speed governors are regarded as primary control elements in an LFC system.

With an increase in the system size due to interconnections, in normal cases, the frequency variations become very less and LFC assumes importance. However, the role of speed governors in rapid control of frequency can be underestimated.

The automatic voltage control or Q–V control is achieved by an excitation control mechanism. The main and important objective of an excitation system is to control the field current of the synchronous machine. The field current is controlled so as to regulate the generating voltage of the machine.

As the field circuit time constant is high (of the order of a few seconds), the fast control of the field current requires 'field forcing'. Thus, the field excited should have a high ceiling voltage, which enables it to operate transiently with voltage levels that are three to four times the normal voltages. The rate of change of voltage should also be fast.

The excitation systems of synchronous machines have an extreme effect on system stability and when evaluated on the basis of an increased power carrying per increase in the system cost, they are by far the most economical source of increased stability limits.

The excitation system often contains other features such as voltage dip compensation to compensate for the voltage drop in some impedance between the generator and the rest of the network.

The functioning of LFC and automatic voltage control loops is presented in detail in Unit-VII (LFC-II).

In this unit, the modeling of speed-governing systems for steam turbines and hydro-turbines is discussed.

The effect of varying excitations on a synchronous generator, methods of providing excitation, and their block diagram representation and modeling are also discussed.

12.2 MODELING OF SPEED-GOVERNING SYSTEMS

According to the principle of control, the speed-governing systems are mainly classified into two categories, for both steam and hydraulic turbines. They are:

1. Mechanical–hydraulic-controlled and

2. Electro–hydraulic-controlled

In both these types, hydraulic servomotors are used for positioning the valve or gate, controlling the steam or the water flow.

12.3 FOR STEAM TURBINES

In this section, we shall discuss mechanical–hydraulic-controlled speed-governing system, electro–hydraulic-controlled speed-governing systems, and general model for speed-governing systems for steam turbines in detail.

12.3.1 Mechanical–hydraulic-controlled speed-governing systems

For a steam turbine, the mechanical–hydraulic-controlled speed-governing system consists of a speed governor, a speed relay, hydraulic servomotor, and governor-controlled valves.

The functional block diagram of a mechanical–hydraulic-controlled speed-governing system is shown in Fig. 12.1.

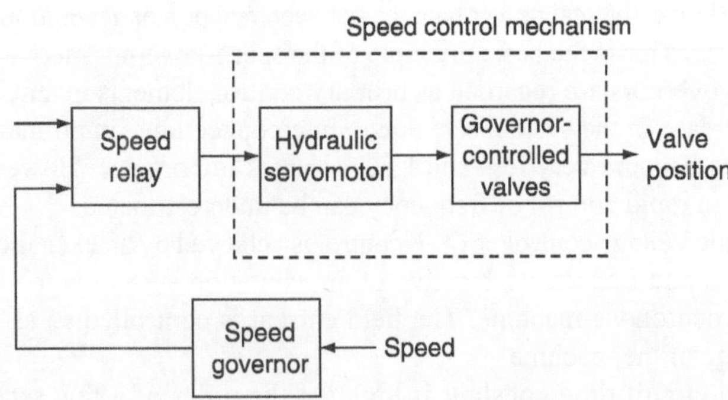

FIG. 12.1   *Functional block diagram*

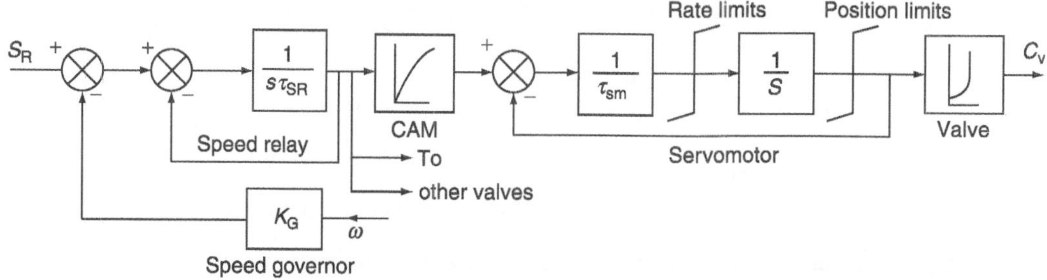

FIG. 12.2   *Approximate non-linear model representation with limits*

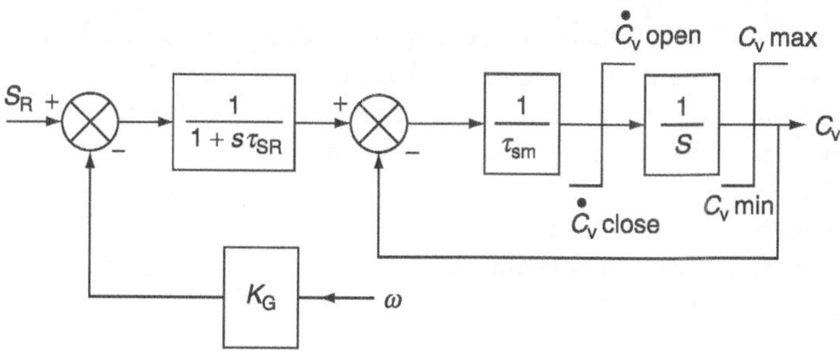

FIG. 12.3   *Simplified diagram of Fig. 12.2*

The approximate non-linear mathematical model can be represented by the block diagram shown in Figs. 12. 2 and 12.3.

K_G is the gain of speed governor, which is the reciprocal of regulation or droop. It represents a position of an assumed linear instantaneous indication of speed produced by the speed governor.

Governor speed-changer position provides the speed regulation (SR) signal and it is determined by a system of automatic generation control.

The signal SR represents a composite load and speed reference. It is assumed to be constant over the interval of a stability study.

τ_{SR} is the time constant of speed relay. The speed relay is represented as an integrator and is provided as direct feedback.

The non-linear property of the valve is compensated by means of providing a non-linear CAM in between the speed relay, and the hydraulic servomotor.

The servomotor controls the valve's movement and is represented as an integrator with time constant τ_{SM} and is provided as direct feedback. Rate limiting of the servomotor may occur for large, rapid-speed deviations, and rate limits that are shown at the input to the integrator. The position limits that are indicated correspond to wide-open valves or the setting of a load limiter.

Generally, the non-linearities present in a speed control mechanism are neglected in the study of power system operation and controlling except for rate limits and the limits on valve position.

The typical parameters for a mechanical–hydraulic system are:

$K_G = 20.0$

$\tau_{SR} = 0.1$ s $=$ speed relay time constant

$\tau_{SM} = 0.2–0.3$ s $=$ valve positioning servomotor time constant

Valve or gate servorate limits $\begin{cases} \dot{C}_{Vopen} = 0.1 \text{ p.u. per sec. per valve} \\ \dot{C}_{Vclose} = 1.0 \text{ p.u. per sec. per valve} \end{cases}$

12.3.2 Electro–hydraulic-controlled speed-governing systems

In this type of speed-governing systems, the mechanical components in the lower power portions are replaced by the static electronic circuits and thus provides more flexibility.

The functional block diagram representation of an electro–hydraulic speed-governing system is shown in Fig. 12.4.

The linearity of the system can be improved compared to mechanical–hydraulic-controlled system by means of providing feedback loops of steam flow and the servomotor.

The approximate mathematical model for a general EHC system is shown in Fig. 12. 5.

The typical parameters for this block diagram are:

$K_G = 20.0$

$K_P = 3.0$ with steam flow feedback

    $= 1.0$ without steam flow feedback

$\tau_{SM} = 0.1$ s

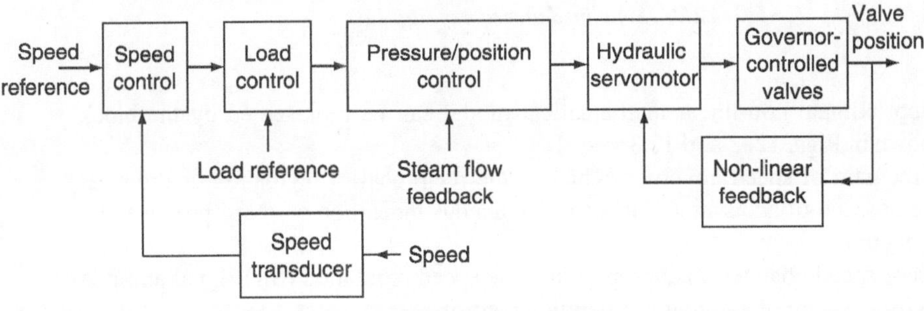

FIG. 12.4   *Functional block diagram*

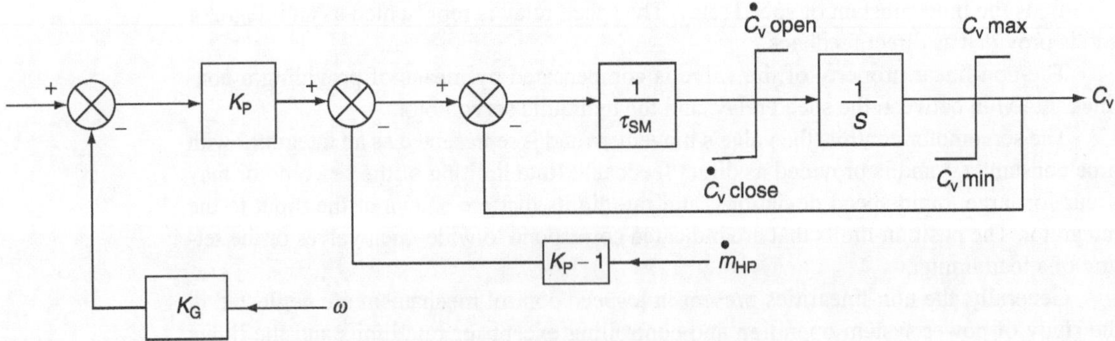

FIG. 12.5   *Block diagram for approximate mathematical model*

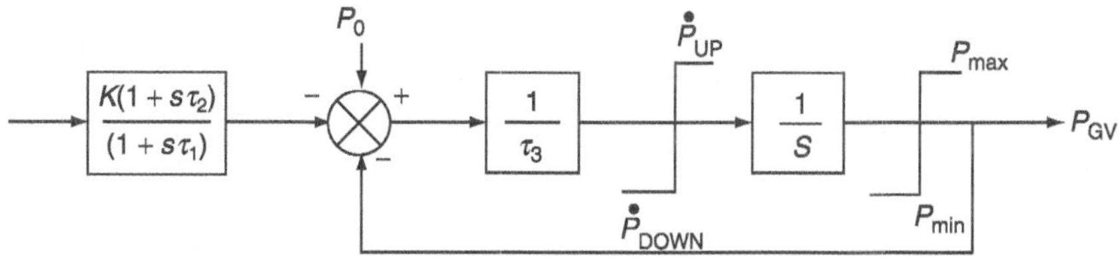

FIG. 12.6 *General model of a speed-governing system for steam turbines*

$$\text{Valve or gate servorate limits} \begin{cases} \dot{C}_{\text{Vopen}} = 0.1 \text{ p.u. per sec. per valve} \\[2mm] \dot{C}_{\text{Vclose}} = 0.1 \text{ p.u. per sec. per valve} \end{cases}$$

12.3.3 General model for speed-governing systems

A simplified, general model of speed-governing systems for steam turbines is shown in Fig. 12.6.

By the proper parameter selection, this general model represents either a mechanical–hydraulic system or an electro–hydraulic system.

This model shows the load reference as an initial power P_0. This initial value is combined with the increments due to speed deviation to obtain total power P_{GV}, subject to the time lag τ_3 introduced by the servomotor mechanism.

The typical values of time constants are:

For a mechanical–hydraulic system:

$\tau_1 = 0.2 - 0.3$ s
$\tau_2 = 0$
$\tau_3 = 0.1$ s

For an electro–hydraulic system:

$\tau_1 = \tau_2$
$\tau_3 = 0.025 - 0.15$ s

Note that when $\tau_1 = \tau_2$, the value of τ_1 or τ_2 has no effect, as there is pole-zero cancellation.

The rate limits are nominally 0.1 p.u. per second. The nominal value of $k = 100/$ (% steady-state SR).

12.4 FOR HYDRO-TURBINES

In this section, mechanical–hydraulic-controlled speed-governing systems, general model for a hydraulic turbine speed-governing system, and EHC-controlled speed-governing systems are discussed in detail.

12.4.1 Mechanical–hydraulic-controlled speed-governing systems

It consists of a speed governor, a unit of pilot valve and servomotor, a unit of distributor valve, and gate servomotor and governor-controlled gates.

The functional block diagram of a mechanical–hydraulic-controlled speed-governing system is shown in Fig. 12.7.

The speed-governing requirements for hydro-turbines are strongly influenced by the effects of water inertia.

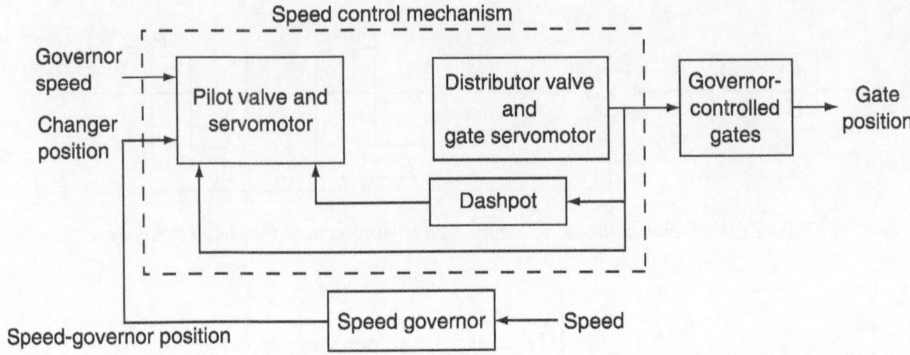

FIG. 12.7   *Functional block diagram representation of mechanical–hydraulic-controlled speed-governing system*

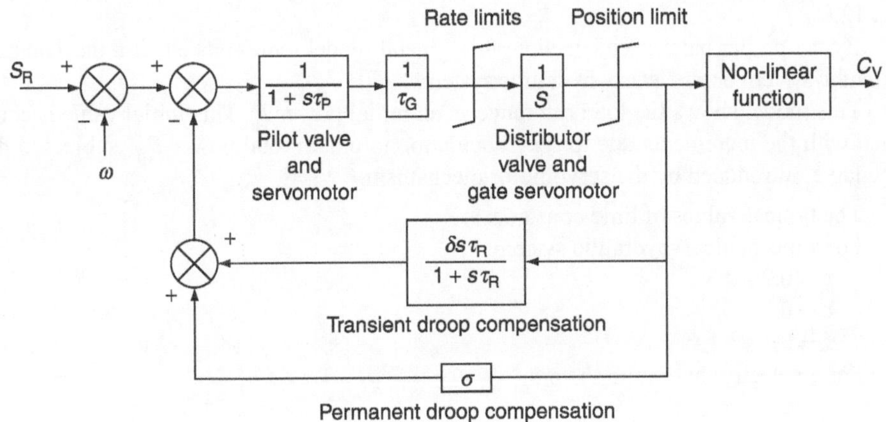

FIG. 12.8   *Block diagram for approximate non-linear model*

To achieve the stable performance of a speed-governing system, the dashpot feedback is required.

An approximate non-linear model for the above system is shown in Fig. 12.8.

The gate servomotor may be rate-limited for large rapid-speed excursions. However, transient droop feedback reduces the likelihood rate-limiting in stability analysis. Position limits exist corresponding to the extremes of gate opening.

The typical parameters of a speed-governing system for hydro-turbines and their values and their ranges are given in Table 12.1, where τ_R is the time constant of dashpot, τ_G the gate time constant of gate servomotor, τ_P the time constant of pilot valve, δ the transient speed droop coefficient, and σ the permanent speed droop coefficient.

Typically, τ_R and δ are computed as

$$\tau_R = 5\tau_w$$

TABLE 12.1   *Typical parameters of a speed-governing system for hydro-turbines*

Parameters	Typical value	Range
τ_R	5.00	2.5–25.0
τ_G	0.20	0.2–0.40
τ_P	0.04	0.03–0.05
δ	0.30	0.2–1.00
σ	0.05	0.03–0.06

and $\;\delta = 2.5\,\dfrac{\tau_w}{2H}$

where τ_w is the water starting time and H the turbine-generator inertia constant.

12.4.1.1 General model for hydraulic turbine speed-governing system

The general model for a hydraulic turbine speed-governing system is shown in Fig. 12.9.

$$\text{Let}\quad \tau_A = \left(\frac{1}{\sigma}\right)\tau_R \tau_G$$

$$\tau_B = \left(\frac{1}{\sigma}\right)(\sigma + \delta)\tau_R + \tau_G$$

Then τ_1 and τ_3 of Fig. 12.9 can be expressed approximately as

$$\tau_1, \tau_3 = \frac{\tau_B}{2} \pm \sqrt{\frac{\tau_B^{\,2}}{2} - \tau_A}$$

Also from Fig. 12.9, $K = \dfrac{1}{\sigma}, \tau_2 = 0$

P_0 is the initial power (load reference determined from automatic generation control).

P_{GV} is the output of the governor and is expressed as power reference in p.u. It is also to be noted that K is the reciprocal of σ (steady-state SR in p.u.).

12.4.2 Electric–hydraulic-controlled speed-governing system

The low-power functions associated with speed sensing and droop compensation in a modern speed-governing system for hydro-turbines can be performed by an electronic apparatus, which results in the better performance and greater flexibility in both dead band and dead time. For interconnected system operation, however, the dynamic performance

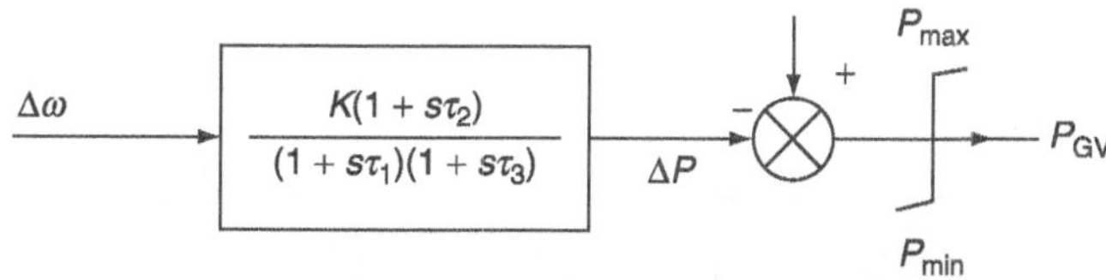

FIG. 12.9   *General model for a speed-governing system for hydro-turbines*

of the electric governor is necessarily adjusted to be essentially the same as that for the mechanical governor, so that a separate model is not needed.

12.5 MODELING WITH LIMITS

There are two types of limiters that are different in terms of behavior:
   (i) Wind-up limiter.

   (ii) Non-wind-up limiter.

12.5.1 Wind-up limiter

The block diagram representation of a wind-up limiter is shown in Fig. 12.10.

In this case of limiter, the output variable (y) of the transfer function $G(s)$ is not limited and is free to vary. Hence, the limiter can be treated as a separate block whose input is 'y' and the output is 'z'.

If $G(s) = \dfrac{1}{1+s\tau}$, the equations with the wind-up limiter are:

$$\frac{dy}{dt} = \frac{u-y}{\tau}$$

If $L \leq y \leq H$, then $z = y$

$y > H$,   then $z = H$

$y < L$,   then $z = L$

where L is the lower limit of output z and H the upper limit of output z.

12.5.2 Non-wind-up limiter

In this case, the output of the transfer function $G(s)$ is limited and there is no separate block for the limiter.

The equations are:

$f = (u-y)\,\tau$

If $y = H$ and $f > 0$,

$y = L$ and $F < 0$

then $\dfrac{dy}{dt} = 0$

Otherwise, $\dfrac{dy}{dt} = f$

and $L \leq y \leq H$

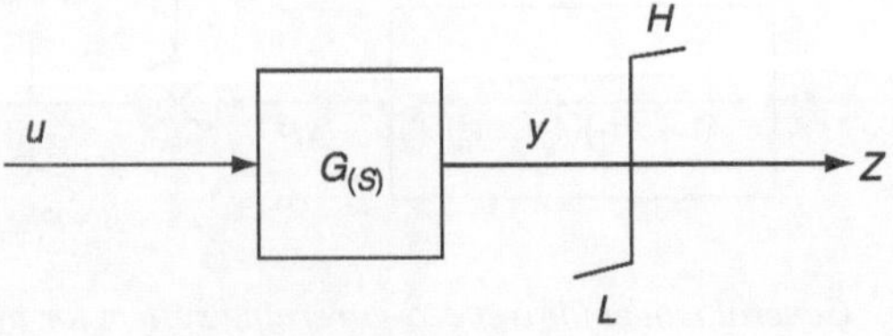

FIG. 12.10   *Block diagram representation of a wind-up limiter*

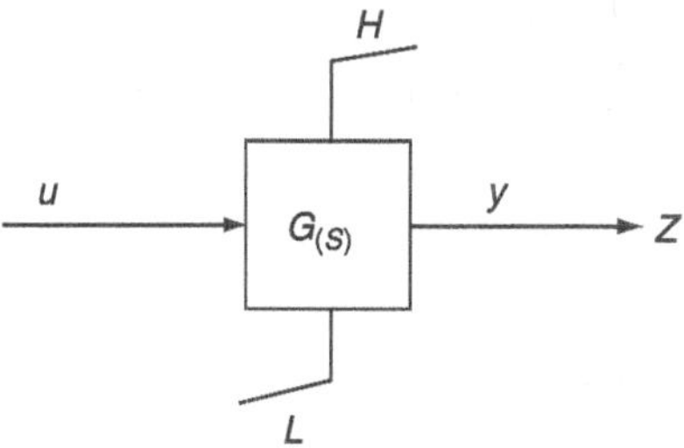

FIG. 12.11 *Block diagram representation of a non-wind-up limiter*

The block diagram representation of a non-wind-up limiter is shown in Fig. 12.11.

Note:

- As the output z of the limiter does not change until y comes within the limits, the wind-up limiter can change in terms of slow response.

- Generally, all integrator blocks have non-wind-up limits.

12.6 MODELING OF A STEAM-GOVERNOR TURBINE SYSTEM

While modeling the steam generators, at every instant the boiler controls and on-line frequency control equipment are to be ignored due to their slower operations.

A boiler contains a certain amount of heat stored in its hot metal and this is usually sufficient to guarantee that the demands for extra steam during system disturbances can be met.

During long-term operations, we must consider the rate of deterioration of steam conditions as a boiler having sufficient capacity of producing indefinitely only a given amount of extra energy at each level of output. Demands for the larger increase will be met for a short time (from 30 s to 5 min); but after that, the steam conditions will deteriorate and the turbine output will decline. It is extremely difficult to examine this problem rigorously at present because boiler turbine models are not comprehensive enough.

12.6.1 Reheat system unit

The basic components of a reheat system unit are shown in Fig. 12.12.

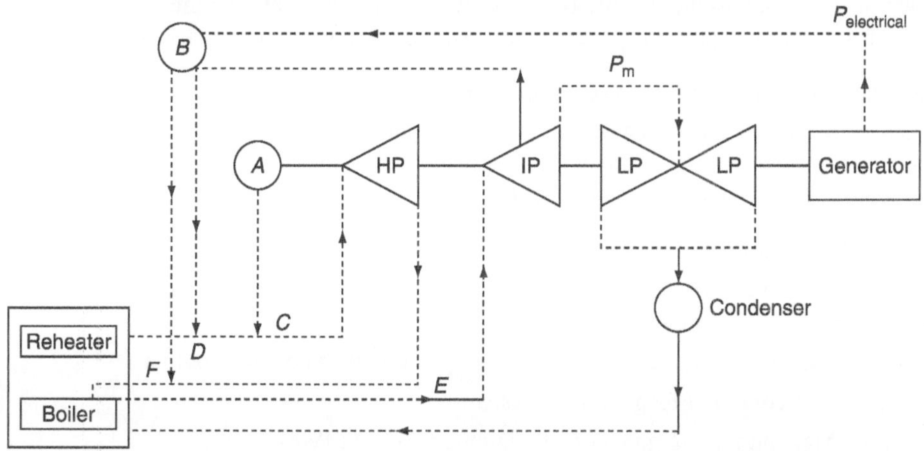

FIG. 12.12 *Basic components of a reheat system unit*

  A is a primary governing system

  B is an anticipatory governor system

  C is the main governing valve or throttle blade

  D is the combined stop and emergency valve

  E is the interceptor governor valve

  F is the combined stop and emergency valve

12.6.1.1 *Primary governing system*

It responds to the speed of the main shaft. It controls either the main governor valve or throttle blades.

12.6.1.2 *Secondary governing system*

The interceptor governing system will act as a secondary governing system and it responds to the frequency of turbines. It controls the interceptor valves between the HP stage and the reheater. It is usually set so that the interceptor valve is closed and it is about 25% to 50% open before the main governing valves commence to open. Consequently, this governing system is usually ignored.

12.6.1.3 *Anticipatory governing system*

It responds to the accelerating power of the unit and it is usually not set to operate if either:

 (i) The generator output is more than a certain value (i.e., 25% of maximum output) or

 (ii) The turbine mechanical power output (P_m) is less than a certain value (i.e., 80% of the maximum capacity).

For the activation of a governor, both these conditions should be violated.

This governing system is activated only when the unit suffers loss of a large percentage of its load and on sensing this condition, the emergency stop valves are closed very rapidly to prevent dangerous overspeed. The emergency stop valves are located very adjacent to the main governing valves.

A reset time delay is included so that when both electrical and mechanical powers revert to within settings, the emergency stop valves will open after a certain time.

This governing system is generally applied only on some modern large steam units.

12.6.1.4 *Emergency overspeed governor trip*

When the shaft velocity exceeds a pre-set value, then this governor will close the combined stop and emergency valves and shut the set down. Starting up is a lengthy process, and usually this set would not figure any further in the stability calculations.

12.6.2 Block diagram representation

The block diagram representation of modeling a reheat system unit with a reduction in elements of Fig. 12.12 is shown in Fig. 12.13.

  $r=$ Steady-state droop system setting in rad/s/MW turbine power output

  $\tau_g=$ Time constant of a governing system

  $R^-=$ Maximum closing rate of a governing valve in MW/s

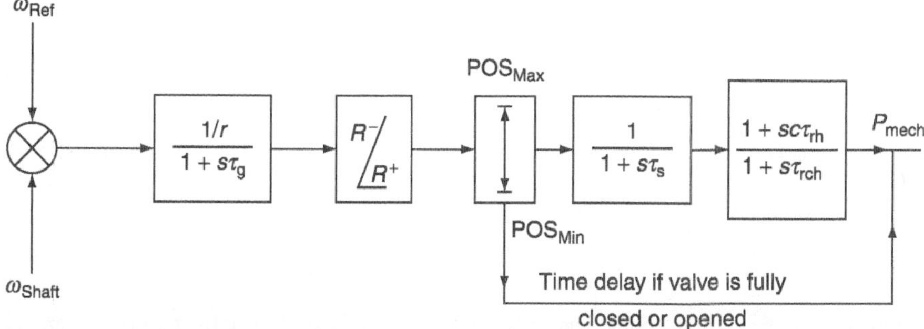

FIG. 12.13   *Block diagram representation of modeling a simple reheat steam turbine unit*

R^+ = Maximum opening rate of the governing valve in MW/s

POS_{max} = Maximum power output of the turbine in MW (maximum governor valve opening)

POS_{min} = Minimum power output of the turbine in MW (governing valve may be closed)

τ_s = Equivalent time constant of steam entrained in the turbine HP stage

τ_{rh} = Equivalent time constant of the reheater and the associated piping

ω_{ref} = Reference speed setting of the governor in rad/s

ω_{shaft} = Actual angular rotor velocity in rad/s

Delays and dead bands are present in the operations of:

(a)   The speed-sensing mechanism, friction, and blacklash.

(b)   Overcapping of oil ports in the servosystem as well as friction.

(c)   Friction in the main governing valve.

12.6.3 Transfer function of the steam-governor turbine modeling

Let GP_1 = Power developed in the turbine at the HP stage

τ_h = Time constant associated with entrained steam in the HP stage

GP_2 = Power developed in the subsequent stage of the turbine

τ_r = Time constant associated with entrained steam in the reheater and connected pipe work

τ_I = Time constant associated with entrained steam in IP and LP stages of the turbine.

The expression for the turbine shaft power P_t as a function of the governor valve opening G is:

$$P_t = \left(\frac{P_1}{1+s\tau_h} + \frac{P_2}{(1+s\tau_h)(1+s\tau_r)(1+s\tau_I)} \right) G$$

$$= \left(\frac{P_1(1+s(\tau_r+\tau_I)+s^2\tau_r\tau_I)+P_2}{(1+s\tau_h)(1+s\tau_r)(1+s\tau_i)} \right) G \tag{12.1}$$

Since the reheater time constant is lower,

$$P_t = \frac{1 + sC\tau_r}{(1 + s\tau_h)(1 + s\tau_r)}(p_1 + p_2)G \tag{12.2}$$

where C is the fraction of power developed in HP stage. The turbine power expression becomes

$$P_t = \frac{(1 + sC\tau_r)G}{(1 + s\tau_h)(1 + s\tau_r)} \quad \text{p.u.} \tag{12.3}$$

For representing a non-reheat system of turbine, simply replace τ_r by τ_I in Equation (12.3) and we get

$$P_{t_{\text{non-linear}}} = \frac{1 + sC\tau_I}{(1 + s\tau_h)(1 + s\tau_I)}$$

12.7 MODELING OF A HYDRO-TURBINE-SPEED GOVERNOR

The block diagram representation of a simple, general hydro-turbine-speed-governor modeling is shown in Fig. 12.14.

$r =$ Steady-state droop setting in rad/s/MW turbine output power

$R =$ Transient droop setting in rad/s/MW turbine output power

$\tau_r =$ Recovery time constant of temperature droop dashpot

$\tau_g =$ Equivalent governor system time constant

$R^- =$ Maximum closing rate of the governor valve in MW/s

$R^+ =$ Maximum opening rate of the governor valve in MW/s

$\text{POS}_{\text{max}} =$ Maximum power output of the turbine in MW (maximum governor valve opening)

$\text{POS}_{\text{min}} =$ Minimum power output of the turbine in MW (usually the governor valve is fully closed)

The above modeling is based on some assumptions according to Kirchmayer:

(i)   Neglecting dead band, delays, and non-linear performance in the governing system.

(ii) Neglecting the variation in head of the set with daily use (or seasonal use).

(iii) Assuming a constant equivalent water starting time constant.

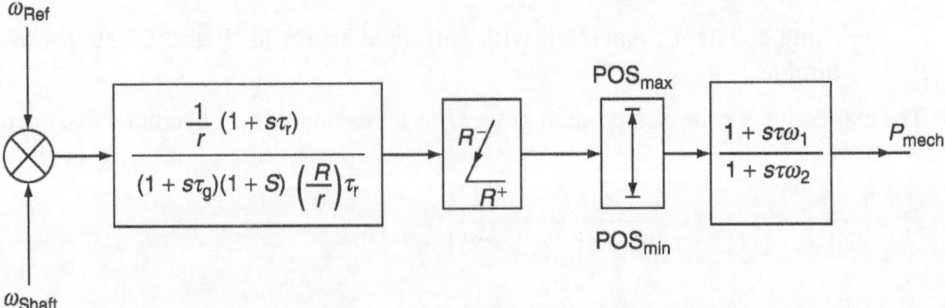

FIG. 12.14 *Block diagram representation of a hydraulic turbine-speed governor*

A transient droop setting 'r' and dashpot (i.e., damping) recovery time constant 'τ' are quite important in most stability studies, as the steady time is usually too short for the effective droop to reduce the steady-state value.

Representation of the water column inertia is important as there is an initial tendency for the turbine torque to change in the opposite direction to that finally produced when there is a change in the wicket gate in the case of a reaction turbine or an orifice opening in the case of an impulse turbine.

12.8 EXCITATION SYSTEMS

The excitation system consists of an exciter and an automatic voltage regulator (AVR). An exciter provides the required field current to the rotor winding of the alternator. The simplest form of an excitation system is an exciter only. When the task of the system becomes maintaining the constant terminal voltage of an alternator during variable load conditions, it incorporates the voltage regulator.

The voltage regulator senses the requirement from the terminal voltage of the alternator and actuates the exciter for the necessary increasing or decreasing of the voltage across the alternator field.

An excitation system with better reliabilities is preferable, even if the initial cost is more because of the fact that the cost of an excitation system is very small as compared to the cost of the alternator.

12.9 EFFECT OF VARYING EXCITATION OF A SYNCHRONOUS GENERATOR

Consider a synchronous generator supplying constant power to an infinite bus through a transmission line of reactance 'X' Ω as shown in Fig. 12.15.

The output power of a synchronous generator is expressed as:

$$P_G = |V|\,|I| \cos \phi \tag{12.4}$$

where $|V|$ is the magnitude of terminal voltage, $|I|$ the magnitude of current, and $\cos \phi$ the p.f. It may be expressed in terms of torque angle δ as

$$P_G = \frac{|E||V|}{X} \sin \delta \tag{12.5}$$

where $|E|$ is the magnitude of excitation voltage, $|V|$ the magnitude of voltage at the bus, and δ the torque angle.

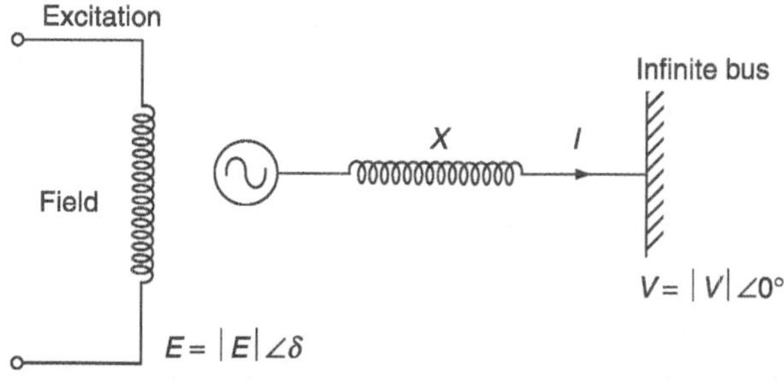

FIG. 12.15 *Synchronous generator connected to an infinite bus*

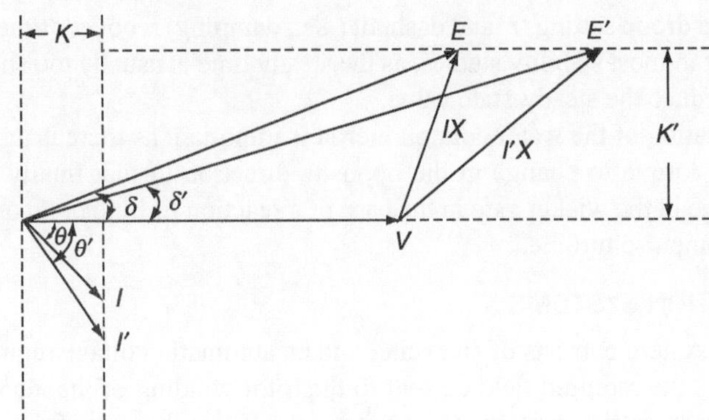

FIG. 12.16   *Effect of varying excitation of a synchronous generator*

The power output P_G and voltage magnitude $|V|$ at an infinite bus are constant; therefore, from Equations (12.4) and (12.5), we have

$$|I| \cos \phi = K \tag{12.6}$$

$$|E| \sin \delta = K' \quad [\because X \text{ is also a constant for this problem}] \tag{12.7}$$

where K and K' are constants.

Equations (12.6) and (12.7) are clearly explained by the phasor diagram given in Fig. 12.16.

According to the phasor diagram shown in Fig. 12.16, for the variation of excitation, the tip of the excitation voltage vector 'E' is restricted to move along the horizontal dotted line, and the tip of the current vector 'I' is restricted to move along the vertical dotted line.

It is observed from Fig. 12.16 that when the excitation increases the torque angle 'δ' reduces (from δ to δ'), the current increases, and power angle increases from ϕ to ϕ' and hence becomes more lagging with respect to the terminal voltage 'V'.

Hence, the torque angle 'δ' is reduced with an increase in excitation, which results in an increase in stiffness of the machine, i.e., the couplings of the generator rotor and rotating armature flux become more tight. In other words, with the increase in excitation, the stability of the machine will become enhanced.

12.9.1 Explanation

For the increment in excitation voltage, the torque angle δ reduces.

Let us assume that a cylindrical rotor (wound rotor) synchronous generator connected to an infinite bus is initially operating at torque angle δ_0 and supplying a power P_G^o.

The generator output is equal to the turbine power, $P_G^o = P_T^o$.

Now, draw the power angle characteristics of the generator as shown in Fig. 12.17(a). We have the power output of the generator as:

$$P_G = \frac{|E||V|}{X} \sin \delta$$

For a stable region, $\dfrac{\mathrm{d}P_G}{\mathrm{d}\delta} > 0$

At $\delta = 90°$, $\dfrac{\mathrm{d}P_G}{\mathrm{d}\delta} = 0$ $\quad [\because \sin\delta = 0]$

For an unstable region, $\dfrac{\mathrm{d}P_G}{\mathrm{d}\delta} < 0$

With the decrease in excitation, the torque angle increases, and hence the stiffness of the machine decreases.

From the power angle characteristics shown in Fig. 12.17(a), a graph between $|E|$ and δ can be drawn as shown in Fig. 12.17(b).

It is observed that by decreasing the excitation, point D is reached and the instability will take place at $\dfrac{\mathrm{d}\delta}{\mathrm{d}|E|} \to -\infty$. The occurrence of instability is shown as a drift in the

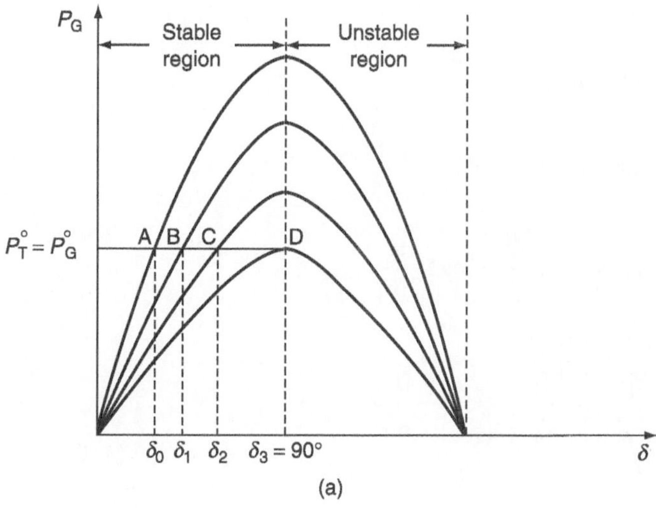

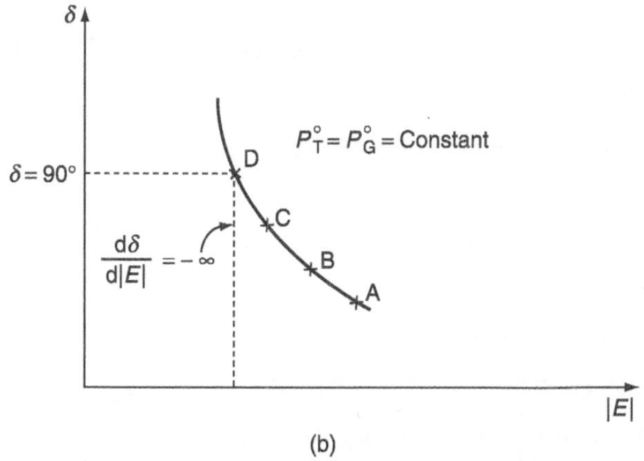

FIG. 12.17 *(a) Power angle characteristics of a synchronous generator at different excitations; (b) the value of δ as a function of $|E|$*

curve. At any point on the lower portion of the curve, $\delta = f\left(|E|\right)$, the stable state is maintained since it corresponds to the point $\dfrac{dP_G}{d\delta} > 0$ on $P_G = f(\delta)$ curve.

$\therefore$ The stability criterion of the system can be mathematically formulated as

$$\frac{dP_G}{d\delta} < 0$$

And its critical point is given by $\dfrac{d|E|}{d\delta} = 0$.

$$P_G = \frac{|E||V|}{X}\sin\delta$$

Differentiating the above equation with respect to $|E|$, we get

$$\frac{dP_G}{d|E|} = \frac{|V|}{X}\sin\delta + \frac{|E||V|}{X}\cos\delta\,\frac{d\delta}{d|E|} \tag{12.8}$$

For $P_G = P_G^0 = \text{constant}$, $\dfrac{dP_G}{d\delta} = 0$

$$\therefore 0 = \frac{|V|}{X}\sin\delta + \frac{|E||V|}{X}\cos\delta\,\frac{d\delta}{d|E|}$$

$$= \frac{|V|}{X}\sin\delta = -\frac{|E||V|}{X}\cos\delta\,\frac{d\delta}{d|E|}$$

$$= \frac{d\delta}{d|E|} = -\frac{\tan\delta}{|E|}$$

Instability will occur when $\delta \to 90°$,

i.e., $\dfrac{d\delta}{d|E|} \to -\infty$

From Fig. 12.17, it is concluded that the steady-state stability of the synchronous generator is improved by increasing its excitation.

12.9.2 Limitations of increase in excitation

The increase in excitation is limited by the following factors:

- Maximum output voltage of the exciter supplying the field current.
- Resistance of the field circuit.
- Saturation of the magnetic circuit and rotor heating.

12.10 METHODS OF PROVIDING EXCITATION

The excitation is provided by the following two methods:

1. Common excitation bus method.
2. Individual excitation method

12.10.1 Common excitation bus method

It is also known as the *centralized excitation method*. In this method, two or more number of exciters feed a common bus, which supplies an excitation to the fields of all generators in the plant.

12.10.2 Individual excitation method

It is also known as the *unit-exciter method*. In this method, each generator is fed from its own exciter, which is usually direct connected to the generator shaft, but sometimes it is driven by a motor or a small prime mover or both.

The individual excitation (or) *unit-exciter method* is more preferable because a fault in any one exciter affects the entire excitation system.

12.10.2.1 *Merits of individual excitation methods*

(a) **Simplicity:** Since each alternator has its own exciter, this method of excitation results in a simple layout of the station. The exciters are so selected according to the requirement of individual generators that the main field rheostats and high-capacity switch gear are not required, which are necessary in the case of the common bus excitation method.

(b) **Less ohmic losses:** The ohmic losses are very less because no rheostats are required in the generator field circuit, and the exciter field rheostats are operated at a much lower power.

(c) **Higher reliability:** As any fault that occurs in exciter affects only the generator to which it is connected, the *unit-exciter method* has higher reliability than common exciter method.

(d) **Incorporation automatic regulators:** The AVRs are incorporated in an individual (or) *unit excitation system* for reliable sharing of reactive power to maintain constant terminal voltage while the generators are running in parallel.

(e) **Less maintenance:** Since the *unit excitation system* has no main field rheostats and high-capacity switchgear, the individual excitation system requires less maintenance and due to this it has less maintenance cost.

It is important to note that an excitation system with better reliability is preferred even though its initial cost is more because of the fact that the cost of an excitation system is very less as compared to the cost of a generator.

12.10.3 Block diagram representation of structure of a general excitation system

A block diagram representation of the structure of a general excitation system is shown in Fig. 12.18.

The main components present in the block diagram are:

12.10.3.1 *Synchronous generator*

It may be the type of high-speed turbo-alternator run by a steam turbine or a low-speed AC generator run by a hydro-turbine. With the help of an excitation system, the terminal voltage of an alternator or a synchronous generator should be maintained constant during variable load situations.

12.10.3.2 *Exciter*

It supplies the field current to the rotor field circuit of the synchronous generator. It may either be a self-excited type or separately exciter type of DC generator.

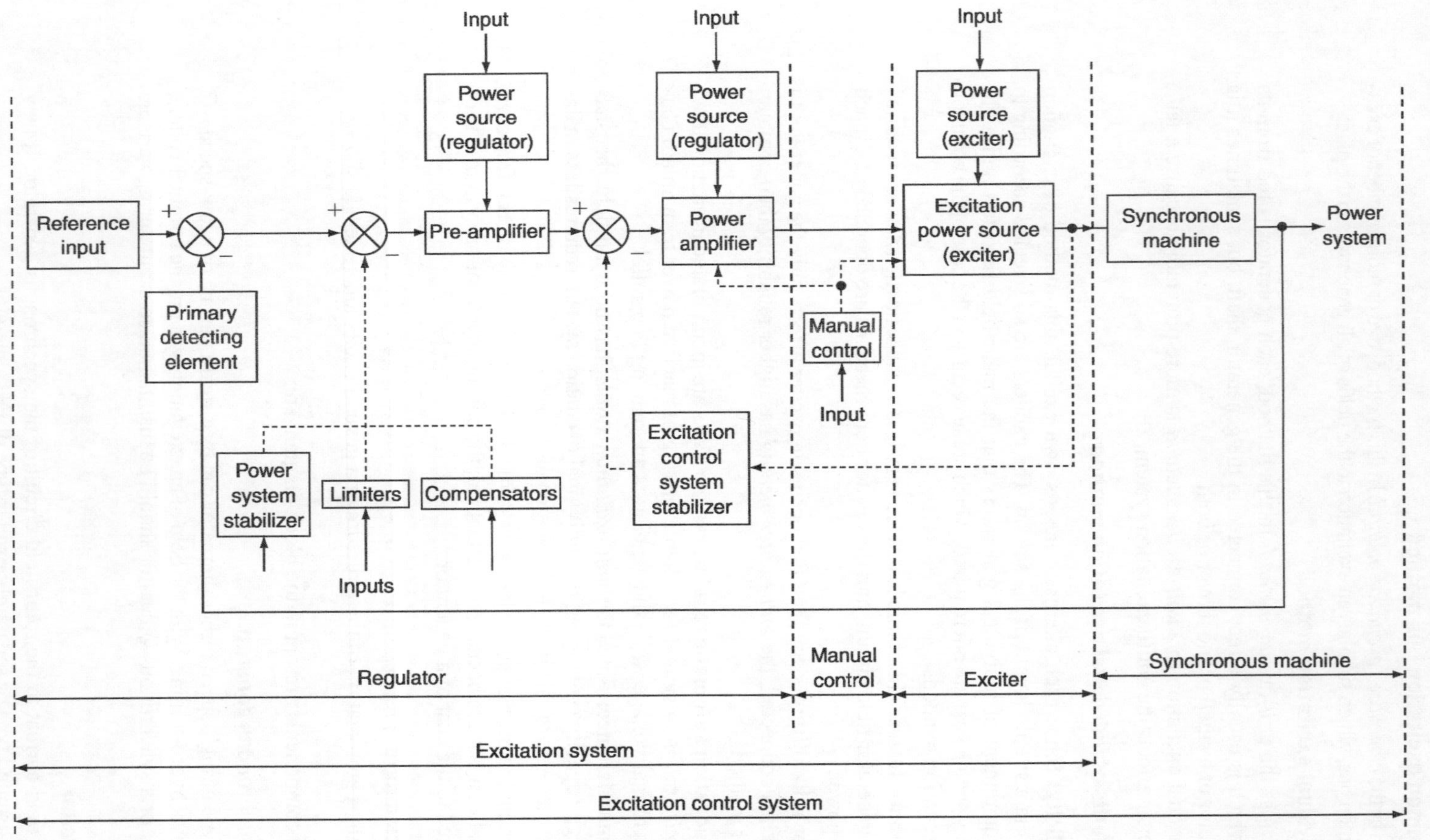

FIG. 12.18   *Block diagram representation of a general excitation system*

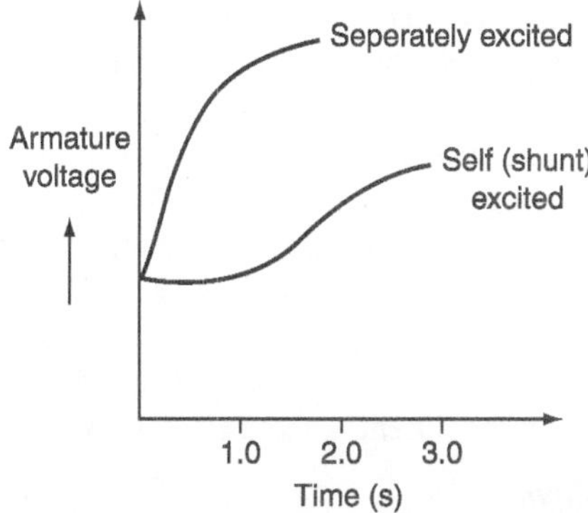

FIG. 12.19   *Response of an exciter when separately excited and self-excited*

In a self-excited exciter, a few turns are added for compounding and inter-poles are used. In separately excited exciters, an exciter field is supplied from a small DC generator known as the *pilot exciter*. A *pilot exciter* is a level compound generator and maintains constant voltage excitation for the main exciter.

The response of an exciter when compared to separately excited with that when self-excited is shown in Fig. 12.19.

Usually, the separately excited exciter, known as main exciter, is provided with two or more than two field windings, as shown in Fig. 12.20. Due to this arrangement of field, an easier automatic voltage regulation is permitted.

The voltage of the main exciter should be controlled from zero to ceiling voltage, the maximum voltage that may be attained by the exciter under specified conditions, to obtain rapid correction of exciter voltage after disturbance or fault. The faults or system disturbances cause an AVR to force an excitation up. After post-fault, rapid reduction of field is necessary to adjust the excitation to the correct value. This is easily achieved with a negative field. The main positive field is arranged in two parallel sections with rheostats for adjusting the field currents as required.

Hence, the positive and negative field windings of the main exciter with the adjustments of currents according to the load on the exciter maintain the exciter voltage and excitation as required.

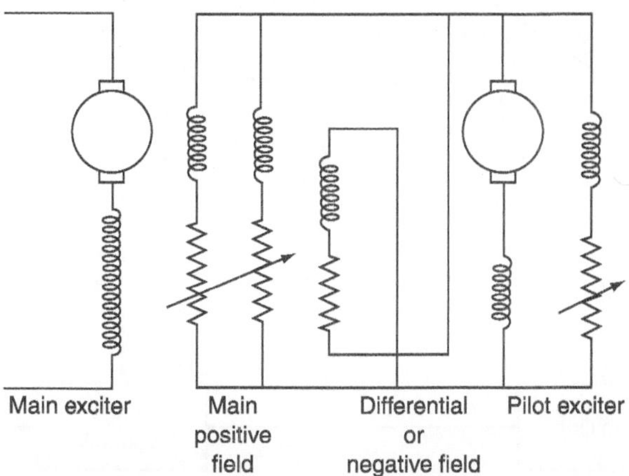

FIG. 12.20   *Exciter field arrangements*

Due to the several parallel connected field windings, the fast response of the exciter is achieved, because of low time constant of the whole field circuit.

For small-sized turbo-generators, the exciters are usually directly coupled to the generator shaft. For medium-and large-sized turbo-generators, the exciters are coupled to the main shaft through the gear and are generally driven at 1,000 rpm.

For smaller generators (i.e., rated up to 25 MVA or so), self-excited exciters must be used and for large-sized generators of above 25 MVA, separately excited exciters are used. The exciter voltage of the main exciter is usually 230 V. In some cases, a nominal voltage of 440 V is used. The main exciter load in the resistance is the alternator field winding and this is generally between 0.25 and 1.0 Ω. The rotor current is about 10 A per MVA of alternator rating.

12.10.3.3 *Use of amplidyne*

In some cases, the DC excitation system is equipped with an amplidyne. An amplidyne provides large currents to the field winding of the main exciter.

It is a high-response cross-field generator and has a number of control windings, which can be supplied from the pilot exciter and a number of feedback circuits of an AVR and magnetic amplifier, etc. for control purposes. An amplidyne has a very high amplification factor of 10^6 or even more and needs very small control power.

12.10.3.4 *AVR*

An AVR in conjunction with the exciter tries to maintain constant terminal voltage of on an AC generator. The voltage regulator, in fact, couples the output variables of the synchronous generator to the input of the exciter through feedback and forwarding elements for the purpose of regulating the synchronous machine output variables. Thus, the voltage regulator may be assumed to consist of an error detector, pre-amplifier, power amplifier, stabilizers, compensators, auxiliary inputs, and limiters. The voltage regulator is treated as the heart of an excitation system. Exciter and regulator constitute an excitation system. Exciter, regulator, and synchronous generator constitute a system known as the excitation control system.

12.11 EXCITATION CONTROL SCHEME

A typical excitation control scheme is shown in Fig. 12.21.

The field winding of an alternator is connected to the exciter. The alternator terminal voltage is rectified by means of a potential transformer (PT) and rectifier, and is fed to a voltage

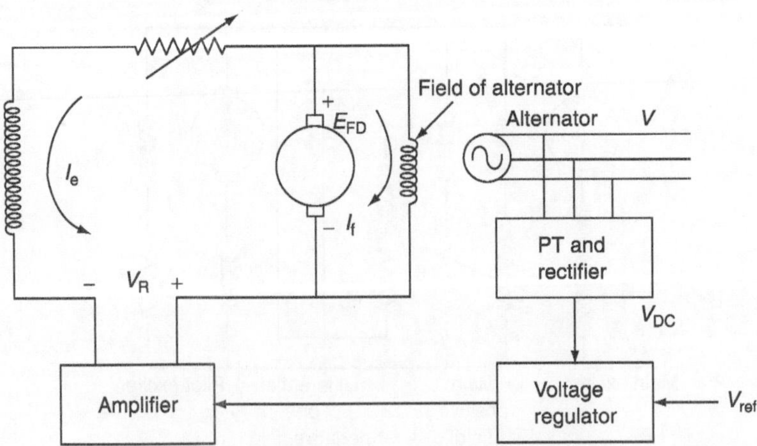

FIG. 12.21   *A typical excitation control scheme*

regulator. A voltage regulator compares the rectified output voltage with a reference voltage V_{ref}. The error signal output $V_e = |V_{ref} - V_{dc}|$ from the voltage regulator is amplified by an amplifier and the amplifier output voltage is fed to the exciter field winding.

There is no error signal output from the regulator and the field winding current of exciter I_e is constant when the output voltage (terminal voltage) of an alternator is at a nominal value.

When the load on the alternator varies, the terminal voltage also varies. Hence, the error signal can be produced by the regulator, amplified, and fed to the field winding of the exciter. The field winding current of the exciter is varied and hence the terminal voltage reaches the required level.

12.12 EXCITATION SYSTEMS—CLASSIFICATION

The excitation systems are broadly classified into the following categories:

1. DC excitation system.
2. AC excitation system.
3. Static excitation system.

12.12.1 DC excitation system

It consists of different configurations of rotating exciters like:

(a) Self-excited exciter with a direct-acting rheostatic-type voltage regulator.

(b) Main and pilot exciters with an indirect-acting rheostatic-type voltage regulator.

(c) Main exciter, amplidyne, and static voltage regulator.

(d) Main exciter, magnetic amplifier, and static voltage regulator.

The main drawbacks of a DC excitation system are:

- Complexity is more due to rotating exciters, voltage regulators, and moving contacts like slip rings and brushes.

- Time constants of exciter, voltage regulator, amplidyne, and magnetic amplifier are large (about 3 s).

- Difficulties of commutation.

- Smoothless operation needs continuous maintenance.

- Reliability is less.

- Noise level is more due to rotating exciters.

12.12.2 AC excitation system

It consists of an AC generator and a thyristor (rectifier) bridge circuit directly connected to the alternator shaft. The main advantage of this method of excitation is that the moving contacts such as slip rings and brushes are completely eliminated thus offering smooth and maintenance-free operation. Such a system is known as a brushless excitation system. In this system, there are no commutation problems.

12.12.3 Static excitation system

It consists of a step-down transformer and a rectifier system using mercury arc rectifiers of silicon-controlled rectifiers (SCRs). The rotating amplifiers and rotating exciters are replaced by the static devices of a rectifier system.

The advantages of a static excitation system are:

(i) Noise-free operation in the plant is obtained as the rotating exciters are replaced with static devices of rectifiers.

(ii) Since the static excitation equipment may be mounted or placed separately at a convenient place, the complexity of the excitation system is reduced.

(iii) Due to static devices, the overall length of the generator shaft is reduced, which simplifies the torsion problem and the problem of critical speed. The generator rotor is easily withdrawn for maintenance purpose.

(iv) High reliability compared to other excitation systems because of having more reliable static devices and

(v) Static devices are provided with low-speed hydro-alternators, where large-sized rotating exciters are needed.

12.13 VARIOUS COMPONENTS AND THEIR TRANSFER FUNCTIONS OF EXCITATION SYSTEMS

In this section, we shall discuss PT and rectifier, voltage comparators, and amplifiers in detail.

12.13.1 PT and rectifier

One possible arrangement of a PT and a rectifier is shown in Fig. 12.22. The terminal voltage of the alternator is stepped down by the PT and rectified to form V_{DC}, which is proportional to the average RMS value of the terminal voltage V_t.

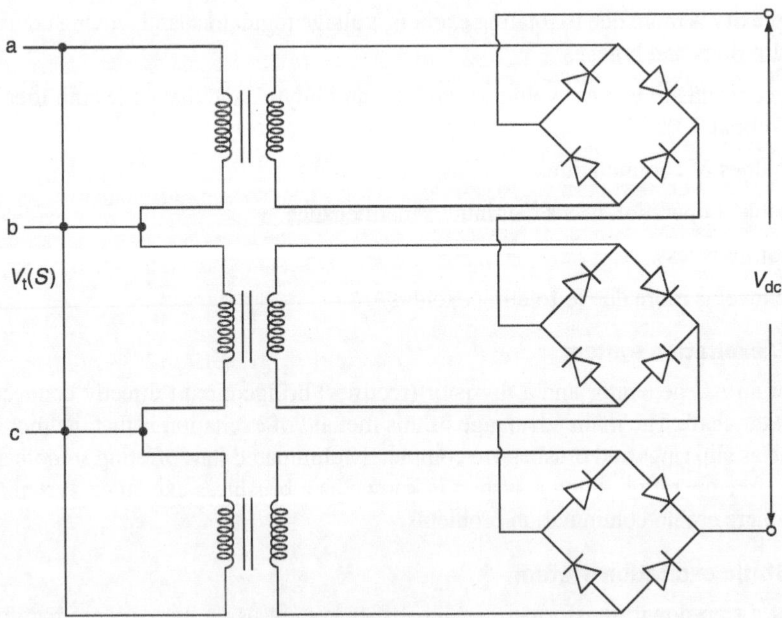

FIG. 12.22 *Connections of PT and rectifier*

Transfer function of the arrangement is represented as:

$$\frac{V_{dc}(s)}{V_t(s)} = \frac{K_R}{1+S\tau_R}$$

where K_R is the proportionality constant (or) gain of the PT and rectifier assembly and τ_R the time constant of the assembly due to filtering in the assembly arrangement.

12.13.2 Voltage comparator

A voltage comparator compares the rectified DC voltage of the generator with a reference voltage V_{ref} and produces an output in the form of an error signal V_e. Figure 12.23 shows an electronic difference amplifier as a comparator.

The output voltage or an error voltage V_e is expressed as

$$V_e(s) = K(V_{ref}(s) - V_{dc}(s))$$

12.13.3 Amplifiers

Among the various types of amplifiers used in an excitation system, the amplidyne and magnetic amplifier have high amplification factors.

12.13.3.1 Amplidyne

Basically, a cross-field DC generator is operated as an amplidyne. An amplidyne configuration is shown in Fig. 12.24.

The operation of an amplidyne consists of two stages of amplification.

First stage of amplification

An amplidyne consists of brushes along the d-axis and the q-axis. The brushes along the q-axis are short-circuited. As the armature resistance is very small, a small amount of field m.m.f. results in a large q-axis current. This production of the large q-axis current is treated as the first stage of amplification.

Second stage of amplification

The large q-axis current produces a flux in time and space phase with itself. Corresponding to this q-axis flux, the voltage will be developed across the d-axis brushes. The development of voltage across d-axis brushes due to the q-axis current is termed as the second stage of amplification.

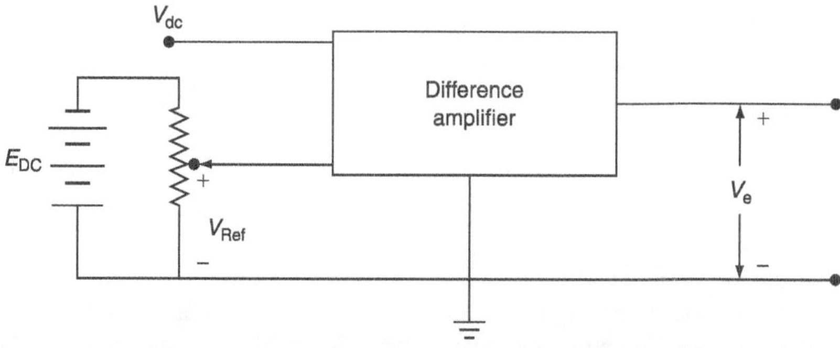

FIG. 12.23   *Electronic difference amplifier as a comparator*

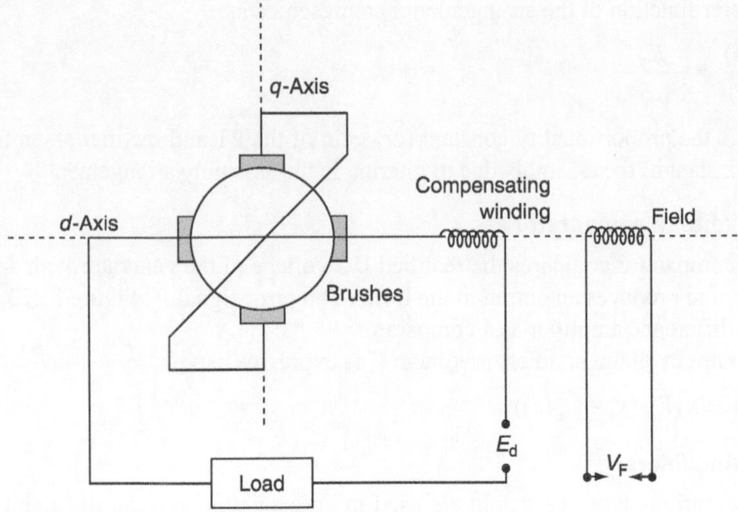

FIG. 12.24   *Amplidyne configuration*

The *d*-axis brushes are connected to the load along with a compensating winding, which provides no resultant excitation due to load current since it may reduce the original field excitation. (It has a number of control windings supplied from the pilot exciter and has a number of feedback circuits of AVR and magnetic amplifier, etc. for control purposes.)

The advantages of the second stage of amplification are as follows:

- Power required for control purpose is very small.

- As the response time is very less, it has a fast response.

- Amplification factor is of 10^6 or even more.

The transfer function of an amblidyne under no-load condition can be expressed as

$$\frac{E_d(s)}{V_f(s)} = \frac{K_A}{(1+s\tau_a)(1+s\tau_f)}$$

where τ_a is the time constant of armature, τ_f the time constant of field, and K_A the voltage amplification factor.

12.13.3.2 *Magnetic amplifier*

Magnetic amplifier configuration is shown is Fig. 12.25. It consists of a saturable core, control winding, and a rectifier circuit.

The saturable core reactor can be designed so that when no DC current is flowing through the DC control winding, the inductive reactance of AC coils is very high and limits the flow of AC current to a small value.

In a magnetic amplifier when large DC current flows through the control winding, the core gets saturated. This results in the decrement in permeability and hence the reactance of AC coils decreases. Therefore, more AC current flows. This AC current is rectified and fed to the load.

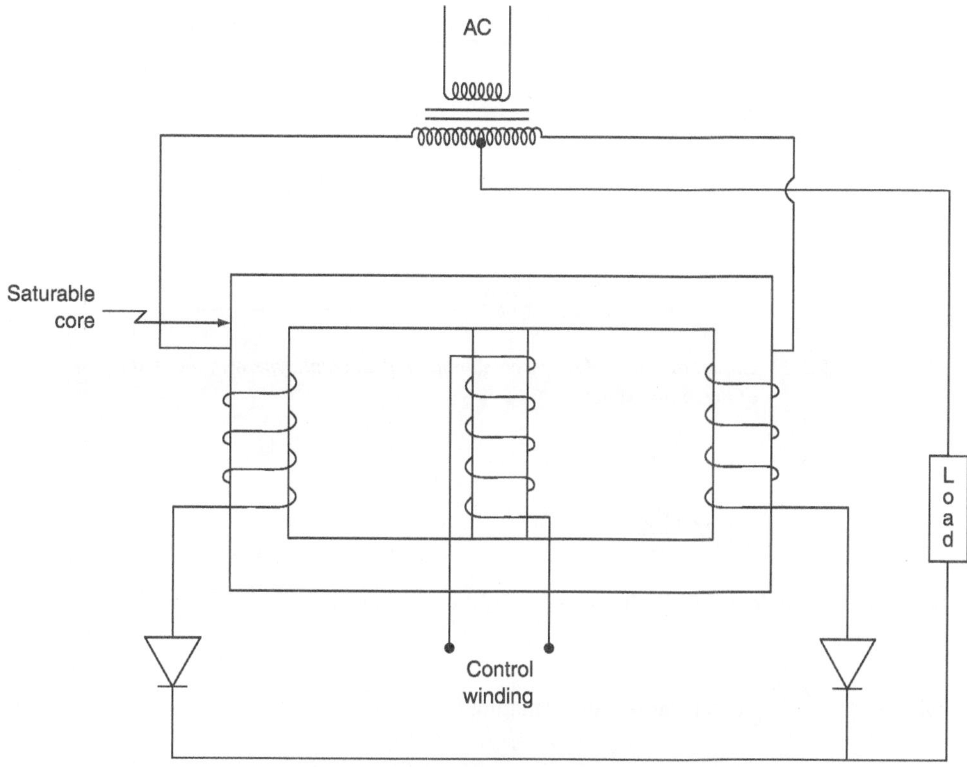

FIG. 12.25   *Configuration of a magnetic amplifier*

The controlling of a large output current by means of a small control current is the main principle of a magnetic amplifier.

The transfer function of a magnetic amplifier can be expressed as

$$\frac{V_R(s)}{V_e(s)} = \frac{K_A}{1 + s\tau_A}$$

where V_e is an error signal input applied to control winding, V_R is the output voltage and is governed by the limits, i.e., $V_{R\,min} \leq V_R \leq V_{R\,max}$, K_A is the amplification factor, and τ_A is the time constant amplifier.

12.14 SELF-EXCITED EXCITER AND AMPLIDYNE

The amplidyne is connected in series with the shunt field of the main exciter as shown in Fig. 12.26.

Let V_R be the armature emf of the amplidyne,

e_{fd} the armature emf of main exciter,

N_e the number of field turns of main exciter under no-load condition,

ϕ_e the flux of the main exciter under no-load condition,

r_e the field circuit resistance of the exciter under no-load condition, and

i_e the field current of the exciter under no-load condition.

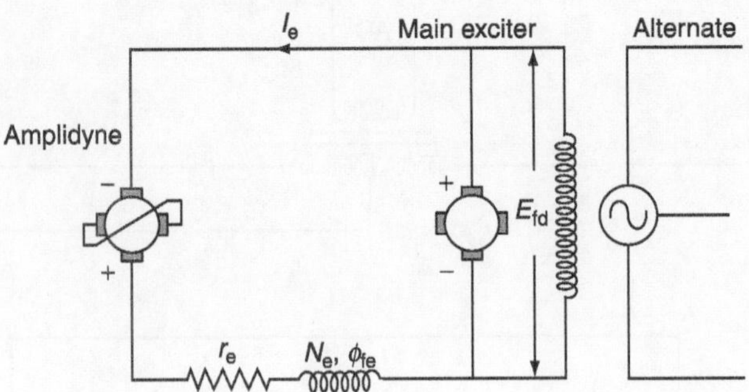

FIG. 12.26   *Circuit diagram of series combination of an amplidyne with the shunt field of the main exciter*

For an exciter field circuit,

$$V_R + e_{fd} = i_e r_e + N_e \frac{d\phi_{fe}}{dt}$$

Since e_{fd} is a function of ϕ_e, effective flux of the main exciter

$$e_{fd} = K\phi_e$$

where $K = \dfrac{ZnP}{\alpha}$, a constant for the armature

Difference of ϕ_{fe} and ϕ_e is an account of leakages flux ϕ_1, proportional to field current i_e, and may be written as

$$\phi_{fe} = \phi_e + \phi_1$$

and $\phi_1 = C_1 \phi_e$

where C_1 is the proportionality constant:

$$\phi_{fe} = \phi_e + C_1 \phi_e$$
$$= (1 + C_1)\, \phi_e$$
$$\phi_{fe} = \sigma \phi_e$$

where σ is known as the coefficient of dispersion having a value in the range of 1.1–1.2.

$$\therefore V_R + e_{fd} = i_e r_e + \frac{N_e \sigma}{K} \frac{d(e_{fd})}{dt}$$

$$\Rightarrow V_R + e_{fd} = i_e r_e + \tau_E \frac{d(e_{fd})}{dt}$$

where τ_E is known as time constant of the exciter.

However, the effect of saturation of the exciter voltage e_{fd} is taken into account while solving the above equation.

The exciter characteristics are shown in Fig. 12.27.

It is evident that the saturation of the exciter S_E is a non-linear function of the exciter voltage e_{fd} and is given as

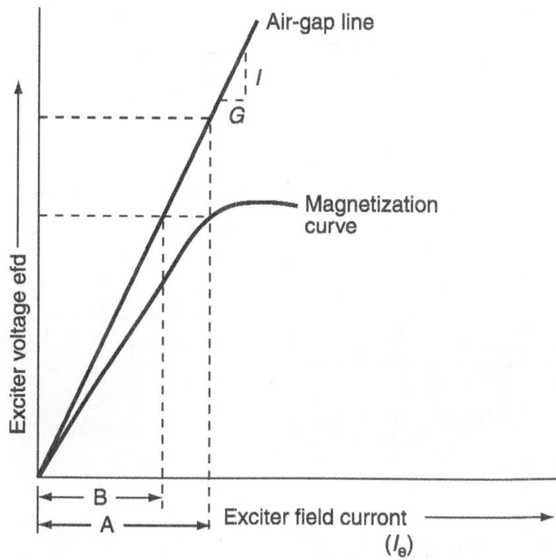

FIG. 12.27   *Exciter characteristics*

$$S_E = \frac{(A - B)}{B}$$

From the above, we can write $B = \dfrac{A}{1 + S_E}$

If the slope of the air-gap line is $\dfrac{1}{G}$

$$\therefore e_{fd} G = \frac{i_e}{1 + S_E}$$

or $i_e = G(1 + S_E) e_{fd}$

Substituting i_e in equation $V_R + e_{fd} = i_e r_e + \tau_E \dfrac{d(e_{fd})}{dt}$, we have

$$\Rightarrow V_R + e_{fd} = r_e(G(1 + S_E)e_{fd}) + \tau_E \frac{de_{fd}}{dt}$$

Taking Laplace transform of the above equation, we get

$$V_R(s) + E_{fd}(s) = r_e G(1 + S_E)E_{fd}(s) + s\tau_E E_{fd}(s)$$

From the above, we can get

$$E_{fd}(s) = \frac{V_R(s) - S_E E_{fd}(s)}{K_E + s\tau_E}$$

where $K_E = r_e G - 1$

12.15 DEVELOPMENT OF EXCITATION SYSTEM BLOCK DIAGRAM

The simplified diagram of an excitation system with fundamental components is as shown in Fig. 12.28.

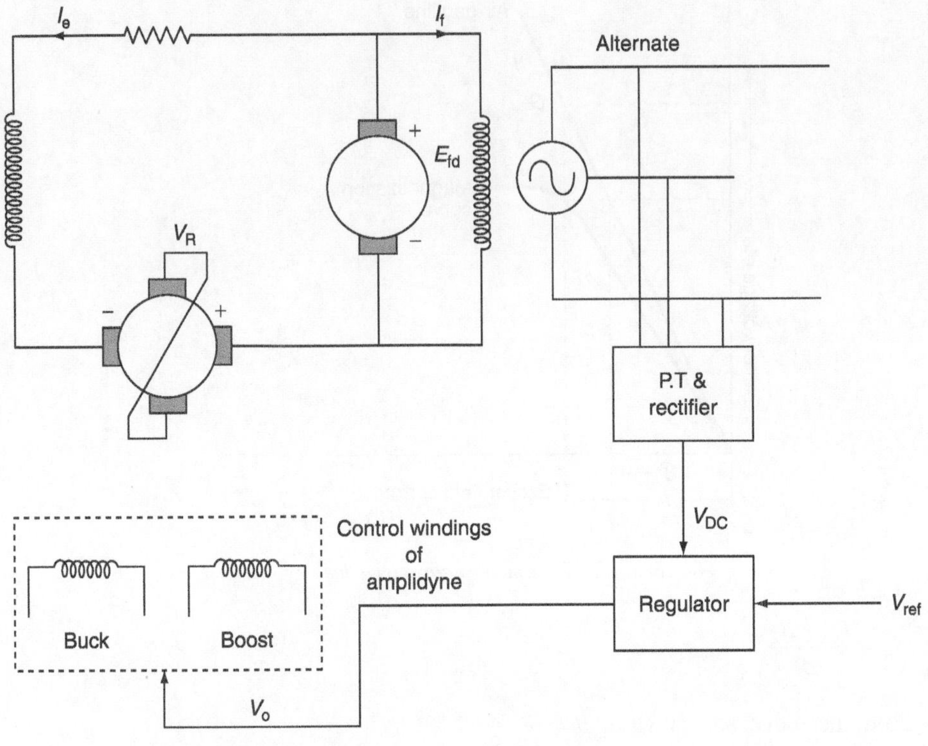

FIG. 12.28   *Simplified diagram of a buck–boost excitation system*

For the complete analysis of the excitation system, it is necessary to develop the transfer function of each component and then the transfer function of an overall excitation system.

Transfer function of PT and rectifier is $\dfrac{V_{DC}(s)}{V_t(s)} = \dfrac{K_R}{1 + s\tau_R}$

Transfer function of an amplifier is $\dfrac{V_R(s)}{V_e(s)} = \dfrac{K_A}{1 + s\tau_A}$ and

Transfer function of the exciter is $\dfrac{E_{fd}(s)}{V_R(s)}$

Here, $E_{fd}(s) = \dfrac{V_R(s) - S_E E_{fd}(s)}{K_E + s\tau_E}$

If the saturation is neglected i.e., $S_E = 0$, $E_{fd} = \dfrac{V_R(s)}{K_E + s\tau_E}$

$\therefore$ Transfer function of the exciter can be written as $\dfrac{E_{fd}(s)}{V_R(s)} = \dfrac{1}{K_E + s\tau_E}$.

12.15.1 Transfer function of the stabilizing transformer

An equivalent circuit of a stabilizing transformer is shown in Fig. 12.29.

The excitation system that was described earlier has a dynamic response that is prone to excessive overshoot and stability problems. These problems are overcome by adding

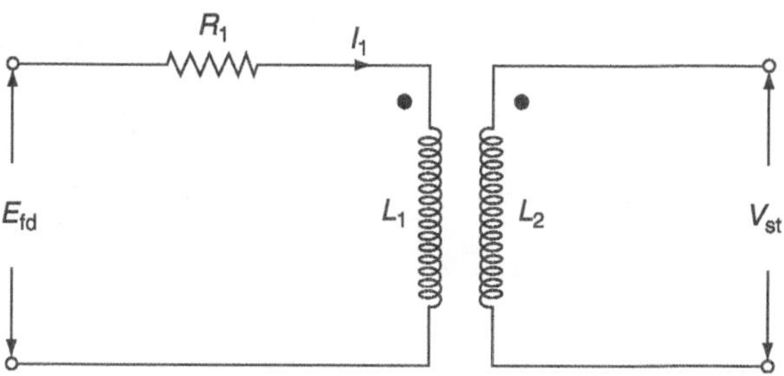

FIG. 12.29   *Equivalent circuit of a stabilizing transformer*

a stabilizing transformer. For the stabilizing transformer, the input becomes E_{fd} and the output is V_{ST}.

The output V_{ST} is subtracted from V, i.e., $V - V_{ST}$ to provide the input to the amplifier:

Input, $e_{fd} = R_1 i_1 + L_1 \dfrac{di_1}{dt}$

In Laplace transform,

$E_{fd}(s) = (R_1 + sL_1)\, I_1(s)$

and output $V_{ST} = M \dfrac{di_1}{dt}$

i.e., $V_{ST}(s) = sMI_1(s)$

$\therefore$ Transfer function of a stabilizing transformer is

$$\frac{V_{ST}(s)}{E_{fd}(s)} = \frac{sMI_1(s)}{(R_1 + sL_1)I_1(s)}$$

$$\Rightarrow \frac{V_{ST}(s)}{E_{fd}(s)} = \frac{sK_F}{1 + s\tau_F}$$

where $K_F = \dfrac{M}{R_1} =$ Transformer gain

and $\tau_F = \dfrac{L_1}{R_1} =$ Transformer time constant.

12.15.2 Transfer function of synchronous generator

$$\frac{V_t(s)}{E_{fd}(s)} = \frac{K_G}{1 + s\tau_G}$$

where K_G is the gain of the generator and τ_G the time constant of a rotor field.

12.15.3 IEEE type-1 excitation system

Most of the excitation systems are modeled based on IEEE type-1 excitation system, which was produced by the report of the first IEEE committee in 1968. The complete block diagram of an IEEE type-1 excitation system is as shown in Fig. 12.30 by interconnecting all the components in the forward path and the feedback control loop.

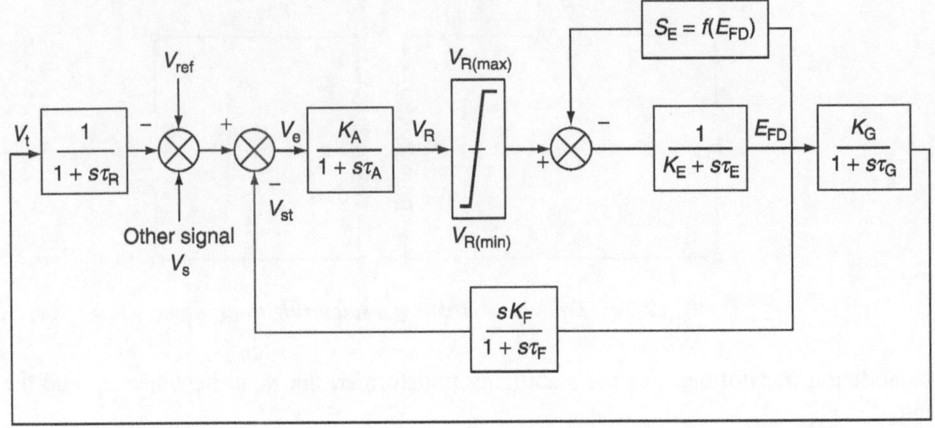

FIG. 12.30   *IEEE type-1 excitation system*

E_{FD} = Exciter output voltage (applied to generator field)

I_{FD} = Generator field current

I_{T} = Generator field terminal current

K_{A} = Regulator gain

K_{E} = Exciter constant related to self-excited field

K_{F} = Exciter saturation function

τ_{A} = Regulator amplifier time constant

τ_{E} = Exciter time constant

τ_{F} = Regulated stabilizing circuit time constant (τ_{F1} and τ_{F2})

τ_{R} = Regulated input filter time constant

V_{R} = Regulator output voltage

V_{t} = Terminal voltage of the generator applied to the regulator input

K_{G} = Gain of the generator

τ_{G} = Time constant of the generator rotor field

When the generator is operating at an equilibrium state, i.e., at rated voltage, the voltage of the rotating amplifier V_{R} becomes zero. If the generator load is increased such that the sensing circuit shown in Fig. 12.30 detects this fall in terminal voltage, it causes the amplidyne to increase the field current I_{e} in the exciter field. Hence, the exciter voltage increases and in turn increases the generator field current I_{f} that ultimately should rise the terminal voltage of generator, V_{t}.

Under steady-state conditions, $E = E_{fd}$.

Under transient conditions, any mismatch between E and E_{fd} will cause the voltage E_{q}' to change after some delay.

Mathematically, $\dfrac{dE_{q}'}{dt} = \dfrac{1}{\tau_{do}'}(E_{fd} - E)$

where τ_{do}' is the open-circuit generator, direct axis transient time constant.

12.15.4 Transfer function of overall excitation system

The transfer function of an overall excitation system, shown in Fig. 12.31, can be obtained by either using the block diagram reduction technique or the signal flow graph method.

First, neglect the effect of saturation:

i.e., $S_E = 0$

and remove the stabilizing transformer from the block diagram shown in Fig. 12.30.

$\therefore$ The transfer function of the system is of the form:

$$\frac{V_t(s)}{V_{ref}(s)} = \frac{G(s)}{1 + G(s)H(s)}$$

where $G(s) = \dfrac{K_A K_G}{(1 + s\tau_A)(K_E + s\tau_E)(1 + s\tau_G)}$ and is known as the feed-forward transfer

function and $H(s) = \dfrac{K_R}{1 + s\tau_R}$ is known as the feedback transfer function.

$\therefore$ The transfer function of the excitation system is expressed as

$$\frac{V_t(s)}{V_{ref}(s)} = \frac{K_A K_G (1 + s\tau_R)}{(1 + s\tau_A)(K_E + s\tau_E)(1 + s\tau_G)(1 + s\tau_R) + K_A K_G K_R}$$

In the transfer function $\dfrac{1}{1 + s\tau_R}$, τ_R is a simple time constant representing regulator input filtering. It is very small and may be considered to be zero for many systems.

The first summing point compares the regulator reference with the output of the input filter to determine the voltage error input to the regulator amplifier.

The AVR usually comprises several control loops and a simple reduction is necessary to the form $\dfrac{K_A}{1 + s\tau_A}$. Voltage regulator gain (K_A) has an important effect on power system performance while the time constant τ_A has a much smaller influence owing to large τ_e in series.

Because of high gain in an excitation system (100–400), errors in forward-path gain K_A are more important than errors in most other parameters (including generator and network).

The second summing point combines the voltage error input with the excitation damping loop signal. K_A and τ_A represent the main regulator gain and its transfer function. The minimum and maximum limits of the regulator are imposed so that large input error signals may not produce a negative output, which exceeds the practical limit.

The next summing point subtracts a signal that represents the saturation function $S_E = f(E_{FD})$ of the exciter. That is, the exciter output voltage (or generator field voltage E_{FD}) is multiplied by a non-linear saturation function and subtracted from the regulator output

signal. The resultant is applied to the exciter transfer function $\left(\dfrac{1}{K_E + s\tau_E}\right)$.

Major loop damping is provided by the feedback transfer function, $\dfrac{sK_F}{1 + s\tau_F}$ from the exciter output E_{FD} to the first summing point.

If the stabilizing loop is omitted, the exciter system and the main generator will be unstable for most practical values of K_A. It can only be omitted when there are additional input signals to the excitation system such as frequency derivatives, etc.

The useful value of K_F is from 0.1 to 0.15 and τ_F varies in the range 0.5–2.0 s.

$V_{ref}=$ Regulator reference voltage setting
$V_{RH}=$ Field rheostat setting
$V_T=$ Generator terminal voltage
$\Delta V_T=$ Generator terminal voltage error

Note that there is an interrelation between the exciter ceiling voltage $E_{FD_{max}}$, regulator ceiling $E_{R_{max}}$, exciter saturation, S_E and K_E.

Under steady-state condition,

$$V_R-(K_E+S_E)\,E_{FD}=0; \quad E_{FD_{min}} \leq E_{FD} \leq E_{FD_{max}}$$

At the ceiling or $E_{FD}\,\alpha\,E_{FD_{max}}$, the above equation becomes

$$V_{R_{max}}-(K_E+S_{E_{max}})E_{FD_{max}}=0$$

The exciter saturation function is defined as the multiplier of the exciter output E_{FD} to represent the increase in exciter excitation requirement because of saturation.

The exciter time constant τ_e is a dominant time constant in the excitation system. If it is not possible to obtain data for the main exciter saturation function, then a useful approximation is to increase τ_e by 20% and decrease the exciter ceiling voltage by 20%.

12.16 GENERAL FUNCTIONAL BLOCK DIAGRAM OF AN EXCITATION SYSTEM

The general functional block diagram of an excitation system is shown in Fig. 12.31.

12.16.1 Terminal voltage transducer and load compensation

The terminal voltage of the alternator is sensed and rectified into a DC voltage by means of a terminal voltage transducer.

The load compensation synthesizes a voltage, which differs from the terminal voltage by the voltage drop in an impedance $(R_{LC}+jX_{LC})$. Both voltage and current phasors must be used in computing the compensating voltage V_C.

12.16.1.1 *Objectives of load compensation*

- Sharing of reactive power among the units, which are bussed together with zero impedance between them. For this, R_{LC} and X_{LC} are positive and the voltage is regulated at a point internal to the generator.

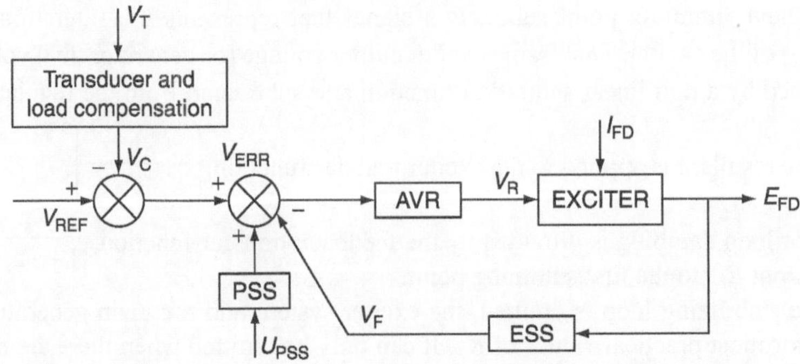

FIG. 12.31 *Functional block diagram of an excitation control system*

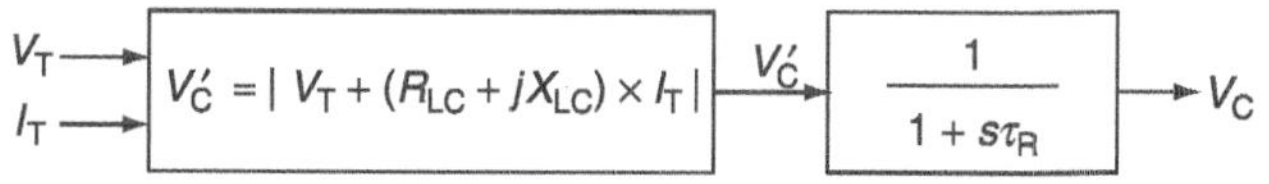

$$V_C' = |\, V_T + (R_{LC} + jX_{LC}) \times I_T \,|$$

$$\frac{1}{1 + s\tau_R}$$

FIG. 12.32   *Modeling of transducer and load compensation*

- When the units are operating in parallel through unit transformers, it is desirable to regulate voltage at a point beyond the machine terminals to compensate for a portion of transformer impedance. For this case, both R_{LC} and X_{LC} are negative values. R_{LC} is neglected in most of the cases.

The modeling of terminal voltage transducers and load compensation is as shown in Fig. 12.32.

The voltage transducer is usually modeled as a single time constant τ_R and it is very small and assumed to be zero for simplicity in many cases.

12.16.2 Exciters and voltage regulators

The AVRs of modern type are continuously acting electronic regulators with high gain and lower time constants.

12.16.2.1 *Types of exciters*

The types of exciters are shown in Fig. 12.33(a). The block diagram representation of an exciter and a regulator is shown in Fig. 12.33(b).

In Fig. 12.33(b), V_R is the output of the regulator, which is limited:

τ_A—single time constant of regulator

K_A—positive gain

Saturation function $S_E = f(E_{FD})$ represents the saturation of the exciter.

Note: The limits on V_R can be found from steady-state equation:

$$V_R - (K_E + S_E)\, E_{RD} = 0$$

This implies limits on E_{FD} such that:

$$E_{FD_{min}} \leq E_{FD} \leq E_{FD_{max}}$$

With the specification of parameters, $K_E = 1$, $\tau_E = 0$, $S_E = 0$, and $V_{R_{max}} = K_P V_T$, IEEE type-1 system represents the static excitation system with potential source-controlled rectifier type.

12.16.3 Excitation system stabilizer and transient gain reduction

This system is used to increase the stability region of operation of the excitation system and also permit higher regulator gains.

The feedback transfer function of the ESS is shown in Fig. 12.34.

The ESS is realized by an ideal transformer whose secondary is connected to high impedance as shown in Fig. 12.35.

The turns ratio of the transformer (n) and the time constant of the circuit $\left(\dfrac{L}{R}\right)$ determine K_F and τ_F such as $\dfrac{nL}{R} = K_F$ and $\dfrac{L}{R} = \tau_F$

τ_F is usually 1 s.

A series-connected load or lag circuit can also be used instead of feedback compensation circuit for ESS as shown in Fig. 12.36.

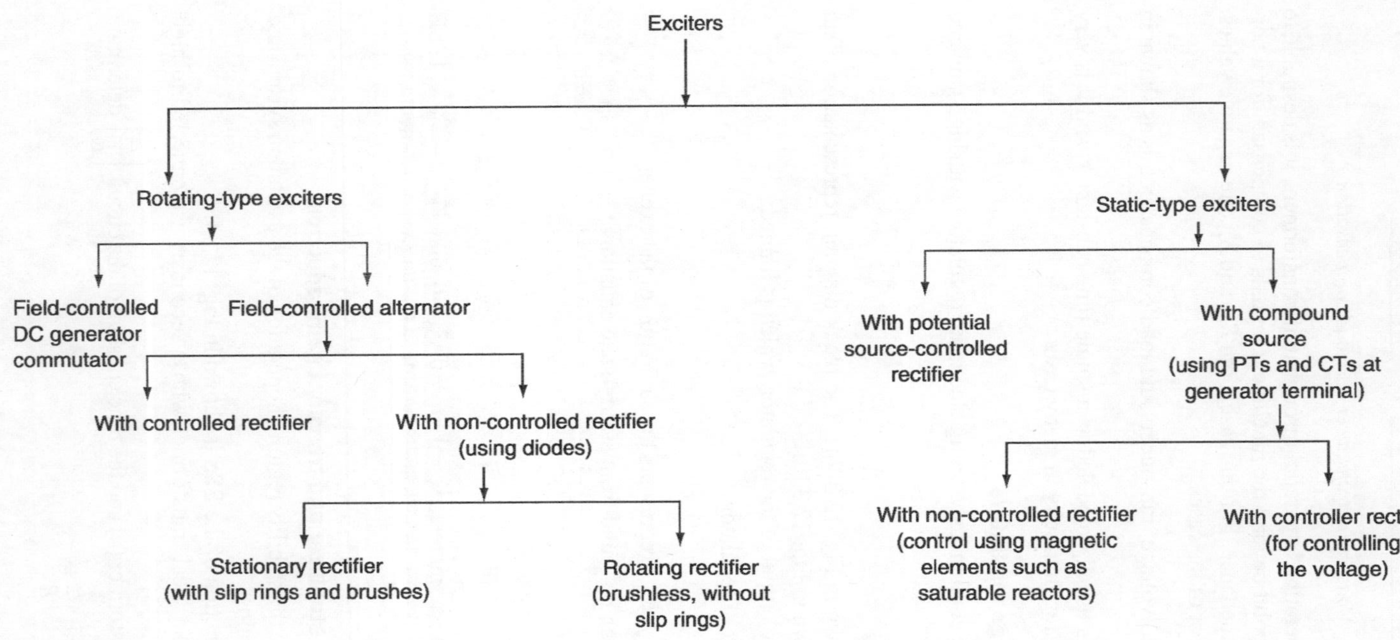

FIG. 12.33   (*Continued on next page*)

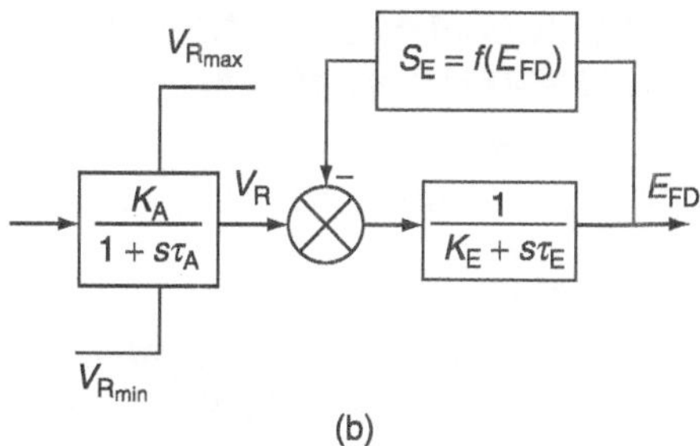

(b)

FIG. 12.33   *(a) Classification of exciters; (b) block diagram representation of exciter and voltage regulator*

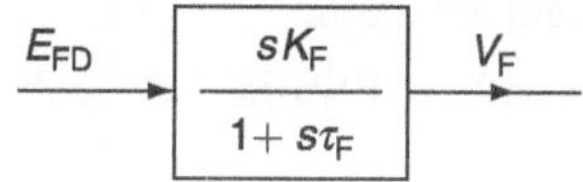

FIG. 12.34   *ESS transfer function*

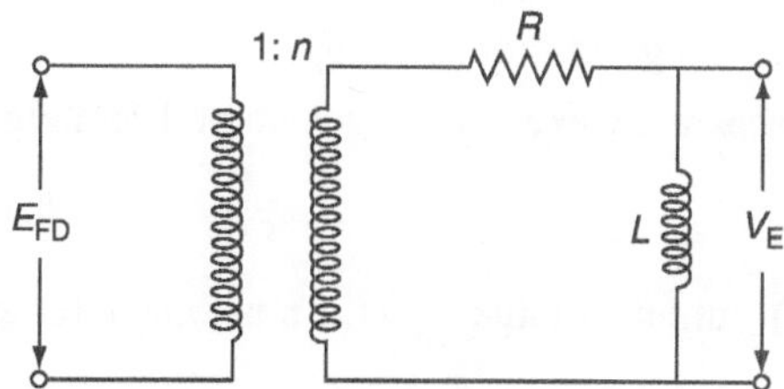

FIG. 12.35   *Realization of ESS*

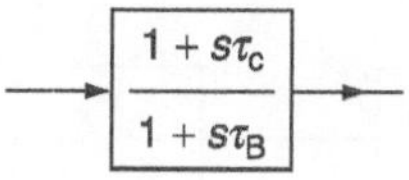

FIG. 12.36   *TGR*

Here, $\tau_C < \tau_B$ and stabilization is termed as TGR. Reducing the transient gain (or gain at higher frequencies), thereby minimizing the negative contribution of the regulator to system damping, is the main objective of TGR. The TGR may not be required, if power system stabilizer (PSS) is specifically used to enhance system damping.

Usually, TGR factor $\dfrac{\tau_B}{\tau_C} = 10$.

12.16.4 Power system stabilizer

During the transient disturbance, the rotor oscillations of frequency 0.2–2.0 Hz are damped out by providing the PSSs. The damping of rotor oscillatations can be impaired by the provision of high-gain AVR, particularly at high loading conditions when a generator is connected through a high external impedance (due to weak transmission network).

The input signal to PSS is derived from speed or frequency or accelerating power or a combination of these signals.

The output of PSS, V_s, is added to the terminal voltage error signal.

12.17 STANDARD BLOCK DIAGRAM REPRESENTATIONS OF DIFFERENT EXCITATION SYSTEMS

The standard block diagrams of different excitation systems based on supply were published by the second IEEE committee report in the year 1981.

12.17.1 DC excitation system

Figure 12.37 shows the type DC-1 excitation system. It consists of a DC commutator exciter with a continuously acting voltage regulator. This is similar to the IEEE type-1 excitation system.

The TGR can be represented by the transfer function $\left(\dfrac{1+s\tau_c}{1+s\tau_E}\right)$ with $\tau_B > \tau_C$. It has the similar function as ESS in the feedback path.

Either TGR in the forward path or ESS in the feedback path is shown in the block diagram representation.

With $\tau_B = \tau_C$, the TGR can be avoided and similarly with $K_F = 0$, ESS can be neglected.

12.17.1.1 *Derivation of transfer function*

(i) *For separately excited DC generator (exciter)*

Figure 12.38 shows the separately excited DC generator. From Fig. 12.38,

$$E_S = I_f R_f + L_f \frac{dI_f}{dt}$$

The generator (exciter) output voltage E_g is a non-linear function of I_f as shown in Fig. 12.39.

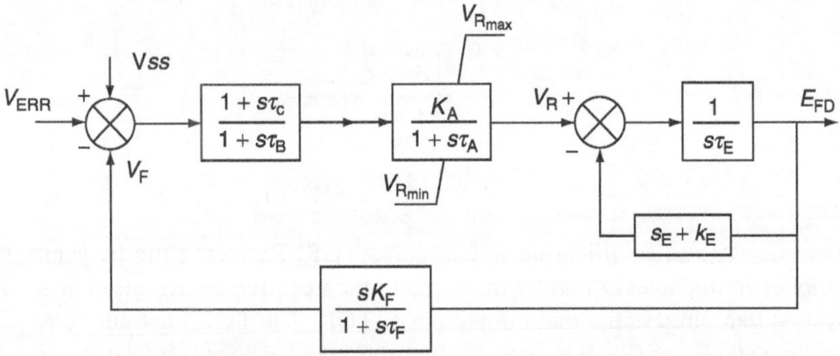

FIG. 12.37   *Type DC 1-DC commutator exciter*

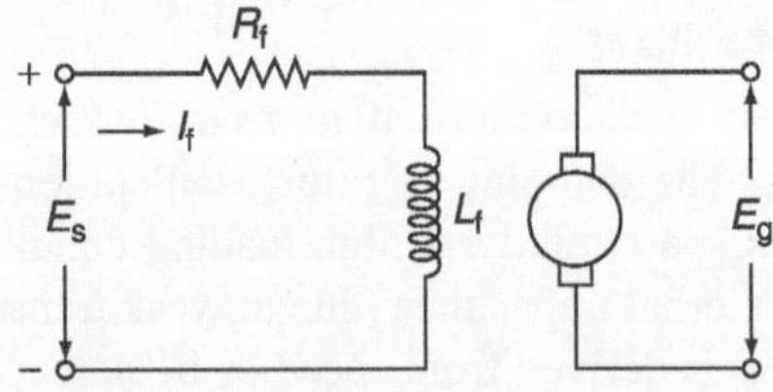

FIG. 12.38   *Separately excited DC generator*

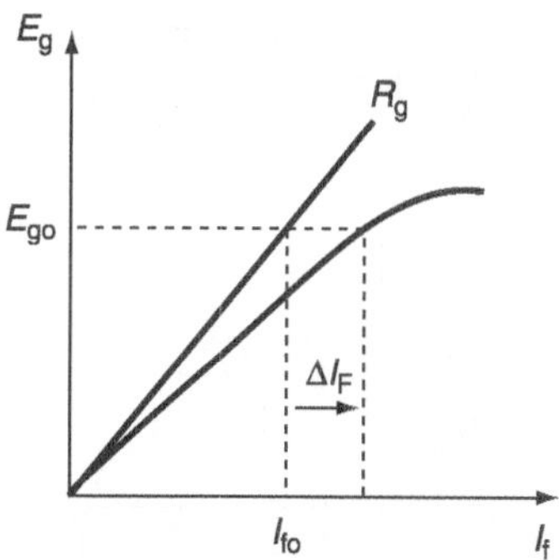

FIG. 12.39 *Exciter load saturation curve*

Assume the speed of the exciter to be constant. From Fig. 12.39, we have the following:

$$I_f = \frac{E_g}{R_g} + \Delta I_f$$

$$\Delta I_f = S_e E_g$$

where R_g is the slope of the saturation curve near $E_g = 0$. Express I_f in p.u. of I_{fb}:

$$I_{fb} = \frac{E_{gb}}{R_g}$$

where E_{gb} is the rated voltage that is defined as the voltage, which produces rated open-circuit voltage in the generator-neglecting saturation:

$$\bar{I}_f = \bar{E}_g + S'_E \bar{E}_g \qquad \text{(in p.u.)}$$

$$\therefore E_s = \bar{I}_f \frac{R_f}{R_g} + \frac{K_f}{R_g} \frac{d\bar{E}_g}{dt}$$

where $S'_E = R_g S_e$

$$K_f = L_f \frac{d\bar{I}_f}{d\bar{E}_g}$$

Since $\bar{E}_g = V_R$

The block diagram of a separately excited generator is shown in Fig. 12.40.

(ii) Self-excited DC generators

The schematic diagram of a self-excited exciter is shown in Fig. 12.41.

E_a represents the voltage of the amplifier in series with the exciter shunt field.

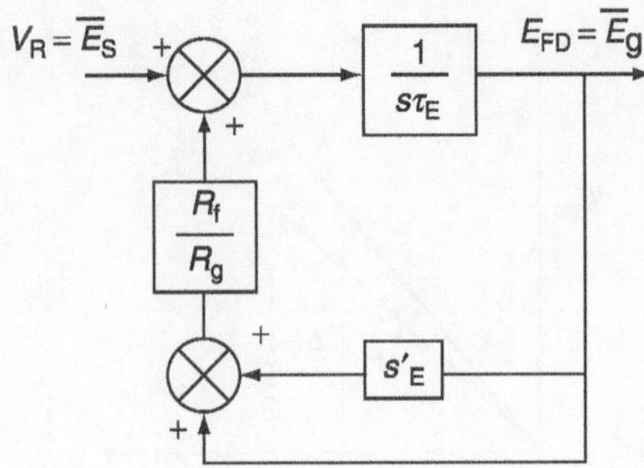

FIG. 12.40   *Block diagram of a separately excited DC generator*

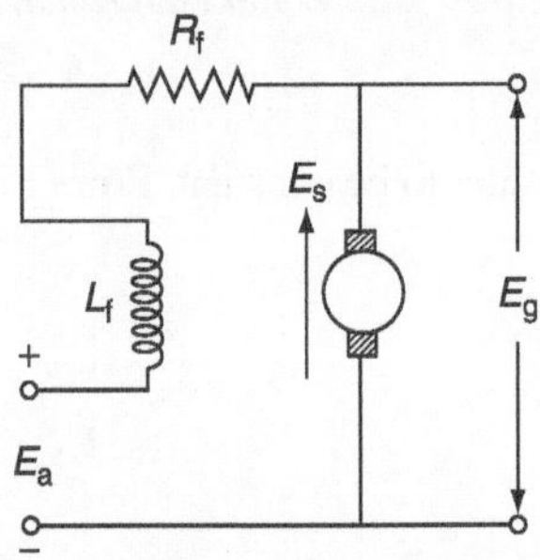

FIG. 12.41   *Schematic diagram of a self-excited exciter*

$$\therefore \ \overline{E}_s = \overline{E}_g + \overline{E}_a$$

The block diagram of Fig. 12.41 with $E_a = V_R$ can be reduced such that $K'_E = \dfrac{R_f}{R_g} - 1$.

The R_f is periodically adjusted to maintain $V_R = 0$ in the steady state, for this $K_E = -S_{E_0}$ where S_{E_0} is the value of saturation function S_E at the initial operating point and K_E is generally negative for a self-excited exciter.

12.17.2 AC excitation system

The block diagram of a type AC-1 excitation system is shown in Fig. 12.42. This represents the field-controlled alternator rectifier with non-controlled rectifier-type AC excitation system.

The term $K_D I_{FD}$ represents armature reaction of the alternator and F_{EX} represents rectifier regulation.

Constant K_D is a function of the synchronous alternator, and transient reactance constant K_C is a function of the commutating reactance.

The function F_{BX} is given as

$$F_{BX} = 1 - \frac{I_N}{\sqrt{3}} \quad \text{if } I_N \leq \frac{\sqrt{3}}{4}$$

$$= \sqrt{\frac{3}{4} - I_N^2} \quad \text{if } \frac{\sqrt{3}}{4} < I_N < \frac{3}{4}$$

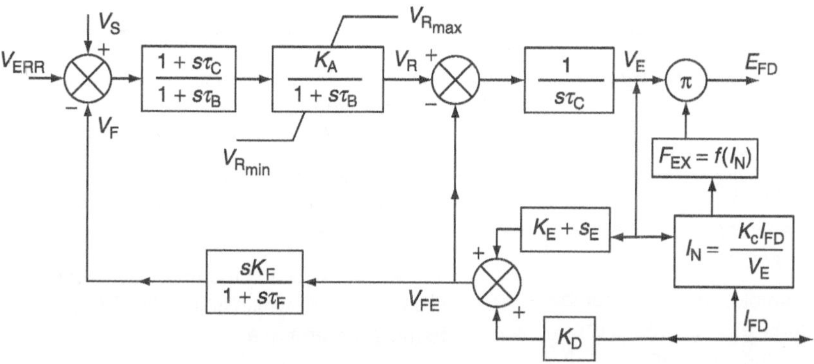

FIG. 12.42 *Block diagram of type AC-1 excitation system*

$$= \sqrt{3}(1 - I_N) \quad \text{if } I_N \geq \frac{3}{4}$$

The signal V_{FE} is proportional to the exciter field current and is used as an input to ESS.

12.17.3 Static excitation system

There are two types of static excitation systems:

(i) *With a potential source-controlled rectifier*—In this, the excitation power is supplied through a PT connected to generator terminals.

(ii) *With a compound source-controlled rectifier*—In this, both current transformer (CT) and PT are used at generator terminals.

The block diagram of the potential source-controlled rectifier excitation system is shown in Fig. 12.43.

In this block diagram, the internal limiter following the summing junction can be neglected, but field voltage limits that are dependent on both V_T and I_{FD} must be considered.

For transformer-fed systems, K_C is small and can be neglected.

In these systems, transformers are used to convert voltage (and also current in compounded systems) to the required level of field voltage with the aid of controlled or uncontrolled rectifiers. As the exciter ceiling voltage tends to be high in static exciters, field current limiters are used to protect the exciter and field circuit.

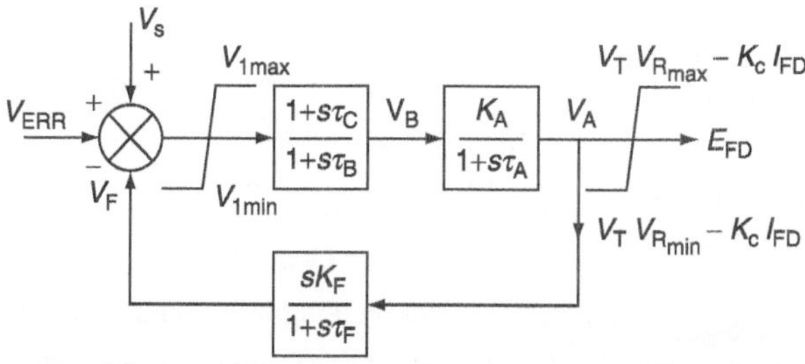

FIG. 12.43 *Block diagram-type ST1-potential source-controlled rectifier excitation system*

KEY NOTES

- According to the control, speed-governing systems are classified as:
  - (i) Mechanical–hydraulic-controlled.
  - (ii) Electro–hydraulic-controlled.
- The significance of rate and position limits in a speed-governing system are:
  - (i) Rate limiting of servomotor may occur for large, rapid-speed deviations, and rate limits are shown at the input to the integrator.
  - (ii) Position limits are indicated that correspond to wide-open valves or the setting of a load limiter.
- For wind-up limiter, the output variable of the transfer function $G(s)$ is not limited and is free to vary. Hence, the wind-up can be treated as a separate block in the modeling of a speed-governing system.
- For non-wind-up limiter, the output variable of the transfer function $G(s)$ is limited and there is no separate block in the modeling of a speed-governing system.

- Secondary governing system responds to the frequency of turbines and it controls the interceptor valves between the HP state and the reheater.
- Exciter provides the required field current to the rotor winding of a synchronous generator. It may either be self-excited type or separately excited type.
- In the unit excitation method, each generator is fed from its exciter, which is usually directly connected to the generator shaft.
- Amplidyne is a high-response cross-field generator, which has a number of control windings that can be supplied from a pilot exciter and a number of feedback circuits of AVR and magnetic amplifier for control purposes.
- An AC excitation system consists of an AC generator and a thyristor bridge circuit directly connected to the generator shaft
- By providing PSS, the rotor oscillations are damped out during the transient disturbance.

SHORT QUESTIONS AND ANSWERS

(1) What is the classification of speed-governing systems according to the control?

- Mechanical–hydraulic-controlled.
- Electro–hydraulic-controlled.

(2) What is the function of hydraulic servomotors used in mechanical–hydraulic-controlled and electro–hydraulic-controlled speed-governing systems?

For positioning valve or gate controlling system or water flow.

(3) What are the components of mechanical–hydraulic-controlled speed-governing systems used for steam turbines?

Speed governor, speed relay, hydraulic servomotor, and speed-governor-controlled systems.

(4) Why do the electro–hydraulic-controlled speed-governing systems provide more flexibility than mechanical–hydraulic-controlled speed-governing systems?

In electro–hydraulic-controlled speed-governing systems, mechanical components in the lower power portions are replaced by the static electronic circuits.

(5) How does the linearity of electro–hydraulic-controlled speed-governing systems improve?

By providing feedback loop of steam flow and the servomotor.

(6) What are the components of mechanical–hydraulic-controlled speed-governing systems for hydro-turbines?

A speed governor, a pilot valve, and servomotor, a distributor valve and gate servomotor, and governor-controlled gates.

(7) What is required to achieve the stable performance of speed-governing system for hydro-turbines?

Dashpot feedback is required to achieve the stable performance of speed-governing system for hydro-turbines.

(8) How is the speed relay represented in the approximate non-linear modeling of a speed-governing system?

As an integrator and provided as a direct feedback.

(9) How is the non-linear property of the speed-governing valve compensated?

By providing a non-linear CAM in between the speed relay and the hydraulic servomotor.

(10) What is the significance of a servomotor in the speed-governing system?

The servomotor controls the movements of valves and it is represented as an integrator with time constant τ_{sm} and it provides as direct feedback.

(11) What is the significance of rate limits and position limits in approximate non-linear modeling of a speed-governing system?

- Rate limiting of servomotor may occur for large, rapid-speed deviations, and rate limits are shown at the input to the integrator.

- Position limits are indicated as there corresponding to wide-open valves or the setting of a load limiter.

(12) What do you mean by wind-up limiter and non-wind-up limiter?

In the case of a wind-up limiter, the output variable of the transfer function $G(s)$ is not limited and is free to vary. Hence, the wind-up can be treated as a separate block in the modeling of a speed-governing system.

In the case of a non-wind-up limiter, the output variable of the transfer function $G(s)$ is limited and there is no separate block in the modeling of the speed-governing system.

(13) When modeling steam-turbine generators, which equipment is to be ignored at every instant?

The boiler controls and on-line frequency control equipment should be ignored at every instant due to their slower operations.

(14) What are the parts of a speed-governing system in a reheat system unit?

Primary governing system, secondary governing system, and anticipatory governing system are the parts of a speed-governing system.

(15) What is the primary governing system of a reheat system unit?

Primary governing system responds to the speed of a main shaft. It controls the main governor valve or throttle blades.

(16) What is the secondary governing system of a reheat system unit?

Secondary governing system responds to the frequency of turbines and it controls the

interceptor valves between the HP state and the reheater. The interceptor governing system will act as the secondary governing system.

(17) What is anticipatory governing system of a reheat system unit?

An anticipatory governing system responds to the accelerating power and is usually not set to operate if either the generator output is more than a certain value (25% of maximum output) or the turbine mechanical power output is less than a certain value (i.e., 80% of maximum capacity).

(18) When will the anticipatory governing system be activated?

Only when the reheat system unit suffers loss of a large percentage of its load and on serving this condition, the emergency stop valves are closed very rapidly to prevent dangerous overspeeds.

(19) When will the emergency overspeed governor trip?

When the velocity of shaft exceeds a pre-set value, then emergency overspeed governor will close the combined stop and emergency valves and shut the set down.

(20) On what assumptions is the modeling of a hydro-turbine based?

According to Kirchmayer, the modeling of single, general hydro-turbines is based on the following assumptions:

- Neglecting dead band, delays and non-linear performance in the governing systems.

- Neglecting the variations in head of the set of hydro-turbine unit with daily use or seasonal use.

- Assuming a constant equipment water starting time (τ_w).

(21) What is the function of an exciter in an excitation system?

It provides the required field current to the rotor winding of a synchronous generator. It may be either self-excited type or separately excited type.

(22) What are the effects of increase in excitation?

- The torque angle δ reduces and

- The current increases and the power angle also increases and hence becomes more lagging with respect to terminal voltage.

(23) What is the effect of increase in excitation on stability of the synchronous machine?

When the excitation increases, the torque angle δ reduces, which results in an increase in stiffness of the machine. In other words, with an increase in excitation, the stability of the machine will improve.

(24) What are the factors by which the increase in excitation is limited?

- Maximum output voltage of the exciter.
- Resistance of the field current.
- Saturation of the magnetic circuit.
- Rotor heating.

(25) What are the methods of providing excitation?

- Common or centralized excitation method.
- Individual or unit excitation method.

(26) What do you mean by individual or unit excitation method?

In individual or unit excitation method, each generator is fed from its own exciter, which is usually directly connected to the generator shaft.

(27) What is the meaning of common or centralized excitation method?

In common or centralized excitation method, two or more number of exciters feed a common bus, which supplies excitation to the fields of all generators in the plant.

(28) Which method of providing excitation is more preferable?

Unit exciter or common excitation method is more preferable.

(29) What are the merits of individual or unit excitation method?

Simplicity, less ohmic losses, higher reliability, less maintenance, and the possibility of incorporation of automatic regulators.

(30) Why are automatic regulators incorporated in individual or unit excitation method?

For the reliable sharing of reactive power to maintain constant terminal voltage while generators are running in parallel.

(31) What is pilot exciter?

In separately excited-type exciters, exciter field is supplied from a small DC generator known as a pilot exciter, which is a level compound generator and maintains constant voltage excitation for the main exciter.

(32) What is the use of an amplidyne in a DC excitation system?

An amplidyne provides large currents to the field winding of a main exciter.

(33) What is an amplidyne?

Amplidyne is high-response cross-field generator, which has a number of control windings that can be supplied from pilot exciter and a number of feedback circuits of an AVR and a magnetic amplifier for control purposes.

(34) What is an AVR? What are its components?

The AVR in conjunction with the excitation tries to maintain a constant terminal voltage of a synchronous generator. It consists of an error detector, pre-amplifier, power amplifier, stabilizer, compensators, auxiliary inputs, and limiters.

(35) Which is treated as the heart of an excited system?

The heart of an excitation system is the AVR.

(36) What is an excitation control system?

An exciter, voltage regulator, and synchronous generator constitute a system known as an excitation control system.

(37) What are the classifications of excitation systems?

- DC excitation system.
- AC excitation system.
- Static excitation system.

(38) What are the drawbacks of a DC excited system?

- More complexity.
- Larger time constants of an exciter, voltage regulator and amplidyne, and magnetic amplifier.
- Difficulties of commutation.
- Smoothless operation.
- Less reliability.
- More noise level due to rotating exciter.

(39) What is an AC excitation system?

An AC excitation system consists of an AC generator and a thyristor bridge circuit directly connected to the generator shaft.

(40) What are the merits of an AC excitation system?

- Moving contacts are completely eliminated.
- Offering smooth and maintenance-free operation.
- There are no commutation problems.

(41) Which system of excitation is brushless excitation?

An AC excitation system is a brushless excitation system.

(42) What are the merits of static excitation systems?

- Noise-free operation.
- Less complexity.
- High reliability compared to other excitation systems.
- The overall length of the generator shaft is reduced, which simplifies torsion problem and critical speed operation.

(43) Write the transfer function of a potential transformer and rectifier of an excitation system.

$$\frac{V_{dc}(s)}{V_E(s)} = \frac{K_r}{1 + s\tau_r}.$$

where K_r is the gain of PT and rectifier assembly

τ_r is the time constant of the assembly

(44) What are the stages of operation of an amplidyne?

First stage of amplification and second stage of amplification are the stages of operation of an amplidyne.

(45) What is the first stage of amplification for the operation of an amplidyne?

An amplidyne consists of brushes along the q-axis and the d-axis. The brushes along the q-axis are short-circuited. As the armature resistance is small, a small amount of field m.m.f. results in a large q-axis current. This is the first stage of amplification.

(46) What is the second stage of amplification of operation of an amplidyne?

The development of voltage across d-axis brushes due to q-axis current is termed as the second stage of amplification.

(47) Write the transfer function of an amplidyne under no-load condition.

$$\frac{E_d(s)}{V_f(s)} = \frac{K_A}{(1 + s\tau_a)(1 + s\tau_f)}$$

where τ_a is the time constant of the armature, τ_f the time constant of field, and K_A the voltage amplification factor.

(48) What are the components of a magnetic amplifier?

A saturable core, control winding, and a rectifier circuit are the components of a magnetic amplifier.

(49) What is the main principle of a magnetic amplifier?

The controlling of a large output current by means of small control current is the main principle of a magnetic amplifier.

(50) Write the transfer function of a magnetic amplifier.

$$\frac{V_r(s)}{V_e(s)} = \frac{K_A}{(1 + s\tau_A)}.$$

where τ_A is the time constant of the magnetic amplifier, V_r the output voltage, K_A the voltage amplification factor, and V_e the error signal input.

(51) Write the transfer function of an exciter.

$$\frac{E_{fd}(s)}{V_R(s)} = \frac{1}{(K_E + s\tau_E)}.$$

where τ_E is the time constant of the exciter, K_E the gain of the exciter, V_E the armature emf of an amplidyne, E_{fd} the armature emf of the main exciter.

(52) What is the function of a stabilizing transformer?

The excitation system has dynamic response which is prone to excessive overshoot and stability problems. These problems are overcome by adding a stabilizing transformer to the excitation system.

(53) What is the function of a terminal voltage transducer represented in a functional block diagram of an excitation system?

The terminal voltage of the alternator is sensed and rectified into a proportionate DC signal by using a terminal voltage transducer.

(54) What is the function of load compensation block in an excitation system block diagram?

Load compensation synthesizes a voltage that differs from the terminal voltage by the voltage drop in the impedance.

(55) What is the significance of a saturation function S_E, which is represented in an excitation system functional block diagram?

The saturation function $S_E = f(E_{fd})$ represents the saturation of the exciter.

(56) What are the main classifications of an exciter?

Rotating-type and static-type exciters are the main classifications of an exciter.

(57) What is the function of an excitation system stabilizer transient gain regulator block?

To increase the stability region of operation of the excitation system and also to permit a higher regulating gain.

(58) How is the excitation system stabilizer realized?

An excitation system stabilizer is realized by an ideal transformer whose secondary is connected to a high impedance.

(59) What is the function of PSS?

During the transient disturbance, the rotor oscillations (of frequency 0.2–2 Hz) are damped out by providing the PSS.

(60) What is a potential source-controlled-type static excitations system?

In a potential source-controlled-type static excitations system, the excitation power specified is supplied through a PT connected to generator terminals.

(61) What is a static excitation system with a compound source-controlled rectifier?

The excitation power is supplied through both CTs and PTs connected to generator terminals.

(62) What is the input signal to PSS?

The input signal to PSS is derived from speed or frequency or oscillating power or a combination of these signals.

MULTIPLE-CHOICE QUESTIONS

(1) Hydraulic servomotors are used in ————— type of speed-governing systems.

(a) Mechanical–hydraulic-controlled.

(b) Electro–hydraulic-controlled.

(c) Either (a) or (b).

(d) Both (a) and (b).

(2) In hydraulic-controlled speed-governing systems, the hydraulic servomotors are used for:

(a) Positioning the valve or gate, controlling steam or water flow.

(b) Removing the valve or gate, controlling steam or water flow.

(c) Improving the water head.

(d) Improving the steam pressure and temperature.

(3) For a steam turbine, the mechanical–hydraulic-controlled speed-governing system consists of which of the following?

(i) A speed governor.

(ii) A speed relay.

(iii) A hydraulic servomotor.

(iv) Governor-controlled valves

   (a) (i) and (iv)   (b) (iii) and (iv)

   (c) All except (ii) (d) All of these.

(4) In the approximate non-linear mathematical model of a mechanical–hydraulic-controlled speed-governing system, the term K_G represents the gain of speed-governor system, which is ___________.

(a) The regulation or droop of characteristics.

(b) The reciprocal of regulation or droop of characteristics.

(c) Not the function of regulation or droop.

(d) None of these.

(5) The gain of a speed-governor K_G represents:

(a) A position of an assumed linear instantaneous indication of a speed produced by the speed governor.

(b) The regulation or droop of speed-governor characteristics.

(c) The governor speed-changer position.

(d) None of these.

(6) The governor speed-changer position provides the relay signal and is determined by:

(a) A system of speed governing.

(b) A system of automatic generation control.

(c) A system of hydraulic servomotor control.

(d) None of these.

(7) The speed relay signal in mechanical–hydraulic speed-governing system represents a composite load and speed reference and is assumed ___________ over the interval of a stability study.

(a) Variable.

(b) Constant.

(c) Either (a) or (b).

(d) None of these.

(8) The speed relay in a mechanical–hydraulic speed-governing system is represented as:

(a) An integrator.

(b) A differentiator.

(c) An amplifier.

(d) None of these.

(9) The speed relay in a mechanical–hydraulic speed-governing system provides:

(a) An indirect feedback.

(b) A direct feedback.

(c) No feedback.

(d) None of these.

(10) The non-linearity property of the valve is compensated by means of providing:

(a) A linear CAM.

(b) A non-linear CAM.

(c) A speed relay.

(d) A hydraulic servomotor.

(11) A non-linear CAM is provided to compensate the non-linear property of the valve in between:

(a) The hydraulic servomotor and governor-controlled valve.

(b) The speed governor and the speed relay.

(c) The speed relay and the hydraulic servomotor.

(d) The hydraulic servomotor and the speed governor.

(12) The hydraulic servomotor:

(a) Controls the moments of valves.

(b) Is represented as an integrator in the approximate linear model.

(c) For which the rate timing may occur for large- and rapid-speed deviations.

(d) All of these.

(13) The hydraulic servomotor control provides:

(a) A direct feedback.

(b) An indirect feedback.

(c) No feedback.

(d) None of these.

(14) The position limits of the hydraulic servomotor that are indicated correspond to:

(a) Wide-open valves.

(b) The setting of a load limiter.

(c) Either (a) or (b).

(d) None of these.

(15) The non-linearities present in speed control mechanism are not neglected for:

(a) Rate limits of servomotor.

(b) Position limits of valve.

(c) Study of power system components.

(d) Both (a) and (b).

(16) The mechanical components in the lower power portions are replaced by the static electronic circuits in:

(a) Mechanical–hydraulic speed-governing system.

(b) Electro–hydraulic speed-governing system.

(c) Both (a) and (b).

(d) None of these.

(17) The flexibility is more in which type of speed-governing system?

(a) Mechanical–hydraulic speed-governing system.

(b) Electro–hydraulic speed-governing system.

(c) Both (a) and (b).

(d) None of these.

(18) The linearity of the electro–hydraulic-controlled type speed-governing system can be improved by means of:

(a) Providing speed relay and hydraulic servomotor.

(b) Providing excitation control signals.

(c) Providing feedback loops of steam flow and servomotors.

(d) Providing linear components.

(19) For hydraulic turbines, the mechanical–hydraulic-controlled speed-governing system consists of:

(i) A speed governor.

(ii) A pilot valve and servomotor.

(iii) A distribution valve and gate servomotor.

(iv) Governor-controlled gates

    (a) (i) and (ii)   (b) (i) and (iii)

    (c) (i) and (iv)   (d) All of these.

(20) The dashpot feedback system is required to achieve the stable performance of a speed-governing system of:

    (a) Hydro-turbines.

    (b) Steam turbines.

    (c) Both (a) and (b).

    (d) Either (a) or (b).

(21) In the speed-governing systems, the gate servomotor rate is limited for:

    (a) Large, rapid-speed excursions.

    (b) Extremes of gate opening.

    (c) Either (a) or (b).

    (d) Both (a) and (b).

(22) In speed-governing systems, the position limits exist corresponding to:

    (a) Large, rapid-speed excursions.

    (b) Extremes of gate opening.

    (c) Either (a) or (b).

    (d) Both (a) and (b).

(23) The speed-governing requirements for hydro-turbines are strongly influenced by the effects of:

    (a) The position of penstock.

    (b) Head of water.

    (c) Water inertia.

    (d) All of these.

(24) According to the behavior, the output variable of the transfer function is not limited and is free to vary in the case of:

    (a) Wind-up limiter.

    (b) Non-wind-up limiter.

    (c) Rate limiter.

    (d) Position limiter.

(25) A separate block is needed to represent in the block diagram in the case of:

    (a) Wind-up limiter.

    (b) Non-wind-up limiter.

    (c) Rate limiter.

    (d) Position limiter.

(26) According to the behavior, the output variable of the transfer function is limited and no separate block is needed for the limiter in the case of:

    (a) Wind-up limiter.

    (b) Non-wind-up limiter.

    (c) Rate limiter.

    (d) Position limiter.

(27) Generally, the integrator blocks have:

    (a) Wind-up limiter.

    (b) Non-wind-up limiter.

    (c) Rate limiter.

    (d) Position limiter.

(28) While modeling steam generators, the following equipment is ignored at every instant due to their slow operations:

    (a) The boiler controls equipment.

    (b) Online frequency controls equipment.

    (c) Both (a) and (b).

    (d) Either (a) or (b).

(29) Primary governing system of a reheat system unit responds to the:

    (a) Speed of the main shaft.

    (b) Frequency of the turbine.

    (c) Either (a) or (b).

    (d) Both (a) and (b).

(30) The secondary governing system of a reheat unit responds to:

    (a) Speed of the main shaft.

    (b) Frequency of the turbine.

    (c) Either (a) or (b).

    (d) Both (a) and (b).

(31) The governing system that controls either main governor valve or throttle blades is:

    (a) Primary governing system.

    (b) Secondary governing system.

    (c) Anticipatory governing system.

    (d) None of these.

(32) The governing system that controls the interceptor valves between the HP state and the reheater is:

(a) Primary governing system.

(b) Secondary governing system.

(c) Anticipatory governing system.

(d) None of these.

(33) Anticipatory governing system of a reheat unit responds to:

(a) Speed of the main shaft.

(b) Frequency of the turbine.

(c) The accelerating power of the unit.

(d) None of these.

(34) The anticipatory governing system is usually set not to operate if:

(a) The governor output is more than a certain value.

(b) The turbine mechanical power output is less than a certain value.

(c) Either (a) or (b).

(d) Both (a) and (b).

(35) The speed-governing system in which, on sensing the loss of a large percentage of its load, the emergency stop valves are closed very rapidly to prevent dangerous overspeed:

(a) Primary governing system.

(b) Secondary governing system.

(c) Anticipatory governing system.

(d) None of these.

(36) In an anticipatory speed-governing system, the emergency stop valves are located very adjacent to:

(a) The servomotor.

(b) The main governing valves.

(c) The speed relays.

(d) All of these.

(37) Emergency overspeed governor will trip when:

(a) The velocity of the shaft exceeds a pre-set value.

(b) The frequency of the system is maintained constant.

(c) The overall efficiency of the speed-governing system and turbine reduces.

(d) The velocity of shaft exceeds the frequency.

(38) The interceptor governing system of a reheat unit will act as:

(a) Primary governing system.

(b) Secondary governing system.

(c) Anticipatory governing system.

(d) None of these.

(39) The simple modeling of a hydro-turbine unit is based on the assumption according to Kirchmayer:

(a) Neglecting dead band, delays.

(b) Neglecting the variation in head of water.

(c) Both (a) and (b).

(d) None of these.

(40) In modeling a hydro-turbine unit, which of the following is important?

(a) Representation of the water column criteria.

(b) Representation of the water head.

(c) Representation of the speed.

(d) All of these.

(41) When the excitation increases, the torque angle 'δ'____________.

(a) Increases.

(b) Reduces.

(c) No effect.

(d) None of these.

(42) When the excitation increases, the current ____________ and the power angle ____________.

(a) Increases, increases.

(b) Decreases, increases.

(c) Increases, decreases.

(d) Decreases, decreases.

(43) When the excitation increases, the power angle becomes ____________ with respect to terminal voltage.

(a) More leading.

(b) More lagging.

(c) Zero.

d) 90°.

(44) With an increase in excitation,

(a) The torque angle δ reduces.

(b) The stiffness of the machine increases.

(c) The coupling of generator and rotating armature flux becomes more tight.

(d) All of these.

(45) With an increase in excitation,

(a) The stability of the machine will improve.

(b) The stability of the machine will decrease.

(c) There is no effect on the stability of the machine.

(d) None of these.

(46) The increase in excitation is limited by which of the following factors?

(i) Resistance of field circuit.

(ii) Saturation of magnetic circuit.

(iii) Rotor heating.

(iv) Maximum output voltage of excitation

    (a) (i) and (ii)   (b) All except (iii)

    (c) All except (i)   (d) All of these.

(47) The excitation system consists of:

(a) An exciter.

(b) An AVR.

(c) Both (a) and (b).

(d) None of these.

(48) In common excitation bus method,

(a) Two or more number of exciters feed a common bus.

(b) Each generator is fed from its own exciter.

(c) Either (a) or (b).

(d) None of these.

(49) In the individual excitation method,

(a) Two or more number of exciters feed a common bus.

(b) Each generator is fed from its own exciter.

(c) Either (a) or (b).

(d) None of these.

(50) Unit-exciter method is nothing but:

(a) Common excitation bus method.

(b) Centralized excitation method.

(c) Individual excitation method.

(d) All of these.

(51) Which of the following methods is more preferable?

(a) Common excitation bus method.

(b) Centralized excitation method.

(c) Individual excitation method.

(d) All of these.

(52) The merits of a unit exciter are:

(a) Simplicity.

(b) Less maintenance.

(c) Less ohmic loss and high reliability.

(d) All of these.

(53) The function of exciter is the structure of excitation:

(a) To supply terminal voltage to the rotor circuit.

(b) To supply current to the rotor field circuit of a synchronous generator.

(c) To supply current to the stator circuit of a synchronous generator.

(d) All of these.

(54) A pilot exciter is:

(a) A level compound small DC generator.

(b) A small servotype synchronous generator.

(c) A main synchronous generator.

(d) A main exciter.

(55) The function of a pilot exciter is:

(a) To supply current to the rotor circuit.

(b) To maintain constant voltage excitation for the main exciter.

(c) To supply variable excitation for the main exciter.

(d) None of these.

(56) The main exciter is:

(a) A level compound small generator.

(b) A main synchronous generator.

(c) A separately excited exciter.

(d) A pilot executer.

(57) The fast response of the exciter is obtained due to:

(a) Several series-connected field windings.

(b) Several parallel-connected field windings.

(c) Combination of series-connected and parallel-connected field windings.

(d) None of these.

(58) The function of an amplidyne is:

(a) To provide the constant excitation to a synchronous generator.

(b) To provide large currents to the field windings of a main exciter.

(c) To provide supply to the synchronous machine.

(d) None of these.

(59) Which of the following is correct regarding the amplidyne?

(a) Amplidyne is a high response cross-field generator.

(b) Amplidyne has a number of control windings supplied from pilot exciter.

(c) Amplidyne has a number of feedback circuits of an AVR and magnetic amplifier.

(d) All of these.

(60) Which is treated as the heart of the excited system?

(a) Main exciter.

(b) Pilot exciter.

(c) Rotor field exciter.

(d) AVR.

(61) Excitation field control system is the system that consists of:

(a) Exciter and regulator.

(b) Exciter and field system.

(c) Exciter regulator and synchronous generator.

(d) None of these.

(62) The drawback of DC excitation system is:

(a) More complexity.

(b) Larger time constants.

(c) Less reliability.

(d) More noise level.

(e) None of these.

(63) The main advantage of an AC excitation system is:

(a) Moving contacts are completely eliminated.

(b) Smooth operation.

(c) Maintenance-free operation.

(d) All of these.

(64) The brushless excitation system is:

(a) DC excitation system.

(b) AC excitation system.

(c) Static excitation system.

(d) None of these.

(65) No commutation problems occur in:

(a) DC excitation system.

(b) AC excitation system.

(c) Static excitation system.

(d) None of these.

(66) The advantage of a static excitation system is:

(a) Noise-free operation.

(b) High reliability due to more reliable static devices.

(c) Overall length of the generator shaft is reduced, which simplifies the torsion and critical speed problems.

(d) All of these.

(67) In the first stage of amplification of operation of amplitude,

(i) A small amount of field m.m.f. results in large q-axis current.

(ii) The voltage will be developed across d-axis brushes due to q-axis current.

(iii) The brushes in q-axis are short-circuited

(a) (i) and (iii)   (b) (ii) and (iii)

(c) Only (i)   (d) All of these.

(68) In the second stage of amplification of operation of amplidyne, which of the following occurs?

(i) A small amount of field m.m.f. results in large q-axis current.

(ii) The voltage will be developed across d-axis brushes due to q-axis current.

(iii) The d-axis brushes are connected to load along with a compensating winding

(a) (i) and (iii)   (b) (ii) and (iii)

(c) Only (i)   (d) All of these.

(69) The main principle of a magnetic amplifier is:

(a) The magnetic core gets saturated when large AC current flows through control winding and results in the decrement of permeability.

(b) Reactance of AC coils increases due to the decrement in permeability.

(c) The controlling of a large output current by means of a small control current.

(d) None of these.

(70) The amplidyne is connected in ___________ with the shunt field of the main exciter.

(a) Series.

(b) Parallel.

(c) Series for some time and parallel for some time.

(d) None of these.

(71) Advantage of a stabilizing transformer is:

(a) The problem of dynamic response that is prone to exclusive overshoot is overcome.

(b) Stability problems are overcome.

(c) Both (a) and (b).

(d) None of these.

(72) Most of the excitation systems are modeled based on:

(a) AC excitation system.

(b) DC excitation system.

(c) Static excitation system.

(d) IEEE type-1 excitation system.

(73) The objective of load compensation is:

(a) Sharing of reactive power among the units, which are bussed together with zero impedance between them.

(b) When the units are operating in parallel through a unit transformer, it is desirable to regulate the voltage at a point beyond the machine terminals to compensate for a portion of transformer impedance.

(c) Both (a) and (b).

(d) None of these.

(74) The function of terminal voltage transducer is:

(a) To sense the terminal voltage of an alternator and rectify it into a proportional DC voltage.

(b) To synchronize a voltage that differs from the terminal voltage by the voltage drop.

(c) Both (a) and (b).

(d) None of these.

(75) The function of an excitation system stabilizer and transient gain regulator is to ___________ the stability and permit ___________ regulator gains

(a) Increase, lower.

(b) Decrease, lower.

(c) Increase, higher.

(d) Decrease, higher.

(76) Excitation system stabilizer is realized by:

(a) A practical transformer whose secondary is connected to a high impedance.

(b) A practical transformer whose secondary is connected to a low impedance.

(c) An ideal transformer whose secondary is connected to a low impedance.

(d) An ideal transformer whose secondary is connected to a high impedance.

(77) Reducing the transient gain or gain at higher frequency, thereby minimizing the negative contribution of the regulator to system damping is the main aim of:

(a) Power system stabilizer (PSS).

(b) Excitation system stabilizer (ESS).

(c) Transient gain regulator (TGR).

(d) Main exciter.

(78) During the transient disturbance, the rotor oscillations of frequency 0.2–2 Hz are damped out by providing ___________.

(a) Power system stabilizer.

(b) Excitation system stabilizer.

(c) Transient gain regulator.

(d) Main exciter.

(79) The input signal to power system stabilizer is derived from:

(i) Speed.

(ii) Frequency.

(iii) Accelerating power.

(iv) Combination of signals of (i), (ii), and (iii)

(a) (i) and (iii)   (b) (ii) and (iii)

(c) Only (iv)   (d) Any of the above.

(80) In a potential source-controlled rectifier type of static excitation system,

(a) The excitation power is supplied through a PT connected to generator terminals.

(b) The excitation power is supplied through both PT and CT connected to generator terminals.

(c) The excitation power is supplied without connecting PT and CT connected to generator terminals.

(d) None of these.

(81) In a compound source-controlled rectifier type of static excitation system,

(a) The excitation power is supplied through a PT connected to generator terminals.

(b) The excitation power is supplied through both PT and CT connected to generator terminals.

(c) The excitation power is supplied without connecting PT and CT connected to generator terminals.

(d) None of these.

(82) The output of a power system stabilizer is:

(a) V_s is added to the terminal voltage.

(b) V_s is added to the terminal voltage error signal.

(c) V_s is subtracted from the terminal voltage error signal.

(d) None of these.

REVIEW QUESTIONS

(1) Discuss the mechanical–hydraulic control and electro–hydraulic control speed-governing system of steam turbines.

(2) Discuss the mechanical–hydraulic control and electro–hydraulic control speed-governing system of hydraulic turbines.

(3) Explain the different types of limiters and their role in speed-governing system modeling.

(4) Explain the effect of varying excitation of a synchronous generator.

(5) Explain the methods of providing excitation systems.

(6) Explain the various components of a block diagram representation of a general excitation system.

(7) Explain the classification of excitation systems.

(8) Derive the transfer function of an overall excitation system.

13 Power System Security and State Estimation

OBJECTIVES

After reading this chapter, you should be able to:

- know the meaning of security control system and its importance
- discuss the applications of planning of security analysis
- develop the mathematical modeling of security-constrained optimization problem
- study the various techniques used for steady-state and transient-state security analysis
- know the need of state estimation in power systems
- discuss the applications of state estimation

13.1 INTRODUCTION

The concept of control is fundamental to the proper functioning of any system. Irrespective of whether it is an engineering system or an economic system or a social system, it is essential to exert some kind of control, such as quality control, inventory control, or population control, to achieve certain objectives like better quality of output or better economics, etc. It is only natural that power system, which is one of the most complex man-made systems, calls for the implementation of a number of controls for satisfactory operation. Power system control has gone through a lot of changes over the past three decades. Beginning with simple governor control at the machine level, it has now grown into a sophisticated multi-level control needing, a real-time computer process, and system-wide instrumentation.

The ultimate objective of power system control is to maintain continuous electric supply of acceptable quality by taking suitable measures against the various disturbances that occur in the system. These disturbances can be classified into two major heads, namely, small-scale disturbances and large-scale disturbances. Small-scale disturbances comprise slowly varying small-magnitude changes occurring in the active and reactive demands of the system. Large-scale disturbances are sudden, large-magnitude changes in system operating conditions such as faults on transmission network, tripping of a large generating unit or sudden connection or removal of large blocks of demand. While the small-scale disturbances can be overcome by regulatory controls using governors and exciters, the large-scale disturbances can only be overcome by proper planning and adopting emergency switching controls.

13.2 THE CONCEPT OF SYSTEM SECURITY

'Security control' or a 'security control system' may be defined as a system of integrated automatic and manual controls for the maintenance of electric power service under all conditions of operation. It may be noted from this definition that security control is a significant departure from the conventional generation control or supervisory control systems. First, the proper integration of all the necessary automatic and manual control functions requires a total systems approach with the human operator being an integral part of the control system design. Second, the mission of security control is all-encompassing, recognizing that control decisions by the main computer system must be made not only when the power system is operating normally but also when it is operating under abnormal conditions. As power systems have become more tightly coupled, the problem of making the operating decisions under varying conditions has become extremely difficult.

To keep the system always secure, it is necessary to perform a number of security-related studies, which can be grouped into three major areas, namely, long-term planning, operational planning, and on-line operation.

Certain significant applications in each of these areas are listed as follows:

13.2.1 Long-term planning

- Evaluation of generation capacity requirements.
- Evaluation of interconnected system power exchange capabilities.
- Evaluation of transmission system adequacy.

13.2.2 Operational planning

- Determination of spinning reserve requirements in the unit commitment process.
- Scheduling of hourly generation as well as interchange scheduling among neighboring systems.
- Outage dispatching of transmission lines and transformers for maintenance and system operation.

13.2.3 On-line operation

- Monitoring and estimation of the operating state of the system.
- Evaluation of steady-state, transient, and dynamic securities.
- Quantitative assessment of security indices.
- Security enhancement through constrained optimization.

13.3 SECURITY ANALYSIS

Security analysis is the determination of the a security of the system based on a next-contingency set. This involves verifying the existence and normalcy of the post-contingency states. If all the post-contingency states exist and are found to be normal, the state is secure. On the other hand, the non-existence of even one of the post-contingency states or emergency nature of an existing post-contingency state indicates that the current state is insecure.

Though it may be theoretically possible to conduct a security analysis for both the steady-state emergency and dynamic instability, the trend has been to have a separate analysis for each of these two types of emergency. The main reason for this is the extreme difficulty in implementing a dynamic-security analysis with the present methods of stability analysis. On the other hand, for the steady-state security (SSS) analysis, several approaches are possible and are in use. Basically, these approaches start with a knowledge of the present state of the system as obtained from the security monitoring function. The system is then tested for various next-contingencies by, in effect, solving for the changes in the system conditions for a given contingency and checking the new values against the operating constraints.

'Transient security analysis' refers to an on-line procedure whose objective is to determine whether or not a postulated disturbance will cause transient instability of the power system. A transient instability condition implies the loss of synchronism or oscillations, which increase in amplitude, leading to cascading outages and subsequent system break-up. As against the SSS analysis where the next-contingencies to be considered are only outages of lines/transformers or generators, in the case of transient security analysis, a much wider range of possible contingencies must be considered such as:

- Single-phase, two-phase, and three-phase fault conditions.
- Faults with or without reclosing.
- Proper operation or failure of protective relays.
- Circuit breaker operation or failure to clear the fault.
- Loss of generation or a large block of load.

Direct methods for transient security analysis have been suggested, but none of these have yet passed the experimental stage. The current industry practice is to express the security constraints associated with transient stability as steady-state operating limits on power transfer or phase-angle difference across selected transmission lines. The general approach for imposing transient security constraints on an operating power system consists of the following steps:

(i) Perform extensive off-line transient stability studies for a range of operating conditions and postulated contingencies.

(ii) On the basis of these studies and pre-determined reliability criteria (e.g., the system must withstand three-phase faults with normal clearing), establish steady-state operating limits for line power flows or line phase-angle differences.

(iii) Operate the system within the constraints determined in the previous step.

Transient stability analysis of a large system, though done using a crude machine model of constant emf behind transient reactance, requires quite a lot of computer time because of the large number of differential algebraic equations involved. When one goes for a model of higher degree of complexity, which may include an excitation system model, detailed generator electrical model, governor control model, and turbine model and if one considers simulation for each of the contingencies in the next-contingency set, then the computation time needed becomes prohibitive. In recent years, considerable amount of research has been devoted to developing efficient and effective techniques for on-line transient stability analysis. The suggested techniques can be classified according to the following basic approaches:

(i) Digital simulation.

(ii) Hybrid computer simulation.

(iii) Lyapunov methods.

(iv) Pattern recognition.

13.3.1 Digital simulation

Digital simulation techniques, though very adaptable and flexible, are slow in speed and hence it appears that they will use on-line analysis in a complementary role with faster, but possibly less accurate, techniques. Recent advances in digital methods have been directed toward implicit integration techniques and the simultaneous solution of the whole set of differential-algebraic equations using sparsity techniques. It is claimed that a method like 'variable integration step transient analysis' (VISTA) can reduce the simulation time as much as five times as compared with the conventional explicit integration methods.

13.3.2 Hybrid computer simulation

Hybrid computer simulation of a transient stability problem could be made many times faster than real time. Although hybrid computers have been able to provide matchless solution speeds, their application to power system operation is limited by disadvantages such as very large initial investment in the case of large systems; applicable only to a limited number of select on-line computation functions and limited flexibility due to the normal patching of the analog computers.

13.3.3 Lyapunov methods

The second method of Lyapunov has received a considerable amount of attention for determining power system transient stability, particularly for on-line application. This method involves the derivation of a scalar Lyapunov $V(X)$, where X is the dynamic-state vector of the system set of differential equations, which has the following properties:

$$V(0) = 0, \quad \text{i.e., } X = 0 \text{ is the equilibrium state}$$

$$V(X) > 0, \quad X \in \Omega, \quad X \neq 0$$

$$V(X) \leq 0, \quad X \in \Omega$$

where Ω is a region around the stable point $X = 0$, which is called the region of stability. While this method can offer considerable gain in computational speed, the drawbacks of this method as follows are:

- Too conservative, especially for systems with more than three or four machines.
- Computational requirements have made the study of large-scale power systems infeasible.
- Requires a simplified system model.

The last limitation is not as severe as the first two since much useful information can be obtained from analytical studies with the simplified models. Very recent developments indicate that a breakthrough in overcoming the first two problems is now possible. First, an efficient method of calculating the unstable equilibrium point has been developed using a modified Newton–Raphson load flow. Second, there is now an increased amount of awareness as to why the second method of Lyapunov is highly conservative.

13.3.4 Pattern recognition

Pattern recognition is another approach aimed at overcoming the high computational requirements of on-line transient stability studies. A large number of off-line stability studies are performed to form a 'training set' and certain important features are selected. An on-line classifier compares the actual operating conditions with the training set and, on the basis of this comparison, classifies the existing state as either secure or insecure. This method is very appealing for on-line assessment because of its tremendous speed and the minimum on-line data that it requires.

However, the disadvantages of this method are:

(i) The accuracy of the classification method is not as good as that of the direct solution methods since it is basically an interpolation technique.

(ii) A very large number of samples (and hence simulation) may be required for the formation of an adequate training set.

(iii) It has difficulty in handling abnormal conditions, which may arise due to unusual load patterns and/or network configurations.

If the duration of the analysis to be conducted is longer than 1–3 s, the dynamics of the boilers, turbines, and other power plant components cannot be ignored. In addition to this, the dynamics of AGC and SVC should be taken into account along with the control action of impedance and under-frequency load-shedding relays. As a result, the effect of a fault-initiated disturbance may continue past the transient stability phase to the so-called long-term dynamic stability phase, which can be of the order of 10–20 min or more. The objectives of a long-term dynamic response assessment are:

(i) Evaluation of dynamic reserve response characteristics including the distribution of reserves and effect of fast-starting units.

(ii) Evaluation of emergency control strategies like load-shedding by under-frequency relays, fast valuing, dynamic braking, and others.

These objectives fall primarily under system planning, control system design, as well as post-disturbance analysis. However, the operating implications cannot be neglected, in view of the fact that serious blackouts that have occurred over the past 15–20 years were generally the result of long-term instability and sequences of cascading events.

13.4 SECURITY ENHANCEMENT

Security enhancement is a logical adjunct to security analysis and it involves on-line decisions aimed at improving (or maintaining) the level of security of a power system in operation. Security enhancement includes a collection of control actions, each aimed at the elimination of security constraint violations. These controls may be classified as:

(i) Preventive controls in the normal operating state, when on-line security analysis has detected an insecure condition with respect to a postulated next-contingency.

(ii) Correctable emergency controls (simply called 'corrective controls') in an emergency state, when an out-of-bound operating condition already exists but may be tolerated for a limited time period.

In either case, the primary objective is to find feasible and practical ways to remedy a potentially dangerous operating condition once the security analysis program reveals the existence of such a condition.

Security enhancement implies the utilization of available generation and transmission capacity to improve the security of a power system. There are five generic approaches to the use of available system resources for security enhancement, namely:

(i) Manipulation of real-power flows in certain parts of the system through rescheduling of generation along with other control variables such as phase-shifter ratios.

(ii) Manipulation of reactive-power flows in the system to maintain a good 'voltage profile' through excitation control of generators along with other control variables such as shunt capacitor or reactor switching, off-nominal tap ratios of transformers, etc.

(iii) Utilizing heat capacity of components like transformers and underground cables to permit short-term overloading of certain pieces of equipment.

(iv) Changing the network topology via switching actions.

(v) Modifying the settings of protective relays or control logic.

All the five options given above involve some trade-offs between the economy and the security of power system operation. For example, generation shifting or rescheduling power transactions usually result in higher operating costs. Hence, for those preventive control actions that drastically affect the economy of operation, the operator may decide not to execute the recommended control actions until the postulated contingency actually takes place, depending on the general operating philosophy of the particular system and the nature of the predicated constraint violations.

Security-constrained optimization may be used as a convenient framework for discussing approaches to system security enhancement, especially for SSS. The constrained optimization problem of obtaining the 'best' operating condition that satisfies not only the load constraints and the operating constraints but also the security constraints may be stated as follows:

Minimize

$f(X, U)$ objective function

Subject to:

$G(X, U) = 0$, load constraints

$H(X, U) \geq 0$, operating constraints

$S(X, U) \leq 0$, security constraints

where f is a scalar-valued function.

The security constraints reflect all the operating and load constraints associated with the postulated post-contingency states; these 'logical constraints' can be rigorously formulated and expressed as a set of inequality constraints as indicated above. These functional constraints, too large in number, make the problem very complex. Two non-linear programming techniques, namely the penalty function technique and the generalized reduced gradient technique, have been identified as most suitable for solving the constrained optimization problem. For a quick on-line solution, the dual linear programming technique using the linear model as well as the successive linear programming technique using linearized models have been found to be most useful.

Only a limited amount of research has been directed at the development of control algorithms for transient security enhancement. Since on-line implementation of control algorithms to enhance system security is very difficult to achieve, one can consider the intermediate step of computing and presenting suitable security indices to the operators who will in turn take control decisions. A number of security indices, both for SSS as well as for transient security, have been proposed along with a suitable technique of obtaining them from the on-line security analysis.

13.5 SSS ANALYSIS

Though SSS analysis is only a part of the overall security assessment process, its importance should not be underestimated. The reasons for its prominence are: first, it is the only simple simulation process that can be implemented on-line. It should be noted that at this time of writing when on-line security analysis is not yet in common use, except for a few pioneering applications, the SSS analysis is either in the early stages of on-line implementation or planned for several new energy control centers in developed countries. Second, it is advantageous to know whether (or not) the post-contingency state of the system would be acceptable from steady-state considerations, even before investigating the transient and dynamic performance. Third, an approximate check of transient stability could also be incorporated by imposing on the post-contingency steady states, appropriate power flow, or other constraints derived from off-line transient stability studies. Hence, it is logical to devote more attention on the various aspects of the SSS analysis.

The objective of the SSS analysis is to determine whether, following a postulated disturbance, there exists a new steady-state operating point where the perturbed power system will settle after the post-fault dynamic oscillations have been damped out. An on-line algorithm simulates the predicted steady-state conditions for a specified set of next-contingencies and checks for operating constraint violations. If the normal system fails to pass any one of the contingency tests, it is declared to be 'steady-state insecure' and the particular contingencies with the attendant limit excursions are noted. More precisely, SSS is defined as the ability of the system to operate steady-state-wise within the specified limits of safety and supply quality following a contingency, in the time period after the fast-acting automatic control devices have restored the system load balance, but before the slow-acting controls, e.g., transformer tapings and human decisions, have responded.

13.5.1 Requirements of an SSS assessor

The SSS assessor is defined as an on-line process using real-time data for conducting SSS analysis on the current state of the system. Each contingency is solved approximately as a steady-state AC power flow problem. Except for simulated outages, the network is the same as the actual operating system and the bus power injection (defined as generation minus load) schedule corresponds to the currently estimated state of the system. The results of each solution are checked against pre-determined constraints. If a contingency causes a constraint violation or if a solution for a contingency is impossible, this information is transferred from the SSS assessor to another function in the control center in which appropriate control actions will be taken to enhance the security of the system.

The SSS assessor will be one of several interrelated programs in an automated dispatch center. The ways in which it can be integrated with other functions are not considered here. However, a simplified schematic diagram for the flow of information is shown in Fig. 13.1. The most critical input is the state vector from the state estimator. This estimate is transmitted in the form of the state vector V consisting of complex voltages at each node of the monitored system.

Other essential inputs are explained as follows.

13.5.1.1 Network data

The passive network is modeled by the bus admittance matrix $[\bar{Y}]$, which is developed from a detailed list of basic network components including transmission lines, transformers, capacitors, and reactors. It is essential to have real-time information on the status of these components at the beginning of each solution cycle. A solution cycle is the solution and checking of results for all contingent outages in a specified contingency list. The change of status of every network component is transmitted to the control computer and whenever a component is switched in or out, its effect is reflected by a change in the admittance matrix.

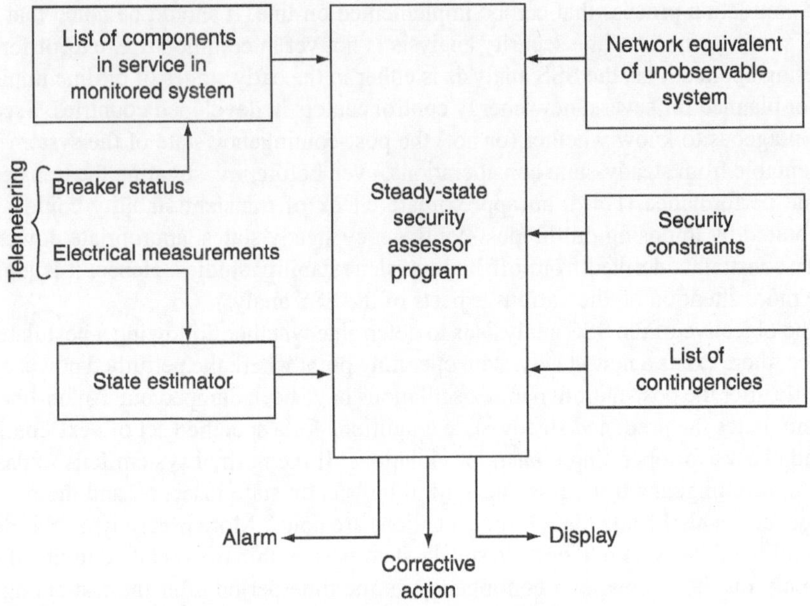

FIG. 13.1　*Flow of information in a security assessor*

13.5.1.2 *Bus power injections*

For a line outage, the injections should correspond to the actual state of the system. The injection schedule is computed once at the beginning of each solution cycle based on the network admittances and the state vector. The injection at every bus k is computed as

$$P_k + jQ_k = \overline{V}_k \sum_{m \varepsilon \alpha_k} \overline{Y}^*_{km} \overline{V}^*_m \tag{13.1}$$

where P_k and Q_k are the real and reactive powers, $\overline{V}_k$ and $\overline{V}_m$ are elements of the state vector $\overline{V}$, $\overline{Y}_{km}$ is an element of the bus admittance matrix $[\overline{Y}]$, and α_k is the set of all nodes adjacent to node k.

For physical as well as mathematical reasons, it is necessary to fix the voltage angle at the slack bus and allow the variation in losses to be supplied by the injection at this bus.

13.5.1.3 *Security constraints*

The constraints are transmission line power flows, bus voltages, and reactive limits. These constraints may originate from customer requirements, relay settings, insulation levels, equipment ratings, or other sources. Constraints can also be established by off-line simulation studies. Line flow constraints are usually expressed either in terms of maximum continuous current or power ratings (normally for shorter lines), or in terms of the allowable maximum steady-state phase-single differences between connected buses (normally for longer lines). As stated earlier, SSS constraints can be derived to suit transient stability requirements. However, these constraints are difficult to specify, since the transient stability properties of a line depend on: the generation/load pattern throughout the entire network, the precise nature of the contingency, the configuration of the post-fault system, etc.

In some systems, it may be desirable to alter the constraints according to the state of the system, with different constraints applying under different operating conditions or contingencies. Establishing appropriate constraints for the SSS assessor is an important sub-problem that requires more investigation.

13.5.1.4 *Contingency list*

For the purpose of SSS analysis, the following contingencies should be considered:

   (i)   Loss of a generating unit.

   (ii)   Sudden loss of a load.

   (iii)   Sudden change in flow in an inter-tie.

   (iv)   Outage of a transmission line.

   (v)   Outage of a transformer.

   (vi)   Outage of a shunt capacitor or reactor.

These outages can be grouped into two categories: 'network outage' and 'power outage'. A network outage involves only changes in the network admittance parameters and includes items (iv) to (vi) given above. A power outage involves only changes in bus power injections and hence includes items (i) to (iii) given above.

The usual practice in SSS analysis is to assume that the network configuration and the injection schedule at the contingency state remain the same as in the base case state except for the simulated outages. However, in dealing with power outages that involve the loss of certain generating units, the injection schedule at the contingency state should take into account the redistribution of lost generation to the remaining generators in service. This may be done with the help of a generation allocation function. Since the analysis is concerned with the new study after the outage transients have settled, the generation allocation will be determined by the natural governor characteristics of the available units in the system. However, if it is required to check the power flows immediately after the outage, then all the remaining generators in the system will share, temporarily, the lost generation in proportion to their inertias.

13.6 TRANSIENT SECURITY ANALYSIS

In recent years, a considerable amount of research has been devoted to developing efficient and effective techniques for on-line transient stability analysis. Transient stability assessment consists of determining if the system's oscillations following a short-circuit fault will cause loss of synchronism among generators. The primary physical phenomenon involved here is that of inertial interaction among the generators as governed by the transmission network and busloads. This phenomenon is of short duration (1–3 s) in general. For longer durations, the dynamics of boilers, turbines, and other power plant components cannot be ignored. The suggested techniques to solve transient stability problems can be classified according to the following basic approaches:

   (i)   Digital simulation.

   (ii)   Pattern recognition.

   (iii)   Lyapunov method.

   (iv)   Hybrid computer simulation.

13.6.1 Digital simulation

Several numerical integration approaches have been proposed and used. As in all integration schemes, the usual limiting factor is the smallest time constant of the system, which is normally caused by synchronizing oscillations. The use of implicit predictor–corrector methods has generally allowed larger step sizes while maintaining a high level of numerical stability. Normally, the transient stability program will alternate between an integration step and a load flow solution to solve the network equations. Thus, sparse matrix methods can be quite effective and useful in this context.

13.6.2 Pattern recognition

It is another approach aimed at overcoming the high computational requirements of on-line transient stability solutions. A large number of off-line stability studies are performed to form a 'training set' and certain important features are selected. An on-line classifier compares the actual operating conditions with the training set and, on the basis of this comparison, classifies the existing system as either secure or insecure. Consequently, the bulk of the computation load is transferred to the off-line studies' time frame. This method leads to the generation of a function known as a 'security function', which is used to assess the security of the system.

13.6.3 Lyapunov method

The second method of Lyapunov has received considerable attention for determining power system transient stability, particularly for on-line application. However, the results of this research have been of little practical value to date, due to three basic problems. The classical Lyapunov method yields sufficient but not necessary conditions for stability; these conditions are discussed in detail in the following sections.

13.7 STATE ESTIMATION

The state of a power system is defined in terms of the voltage magnitude and phase angle of every bus in the power system. The state estimation plays a very vital role in power system operation, monitoring and control in terms of avoiding system failures and regional blackouts. The main objective of state estimation is to obtain the best possible values of the magnitudes of bus voltages and their angles and it requires the measurement of electrical quantities, such as real and reactive-power flows in transmission lines and real and reactive-power injections at the buses.

State estimation is an available data processing scheme to find the best state vectors, using the weighted least square method to fit a scatter of data. In order to obtain a higher degree of accuracy of the solution of the state estimation technique, two modifications are introduced. First, it is recognized that the numerical values of the available data to be processes for the state estimation are generally noisy due to the presence of errors. Second, it is noted that there are a large number of variables in the system (active and reactive-power line flows), which can be measured but not utilized in the load flow analysis. Thus, the process involves imperfect measurements that are redundant and the process of system state estimation is based on a statistical criterion that estimates the true values of the state variables either to minimize or maximize the selected criterion. A commonly used criterion is that of minimizing the sum of the squares of the differences between the estimated and measured true values of a function.

All the system information is collected by the centralized automation control of power system dispatch through remote terminal units (RTUs). The RTUs sample the analog variables and convert them into a digital form. These digital signals are interrogated periodically for the latest values and are transmitted by telephone and microwave communication link to the control center.

The control center operation must depend on measurements that are incomplete, inaccurate, delayed, and unreliable. The state estimation technique is used to process all the available data and hence the best possible estimate of the true value of the system is found.

13.7.1 State estimator

It processes real-time system data, which is redundant and computes the magnitudes of bus voltages and bus voltage phase angles with the help of a computer program. The inputs to an estimator are imperfect (noisy) power system measurements. It is designed to give the best estimate of system state variables (i.e., bus voltage magnitudes and phase angles).

The state estimator detects bad or inaccurate data by using statistical techniques. For this, state estimators are designed such that they have well-defined error limits and are based on the number, types, and accuracy of measurements.

The state estimator approximates the power flows and voltages at a bus whose measurements are not available because of RTU failure or breakdown of telephone or a communication link. Under such a condition, the state estimator is required to make available a set of measurements to replace missing or defective data.

13.7.2 Static-state estimation

There are two different modes of state estimation as applied to power systems:

    (i)   Static-state estimation.

    (ii) Dynamic-state estimation.

Static-state estimation pertains to the estimation of a system state frozen at a particular point in time. Figuratively speaking, it is a snapshot of the system. In the steady-state operation of a system (e.g., the sudden opening of one of the phases of a transmission line is reflected in the power flow in the two healthy phases much lesser than the average power flow indicated by the last-state estimation), the state estimator is required to detect a change in network configuration and convey a signal indicating the change in circuit configuration and to prepare the operator for corrective action on the first data scan. On the other hand, dynamic-state estimation is a continuous process, which takes into account the dynamics of the system and gives an estimate of the system state as it evolves in time. At the present moment, most of the state estimators in power systems, which are operational, belong to the first category.

On the face of it, it may appear as if there is not much of a difference between load flow calculations and static-state estimation. But, this is a superficial point of view. In load flow studies, it is taken for granted that the date on which calculations are based are absolutely free from error. On the other hand, in state-estimation methods, accuracy of measurement on modeling errors are taken into account by ensuring redundancy of input data. This means that the number of input data 'm' on which calculations are based are much more than the number of unknown variables 'n' whose knowledge completely specifies the system. The more the redundancy, the better it is from an estimation point of view. But redundancy has a price to pay in terms of installation of additional measuring equipment and communication facilities.

13.7.3 Modeling of uncertainty

From a mathematical viewpoint, the simplest way of describing a random vector 'v' is by assigning a Gaussian distribution to it. The probability density function for 'v' is then given by

$$p(v) = \frac{1}{\left[(2\pi)^m |R|\right]^{1/2}} \exp\left[-\frac{1}{2} v\, R^{-1} v\right] \tag{13.2}$$

Here, the expected value of v is assumed to be zero and R denotes the covariance matrix of v. The random vector v represents the following errors:

(i) Instrumentation errors (meter errors, incomplete instrumentation, and bad data).

(ii) Operational uncertainties (unexpected system changes, measurement delay).

(iii) Incompleteness of the mathematical model (modeling errors, inaccuracy in network parameters).

13.7.4 Some basic facts of state estimation

There are three important quantities of interest in state estimation. They are:

(i) The variable to be estimated.

(ii) The observations.

(iii) The mathematical model showing how the observations are related to the variables of interest (which are to be estimated) and the ever-present uncertainties.

The variables to be estimated are the state variables x, the observations are represented by z, and the mathematical model is given by

$$z = h(x) + v \tag{13.3}$$

In Equation (13.3), 'h' represents a known non-linear relation connecting z and x. For pedagogical reasons, the above quantities are represented specifically as

$x_{\text{true}} = $ True value of state x

$z_{\text{actual}} = $ Actual value of observation

$v_{\text{actual}} = $ Actual value of observation uncertainty

Further, for simplicity of explanation, let us assume that the non-linear relation in Equation (13.3) is replaced by a linear relation viz.,

$$z_{\text{actual}} = h(x)_{\text{true}} + v_{\text{actual}} \tag{13.4}$$

where

$$h = \frac{\partial h(x)}{\partial x}$$

In Equation (13.4), we know that

$$\left.\begin{array}{c} z_{\text{actual}} \\ h \end{array}\right\} \text{Numerical values known}$$

$$\left.\begin{array}{c} x_{\text{true}} \\ v_{\text{actual}} \end{array}\right\} \text{Numerical values not known}$$

We note that even though x_{true} and v_{actual} are not known, the mathematical model conveys some information on their values, i.e., there is a model for their uncertainty. Now define:

$\hat{x}$: estimate of value x_{true}

The estimate $\hat{x}$ depends on the value z and the mathematical model (and the uncertainty models for x_{true} and v_{actual}). Usually, it is desirable to view the estimate $\hat{x}$ as some specified function of the observation z_{actual}. This function is called an estimator. The nature of this estimator can be determined from h and the models of x_{true} and v_{actual}. It can therefore be specified before observations are actually made. The estimator for linear systems is often a linear matrix operator W.

Thus, $\qquad \hat{x} : W \, z_{actual}$ $\hfill$ (13.5)

In general, $\hat{x}$ is not equal to x_{true}. Hence, the first problem is to choose the best estimator (the best W), which minimizes, in some sense, the error $(x_{true} - \hat{x})$. Assuming that such a W has been chosen, the second problem is to determine how close $\hat{x}$ is to x_{true}. Since the numerical value of the error $(x_{true} - \hat{x})$ is not known, the problem is to develop an uncertainty model for the same. The uncertainty in $(x_{true} - \hat{x})$ depends upon h, the uncertainty in x_{true} and v_{actual} and of course the estimator W. Hence, in general terms, the basic estimation problem involves the following steps:

(i) Find the estimator W

z_{actual} ⟶ | Estimator | ⟶ Estimate $\hat{x}$

such that $\hat{x}$ is as close to x_{true} as possible.

(ii) Determine the model for the uncertainty in $(x_{ture} - \hat{x})$. This model depends on the chosen W.

There are two models for the uncertainty x_{true} and they are:

(i) *A priori* **model:** The model for x_{true}, which models the uncertainty before the observation is made.

(ii) *A posteriori* **model:** It is the model for $(x_{true} - \hat{x})$, which models the uncertainty in x_{true} after the observation has been made and processed to yield $\hat{x}$.

The choice of estimator such as W depends on the *a priori* model. The *a posteriori* model depends on which estimator W is chosen. In what is to follow, we drop the notation x_{true} and z_{actual} in favor of x and z.

There are many ways of modeling the uncertainty of x and v. Some of the more important ways are:

(i) Bayesian model $\qquad\qquad$: x and v are random vectors.

(ii) Fisher model $\qquad\qquad\quad$: x is completely unknown; 'v' is random vector.

(iii) Weighted least squares $\quad$: No models for x and v.

(iv) Unknown but bounded $\quad$: x and v are constrained to lie in specified sets.

13.7.5 Least squares estimation

Consider the relation:

$$z = h(x) + v$$

or $\quad v = [z - h(x)]$ $\hfill$ (13.6)

and from Equation (13.2), $p(v) = \dfrac{1}{\left[(2\pi)^m |R|^{1/2}\right]} \exp\left[-\frac{1}{2} v' R^{-1} v\right]$

The optimal estimate $\hat{x}$ is given by that value of x for which the scalar function of the weighted squares:

$$J = v'R^{-1}v = [z - h(x)]'R^{-1}[z - h(\mathrm{x})] \tag{13.7}$$

has a minimum value. The weighting matrix R^{-1} is the inverse of the covariance matrix of the observation noise v.

Applying the first-order necessary conditions for minimizing J, we have

$$g(x) = \frac{\partial J}{\partial x} = -2H'R^{-1}\left[z - h(x)\right] = 0 \tag{13.8}$$

The second partial derivative of J with resect to 'x' viz., $\dfrac{\partial^2 J}{\partial x^2}$ is a matrix known as the *Hessian matrix* and is denoted here by $G(x)$:

$$\therefore G(x) = \frac{\partial^2 J}{\partial x^2} \tag{13.9}$$

The second-order sufficiency condition demands that $G(x)$ be positive, definite at the minimum.

As usual in such problems, we follow the iterative procedure to successively close in on the minimum point $\hat{x}$, which in this case, is the least square estimate.

Therefore, assume the iterative form:

$$x_{k+1} = x_k - A_k g(x_k) \tag{13.10}$$

As k tends to infinity, hopefully $x_k \to x_{k+1}$ and $A_k g(x_k) \to 0$, which implies for non-singular A_k that $g(x_k) = 0$. This is precisely the condition to be satisfied by Equation (13.8) and hence the desired result is obtained.

There are several methods to arrive at the matrix A_k. A_k is a scalar multiple of unit matrix in the steepest descent method; it is the inverse of the Hessian matrix $G(x_k)$ in Newton's method. It is possible to choose A_k by taking Taylor's series expansion of $h(x)$ about a initial point x_0:

$$\text{i.e.,} \quad h(x) = h(x_0) + h(x_0)\,(x - x_0) + \text{higher order terms} \tag{13.11}$$

Substituting this approximate value from Equation (13.11) after neglecting higher order terms in the objective function J given by Equation (13.7), we get

$$J_1 = [z - h(\mathrm{x}_0) - H(x_0)\,(x - x_0)]' \quad R^{-1}\,[z - h(x_0) - H(x_0)\,(x - x_0)]$$

Hence, $\dfrac{\partial J_1}{\partial x} = -2H'R^{-1}\left[z - h(x_0) - h(x - x_0)\right] \tag{13.12}$

$$H'\,R^{-1}\,[z - h(x_0) - H\,\Delta x]$$

where $\quad \Delta x = (x - x_0)$

Using the optimality condition $\dfrac{\partial J_1}{\partial_x} = 0$, we get

$$h' \, R^{-1} \, [z - h(x_0) - h \, \Delta x] = 0$$

Hence, $\Delta x = [H' \, R^{-1} \, H]^{-1} \, H' R^{-1} \, [z - h(x_0)]$ (13.13)

The vector $x = (x_0 + \Delta x)$ yields the absolute minimum of J_1, but does not yield the minimum for the function J. This calls for further iterations till the value $|x_k - x_{k+1}|$ is within prescribed bounds.

Specifically, $x_{k+1} = x_k + \Delta x_k$

$$= x_k \, [H'R^{-1} \, H^{-1}] \, H'R^- \, [z - h(x_k)] + x_k \qquad (13.14)$$

But by Equation (13.10), $H'R^{-1}\left[z - h(x_k)\right] = -\dfrac{1}{2} g(x)$

We may also identify $\dfrac{1}{2} H'R^{-1}$ with A_k of Equation (13.10). Hence, $x_{k+1} = x_k - A_k g(x_k)$, which is the general form originally postulated.

To start with, we assume a suitable value for x_0. This may be obtained either from a previous load flow study or may be arbitrarily chosen, e.g., choose $V_i = e_i + j f_i$ with $e_i = 1$ and $f_i = 0$ for all i ranging from 1 to N. The algorithm given by Equation (13.12) is not an easily implemented table for the following two reasons:

(i) The Jacobian H has to be evaluated for every iteration.

(ii) Each iteration requires a matrix inversion.

For example, considerable simplification may be achieved if the matrix $[H'R^{-1}H]^{-1}$ of Equation (13.14) is evaluated only once for the initial state x_0.

Let $\left[H'R^{-1}H\right]^{-1}_{x-x_0} = P_0$ (13.15)

Then Equation (13.14) becomes

$$x_{k+1} = x_k + P_0 \, H_0 \, R^{-1} \, [z - h(x_k)] \qquad (13.16)$$

This simplification, no doubt, reduces the convergence speed as compared to Equation (13.14) but this is offset by the greatly reduced computing time.

13.7.6 Applications of state estimation

Static-state estimation may be successfully used in estimating the status of the circuit breakers and other switches in the system. In a complex power system, the network topology continuously changes. The data regarding the wrong information of switch positions may be easily checked by comparing estimation runs obtained at different instants. It is also possible to decide on the quantum of additional instrumentation by merely comparing the minimum values of the objective function $J(x)$ for different instrumentation configurations, the uses to which state estimation may be:

(i) Data processing and display [bad data detection, sampling rate].

(ii) Security monitoring [overload limits, rescheduling, switching, and load shedding].

(iii) Optimal control [load frequency control (LFC), economic load dispatch].

KEY NOTES

- 'Security control' or a 'security control system' may be defined as a system of integrated automatic and manual controls for the maintenance of electric power service under all conditions of operation.

- To keep the system always secure, it is necessary to perform a number of security-related studies, which can be grouped into three major areas, namely: long-term planning, operational planning, and on-line operation.

- Security analysis is the determination of the security of the system based on a next-contingency set. This involves verifying the existence and normalcy of the post-contingency states.

- The possible contingencies considered in transient security analysis are:

  (i) Single-phase, two-phase, and three-phase fault conditions.

  (ii) Faults with or without reclosing.

  (iii) Proper operation or failure of protective relays.

  (iv) Circuit breaker operation or failure to clear the fault.

  (v) Loss of generation or a large block of load.

- Transient stability analysis techniques are based on:

  (i) Digital simulation.

  (ii) Hybrid computer simulation.

  (iii) Lyapunov methods.

  (iv) Pattern recognition.

- The objective of an SSS analysis is to determine whether, following a postulated disturbance, there exists a new steady-state operating point where the perturbed power system will settle after the post-fault dynamic oscillations have been damped out.

- The main objective of state estimation is to obtain the best possible values of the magnitudes of bus voltages and their angles and it requires the measurement of electrical quantities, such as real and reactive-power flows in transmission lines and real and reactive-power injections at the buses.

- Functions of a state estimator are:

  (i) It processes real-time system data, which are redundant and compute the magnitudes of bus voltages and bus voltage phase angles with the help of a computer program.

  (ii) It detects bad or inaccurate data by using statistical techniques.

- The applications of state estimation are:

  (i) Data processing and display.

  (ii) Security monitoring.

  (iii) Optimal control.

SHORT QUESTIONS AND ANSWERS

(1) How is the security control system defined?

'Security control' or a 'security control system' may be defined as a system of integrated automatic and manual controls for the maintenance of electric power service under all conditions of operation.

(2) How is the security control considered as a significance departure from conventional generation control or supervisory control?

First, the proper integration of all the necessary automatic and manual control functions requires a total systems approach with the human operator being an integral part of the control system design. Second, the mission of security control is all-encompassing, recognizing that control decisions by the human computer system must be made not only when the power system is operating normally but also when it is operating under abnormal conditions.

(3) **What are the three major areas of security-related studies?**

To keep the system always secure, it is necessary to perform a number of security-related studies, which can be grouped into three major areas, namely long-term planning, operational planning, and on-line operation.

(4) **What are the applications of long-term planning?**

The applications of long-term planning are:

- Evaluation of generation capacity requirements
- Evaluation of interconnected system power exchange capabilities.
- Evaluation of transmission system adequacy.

(5) **What are the applications of operational planning?**

The applications of operational planning are:

- Determination of spinning reserve requirements in the unit commitment process.
- Scheduling of hourly generation as well as interchange scheduling among neighboring systems.
- Outage dispatching of transmission lines and transformers for maintenance and system operation.

(6) **What are the applications of on-line planning?**

The applications of on-line planning are as follows:

- Monitoring and estimation of the operating state of the system.
- Evaluation of steady-state, transient, and dynamic securities.
- Quantitative assessment of security indices.
- Security enhancement through constrained optimization.

(7) **What is security analysis?**

Security analysis is the determination of the security of the system based on a next-contingency set. This involves verifying the existence and normalcy of the post-contingency states.

(8) **What indicates the insecurity of a current state?**

The non-existence of even one of the post-contingency states or emergency nature of an existing post-contingency state indicates that the current state is insecure.

(9) **What is the objective of transient security analysis?**

'Transient security analysis' refers to an on-line procedure whose objective is to determine whether or not a postulated disturbance will cause transient instability of the power system.

(10) **What are the possible contingencies considered in transient security analysis?**

The possible contingencies considered in transient security analysis are:

- Single-phase, two-phase, and three-phase fault conditions.
- Faults with or without reclosing.
- Proper operation or failure of protective relays.
- Circuit breaker operation or failure to clear the fault.
- Loss of generation or a large block of load.

(11) **What are the steps of general approach for importing transient security constraints on an operating power system?**

The general approach for imposing transient security constraints on an operating power system consists of the following steps:

(i) Perform extensive off-line transient stability studies for a range of operating conditions and postulated contingencies.

(ii) On the basis of these studies and pre-determined reliability criteria (e.g., the system must withstand three-phase faults with normal clearing), establish steady-state operating limits for line power flows or line phase-angle differences.

(iii) Operate the system within the constraints determined in the previous step.

(12) What are the suggested techniques to be carried out in the transient stability analysis?

The suggested techniques can be classified according to the following basic approaches:

(i) Digital simulation.

(ii) Hybrid computer simulation.

(iii) Lyapunov methods.

(iv) Pattern recognition.

(13) What is security enhancement?

Security enhancement is a logical adjunct to security analysis and it involves on-line decisions aimed at improving (or maintaining) the level of security of a power system in operation. Security enhancement includes a collection of control actions each aimed at the elimination of security constraint violations.

(14) What are the two controls used for security enhancement?

The controls used for security enhancement are classified as:

(i) Preventive controls in the normal operating state, when on-line security analysis has detected an insecure condition with respect to a postulated next-contingency.

(ii) Correctable emergency controls (simply called 'corrective controls') in an emergency state, when an out-of-bound operating condition already exists but may be tolerated for a limited time period.

(15) What are the techniques used for solving the security-constrained optimization problem?

Two non-linear programming techniques, namely the penalty function technique and the generalized reduced gradient technique have been identified as the most suitable ones for solving the constrained optimization problem. For a quick on-line solution, the dual linear programming technique using linear model as well as the successive linear programming technique using linearized models have been found to be most useful.

(16) Define SSS.

SSS is defined as the ability of the system to operate steady-state-wise within the specified limits of safety and supply quality following a contingency, in the time period after the fast-acting automatic control devices have restored the system load balance, but before the slow-acting controls, e.g., transformer tapings and human decisions, have responded.

(17) What are the objectives of SSS analysis?

The objective of SSS analysis is to determine whether, following a postulated disturbance, there exists a new steady-state operating point where the perturbed power system will settle after the post-fault dynamic oscillations have been damped out.

(18) What are the security constraints?

The constraints are transmission line power flows, bus voltages, and reactive limits.

(19) What are the contingencies that should be considered for SSS analysis?

For the purpose of SSS analysis, the following contingencies should be considered:

(i) Loss of a generating unit.

(ii) Sudden loss of a load.

(iii) Sudden change in flow in an inter-tie.

(iv) Outage of a transmission line.

(v) Outage of a transformer.

(vi) Outage of a shunt capacitor or reactor.

(20) What is the main objective of state estimation?

The main objective of state estimation is to obtain the best possible values of the magnitudes of bus voltages and their angles and it requires the measurement of electrical quantities, such as real and reactive-power flows in transmission lines and real and reactive-power injections at the buses.

(21) What are the two modifications introduced to obtain a higher degree of accuracy of the solution to the state estimation technique?

In order to obtain a higher degree of accuracy of the solution to the state estimation technique, two modifications are introduced. First, it is recognized that the numerical values of the available data to be processed for the state estimation are generally noisy due to the presence of errors. Second, it is noted that there are a large number of variables in the system (active and reactive-power line flows), which can be measured but not utilized in the load flow analysis.

(22) What is the function of a state estimator?

- It processes real-time system data, which are redundant and compute the magnitudes of bus voltages and bus voltage phase angles with the help of a computer program.

- It detects bad or inaccurate data by using statistical techniques.

(23) What do you mean by static-state and dynamic-state-estimation modes?

Static-state estimation pertains to the estimation of a system state frozen at a particular point in time.

Dynamic-state estimation is a continuous process, which takes into account the dynamics of the system and gives an estimate of the system state as it evolves in time.

(24) What are the applications of state estimation?

The applications of state estimation are:

(i)   Data processing and display.

(ii) Security monitoring.

(iii) Optimal control.

MULTIPLE-CHOICE QUESTIONS

(1) Security control system is a system of:

(a) Manual control.

(b) Integrated automatic control.

(c) Conventional generation control.

(d) Both (a) and (b).

(2) Evaluation of generation capacity requirements is a:

(a) Long-term planning of system security.

(b) Operational planning of system security.

(c) On-line operation application of system security.

(d) All of these.

(3) The operational planning of system security control includes:

(a) Spinning reserve requirements determination.

(b) Scheduling of hourly generation as well as interchange scheduling.

(c) Outage dispatching of transmission lines and transformers.

(d) All of these.

(4) The monitoring and estimation of operating state of the system and evaluation of SSS state, transient, and dynamic securities are the applications of:

(a) On-line operation of security control system.

(b) Operational planning of security control system.

(c) Long-term planning of security control system.

(d) All of these.

(5) Security analysis is the determination of the security of a system.

(a) Based on a next-contingency set.

(b) Involves verifying the existence of post-contingency states.

(c) Involves verifying the normalcy of post-contingency states.

(d) All of these.

(6) Non-existence of even one of the post-contingency states or emergency nature of an existing post-contingency state indicates:

(a) Security of current state.

(b) Security of previous state.

(c) Insecurity of current state.

(d) Insecurity of previous state.

(7) In SSS analysis, the next contingencies to be considered are:

(a) Outages of lines or transformers or generators.

(b) Faults with or without reclosing.

(c) Circuit breaker operation or failure to clear the fault.

(d) Loss of generation.

(8) Security enhancement involves:

(a) On-line decisions aimed at maintaining the level of security.

(b) A collection of control actions aimed at the elimination of security constraint violations.

(c) Failure of even one of post-contingencies.

(d) Both (a) and (b).

(9) For getting quick on-line solution to a security-constrained optimization problem, the technique used is:

(a) Dual linear programming technique using linearized model.

(b) Successive linear programming technique using linearized model.

(c) Both (a) and (b).

(d) None of these.

(10) If the normal system fails to pass any one of the contingency tests, it is declared to be:

(a) Steady-state secure.

(b) Steady-state insecure.

(c) Transient-state secure.

(d) Transient-state insecure.

(11) The SSS assessor is an on-line process using real-time data for conducting SSS analysis on:

(a) The previous state of the system.

(b) The current state of the system.

(c) The post-state of the system.

(d) All of these.

(12) A network outage involves:

(a) Only changes in the network admittance parameters.

(b) Outages of transmission line or transformer or shunt capacitor or reactor.

(c) Only changes in bus power injections.

(d) Both (a) and (b).

(13) A power outage involves:

(a) Only changes in network admittance parameters.

(b) Only change in bus power injections.

(c) Loss of a generating unit or sudden loss of load.

(d) Both (b) and (c).

(14) The main objective of state estimation is:

(a) To obtain the best values of the magnitudes of bus voltages and angles.

(b) To maintain constant frequency.

(c) To reduce the load levels.

(d) To increase the power generation capacity.

(15) State estimation process requires the measurement of:

(a) Real and reactive-power flows in transmission lines.

(b) Real and reactive-power injections at the buses.

(c) Only reactive power absorbed by load.

(d) Both (a) and (b).

(16) State estimation is:

(a) An available data-sharing scheme.

(b) An available data-measuring scheme.

(c) An available data-processing scheme.

(d) An available data-sending scheme.

(17) State estimation scheme uses:

(a) Lagrangian function method.

(b) Negative gradient method.

(c) Lyapunov method.

(d) Weighted least square method.

(18) In the state estimation scheme, all the system information is collected by the centralized automation control of power system dispatch through:

(a) Remote terminal units.

(b) Transmitters.

(c) Digital signal processors.

(d) All of these.

(19) The inputs to state estimation are:

(a) Perfect power system measurements.

(b) Imperfect power system measurements.

(c) Depends on load connected to power system.

(d) All of these.

(20) Most of the state estimators in power systems at present belong to:

(a) Static-state estimators.

(b) Dynamic-state estimators.

(c) Either (a) or (b).

(d) Both (a) and (b).

REVIEW QUESTIONS

(1) Explain the concept of system security.

(2) Discuss the significance applications of system security.

(3) Explain the techniques used for transient security analysis.

(4) Explain the security enhancement.

(5) Explain the mathematical modeling of security-constrained optimization problem.

(6) Explain the SSS analysis.

(7) Discuss the need of state estimation.

(8) Explain the function of a state estimator.

(9) Discuss the difference between static-state estimation and dynamic-state estimation.

(10) Explain the least square estimation process.

(11) Explain the applications of state estimation process.

Appendix A
Answers to Multiple-choice Questions

CHAPTER 1

(1) a	(12) b	(23) b	(34) b
(2) c	(13) c	(24) a	(35) d
(3) b	(14) a	(25) d	(36) a
(4) a	(15) a	(26) c	(37) a
(5) a	(16) c	(27) c	(38) c
(6) d	(17) b	(28) c	(39) c
(7) b	(18) d	(29) a	(40) d
(8) b	(19) b	(30) d	(41) d
(9) e	(20) a	(31) c	(42) d
(10) c	(21) c	(32) c	
(11) a	(22) d	(33) a	

CHAPTER 2

(1) a	(10) a	(19) a	(28) a
(2) a	(11) a	(20) d	(29) c
(3) b	(12) a	(21) c	(30) d
(4) a	(13) c	(22) d	(31) d
(5) c	(14) d	(23) b	(32) a
(6) d	(15) c	(24) b	(33) a
(7) d	(16) b	(25) c	
(8) d	(17) c	(26) b	
(9) c	(18) d	(27) a	

CHAPTER 3

(1) c	(10) d	(19) b	(28) c
(2) b	(11) a	(20) a	(29) c
(3) d	(12) a	(21) c	(30) d
(4) d	(13) d	(22) a	(31) c
(5) a	(14) a	(23) c	(32) c
(6) c	(15) a	(24) b	(33) d
(7) c	(16) b	(25) a	(34) d
(8) c	(17) c	(26) a	
(9) a	(18) d	(27) d	

CHAPTER 4

(1) c	(14) d	(27) b	(40) b
(2) c	(15) d	(28) a	(41) a
(3) a	(16) a	(29) b	(42) b
(4) d	(17) c	(30) a	(43) b
(5) a	(18) a	(31) a	(44) a
(6) b	(19) a	(32) b	(45) b
(7) a	(20) b	(33) c	(46) b
(8) d	(21) b	(34) a	(47) b
(9) d	(22) d	(35) c	(48) b
(10) a	(23) b	(36) a	(49) a
(11) d	(24) a	(37) d	
(12) d	(25) b	(38) b	
(13) d	(26) a	(39) b	

CHAPTER 5

(1) d	(9) b	(17) b	(25) b
(2) c	(10) a	(18) a	(26) c
(3) b	(11) c	(19) Rigid, soft	(27) b
(4) a	(12) a	(20) b	(28) d
(5) b	(13) d	(21) a	(29) a
(6) c	(14) a	(22) b	(30) d
(7) d	(15) a	(23) c	
(8) b	(16) c	(24) b	

CHAPTER 6

(1) a	(12) a	(23) a	(34) d
(2) b	(13) d	(24) a	(35) b
(3) a	(14) d	(25) b	(36) b
(4) b	(15) c	(26) c	(37) a
(5) b	(16) d	(27) b	(38) a
(6) c	(17) a	(28) c	(39) d
(7) a	(18) b	(29) d	(40) d
(8) c	(19) b	(30) c	(41) c
(9) d	(20) c	(31) a	
(10) c	(21) c	(32) b	
(11) d	(22) d	(33) d	

CHAPTER 7

(1) a	(14) b	(27) d	(40) b
(2) c	(15) b	(28) d	(41) c
(3) c	(16) b	(29) a	(42) c
(4) a	(17) b	(30) a	(43) c
(5) d	(18) c	(31) c	(44) d
(6) b	(19) b	(32) d	(45) a
(7) c	(20) c	(33) c	(46) a
(8) c	(21) a	(34) b	(47) b
(9) b	(22) b	(35) a	(48) a
(10) c	(23) a	(36) a	(49) d
(11) c	(24) c	(37) b	(50) c
(12) c	(25) b	(38) d	
(13) c	(26) d	(39) c	

CHAPTER 8

(1) b	(6) a	(11) c	(16) b
(2) c	(7) d	(12) b	(17) c
(3) b	(8) c	(13) c	(18) a
(4) b	(9) d	(14) a	(19) b
(5) c	(10) d	(15) a	(20) c

CHAPTER 9

(1) a	(17) c	(33) c	(49) d
(2) a	(18) a	(34) c	(50) d
(3) b	(19) a	(35) c	(51) a
(4) b	(20) a	(36) c	(52) a
(5) b	(21) d	(37) c	(53) b
(6) b	(22) a	(38) c	(54) b
(7) b	(23) b	(39) a	(55) b
(8) b	(24) b	(40) c	(56) a
(9) d	(25) d	(41) c	(57) d
(10) a	(26) c	(42) c	(58) a
(11) a	(27) d	(43) c	(59) c
(12) c	(28) b	(44) d	(60) c
(13) a	(29) c	(45) c	(61) c
(14) d	(30) c	(46) b	(62) d
(15) c	(31) c	(47) b	(63) c
(16) b	(32) c	(48) d	(64) a

CHAPTER 10

(1) a	(8) a	(15) b	(22) d
(2) c	(9) b	(16) a	(23) c
(3) b	(10) d	(17) d	(24) d
(4) d	(11) a	(18) c	(25) a
(5) a	(12) b	(19) b	
(6) d	(13) a	(20) a	
(7) a	(14) a	(21) c	

CHAPTER 11

(1) b	(8) d	(13) b	(20) d
(2) c	(9) d	(14) a	(21) c
(3) c	(10) (i)-(c), (ii)-(b), (iii)-(a), and (iv)-(d)	(15) b	(22) d
(4) a		(16) a	(23) a
(5) c	(11) a	(17) d	(24) b
(6) a	(12) a	(18) c	(25) a
(7) d		(19) c	(26) a

(27) d	(31) b	(35) d	(39) a
(28) d	(32) a	(36) c	(40) b
(29) a	(33) d	(37) a	(41) c
(30) a	(34) b	(38) d	

CHAPTER 12

(1) d	(22) b	(43) b	(64) b
(2) a	(23) c	(44) d	(65) b
(3) d	(24) a	(45) a	(66) d
(4) b	(25) a	(46) d	(67) a
(5) a	(26) b	(47) c	(68) b
(6) b	(27) b	(48) a	(69) c
(7) b	(28) c	(49) b	(70) a
(8) a	(29) a	(50) c	(71) c
(9) b	(30) b	(51) c	(72) d
(10) b	(31) a	(52) d	(73) c
(11) c	(32) b	(53) b	(74) a
(12) d	(33) c	(54) a	(75) b
(13) a	(34) c	(55) b	(76) d
(14) c	(35) c	(56) c	(77) c
(15) d	(36) b	(57) b	(78) a
(16) b	(37) a	(58) b	(79) d
(17) b	(38) b	(59) d	(80) a
(18) c	(39) c	(60) d	(81) b
(19) d	(40) a	(61) c	(82) b
(20) a	(41) b	(62) e	
(21) a	(42) a	(63) d	

CHAPTER 13

(1) d	(6) c	(11) b	(16) c
(2) a	(7) a	(12) d	(17) d
(3) d	(8) d	(13) d	(18) a
(4) a	(9) c	(14) a	(19) b
(5) d	(10) a	(15) d	(20) a

Index